METHODS IN M

Series Editor
John M. Walker
School of Life Sciences
University of Hertfordshire
Hatfield, Hertfordshire, AL10 9AB, UK

For further volumes:
http://www.springer.com/series/7651

Lectins

Methods and Protocols

Edited by

Jun Hirabayashi

National Institute of Advanced Industrial Science and Technology, Tsukuba, Japan

Editor
Jun Hirabayashi
National Institute of Advanced Industrial
Science and Technology
Tsukuba, Japan

ISSN 1064-3745 ISSN 1940-6029 (electronic)
ISBN 978-1-4939-5466-7 ISBN 978-1-4939-1292-6 (eBook)
DOI 10.1007/978-1-4939-1292-6
Springer New York Heidelberg Dordrecht London

Printed on acid-free paper

Humana Press is a brand of Springer
Springer is part of Springer Science+Business Media (www.springer.com)

Preface

Lectins, a wide range of carbohydrate-binding proteins, have an extraordinarily long history of investigations, with its origin in the discovery of a plant toxin, "ricin," in 1888 by Peter Hermann Stillmark in Russia. Since then, a number of plant lectins have been identified with their detailed biochemical properties and biological implications in relevance to interaction with animal cell surface components (glycoconjugates). They have been used as useful tools in cell biology, basic biochemistry, glycan/glycoprotein separation as well as clinical applications as cell stimulants (e.g., mitogen). However, coming to the era of post-genome science with modern biotechnology, various new fields of lectin technology have emerged; these are represented by exploration of animal lectins including those of human membrane-integrated type. Their functional analyses necessitated development of new assay systems using live cells, while advanced microarray technology enabled a direct glycan profiling without liberation of glycans from proteins. The latter approach established an old but new principle to analyze, or "decode," extremely diverse and heterogeneous features of glycans under the concept "cellular glycome."

The edition "Lectins" thereby summarizes not only classic lectin technologies, which are still important and more accessible (*Part I: Standard Techniques*, comprising 19 chapters) but also a series of advanced techniques with high throughputs and sensitivities (*Part II: Emerging Techniques for Lectin-Based Glycomics*, comprising 9 chapters). The former part includes conventional hemagglutination assay, serial lectin-affinity chromatography procedure, lectin-probed western blot and histochemical analyses, quantitative interaction analyses based on equilibrium dialysis, isothermal calorimetry, surface plasmon resonance, etc. Line-up of these comparative methods will provide readers solid criteria for selection and optimization of their researches. The latter part is the latest collection of the most advanced techniques to analyze complex feature of glycans. Actually, lectin microarray represents an advanced method for glycomics, which was investigated in the twenty-first century, enabling an alternative, so-far inaccessible approach to glycoprotein without liberation of glycans from the core protein.

The edition also covers variable techniques for elucidating functions of endogenous animal lectins (*Part III: Techniques for Elucidating Functions of Endogenous Animal Lectins*, comprising 8 chapters). Necessity of this approach is emphasized, because most of the animal lectins, e.g., human lectins, are relatively unstable and show weaker binding activity compared with conventional plant lectins. Moreover, many animal lectins of interest are expressed in complex on the cell surface. For such "difficult" lectins, many effective methods including various cell modifying technologies are described. These approaches have now been extensively utilized and seem essential to elucidate glycan/lectin functions in the context of animal cells. In *Part IV (Structural Biology and Engineering of Lectins*, comprising 8 chapters), more challenging methodologies are described, which include advanced ones to "engineer" novel lectins by evolutionary concepts. This approach is in particular important, because it is generally difficult to prepare anti-carbohydrate antibodies, where the epitope structures are common between the species. Of course, for this realization, detailed information of three-dimensional structures of extensive lectins is necessary, for which several useful approaches are also described in this Part.

Although all of these features based on standard chapters of Methods and Protocols are sufficient to learn both necessary and more advanced techniques for lectins, the present edition also provides as many as 9 Overviews, each of which reviews the relevant field from a historical viewpoint referring to the "state-of-the arts" in front of each Part. Such introduction will guide readers more smoothly to the main part of Methods and Protocols, and make it more and more attractive. Another topic of this edition is two special equipments of "Comprehensive Lists" (*Part V*). However, these are not usual appendixes because they are edited based on the concept that this edition be not just for memory but for continuous use for not only lectin specialists but also non-experts including both young scientists and those who "happen" to meet necessity to use lectins. Lastly, as the editor of Lectins, I believe that researchers concerned with life sciences will meet every necessity to use lectins if they target glycans. Visiting old, learn new.

Tsukuba, Japan ***Jun Hirabayashi***

Contents

PART III TECHNIQUES FOR ELUCIDATING FUNCTIONS OF ENDOGENOUS ANIMAL LECTINS

PART IV STRUCTURAL BIOLOGY AND ENGINEERING OF LECTINS

Contributors

YOSHIHIRO AKIMOTO • *Department of Anatomy, Kyorin University School of Medicine, Mitaka, Japan*

MAHO AMANO • *Hokkaido University, Sapporo, Japan*

TOMAS BERTOK • *Slovak Academy of Sciences, Bratislava, Slovakia*

JULIE BOUCKAERT • *Université de Lille 1, Villeneuve d'Ascq, France*

KIMIE DATE • *Graduate School of Humanities and Sciences and Glycoscience Institute, Ochanomizu University, Tokyo, Japan*

ZUI FUJIMOTO • *Biomolecular Research Unit, National Institute of Agrobiological Sciences, Tsukuba, Japan*

JUN-ICHI FURUKAWA • *Hokkaido University, Sapporo, Japan*

PETER GEMEINER • *Slovak Academy of Sciences, Bratislava, Slovakia*

TOHRU GONOI • *Medical Mycology Research Center, Chiba University, Chiba, Japan*

TOMOMITSU HATAKEYAMA • *Nagasaki University, Nagasaki, Japan*

HIKARU HEMMI • *National Food Research Institute, National Agriculture and Food Research Organization (NARO), Tsukuba, Japan*

HIROSHI HINOU • *Hokkaido University, Sapporo, Japan*

JUN HIRABAYASHI • *Research Center for Stem Cell Engineering, National Institute of Advanced Industrial Science and Technology (AIST), Tsukuba, Japan*

MAKOTO HIRAYAMA • *Graduate School of Biosphere Science, Hiroshima University, Higashi-Hiroshima, Japan*

KANJI HORI • *Graduate School of Biosphere Science, Hiroshima University, Higashi-Hiroshima, Japan*

DAN HU • *Institute of Traditional Chinese Medicine and Natural Products, Jinan University, Guangzhou, People's Republic of China*

YUZURU ITO • *Research Center for Stem Cell Engineering, National Institute of Advanced Industrial Science and Technology (AIST), Tsukuba, Japan*

KAZUAKI KAKEHI • *Department of Pharmaceutical Sciences, School of pharmacy, Kinki University, Higashi-osaka, Japan*

AKIHIKO KAMEYAMA • *Bioproduction Research Institute, National Institute of Advanced Industrial Science and Technology (AIST), Tsukuba, Japan*

KEN-ICHI KASAI • *Department of Biological Chemistry, School of Pharmaceutical Sciences, Teikyo University, Tokyo, Japan*

RYUICHI KATO • *High Energy Accelerator Research Organization (KEK), Tsukuba, Japan*

HIROKAZU KAWAGISHI • *Shizuoka University, Shizuoka, Japan*

HAYATO KAWAKAMI • *Department of Anatomy, Kyorin University School of Medicine, Mitaka, Japan*

TOSHISUKE KAWASAKI • *Research Center for Glycobiotechnology, Ritsumeikan University, Shiga, Japan*

NORIHITO KAWASAKI • *Food & Health Programme, Institute of Food Research, Norwich, UK*

HIROTO KAWASHIMA • *Department of Biochemistry, School of Pharmacy and Pharmaceutical Sciences, Hoshi University, Tokyo, Japan*

MITSUHIRO KINOSHITA • *Department of Pharmaceutical Sciences, School of pharmacy, Kinki University, Higashi-osaka, Japan*
YUKA KOBAYASHI • *J-Oil Mills, Inc., Yokohama, Japan*
KYOKO KOJIMA-AIKAWA • *Ochanomizu University, Tokyo, Japan*
ATSUSHI KUNO • *Research Center for Medical Glycoscience, National Institute of Advanced Industrial Science and Technology (AIST), Tsukuba, Japan*
HISAYOSHI MAKYIO • *High Energy Accelerator Research Organization (KEK), Tsukuba, Japan*
ATSUSHI MATSUDA • *Research Center for Medical Glycoscience, National Institute of Advanced Industrial Science and Technology (AIST), Tsukuba, Japan*
YU-KI MATSUNO • *Bioproduction Research Institute, National Institute of Advanced Industrial Science and Technology (AIST), Tsukuba, Japan*
ICHIRO MATSUO • *Division of Molecular Science, Faculty of Science and Technology, Gunma University, Kiryu, Japan*
RISHO MIYOSHI • *Hokkaido University, Sapporo, Japan*
EIJI MIYOSHI • *Department of Molecular Biochemistry and Clinical Investigation, Osaka University Graduate School of Medicine, Suita, Japan*
MAMORU MIZUNO • *Laboratory of Glyco-organic Chemistry, The Noguchi Institute, Tokyo, Japan*
KENTA MORIWAKI • *Department of Molecular Biochemistry and Clinical Investigation, Osaka University Graduate School of Medicine, Suita, Japan*
JOZEF NAHALKA • *Slovak Academy of Sciences, Bratislava, Slovakia*
YUKO NAITO-MATSUI • *Department of Cellular and Molecular Medicine, University of California, San Diego, CA, USA*
HISASHI NARIMATSU • *Research Center for Medical Glycoscience, National Institute of Advanced Industrial Science and Technology (AIST), Tsukuba, Japan*
SHIN-ICHIRO NISHIMURA • *Hokkaido University, Sapporo, Japan*
MOTOHIRO NONAKA • *Sanford-Burnham Medical Research Institute, La Jolla, CA, USA*
HARUKO OGAWA • *Graduate School of Humanities and Sciences and Glycoscience Institute, Ochanomizu University, Tokyo, Japan*
TOMOHISA OGAWA • *Tohoku University, Sendai, Japan*
TAKASHI OHKURA • *Department of Reproductive Biology, National Center for Child Health and Development, Tokyo, Japan*
YASUKO ONUMA • *Research Center for Stem Cell Engineering, National Institute of Advanced Industrial Science and Technology (AIST), Tsukuba, Japan*
TOMOYUKI SAKO • *Yakult Europe B.V., Almere, The Netherlands*
KOTONE SANO • *Graduate School of Humanities and Sciences and Glycoscience Institute, Ochanomizu University, Tokyo, Japan*
CHIHIRO SATO • *Bioscience and Biotechnolgy Center, Nagoya University, Nagoya, Japan*
TAKESHI SATO • *Laboratory of Glycobiology, Department of Bioengineering, Nagaoka University of Technology, Nagaoka, Japan*
AZUSA SHIBAZAKI • *Laboratory for Inflammatory Regulation, RIKEN Research Center for Integrative Medical Science (IMS-RCAI), Yokohama, Japan*
YASURO SHINOHARA • *Hokkaido University, Sapporo, Japan*
TSUYOSHI SHIRAI • *Bioinfomatic Research Division, Nagahama Institute of Bio-Science and Technology, Japan Science and Technology Agency, Nagahama, Japan*
NONGLUK SRIWILAIJAROEN • *Department of Preclinical Sciences, Faculty of Medicine, Thammasat University, Pathumthani, Thailand; Health Science Hills, College of Life and Health Sciences, Chubu University, Kasugai, Japan*

YASUO SUZUKI • *Health Science Hills, College of Life and Health Sciences, Chubu University, Kasugai, Japan*
YOICHI TAKEDA • *ERATO, Ito Glycotrilogy Project, Japan Science and Technology Agency (JST), Wako, Japan*
HIROMU TAKEMATSU • *Graduate School of Medicine, Kyoto University, Kyoto, Japan*
BINBIN TAN • *Ministry of Education Key Laboratory of Systems Biomedicine, Shanghai Center for Systems Biomedicine, Shanghai Jiao Tong University, Shanghai, China*
HIROAKI TATENO • *Research Center for Stem Cell Engineering, National Institute of Advanced Industrial Science and Technology (AIST), Tsukuba, Japan*
JAN TKAC • *Slovak Academy of Sciences, Bratislava, Slovakia*
SACHIKO UNNO • *Research Center for Medical Glycoscience, National Institute of Advanced Industrial Science and Technology (AIST), Tsukuba, Japan*
ELS J.M. VAN DAMME • *Ghent University, Ghent, Belgium*
STEPHEN D. WEEKS • *Pharmaceutical and Pharmacological Sciences, KU Leuven, Leuven, Belgium*
KAZUO YAMAMOTO • *The University of Tokyo, Chiba, Japan*
KATSUKO YAMASHITA • *Department of Histology and Cell Biology, School of Medicine, Yokohama City University, Yokohama, Japan*
EMI YASUDA • *Yakult Central Institute for Microbiological Research, Kunitachi, Japan*

Part I

Standard Techniques

Chapter 1

History of Plant Lectin Research

Els J.M. Van Damme

Abstract

Numerous plant species are known to express one or more lectins or proteins containing a lectin domain, enabling these proteins to select and bind specific carbohydrate structures. The group of plant lectins is quite heterogeneous since lectins differ in their molecular structure, specificity for certain carbohydrate structures, and biological activities resulting therefrom. This chapter presents a short historical overview on how plant lectin research has evolved over the years from a discipline aiming merely at the purification and characterization of plant lectins towards the application of plant lectins as tools in glycobiology.

Key words Agglutinin, Carbohydrate, Carbohydrate recognition domain, Classification, Evolution, Lectin, Lectin domain, Specificity, Sugar binding

1 Short Historical Overview

Plant lectins, also called (phyto)hemagglutinins, have a very long history. It is generally accepted that the first lectin was reported by Peter Hermann Stillmark in 1888 [1]. As part of his doctoral thesis Stillmark isolated ricin from the seeds of castor bean (*Ricinus communis*), and described it as an extremely toxic protein. Stillmark also tested the reactivity of partially purified protein extracts towards red blood cells and observed some clumping of the cells. This phenomenon was analyzed in more detail in the years to come. In 1898 Elfstrand described the activity of proteins to clump or agglutinate erythrocytes, which led to the introduction of the term "agglutinin" [2]. Soon after the discovery of ricin several toxic proteins have been identified. However, as more and more substances were purified it became obvious that not all proteins with agglutinating activity are toxic. Landsteiner and Raubitschek (1907) presented evidence for nontoxic lectins from legume seeds, in particular *Phaseolus vulgaris* (bean), *Pisum sativum* (pea), *Lens culinaris* (lentil), and *Vicia sativa* (vetch) [3].

Though the ability of some lectins to agglutinate red blood cells was well known since the early 1900s, it was only in the 1950s

Jun Hirabayashi (ed.), *Lectins: Methods and Protocols*, Methods in Molecular Biology, vol. 1200,
DOI 10.1007/978-1-4939-1292-6_1,

that researchers discovered the reasons for this agglutination reaction. Already in 1936 Sumner and Howell reported that cane sugar could inhibit the agglutination activity of concanavalin A [4], but it took until 1952 when Watkins and Morgan first demonstrated the link between the agglutination activity of lectins and their ability to recognize the carbohydrate structures present on the surface of the erythrocytes, thus proving that lectins possess carbohydrate-binding activity [5].

Since the discovery of the first plant lectin more than 100 years ago, a steadily growing number of plant lectins has been reported in literature [6–10]. Owing to their abundance in legume seeds, lectin research has focused on legume lectins for several years [11, 12]. Consequently, concanavalin A was the first lectin to be purified from the seeds of jack bean (*Canavalia ensiformis*) [4]. In the 1980s plant lectin research turned to storage tissues (especially bark, rhizomes, bulbs, and corms) as a rich source of carbohydrate-binding proteins [7]. The extension from seeds to other plant tissues enabled lectinology to progress since the purification of these lectins resulted in the discovery of more lectins with a large variety of molecular structures and carbohydrate-binding properties.

In the early days of lectinology the discipline was hampered by the lack of tools to detect and/or purify particular carbohydrate-binding proteins. For a long time agglutination assays using red blood cells from different animal species have been the method of choice to search for new multivalent lectins. In combination with hapten inhibition assays these assays were an easy tool to get some information on the carbohydrate-binding properties of the lectin under study. Surely the elaboration of affinity chromatography techniques greatly facilitated the purification of numerous plant lectins from different species and tissues. Starting from the 1970s most lectin purification protocols included one or more affinity chromatography steps taking advantage of the carbohydrate-binding specificity of the lectins.

In the 1980s the focus of the lectin research shifted towards a molecular characterization, including the determination of the amino acid sequences of lectin polypeptides and their three-dimensional structure [13]. Concanavalin A was the first lectin for which a complete protein sequence and three-dimensional structure became available [14, 15]. In 1983 the first lectin was cloned, in particular the soybean lectin [16]. The introduction of advanced biochemical and molecular techniques ultimately led to a better understanding of lectin activity at the molecular level [9].

The availability of purified lectin samples allowed to study both the physicochemical properties as well as the biological activity of the proteins in more detail. Over the years substantial progress was made in the technology available to, e.g., determine the molecular structure of the protein and its carbohydrate-binding specificity. The advances in biochemistry and molecular biology also greatly determined the progress made at the level of structural analyses of the protein since the availability of sequence information, either from protein sequencing or molecular cloning of cDNAs encoding lectins, was a prerequisite to unravel the three-dimensional structure of the lectin and its carbohydrate-binding site [9, 10, 17].

The multitude of sequence information that became available in the last decade from the proteomics field, but also from data obtained through genomics and transcriptomics, enhanced our knowledge on the distribution of lectins and their evolutionary relationships, but also yielded important information with respect to when and where lectins are expressed, in turn allowing to gain better insight into the physiological role of plant lectins [9].

2 Definition

The term “lectin” (derived from the Latin *legere*, which means to select, to pick out, to choose) was first introduced by Boyd and Shapleigh in 1954 [18] to refer to the fact that lectins can recognize and bind specific carbohydrate structures. Furthermore some plant lectins have the ability to distinguish between erythrocytes of different blood types, and thus can select certain cells. However, this selectivity is not a general characteristic of plant lectins.

Over the years numerous definitions have been proposed to describe the group of plant lectins and their biological activity. In the early days of lectinology the definition focused on the ability of plant lectins to agglutinate red blood cells, and thus accidentally a selection was made for those proteins containing two or more carbohydrate-binding sites, a prerequisite to form a network of erythrocytes linked through lectin molecules. With the expansion of our knowledge on plant lectins the definition of plant lectins had to be adapted several times for different reasons [19–23]. For instance, it was shown that some proteins contain only a single carbohydrate-binding site, whereas other lectins show only poor agglutination activity. Furthermore it became clear that some lectins are chimeric molecules consisting of multiple protein domains, only one of which has lectin activity. All these elements urged to adapt and expand the definition of plant lectins, finally leading to a definition that is far less restrictive than the early definitions. Taking into consideration most of the issues raised above Peumans and Van Damme [23] proposed to define plant lectins as “all plant proteins possessing at least one non-catalytic domain, which binds reversibly to a specific mono- or oligosaccharide.” This definition comprises a broad range of proteins or protein domains which all have the ability to select for and bind to specific carbohydrate structures through their carbohydrate-binding site. Although our knowledge on the carbohydrate-binding properties of different lectins has still improved considerably over the last decade this definition is still accepted to date.

3 Lectins and Lectin Domains

As indicated above, lectin research originally focussed on the purification of abundant proteins with two or more carbohydrate-binding sites, allowing them to agglutinate cells. These lectins can be referred to as hololectins that consist solely of carbohydrate-binding domains. However, with the advent of genomics and proteomics a lot of sequence information became available. Genome sequence analyses combined with transcriptome analyses provided ample evidence for the occurrence of numerous genes with one or more lectin domain(s) linked to unrelated domain(s) [9, 10, 17]. Interestingly, these chimerolectins are more wide-spread in plants than the hololectins. Due to the complex protein architecture it is more appropriate to think in terms of carbohydrate recognition domains, as part of more complex proteins rather than "simple" lectins composed of carbohydrate-binding domains only. Evidence is accumulating that through evolution these lectin domains have been used as building blocks to create new chimeric proteins with multiple domains and with multiple activities. Similar to the animal field the lectin domains in plants are now considered as functional units within multidomain proteins [24]. This idea nicely fits with the idea of lectins being recognition or signaling molecules.

4 Classification of Plant Lectins

Over the years hundreds of plant lectins have been reported in literature and many of them have been characterized in some detail. Furthermore genome/transcriptome analyses revealed that most lectin families are far more widespread than can be inferred from the work with the purified proteins. By now it is generally accepted that plant lectins are omnipresent in the plant kingdom [7–9, 25].

From the early days of lectin research it was clear that the group of lectins was quite heterogeneous in that, e.g., different legume lectins recognized different carbohydrates [11, 12]. Taking into account that in the 1970s and 1980s the focus in lectinology was mainly on protein characterization and determination of their sugar-binding properties it seemed logic to try and set up a classification system of lectins based on the differences observed in sugar specificity. Since the specificity of a lectin was often defined in terms of its reactivity towards monosaccharides, different specificity groups were created: mannose-binding lectins, galactose-binding lectins, chitin-binding lectins, ... Although this classification was of some practical use, it turned out very soon that this classification system was difficult to maintain. Furthermore the system did not take into account any of the molecular and taxonomical relationships among the plant species.

Since the classification of all carbohydrate-binding proteins into one large family relies solely on the ability of the proteins to bind sugars it is rather artificial. Furthermore it does not take into account evolutionary or sequence relationships among the proteins. As soon as more information at molecular level became available, such as sequence information of lectin polypeptides, attempts were made to classify the whole group of lectins based on sequence relationships. Based on a comprehensive analysis of plant lectin sequences a system was elaborated in 1998, whereby the majority of all plant lectins were classified into seven families of structurally and evolutionarily related proteins [7, 9]. This classification system proved to be useful but required several updates as new sequence information became available.

A careful analysis of the sequences available today combined with relevant data from plant genome/transcriptome analyses shows that all plant lectins for which sufficient sequence information is available can be classified into 12 families, based on the sequence of the lectin polypeptides and the structure of their carbohydrate recognition domains (CRDs) (Table 1). In modern lectinology these CRDs are the basis for the classification of the whole group of carbohydrate-binding proteins. Each CRD is characterized by its own amino acid sequence, typical folding of the lectin polypeptide, and the structure of the binding site. In principle, each lectin family comprises all proteins possessing a domain that is evolutionarily related—in terms of sequence similarity—to the characteristic CRD. Different CRDs differ from each other in their sequences but can show reactivity towards similar carbohydrate structures, indicating that specificity is not linked to the occurrence of one particular CRD [9, 10].

5 New Insights into Plant Lectin Research

Over the years many plant lectins have been purified and characterized from wild and cultivated species, including several crops, such as potato, tomato, bean, rice, wheat, …. The introduction of more advanced technologies to study these proteins resulted in new insights related to the evolution and physiological importance of lectin domains in plants.

5.1 Structural Basis of Lectin–Carbohydrate Interactions

Numerous plant lectins have been studied in great detail to unravel their carbohydrate-binding specificity. For a long time these specificity studies were performed using an indirect approach involving assays to check what sugars or glycans were able to inhibit the agglutination, precipitation, or binding reactions.

The interaction of plant lectins with simple monosaccharides has been recognized since the early days of lectinology. However, one can question whether this low-affinity interaction ($K_d = 10^{-3}$–10^{-4} M)

Table 1
Overview of the 12 plant lectin domains

Lectin domain	Protein characteristics/ 3D structure	Specificity	Localization in plant cell
Agaricus bisporus homolog	Homodimer β-Sandwich	T antigen *N*-glycans	Nucleus, cytoplasm
Amaranthin domain	Homodimer β-Trefoil	GalNAc T antigen	Nucleus, cytoplasm
Homolog of class V chitinases	Homodimer TIM barrel	High-mannose *N*-glycans	Vacuole
Cyanovirin domain	Homodimer Triple-stranded β-sheet and a β-hairpin	High-mannose *N*-glycans	Vacuole
Euonymus europaeus lectin domain	Homodimer Structure unknown	Galactosides, high-mannose *N*-glycans	Nucleus, cytoplasm Vacuole?
Galanthus nivalis agglutinin domain	Different oligomerization states β-Barrel	Mannose, oligomannosides High-mannose *N*-glycans, Complex *N*-glycans	Vacuole Nucleus, cytoplasm
Hevein domain	Different oligomerization states Hevein domain	GlcNAc $(GlcNAc)_n$	Vacuole, cell wall
Jacalin domain	Different oligomerization states β-Prism	Mannose-specific subgroup Galactose-specific subgroup	Nucleus, cytoplasm Vacuole
Legume domain	Different oligomerization states β-Sandwich	Man/Glc, Gal/GalNAc, $(GlcNAc)_n$, fucose, Siaα2,3Gal/GalNAc, complex *N*-glycans	Vacuole Cytoplasm?
Lys M domain	Different oligomerization states β-α-α-β-Structure	$(GlcNAc)_n$	Vacuole Nucleus, cytoplasm
Nictaba-like domain	Homodimer Structure unknown	$(GlcNAc)_n$ High-mannose *N*-glycans Complex *N*-glycans	Nucleus, cytoplasm
Ricin-B domain	Different oligomerization states β-Trefoil	Gal/GalNAc Siaα2,6Gal/GalNAc	Vacuole Nucleus, cytoplasm

has any physiological relevance taken into account that the lectins show a much higher affinity ($K_d = 10^{-6}$-10^{-8} M) towards oligosaccharides and more complex glycans. The binding of these more complex carbohydrate structures into the lectin-binding site was proven by co-crystallization of the lectins and their complementary glycan structures, and demonstrated the occurrence of what is now referred to as "an extended binding site," allowing anchoring of the elongated carbohydrate structure through hydrogen bonds and van der Waals interactions to the core of the site as well as to different amino acids located in the vicinity of the primary lectin-binding site [10].

A major drawback of the carbohydrate-binding assays was the need for large amounts of the pure protein as well as of the (rare and expensive) carbohydrates. Our knowledge of the carbohydrate-binding specificity of plant lectins has benefited greatly from the introduction of high-performing techniques that enable screening of large collections of carbohydrates and more complex glycans. Frontal affinity chromatography is a biophysical method that allows the study of molecular interactions between immobilized lectins and complex carbohydrate structures in a flow-based system, and permits the thermodynamic and kinetic characterization of the interaction [26]. Glycan microarrays are a high-throughput method to determine the specificity of glycan-binding proteins [27, 28]. Therefore a labeled lectin is allowed to interact with hundreds of glycan structures immobilized on the array. A major advantage of this technology is that only small amounts of protein are needed.

Without doubt the new advanced technologies to study the carbohydrate recognition of lectins allowed refining our ideas with respect to the specificity of plant carbohydrate-binding proteins. Different carbohydrate or glycan microarrays have been developed which allow a rapid and comprehensive screening of carbohydrate-binding proteins for their interaction with a large set of carbohydrate structures [28]. In addition, the glycan array technology also allows quantification of the relative affinity of carbohydrate-binding proteins to glycans [28, 29].

Glycan array analyses of plant lectins allowed refining the specificity of many plant lectins and confirmed the preferential binding of most CRDs to oligosaccharides and glycans rather than to monosaccharides, which contributed greatly to our knowledge on the biological activity of plant lectins and urged to adapt ideas with respect to the physiological role of plant lectins and their application in glycoconjugate research [30].

5.2 Plant Lectins and Their Role in the Plant: Defense Versus Signaling Molecules

For practical reasons research on plant lectins was concentrated on those lectins that are abundant in seeds or vegetative storage tissues for a long time [7]. In the last two decades more attention was given to plant lectins that are weakly expressed in non-storage tissues (such as leaves, roots, flowers). This change in interest together with the availability of more advanced technologies that allow a study of low-abundance proteins resulted in the discovery of several novel lectins that are not constitutively expressed but are induced by a specific treatment with a biotic/abiotic stress factor

[25, 31, 32]. The discovery of these low-abundance lectins was often due to the identification of a lectin motif in a gene sequence after which researchers started to look for proteins with carbohydrate-binding activity.

The first class of abundant proteins is often referred to as the group of "classical lectins." Many of these lectins specifically recognize typical animal glycans (e.g., sialic acid and GalNAc containing *O*- and *N*-linked oligosaccharides) that are abundantly present on the surface of the epithelial cells exposed along the intestinal tract of higher and lower animals [33]. Since these glycans are accessible for dietary proteins they represent potential binding sites for dietary plant lectins. If binding to these receptors results in an adverse effect the lectin will exert harmful or toxic effects. Feeding trials with insects and higher animals confirmed that some plant lectins provoke toxic effects ranging from a slight discomfort to a deadly intoxication [34–36], which leaves little doubt that lectins play a role in plant defense against insects and/or predating animals. To reconcile the carbohydrate-binding activity of these lectins with the high concentration of these proteins in the plant the concept was developed that these lectins are a class of a specific constitutively expressed defense proteins that help the plant to cope with attacks by predators, such as phytophagous invertebrates and/or herbivorous animals. It was suggested that these lectins combine a defense-related role with a function as a storage protein and whenever appropriate can be recruited by the plant for defense purposes [23].

During the last decade it was unambiguously shown that next to the classical lectins plants also synthesize minute amounts of carbohydrate-binding proteins upon exposure to stress situations like drought, high salt, hormone treatment, pathogen attack, or insect herbivory [25, 31, 32]. Since lectin activity cannot be detected under normal growth conditions but is clearly upregulated after stress application these lectins are referred to as "inducible" lectins. In contrast to the classical lectins that accumulate in the vacuole or the cell wall, these inducible lectins locate to the nucleus and/or the cytoplasm of plant cells. Based on these observations the concept was developed that lectin-mediated protein–carbohydrate interactions in the cytoplasm and the nucleus play an important role in the stress physiology of the plant cell [9, 30, 31]. At present at least six CRDs have been identified within the group of nucleocytoplasmic plant lectins (Table 1) [9, 25].

Bioinformatics studies on the sequences of all available databases have shown that some of the carbohydrate-binding motifs are widespread (ranging from plants to animals, fungi, and bacteria), whereas others are confined to certain plant families [9]. It should be mentioned however that in all cases the lectin concentrations in the plant tissues remain low. Therefore, it has been suggested that these cytoplasmic lectins are involved in signaling reactions in plant cells/tissues or between plants and other organisms [9, 25, 31, 32].

6 Plant Lectins as Research Tools: Exploiting the Biological Activity of Plant Lectins

In the last two decades plant lectin research shifted from an analytical study of lectins towards a more functional study, aiming at deciphering the physiological importance of lectins in the plant. Significant progress has been made in understanding the biological activity of different lectins. It is generally accepted that these biological properties of lectins will be of utmost importance for their physiological role. Furthermore understanding the biological activities of plant lectins has also enabled to develop some plant lectins as biotechnological tools for biomedical and agronomical applications [37, 38].

Plant lectins made an essential contribution to the field of modern glycobiology and are an important tool that can help to study glycoconjugates in solution and on cell surfaces. As soon as it was resolved that the agglutinating activity of lectins was due to their specific sugar-binding activity, the proteins were studied mainly because of the interaction of the protein with carbohydrate structures. Consequently lectins are interesting tools to study protein carbohydrate interactions playing an important role in, e.g., host-pathogen interaction(s), development, cell–cell communication, and cell signaling. For this purpose protocols were established to purify plant lectins on a large scale and make them commercially available. For instance concanavalin A is now the most used lectin for characterization and purification of high-mannose *N*-glycan-containing glycoconjugates and the detection of these carbohydrate structures on biomolecules and cells [38, 39].

The specific binding of a lectin to a carbohydrate structure can be exploited in lectin affinity chromatography [40]. This technique allows selecting for, e.g., glycosylated proteins from tissue or cell extracts and enables the use of lectins as tools for the separation and structural analysis of glycoproteins and oligosaccharides. The combination of affinity supports with immobilized lectins with different carbohydrate-binding properties allows the fractionation of oligosaccharides into structurally related subsets and can help to determine the glycan profile for different cells/tissues. Alternatively lectin arrays can be used to determine carbohydrate expression on purified proteins or on cells. Furthermore, the lectin microarray technology has been developed into a sensitive tool for high-throughput analysis of, e.g., tumor-associated changes in glycosylation [39].

References

1. Stillmark H (1888) Über Ricin ein giftiges Ferment aus den Samen von *Ricinus communis* L. und einige anderen Euphorbiaceen. Inaugural dissertation Dorpat, Tartu
2. Elfstrand M (1898) Über blutkörperchenagglutinierende Eiweisse. In: Kobert R (ed) Görberdorfer veröffentlichungen a Band I. Enke, Stuttgart, Germany, pp 1–159
3. Landsteiner K, Raubitschek H (1907) Beobachtungen über hämolyse und hämagglutination. Zentralbl Bakteriol Parasitenk Infektionskr Hyg Abt 1: Orig 45, 660–667
4. Sumner JB, Howell SF (1936) The identification of the hemagglutinin of the Jack bean with concanavalin A. J Bacteriol 32:227–237
5. Watkins WM, Morgan WTJ (1952) Neutralization of the anti-H agglutinin in eel by simple sugars. Nature 169:825–826
6. Etzler ME (1986) Distribution and function of plant lectins. In: Liener IE, Sharon N, Goldstein IJ (eds) The lectins: properties, functions, and applications in biology and medicine. Academic, Orlando, FL, pp 371–435
7. Van Damme EJM, Peumans WJ, Barre A, Rougé P (1998) Plant lectins: a composite of several distinct families of structurally and evolutionary related proteins with diverse biological roles. Crit Rev Plant Sci 17:575–692
8. Van Damme EJM, Peumans WJ, Pusztai A, Bardocz S (1998) Handbook of plant lectins: properties and biomedical applications. Wiley, Chichester
9. Van Damme EJM, Lannoo N, Peumans WJ (2008) Plant lectins. Adv Bot Res 48:107–209
10. Van Damme EJM, Rougé P, Peumans WJ (2007) Carbohydrate-protein interactions: plant lectins. In: Kamerling JP, Boons GJ, Lee YC, Suzuki A, Taniguchi N, Voragen AJG (eds) Comprehensive glycoscience – from chemistry to systems biology, vol 3. Elsevier, Oxford, pp 563–599
11. Sharon N, Lis H (1990) Legume lectins – a large family of homologous proteins. FASEB J 4:3198–3208
12. Strosberg AD, Buffard D, Lauwereys M, Foriers A (1986) Legume lectins: a large family of homologous proteins. In: Liener IE, Sharon N, Goldstein IJ (eds) The lectins, properties, functions, and applications in biology and medicine. Academic, Orlando, FL, pp 249–264
13. Etzler ME (1985) Plant lectins: molecular and biological aspects. Annu Rev Plant Physiol Plant Mol Biol 36:209–234
14. Edelman GM, Cunningham BA, Reeke GN Jr, Becker JW, Waxdal MJ, Wang JL (1972) The covalent and three-dimensional structure of concanavalin A. Proc Natl Acad Sci U S A 69:2580–2584
15. Hardman KD, Ainsworth CF (1972) Structure of concanavalin A at 2.4 Å resolution. Biochemistry 11:4910–4919
16. Vodkin LO, Rhodes PR, Goldberg RB (1983) A lectin gene insertion has the structural features of a transposable element. Cell 34: 1023–1031
17. Van Damme EJM, Fouquaert E, Lannoo N, Vandenborre G, Schouppe D, Peumans WJ (2011) Novel concepts about the role of lectins in the plant cell. In: Wu AM (ed) The molecular immunology of complex carbohydrates-3. Springer, New York, NY, pp 295–324. ISBN 978-1-4419-7876-9
18. Boyd WC, Shapleigh E (1954) Specific precipitating activity of plant agglutinins (lectins). Science 119:419
19. Dixon HBF (1981) Defining a lectin. Nature 338:192
20. Kocourek J, Horejsi V (1983) A note on the recent discussion on definition of the term 'lectin'. In: Bøg-Hansen TC, Spengler GA (eds) Lectins: Biology, Biochemistry and Clinical Biochemistry, vol 3, Walter de Gruyter. Berlin, Germany, pp 3–6
21. Goldstein IJ, Hughes RC, Monsigny M, Osawa T, Sharon N (1980) What should be called a lectin? Nature 285:66
22. Barondes SH (1988) Bifunctional properties of lectins: lectins redefined. Trends Biochem Sci 13:480–482
23. Peumans WJ, Van Damme EJM (1995) Lectins as plant defense proteins. Plant Physiol 109: 347–352
24. Taylor ME, Drickamer K (2011) Introduction to glycobiology, vol 3. Oxford University Press, London, p 283. ISBN 978-0-19-956911-3
25. Lannoo N, Van Damme EJM (2010) Nucleocytoplasmic plant lectins. Biochim Biophys Acta Gen Subj 1800:190–201
26. Hirabayashi J (2008) Concept, strategy and realization of lectin-based glycan profiling. J Biochem 144:139–147
27. Blixt O, Head S, Mondala T, Scanlan C, Huflejt ME, Alvarez R, Bryan MC, Fazio F, Calarese D, Stevens J, Razi N, Stevens DJ, Skehel JJ, van Die I, Burton DR, Wilson IA, Cummings R, Bovin N, Wong C-H, Paulson JC (2004) Printed covalent glycan array for ligand profiling of diverse glycan binding proteins. Proc Natl Acad Sci U S A 101: 17033–17038

28. Taylor ME, Drickamer K (2009) Structural insights into what glycan arrays tell us about how glycan-binding proteins interact with their ligands. Glycobiology 19:1155–1162
29. Van Damme EJM, Smith DF, Cummings R, Peumans WJ (2011) Glycan arrays to decipher the specificity of plant lectins. In: Wu AM (ed) The molecular immunology of complex carbohydrates-3. Springer, New York, NY, pp 841–854. ISBN ISBN: 978-1-4419-7876-9
30. Smith DF, Song X, Cummings RD (2010) Use of glycan microarrays to explore specificity of glycan-binding proteins. Methods Enzymol 480:417–444
31. Van Damme EJM, Barre A, Rougé P, Peumans WJ (2004) Cytoplasmic/nuclear plant lectins: a new story. Trends Plant Sci 9:484–489
32. Van Damme EJM, Lannoo N, Fouquaert E, Peumans WJ (2004) The identification of inducible cytoplasmic/nuclear carbohydrate-binding proteins urges to develop novel concepts about the role of plant lectins. Glycoconj J 20:449–460
33. Peumans WJ, Barre A, Hao Q, Rougé P, Van Damme EJM (2000) Higher plants developed structurally different motifs to recognize foreign glycans. Trends Glycosci Glycotechnol 12:83–101
34. Pusztai A, Bardocz S (1996) Biological effects of plant lectins on the gastrointestinal tract: metabolic consequences and applications. Trends Glycosci Glycotechnol 8:149–165
35. Van Damme EJM (2008) Plant lectins as part of the plant defence system against insects. In: Schaller A (ed) Induced plant resistance to herbivory. Springer, Dordrecht, pp 285–307
36. Vandenborre G, Smagghe G, Van Damme EJM (2011) Plant lectins as defense proteins against phytophagous insects. Phytochemistry 72:1538–1550
37. Hirabayashi J (2004) Lectin-based structural glycomics: glycoproteomics and glycan profiling. Glycoconj J 21:35–40
38. Paulson JC, Blixt O, Collins BE (2006) Sweet spots in functional glycomics. Nat Chem Biol 2:238–248
39. Fry S, Afrough B, Leathem A, Dwek M (2012) Lectin array-based strategies for identifying metastasis-associated changes in glycosylation. Methods Mol Biol 878:267–272
40. Van Damme EJM (2011) Lectins as tools to select for glycosylated proteins. In: Vandekerckhove J, Gevaert K (eds) Gel-free proteomics, methods and protocols. Methods Mol Biol 753, 289–297

Chapter 2

Fungal Lectins: A Growing Family

Yuka Kobayashi and Hirokazu Kawagishi

Abstract

Fungi are members of a large group of eukaryotic organisms that include yeasts and molds, as well as the most familiar member, mushrooms. Fungal lectins with unique specificity and structures have been discovered. In general, fungal lectins are classified into specific families based on their amino acid sequences and three-dimensional structures. In this chapter, we provide an overview of the approximately 80 types of mushroom and fungal lectins that have been isolated and studied to date. In particular, we have focused on ten fungal lectins (*Agaricus bisporus, Agrocybe cylindracea, Aleuria aurantia, Aspergillus oryzae, Clitocybe nebularis, Marasmius oreades, Psathyrella velutina, Rhizopus stolonifer, Pholiota squarrosa, Polyporus squamosus*), many of which are commercially available and their properties, sugar-binding specificities, structural grouping into families, and applications for biological research being described. The sialic acid-specific lectins (*Agrocybe cylindracea* and *Polyporus squamosus*) and fucose-specific lectins (*Aleuria aurantia, Aspergillus oryzae, Rhizopus stolonifer*, and *Pholiota squarrosa*) each showed potential for use in identifying sialic acid glycoconjugates and fucose glycoconjugates. Although not much is currently known about fungal lectins compared to animal and plant lectins, the knowledge accumulated thus far shows great promise for several applications in the fields of taxonomy, biomedicine, and molecular and cellular biology.

Key words Lectin, Agglutinin, Microorganisms, Fungi, Mushroom, Sialic acid, Fucose

1 Introduction

All organisms produce lectins, which serve various biological purposes and can recognize specific carbohydrates in a non-catalytic, specific, and reversible manner. Plant lectins have been studied extensively and have been used as tools in glycobiology and biomedical research [1]. Animal lectins have also been studied and their roles and functions have been elucidated [2]. Recently, algal lectins, which have unique structures and specificities, have been identified. Meanwhile, of fungal lectins identified, 82 % were from mushrooms, 15 % from microfungi (molds), and 3 % from yeasts [3–5].

Fungi are members of a large group of eukaryotic microorganisms that includes yeasts and molds, as well as the following mushroom taxa: the subkingdom Dikarya (phyla Ascomycota

Jun Hirabayashi (ed.), *Lectins: Methods and Protocols*, Methods in Molecular Biology, vol. 1200,
DOI 10.1007/978-1-4939-1292-6_2, © Springer Science+Business Media New York 2014

ORIGIN	ROLES	RESERCH AND BIOTECHNOLOGICAL APPLICATION
◆ **Mushroom** ◆ **Microfungi** ◆ **Yeast**	✧ storage protein ✧ growth ✧ morphogenesis ✧ parasitism ✧ infections ✧ molecular recognition ✧ defense cell flocculation ✧ mating process	➢ Glycoproteins purification ➢ Carbohydrates purification ➢ Glycomics studies ➢ Biomarkers ➢ Cancer research ➢ Antiviral ➢ Insecticide/Vermicide ➢ Targeted drug delivery

Fig. 1 Origin, roles, and applications of fungal lectins [5]

and Basidiomycota); the phyla Glomeromycota, Zygomycota, Blastocladiomycota, Neocallimastigomycota, and Microsporidia; and the division Chytridiomycota. In recent years, lectins from mushrooms, microfungi (molds), and yeasts have attracted much research interest (Fig. 1). The biological roles of fungi lectins are estimated to be related to storage, growth, morphogenesis, parasitism, infections, molecular recognition, defense, cell flocculation, and mating.

Here, we provide an overview of the recent knowledge attained on the properties of fungal lectins, particularly those of mushrooms.

2 Fungi as a Source of Lectins

In an attempt to locate lectins in microfungi, many studies have focused on their mycelium. However, lectin activity has also been detected in the culture filtrate of some fungal species, including *Neurospora sitophila, Aspergillus fumigatus, A. nidulans, Fusarium sp., Mucor javanicum, M. rouxianus*, and *Penicillium chrysogenum*. In addition, some reports have also revealed the presence of lectins in the spores. On the other hand, a thorough investigation of the occurrence of lectins in 25 species of *Aspergillus* showed that, in 13 species, lectin activity is restricted only to the mycelium, whereas none of the other species displayed lectin activity either in culture filtrates or mycelia [6].

Lectins have also been isolated also from the fruiting bodies of numerous species of macromycetes. Fruiting bodies of spore-producing fungi are generally referred to as mushrooms, whose species number is estimated at more than 100,000, although only 10 % of them have been named to date [7]. Therefore, although mushrooms are a vast fungal group, they are a largely untapped source of potentially new and powerful glycobiological products. Fungal lectins are known to play roles in several physiological events. These include the formation of primordial fruiting bodies, the creation of mycelium structures to ease the penetration of parasitic fungi into host organisms, and the identification of appropriate partners during the early stage of mycorrhization [5]. The fruiting bodies have been being studied as potential sources of lectins with novel carbohydrate-binding activity. Table 1 shows the properties of the lectins isolated from mushrooms and some other fungi thus far. The mushroom lectins have found applications in the isolation and structural determination of cell glycoconjugates, and for monitoring the changes that occur on the surface of the cell membranes at various physiological or pathological stages. In addition, the lectins are useful for embryological, microbiological, and taxonomic research. In some cases, mushroom lectins display immunological and antitumor activity.

Although reports on lectins from mushrooms are relatively scarce, these proteins are valuable tools to study the glycobiology of diseases such as cancer and for biomedical research. In this section, the properties, sugar-binding specificities, and examples of applications of some fungal- mostly mushroom-lectins are reviewed in detail.

2.1 *Agaricus bisporus* Lectin (ABL)

The common edible mushroom *Agaricus bisporus* is so ubiquitous that its common name is simply "mushroom." *Agaricus bisporus* is the most highly produced mushroom in the world. The molecular weight of a lectin from the mushroom, *A. bisporus* lectin (ABL), 58.5 kDa, was determined by gel filtration analysis. ABL has an affinity for *N*-acetylglucosamine (GlcNAc) and galactose (Gal) and strongly binds to the T antigen (Galβ1-3GalNAcα1-Ser/Thr). It has two distinct binding sites per monomer that recognize the different configurations of a single epimeric hydroxyl, corresponding to the T-antigen- and GlcNAc-binding sites. ABL is commercially available and has been employed for various purposes, including cancer research and the detection and isolation of the T antigen [8–10]

2.2 *Agrocybe cylindracea* Galectin (ACG)

Agrocybe species contain lectins specific for galactose. The result of polyacrylamide gel electrophoresis and gel filtration indicated that a lectin from *A. cylindracea* is a dimer consisted of the same subunit of 15 kDa. This lectin exhibits broad binding specificity for β-galactose-containing glycans. Its complete amino acid sequence was determined by fragmenting the protein into peptides by chemical cleavage and enzymatic hydrolysis. The sequence of this lectin has 19.1 % identity with that of human galectin-1. Seven residues, which are commonly found in the carbohydrate-recognizing domain (CRD) of galectins, are conserved in its sequence. This lectin is a member of the galectin family and therefore is designated as ACG (*A. cylindracea* galectin). Hemagglutination inhibition tests showed that the affinity of ACG is weak for sialic acid and lactose, whereas it is high for trisaccharides containing *N*-acetylneuraminyl galactose (NeuAcα2-3Galβ) residue. ACG strongly interacts with glycoconjugates containing NeuAcα2-3Galβ1-3GlcNAc-/GalNAc

Table 1
Properties of lectins isolated from fungi (mushroom and the others)

Origin		Blood-group specificity	Saccharide specificity	MW		Ref.
				Native	Subunit	
Mushroom						
Agaricus arvensis			Inulin	30	15	[44]
Agaricus bisporus (ABA)		No	βGal, GalNAc	64	16	[10, 45]
Agaricus blazei (ABL)		No	–	64	16	[46]
Agaricus campestris		No	–	64	16	[47]
Agaricus edulis	I	No	–	60	14	
	II	No	–	32	14	[48]
Agrocybe aegerita (AAL)		No	Gal	44	22	[49]
Agrocybe aegerita	I	No	(Bovine mucin)	32	15.8	[50]
	II	No	GlcNAc	43		[51]
Agrocybe cylindracea (ACG)			Gal, NeuAcα2-3	31	15, 16	[15, 52]
Aleuria aurantia (AAL)		No	L-Fuc	72	36	[17, 18, 20]
Amanita pantherina (APL)		A > B,O	GlcNAcβ1-4Manβ-*p*NP	43	22	[53]
Amanita virosa						[54]
Auricularia polytricha		No	Lac, Gal	23	23	[55]
Boletopsis leucomelas (BLL)		–		15	15	[56]
Boletus edulis (BEL)			Galβ1-3GalNAc	60	15	[57]
Boletus venenatus (BVL)		No	–	11	33	[58]
Boletus satanas			Gal	63	63	[59–61]

Chlorophyllum molybdites (CML)		A, O > B	NeuGc, GalNAc	32	16	[62]
Ciborinia camelliae			GalNAc		17	[63]
Clavaria purpurea (CpL)			Gal	32	15	[64]
Clitocybe nebularis (CNL)		No	Lac, GalNAc	70	19,14.5	[25]
Coprinus cinereus	(CGLI)		Galβ	16.4	16.4	
	(CGLII)		Galβ	16.7	16.7	[65]
	(CGLIII)		GlcNAc			[66]
Cordyceps militaris (CML)		No	(Sialoglycoproteins)	31	31	[67]
Flammulina velutipes		No	–	20	12, 8	
			–	12.7	12.7	[68, 69]
Fomes fomentarius		B > O > A	GalNAc, raffinose	60	35, 21	[25]
Ganoderma lucidum (GSL)			(Porcine mucin)		24	[70]
				114		[71]
Grifola frondosa		No	GalNAc	>100	66, 33	[72]
Hericium erinaceus	(HEL)	A, O > B	NeuGc	54	16	[73]
	(HEA)		Inulin	51	51	[74]
Hygrophorus russula (HRL)		A, O > B	Manα1-6	64	18.5	[75, 76]
Inocybe umbrinella			Lac, Gal	17		[77]
Ischnoderma resinosum (IRA)		B > A, O	Lactulose, Gal	32	16	[46]
Lactarius flavidulus (LFL)			*p*NPGlc	60	30	[78]
Laccaria amethystine	II	A > O	Lac, GalNAc	19	17.5	
	I	O > A	L-Fuc	1	16	[79]
Lactarius deliciosus (LDL)		No	Galβ1-3GalNAc	37	18, 19	[80]

(continued)

Table 1 (continued)

Origin		Blood-group specificity	Saccharide specificity	MW		Ref.
				Native	Subunit	
Lactarius deterrimus (LDetL)		No	Galβ1-3GalNAc	37	18	[81]
Lactarius flavidulus (LFL)		No	Galβ	30	15	[78]
Lactarius lignyotus		No	–	>100	22	[82]
Lactarius salmonicolor			Galβ1-3GalNAc		37	[83, 84]
Laetiporus sulphureus		No	LacNAc		190	[85]
Laetiporus sulphureus (LSL)			Galβ1-4GlcNAc	140	35	[86]
Lentinus edodes	L1	No	Gal,Lac		43	
	L2	No	Gal,Lac		37	[87]
Lyophyllum decastes			Galα1-4Gal	20	10	[88]
Lyophyllum shimeji			–		30	[89]
Marasmius oreades	(MOA)	B>O>A	Galα1-3GalNAc	50	33, 23	[26–33]
	(MOL)		Man			[90]
Melastiza chateri			L-Fuc		40	[91]
Mycoleptodonoides aitchisonii		No	–	64	16	[92]
Oudemansiella platyphylla (OPL)			βGalNAc			[93]
Paecilomyces Japonica (PJA)		ABO	Sialic acid		16	[94]
Paxillus atrotomentosus			–		40	[95]
Paxillus involutus			*p*NPGal	28	7	[96]
Phallus impudicus			–		75	[97]

Pholiota adiposa		*p*NPGal	32	16	[98]
Pholiota aurivella (PAA)	No	–	>100	18	[99]
Pholiota squarrosa (PhoSL)	No	Fucα1-6GlcNAc	14	4.5	[42]
Pleurotus cornucopiae-a	–	–	32	16	
-b,c	–	–	31	16 and 15	[100–102]
Pleurotus eous (PEL)				16	[103]
Pleurotus citrinopileatus		Insulin	64.8	32.4	[104]
Pleurotus florida (PFL)					[105]
Pleurotus ostreatus (POL)	O > A, B	GalNAc	80	40	[106, 107]
Pleurocybella porrigens (PPL)		Gal, GalNAc	14	56	[108]
Pleurotus tuber-regium		GlcNAc		32	[109]
Polyporus adusta		Turanose	24	12	[110]
Polyporus squamosus (PSL)	No	Neu5Acα2-6Galβ1-4Glc	56	28	[43]
Psathyrella asperospora	No	GlcNAc	42	42	[111]
Psathyrella velutina (PVL)	No	GlcNAc	40	40	[37, 38, 112]
Psilocybe barrerae		Gal		15	[113]
Rigidoporus lignosus	–			150	[114]
Russula delica		Inulin, *p*NPβGal	60	30	[115]
Schizophyllum commune (SCL)		Lac		64	
		GalNAc		31.5	[116, 117]
Tricholoma mongolicum I, II		Lac	37	17.5	[118]
Volvariella volvacea (VVL)	AB > A, B, O	–	26	13	[119]
Xerocomus chrysenteron (XCL)	No	–	22	17	[82]

(continued)

Table 1 (continued)

Origin	Blood-group specificity	Saccharide specificity	MW Native	MW Subunit	Ref.
Xylaria hypoxylon		Xyl	28.8	14.4	[120]
Other fungus					
Aspergillus fumigatus (AFL)		Sia	32		[121]
		L-Fuc			[122, 123]
Aspergillus oryzae lectin (AOL)		L-Fuc	70	35	[21–24]
Arthrobotrys oligospora		(Asialofetuin, frsuin, mucin)	36	15	[124]
Beauveria bassiana (BBL)	No	Galβ1-3GalNAc		15	[125]
Fusarium solani		Galβ1-3GalNAc, LacNAc	26		[125, 126]
Kluyveromyces bulgaricus (Kb-CWL I)		GlcNAc	38	18.9	[127]
(Kb-CWL II)		Gal	150	18.9	
Kluyveromyces bulgaricus		GlcNAc	61		[128]
		Gal	65		
Macrophomina phaseolina		NeuAc, LacNAc	34		[129]
Penicillium chrysogenum (PeCL)		Man			[130]
Peziza sylvestris		Arabinose	20		[131]
Punctularia atropurpurascens		GlcNAc	67		[132]
Rhizoctonia bataticola (RBL)	No		44	11	[133]
Rhizopus stolonifer		L-Fuc	28	4.5	[41]
Sclerotium rolfsii (SRL)		Galβ1-3GalNAcα		45	[134–136]

pNP p-nitrophenyl

and has a strong preference for *N*-acetylneuraminyl lactose (NeuAcα2-3Lac). Interestingly, the glycan-binding specificity of this lectin has been engineered toward α2-3-linked sialic acid by saturation mutagenesis of ACG. The ACG mutant E86D, in which Glu86 was substituted with Asp, exhibits a rather strict binding to the α2-3disialo biantennary *N*-linked glycans. The ACG mutant N46A had enhanced affinity toward blood group A tetraose (type 2), A hexaose (type 1), and Forssman pentasaccharide [11–16].

2.3 *Aleuria aurantia* Lectin (AAL)

The orange peel fungus (*Aleuria aurantia*) is a widespread ascomycete fungus belonging to the order of Pezizales. The lectin purified from this organism was designated as AAL (*A. aurantia* lectin). This lectin is a dimer of two identical subunits of approximately 36 kDa. Isoelectric focusing measurement of AAL showed that its isoelectric point (pI) is 4.0. AAL binds preferentially to fucose linked to α1-6 *N*-acetylglucosamine or to fucose linked to α1-3 *N*-acetyl-lactosamine residues. AAL is a commercially available lectin that is commonly used for detection and separation of L-Fuc and its derivatives as well as fucose-oligosaccharides, and is widely used to estimate L-Fuc stereoisomeric ratios [17–20].

2.4 *Aspergillus oryzae* Lectin (AOL)

In Japan, *Aspergillus oryzae* is used to produce various fermented foods such as miso, soy sauce, and brew. *A. oryzae* lectin (AOL) binds to α1-6 fucose more clearly than AAL. The subtle specificities of both the lectins for fucose linked to oligosaccharides through the 2-, 3-, 4-, or 6-position were determined by frontal affinity chromatography (FAC). AOL showed a similar specificity to AAL with respect to its high affinity for α1-6-fucosylated oligosaccharides (Fucα1-2Gal, Fucα1-3GlcNAc, and Fucα1-4GlcNAc), and fucose attaching to the reducing terminal core, GlcNAc. However, AOL showed twofold higher affinity constants for α1-6-fucosylated oligosaccharides than AAL and only AAL recognized α1-3-fucosylated GlcNAc at the reducing terminal. The difference of oligosaccharide specificities between AAL and AOL has also been studied by using FAC and surface plasmon resonance (SPR), and fucosylation levels in embryo fibroblasts of wild-type and Fut8-knockout mice were compared with each other using AOL and AAL [21–24].

2.5 *Clitocybe nebularis* Lectin (CNL)

Clitocybe nebularis, commonly known as the clouded agaric or cloud funnel, is an abundant gilled fungus, which appears both in conifer-dominated forests and broad-leaved woodlands throughout the world. It is edible but also occasionally causes gastric upset in many individuals. The isolated *C. nebularis* lectin (CNL) was a homodimeric lectin with 15.9 kDa subunits and agglutinated human group A, followed by B, O, and bovine erythrocytes. This hemagglutination was inhibited by asialofetuin and lactose. Glycan microarray analysis revealed that the lectin recognized the human blood group A determinant, that is, GalNAcα1-3(Fucα1-2)Galβ-containing carbohydrates, and GalNAcβ1-4GlcNAc (*N,N′*-diacetyllactosediamine). The lectin exerted antiproliferative activity specific to human leukemic T cells. In addition, the bivalent carbohydrate-binding property of CNL was essential for its activity. CNL is a ricin B-like lectin [25].

2.6 *Marasmius oreades* Agglutinin (MOA)

The Scotch bonnet (*Marasmius oreades*) is known as a fairy ring-forming mushroom. The lectin isolated from this organism (*Marasmius oreades* agglutinin, MOA) is a 30 kDa dimer composed of two subunits of approximately 15 kDa and is highly specific for Galα1-3Gal, Galα1-3Galβ1-4GlcNAc, and the blood group B determinant. MOA is an ideal reagent for detection and identification of Galα1-3Gal and Galα1-3Galβ1-4GlcNAc/Glc epitopes present on the chain ends of glycoproteins and glycolipids. It is much more specific for the above di- and trisaccharides than the plant isolectin from *Griffonia simplicifolia* (GSL-I-B4), which recognizes α-Gal end groups in any linkage [26–33].

2.7 *Psathyrella velutina* Lectin (PVL)

Psathyrella velutina lectin (PVL), whose monomeric molecular mass is 40 kDa, is specific for heparin/pectin and *N*-acetylglucosamine/*N*-acetylneuraminic acid. PVL binds well to oligosaccharides bearing nonreducing terminal β-GlcNAc-linked 1-6 or 1-3 but poorly to those having 1-4 linkages such as *N*-acetylated chito-oligosaccharides. It also binds to subterminal GlcNAc moiety when it is substituted at the C-6 position but does not interact with the moiety when substituted at either C-3 or C-4. In addition, PVL was found to exhibit multispecificity to acidic polysaccharides and sulfatides. The binding between PVL and heparin/pectin occurs at a site different from the GlcNAc/NeuAc-specific site. The binding activity to heparin is not inhibited by GlcNAc and the binding of heparin and GlcNAc shows different pH dependencies (the maximum binding to heparin at pH 7 and the maximum binding to GlcNAc at pH 3–4) [34–40].

2.8 *Rhizopus stolonifer* Lectin (RSL)

Rhizopus stolonifer (black bread mold) is a widely distributed thread-like mucoralean mold commonly found on bread surfaces, from which it obtains nutrients. Agglutinating activity was detected in the extract of the mycelium-forming spores cultured on agar plates but not in liquid medium. The *N*-terminal amino acid sequence of *Rhizopus stolonifer* lectin (RSL) corresponds to NH_2.IDPVNVKKLQCDGDTYKCTADLDFGDGR. The sugar-binding specificity of RSL was investigated by hemagglutination inhibition assay and capillary affinity electrophoresis analysis. The results showed that L-Fuc, fucoidan, fetuin, porcine stomach mucin, and thyroglobulin inhibited RSL-mediated hemagglutination. The most potent inhibitor of RSL is $Man_3GlcNAc_2Fuc$, which is the core structure of *N*-glycan with an L-fucose residue attached to

GlcNAc at the reducing terminus via an α1-6 linkage. *N*-glycans without α1-6-linked fucose residues show 100-fold weaker specificity for RSL than those with α1-6-linked fucose. Oligosaccharides with α1-2, -3, and -4 fucose showed no binding specificity in the assays [41].

2.9 Pholiota squarrosa Lectin (PhoSL)

Pholiota is a genus of gilled agarics that is characterized by brown spores, and produces a unique lectin. The lectin purified from *Pholiota squarrosa* (designated PhoSL) has been analyzed by sodium dodecyl sulfate-polyacrylamide gel electrophoresis, matrix-assisted laser desorption/ionization–time-of-flight mass spectrometry, and *N*-terminal amino acid sequencing. These analyses indicated that PhoSL has a molecular mass of 4.5 kDa and consists of 40 amino acids (NH_2-APVPVTKLVCDGDTYKCTAYLDFGDGRWVAQW DTNVFHTG-OH). Isoelectric focusing of PhoSL showed bands with a pI near 4.0. The activity of PhoSL was found to be stable between pH 2.0 and 11.0 and at temperatures ranging from 0 to 100 °C for incubation times of 30 min. In sugar-binding assays, PhoSL bound only to core α1-6-fucosylated *N*-glycans and not to other types of fucosylated oligosaccharides, such as α1-2, α1-3, and α1-4-fucosylated glycans. Furthermore, PhoSL bound to α1-6-fucosylated α-fetoprotein (AFP) but not to non-fucosylated AFP. PhoSL has been used to discriminate the differential expression of α1-6 fucosylation between primary and metastatic colon cancer tissues. Thus, PhoSL appears to be a promising tool for analyzing the biological functions of α1-6 fucosylations and for evaluating Fucα1-6 oligosaccharides as cancer biomarkers [42].

2.10 Polyporus squamosus Lectin (PSL)

A lectin of *Polyporus squamosus*, designated as PSL, is composed of two identical 28 kDa subunits associated by non-covalent bonds. cDNA cloning revealed that PSL contains a ricin B chain-like (QXW)3 domain at its N-terminus that is composed of three homologous subdomains (α, β, and γ). A recombinant PSL was expressed in *Escherichia coli* as a fully active, soluble form. PSL was observed to agglutinate human A, B, and O and rabbit red blood cells but was precipitated only with human α2-macroglobulin. Although PSL bound to β-D-galactosides, it had an extended carbohydrate-combining site that exhibited its highest specificity and affinity for the nonreducing terminal Neu5Acα2,6Galβ1,4Glc/GlcNAc (6′-sialylated type II chain) of *N*-glycans, which showed 2,000-fold stronger affinity compared to that for galactose. The strict specificity of PSL for α2-6-linked sialic acid renders this lectin a valuable tool for glycobiological studies in biomedical and cancer research. It does not recognize the Neu5Acα2-6GalNAc moiety present in ovine submaxillary mucin, in contrast to the *Sambucus nigra* agglutinin (SNA-I), which requires only the disaccharide group. This indicates the necessity of three structural features for the interaction of PSL with its substrates [43].

3 Comparison of the Sugar-Binding Specificities of Fungal Lectins

3.1 Fungal Lectins (ACG and PSL) for Sialic Acid Detection

Figure 2 summarizes the preferred glycan structure (epitope) of each fungal lectin for detection of sialic acid. The affinity of *Agrocybe cylindracea* galectin (ACG) for Neu5Acα2-3Gal is similar to those of *Maackia amurensis* leukoagglutinin (MAM/MAL-I) and *Maackia amurensis* hemagglutinin (MAH/MAL-II), and ACG has weak affinity for Lac and Gal. These three lectins do not bind to α2-6-linked NeuAc. However, the recognized epitopes slightly differ among these lectins. MAM binds to glycoconjugates with the Siaα2-3Galβ1-4GlcNAc epitope and does not bind to glycolipid-type glycans (Siaα2-3Galβ1-3GalNAc), indicating that MAM requires lactosamine moiety for the binding. In contrast, the *O*-linked oligosaccharides of glycoproteins, Siaα2-3Galβ1-3GalNAc, are recognized by MAH. ACG recognizes both of the epitope. These strict different binding specificities of MAM (MAL-I) and

Epitope	Siaα2-3 (N-linked)	Siaα2-3 (O-linked)	Sia2-6 (N-linked)	Siaα2-6 (Glycolipid)	Lactose
Structure	α2-3 β1-4	α2-3 β1-3	α2-6 β1-4	α2-6 β1-3	β1-4
Glyco-biological Roles		➢ Malignant Transform ation and Mucins		➢ canceration (sualyl-Tn pitope)	
Probe Lectins					
ACG	●	●			○
MAM/MAL-I	●				
MAH/MAL-II		●			
SSA			●	●	
SNA-I			●	●	
PSL			●	●	○
mutant ACG (E86D)	●				

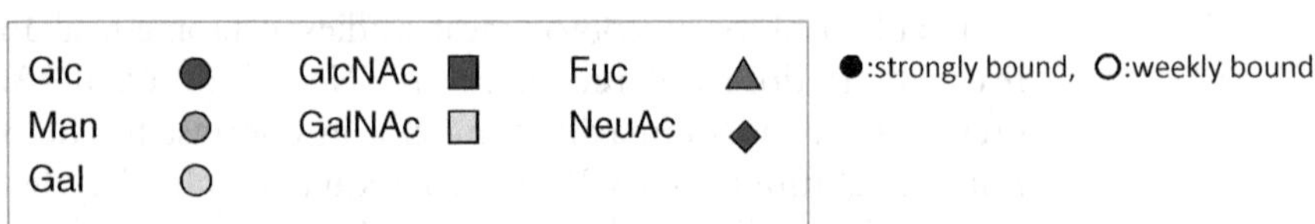

Fig. 2 Sialic acid-specific fungal lectins for detection tool

MAH (MAL-II) are useful for distinction between *N*-linked glycans (*N*-glycans) and *O*-linked glycans (*O*-glycans). ACG can be used for detection of both. In addition, genetic engineering of ACG to establish the E86D mutation in its glycan-binding site resulted in a large decrease in the binding affinity for asialo complex-type *N*-glycans, sialo complex-type *N*-glycans (except for the α2-3-disialo biantennary *N*-linked glycan), glycolipid-type glycans, and other glycans. Thus, E86D exhibited rather strict binding to the α2-3-disialo biantennary *N*-linked glycans. The binding properties of MAM (MAL-I) are similar to those of the mutant ACG (E86D) but MAM (MAM-I) has higher binding affinity for mono-terminal α2-3-Gal-linked glycans (*see* Chapter 45 by Y. Kobayashi et al.) [11–16].

Polyporus squamosus lectin (PSL) has affinity for Neu5Acα2-6Gal-GlcNAc in a similar manner to *Sambucus nigra* agglutinin (SNA-I) and *Sambucus sieboldiana* agglutinin (SSA), but also displays weak affinity for Lac. Sialic acid-specific lectins have been widely used for studies of several diseases, including cancer and Alzheimer's disease, as well as for the comparative study of influenza virus hemagglutinin and tissue staining. Further investigation of fungal sialic acid-specific lectins might provide great insights into these diseases.

3.2 Fungal Lectins (AAL, AOL, and PhoSL) for Fucose Detection

Figure 3 shows the preferred glycan structure (epitope) of each lectin for detection of L-fucose. *Aleuria aurantia* lectin (AAL) and *Aspergillus oryzae* lectin (AOL), which have affinity for Fucα1-2Gal, Fucα1-3GlcNAc, Fucα1-4GlcNAc, and Fucα1-5GlcNAc, have been used as probes for both fucosylated glycolipids and glycoproteins (*O*- and *N*-linked to sugars). Thus, these lectins could be used to detect the outer and core fucose chains. The plant lectin *Ulex europaeus* agglutinin (UEA-I) binds to many glycoproteins and glycolipids containing L-fucose residues, such as ABO blood group glycoconjugates. This lectin preferentially binds to blood group O (type 2H). Although *Lotus tetragonolobus* lectin (Lotus/LTL) has loosely similar binding specificity to UEA-I, their binding affinities and some specificities for oligosaccharides are markedly different from each other. Lotus lectin has high affinity toward saccharides with Le^Y (Fucα1-2Galβ1-4(Fucα1-3)GlcNAc) and weak affinity toward saccharides with Le^x (Galβ1-4(Fucα1-3)GlcNAc), but does not have affinity toward saccharides with Le^a (Galβ1-3(Fucα1-4) GlcNAc) and Le^b(Fucα1-2Galβ1-3(Fucα1-4)GlcNAc). UEA-I preferentially binds to blood group O (Fucα1-2Galβ1-4GlcNAc-) cells and weakly binds to Le^Y (Fucα1-2Galβ1-4(Fucα1-3)GlcNAc) (*see* Chapter 45 by Y. Kobayashi et al.).

The plant lectin, *Lens culinaris* lectin (LCA), is a useful probe for the detection of core-fucosylated biantennary *N*-glycan and also binds to D-mannose. This specificity has been exploited to diagnose hepatocellular carcinoma, by determining the content

Epitope	Fucα1-3	Fucα1-4	Fucα1-6 (coreFuc)
Structure	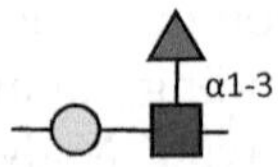 α1-3	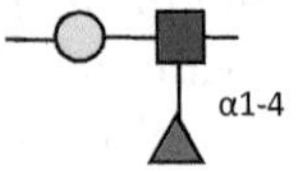 α1-4	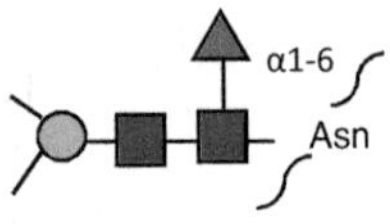 α1-6 Asn
Roles	➢ cell adhesion (L-selectin) ➢ tumor marker (CA19-9)		➢ functional regulation ➢ cell adhesion (integrin) ➢ tumor marker (AFP-L3)
Probe Lectins			
AAL	●	●	●
AOL	●	●	●
PhoSL			●
RSL			●
LCA			●
UEA-I	○		
Lotus/LTL	○		

● :strongly bound, ○:weekly bound

Fig. 3 Fucose-specific fungal lectins for detection tool

(%) of AFP-L3 in a sample. *Pholiota squarrosa* lectin (PhoSL) does not bind to L-fucose α1-3 and α1-4 sugar chains and strictly recognizes L-fucose α1-6 chains with a high association constant compared to those of AAL, AOL, LCA, and *Pisum sativum* agglutinin (PSA), which is well known for its affinity for L-fucose α1-6. RSL has also been isolated as a core fucose-specific lectin. RSL has high affinity toward saccharides with Fuc α1-6 and weak affinity toward saccharides with Fucα1-2, Fucα1-3, and Fuc α1-4. The specificity for the core fucose of PhoSL and RSL makes a significant contribution to exclusive detection of the core, compared with other fucose-binding lectins such as AOL, AAL, and LCA. On the other hand, AOL, AAL, or LCA is valuable toward detection of other fucose-containing sugar chains like Lex (x: upper).

The fucose-specific plant and fungal lectins summarized in this chapter may prove to be useful for the detection of AFP-L3 and other new biomarkers, and for determining the physiological functions of oligosaccharides.

4 Classification of Fungal Lectins

The amino acid sequences of several hundreds of lectins have now been established, and many of their three-dimensional structures have also been elucidated. Recently, a classification system was proposed, which subdivided these lectins into 48 families of structurally and evolutionarily related proteins (*see* Chapter 46 by Z. Fujimoto et al.). Several fungal lectins are classified into the same family as animal and plant lectins such as galectin, ricin B-chain like (R type), Jacalin related, and monocot lectin families, which comprise the majority of all currently known animal and plant lectins (Table 2). However, some fungal lections cannot be classified into the known families because of no significant sequence similarity to any lectins from animals and plants. Therefore, recently three fruit-body lectin families, ABA-like lectins, AAL-like, and PVL-like families, have been established based on their sequence identities. Although the classification of fungal lectins is less clear compared with those of animals and plants, in this section, we summarize the classification of the major fungal lectin families and propose a new family of mushroom lectins.

4.1 Galectin, Ricin B-Chain-Like (R-Type), Jacalin-Related, and Monocot Families

The classification of fungal lectins is not as clear as those of animals and plants. In general, the amino acid sequences of proteins are used to determine evolutionary relationships and to assign reactivity. However, several fungal lectins have been classified into the same family as animal and plant lectins. For example, *Agrocybe aegerita* lectin (AAA) and *Agrocybe cylindracea* lectin (ACG) have been assigned to the galectin family along with animal lectins, whereas *Pleurocybella porrigens* lectin (PPL), *Polyporus squamosus* lectin (PSL), and *Marasmius oreades* agglutinin (MOA) have been classified as R-type lectins along with plant lectins. There are many other galactose-binding lectins found from mushrooms, but their structures have not yet been elucidated.

Hygrophorus russula lectin (HRL) and *Grifola frondosa* lectin (GFL) have been assigned into the Jacalin-related family. The hemagglutinating activity of HRL was inhibited by mannose, whereas GFL is not inhibited by any monosaccharides and only inhibited by porcine stomach mucin.

The ricin B-chain-like lectin, *Marasmius oreades* lectin (MOA), was investigated for its molecular structure and carbohydrate-binding properties. However, another lectin, mannose-recognizing lectin (MOL), was purified from the same mushroom grown in Japan as MOA, which was reported to be a *Galanthus nivalis* agglutinin (GNA)-like lectin.

Table 2
Family and carbohydrate specificity of lectin in selected fungi

Nomenclature (origin)	Monosaccharide specificity	Preferred glycan structure (terminal epitope)	Ref.
Galectin family (*see* Chapter 46, No. 2)			
AAL (mushroom, *Agrocybe aegerita*)	Gal	Galβ1–4GlcNAc	[137, 138]
ACG (mushroom, *Agrocybe cylindracea*)	Gal	Siaα2–3Galβ1–4GlcNAc	[11, 19]
CCG (mushroom, *Coprinus cinereus*)	Gal	Galβ1–4GlcNAc	[66]
Ricin B-chain-like (R-type) family (*see* Chapter 46, No. 9)			
CNL (mushroom, *Clitocybe nebularis*)	Gal	Galβ1–4GlcNAc	[66]
PPL (mushroom, *Pleurocybella porrigens*)	Gal	GalNAc	[108]
PSL (mushroom, *Polyporus squamosus*)	Gal	Siaα2–6Gal	[139]
MOA (mushroom, *Marasmius oreades*)	Gal	Galα1-3Gal Galα1-3Galβ1-4GlcNAc/Glc	[26]
MPL (mushroom, *Macrolepiota procera*)	–	–	
Jacalin-related family (*see* Chapter 46, No. 22)			
GFL (mushroom, *Grifola frondosa*)	–	–	[70]
HRL (mushroom, *Hygrophorus russula*)	Man	Manα1–6Man	[75, 76]
Monocot (GNA-related) family (*see* Chapter 46, No. 31)			
MOL (mushroom, *Marasmius oreades*)	Man	Complex-type *N*-glycans	[90]
Fungal fruit-body ABL-like (α/β-sandwich Actinoporin-like) family (*see* Chapter 46, No. 31)			
ABA (mushroom, *Agaricus bisporus*)	Gal, GlcNAc	Galβ1–3GalNAc, GlcNAc	[140, 141]
BEL (mushroom, *Boletus edulis lectin*)	Gal, GlcNAc	Galβ1–3GalNAc	[142]
SRL (fungus, *Sclerotium rolfsii*)	GalNAc	Galβ1,3GalNAc	[136]
XCL (mushroom, *Xerocomus chrysenteron*)	GalNAc	Galβ1,3GalNAc	[143]
Fungal fruit-body PVL-like (7-bladed β-propeller) family (*see* Chapter 46, No. 33)			
PVL (mushroom, *Psathyrella velutina* lectin)	GlcNAc	GlcNAcβ1–4GalNAc	[34]

(continued)

Table 2 (continued)

Nomenclature (origin)	Monosaccharide specificity	Preferred glycan structure (terminal epitope)	Ref.
Fungal fruit-body fucose lectin AAL-like (six-bladed β-propeller) family (*see* Chapter 46, No. 34)			
AAL (mushroom, *Aleuria aurantia*)	Fuc	Fucα1-6GlcNAc, Fucα1-3(Galβ1-4)GlcNAc (Le^x)	[34, 144–146]
AFL (fungus, *Aspergillus fumigatus*)			[123]
AOL (fungus, *Aspergillus oryzae*)	Fuc	Fucα6GlcNAc(coreFuc) Fuca2Galβ4GlcNAc(H2)	[22]
Fungal fruit-body PhoSL-like family			
PhoSL (mushroom, *Pholiota squarrosa*)	Fuc	Fucα6GlcNAc (core Fuc)	[42]
RSL (fungus, *Rhizopus stolonifer*)	Fuc	Fucα6GlcNAc (core Fuc)	[41]
Fungal fruit-body LSL family			
LSL (mushroom, *Laetiporus sulphureus*)	LacNAc	Galβ1–4GlcNAcβ1–3Galβ1–4Glc	[86, 147–149]

4.2 Fungal Fruiting Bodies (ABL-Like) Family

Agaricus bisporus lectin (ABL) is a homotetramer protein with each monomer organized into an α/β-sandwich structure and a novel fold consisting of two β-sheets connected by a helix-loop-helix motif. This lectin has two distinct binding sites per monomer that recognize the different configurations of a single epimeric hydroxyl. This lectin family, fungal fruiting bodies (ABL-like) include *Xerocomus chrysenteron* lectin (XCL), *Sclerotium rolfsii* lectin (SRL), ABL, and *Boletus edulis* lectin (BEL). In addition, it has been suggested that XCL and ABL may have similar binding mechanism.

4.3 Fuc-Lectin (AAL-Like) Family

Fungal Fuc-specific lectins, such as *Aleuria aurantia* lectin (AAL), constitute a particular family of lectins that specifically recognize fucosylated glycans. AAL is a dimeric protein with each monomer being organized into a six-bladed β-propeller and a small antiparallel two-stranded β-sheet. The β-propeller fold is important for fucose recognition; five binding pockets are located between the propeller blades. The small β-sheet is involved in the dimerization process. *Aspergillus oryzae* lectin (AOL) belongs to this family, along with two lectins isolated from the bacteria *Ralstonia solanacearum* and *Burkholderia ambifaria*.

4.4 PhoSL-Like Family

Most lectins are multimeric, consisting of non-covalently associated subunits. This multimeric structure confers lectins their ability to agglutinate cells or form precipitates with glycoconjugates in a manner similar to antigen-antibody interactions. This unique group of proteins has provided powerful tools to explore a myriad of biological structures and processes. However, the mushroom *Pholiota squarrosa* produces a Fucα1-6GlcNAc (core Fuc)-binding lectin, PhoSL, a trimer protein of small subunits with 40 amino acids, and a chemically synthesized PhoSL peptide (4.5 kDa) corresponding to the determined sequence of the subunit exhibiting identical binding specificity to native PhoSL. In our preliminary investigation, some Strophariaceae mushrooms were found also to contain the same Fucα1-6GlcNAc (core Fuc)-specific lectins. The amino acidic sequences of PhoSL and *Rhizopus stolonifer* lectin (RSL) share 85 % homology. The structural analysis of these lectins has also been performed, and the results suggest that they do not belong to any known families. Therefore, here we propose a novel family, PhoSL-like family.

In the near future, the structures of many fungal lectins will be elucidated, thus enabling the lectins to classify into known families or novel families.

References

1. Damme EJMV, Peumans WJ, Pusztai A et al. (1998) In: Handbook of Plant Lectins: Properties and Biomedical Applications
2. Kilpatrick DC (2002) Animal lectins: a historical introduction and overview. Biochim Biophys Acta 1572:187–197
3. Diaz EM, Vicente-Manzanares M, Sacristan M et al (2011) Fungal lectin of *Peltigera canina* induces chemotropism of compatible Nostoc cells by constriction-relaxation pulses of cyanobiont cytoskeleton. Plant Signal Behav 6:1525–1536
4. Khan F, Khan M (2011) Fungal lectins: current molecular and biochemical perspectives. Int J Biol Chem 5:1–20
5. Varrot A, Basheer SM, Imberty A (2013) Fungal lectins: structure, function and potential applications. Curr Opin Struct Biol 23: 678–685
6. Singh RS, Bhari R, Kaur HP (2011) Current trends of lectins from microfungi. Crit Rev Biotechnol 31:193–210
7. Kobayashi Y, Ishizaki T, Kawagishi H (2004) Screening for lectins in wild and cultivated mushrooms from Japan and their sugar-binding specific cities. Int J Med Mush 6:113–125
8. Kaifu R, Osawa T (1979) Syntheses of *O*-beta-D-galactopyranosyl-(1 leads to 3)-0-(2-acetamido-2-deoxy-alpha(and -beta)-D-galactopyranosyl)-N-tosyl-L-serine and their interaction with D-galactose-binding lectins. Carbohydr Res 69:79–88
9. Nakamura-Tsuruta S, Kominami J, Kuno A et al (2006) Evidence that *Agaricus bisporus* agglutinin (ABA) has dual sugar-binding specificity. Biochem Biophys Res Commun 347:215–220
10. Sueyoshi S, Tsuji T, Osawa T (1985) Purification and characterization of four isolectins of mushroom (*Agaricus bisporus*). Biol Chem Hoppe Seyler 366:213–221
11. Ban M, Yoon HJ, Demirkan E et al (2005) Structural basis of a fungal galectin from *Agrocybe cylindracea* for recognizing sialoconjugate. J Mol Biol 351:695–706
12. Imamura K, Takeuchi H, Yabe R et al (2011) Engineering of the glycan-binding specificity of *Agrocybe cylindracea* galectin towards alpha(2,3)-linked sialic acid by saturation mutagenesis. J Biochem 150:545–552
13. Liu C, Zhao X, Xu XC et al (2008) Hemagglutinating activity and conformation of a lactose-binding lectin from mushroom

Agrocybe cylindracea. Int J Biol Macromol 42:138–144

14. Ngai PH, Zhao Z, Ng TB (2005) Agrocybin, an antifungal peptide from the edible mushroom *Agrocybe cylindracea*. Peptides 26: 191–196
15. Wang H, Ng TB, Liu Q (2002) Isolation of a new heterodimeric lectin with mitogenic activity from fruiting bodies of the mushroom *Agrocybe cylindracea*. Life Sci 70:877–885
16. Yagi F, Hiroyama H, Kodama S (2001) *Agrocybe cylindracea* lectin is a member of the galectin family. Glycoconj J 18:745–749
17. Debray H, Montreuil J (1989) *Aleuria aurantia* agglutinin. A new isolation procedure and further study of its specificity towards various glycopeptides and oligosaccharide. Carbohydr Res 185:15–26
18. Kochibe N, Furukawa K (1980) Purification and properties of a novel fucose-specific hemagglutinin of *Aleuria aurantia*. Biochemistry 19:2841–2846
19. Kuwabara N, Hu D, Tateno H et al (2013) Conformational change of a unique sequence in a fungal galectin from *Agrocybe cylindracea* controls glycan ligand-binding specificity. FEBS Lett 587:3620–3625
20. Ticha M, Dudova V, Kocourek J (1985) Studies on lectins. Lectins, Biol Biochem Clin Biochem 4:491
21. Ishida H, Hata Y, Kawato A et al (2004) Isolation of a novel promoter for efficient protein production in *Aspergillus oryzae*. Biosci Biotechnol Biochem 68:1849–1857
22. Ishida H, Moritani T, Hata Y et al (2002) Molecular cloning and overexpression of fleA gene encoding a fucose-specific lectin of *Aspergillus oryzae*. Biosci Biotechnol Biochem 66:1002–1008
23. Matsumura K, Higashida K, Hata Y et al (2009) Comparative analysis of oligosaccharide specificities of fucose-specific lectins from *Aspergillus oryzae* and *Aleuria aurantia* using frontal affinity chromatography. Anal Biochem 386:217–221
24. Matsumura K, Higashida K, Ishida H et al (2007) Carbohydrate binding specificity of a fucose-specific lectin from *Aspergillus oryzae*: a novel probe for core fucose. J Biol Chem 282:15700–15708
25. Horejsi V, Kocourek J (1978) Studies on lectins XXXVI. Properties of some lectins prepared by affinity chromatography on O-glycosyl polyacrylamide gels. Biochim Biophys Acta 538:299–315
26. Grahn E, Askarieh G, Holmner A et al (2007) Crystal structure of the *Marasmius oreades* mushroom lectin in complex with a xenotransplantation epitope. J Mol Biol 369:710–721
27. Grahn E, Holmner A, Cronet C et al (2004) Crystallization and preliminary X-ray crystallographic studies of a lectin from the mushroom *Marasmius oreades*. Acta Crystallogr D Biol Crystallogr 60:2038–2039
28. Grahn EM, Winter HC, Tateno H et al (2009) Structural characterization of a lectin from the mushroom *Marasmius oreades* in complex with the blood group B trisaccharide and calcium. J Mol Biol 390:457–466
29. Kirkeby S, Winter HC, Goldstein IJ (2004) Comparison of the binding properties of the mushroom *Marasmius oreades* lectin and *Griffonia simplicifolia* I-B isolectin to alpha-galactosyl carbohydrate antigens in the surface phase. Xenotransplantation 11:254–261
30. Loganathan D, Winter HC, Judd WJ et al (2003) Immobilized *Marasmius oreades* agglutinin: use for binding and isolation of glycoproteins containing the xenotransplantation or human type B epitopes. Glycobiology 13:955–960
31. Rempel BP, Winter HC, Goldstein IJ et al (2002) Characterization of the recognition of blood group B trisaccharide derivatives by the lectin from *Marasmius oreades* using frontal affinity chromatography-mass spectrometry. Glycoconj J 19:175–180
32. Tateno H, Goldstein IJ (2004) Partial identification of carbohydrate-binding sites of a Galalpha1,3Galbeta1,4GlcNAc-specific lectin from the mushroom *Marasmius oreades* by site-directed mutagenesis. Arch Biochem Biophys 427:101–109
33. Teneberg S, Alsen B, Angstrom J et al (2003) Studies on Galalpha3-binding proteins: comparison of the glycosphingolipid binding specificities of *Marasmius oreades* lectin and *Euonymus europaeus* lectin. Glycobiology 13:479–486
34. Cioci G, Mitchell EP, Chazalet V et al (2006) Beta-propeller crystal structure of *Psathyrella velutina* lectin: an integrin-like fungal protein interacting with monosaccharides and calcium. J Mol Biol 357:1575–1591
35. Endo T, Ohbayashi H, Kanazawa K et al (1992) Carbohydrate binding specificity of immobilized *Psathyrella velutina* lectin. J Biol Chem 267:707–713
36. Kobata A, Kochibe N, Endo T (1994) Affinity chromatography of oligosaccharides on *Psathyrella velutina* lectin column. Methods Enzymol 247:228–237
37. Kochibe N, Matta KL (1989) Purification and properties of an *N*-acetylglucosamine-specific

lectin from *Psathyrella velutina* mushroom. J Biol Chem 264:173–177

38. Ueda H, Kojima K, Saitoh T et al (1999) Interaction of a lectin from *Psathyrella velutina* mushroom with *N*-acetylneuraminic acid. FEBS Lett 448:75–80
39. Ueda H, Saitoh T, Kojima K et al (1999) Multi-specificity of a *Psathyrella velutina* mushroom lectin: heparin/pectin binding occurs at a site different from the *N*-acetylglucosamine/*N*-acetylneuraminic acid-specific site. J Biochem 126:530–537
40. Ueda H, Takahashi N, Ogawa H (2003) *Psathyrella velutina* lectin as a specific probe for *N*-acetylneuraminic acid in glycoconjugates. Methods Enzymol 363:77–90
41. Oda Y, Senaha T, Matsuno Y et al (2003) A new fungal lectin recognizing alpha(1-6)-linked fucose in the *N*-glycan. J Biol Chem 278:32439–32447
42. Kobayashi Y, Tateno H, Dohra H et al (2012) A novel core fucose-specific lectin from the mushroom *Pholiota squarrosa*. J Biol Chem 287:33973–33982
43. Mo H, Winter HC, Goldstein IJ (2000) Purification and characterization of a Neu5Aca2——6Galb1——4Glc/GlcNAc-specific lectin from the fruiting body of the polypore mushroom *Polyporus squamosus*. J Biol Chem 275:10623–10639
44. Zhao JK, Zhao YC, Li SH et al (2011) Isolation and characterization of a novel thermostable lectin from the wild edible mushroom *Agaricus arvensis*. J Basic Microbiol 51:304–311
45. Presant CA, Kornfeld S (1972) Characterization of the cell receptor for the *Agaricus bisporus* hemagglutinin. J Biol Chem 247:6937–6945
46. Kawagishi H, Nomura A, Yumen T et al (1988) Isolation and properties of a lectin from the fruiting bodies of Agaricus blazei. Carbohydr Res 183(1):150–154
47. Sage HJ, Vazquez JJ (1967) Studies on a hemagglutinin from the mushroom *Agaricus campestris*. J Biol Chem 242:120–125
48. Eifler R, Ziska P (1980) The lectins from Agaricus edulis. Isolation and characterization. Experientia 36:1285–1286
49. Ticha M, Dudova V, Kocourek J (1985) Lectins. Biol Biochem Clin Biochem 4:491
50. Sun H, Zhao CG, Tong X et al (2003) A lectin with mycelia differentiation and antiphytovirus activities from the edible mushroom *Agrocybe aegerita*. J Biochem Mol Biol 36:214–222
51. Jiang S, Chen Y, Wang M et al (2012) A novel lectin from *Agrocybe aegerita* shows high binding selectivity for terminal *N*-acetylglucosamine. Biochem J 443:369–378
52. Yagi F, Miyamoto M, Abe T et al (1997) Purification and carbohydrate-binding specificity of *Agrocybe cylindracea* lectin. Glycoconj J 14:281–288
53. Zhuang C, Murata T, Usui T et al (1996) Purification and characterization of a lectin from the toxic mushroom *Amanita pantherina*. Biochim Biophys Acta 1291:40–44
54. Antonyuk VO, Klyuchivska OY, Stoika RS (2010) Cytotoxic proteins of *Amanita virosa* Secr. mushroom: purification, characteristics and action towards mammalian cells. Toxicon 55:1297–1305
55. Yagi F, Tadera K (1988) Purification and characterization of lectin from *Auricularia polytricha*. Agric Biol Chem 52:2077–2079
56. Koyama Y, Katsuno Y, Miyoshi NHS, Mita T, Muto H, Isemura S, Aoyagi Y, Isemura M (2002) Apoptosis induction by lectin isolated from the mushroom *Boletopsis leucomelas* in U937 cells. Biosci Biotechnol Biochem 66: 784–789
57. Castillo C, Lara B, Cruz MJ et al (2013) Protein identification of two allergens of *Boletus edulis* causing occupational asthma. Am J Respir Crit Care Med 187:1146–1148
58. Horibe M, Kobayashi Y, Dohra H et al (2010) Toxic isolectins from the mushroom *Boletus venenatus*. Phytochemistry 71:648–657
59. Kretz O, Creppy EE, Dirheimer G (1991) Characterization of bolesatine, a toxic protein from the mushroom *Boletus satanas* Lenz and its effects on kidney cells. Toxicology 66:213–224
60. Kretz O, Reinbolt J, Creppy EE, Dirheimer G (1992) Properties of bolesatine, a translational inhibitor from *Boletus satanas* Lenz. Amino-terminal sequence determination and inhibition of rat mitochondrial protein synthesis. Toxicol Lett 64–65
61. Licastro F, Morini MC, Kretz O, Dirheimer G, Creppy EE, Stirpe FR, Articles L (1993) Mitogenic activity and immunological properties of bolesatine, a lectin isolated from the mushroom *Boletus satanas* Lenz. Int J Biochem 25:789–792
62. Kobayashi Y, Kobayashi K, Umehara K et al (2004) Purification, characterization, and sugar binding specificity of an *N*-glycolylneuraminic acid-specific lectin from the mushroom *Chlorophyllum molybdites*. J Biol Chem 279:53048–53055

63. Otta Y, Amano K, Nishiyama K, Ando A, Ogawa S, Nagata Y (2002) Purification and properties of a lectin from ascomycete mushroom, *Ciborinia camelliae.* Phytochemistry 60:103–107
64. Lyimo B, Funakuma N, Minami Y et al (2012) Characterization of a new alpha-galactosyl-binding lectin from the mushroom *Clavaria purpurea.* Biosci Biotechnol Biochem 76:336–342
65. Cooper DN, Boulianne RP, Charlton S, Farrell EM, Sucher A, Lu BC (1997) Fungal galectins, sequence and specificity of two isolectins from *Coprinus cinereus.* J Biol Chem 272:1514–1521
66. Walti MA, Walser PJ, Thore S et al (2008) Structural basis for chitotetraose coordination by CGL3, a novel galectin-related protein from *Coprinopsis cinerea.* J Mol Biol 379: 146–159
67. Jung EC, Kim KD, Bae CH et al (2007) A mushroom lectin from ascomycete *Cordyceps militaris.* Biochim Biophys Acta 1770: 833–838
68. Ko J, Hsu CI, Lin RH, Kao CL, Lin JY (1995) A new fungal immunomodulatory protein, FIP-fve isolated from the edible mushroom, *Flammulina velutipes* and its complete amino acid sequence. Eur J Biochem 228:244–249
69. Tsuda M (1979) Purification and characterization of a lectin from the mushroom *Flammulina velutipes.* J Biochem 86:1463–1468
70. Nagata Y, Yamashita M, Honda H et al (2005) Characterization, occurrence, and molecular cloning of a lectin from *Grifola frondosa:* jacalin-related lectin of fungal origin. Biosci Biotechnol Biochem 69:2374–2380
71. Thakur A, Rana M, Lakhanpal TN et al (2007) Purification and characterization of lectin from fruiting body of *Ganoderma lucidum*: lectin from *Ganoderma lucidum.* Biochim Biophys Acta 1770:1404–1412
72. Kawagishi H, Nomura A, Mizuno T et al (1990) Isolation and characterization of a lectin from *Grifola frondosa* fruiting bodies. Biochim Biophys Acta 1034:247–252
73. Kawagishi H, Mori H, Uno A et al (1994) A sialic acid-binding lectin from the mushroom *Hericium erinaceum.* FEBS Lett 340:56–58
74. Li Y, Zhang G, Ng TB et al (2010) A novel lectin with antiproliferative and HIV-1 reverse transcriptase inhibitory activities from dried fruiting bodies of the monkey head mushroom *Hericium erinaceum.* J Biomed Biotechnol 2010:716515
75. Imai Y, Hirono S, Matsuba H et al (2012) Degradation of target oligosaccharides by anthraquinone-lectin hybrids with light switching. Chem Asian J 7:97–104
76. Suzuki T, Sugiyama K, Hirai H et al (2012) Mannose-specific lectin from the mushroom *Hygrophorus russula.* Glycobiology 22:616–629
77. Zhao JK, Wang HX, Ng TB (2009) Purification and characterization of a novel lectin from the toxic wild mushroom *Inocybe umbrinella.* Toxicon 53:360–366
78. Wu Y, Wang H, Ng TB (2011) Purification and characterization of a lectin with antiproliferative activity toward cancer cells from the dried fruit bodies of *Lactarius flavidulus.* Carbohydr Res 346:2576–2581
79. Guillot J, Genaud L, Gueugnot J et al (1983) Purification and properties of two hemagglutins of the mushroom *Laccaria amethystina.* Biochemistry 22:5365–5369
80. Guillot JGM, Damez M, Dusser M (1991) Isolation and characterization of a lectin from the mushroom, *Lactarius deliciosus.* J Biochem 109:840–845
81. Giollant M, Guillot J, Damez M, Dusser M, Didier P, Didier E (1993) Characterization of a lectin from *Lactarius deterrimus* (research on the possible involvement of the fungal lectin in recognition between mushroom and spruce during the early stages of mycorrhizae formation). Plant Physiol 101:513–522
82. Sychrova H, Tichá M, Kocourek J (1985) Studies on lectins. LIX. Isolation and properties of lectins from fruiting bodies of *Xerocomus chrysenteron* and *Lactarius lignyotus.* Can J Biochem Cell Biol 63:700–704
83. Giollant M (1991) Les lectines deslactaires du groupe Dapetes: (*L. deliciosus, L. deterrimus, L. salmonicolor*). Purification, etude biochimique et specificite. Intervention des lectines dans les phenomenenes de reconnaisance molecularire au cours des evenements precoces de la mycorhization avec les coniferes associes. These de doctorat en pharmacie, Clemont-Ferrand
84. Jeune KH, Moon IJ, Kim MK et al (1990) Studies on lectins from Korean higher fungi; ‡W. A mitogenic lectin from the mushroom *Lentinus edodes.* Planta Med 56:592
85. Konska G, Guillot J, Dusser M et al (1994) Isolation and characterization of an *N*-acetyllactosamine-binding lectin from the mushroom *Laetiporus sulphureus.* J Biochem 16:519–523
86. Tateno H, Goldstein IJ (2003) Molecular cloning, expression, and characterization of novel

hemolytic lectins from the mushroom *Laetiporus sulphureus*, which show homology to bacterial toxins. J Biol Chem 278:40455–40463

87. Tsivileva OM, Nikitina VE, Makarov OE (2008) Characterization of an extracellular glycolipid from *Lentinus edodes* (Berk.) Sing [Lentinula edodes (Berk.) Pegler]. Mikrobiologiia 77:490–495
88. Goldstein IJ, Winter HC, Aurandt J et al (2007) A new alpha-galactosyl-binding protein from the mushroom *Lyophyllum decastes*. Arch Biochem Biophys 467:268–274
89. Ng TB, Lam YW (2002) Isolation of a novel agglutinin with complex carbohydrate binding specificity from fresh fruiting bodies of the edible mushroom *Lyophyllum shimeji*. Biochem Biophys Res Commun 290:563–568
90. Shimokawa M, Fukudome A, Yamashita R et al (2012) Characterization and cloning of GNA-like lectin from the mushroom *Marasmius oreades*. Glycoconj J 29:457–465
91. Ogawa S, Otta Y, Ando A, Nagata Y (2001) A lectin from an ascomycete mushroom, Melastiza chateri: no synthesis of the lectin in mycelial isolate. Biosci Biotechnol Biochem 65:686–689
92. Kawagishi H, Takagi J, Taira T et al (2001) Purification and characterization of a lectin from the mushroom *Mycoleptodonoides aitchisonii*. Phytochemistry 56:53–58
93. Matsumoto H, Natsume A, Ueda H et al (2001) Screening of a unique lectin from 16 cultivable mushrooms with hybrid glycoprotein and neoproteoglycan probes and purification of a novel *N*-acetylglucosamine-specific lectin from *Oudemansiella platyphylla* fruiting body. Biochim Biophys Acta 1526:37–43
94. Park JH, Ryu CS, Kim HN et al (2004) A sialic acid-specific lectin from the mushroom *Paecilomyces Japonica* that exhibits hemagglutination activity and cytotoxicity. Protein Pept Lett 11:563–569
95. Guillot J, Konska G (1997) Lectins in higher fungi. Biochem Syst Ecol 25:203–230
96. Wang SX, Zhang GQ, Zhao S et al (2013) Purification and characterization of a novel lectin with antiphytovirus activities from the wild mushroom *Paxillus involutus*. Protein Pept Lett 20:767–774
97. Entlicher G, Jesensk K, Jarosova-Dejlova L et al (1985) Isolation and characterization of a lectin from the shinkhorn mushroom (*Phallus impudicus* L. ex Pers). Lectins: Biol Biochem Clin Biochem 4:491
98. Zhang GQ, Sun J, Wang HX et al (2009) A novel lectin with antiproliferative activity from the medicinal mushroom *Pholiota adiposa*. Acta Biochim Pol 56:415–421
99. Kawagishi H, Abe Y, Nagata T et al (1991) A lectin from the mushroom *Pholiota aurivella*. Agric Biol Chem 55:2485–2489
100. Oguri S, Nagata Y (1994) Complete amino acid sequence of a lectin-related 16.5 kDa protein isolated from fruit bodies of a lectin-deficient strain of *Pleurotus cornucopiae*. Biosci Biotechnol Biochem 58:507–511
101. Oguri S, Yoshida M, Nagata Y (1994) Isolation, crystallization, and characterization of a 16.5 kDa protein from fruit bodies of a lectin-deficient strain of *Pleurotus cornucopiae*. Biosci Biotechnol Biochem 58:502–506
102. Yoshida M, Kato S, Oguri S et al (1994) Purification and properties of lectins from a mushroom, *Pleurotus cornucopiae*. Biosci Biotechnol Biochem 58:498–501
103. Mahajan RG, Patil SI, Mohan DR et al (2002) *Pleurotus Eous* mushroom lectin (PEL) with mixed carbohydrate inhibition and antiproliferative activity on tumor cell lines. J Biochem Mol Biol Biophys 6:341–345
104. Li YR, Liu QH, Wang HX et al (2008) A novel lectin with potent antitumor, mitogenic and HIV-1 reverse transcriptase inhibitory activities from the edible mushroom *Pleurotus citrinopileatus*. Biochim Biophys Acta 1780:51–57
105. Bera AK, Rana T, Das S et al (2011) Mitigation of arsenic-mediated renal oxidative stress in rat by *Pleurotus florida* lectin. Hum Exp Toxicol 30:940–951
106. Conrad F, Rudiger H (1994) The lectin from *Pleurotus ostreatus*: purification, characterization and Interaction with a phosphatase. Phytochemistry 36:277–283
107. Kawagishi H, Suzuki H, Watanabe H et al (2000) A lectin from an edible mushroom *Pleurotus ostreatus* as a food intake-suppressing substance. Biochim Biophys Acta 1474: 299–308
108. Suzuki T, Amano Y, Fujita M et al (2009) Purification, characterization, and cDNA cloning of a lectin from the mushroom *Pleurocybella porrigens*. Biosci Biotechnol Biochem 73:702–709
109. Wang H, Ng T (2003) Isolation of a novel *N*-acetylglucosamine-specific lectin from fresh sclerotia of the edible mushroom *Pleurotus tuber-regium*. Protein Expr Purif 29:156–160
110. Wang H, Ng TB, Liu Q (2003) A novel lectin from the wild mushroom *Polyporus adusta*. Biochem Biophys Res Commun 307:535–539
111. Rouf R, Stephens AS, Spaan L et al (2014) G/M cell cycle arrest by an *N*-acetyl-D-

glucosamine specific lectin from *Psathyrella asperospora*. Glycoconj J 31:61–70

112. Ueda H, Matsumoto H, Takahashi N et al (2002) *Psathyrella velutina* mushroom lectin exhibits high affinity toward sialoglycoproteins possessing terminal *N*-acetylneuraminic acid alpha 2,3-linked to penultimate galactose residues of trisialyl *N*-glycans. Comparison with other sialic acid-specific lectins. J Biol Chem 277:24916–24925
113. Hernandez E, Ortiz R, Lopez F et al (1993) Purification and characterization of galactose-specific lectin from *Psilocybe barrerae*. Phytochemistry 32:1209–1211
114. Richard, T (1995) Contribution a l'etude de la lectine champignon basidomy *Rigidoporus lignosus*. Purification, proprietes physicochimiques et location de la lectine et de ses sites d'alfinite. These de doctorat en science
115. Zhao S, Zhao Y, Li S et al (2010) A novel lectin with highly potent antiproliferative and HIV-1 reverse transcriptase inhibitory activities from the edible wild mushroom *Russula delica*. Glycoconj J 27:259–265
116. Chumkhunthod P, Rodtong S, Lambert SJ et al (2006) Purification and characterization of an *N*-acetyl-D-galactosamine-specific lectin from the edible mushroom Schizophyllum commune. Biochim Biophys Acta 1760: 326–332
117. Han CH, Liu QH, Ng TB et al (2005) A novel homodimeric lactose-binding lectin from the edible split gill medicinal mushroom *Schizophyllum commune*. Biochem Biophys Res Commun 336:252–257
118. Wang HX, Ng TB, Liu WK, Ooi VE, Chang ST (1995) Isolation and characterization of two distinct lectins with antiproliferative activity from the cultured mycelium of the edible mushroom *Tricholoma mongolicum*. Int J Pept Protein Res 46:508–513
119. Lin J, Chou T (1984) Isolation and characterization of a lectin from edible mushroom, *Volvariella volvacea*. J Biochem 96:35–40
120. Liu Q, Wang H, Ng TB (2006) First report of a xylose-specific lectin with potent hemagglutinating, antiproliferative and anti-mitogenic activities from a wild ascomycete mushroom. Biochim Biophys Acta 1760:1914–1919
121. Tronchin G, Esnault K, Sanchez M et al (2002) Purification and partial characterization of a 32-kilodalton sialic acid-specific lectin from *Aspergillus fumigatus*. Infect Immun 70:6891–6895
122. Garcia ML, Herreras JM, Dios E et al (2002) Evaluation of lectin staining in the diagnosis of fungal keratitis in an experimental rabbit model. Mol Vis 8:10–16
123. Houser J, Komarek J, Kostlanova N et al (2013) A soluble fucose-specific lectin from *Aspergillus fumigatus* conidia - structure, specificity and possible role in fungal pathogenicity. PLoS One 8:e83077
124. Rosen S, Bergstrom J, Karlsson KA et al (1996) A multispecific saline-soluble lectin from the parasitic fungus *Arthrobotrys oligospora*. Similarities in the binding specificities compared with a lectin from the mushroom agaricus bisporus. Eur J Biochem 238:830–837
125. Kossowska B, Lamer-Zarawska E, Olczak M et al (1999) Lectin from *Beauveria bassiana* mycelium recognizes Thomsen-Friedenreich antigen and related structures. Comp Biochem Physiol B Biochem Mol Biol 123:23–31
126. Khan F, Ahmad A, Khan MI (2007) Purification and characterization of a lectin from endophytic fungus *Fusarium solani* having complex sugar specificity. Arch Biochem Biophys 457:243–251
127. al-Mahmood S, Colin S, Bonaly R (1991) *Kluyveromyces bulgaricus* yeast lectins. Isolation of two galactose-specific lectin forms from the yeast cell wall. J Biol Chem 266, 20882–20887
128. Al-Mahmood S, Giummelly P, Bonaly R et al (1988) *Kluyveromyces bulgaricus* yeast lectins. Isolation of N-acetylglucosamine and galactose-specific lectins: their relation with flocculation. J Biol Chem 263:3930–3934
129. Bhowal J, Guha AK, Chatterjee BP (2005) Purification and molecular characterization of a sialic acid specific lectin from the phytopathogenic fungus *Macrophomina phaseolina*. Carbohydr Res 340:1973–1982
130. Francis F, Jaber K, Colinet F et al (2011) Purification of a new fungal mannose-specific lectin from *Penicillium chrysogenum* and its aphicidal properties. Fungal Biol 115: 1093–1099
131. Wang H, Ng TB (2005) First report of an arabinose-specific fungal lectin. Biochem Biophys Res Commun 337:621–625
132. Albores S, Mora P, Cerdeiras MP et al (2014) Screening for lectins from basidiomycetes and isolation of *Punctularia atropurpurascens* lectin. J Basic Microbiol 54:89–96
133. Nagre NN, Chachadi VB, Sundaram PM et al (2010) A potent mitogenic lectin from the mycelia of a phytopathogenic fungus, *Rhizoctonia bataticola*, with complex sugar specificity and cytotoxic effect on human ovarian cancer cells. Glycoconj J 27:375–386
134. Chachadi VB, Inamdar SR, Yu LG et al (2011) Exquisite binding specificity of *Sclerotium rolfsii* lectin toward TF-related *O*-linked mucin-type glycans. Glycoconj J 28:49–56

135. Inbar J, Chet I (1994) A newly isolated lectin from the plant pathogenic fungus *Sclerotium rolfsii:* purification, characterization and role in mycoparasitism. Microbiology 140(Pt 3): 651–657
136. Leonidas DD, Swamy BM, Hatzopoulos GN et al (2007) Structural basis for the carbohydrate recognition of the *Sclerotium rolfsii* lectin. J Mol Biol 368:1145–1161
137. Feng L, Sun H, Zhang Y et al (2010) Structural insights into the recognition mechanism between an antitumor galectin AAL and the Thomsen-Friedenreich antigen. FASEB J 24:3861–3868
138. Yang N, Li DF, Feng L et al (2009) Structural basis for the tumor cell apoptosis-inducing activity of an antitumor lectin from the edible mushroom *Agrocybe aegerita.* J Mol Biol 387:694–705
139. Kadirvelraj R, Grant OC, Goldstein IJ et al (2011) Structure and binding analysis of *Polyporus squamosus* lectin in complex with the Neu5Ac{alpha}2-6Gal{beta}1-4GlcNAc human-type influenza receptor. Glycobiology 21:973–984
140. Carrizo ME, Capaldi S, Perduca M et al (2005) The antineoplastic lectin of the common edible mushroom (*Agaricus bisporus*) has two binding sites, each specific for a different configuration at a single epimeric hydroxyl. J Biol Chem 280:10614–10623
141. Ismaya WT, Rozeboom HJ, Weijn A et al (2011) Crystal structure of *Agaricus bisporus* mushroom tyrosinase: identity of the tetramer subunits and interaction with tropolone. Biochemistry 50:5477–5486
142. Bovi M, Carrizo ME, Capaldi S et al (2011) Structure of a lectin with antitumoral properties in king bolete (*Boletus edulis*) mushrooms. Glycobiology 21:1000–1009
143. Birck C, Damian L, Marty-Detraves C et al (2004) A new lectin family with structure similarity to actinoporins revealed by the crystal structure of *Xerocomus chrysenteron* lectin XCL. J Mol Biol 344:1409–1420
144. Fujihashi M, Peapus DH, Kamiya N et al (2003) Crystal structure of fucose-specific lectin from *Aleuria aurantia* binding ligands at three of its five sugar recognition sites. Biochemistry 42:11093–11099
145. Fujihashi M, Peapus DH, Nakajima E et al (2003) X-ray crystallographic characterization and phasing of a fucose-specific lectin from *Aleuria aurantia.* Acta Crystallogr D Biol Crystallogr 59:378–380
146. Wimmerova M, Mitchell E, Sanchez JF et al (2003) Crystal structure of fungal lectin: six-bladed beta-propeller fold and novel fucose recognition mode for *Aleuria aurantia* lectin. J Biol Chem 278:27059–27067
147. Angulo I, Acebron I, de las Rivas B et al. (2011) High-resolution structural insights on the sugar-recognition and fusion tag properties of a versatile beta-trefoil lectin domain from the mushroom *Laetiporus sulphureus.* Glycobiology 21, 1349–1361
148. Mancheno JM, Tateno H, Goldstein IJ et al (2004) Crystallization and preliminary crystallographic analysis of a novel haemolytic lectin from the mushroom *Laetiporus sulphureus.* Acta Crystallogr D Biol Crystallogr 60:1139–1141
149. Mancheno JM, Tateno H, Goldstein IJ et al (2005) Structural analysis of the *Laetiporus sulphureus* hemolytic pore-forming lectin in complex with sugars. J Biol Chem 280: 17251–17259

Chapter 3

The "White Kidney Bean Incident" in Japan

Haruko Ogawa and Kimie Date

Abstract

Lectin poisoning occurred in Japan in 2006 after a TV broadcast that introduced a new diet of eating staple foods with powdered toasted white kidney beans, seeds of *Phaseolus vulgaris*. Although the method is based on the action of a heat-stable α-amylase inhibitor in the beans, phaseolamin, more than 1,000 viewers who tried the method suffered from acute intestinal symptoms and 100 people were hospitalized. Lectins in the white kidney beans were suspected to be the cause of the trouble. We were asked to investigate the lectin activity remaining in the beans after the heat treatment recommended on the TV program. The test suggested that the heat treatment was insufficient to inactivate the lectin activity, which, combined with our ignorance of carbohydrate signaling in the intestine, was the cause of the problem.

Key words White kidney bean diet, Phaseolamin, Mass poisoning, Roles of intestinal glycan ligands, Lectin as antinutrients, Intestinal glycobiology

1 Background

To lose weight quickly, most people decrease their intake of foods containing carbohydrates and lipids. However, they soon notice that maintaining the diet is much harder than they expected. Therefore, it is no wonder that so many dietary methods, drugs, and apparatuses for slimming down are advertised, especially in advanced nations where food is plentiful. However, how many of these methods are scientifically supported?

Lectins, enzyme inhibitors of animal digestive enzymes, tannins, and phytates have been called "antinutrients" because they decrease the absorption of nutrients. These antinutrients are classified as self-defense proteins in plants, and most legume lectins are considered to cause intestinal inflammation in animals when consumed without heat treatment.

Red kidney beans (*Phaseolus vulgaris*) are known to contain five tetramer lectins (designated as "PHA" from phytohemagglutinin) consisting of combinations of E (erythroagglutinin) and L (leukoagglutinin) subunits: E_4, E_3L_1, E_2L_2, E_1L_3, and L_4, of which

Jun Hirabayashi (ed.), *Lectins: Methods and Protocols*, Methods in Molecular Biology, vol. 1200,
DOI 10.1007/978-1-4939-1292-6_3,

E-PHA (or PHA-E4) and L-PHA (or PHA-L4) are the most commonly used in structural carbohydrate chemistry owing to their unique specificities for multiantennary complex-type *N*-glycans [1]. White kidney beans also contain two α-amylase inhibitors: one is the heat-labile α-AI-u [2] and the other is the heat-stable phaseolamin or α-AI-s [3]. A TV program on a novel dietary method depending on the action of phaseolamin by heat treatment to inactivate the endogenous lectins of white kidney beans was broadcast in 2006 in Japan.

2 White Kidney Bean Diet and Response of Viewers

One Saturday evening in a holiday week, a weekly TV program promoting health and beauty introduced a new method to lose weight utilizing a common white kidney bean (*P. vulgaris*). The method was simple: toasting white kidney beans for 3 min in a frying pan, grinding the beans to a powder using a coffee mill, and then dusting the bean powder on cooked rice. Starting the next evening, complaints from viewers of serious intestinal symptoms, such as severe diarrhea and vomiting, were sent to the television station. Two days after the broadcast, the television station officially declared that the symptoms were due to the TV program and warned viewers not to go on a diet using white kidney beans. Finally, more than 1,000 people, including 122 patients who were hospitalized for treatment, and 467 people who consulted a doctor, reported intestinal troubles including vomiting and/or diarrhea to the TV station.

3 Investigation of the Cause

The television station asked the Glycoscience Institute of Ochanomizu University to identify the component that caused the intestinal symptoms. Because it was suspected that a kidney bean lectin (PHA) that remained active after inadequate heat treatment was a causative factor in the intestinal symptoms, our mission was to elucidate whether and how much PHAs remained active after the heat treatment demonstrated by the TV program.

3.1 Stability of PHA Activities in Heat Treatment of White Kidney Beans

The samples examined were commercial white kidney beans from two typical cultivars in Japan, Shirohana and Oofuku, each purchased from three different dealers, plus toasted beans supplied by two patients. The raw kidney beans were toasted for 2–20 min in a frying pan (Figs. 1 and 2) and ground in a coffee mill as demonstrated on the television program. The powder was extracted with 9 vol. (v/w) of phosphate-buffered saline (PBS) with stirring

Fig. 1 Heat treatments of white kidney beans by toasting (*left*) and boiling in hot water (*right*)

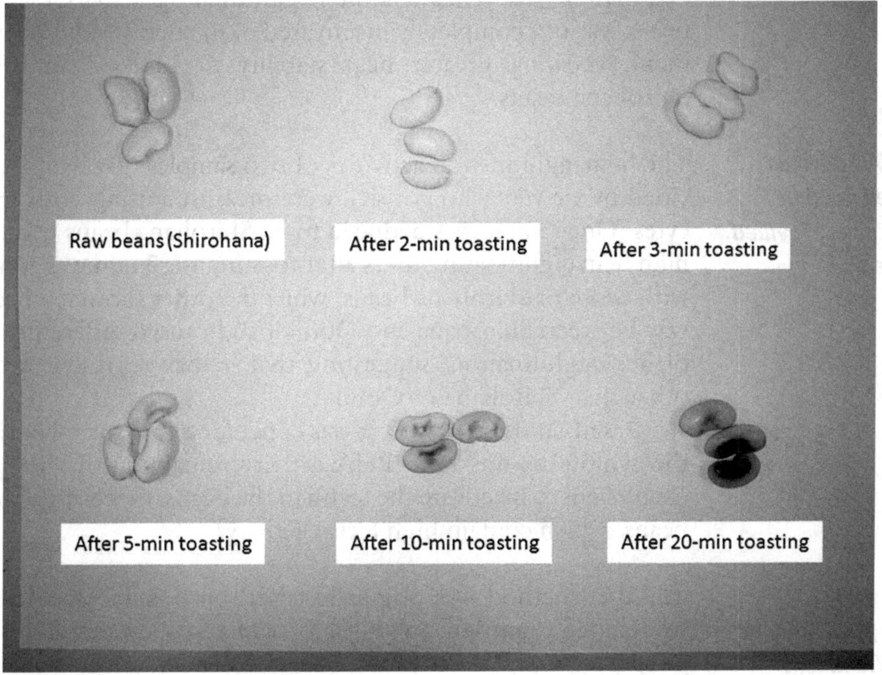

Fig. 2 White kidney beans and heat treatment

overnight at 4 °C. Alternatively, the raw beans were soaked in cold water overnight, and then boiled in hot water for 3–60 min as kidney beans are traditionally cooked in Japan; then, they were mashed using a mortar and pestle and extracted with PBS. After centrifuging the extract, the hemagglutinating activity of each supernatant was measured (*see* Chapter 4).

The PHA activity was measured by a hemagglutination test using guinea pig and human erythrocytes. The hemagglutinating activities of the samples from different dealers were not significantly different. All extracts of Shirohana beans showed very strong hemagglutinating activity for guinea pig erythrocytes, more than 2,000 times that of Oofuku beans. The hemagglutination titer of Shirohana beans was not changed by toasting for 3 min, but it did start to decrease after 5 min, and was almost lost (<1 %) after 20 min, indicating that the lectin activity is still present after the 3-min heat treatment. On the other hand, the hemagglutinating activity of Shirohana beans was inactivated to less than 5 % by boiling in hot water for 3 min and was completely lost after 10 min. In contrast, extracts of the Oofuku beans obtained from three manufacturers did not show detectable hemagglutinating activity toward guinea pig erythrocytes and showed very low hemagglutinating activities toward trypsin-treated human type A erythrocytes (titer $\leq$8). Thus, the hemagglutinating activity of Oofuku beans was not completely inactivated even after 20-min heat treatment, showing greater heat stability of PHAs than those of Shirohana beans.

3.2 PHA Activities Remained in White Kidney Beans Provided by Viewers

The hemagglutination activities of two samples that had been provided by viewers who got sick were measured using both erythrocytes. One sample was deduced to be Shirohana beans because the high hemagglutinating titers after toasting for 3 min was consistent with those of Shirohana beans, while the other showed a titer halfway between Shirohana and Oofuku and caused different patterns of hemagglutination, suggesting that it may represent a cultivar other than Shirohana or Oofuku.

Based on these results, it was concluded that the main cause of the symptoms was high PHA activity remaining due to heating insufficient to inactivate the lectin in the beans, especially Shirohana beans, which contain high lectin activity.

3.3 Stability of α-Amylase Inhibitor in Heat Treatment

The diet method was originally based on possible activity of the heat-stable α-amylase inhibitor of white kidney beans, phaseolamin. We examined whether the heat treatment affected the inhibitory activity of phaseolamin or not. Under the experimental condition, the starch-degrading activity of porcine pancreatic α-amylase (PPA) was suppressed to 71–72 % in the presence of extracts of both Oofuku and Shirohana white kidney beans, and the inhibition activities of PPA were initially decreased by 14–21 % within 3 min by toasting or boiling the beans, but thereafter, the inhibition activities were unchanged until they had been toasted or boiled for 20 min. The initial decrease of the inhibitor activity is attributable to inactivation of the α-AI-u in kidney beans coexisting with phaseolamin [2, 4], and the stability of phaseolamin was confirmed by a 20-min heat treatment.

4 A Possible Mechanism of the Reaction

The intestinal luminal brush-border membranes of mammals are heavily glycosylated with *N*- and *O*-linked oligosaccharides produced by enterocytes and goblet cells. Because the glycans of most membrane proteins including transporters, brush-border enzymes, and hormone and growth factor receptors are known to function as reservoirs of endogenous growth factors, hormones, and bacteria, exogenous lectins or other carbohydrate-binding proteins may interfere with signaling events biologically important for maintaining health [5, 6].

4.1 Effects of PHAs Suggested by Small Animal Studies

Small animal studies have shown that PHAs can damage the gut, which may lead to various nutritional disorders [7–9], partly due to their relatively high resistance to degradation by digestive enzymes. More than 90 % of PHAs are stable and functionally active in the small intestine of rats [10]. Therefore, PHAs that bind to the glycans of rat intestinal cells in vivo disrupt the brush borders of duodenal and jejunal enterocytes [11–13]. Furthermore, PHAs change the intestinal microbial flora [14], hormone balance [15], and metabolic reactions [16] in rats, and consequently, acute intestinal problems may result [5]. Based on these reports, it is considered that when significant amounts of active PHAs are ingested, they react with the glycans of intestinal mucous membranes to cause acute mucosal disarrangement and impede functions of the gut.

4.2 Roles of Glycans in Nutrient Transportation in the Intestine

It was reported that glycosylation of transporters in the intestinal brush-border membrane is important for absorption of cholesterol [17], multiple vitamins [18], iron [19], and anions [17, 18, 20]. Exogenous lectins and carbohydrate-binding proteins may change the activities of these transporters and the balance of metabolic reactions by interacting with the glycans of the transporters. Because the transporters play essential roles in intestinal nutrient absorption, oral ingestion of a lectin in excess amounts may cause intestinal disorder.

4.3 Emerging Roles of Intestinal Glycans in Digestive Enzymes

Both porcine and human pancreatic α-amylases exhibit carbohydrate-specific binding activity [21, 22]. They bind to *N*-glycosylated glycoproteins of the duodenal brush-border membrane to control starch degradation and glucose transport [22]. Moreover, bovine and human trypsin and trypsinogen exhibit a high affinity to carbohydrates [23], and most pancreatic exocrine enzymes have a unique carbohydrate-binding specificity (our unpublished results). It is assumed that the carbohydrate-specific interactions between pancreatic enzymes and glycoligands in the luminal mucus at the brush border of enterocytes play modulatory roles in the metabolic reactions that maintain biological homeostasis, though not all of the roles have been clarified.

4.4 A Possible Mechanism of Reaction from Accumulated Knowledge

As described in Subheadings 4.1–4.3, the luminal side of the intestinal tract is not only the site of food digestion and absorption but also a hot spot of stimuli input by exocrine proteins that have carbohydrate-binding properties and exogenously taken-up lectins via common carbohydrate ligands. In this context, the incident described in this chapter was a disturbance of a modulation system that is normally performed by endogenous glycan-binding proteins related to digestive functions, caused by an exogenous white kidney bean lectin crossing the brush-border membrane.

5 Conclusions: What We Learned from the Incident

The mass poisoning by a legume lectin taught us that the signal input of a plant lectin through intestinal carbohydrates can induce acute stress to human digestive organs. "Do not eat raw beans" is a traditional admonition in Japan to prevent intestinal problems. However, we did not know the precise mechanism that induced the inflammation in the human intestine, especially molecular level responses caused by interactions between intestinal glycan ligands and carbohydrate-binding proteins of both exogenous and endogenous origins. The incident demonstrates the significance of glycobiological research on intestinal glycan receptors and carbohydrate-binding proteins to understand reactions that are ordinarily carried out at the brush-border membrane in order to maintain nutritional homeostasis through the digestion and absorption systems. This research will contribute basic knowledge to overcome several worldwide problems including metabolic syndromes and starvation.

References

1. Cummings RD, Etzler ME (2009) Chapter 45: antibodies and lectins in glycan analysis. In: C.R. Varki A, Cummings RD, Esko JD, Freeze H, Stanley P, Bertozzi C, Hart G, Etzler M (eds) Essentials of glycobiology, 2nd edn. Cold Spring Harbor Laboratory Press, Cold Spring Harbor, NY
2. Yamaguchi H (1993) Isolation and characterization of the subunits of a heat-labile alpha-amylase inhibitor from Phaseolus vulgaris white kidney bean. Biosci Biotechnol Biochem 57:297–302
3. Marshall JJ, Lauda CM (1975) Purification and properties of phaseolamin, an inhibitor of alpha-amylase, from the kidney bean, *Phaseolus vulgaris*. J Biol Chem 250:8030–8037
4. Wato S, Kamei K, Arakawa T, Philo JS, Wen J, Hara S, Yamaguchi H (2000) A chimera-like alpha-amylase inhibitor suggesting the evolution of *Phaseolus vulgaris* alpha-amylase inhibitor. J Biochem 128:139–144
5. Pusztai A, Bardocz S, Ewen SW (2008) Uses of plant lectins in bioscience and biomedicine. Front Biosci 13:1130–1140
6. Pusztai A, Ewen SW, Grant G, Peumans WJ, Van Damme EJ, Coates ME, Bardocz S (1995) Lectins and also bacteria modify the glycosylation of gut surface receptors in the rat. Glycoconj J 12:22–35
7. Banwell JG, Boldt DH, Meyers J, Weber FL Jr (1983) Phytohemagglutinin derived from red kidney bean (*Phaseolus vulgaris*): a cause for intestinal malabsorption associated with bacterial overgrowth in the rat. Gastroenterology 84:506–515
8. Pusztai A (1996) Characteristics and consequences of interactions of lectins with the intestinal mucosa. Arch Latinoam Nutr 44:10S–15S

9. Van Damme EJM, Peumans WJ, Pusztai A, Bardocz S (1998) Handbook of plant lectins: properties and biomedical applications. *Phaseolus vulgaris* lectin. Wiley, Chichester, UK
10. Pusztai A, Bardocz S (1996) Biological effects of plant lectins on the gastrointestinal tract: metabolic consequences and applications. Trends Glycosci Glycotechnol 8:149–165
11. Pusztai A, Clarke EM, King TP (1979) The nutritional toxicity of *Phaseolus vulgaris* lectins. Proc Nutr Soc 38:115–120
12. King TP, Pusztai A, Clarke EM (1980) Immunocytochemical localization of ingested kidney bean (*Phaseolus vulgaris*) lectins in rat gut. Histochem J 12:201–208
13. Rossi MA, Mancini Filho J, Lajolo FM (1984) Jejunal ultrastructural changes induced by kidney bean (*Phaseolus vulgaris*) lectins in rats. Br J Exp Pathol 65:117–123
14. Banwell JG, Howard R, Cooper D, Costerton JW (1985) Intestinal microbial flora after feeding phytohemagglutinin lectins (*Phaseolus vulgaris*) to rats. Appl Environ Microbiol 50:68–80
15. Jordinson M, Goodlad RA, Brynes A, Bliss P, Ghatei MA, Bloom SR, Fitzgerald A, Grant G, Bardocz S, Pusztai A, Pignatelli M, Calam J (1999) Gastrointestinal responses to a panel of lectins in rats maintained on total parenteral nutrition. Am J Physiol 276:G1235–G1242
16. Baintner K, Kiss P, Bardocz S, Pusztai A (2004) Effect of orally administered plant lectins on intestinal liquor accumulation and amylase activity in rats. Acta Physiol Hung 91:73–81
17. Wang LJ, Wang J, Li N, Ge L, Li BL, Song BL (2011) Molecular characterization of the NPC1L1 variants identified from cholesterol low absorbers. J Biol Chem 286:7397–7408
18. Ghosal A, Subramanian VS, Said HM (2011) Role of the putative N-glycosylation and PKC-phosphorylation sites of the human sodium-dependent multivitamin transporter (hSMVT) in function and regulation. Biochim Biophys Acta 1808:2073–2080
19. Zhao N, Enns CA (2013) *N*-linked glycosylation is required for transferrin-induced stabilization of transferrin receptor 2, but not for transferrin binding or trafficking to the Cell Surface. Biochemistry 52:3310–3319
20. Hayashi H, Yamashita Y (2012) Role of *N*-glycosylation in cell surface expression and protection against proteolysis of the intestinal anion exchanger SLC26A3. Am J Physiol Cell Physiol 302:C781–C795
21. Matsushita H, Takenaka M, Ogawa H (2002) Porcine pancreatic alpha-amylase shows binding activity toward *N*-linked oligosaccharides of glycoproteins. J Biol Chem 277: 4680–4686
22. Asanuma-Date K, Hirano Y, Le N, Sano K, Kawasaki N, Hashii N, Hiruta Y, Nakayama K, Umemura M, Ishikawa K, Sakagami H, Ogawa H (2012) Functional regulation of sugar assimilation by *N*-glycan-specific interaction of pancreatic alpha-amylase with glycoproteins of duodenal brush border membrane. J Biol Chem 287:23104–23118
23. Takekawa H, Ina C, Sato R, Toma K, Ogawa H (2006) Novel carbohydrate-binding activity of pancreatic trypsins to N-linked glycans of glycoproteins. J Biol Chem 281:8528–8538

Chapter 4

Hemagglutination (Inhibition) Assay

Kotone Sano and Haruko Ogawa

Abstract

The hemagglutination assay is a simple and easy method to obtain semi-quantitative data on the sugar binding and specificity of a lectin. An active lectin agglutinates erythrocytes by recognizing a carbohydrate on the cell surface and forming a cross-linked network in suspension. By serially diluting the lectin in a 96-well microtiter plate and adding a constant quantity of erythrocytes, the lectin activity can be estimated.

Key words Hemagglutination, Hemagglutinin, Agglutinin, Carbohydrate recognition, Erythrocyte

1 Introduction

Lectins are also called agglutinins because they cause agglutination by binding to the cell surface. Lectins were first described in 1888 by Stillmark who was working with an extract from beans of the castor tree, *Ricinus communis*. He first reported in his doctoral thesis that the castor bean extract contains a lectin, which he called hemagglutinin, because it caused hemagglutination [1]. In 1919, James B. Sumner crystallized a pure hemagglutinin from jack bean (*Canavalia ensiformis*) for the first time and named it concanavalin A [2]. Nearly two decades later, Sumner and Howell reported that concanavalin A agglutinated cells such as erythrocytes and yeasts and also precipitated glycogen in solution [3]. They further showed that hemagglutination by concanavalin A was inhibited by sucrose, demonstrating the sugar-binding specificity of lectins for the first time. These results suggested that the hemagglutination induced by concanavalin A is a consequence of a reaction of the plant protein with carbohydrates on the surface of the erythrocyte. Moreover, it was reported that lectins distinguish types of cells [4] because lectins bind to a specific kind of sugar, and sugar chains on the cell surface differ according to the cell type and animal species. Here, we describe the hemagglutination assay that can easily detect lectins and screen their sugar-binding specificity.

Jun Hirabayashi (ed.), *Lectins: Methods and Protocols*, Methods in Molecular Biology, vol. 1200,
DOI 10.1007/978-1-4939-1292-6_4, © Springer Science+Business Media New York 2014

2 Materials

Prepare all solutions using analytical grade reagents and store at 4 °C.

1. Fresh rabbit erythrocytes.
2. Phosphate-buffered saline (PBS): 150 mM NaCl, 20 mM Na-phosphate (pH 7.2), or an appropriate buffer (e.g., Tris–HCl containing 5 mM $CaCl_2$ for Ca^{2+}-requiring lectins, or MEPBS: 20 mM PBS (pH 7.2) with 4 mM β-mercaptoethanol, 2 mM EDTA for lectins such as galectins that require a reducing agent).
3. Trypsin (Sigma): store at −20 °C.
4. Glutaraldehyde.
5. Glycine: store at room temperature.
6. Neuraminidase (from *Vibrio cholerae*, Roche).
7. Incubation buffer for neuraminidase: 0.1 M acetate buffer containing 1 mM $CaCl_2$ (pH 5.5).
8. Carbohydrate solution(s) for inhibition assay: dissolve at 10 mg/mL in the same buffer used for the hemagglutination assay.
9. A 96-well microtiter plate, V shape or U shape (Costar) (*see* **Note 1**).

3 Methods

Hemagglutination is commonly assayed by serial dilution using human or rabbit erythrocytes. Occasionally, erythrocytes that have been treated with trypsin or sialidase are used because these procedures change the agglutination activity by exposing or removing the sugar residue(s) that the lectin recognizes (*see* **Note 2**).

3.1 Isolation of Erythrocytes

1. Centrifuge 10 mL of rabbit blood at 500 × *g* for 5 min at 4 °C (*see* **Notes 3** and **4**). The procedures can be scaled down depending on the blood volume.
2. Remove the plasma and white cell ghost layer at the top of the pellet, and resuspend the cells in 50 mL of cold PBS by gently aspirating and expelling them with a pipette, then centrifuge at 500 × *g* for 5 min at 4 °C.
3. Remove the supernatant and repeat **step 2** three times. Store the pelleted erythrocytes at 4 °C until use. Use within 24 h.

3.2 Preparation of Trypsinized Glutaraldehyde-Fixed Erythrocytes

1. Suspend the isolated erythrocytes (ca. 5 mL) in 100 mL of 0.1 % (w/v) trypsin solution in PBS. The procedures can be scaled down depending on the quantity of erythrocytes.
2. Incubate at 37 °C for 1 h with an occasional shake.
3. Resuspend in 50 mL of cold PBS and wash by centrifugation as described in Subheading 3.1.
4. Suspend trypsinized erythrocytes in 25 mL of 1 % glutaraldehyde in PBS (*see* **Note 5**).
5. Incubate at room temperature for 1 h with continuous gentle shaking.
6. Centrifuge at 500 ×*g* for 5 min at 4 °C, remove the supernatant, and wash twice with 25 mL of 0.1 M glycine in PBS.
7. Wash with PBS and store as 10 % (v/v) erythrocyte suspension at 4 °C.

3.3 Preparation of Sialidase-Treated Erythrocytes

1. Mix the suspension of untreated erythrocytes (10 %, v/v) with an equal volume of incubation buffer (pH 5.5) containing neuraminidase (1 unit/mL) (*see* **Notes 6** and **7**).
2. Incubate at 37 °C for 1 h with occasional shaking.
3. Wash with cold PBS and centrifuge at 4 °C as described in Subheading 3.1 and store as a 10 % suspension at 4 °C.

3.4 Measurement of Hemagglutination

The following method is based on the report by Kawsar et al. (*see* ref. 5).

1. Add 25 μL of lectin-containing solutions at physiological salt concentration in the left well of a horizontal row of a microtiter plate.
2. Make a series of twofold dilutions (2^{-1}, 2^{-2}, 2^{-3} ...) of the sample in PBS in the horizontal row. Make a control well using PBS without the lectin.
3. Add 25 μL of PBS to each well.
4. Mix the plate by swirling gently on the top of the bench in a circular or figure "8" motion for 10 s. (For this step, the mix function of the microplate reader can be used.)
5. Add 4 % (v/v) rabbit erythrocyte suspension prepared as above.
6. Mix the plate by gently swirling on the top of the bench in a circular or figure "8" motion for 10 s. Do *not* use the mix function of the microplate reader.
7. Allow the hemagglutination reaction to proceed for 30–60 min at room temperature.

8. Determine whether agglutination has occurred using the naked eye or a microscope. Hemagglutination activity is expressed by a titer defined as the reciprocal of the maximum dilution giving positive hemagglutination. If a lectin solution causes hemagglutination at a maximum dilution of 2^{-9} (=1/512), the titer of the original solution is 512 (*see* **Notes 8** and **9**).

3.5 Measurement of Hemagglutination Inhibition by Carbohydrates

The following method is based on the report by Ueno et al. (*see* ref. 6).

1. Add 25 μL of the carbohydrate solution to the left well of a horizontal row of microtiter plate.
2. Make a series of twofold dilution (2^{-1}, 2^{-2}, 2^{-3} ...) of the carbohydrate solution in PBS in the horizontal line. Make a control well with PBS.
3. Add 25 μL of a fixed titer of lectin solution (titer 2 is most often used) to each well.
4. Mix the contents of the wells and incubate for 30 min at room temperature.
5. Add 4 % (v/v) rabbit erythrocyte suspension prepared above.
6. Mix the plate by gently swirling on the top of the bench in a circular or figure "8" motion. Do *not* use the mix function of the microplate reader.
7. Allow the hemagglutination reaction to proceed for 1–2 h at room temperature.
8. Determine whether agglutination has occurred by naked eye or microscope. The hemagglutination inhibition activity is expressed as the lowest concentration of the sugar solution that completely inhibits hemagglutination.

4 Notes

1. A plate with U-shaped wells is sometimes used instead of one with V-shaped wells for assays using untreated erythrocytes. Choose the type of plate that clearly distinguishes positive and negative agglutination, which may differ depending on the type of erythrocytes.
2. Trypsinized glutaraldehyde-fixed rabbit erythrocytes are often used because the trypsin treatment renders cells more sensitive to agglutination and glutaraldehyde fixation allows cells to be stored for a long period (>1 month).
3. Hemolysis may occur if the centrifuge speed is too high. Centrifugation is usually operated at lower than 2,000 × *g*.

4. Erythrocytes from various species (chicken, guinea pig, goose, horse, cow, sheep, rabbit, and human types A, B, AB, and O) are widely used.
5. Use gloves of the appropriate type and length, eye protection, and a face mask when measuring glutaraldehyde.
6. Neuraminidase is most active at pH 5.0–5.5 (e.g., 0.1 M acetate buffer); it retains about 30–50 % activity at pH 7.0 in a phosphate buffer.
7. Neuraminidase from *Arthrobacter ureafaciens* or *Clostridium perfringens* can also be used.
8. A schematic representation of hemagglutination is shown in Fig. 1.
9. An example of the hemagglutination assay using a 96-well microtiter plate is shown in Fig. 2.

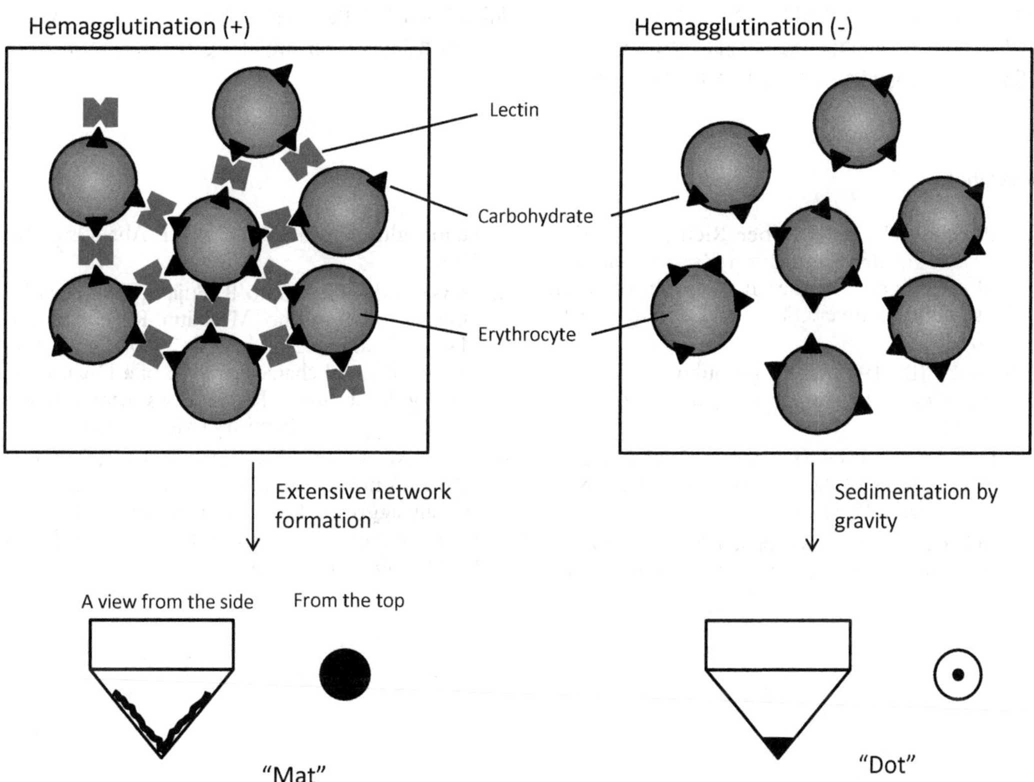

Fig. 1 Schematic representation of hemagglutination. Erythrocytes agglutinate in the presence of a lectin that recognizes sugar chains on the cell surface. They form an apparent "*mat*" on the microtiter plate; if the lectin is inactive or an inhibitor is present, erythrocytes simply form a sediment by gravity and appear as a "*dot*" on the plate

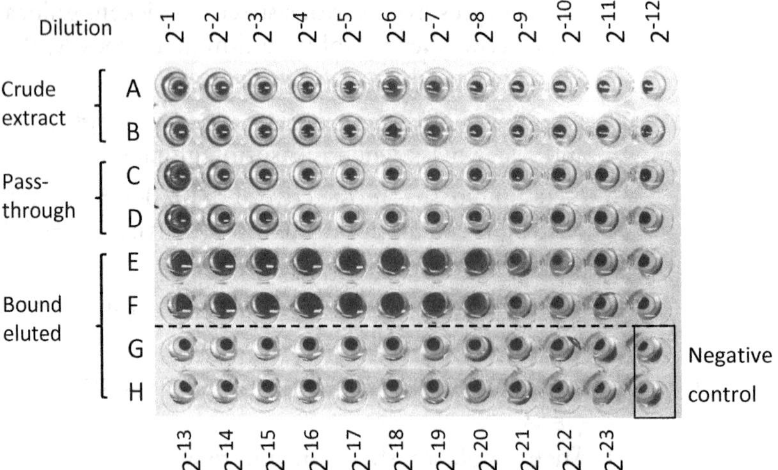

Fig. 2 Typical hemagglutination assay result. Hemagglutination assay of the extract of *Pleurocybella porrigens* and fractions of lectin-affinity chromatography. Twofold dilutions of sample of crude extract (*line A, B*), pass-through (*line C, D*) and bound eluted (*line E–H*) fractions from affinity chromatography were prepared and mixed with trypsin-treated human erythrocytes. Bound eluted fraction causes hemagglutination up to the 2^{-8} dilution. The concentrations of undiluted samples were; crude extract, 2.35 mg/mL; pass-through fraction, 1.58 mg/mL; bound eluted fraction, 0.512 mg/mL

References

1. Stillmark PH (1888) Über Ricin, ein giftiges Ferment aus den Samen von *Ricinus comm.* L. und einigen anderen Euphorbiaceen. Kaiserliche Universität zu Dorpat, Tartu, Estonia
2. Sumner JB (1919) The globulins of the jack bean, *Canavalia ensiformis*. J Biol Chem 37: 137–142
3. Sumner JB, Howell SF (1936) Identification of hemagglutinin of Jack bean with concanavalin A. J Bacteriol 32:227–237
4. Landsteiner K, Raubitschek H (1907) Beobachtungen über Hämolyse und Hämagglutination. Zbl Bakt I Abt Orig 45: 600–607
5. Kawsar SM, Matsumoto R, Fujii Y, Yasumitsu H, Dogasaki C, Hosono M, Nitta K, Hamako J, Matsui T, Kojima N, Ozeki Y (2009) Purification and biochemical characterization of a D-galactose binding lectin from Japanese sea hare (*Aplysia kurodai*) eggs. Biochemistry (Mosc) 74:709–716
6. Ueno M, Ogawa H, Matsumoto I, Seno N (1991) A novel mannose-specific and sugar specifically aggregatable lectin from the bark of the Japanese pagoda tree (*Sophora japonica*). J Biol Chem 266:3146–3153

Chapter 5

Preparation of Affinity Adsorbents and Purification of Lectins from Natural Sources

Kimie Date and Haruko Ogawa

Abstract

Lectins are purified by affinity chromatography to take advantage of their carbohydrate-specific interactions. Highly efficient affinity adsorbents are powerful tools to obtain homogeneous lectins with distinct specificities. Here, we describe three methods to prepare affinity adsorbents by immobilizing carbohydrates or glycoconjugates on agarose gel beads. Because the ligands are immobilized via a stable and nonionic linkage under mild conditions, the adsorbents possess high binding capacity for lectins with low nonspecific adsorption and can withstand repeated use. The procedures require neither specialized techniques and apparatus nor highly toxic compounds. Using these adsorbents, many plant and animal lectins can be purified in a few steps.

Key words Lectin purification, Affinity adsorbent, Affinity chromatography, Immobilization of carbohydrates, Glycamyl agarose, Heparin-agarose, Glycoprotein-agarose

1 Introduction

Lectins are widely distributed among plants, microbes, and animals, and their importance has increased with progress in understanding the functions of carbohydrates in various biological systems. Plant and microbe lectins are widely used to detect and analyze specific carbohydrate structures [1]. Animal lectins are important in recognition of many intra- and/or extracellular events through carbohydrate-specific interactions, as described by N. Kawasaki (*see* Chapter 29). Lectins are purified by affinity chromatography on immobilized carbohydrate adsorbents, and effective and simple procedures to purify the lectins from natural sources are essential. In this chapter, convenient protocols to prepare adsorbents to purify lectins using optional carbohydrate and commercially available agarose gels (Sepharose) are described.

The quality of an affinity adsorbent determines the success or the failure of a purification. Affinity adsorbents for lectins, using

Jun Hirabayashi (ed.), *Lectins: Methods and Protocols*, Methods in Molecular Biology, vol. 1200, DOI 10.1007/978-1-4939-1292-6_5, © Springer Science+Business Media New York 2014

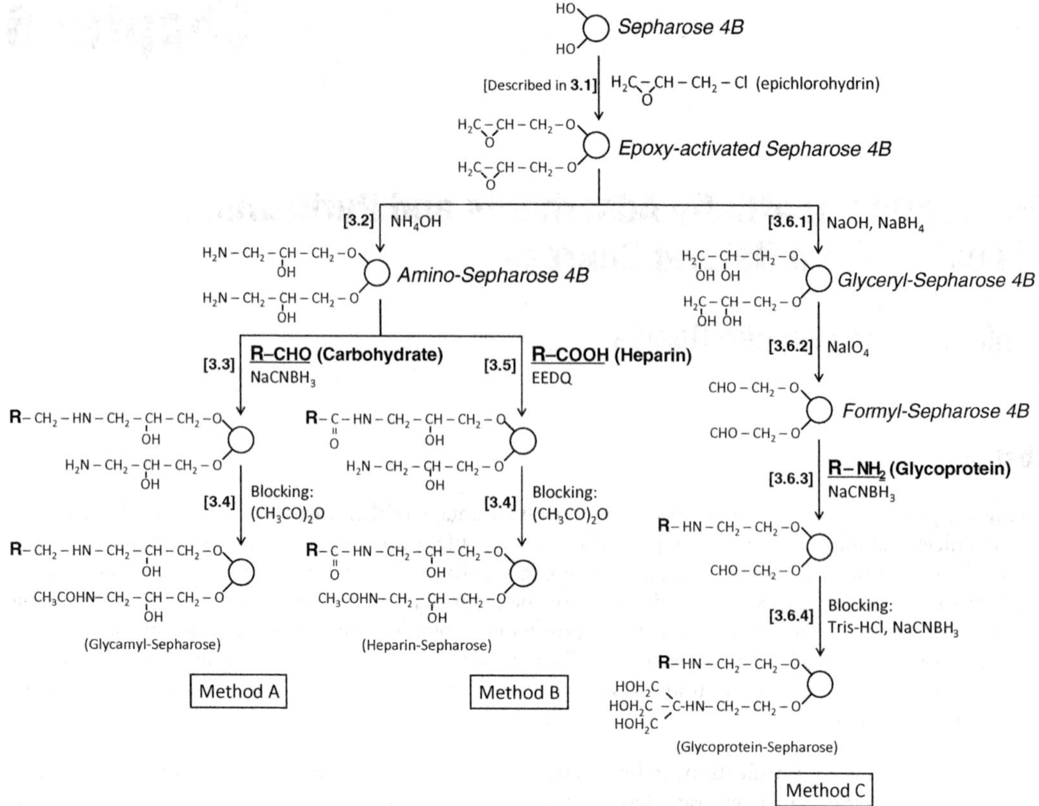

Fig. 1 Preparation of the three kinds of lectin adsorbents described in this chapter

agarose as a carrier, are usually prepared by one of the two methods: the cyanogen bromide (CNBr) method or the epoxy method [2]. To circumvent the disadvantages of the CNBr method, such as low ligand concentration, instability, and nonspecific adsorption [2, 3], we use the epoxy-activation method [4], followed by three immobilization procedures utilizing formyl (-CHO), carboxyl (-COOH), or amino (-NH_2) groups of carbohydrates and glycoprotein ligands (*see* Fig. 1). The prepared adsorbents are stable and have hydrophilic spacers; moreover, these methods for lectin purification are technically feasible in most laboratories without synthetic technologies or special apparatus. By the methods in this chapter, anyone can prepare affinity adsorbents for lectin purification in a few days with surprisingly simple protocols and little cost. The immobilization methods are also applicable to the gels which are suitable for high-performance liquid affinity chromatography (HP-AFC) (*see* **Note 1**).

Although large columns are frequently used for purification in open-column chromatography, the adsorbents we developed achieve the high purification performance with a small bed volume (~6 mL, *see* Table 1) [5–7]. Hundreds of milligrams of plant lectins

Table 1
Binding capacity of plant lectins to various glycamyl-Sepharose through purification procedures

Affinity adsorbent	Column size (cm × cm I.D.)	Column volume (ml)	Lectin	Yield (mg)	Binding capacity (mg/ml gel)
Maltamyl-Sepharose 4B	1.5 × 2.2	5.7	Concanavalin A (ConA)	870	153
Lactamyl-Sepharose 4B	6.3 × 0.86	3.7	*Arachis hypogaea* agglutinin (AHA)	250	67
			Soybean agglutinin (SBA)	100	27
			Ricinus communis agglutinin (RCA)	420	114
$(GlcNAc)_4$-Sepharose 4B	1.4 × 2.0	4.4	Wheat germ agglutinin (WGA)	160	36
GalN-Sepharose 4B (*see* **Note 12**)	5.3 × 0.9	3.4	Soybean agglutinin (SBA)	488	145

Three kinds of glycamyl-Sepharose are prepared as described in Subheadings 3.1–3.4. Adapted from ref. 5 with modifications

have been purified by single-step affinity chromatography using the adsorbents [5]. The adsorbents were successfully used to isolate various lectins and heparin-binding proteins of animal origin by affinity chromatography [5–11].

2 Materials

Prepare all solutions using distilled water and analytical grade reagents. Prepare and store all preparations at room temperature (unless otherwise indicated).

2.1 Epoxy-Activation of Sepharose Gel

This step is a common pre-step for Methods A–C.

1. Sepharose 4B (GE Healthcare): Wash ~20 mL of the resin in a glass filter funnel with 50 volumes of water to remove alcohol and other preservatives in the stock suspension of commercial Sepharose before use (*see* **Note 1**).
2. Sodium borohydride ($NaBH_4$): Store under N_2 in an airtight container. Prepare 0.2 % (w/v) $NaBH_4$ in 2 M NaOH just before use.
3. Epichlorohydrin: A toxic organochlorine compound. Handle under a fume hood.

Hereafter optional

4. 1.3 M sodium thiosulfate ($Na_2S_2O_3$).
5. Phenolphthalein solution: Dissolve 0.1 g phenolphthalein in 100 mL of 90 % (v/v) ethanol.

2.2 Amination of Epoxy-Activated Sepharose Gel

1. Epoxy-activated Sepharose 4B (*see* Subheading 2.1).
2. 25 % (v/v) ammonium hydroxide (concentrated ammonium hydroxide).
3. $NaBH_4$: Store under N_2 in an airtight container.

Hereafter optional

4. 95 % (v/v) ethanol.
5. 2,4,6-Trinitrobenzene sulfonic acid (TNBS).
6. 0.1 M sodium tetraborate decahydrate ($Na_2B_4O_7 \cdot 10H_2O$).

2.3 Immobilization of Free Carbohydrates to Amino-Sepharose by Reductive Amination (Method A)

1. Amino-Sepharose (*see* Subheadings 2.1 and 2.2).
2. Carbohydrates: Maltose, lactose, and heparin, melibiose, and D-galactosamine hydrochloride ($GalNH_2$–HCl).
3. 0.2 M dipotassium hydrogen phosphate (K_2HPO_4).
4. Sodium cyanoborohydride ($NaCNBH_3$): Store under N_2 in an airtight container.

Hereafter optional

5. 95 % (v/v) ethanol.
6. TNBS.
7. $Na_2B_4O_7 \cdot 10H_2O$.

2.4 Blocking of Unreacted Amino Groups in the Gel

1. Carbohydrate or heparin-coupled Sepharose (Subheading 2.3 or Subheading 2.5).
2. 0.2 M sodium acetate.
3. Acetic anhydride.
4. 0.1 M NaOH.
5. Phosphate-buffered saline (PBS): 10 mM phosphate buffer, 0.15 M NaCl, pH 7.0 (*see* **Note 2**).

Hereafter optional

6. 95% (v/v) ethanol.
7. TNBS.
8. $Na_2B_4O_7 \cdot 10H_2O$.

2.5 Immobilization of Heparin to Amino-Sepharose by Amide Formation (Methods B)

1. Amino-Sepharose 4B (*see* Subheadings 2.1 and 2.2).
2. Heparin.
3. *N*-Ethoxycarbonyl-2-ethoxy-1,2-dihydroquinoline (EEDQ).
4. Ethanol (>99.5 %, v/v).
5. 1 M NaCl.
6. 0.2 M sodium acetate.
7. Acetic anhydride.

8. 0.1 M NaOH.
9. PBS (*see* **Note 2**).

Hereafter optional

10. 95% (v/v) ethanol.
11. TNBS.
12. $Na_2B_4O_7 \cdot 10H_2O$.

2.6 Immobilization of Glycoproteins to Formyl-Sepharose by Reductive Amination (Method C)

1. Epoxy-activated Sepharose (*see* Subheading 2.1).
2. $NaBH_4$: Prepare 0.33 mg/mL solution in 0.1 M NaOH immediately before use.
3. Sodium metaperiodate ($NaIO_4$): Store under N_2 in an airtight container. Prepare 0.1 M solution immediately before use and protect from light.
4. Glycoprotein: For example fetuin (lyophilized powder). Prepare 2 mg/mL solution in PBS. Prepare just before use [12, 13].
5. $NaCNBH_3$: Store under N_2 in an airtight container.
6. 1 M Tris–HCl buffer, pH 7.5.
7. PBS (*see* **Note 2**).

2.7 Purification of Lectins by Affinity Chromatography

1. Natural lectin source (e.g., legume seeds or bovine kidney).
2. Acetone or methanol.
3. 0.15 M NaCl.
4. Ammonium sulfate (*see* **Note 3**): Pulverize lump using a mortar and pestle.
5. Affinity adsorbent (*see* Subheadings 2.1–2.6).
6. Equilibrating and washing buffer: For example PBS (*see* **Note 2**).
7. Eluting solution: 0.1–0.2 M specific sugar in washing buffer (*see* **Note 4**).

3 Methods

Carry out all procedures at room temperature unless otherwise specified. Never allow gel to become desiccated. Remove 10–20 mg of the gel after each step and wash to check the progress of each reaction.

3.1 Epoxy Activation of Sepharose Gel

This step is the starting point for the preparation of all the adsorbents described in this chapter. It takes 3 h to carry out the following procedures [4]. It will take further 30 min if the *optional* **step 6** is performed.

1. Place 20 g Sepharose 4B gel (suction-dried, wet gel) in a 200-mL round-bottom flask. Use a flask with at least ten times the volume of the gel to mix enough by shaking and to prevent the gel from spilling out of the flask during the reaction.
2. Add 13 mL of 0.2 % (w/v) $NaBH_4$-2 M NaOH, 3 mL of epichlorohydrin, and 30 mL of water and swirl to mix.
3. Put a double layer of clingfilm over the top of the flask and poke a hole in the film.
4. Incubate at 40 °C for 2 h with vigorous shaking so that the gel does not settle on the bottom of the flask.
5. Transfer the gel to a glass filter funnel and wash with 100 volumes of water. Epoxy-activated Sepharose is unstable and should be reacted successively.

Hereafter optional

6. Check the introduction of epoxy groups to the gel. Using a portion of the reacted gel, check the completion of the reaction before proceeding to the next step. Place 10–20 mg suction-dried gel in a 1.5-mL tube. Add 1 mL of 1.3 M sodium thiosulfate and dissolve, and then add one drop of phenolphthalein solution. The phenolphthalein develops a red color due to OH hydrolysis of epoxy groups if the activation process was successful (*see* **Note 6**).

3.2 Amination of Epoxy-Activated Sepharose Gel

It takes 2 h to carry out the following procedures [14]. It will take further 4 h if the optional **step 6** is performed.

1. Put 20 g of epoxy-activated Sepharose 4B gel (suction-dried, wet gel described in Subheading 3.1) in a 200-mL round-bottom flask.
2. Add 1.5 vol. (30 mL) of 25 % (v/v) ammonium hydroxide and 60 mg of $NaBH_4$ (*see* **Note 7**).
3. Put a double layer of clingfilm over the top of the flask and poke a hole in the film.
4. Incubate the suspension at 40 °C for 1.5 h with vigorous shaking.
5. Transfer the gel to a glass filter funnel and wash extensively with water (1–2 L) until the ammonium odor is gone.

   Amino-Sepharose can be stored in 10 % (v/v) ethanol containing 0.2 % (w/v) NaN_3 at 4 °C for several months.

Hereafter optional

6. Detect amino groups in the gel (TNBS color test): Wash 10–20 mg and suction-dried gel with 10 mL of 95 % (v/v) ethanol and 80 mL of water on a small Buchner funnel (8ϕ). Suspend the gel in 1 mL of 0.1 M $Na_2B_4O_7 \cdot 10H_2O$ containing

15 mg TNBS. Incubate for more than 4 h at room temperature, and then wash the gel with 50 mL of water. An orange color indicates the presence of amino groups in the Sepharose gel (*see* **Note 8**).

3.3 Immobilization of Free Carbohydrates to Amino-Sepharose by Reductive Amination (Method A)

This method immobilizes the reducing terminal group of carbohydrates using to amino-Sepharose by reductive amination (Method A in Fig. 1). The adsorbents are generally called "glycamyl-Sepharose," e.g., lactamyl-Sepharose and maltamyl-Sepharose. Coupling of free carbohydrates with amino-Sepharose takes 2–3 days [5].

1. Put 4 g amino-Sepharose (suction-dried, wet gel described in Subheading 3.2) in a 100-mL round-bottom flask. Add 3 mL of 0.2 M K_2HPO_4 containing carbohydrates, e.g., 104 mg of disaccharides [5], 320 mg of chitooligosaccharides, $GlcNAc_{5-6}$ [15], or 120 mg of heparin [16], mix, and then add 51 mg of $NaCNBH_3$.
2. Put a double layer of clingfilm over the top of the flask and poke a hole in the film.
3. Incubate at 40 °C with vigorous shaking for 2–3 days (*see* **Note 9**).
4. Transfer the gel to a glass filter funnel and wash with 50 volumes of water.

Hereafter optional

5. TNBS color test to confirm the coupling of carbohydrates (*see* **Note 8**): Do the TNBS color test on the suction-dried gel. Confirm that the strong orange color of amino-Sepharose after an epoxy-activation procedure (Subheading 3.2) becomes lighter due to the coupling of carbohydrates.

3.4 Blocking of Unreacted Amino Groups in the Gel

To block the ionic interaction which is exhibited by free amino groups in the gel, unreacted amino groups are blocked by *N*-acetylation. It takes 3 h to carry out the following procedures [14].

1. Put 4 g carbohydrate-coupled Sepharose (suction-dried, wet gel described in Subheading 3.3) in 4 mL of 0.2 M sodium acetate in a 100-mL glass beaker and cool the suspension on ice with gentle stirring using a magnetic stirrer.
2. Add 2 mL of acetic anhydride to the beaker on ice.
3. Put a double layer of clingfilm over the top of the beaker.
4. Incubate for 30 min on ice stirring gently enough not to grind the gel.
5. Add another 2 mL of acetic anhydride, and stir the gel for another 30 min at room temperature.
6. Transfer the gel to a glass filter funnel and wash subsequently with 50 volumes each of water, 0.1 M NaOH, water, and PBS (*see* **Note 2**).

7. Repeat **steps 1–5**. By carrying out *N*-acetylation treatment twice, the remaining amino groups in Sepharose gel are completely blocked.

Hereafter optional

8. Perform TNBS color test (*see* **Note 8**): The gel becomes white after re *N*-acetylation of the remaining amino groups.

To quickly check the binding efficiency of the prepared affinity adsorbent for lectin (glycamyl-Sepharose, and/or glycoprotein-Sepharose), run the mini-column test (*see* **Note 5**).

3.5 Immobilization of Heparin to Amino-Sepharose by Amide Formation (Method B)

Heparin or other glycans with carboxyl groups are immobilized to amino-Sepharose by condensation to form amide groups. Using EEDQ which is nontoxic and less expensive than $NaCNBH_3$ as a condensation agent immobilizes heparin in higher amounts in less time than reductive amination (Method B in Fig. 1) [10].

Epoxy activation and amination procedures are the same as described in Subheadings 3.1 and 3.2, respectively. To carry out the following procedures, it takes 12–13 h.

1. Put 10 g of amino-Sepharose 4B (suction-dried, wet gel described in Subheading 3.2) in 200-mL round-bottomed flask.
2. Add 6 mL of water containing 200 mg of heparin and mix well. Then add 200 mg of EEDQ dissolved in 4 mL of ethanol (*see* **Note 10**).
3. Put a double layer of clingfilm over the top of the flask and poke a hole in the film.
4. Incubate at 40 °C overnight with vigorous shaking.
5. Transfer the gel to a glass filter funnel and wash sequentially with 200 mL of 40 % (v/v) ethanol, 100 % ethanol, 1 M NaCl, and water.

Hereafter optional

6. Perform TNBS color test (*see* **Note 8**): The orange color of amino-Sepharose becomes lighter when heparin is coupled to amino groups of the gel.

The blocking procedure is the same as described in Subheading 3.4.

The immobilized heparin concentration can be measured by quantifying the hexosamine after acid hydrolysis of heparin-Sepharose [16]. The amount of heparin immobilized is 7.2–10 mg/g suction-dried gel by this method [16, 17].

3.6 Immobilization of Glycoproteins to Formyl-Sepharose by Reductive Amination (Method C)

This method immobilizes glycoproteins using their amino groups to formyl-derivatized Sepharose by reductive amination (Method C in Fig. 1) [6, 11–13, 18, 19].

3.6.1 Preparation of Glyceryl-Sepharose (24 h)

1. Put 10 g of epoxy-activated Sepharose (suction-dried, wet gel as described in Subheading 3.1) in a 200-mL round-bottom flask.
2. Add 60 mL of 0.33 mg/mL $NaBH_4$-0.1 M NaOH and mix.
3. Put a double layer of clingfilm over the top of the flask and poke a hole in the film.
4. Incubate the reaction mixture at 40 °C for 24 h with vigorous shaking.
5. Transfer the gel to a glass filter funnel and wash with 1 L of water.

   Glyceryl-Sepharose can be stored in 10 % (v/v) ethanol containing 0.2 % (w/v) NaN_3 at 4 °C for several months.

3.6.2 Preparation of Formyl-Sepharose (1.5 h)

1. Put 10 g of glyceryl-Sepharose (suction-dried, wet gel) in a 200-mL round-bottomed flask.
2. Add 15 mL of 0.1 M $NaIO_4$.
3. Put a double layer of clingfilm over the top of the flask and poke a hole in the film.
4. Incubate the mixture in the dark at 4 °C for 1 h with shaking.
5. Transfer the gel to a glass filter funnel and wash with 1 L of water.

   Formyl-Sepharose is unstable and should be reacted successively.

3.6.3 Coupling of Glycoprotein with Formyl-Sepharose (Overnight to 2 Days)

1. Put 20 mg of fetuin (glycoprotein) in 10 mL PBS and measure the absorbance at 280 nm.
2. Put 10 g of formyl-Sepharose 4B gel (as suction-dried gel) in a 100-mL round-bottom flask.
3. Add 10 mL fetuin solution and mix.
4. Add 120 mg of $NaCNBH_3$ to the flask and mix.
5. Put a double layer of clingfilm over the top of the flask and poke a hole in the film.
6. Incubate at 4 °C overnight to 2 days with gentle shaking.
7. Centrifuge the reaction mixture in a 40-mL centrifuge tube to separate gel (about 1,800×g) for 5 min at 4 °C and transfer the supernatant to a separate tube. Measure the absorbance of the supernatant at A280.
8. Determine glycoprotein immobilized on the gel. Calculate the amount of glycoprotein immobilized to the gel from the difference between the A280 of the supernatants before and after incubation. The amount of immobilized glycoprotein is expressed as moles (or mg) per gram of suction-dried gel.

3.6.4 Blocking of Unreacted Formyl Groups in the Gel (1 h)

1. Put 10 g fetuin-Sepharose from Subheading 3.6.3 (suction-dried gel) in a 200-mL round-bottom flask.
2. Add 40 mL of 1 M Tris–HCl (pH 7.5) containing 120 mg of $NaCNBH_3$ to the flask.
3. Put a double layer of clingfilm over the top of the flask and poke a hole in the film.
4. Incubate at 4 °C for more than 1 h or overnight with shaking.
5. Transfer the gel to a glass filter funnel and wash with 1–2 L of cold PBS (*see* **Note 2**).

To quickly check the binding efficiency of the prepared affinity adsorbent for lectin (glycamyl-Sepharose, and/or glycoprotein-Sepharose), run the mini-column test (*see* **Note 5**).

3.7 Purification of Lectins by Affinity Chromatography

Store all reagents and carry out all procedures at 4 °C. Monitor the fractions eluted from affinity chromatography at A280 nm.

3.7.1 Preparation of Crude Lectins from Seeds (See Ref. 6 on Preparation of Crude Lectins from Bovine Kidney) (1–2 Days)

1. Delipidate plant seeds (soybean, castor bean, jack bean, wheat germ) by homogenizing in cold acetone or methanol using a homogenizer. Dry the seed powder (overnight).
2. Add 3–4 volumes (300–400 mL) of 0.15 M NaCl to 100 g of the seed powder.
3. Stir gently with a magnetic stirrer at 4 °C overnight to extract proteins.
4. Centrifuge the extract to separate insoluble residue (about 10,000 × *g*) at 4 °C for 30 min. Remove the supernatant.
5. Re-extract the precipitate with 200 mL of 0.15 M NaCl for 2 h at 4 °C and centrifuge under the same conditions. Combine the supernatants and discard the precipitate.
6. Add ammonium sulfate powder to the extract to 90 % (v/v) saturation, and then gently stir overnight at 4 °C.
7. Centrifuge the extract at about 12,800 × *g* for 30 min at 4 °C and remove the supernatant.
8. Dissolve the precipitate in a small volume of water. Dialyze against water with frequent changes of the outer solution and lyophilize (crude lectin).

3.7.2 Affinity Chromatography (1–2 Days)

1. Pack the affinity adsorbent, lactamyl-, maltamyl-, $(GlcNAc)_4$-, and GalN-Sepharose 4B, into the empty open column (*see* Table 1) (also *see* **Note 11**).
2. Equilibrate the affinity column with 4–5 column volumes of equilibrating buffer at a slow flow of about 1 drop/20 s (1 mL/10 min).
3. Dissolve the crude lectin preparations [*Arachis hypogaea* agglutinin (AHA), soybean agglutinin (SBA), *Ricinus communis*

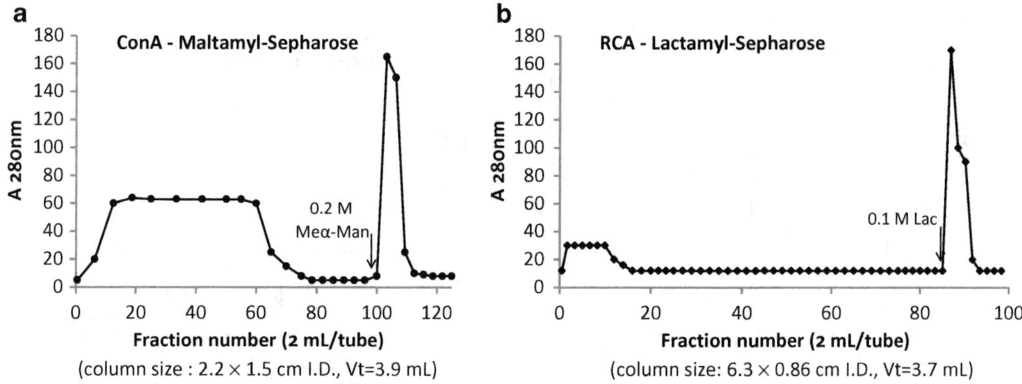

Fig. 2 Purification of Con A on a maltamyl-Sepharose column (**a**) and RCA on a lactamyl-Sepharose column (**b**). Crude lectin solutions are applied on a respective glycamyl-Sepharose 4B column. After the columns are washed with the buffer, lectins are eluted with each specific sugar at the points indicated by arrows in the figures. (**a**) Running buffer: 0.1 M acetate buffer (pH 6.4) containing 1 mM $MgCl_2$, 1 mM $MnCl_2$, and 1 mM $CaCl_2$. Elution is performed with 0.2 M methyl α-D-mannoside in the same buffer. (**b**) Running buffer: PBS. Elution is performed with 0.1 M lactose in PBS. Adapted from ref. 5 with modifications

agglutinin (RCA), concanavalin A (Con A), and wheat germ agglutinin (WGA)] separately in PBS at concentrations of 6–10 % (w/v) (*see* **Note 2**).

4. After removing the insoluble material by centrifugation at about 1,300×*g* for 15 min at 4 °C, load each crude lectin solution onto a separate column.
5. Wash the columns with the washing buffer until the A280 nm of the effluents in a 1-cm cell is less than 0.05.
6. Elute each bound lectin by eluting solution. Collect the effluent in a fraction collector at 1 mL/fraction and measure the effluent at A280 nm. Do not stop or change the flow rate to prevent leakage of the bound protein (examples are shown in Figs. 2 and 3 for animal lectins).
7. Collect peak fractions, dialyze against water, and then lyophilize. Weigh each purified lectin and calculate adsorption capacities of lectin adsorbents (Table 1; also *see* **Note 12**).
8. Check the purity of the lectin by SDS-polyacrylamide gel electrophoresis and measure the activity by hemagglutination assay (*see* Chapter 4).

4 Notes

1. Sepharose is available in three different agarose concentrations, 2 %, 4 %, and 6 %, which are designated Sepharose 2B, Sepharose 4B, and Sepharose 6B, respectively. Because the exclusion limits for proteins and mechanical strength of Sepharose differ by the agarose concentration, Sepharose 4B is mostly commonly recommended for a wide range of proteins.

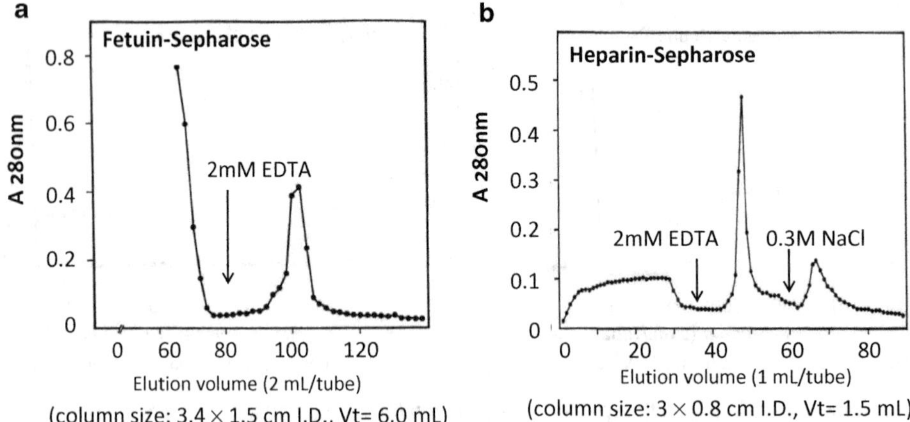

Fig. 3 Chromatography of bovine kidney lectin on (**a**) fetuin- and (**b**) heparin-Sepharose 4B columns. (**a**) Crude extract from bovine kidney dissolved in 5 mM $CaCl_2$-MTBS (5 mM $CaCl_2$, 4 mM 2-mercaptoethanol, 10 mM Tris–HCl (pH 7.5), 150 mM NaCl, 0.5 mM phenylmethanesulfonyl fluoride) is mixed with fetuin-Sepharose 4B gel and incubated for 18 h at 4 °C. The suspension was washed with 5 mM $CaCl_2$-MTBS. The bound protein was eluted with 2 mM EDTA-MTBS. (**b**) The eluted protein fractions from a fetuin-Sepharose column in (**a**) are pooled and applied to a heparin-Sepharose 4B column. The bound protein is eluted with 2 mM EDTA and 0.3 M NaCl by stepwise elution. Adapted from ref. 6 with modifications

The derivatization methods in this chapter are also applicable to other hydroxyl group-possessing bead-formed hydrophilic gels that were developed for gel filtration media and usable for HP-AFC or FPLC, e.g., TSKgel HW-55, -65, G5000PW, and G3000PW (Tosoh Bioscience LLC) [12, 13, 16].

2. An appropriate buffer can be used instead of PBS. If Ca^{2+} is required for the carbohydrate-binding activity of a lectin, use a buffer that is compatible with Ca^{2+}, e.g., 15 mM Tris–HCl (pH 8.5) containing 0.15 M NaCl, 0.1 mM $CaCl_2$, and 0.05 % (w/v) NaN_3 for *Sophora japonica* bark lectin [8], for example, 0.1 M acetate buffer (pH 6.4) containing 1 mM $MgCl_2$, 1 mM $MnCl_2$, and 1 mM $CaCl_2$ for PBS for AHA, SBA, RCA, and WGA [4]. Choose the solution depending on the ion requirement of the lectin. For example, use TBS (pH 7.5) containing 1 mM $CaCl_2$ and 1 mM $MnCl_2$ for crude ConA from mashed commercial jack bean meal or acetone powder of legume seeds. When screening novel lectins from natural sources, we use 0.15 M NaCl [9].
3. Choose the appropriate concentration of ammonium sulfate. For example, fractionation of crude ConA was done at 0–90 % saturation [20], while that of WGA was done at 45 % saturation [21].
4. Use the specific sugar for the lectin except WGA, for example, 0.2 M methyl α-D-mannoside/0.1 M acetate buffer (pH 6.4) containing 1 mM $MgCl_2$, 1 mM $MnCl_2$, and 1 mM $CaCl_2$ for

ConA, 0.1 M lactose/PBS for AHA, SBA, and RCA, or 0.2 M acetic acid for WGA [5]. If the amount of available specific sugar is inadequate, the sugar concentration for elution can be reduced to 50 mM.

5. To quickly check the lectin-binding ability of the affinity adsorbent, run a mini-column test. Put a cotton stopper in the bottom of a capillary pipet, and pour the gel slurry (0.1–0.2 mL) into the mini-column. Apply the crude lectin solution to the mini-column, wash the mini-column with saline, and then apply a suspension of erythrocytes in saline (0.2–0.3 mL). After washing with saline (1–2 mL), judge the binding of the erythrocytes to the mini-column by the red color, which indicates that the ligands are successfully immobilized to the gel.
6. Epoxy groups introduced can be quantified by titrating the OH^- ions with 5 mM hydrochloric acid using phenolphthalein as a pH indicator. Usually 40–50 μmols of epoxy groups per gram of suction-dried gel are introduced.
7. Instead of ammonium hydroxide, hydrazine monohydrate can be used for amination of epoxy-activated Sepharose to obtain hydrazino-Sepharose [12], which reacts more readily with formyl groups than amino-Sepharose. Hydrazino-Sepharose is immobilized on the ligand within 50 h and immobilizes a larger amount (60 μmol per gram of suction-dried gel) than amino-Sepharose [12].
8. Amino groups can be quantified by solubilizing the gel after the TNBS color test. Add 2.5 ml of 50 % (v/v) acetic acid to 10–20 mg of the suction-dried gel and heat at 100 °C for 5 h. After cooling, measure A_{340} of the solution. Determine the trinitrophenylamine concentration in the solution using the absorption coefficient, $\varepsilon = 1.4 \times 10^4\ M^{-1}\ cm^{-1}$. Amino groups in the gel generate equimolar trinitrophenylamine by reacting with TNBS.
9. If the volume of the liquid in the reaction mixture is too small to mix the gel with the reagent, add water so that the gel can move in the reaction mixture.
10. EEDQ is soluble in 100 % (v/v) EtOH but has poor solubility in water.
11. The binding reaction of the lectin to the adsorbent can be done out of the column (batchwise method). Equilibrate the adsorbent with the buffer, add the gel slurry to 10 volumes (v/w) of the crude lectin extract, and incubate overnight at 4 °C with gentle shaking. After the binding reaction, separate the gel by centrifugation at $400 \times g$ for 10 min, wash a few times with the buffer, and then pour into a column [6]. Washing and elution are performed in the same way as described in **step 2**. The batchwise method sometimes gives

higher yield of purified lectin than the column method, especially for the case that the affinity of lectin is low.

12. The purification efficiency varies by the combination of lectin and immobilization mode of carbohydrate. For example, a GalNAc/Gal-specific soybean lectin bound better to the D-GalN-Sepharose which had been prepared by coupling D-GalN to carboxyl-Sepharose [4] than lactamyl-Sepharose (Table 1). In contrast, another GalNAc/Gal-specific lectin in *Clerodendrum trichotomum* fruit did not bind to the D-GalN-Sepharose at all, but bound and sugar-specifically eluted from lactamyl-Sepharose [7].

Acknowledgement

We thank Prof. Isamu Matsumoto for supervision of development of preparation of the adsorbents.

References

1. Roth J (2011) Lectins for histochemical demonstration of glycans. Histochem Cell Biol 136:117–130
2. Lowe CR, Dean PDG (1974) Affinity chromatography. Wiley, London
3. Tercero JC, Diaz-Maurino T (1988) Affinity chromatography of fibrinogen on Lens culinaris agglutinin immobilized on CNBr-activated sepharose: study of the active groups involved in nonspecific adsorption. Anal Biochem 174: 128–136
4. Matsumoto I, Mizuno Y, Seno N (1979) Activation of Sepharose with epichlorohydrin and subsequent immobilization of ligand for affinity adsorbent. J Biochem 85:1091–1098
5. Matsumoto I, Kitagaki H, Akai Y, Ito Y, Seno N (1981) Derivatization of epoxy-activated agarose with various carbohydrates for the preparation of stable and high-capacity affinity adsorbents: their use for affinity chromatography of carbohydrate-binding proteins. Anal Biochem 116:103–110
6. Kojima K, Ogawa HK, Seno N, Matsumoto I (1992) Affinity purification and affinity characterization of carbohydrate-binding proteins in bovine kidney. J Chromatogr 597:323–330
7. Kitagaki H, Seno N, Yamaguchi H, Matsumoto I (1985) Isolation and characterization of a lectin from the fruit of Clerodendron trichotomum. J Biochem 97:791–799
8. Ueno M, Ogawa H, Matsumoto I, Seno N (1991) A novel mannose-specific and sugar specifically aggregatable lectin from the bark of the Japanese pagoda tree (Sophora japonica). J Biol Chem 266:3146–3153
9. Ina C, Sano K, Yamamoto-Takahashi M, Matsushita-Oikawa H, Takekawa H, Takehara Y, Ueda H, Ogawa H (2005) Screening for and purification of novel self-aggregatable lectins reveal a new functional lectin group in the bark of leguminous trees. Biochim Biophys Acta 1726:21–27
10. Nakagawa K, Nakamura K, Haishima Y, Yamagami M, Saito K, Sakagami H, Ogawa H (2009) Pseudoproteoglycan (pseudoPG) probes that simulate PG macromolecular structure for screening and isolation of PG-binding proteins. Glycoconj J 26:1007–1017
11. Matsumoto H, Natsume A, Ueda H, Saitoh T, Ogawa H (2001) Screening of a unique lectin from 16 cultivable mushrooms with hybrid glycoprotein and neoproteoglycan probes and purification of a novel *N*-acetylglucosamine-specific lectin from Oudemansiella platyphylla fruiting body. Biochim Biophys Acta 1526: 37–43
12. Ito Y, Yamasaki Y, Seno N, Matsumoto I (1986) Preparation of high capacity affinity adsorbents using new hydrazino-carriers and their use for low and high performance affinity chromatography of lectins. J Biochem 99: 1267–1272
13. Kanamori A, Seno N, Matsumoto I (1986) Preparation of high-capacity affinity adsorbents

using formyl carriers and their use for low-and high-performance affinity chromatography of trypsin-family proteases. J Chromatogr 363: 231–242

14. Matsumoto I, Seno N, Golovtchenko-Matsumoto AM, Osawa T (1980) Amination and subsequent derivatization of epoxy-activated agarose for the preparation of new affinity adsorbents. J Biochem 87:535–540
15. Ueda H, Kojima K, Saitoh T, Ogawa H (1999) Interaction of a lectin from Psathyrella velutina mushroom with *N*-acetylneuraminic acid. FEBS Lett 448:75–80
16. Sasaki H, Hayashi A, Kitagaki-Ogawa H, Matsumoto I, Seno N (1987) Improved method for the immobilization of heparin. J Chromatogr 400:123–132
17. Kitagaki-Ogawa H, Yatohgo T, Izumi M, Hayashi M, Kashiwagi H, Matsumoto I, Seno N (1990) Diversities in animal vitronectins. Differences in molecular weight, immunoreactivity and carbohydrate chains. Biochim Biophys Acta 1033:49–56
18. Asanuma-Date K, Hirano Y, Le N, Sano K, Kawasaki N, Hashii N, Hiruta Y, Nakayama K, Umemura M, Ishikawa K, Sakagami H, Ogawa H (2012) Functional regulation of sugar assimilation by N-glycan-specific interaction of pancreatic alpha-amylase with glycoproteins of duodenal brush border membrane. J Biol Chem 287:23104–23118
19. Ueda H, Matsumoto H, Takahashi N, Ogawa H (2002) *Psathyrella velutina* mushroom lectin exhibits high affinity toward sialoglycoproteins possessing terminal *N*-acetylneuraminic acid alpha 2,3-linked to penultimate galactose residues of trisialyl *N*-glycans. Comparison with other sialic acid-specific lectins. J Biol Chem 277:24916–24925
20. Agrawal BB, Goldstein IJ (1965) Specific binding of concanavalin A to cross-linked dextran gels. Biochem J 96:23contd–25contd
21. LeVine D, Kaplan MJ, Greenaway PJ (1972) The purification and characterization of wheat-germ agglutinin. Biochem J 129: 847–856

Chapter 6

High-Performance Lectin Affinity Chromatography

Yuka Kobayashi

Abstract

Lectin high-performance liquid chromatography techniques have contributed to the growing interest in glycoproteomics. Affinity chromatography is a very effective method to separate and purify trace amount of biological substances. In this chapter, we describe a basic procedure for separation of glycoproteins using commercially available lectin-HPLC columns. As an example, α-fetoprotein, known as a biomarker of liver cancer, can be separated at the level of their glyco-isomers by using a *Lens culinaris* agglutinin (LCA) column.

Key words Affinity chromatography, High-performance liquid chromatography (HPLC), Isolation, Fractionation, Sugar chains

1 Introduction

The ability of lectins to bind specific sugar chains is the underlying principle that governs lectin affinity chromatography. Many biological substances contain sugar chains in their structures, and lectin affinity chromatography is effective at purifying and concentrating trace amounts of such biological substances. The benefits of lectin affinity chromatography are as follows: (1) It can separate substances (many bioactive substances) that bind to target sugar chains in a single column passage. (2) Target substances can be collected under mild conditions (sugar is primarily used for elution at approximately neutral pH conditions). (3) Glycan structures can be approximated (affinity strength depends on glycan structure).

It is common to perform lectin affinity chromatography by using high-performance liquid chromatography (HPLC) columns in addition to conventional open columns packed with agarose resin. HPLC purification can be performed at a higher flow rate than the latter. Furthermore, it only takes a short time to analyze HPLC data. Finally, even small samples can be analyzed. Diffusion in the column is limited, yielding better separation of the target substances. In addition, the lectin HPLC column method

Jun Hirabayashi (ed.), *Lectins: Methods and Protocols*, Methods in Molecular Biology, vol. 1200,
DOI 10.1007/978-1-4939-1292-6_6, © Springer Science+Business Media New York 2014

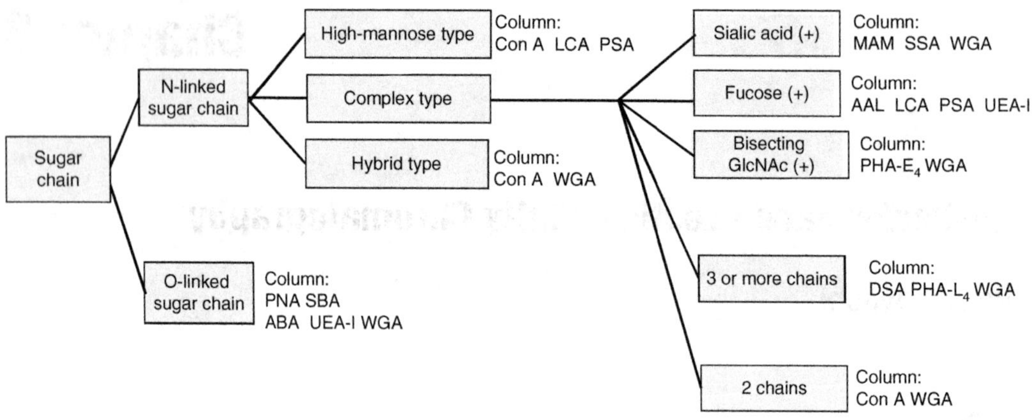

Fig. 1 Selection of lectin column based on sugar chain-lectin interaction

generates highly reproducible separation data, and control standards facilitate easy column monitoring compared to agarose columns. Therefore, lectin HPLC is very effective at analyzing trace quantities of glycoproteomics.

Lectins recognize and bind to specific moieties of complex sugar chain structures. The specificity and strength of the binding depend on the type of lectin used. To purify and analyze a target substance (i.e., oligosaccharides, glycopeptides, and glycoproteins), the type of lectin should be selected based on the structure of the glycan target substance. Figure 1 shows various lectins and sugars to which they bind. If the structure of the sugar chain present in the target substance is already known, the lectin is chosen based on the affinity between the lectin and sugar chain. Because some glycoproteins have multiple glycosylation sites, they may be separated in a different manner from oligosaccharides and glycopeptides. When multiple sugar chains exist in the target substance, the binding between lectins and the target substance often becomes tighter, a phenomenon known as "avidity," which results in poor elution of the target substance from the lectin column used. In such a case, increasing the sugar concentration may improve elution. However, if a higher sugar concentration is not effective, the use of lectins that exhibit weaker affinities is recommended [1–4].

2 Materials

2.1 Lectin HPLC Columns

Choice of column: A series of lectin HPLC columns are commercially available for the analysis and fractionation of sugar chains as well as for glycoprotein purification (*see* Table 1). A packing procedure is unnecessary, because the HPLC column is pre-packed (*see* **Note 1**).

Table 1
Affinity HPLC columns and buffer solutions used for elution

Lectin affinity HPLC columns	Substances used for elution	Concentration
Aleuria aurantia lectin (AAL)	L-Fucose	0.02 M
Artocarpus integrifolia lectin (AIL, Jacalin)	Methyl α-D-galactoside	0.02–0.5 M
Canavalia ensiformis lectin (Con A)	D-Glucose Methyl α-D-glucoside Methyl α-D-mannoside	0.1–0.5 M 0.02–0.5 M 0.02–0.5 M
Datura stramonium agglutinin (DSA)	Chitooligosaccharide Acid	1 % pH 5
Erythrina cristagalli agglutinin (ECA)	Methyl α-D-galactoside	0.02–0.5 M
Galanthus nivalis agglutinin (GNA)	Methyl α-D-mannoside	0.02–0.5 M
Lens culinaris agglutinin (LCA)	D-Glucose Methyl α-D-glucoside Methyl α-D-mannoside	0.2–0.5 M 0.2–0.5 M 0.2–0.5 M
Lotus tetragonolobus lectin (LTL)	L-Fucose	0.02 M
Maackia amurensis mitogen (MAM)	Ethylenediamine	0.05 M
Phaseolus vulgaris agglutinin (PHA-E4)	Potassium borate ($K_2B_4O_7$)	0.1 M (0.2 M)
Phaseolus vulgaris agglutinin (PHA-L4)	Potassium borate ($K_2B_4O_7$)	0.1 M (0.2 M)
Arachis hypogaea (peanut) agglutinin (PNA)	D-Galactose Lactose	0.2 M 0.2 M
Pisum sativum agglutinin (PSA)	Methyl α-D-glucoside Methyl α-D-mannoside	0.2–0.5 M 0.2–0.5 M
Phytolacca americana agglutinin (PWA)	Chitooligosaccharide	1 %
Glycine max (soybean) agglutinin (SBA)	*N*-Acetyl-D-galactosamine	0.2 M
Sambucus sieboldiana agglutinin (SSA)	Lactose	0.2 M
Sambucus nigra agglutinin (SNA)	Lactose	0.2 M
Ulex europaeus agglutinin (UEA)	L-Fucose	0.02 M
Vicia villosa lectin (VVL)	*N*-Acetyl-D-galactosamine	0.02–0.5 M
Triticum vulgare (wheat germ) agglutinin (WGA)	*N*-Acetyl-D-glucosamine Acid (acetic acid)	0.2 M 0.2–0.5 M

2.2 Buffer Solutions Used for Elution

1. Choice of buffer solution: The type of sugar used for elution (hapten sugar) varies depending on the specificity of the lectin. For some lectins, it is actually difficult to get fully effective hapten sugars for various reasons (e.g., price or technical difficulties) (*see* **Note 2**). For example, the hapten sugar for PHA-E4 and L4 is D-GalNAc, which is too expensive for

Table 2
Analysis conditions for typical glycoproteins using Con A, WGA, and LCA-HPLC columns

Column	Con A-HPLC column (4.6 mm I.D. × 150 mm)		
Sample	α1-Acid glycoprotein 1 mg/mL × 20 μL	Ovalbumin	Transferrin
Buffer	A: 50 mM Tris–HCl (pH 7.2) B: 50 mM Methyl α-D-glucoside in buffer A		
Elution method	B conc. 0 → 100 % (60 min; linear gradient)		
Flow rate	0.5 mL/min		
Temperature	25 °C		
Detection	UV-280 nm (AUFS; 0.08)		

Column	WGA-HPLC column (4.6 mm I.D. × 150 mm)		
Sample	α1-Acid glycoprotein 1 mg/mL × 20 μL	Ovalbumin	Thyroglobulin
Buffer	A: 50 mM Tris–H_2SO_4 (pH 7.2) B: 0.2 M GlcNAc in buffer A	A: 0.2 M Sodium acetate B: 0.2 M Acetic acid	
Elution method	B conc. 0 → 100 % (60 min; linear gradient)	B conc. 85 → 100 % (stepwise)	B conc. 0 → 100 % (stepwise)
Flow rate	0.5 mL/min		
Temperature	25 °C		
Detection	UV-280 nm (AUFS; 0.08)		

Column	LCA-HPLC column (4.6 mm I.D. × 150 mm)		
Sample	α-Fetoprotein (from HCC) 1 mg/mL × 20 μL	2 mg/mL × 20 μL	4 mg/mL × 20 μL
Buffer	A:50 mM Tris–H_2SO_4 (pH 7.2) B:0.2 M Methyl α-Glc in buffer A		
Elution method	B conc. 0 % (0–5 min; stepwise) B conc. 100 % (5–20 min; stepwise) B conc. 0 % (20–30 min; stepwise)		
Flow rate	0.5 mL/min		
Temperature	25 °C		
Detection	UV-280 nm (AUFS; 0.08)		

routine use. For such cases, borate is often used as an alternative (*see* **Note 3**).

2. Buffer: Use appropriate HPLC buffer (e.g., for the cases of Con A, WGA, and LCA, *see* Table 2). Buffer should be filtered (0.45 μm filter) and sufficiently degassed before use.

2.3 Samples

Glycoprotein, e.g., α1-acid glycoprotein, ovalbumin, transferrin, thyroglobulin, human α-fetoprotein (AFP) from cord blood or hepatocellular carcinoma (HCC) cells.

3 Methods

3.1 Sample Preparation

1. Mix 2 mg of glycoprotein with 500 μL of preparation buffer (buffer A). Dissolve sample in buffer. Dilutions of samples were made with the buffer at a final concentration of 1–4 mg/mL.
2. Insoluble materials are removed from samples by filtration (0.45 μm filter) and centrifugation (10,000 × *g*, 15 min).

3.2 Lectin Affinity HPLC

1. Connect the column to the HPLC system and let the equilibration buffer flow through the column at a flow rate of 0.5 mL/min. When the gradient is applied, let a buffer for elution (about 5 column volumes) run through the column once and observe the baseline change (e.g., absorbance). Let the equilibration buffer (≥10 column volumes) flow through the column until a stable baseline is obtained.
2. Inject the sample.
3. Initiate the linear gradient elution (e.g., buffer B concentration from 0 to 100 % for 60–120 min). Determine the appropriate conditions for elution.

3.3 Regeneration of Columns and Storage

1. Let initial buffer (≥5 column volumes) run through the column.
2. Fill the column with the packing buffer used during shipment and store it at room temperature.

3.4 Warnings

Procedural requirements are as follows:

pH: pH 5–8 (*see* **Note 4**).

Temperature: 4 °C to room temperature (*see* **Note 5**).

Organic solution: Methanol and ethanol (≤10 %).

Surfactant: The influence of surfactants on lectins varies depending on the surfactant type and concentration (*see* **Note 6**).

(*See* **Note** 7.)

3.5 Application Examples (Con A, WGA, and LCA-HPLC Columns)

1. Connect the lectin column (150 mm × 4.6 mm I.D.) to the HPLC system.
2. Equilibrate the column with Solvent A as explained below.
3. Perform the HPLC analyses according to the conditions described in Table 2.
4. Inject glycoprotein (20 μL) into each column using an autosampling system.

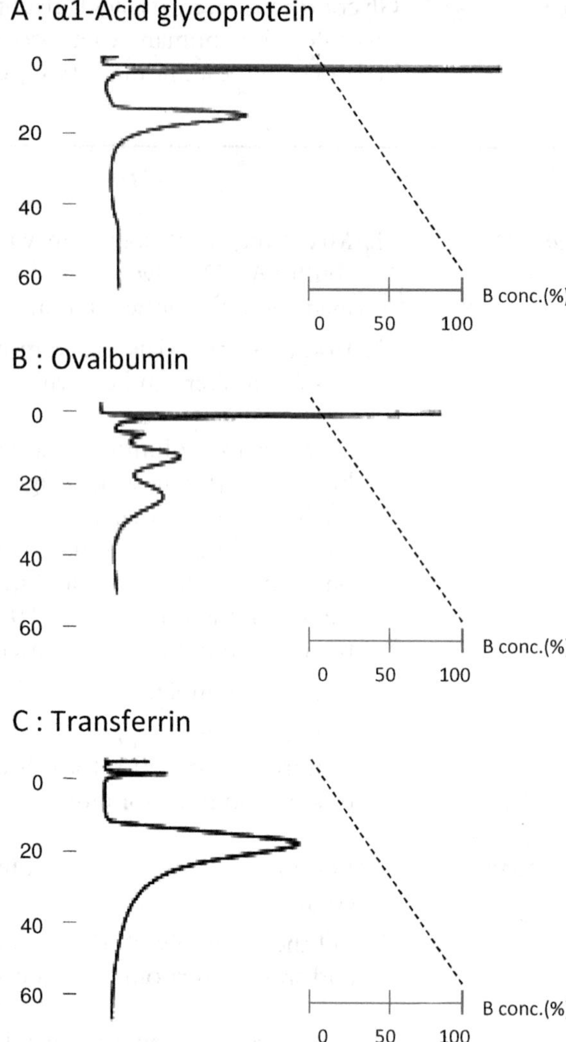

Fig. 2 Separation pattern of glycoprotein on a Con A-HPLC column. Three types of glycoproteins are analyzed using a mannose-specific Con A-HPLC columns

5. Determine UV absorbance at 280 nm. The UV peaks can be identified by comparing the retention time of the sample in each case with its standard retention time (Figs. 2, 3, and 4).

4 Notes

1. Lectin affinity HPLC columns (4.6 I.D. × 50 mm, 4.6 × 150 mm) are produced by J-Oil Mills, Inc. (Tokyo, Japan) (Cosmo bio Co., LTD.). AffiSep® lectin columns (4.6 × 50 mm, 4.6 × 100 mm, 4.6 × 150 mm, 4.6 × 250 mm I.D.)

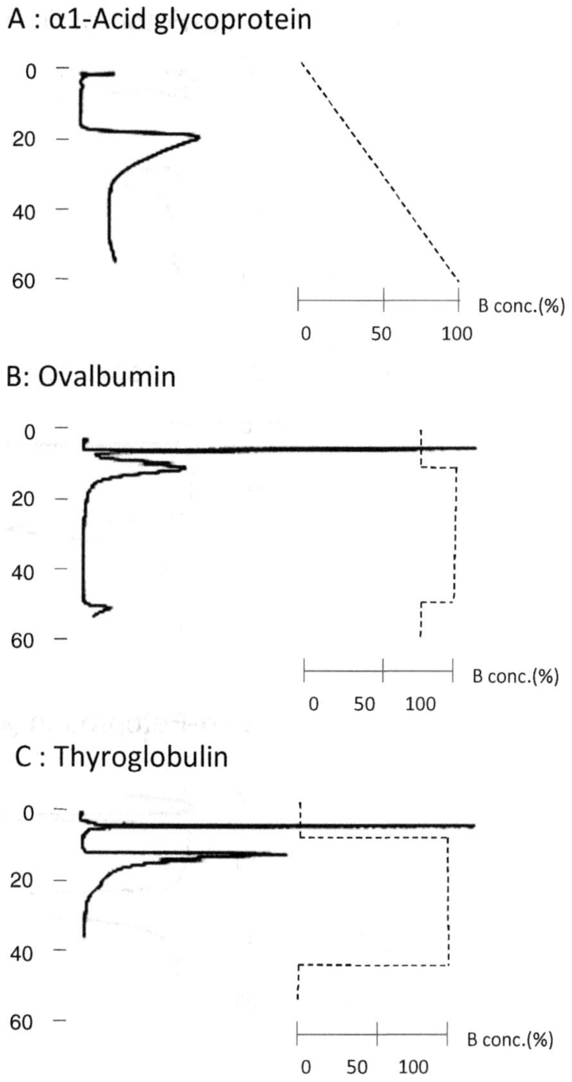

A : α1-Acid glycoprotein

0 20 40 60

B conc.(%) 0 50 100

B: Ovalbumin

0 20 40 60

B conc.(%) 0 50 100

C : Thyroglobulin

0 20 40 60

B conc.(%) 0 50 100

Fig. 3 Separation pattern of glycoproteins on a WGA-HPLC column. Three types of glycoproteins are analyzed using a GlcNAc/sialic acid-specific WGA-HPLC column

are made by GALAB Technologies (Geesthacht, Germany). ProSwift ConA-1S affinity column (5 × 50 mm) is made by Thermo Fisher Scientific Inc. (MA, USA). The column may be stable for approximately 100–500 h.

2. Some lectins, especially legume lectins, need metal ions for binding, so the buffers used for the affinity chromatography on the lectin column must contain 1 mM $CaCl_2$ and $MnCl_2$ (for details, *see* Table 1 in Chapter 45).
3. Binding ability of PHA is inhibited by 0.1 M $K_2B_4O_7$.

A : α-Fetoprotein (1 mg/mL, 20 μL)

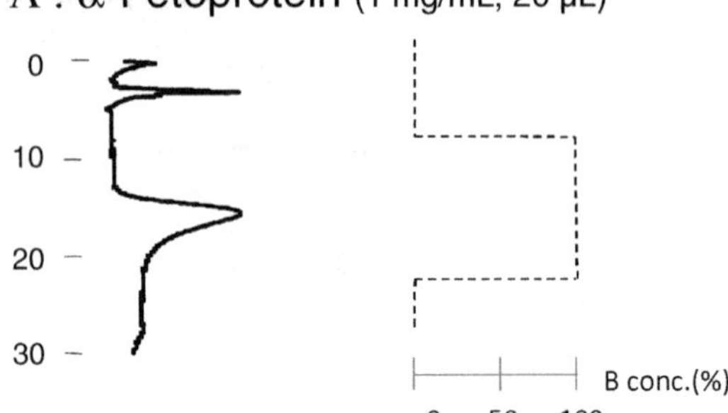

A : α-Fetoprotein (2 mg/mL, 20 μL)

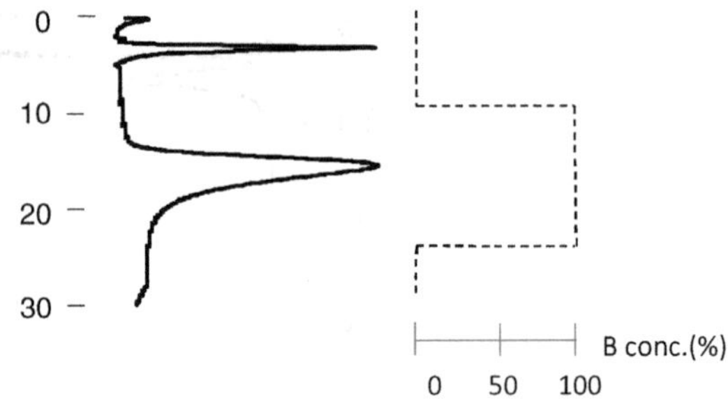

A : α-Fetoprotein (4 mg/mL, 20 μL)

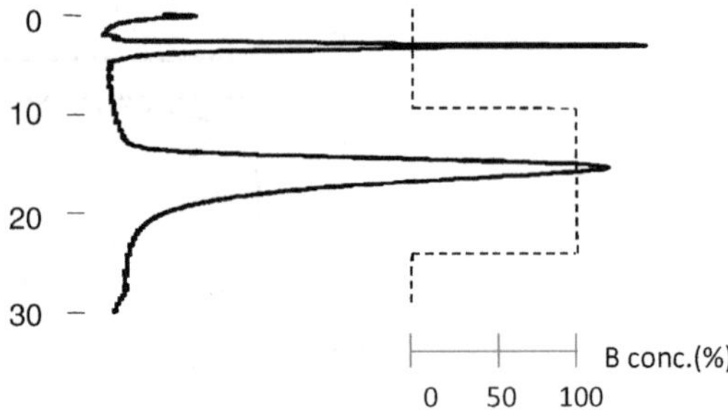

Fig. 4 Separation pattern of AFP (AFP-L1 and L3) on an LCA-HPLC column. AFP from HCC cells is analyzed using an LCA HPLC column. LCA is categorized as a mannose-specific lectin based on monosaccharide inhibition, but it shows particular preference for *N*-linked glycans containing core α1-6 fucose. AFP is separated into two peaks: one of them corresponds to lectin-unbound AFP and the other corresponds to lectin-bound AFP (α1-6 fucosylated AFP). This gives us general information about the analysis of the sugar chain structures of glycoproteins. With lectin HPLC columns, it is possible to analyze the sugar chain structure of glycoproteins (*see* **Note 8**)

4. After using, pH 7–8 should be maintained.
5. Lectin affinity tends to be higher at lower temperatures.
6. After using, the column should be washed with a buffer solution not containing a surfactant.

7. Gel: The gel is a highly hydrophilic, high-molecular-weight polymer. The elution of highly hydrophobic proteins tends to be delayed; however, elution can be expedited with a buffer containing 5 % (v/v) ethanol.

   Column: Corrosion of stainless steel columns during long-term use may be averted by using a buffer devoid of chloride ions [Cl^-]. For long-term storage, fill the column with the packing buffer used during shipment, but do not freeze the column. Connect the column to the HPLC system and let the initial buffer flow through the column at a flow rate of 0.5 mL/min.

8. Typically, the elution buffer concentrations described in this chapter are appropriate for eluting glycoproteins; however, the ideal concentration may change depending on the affinity between the lectins and target substances. For substances that bind weakly to lectin, low concentrations may be sufficient for elution. When broad peaks are observed, an increase in the concentration yields sharper peaks. During experiments, ensure that the concentrations used are appropriate for your purposes.

References

1. Larsson PO, Glad M, Hansson L et al (1983) High-performance liquid affinity chromatography. Adv Chromatogr 21:41–84
2. Green ED, Brodbeck RM, Baenziger JU (1987) Lectin affinity high-performance liquid chromatography: interactions of *N*-glycanase-released oligosaccharides with leukoagglutinating phytohemagglutinin, concanavalin A, *Datura stramonium* agglutinin, and *Vicia villosa* agglutinin. Anal Biochem 167(1):62–75
3. Green ED, Brodbeck RM, Baenziger JU (1987) Lectin affinity high-performance liquid chromatography. Interactions of *N*-glycanase-released oligosaccharides with *Ricinus communis* agglutinin I and *Ricinus communis* agglutinin II. J Biol Chem 262:12030–12039
4. Tokuda M, Kamei M, Yui S et al (1985) Rapid resolution of nucleotide sugars by lectin affinity high-performance liquid chromatography. J Chromatogr 323:434–438

Chapter 7

Determination of Glycan Motifs Using Serial Lectin Affinity Chromatography

Katsuko Yamashita and Takashi Ohkura

Abstract

Serial lectin affinity chromatography is a convenient technique for characterizing glycan motifs (terminal glycan structures) of glycoproteins or released glycans. When these glycoconjugates are applied serially or in parallel to lectin-immobilized columns, information regarding the glycan motifs can be obtained. We demonstrate lectin affinity chromatographic methods for determining *O*-linked glycan structures of MUC1 purified from a breast cancer cell line, YMB-S, *N*-linked glycan structures of serum prostate-specific antigen from prostate cancer, and serum alkaline phosphatases from choriocarcinoma. These lectin-fractionated samples are analyzed quantitatively by measuring radioactivity, antigen contents are analyzed using enzyme-linked immunosorbent assay, and enzymatic activities are assessed.

Key words Lectin, *N*-glycan, *O*-glycan, Fractionation, Glycan motif, ELISA, Glycoprotein, [^{3}H]-oligosaccharide

1 Introduction

Various glycoproteins contain *N*-linked and/or *O*-linked glycans. Their glycan structures change depending on development, differentiation, malignancy, cell type, and species [1]. To elucidate the functional roles of glycoprotein-linked glycans in the human body, the specific profiles of glycan structures must be determined. Nuclear magnetic resonance (NMR) spectrometry is useful for determining anomeric configuration, monosaccharide sequence, and linkage of the purified glycans, although large amounts of samples are required [2]. In contrast, matrix-assisted laser desorption/ionization (MALDI)-time-of-flight-mass spectrometry (TOFMS) is convenient for monosaccharide sequence analysis [3]; however, anomeric configurations cannot be determined, and monosaccharide composition analysis is limited to hexose, hexosamine, and methylpentose. Linkage analysis and monosaccharide determination by methylation analysis are useful in combination with other techniques for structural examination of pure oligosaccharides [4],

Jun Hirabayashi (ed.), *Lectins: Methods and Protocols*, Methods in Molecular Biology, vol. 1200, DOI 10.1007/978-1-4939-1292-6_7,

but the protocol is complicated. Lectin affinity chromatographic analysis has several advantageous characteristics [5, 6]. (1) Glycans or glycoproteins consisting of the specific glycan motifs can be separated by using simple instruments or an automated system. (2) Crude samples such as sera or cell extracts can be used, and the glycan motifs of fractionated target glycoproteins can be monitored by their enzymatic activities or by using an enzyme-linked immunosorbent assay (ELISA). (3) Precise glycan structures can be determined by separating free oligosaccharides or the target glycoproteins using serial or parallel lectin affinity columns, which show different sugar-binding specificities. (4) Glycan motifs of metabolically radiolabeled glycoproteins or released glycans with extremely high sensitivity can be determined by measuring radioactivity. (5) The binding ratio of glycoproteins to lectin affinity columns results in multiplication of the specific motif ratio by the number of glycans per glycoprotein. Accordingly, this method shows higher sensitivity in detecting low amounts of glycan motifs in the glycoproteins.

This chapter describes the precise determination of glycan motifs and the evaluation of invisible glycan changes by malignant transformation of target glycoproteins using lectin affinity chromatography. For example, radiolabeled *O*-linked glycan structures of MUC1 purified from the medium of YMB-S cells can be determined in combination with serial lectin affinity chromatography and substrate specific exo-glycosidase digestion [7]. Another example is prostate-specific antigen (PSA) which possesses one *N*-linked glycan. After PSAs derived from the sera of prostate cancer (PC) patients, benign prostatic hyperplasia (BPH) patients, seminal fluids, and LNCaP prostate cancer cells are fractionated using a set of lectin affinity columns, PSA concentration in each fraction can be quantitatively monitored by ELISA [8]. Furthermore, glycosylphosphatidylinositol (GPI)-anchored alkaline phosphatases (ALPs) produced in the placenta and choriocarcinoma contain two *N*-linked glycans per mole ALP. Serum ALPs of pregnant women and choriocarcinoma patients, ALP released by phosphatidylinositol-specific phospholipase C (PI-PLC) from the choriocarcinoma cell line JEG-3, are fractionated in parallel using a set of lectin affinity chromatography columns. By measuring the enzymatic activities of fractionated ALPs, changes in *N*-glycan structures by malignant transformation can be monitored, even if ALP is not purified [9].

2 Materials

2.1 Preparation of Immobilized Lectin Columns

1. Lectins can be purchased from J-Oil-Mills Corp. (Tokyo, Japan), Sigma-Aldrich (St. Louis, MO, USA), Vector Laboratories (Burlingame, CA, USA), and EY Lab (San Malto, CA, USA) (also, *see* Table 1 in Chapter 45). Lectins are immobilized by

Table 1
Carbohydrate-binding specificities of various lectin-Sepharose 4B column chromatography

Lectin	Binding specificity	References
Ricinus communis agglutinin I (RCA-I)	Strong binding (RCA-I$^+$ fraction, eluted with 10 mM lactose[a]) Galβ1-4GlcNAcβ1-R[b], Galβ1-4(SO_3^--6)GlcNAcβ1-R Weak binding (RCA-I^r fraction) Galβ1-3GalNAc(α1-pNP or $_{OH}$), Galβ1-6 Galβ1-R Galβ1-3(SO_3^--6)GalNAc(α1-pNP or $_{OH}$), Galβ1-3GlcNAcβ1-R	[7, 13]
Trichosantes japonica agglutinin-I (TJA-I)	Strong binding (TJA-I$^+$ fraction, eluted with 0.1 M lactose) Neu5Acα2-6Galβ1-4GlcNAcβ1-R, SO_3^--6Galβ1-4GlcNAcβ1-R	[14]
Maackia amurensis lectin (MAL)	Strong binding (MAL$^+$ fraction, eluted with 0.4 M lactose) SO_3^--3Galβ1-4GlcNAcβ1-R Weak binding (RCA-I^r fraction) Neu5Acα2-3Galβ1-4GlcNAcβ1-R	[7, 15]
Sambucus nigra agglutinin (SNA)	Strong binding (SNA$^+$ fraction, eluted with 0.2 M lactose) Neu5Acα2-6Galβ1-4GlcNAcβ1-R, SO_3^--6Galβ1-4GlcNAcβ1-R SO_3^--6GalNAc(α1-pNP or $_{OH}$), Neu5Acα2-6GalNAc (α1-pNP or $_{OH}$)	[7, 16]
Psathyrella velutina lectin (PVL)	Strong binding (PVL$^+$ fraction, eluted with 0.3 M GlcNAc) GlcNAcβ1-3Galβ1-R, SO_3^--6GlcNAcβ1-3Galβ1-R	[17]
Galectin-3 (Gal-3)	Strong binding (Gal3$^+$ fraction, eluted with 0.1 M lactose) SO_3^--3Galβ1-3GalNAc(α1-pNP or $_{OH}$)	[7, 18]
Wisteria floribunda agglutinin (WFA)	Strong binding (WFA$^+$ fraction, eluted with 0.3 M lactose) GalNAcβ1-4GlcNAcβ1-R Weak binding (WFA^r fraction) GalNAcβ1-3Galβ1-R	[11, 19]
Ulex europaeus agglutinin-I (UEA-I)	Strong binding (UEA-I$^+$ fraction, eluted with 50 mM fucose) Fucα1-2Galβ1-4GlcNAcβ1-R	[20]

(continued)

Table 1
(continued)

Lectin	Binding specificity	References
Lens culinaris agglutinin (LCA)	Strong binding (LCA$^+$ fraction, eluted with 0.2 M MeαMan) R-(Manα1-)$_2$Manβ1-4GlcNAcβ1-4(Fucα1-6)GlcNAcβ (*see* **Note 3**)	[21]
Aleuria aurantia lectin (AAL)	Strong binding (AAL$^+$ fraction, eluted with 5 mM fucose) R-(Fucα1-6)GlcNAc Weak binding (AAL^r fraction) Fucα1-2Galβ1-4GlcNAcβ1-R, Galβ1-4 (Fucα1-3)GlcNAcβ1-R	[22]
Datura stramonium agglutinin (DSA)	Strong binding (DSA$^+$ fraction, eluted with 1 % GlcNAc oligomer) Galβ1-4GlcNAcβ1-6 (Galβ1-4GlcNAcβ1-2)Manα1-R Galβ1-4GlcNAcβ1-6 (Galβ1-4GlcNAcβ1-3) Galβ1-4R (Galβ1-4GlcNAcβ1-3)$_{2\text{-}n}$ Galβ1-4R Weak binding (DSA^r fraction) (*see* **Note 4**) Galβ1-4GlcNAcβ1-4(Galβ1-4GlcNAcβ1-2)Manα-R	[23]
ConcanavalinA (ConA)	Strong binding (ConA^{++} fraction, eluted with 0.2 M Meα-Man) High-mannose type of glycans Hybrid type of glycans Weak binding (ConA$^+$ fraction, eluted with 5 mM Meα-Man) Complex type of bi-antenna glycans	[24]

[a]Haptens in TB (for oligosaccharides) or TBS (for glycoproteins)
[b]R: H or sugar

reacting the amino groups of lectins and cyanogen bromide-activated Sepharose 4B [10] (*see* **Note 1**). CNBr-activated Sepharose 4B can be obtained from GE Healthcare (Little Chalfont, UK).

2. Next, 10 mM Tris–HCl buffer (pH 7.4) containing 0.02 % sodium azide (TB) is used for separating oligosaccharides and TB-0.1 % bovine serum albumin (TB-BSA) is used for fractionation of glycoproteins.
3. Oligosaccharides or glycoproteins bound to the respective lectin columns can be eluted by using TB or TB-BSA containing appropriate haptenic sugars such as 0.2 M (or 5 mM) α-methylmannoside (Meα-Man) for Con A, 50 mM fucose for UEA-I, 0.1 M lactose for TJA-I, 0.2 M lactose for SNA, 0.4 M lactose for MAL, 0.3 M GlcNAc for PVL, 0.3 M lactose for WFA [11] (*see* **Note 2**), 10 mM lactose for RCA-I, 1 % GlcNAc oligomer for DSA, 0.1 M lactose for Gal-3, 0.2 M Meα-Man for LCA (*see* **Note 3**), and 5 mM fucose for AAL.

2.2 Preparation of Human PSA

1. Prepare PSA samples from sera of PC and BPH and from the medium of the prostate cancer cell line LNCaP.
2. Purchase PSA purified from human seminal fluid from Sigma-Aldrich.
3. Purchase LNCaP (RCB2144) cells from RIKEN Bio-Resource Center through the National Bio-Resource Project of the MEXT, Japan. Culture cells in an RPMI 1640 medium (Life Technologies, Carlsbad, CA, USA) supplemented with 10 % fetal calf serum (FCS) [8].

2.3 Preparation of Placental Type of ALPs

1. Prepare ALPs by PI-PLC digestion from the choriocarcinoma cell line JEG-3, serum ALPs of a pregnant woman (24 weeks), and a choriocarcinoma patient.
2. Purchase the choriocarcinoma cell line JEG-3 from the Health Science Research Resources Bank (Osaka, Japan). Culture JEG-3 cells in RPMI 1640 medium (Life Technologies) supplemented with 10 % FCS.
3. Prepare the soluble form of ALP from the microsomes (2 mg protein) of JEG-3 cells by digestion with PI-PLC (100 m-units) in PBS, pH 7.4, at 37 °C for 2 h. Centrifuge the digests at 100,000×*g* for 1 h, and apply to various lectin affinity columns [9].

2.4 Purification of MUC1 from YMB-S Cells and Release of O-Glycans from MUC1

1. Apply 15 L of culture media of breast cancer YMB-S cells to an anti-MUC1 monoclonal antibody KL-6 (Sanko Junyaku Co., Ltd., Tokyo, Japan)-conjugated Sepharose 4B column (1 mg of mAb/mL, 0.9×25 cm, equilibrated with PBS). After extensive washing with PBS, elute bound MUC1 by using 10 mM KH_2PO_4-4 M $MgCl_2$ (pH 2.0), and neutralize the fractions immediately. Dialyze the fractions against water and concentrate [7].

2. Release *O*-linked glycans from MUC1 (2 mg) in a homogenous gas-phase reaction of anhydrous hydrazine at 65 °C for 6 h. Dissolve the residue in a saturated $NaHCO_3$ solution, and *N*-acetylate all free amino groups in the residue using acetic anhydride. After *N*-acetylation, reduce the released oligosaccharides with 3.7 MBq of NaB^3H_4 (American Radiolabeled Chemicals, Inc., St. Louis, MO, USA; 13 GBq/mmol) (yield: 4×10^6 dpm) [7, 12].

3 Methods

3.1 Operation of Immobilized Lectins

1. Pack 1 mL of immobilized lectin-Sepharose 4B (3 mg/mL of gel, 1 mL) in a plastic column (i.d. 0.7 cm × 7 cm long) and place a disc filter on the top of the gel.
2. Precool the column (bed volume, 1 mL) at 4 °C and wash with ten column volumes of TB or TB-BSA.
3. Dissolve glycoproteins, serum, or tritium-labeled glycans in 100 μL of TB or TB-BSA and apply this solution to the column. After standing at 4 °C for 15 min, add 900 μL of TB or TB-BSA to the column and collect the eluent in the tube. Move the column to the next tube and add 1 mL of TB or TB-BSA. Repeat this step three times. Move the column to the next tube, add 1 mL of TB or TB-BSA containing the respective haptenic sugars, collect the eluent in the tube, and repeat this step four times.
4. Take aliquots from each fraction and measure the radioactivity for tritium-labeled glycan released from MUC1, or ELISA for PSA antigen, and the enzymatic activity for ALP. Collect the flow-through fractions 1–3 (−), the retarded fractions 4–5 (r), and the bound fractions 6–8 (+). The carbohydrate-binding specificities in the respective lectin-Sepharose 4B column chromatography used in this chapter are summarized in Table 1 [13–24] (*see* **Notes 3** and **4**).
5. After operation, wash the columns with 20 volumes of TB or TB-BSA. Stock lectins involving divalent ions in the molecules at 4 °C by saturating with TB containing 1 mM $CaCl_2$, 1 mM $MgCl_2$, and 1 mM $MnCl_2$. These lectin-Sepharose columns can be used several hundred times repeatedly, given that proteases in the samples have not contaminated the columns.

3.2 Glycosidase Digestion

1. Digest radioactive oligosaccharides or glycoproteins in the following reaction mixtures (50 μL) at 37 °C for 18 h:

   5 m-units of Siaα2-3Gal-specific sialidase, *Salmonella typhimurium* LT2 (Takara Biochemicals, Kyoto, Japan) in 0.1 M sodium acetate buffer, pH 5.5 [25].

0.1 units of *Arthrobacter ureafaciens* sialidase (Nacalai Tesque, Kyoto, Japan) in sodium acetate buffer, pH 5.0 [26].

50 mU of Fucα1-2Gal-specific *Corynebacterium* α-L-fucosidase (Takara Biochemicals) in 10 mM Tris–sulfate buffer, pH 8.0 [27].

5 mU of *Streptococcus* 6646K β-galactosidase (Seikagaku Corporation, Tokyo, Japan) in 0.1 M citrate-phosphate buffer, pH 5.5 [28].

0.5 U of jack bean β-*N*-acetylhexosaminidase in 0.2 M citrate-phosphate buffer, pH 5.0 [29].

2. Add one drop of toluene to each reaction mixture to prevent bacteria growth. After incubation, heat each reaction mixture at 100 °C for 2 min to stop the reaction.

3.3 Serial Lectin Affinity Chromatography of O-Glycans Released from MUC1

Separate *O*-linked type of NaB^3H_4-reduced glycans into four acidic fractions (A1–A4) by high-voltage paper electrophoresis (PEP) at pH 5.4 (pyridine/acetic acid/water 3:1:387) at a potential of 73 V/cm for 90 min. Next, extract A1, A2, A3, and A4 from the paper using Milli-Q water [7]. Determine the respective structures in combination with serial lectin affinity chromatography (*see* Subheading 3.1), substrate-specific exo-glycosidase digestion (*see* Subheading 3.2), and methanolysis (*see* Fig. 1). Demonstrate structural analysis of A1 (monoacidic extended core 1) and A2 (monoacidic core 1) (*see* **Note 5**). After drying the extracted A1 with water, digest A1 with Siaα2-3Gal-specific *Salmonella* sialidase (*S.* sialidase in Fig. 1) (50 μL, pH 5.5, sodium acetate), and apply to anion exchange LC column, Bio-Scale Mini Unosphere Q cartridge (BIO-RAD Lab, Hercules, CA, USA), at a flow rate of 0.5 mL/min at room temperature over a linear gradient (0–0.5 M of acetate-triethylamine, pH 7.0). Treat an aliquot of the remaining acidic A1a with both Siaα2-6 and 2-3Gal hydrolyzing *Arthrobacter* sialidase (*A.* sialidase in Fig. 1) and all of A1a is resistant (Ala′ in Fig. 1). Convert an aliquot of A1a′ to neutral component (N) using methanolysis (1 mL of 0.05 N HCl/dry methanol, 37 °C for 4 h), which specifically hydrolyzes sulfated oligosaccharides [7]. After the reaction, evaporate the solution until it is dry using a centrifugal concentrator; the residues are freed from HCl by evaporation with 1 mL methanol three times. Apply the remaining sulfated A1a′ sequentially to TJA-I, MAL, RCA-I, and PVL-Sepharose 4B columns. The TJA-I$^+$ component is referred to as A1-II. Sequentially, apply the TJA-I$^-$ component to the MAL column. The MAL$^+$ component is referred to as A1-III, and the MAL$^-$ component binds sequentially to the RCA-I column. Digest the RCA-I$^+$ component with β-galactosidase, and apply this sample to the PVL column. All of the sample binds to the PVL column and is referred to as A1-IV. The structures of these fractionated components of A1 are summarized in Fig. 1.

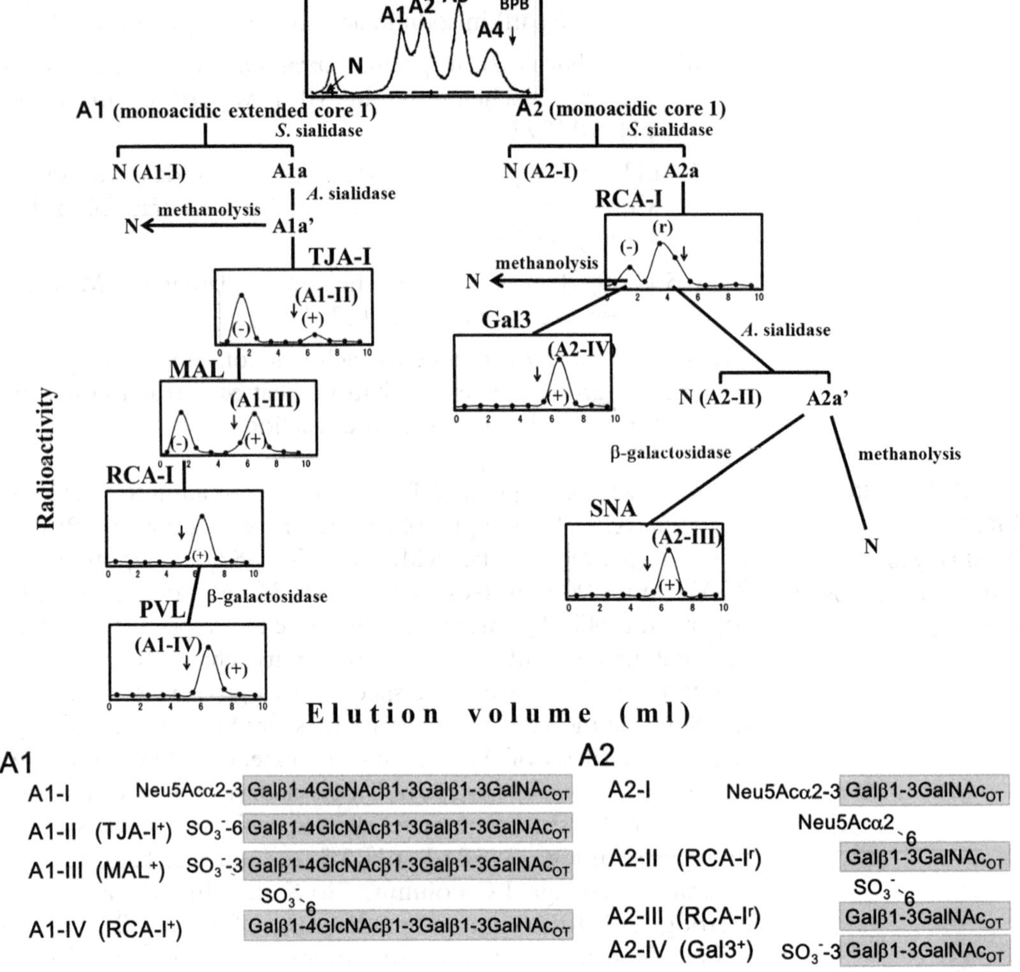

Fig. 1 pH-5.4 paper electrophoresis (PEP) of tritium-labeled *O*-linked oligosaccharides released from MUC1, and elution profiles of serial lectin affinity chromatography of A1 and A2. (*Top*) PEP pattern. (*Middle*) Elution patterns on RCA-I-, TJA-I-, MAL-, PVL-, Gal-3-, and SNA-Sepharose 4B columns of substrate-specific exo-glycosidase-digested A1 and A2 by monitoring radioactivity. *Arrows* show positions where buffers are switched to TB containing various haptens (*see* Table 1). (*Lower*) Possible structures of A1 and A2 predicted from serial lectin-Sepharose 4B column chromatography

Using similar methods, determine the structures of A2. Through digestion with *Salmonella* sialidase, some of the A2 is converted to a neutral component and referred to as A2-I. Apply the remaining acidic fraction A2a to the RCA-I column for separation into RCA-I$^-$ and RCA-I^r fractions. By digestion with *Arthrobacter* sialidase of the RCA-I^r components, some of the RCA-I^r fraction is converted to the neutral component, which is referred to as A2-II. The remaining sialidase-resistant RCA-I^r component is named A2a′. An aliquot of A2a′ is converted to the neutral fraction by methanolysis (N), indicating that this component is sulfated. The A2a′ digested with β-galactosidase binds to the SNA column, and referred to as A2-III. An aliquot of the RCA-I$^-$ component is also converted to the neutral fraction by methanolysis (N), indicating that this component is sulfated. Sequentially, apply this RCA-I$^-$ component to the Gal-3 column. The RCA-I$^-$ component binds to the Gal-3 column, named A2-IV. These fractionated components of A2 are summarized in Fig. 1.

Furthermore, *O*-linked glycans released from metabolic labeled MUC1 can be also determined to be the same structures as NaB^3H_4-reduced *O*-glycans by serial lectin affinity chromatography, although the data are not introduced in this chapter. [^{3}H]-*O*-glycans released from metabolic labeled culture cells (*see* **Note 6**) can be analyzed with 100 times higher sensitivity than NaB^3H_4-reduced glycans and the metabolic labeling method in combination with serial lectin affinity chromatography is a powerful method for determining the glycan profiles of the specified cultured cells.

3.4 Set of Lectin Affinity Chromatography of PSAs in PC and BPH Sera

Prepare PSA samples from PC serum, BPH serum, and LNCaP cell medium [8]. Apply these PSA samples to a set of lectin-Sepharose 4B columns, including LCA, DSA, MAL, TJA-I, UEA-I, and WFA. Resolve 10 μL of PC serum, 50 μL of BPH serum, and 10 μL of LACaP cell medium with 90 μL, 50 μL, and 90 μL of TB-BSA, respectively (*see* **Note 7**). Operate according to Subheading 3.1. Measure the concentration of PSA in each fraction by using the Hybritech-free PSA test kit (Beckman Coulter, Inc., Brea, CA, USA). The elution patterns on the respective lectin affinity columns are shown in Fig. 2. The possible glycan structures of PSAs based on the behavior in lectin affinity columns are also summarized in Fig. 2. These results indicate that WFA and UEA-I can be used to discriminate between BPH and PC.

3.5 Fractionation with a Set of Lectin-Sepharose 4B Columns of Choriocarcinoma ALPs in Sera

Placental type of ALP is a GPI-anchored glycoprotein, which has two *N*-linked glycans at Asn_{122} and Asn_{249}. The sugar chain structures of placental ALP have been reported as follows [30]: Siaα2-3 galactosylated bi-antennary glycan, which binds to MAL-, LCA-, and Con A-Sepharose 4B columns (*see* Fig. 3).

Apply aliquots of serum ALPs from pregnant woman (24 weeks) and choriocarcinoma patients, ALPs released by PI-PLC digestion (100 m-units, 37 °C, 2 h) from JEG-3 cells to a set of lectin affinity columns. Use TJA-I-, UEA-I-, MAL-, LCA-, Con A-, and DSA-Sepharose 4B columns to discriminate the terminal glycan motifs of ALPs between healthy control and malignant transformation. Fractionate each sample according to Subheading 3.1. Aliquots from each fraction should be added to each well of microtiter plates (9018; Costar, Corning, NY, USA) for measurement of the enzymatic activity of ALP using *p*-nitrophenylphosphate. Elution patterns on these six lectin columns and the possible glycan structures are shown in Fig. 3. TJA-I, UEA-I, and DSA can be used to discriminate malignant-ALP and normal-ALP.

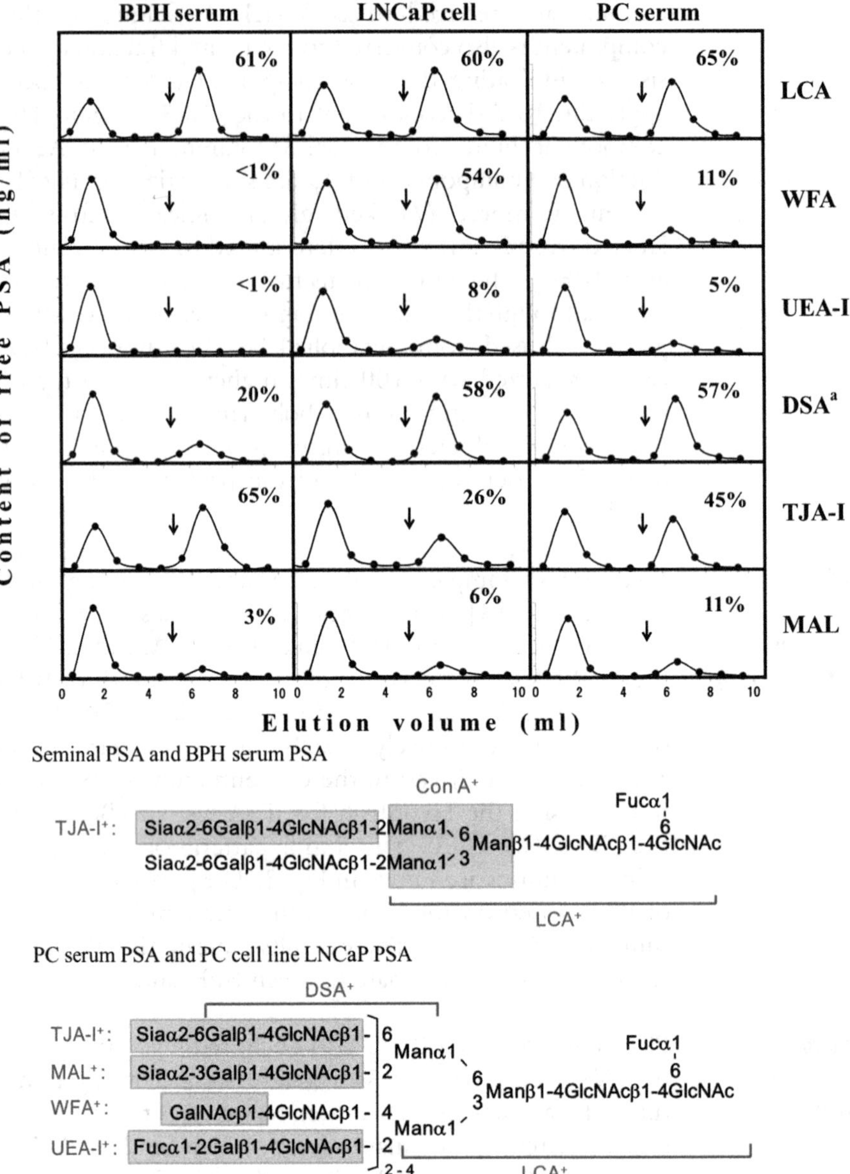

Fig. 2 Lectin-Sepharose 4B column chromatography of PSAs and the possible *N*-glycan structures of PSAs. PSAs are derived from BPH serum, PC patient serum, and PC cell line, LNCaP. (*Upper*) Elution patterns on LCA-, WFA-, UEA-I-, DSA-, TJA-I-, Con A-, and MAL-Sepharose 4B columns by monitoring contents of PSAs using ELISA. DSA[a]: *Arthrobacter* sialidase-digested PSA samples are applied to DSA column. *Arrows* show positions at which buffers are switched to TB-BSA containing various haptens (*see* Table 1). (*Lower*) Structures of *N*-linked glycans of PSAs predicted from elution patterns on various lectin-Sepharose 4B columns

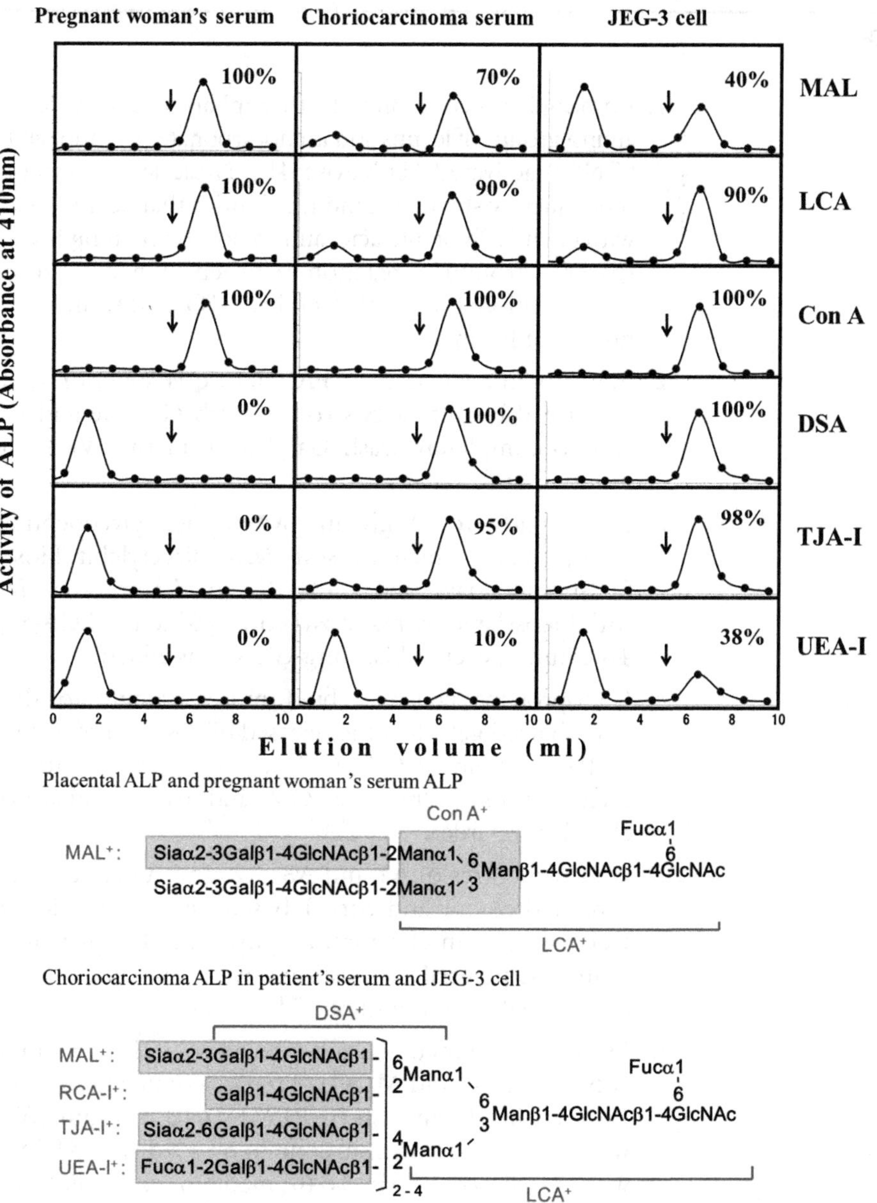

Fig. 3 Elution patterns on lectin-Sepharose 4B column chromatography of human ALPs and the possible *N*-glycan structures of ALPs. ALPs are derived from pregnant woman's serum (24 weeks), choriocarcinoma patient serum, and the choriocarcinoma cells JEG-3. (*Upper*) Elution patterns on MAL-, LCA-, Con A-, DSA-, TJA-I-, and UEA-I-Sepharose 4B columns by monitoring enzymatic activities of ALPs. *Arrows* indicate positions at which buffers are switched to TB-BSA containing various haptens (*see* Table 1). (*Lower*) Structures of major *N*-linked glycans of ALPs are predicted from elution patterns on various lectin-Sepharose 4B columns

4 Notes

1. Immobilize lectins directly to Sepharose 4B by reacting the amino groups of lectins and imidocarbonate of cyanogen bromide (CNBr)-activated Sepharose 4B. These lectin columns show reproducibly stronger binding abilities than lectin-agarose gels with chemically stable acid-amido bonds retaining hydrophobic spacers through a reaction between *N*-hydroxysuccinimide esters (Affi-Gel 10, Affi-Gel 15, BIO-RAD) and the amino groups of lectins.
2. For haptenic sugar to WFA, lactose is superior to GalNAc because WFA interacts strongly with GalNAc and it is difficult to completely wash GalNAc from the WFA-Sepharose 4B column.
3. LCA binds to *N*-glycan pyranoside, glycopeptides, and glycoproteins which possess *N,N'*-diacetylchitobiose. C-2-, C-2-bi-antennary and C-2-, C-2,6-tri-antennary *N*-glycans with fucosylated tri-mannosyl core bind to the LCA-Sepharose 4B column, even if bisecting GlcNAc is added.
4. Glycoproteins generally bind more strongly to the lectin-immobilized gels than the released oligosaccharides. DSA binds to PSA containing C-2, C-2, and 4-substituted tri-antennary glycan, although free C-2, C-2, and 4-tri-antennary oligosaccharide is retarded.
5. Core structures of A1 and A3, A2, and A4 are determined as extended core 1 and core 1 based on the behaviors on Bio-Gel P-4 column chromatography, RCA-I-Sepharose 4B column, sequential digestion with β-galactosidase, and β-*N*-acetylhexosaminidase [7].
6. Purification of metabolically labeled MUC1 is performed similarly to that of cold MUC1. Prepare confluent cells, YMB1-E, cultured in 10-cm dish in RPMI 1640 medium. Wash once with glucose-free medium containing 2 % dialyzed FCS. Replace with 4 mL of RPMI 1640 medium containing 7.4 MBq (200 μCi) of glucosamine–HCl, D-(6-^{3}H) (2.22 TBq/mmol; American Radiolabeled Chemicals, Inc., St. Louis, MO, USA), 1/10 concentration of glucose, and 2 % dialyzed FCS [11]. Incubate at 37 °C for 24 h under a 5 % CO_2 atmosphere. Centrifuge the medium at 10,000 × *g* for 10 min, and collect the supernatant. Remove the lower molecular weights of radioactive components, and purify the sample using a KL-6/mAb-conjugated Sepharose 4B column (0.05 mg mAb/mL, 1 mL bed volume). The radiolabeled *O*-glycans of metabolic labeled MUC1 are released by incubation with 0.05 M KOH/1 M $NaBH_4$ solution at 30 °C for 15 h and are *N*-acetylated. Monosaccharide composition analysis in *O*-linked glycans of

MUC1 obtained by metabolic labeling with [^{3}H]-glucosamine at 37 °C for 24 h shows substantial conversion to tritium-labeled *N*-acetylgalactosamine and galactose, but not to mannose, suggesting that [^{3}H]-glucosamine labeling is effective for analyzing *O*-linked glycans with high sensitivity.

7. The binding capacities of lectins for oligosaccharides are several nmole/mL. Since the content of serum glycoproteins is rather high, quantities of sera applied to the lectin-Sepharose 4B column generally should be kept under 50 μL to prevent overcharging.

Acknowledgements

This work was supported in part by JSPS KAKENHI, Grant Number 24590345, in Japan.

References

1. Valki A (1993) Biological roles of oligosaccharides: all of theories are correct. Glycobiology 3:97–130
2. Vliegenthart JFG, Kamerling JP (2007) 1H-NMR structural-reporter-group concepts in carbohydrate analysis. In: Kamerling JP (ed) Comprehensive glycoscience, vol 2. Elsevier Ltd, Oxford, pp 133–191
3. Dell A, Morris HR (2001) Glycoprotein structure determination by mass spectrometry. Science 291:2351–2356
4. Geyer R, Geyer H (1994) Saccharide linkage analysis using methylation and other techniques. In: Lennarz WJ, Hart GW (eds) Methods in enzymology, vol 230. Academic, San Diego, pp 86–108
5. Cummings RD, Etzler ME (2009) Antibodies and lectins in glycan analysis. In: Varki A, Cummings RD, Esko JD, Freeze H, Hart G, Marth J (eds) Essentials of glycobiology, 2nd edn. Cold Spring Harbor Laboratory Press, New York, Chapter 45
6. Kobata A, Yamashita K (1993) Fractionation of oligosaccharides by serial affinity chromatography with use of immobilized lectin columns. In: Fukuda M, Kobata A (eds) Practical approach-glycoprotein analysis. Oxford University Press, New York, pp 103–125
7. Seko A, Ohkura T, Ideo H et al (2012) Novel *O*-linked glycans containing 6'-sulfo-Gal/GalNAc of MUC1 secreted from human breast cancer YMB-S cells: possible carbohydrate epitopes of KL-6(MUC1) monoclonal antibody. Glycobiology 22:181–195
8. Fukushima K, Satoh T, Baba S et al (2010) α-1,2-Fucosylated and β-*N*-acetylgalactosaminylated prostate-specific antigen as an efficient marker of prostatic cancer. Glycobiology 20:452–460
9. Fukushima K, Hara-Kuge S, Seko A et al (1998) Elevation of α2-6sialyltransferase and α1-2fucosyltransferase in human choriocarcinoma. Cancer Res 58:4301–4306
10. Cuatrecasas P, Anfinsen CB (1971) Affinity chromatography. In: Jakohy WS (ed) Methods in enzymology. Academic, New York, pp 345–378
11. Ohkura T, Seko A, Hara-Kuge S et al (2002) Occurrence of secretory glycoprotein-specific GalNAcβ1-4GlcNAc sequence in *N*-glycans in MDCK cells. J Biochem 132:891–901
12. Ohkura T, Hada T, Higashino K et al (1994) Increase of fucosylated serum cholinesterase in relation to high risk groups for hepatocellular carcinomas. Cancer Res 54:55–61
13. Yamashita K, Umetsu K, Suzuki T et al (1988) Carbohydrate binding specificity of immobilized *Allomyrina dichotoma* lectin II. J Biol Chem 263:17482–17489
14. Yamashita K, Umetsu K, Suzuki T et al (1992) Purification and characterization of a Neu5Acα2→6Galβ1→4GlcNAc and HSO_3^-→6Galβ1→4GlcNAc specific lectin in tuberous roots of *Trichosanthes japonica*. Biochemistry 31:11647–11650
15. Wang WC, Cummings RD (1988) The immobilized leukoagglutinin from the seeds of *Maackia amurensis* binds with high affinity to complex-type Asn-linked oligosaccharides

containing terminal sialic acid-linked α2,3 to penultimate galactose residues. J Biol Chem 263:4576–4585

16. Shibuya N, Goldstein IJ, Broekaert WF et al (1987) The elderberry (*Sambucus nigra* L.) bark lectin recognizes the Neu5Ac(α2-6)Gal/GalNAc sequence. J Biol Chem 262:1596–1601
17. Kochibe N, Matta KL (1989) Purification and properties of an *N*-acetylglucosamine-specific lectin from *Psathyrella velutina* mushroom. J Biol Chem 264:173–177
18. Ideo H, Seko A, Ohkura T et al (2002) High-affinity binding of recombinant human galectin-4 to $SO_3^- \rightarrow 3Gal\beta1 \rightarrow 3GalNAc$ pyranoside. Glycobiology 12:199–208
19. Smith DF, Torres BV (1989) Lectin affinity chromatography of glycolipids and glycolipid-derived oligosaccharides. In: Ginsburg V (ed) Methods in enzymology, vol 179. Academic, New York, pp 30–45
20. Hindgaul O, Norberg T, Le Pendu J et al (1982) Synthesis of type 2 human blood-group antigenic determinants. The H, X and Y haptens and variations of the H type 2 determinants probes for the combining site of the lectin I of *Ulex europaeus.* Carbohydr Res 109:109–142
21. Kornfeld K, Reitman ML, Kornfeld R (1981) The carbohydrate-binding specificity of pea and lentil lectins: fucose is an important determinant. J Biol Chem 256:6633–6640
22. Yamashita K, Kochibe N, Ohkura T et al (1985) Fractionation of L-fucose-containing oligosaccharides on immobilized *Aleuria aurantia* lectin. J Biol Chem 260:4688–4693
23. Yamashita K, Totani K, Ohkura T et al (1987) Carbohydrate binding properties of complex-type oligosaccharides on immobilized *Datura stramonium* lectin. J Biol Chem 262:1602–1607
24. Baenziger JU, Fiete D (1979) Structural determinants of concanavalin A specificity for oligosaccharides. J Biol Chem 254:2400–2407
25. Hoyer LL, Roggentin P, Schauer R et al (1991) Purification and properties of cloned *Salmonella typhimurium* LT2 sialidase with virus: typical kinetic preference for sialyl α2-3 linkages. J Biochem 110:29–41
26. Uchida Y, Tsukada Y, Sugimori T (1974) Production of microbial neuraminidase induced by colominic acid. Biochim Biophys Acta 350:425–431
27. Fukushima K, Hada T, Higashino K et al (1998) Elevated serum levels of *Trichosanthes japonica* agglutinin-I binding alkaline phosphatase in relation to high-risk groups for hepatocellular carcinomas. Clin Cancer Res 4:2771–2777
28. Kiyohara T, Terao T, Shioiri-Nakano K, Osawa T (1976) Purification and characterization of β-*N*-acetylhexosaminidase, and β-galactosidase from Streptococcus 6646K. J Biochem 80:9–17
29. Li YT, Li SC (1972) α-Mannosidase, β-*N*-acetylhexosaminidase, and β-galactosidase from jack bean meal. In: Ginsburg V (ed) Methods in enzymology. Academic, New York, pp 702–713
30. Endo T, Ohbayashi H, Ikehara Y et al (1988) Structural study on the carbohydrate moiety of human placental alkaline phosphatase. J Biochem 103:182–187

Chapter 8

Lectin-Probed Western Blot Analysis

Takeshi Sato

Abstract

Lectin-probed western blot analysis, the so-called lectin blot analysis, is a useful method to yield basic information on the glycan structures of glycoproteins, based on the carbohydrate-binding specificities of lectins. By lectin blot analysis, researchers can directly analyze the glycan structures without releasing the glycans from glycoproteins. Here, the author describes protocols for standard analysis, and applies analysis in combination with glycosidase digestion of blot.

Key words Western blot analysis, Lectin, Carbohydrate-binding specificity, Glycoprotein, Glycosidase

1 Introduction

Diverse glycan structures of glycoproteins have been shown to be involved in a variety of biological events including embryonic development and functional maintenance of tissues and organs [1, 2]. Analysis of such glycan structures is, therefore, essential for studying life sciences by researchers in not only the field of glycobiology but also others. In order to obtain information about the fine structures of glycans released from glycoproteins, the instrumental analyses including high-performance liquid chromatography, nuclear magnetic resonance, and mass spectrometry have already been established. However, researchers need to set up expensive equipment. Furthermore, it takes too long to prepare samples for such analyses and analyze the glycan structures with the aid of experts in most cases.

Lectin blot analysis is a simple and useful method based on western blot analysis, which is a routine experiment for many laboratories [3]. In the case of lectin blot analysis, researchers are not required to release the glycans from glycoproteins, and can obtain information about the glycan structures on their own. The combination of glycosidase digestion and lectin staining provides information regarding the carbohydrate structures at the nonreducing termini, types of *N*-glycan subgroups, and presence or absence of

Jun Hirabayashi (ed.), *Lectins: Methods and Protocols*, Methods in Molecular Biology, vol. 1200,
DOI 10.1007/978-1-4939-1292-6_8,

O-glycans in individual glycoproteins [4–7]. The glycosylation patterns between the control and test group can be compared by lectin blot analysis. For instance, lectin blot analysis has revealed that the expression of β-*N*-acetylgalactosaminylated *N*-glycans is associated with functional development of bovine mammary gland [6] and malignant transformation of human mammary epithelial cells [7], that the treatment with anticancer drug affects the glycosylation of SW480 human colorectal cancer cells [8], and that the β-1,4-galactosylation of *N*-glycans increases depending on cell density of mouse BALB/3T3 fibroblasts [9].

2 Materials

2.1 SDS-Polyacrylamide Gel Components

1. Electrophoresis apparatus with glass plates, casting stand, clamps, and buffer chambers (Nihon Eido Co., Ltd., Tokyo, or Bio-Rad, Hercules, CA).
2. Ammonium persulfate: 10 % solution in distilled water.
3. *N*,*N*,*N*′,*N*′-tetramethylethylenediamine (TEMED).
4. Sodium dodecyl sulfate (SDS): 10 % solution in distilled water.
5. SDS lysis buffer: 62.5 mM Tris–HCl (pH 6.8), 10 % glycerol, 2 % SDS, 5 % β-mercaptoethanol, 0.0125 % bromophenol blue.
6. Resolving gel buffer: 1.5 M Tris–HCl, pH 8.8 (Bio-Rad).
7. Stacking gel buffer: 0.5 M Tris–HCl, pH 6.8 (Bio-Rad).
8. Thirty percent acrylamide/Bis solution (29:1 acrylamide:Bis) (Bio-Rad).
9. SDS-PAGE running buffer: 25 mM Tris (base), 192 mM glycine, 0.1 % SDS.

2.2 Western Blotting Components

1. Mini Trans-Blot Electrophoretic Transfer Cell (Bio-Rad) or equivalent.
2. Western blot transfer buffer: 25 mM Tris (base), 192 mM glycine.
3. Polyvinylidene difluoride (PVDF) membrane (Immobilon-P Transfer Membrane, pore size 0.45 μm; Millipore Corp., Bedford, MA).
4. Mini Trans-Blot Filter Paper (Bio-Rad).
5. Ten millimolar phosphate-buffered saline, pH 7.2 (PBS).
6. PBS containing 0.05 % (v/v) Tween 20 (PBST).
7. Blocking solution: 1 % bovine serum albumin (BSA) (Fraction V, Roche Applied Science, Indianapolis, IN) in PBS.
8. Twenty five millimolar sulfuric acid.

9. Peroxidase-substrate solution: 0.05 % 4-chloro-1-naphthol, 0.03 % hydrogen peroxide, 17 % (v/v) methanol in PBS. Dissolve 15 mg of 4-chloro-1-naphthol in 5 mL of methanol and mix with 25 mL of PBS. Add 10 μL of hydrogen peroxide to the solution and mix just before use.

2.3 Lectins

1. Horseradish peroxidase (HRP)-conjugated lectins (concanavalin A (Con A), *Datura stramonium* agglutinin (DSA), peanut agglutinin (PNA), *Ricinus communis* agglutinin-I (RCA-I), leuko-agglutinating phytohemagglutinin (L-PHA), *Wisteria floribunda* agglutinin (WFA), wheat germ agglutinin (WGA), etc.; Cosmo Bio Co. Ltd., Tokyo, or EY Laboratories, San Mateo, CA).

   Dilute HRP-conjugated lectin to 10 μg/mL with PBST.
2. Biotinylated lectins (*Lycopersicon esculentum* (Tomato) lectin (LEL), *Maackia amurensis* agglutinin (MAA), *Sambucus nigra* agglutinin (SNA), etc.; Vector Laboratories Inc., Burlingame, CA).

Details in the carbohydrate-binding specificities of lectins should be consulted with references [10–13].

2.4 Glycosidases

Conditions of glycosidase digestion (for one sample lane).

1. Recombinant *N*-glycanase (*Flavobacterium meningosepticum*; Roche Applied Science).
   2.0 units/200 μL of 0.1 M phosphate buffer (pH 8.6).
2. Sialidase (*Arthrobacter ureafaciens*; Nacalai Tesque, Inc.).
   0.5 units/200 μL of 0.5 M acetate buffer (pH 5.0).
3. β-Galactosidase (*Diplococcus pneumoniae*; Roche Applied Science).
   30 munits/200 μL of 0.3 M citrate-phosphate buffer (pH 6.0).
4. β-*N*-Acetylhexosaminidase (jack bean; Cosmo Bio Co. Ltd.).
   1.5 units/200 μL of 0.3 M citrate-phosphate buffer (pH 4.0–5.0).

Details in the digestion conditions of other glycosidases should be consulted with ref. 14.

3 Methods

3.1 7.5 % Sodium Dodecyl Sulfate-Polyacrylamide Gel Electrophoresis (SDS-PAGE)

1. Assemble the glass-plate sandwich of the electrophoresis apparatus according to the manufacturer's instructions.
2. Mix 3.03 mL distilled water, 1.58 mL of resolving gel buffer, 1.58 mL of 30 % acrylamide/Bis solution, and 62.5 μL of SDS in a 15 mL centrifuge tube. Add 31.25 μL of ammonium persulfate and 3.15 μL of TEMED, and mix gently.

Pour the gel solution to the sandwich (size: 8.5 cm × 9.5 cm × 1.0 mm), and gently overlay with distilled water. Allow the gel to polymerize.

3. Prepare the stacking gel by mixing 1.52 mL distilled water, 0.63 mL of stacking gel buffer, 0.33 mL of 30 % acrylamide/Bis solution, and 25 μL of SDS in a 15 mL centrifuge tube. Add 15.7 μL of ammonium persulfate and 3.15 μL of TEMED, and mix gently. Overlay the stacking gel on the resolving gel. Insert a 12-well gel comb immediately without introducing air bubbles. Allow the gel to polymerize.
4. Dissolve protein samples in appropriate volume of SDS lysis buffer, and heat them at 100 °C for 5 min. Centrifuge the heated samples at 10,000 × *g* for 10 min. Transfer the supernatant to a new tube.
5. Apply 1–20 μg of sample protein to each well. Subject the gel to SDS-PAGE according to the manufacturer's instructions.
6. After electrophoresis, take the top grass plate off with a spatula. Remove the stacking gel. Transfer the gel carefully to a container with western blot transfer buffer (*see* **Note 1**).
7. Cut a PVDF membrane to the size of the gel and immerse in methanol, and then in western blot transfer buffer until use.
8. Soak fiber pads and filter papers in western blot transfer buffer.

3.2 Electrophoretic Transfer

1. Prepare the gel sandwich according to the manufacturer's instructions. Place the gel holder cassette in a tray that is filled with western blot transfer buffer.
2. On the bottom half of the cassette, place one pre-wetted fiber pad.
3. Place a sheet of pre-wetted filter paper on the fiber pad.
4. Place the gel on the filter paper, and then the pre-wetted PVDF membrane on the gel. Remove all air bubbles.
5. Place a sheet of pre-wetted filter paper on the membrane, and then another pre-wetted fiber pad on the filter paper.
6. Close the cassette tightly.
7. Transfer proteins from gel to the PVDF membrane in western blot transfer buffer at 100 V for 75 min according to the manufacturer's instructions.

3.3 Lectin Staining

1. Block the membrane with 1 % BSA solution at 20 °C for 2 h or at 4 °C overnight.
2. Wash the membrane with PBST for 10 min three times (*see* **Note 2**).

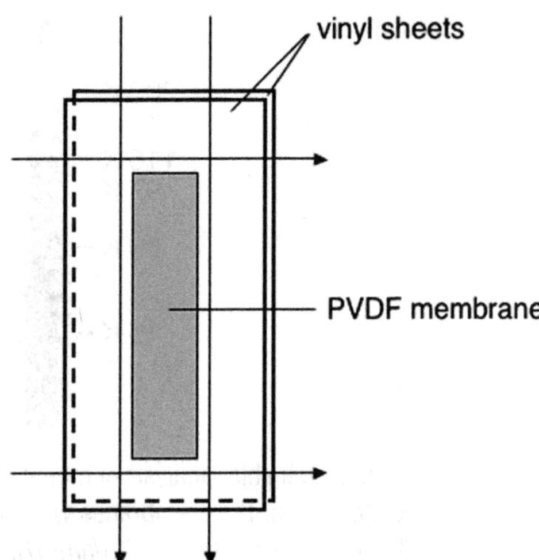

Fig. 1 Sealing of vinyl sheets. Seal two commercial vinyl sheets by Polysealer to fit on the size of a PVDF membrane. The *arrows* indicate the sealed positions by Polysealer

3. Place the membrane between two vinyl sheets followed by sealing the three corners (Fig. 1).
4. Add 200 μL of PBST containing HRP-conjugated lectin into the sealed bag (*see* **Note 3**).
5. Squeeze air bubbles out from the bag with fingers followed by sealing the mouth.
6. Fix the bag on a supporting plate with tape (*see* **Note 4**).
7. Incubate the plate at 4 °C for 1 h (*see* **Note 5**).
8. Open the bag and wash the membrane with 30 mL of PBST for 5 min five times.
9. Place the membrane into 30 mL of peroxidase-substrate solution and let bands develop (*see* **Note 6**) (Fig. 2).
10. Terminate the reaction by washing the membrane several times in distilled water before the background color appears (*see* **Note 7**).

3.4 Treatment of Membrane with Glycosidases

The following steps are added after **step 2** of Subheading 3.3.

1. Equilibrate the membrane in several washes with the buffer to be used for glycosidase treatment.
2. Place the membrane between two vinyl sheets followed by sealing the three corners.
3. Add 200 μL of enzyme solution into the sealed bag (*see* **Note 3**).
4. Squeeze air bubbles out from the bag with fingers followed by sealing the mouth.

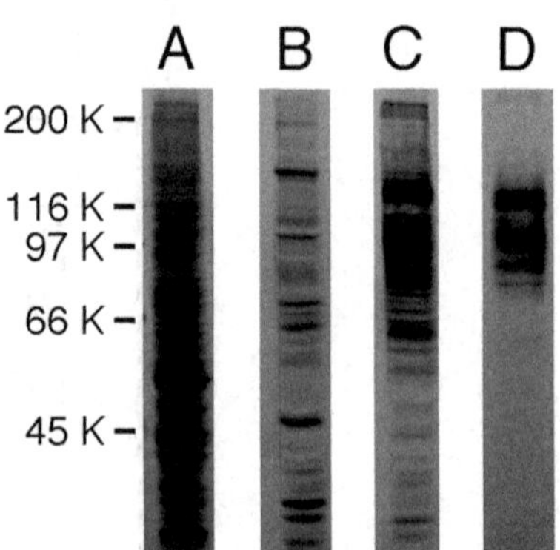

Fig. 2 Lectin blot analysis of membrane glycoproteins from MKN1 human gastric cancer cells. The membranes were incubated with CBB (*lane A*), Con A (*lane B*), RCA-I (*lane C*), or L-PHA (*lane D*), followed by visualization. Note that Con A bound mainly to high-mannose-type oligosaccharides [11], RCA-I interacted with glycans terminating with Galβ1 → 4GlcNAc group [10], and L-PHA specifically bound to highly branched *N*-glycans [12]

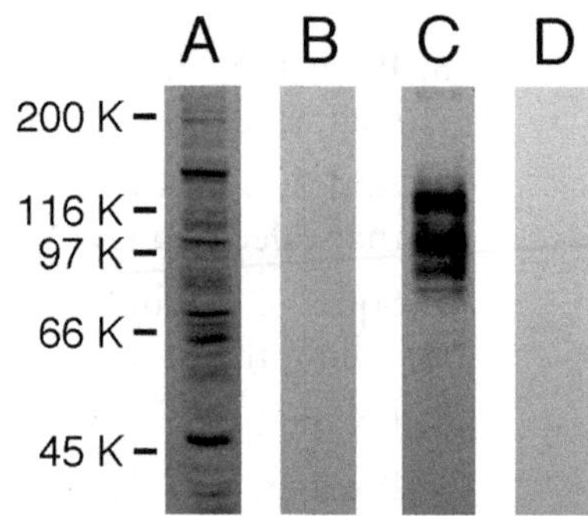

Fig. 3 Lectin blot analysis of membrane glycoproteins from MKN1 cells. The membranes were incubated either with Con A before (*lane A*) and after (*lane B*) treatment with *N*-glycanase or with L-PHA before (*lane C*) and after (*lane D*) treatment with *N*-glycanase, followed by visualization. Note that no lectin binding was observed upon digestion with *N*-glycanase (*lanes B* and *D*)

5. Fix the bag on a supporting plate with tape (*see* **Note 4**).
6. Incubate the plate at 37 °C for 24 h.
7. Open the bag and immerse the membrane again in 1 % BSA solution at 20 °C for 30 min (*see* **Note 8**).
8. Wash the membrane with PBST for 10 min three times.
9. Incubate the membrane with HRP-conjugated lectin at 4 °C for 1 h.

   From this point forward, return to **step 8** of Subheading 3.3.

The results of *N*-glycanase treatment followed by lectin staining are shown in Fig. 3.

4 Notes

1. Use gloves when manipulating gels and PVDF membranes.
2. When using RCA-I, which interacts with glycans terminating with Galβ1→4GlcNAc group [10], the membranes should be treated with 25 mM sulfuric acid at 80 °C for 1 h to remove sialic acid prior to incubation with lectin [8]. The treatment is necessary for enhancement of RCA-I binding because most of the galactose residues on glycoproteins are sialylated in mammalian cells (Fig. 4). After sulfuric acid treatment, the membranes are re-blocked with 1 % BSA solution at 20 °C for 30 min (*see* **Note 8**). Alternatively, sialidase digestion can be conducted instead of the treatment with sulfuric acid.
3. The author usually uses 200 μL of lectin or enzyme solution for a PVDF membrane (size: 1 cm × 7 cm) corresponding to one sample lane of a typical minigel.
4. A Styrofoam plate can be used as a supporting plate. If the membranes are bent during incubation with lectins or enzymes, patterns of band staining with lectins may be unevenly visualized.
5. If researchers use biotinylated lectins, after 1-h incubation at 4 °C with lectin, the membranes are washed as in **step 8** of Subheading 3.3. Then, after 1-h incubation at 4 °C with HRP-conjugated streptavidin, follow **steps 8–10** of Subheading 3.3.
6. In order to detect rare structures and/or small amounts of glycans, chemiluminescence-based technique [e.g., ECL Western Blotting Analysis System (GE Healthcare, Buckinghamshire, UK)] can be used for increasing the sensitivity of detection system.

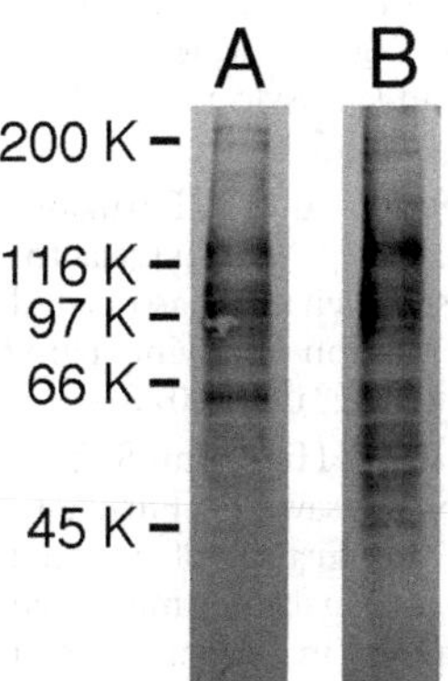

Fig. 4 Lectin blot analysis of membrane glycoproteins from MKN1 cells. The membranes were incubated with RCA-I before (*lane A*) and after (*lane B*) treatment with sulfuric acid, followed by visualization. Note that RCA-I binding was enhanced by the treatment with sulfuric acid (*lane B*)

7. After washing the membranes in distilled water, the membranes should be dried completely, and kept in vinyl sheets without light. Otherwise, the visualized bands will be faint or disappear.
8. Empirically, good and clear patterns of band staining with lectins are obtained with low background.

Acknowledgement

The author thanks Ms. Riho Tange in his laboratory for her excellent technical assistance.

References

1. Varki A, Cummings RD, Esko JD, Freeze HH, Stanley P, Bertozzi CR, Hart GW, Etzler ME (2009) Essentials of glycobiology, 2nd edn. Cold Spring Harbor Laboratory Press, Cold Spring Harbor, NY
2. Furukawa K, Sato T (1999) β-1,4-Galactosylation of *N*-glycans is a complex process. Biochim Biophys Acta 1473:54–66
3. Sato T, Furukawa K, Greenwalt DE, Kobata A (1993) Most bovine milk fat globule membrane glycoproteins contain asparagine-linked sugar chains with GalNAcβ1→4GlcNAc groups. J Biochem 114:890–900
4. Taka J, Sato T, Sakiyama T, Fujisawa H, Furukawa K (1996) Bovine pituitary membrane glycoproteins contain β-N-acetylgalactosaminylated *N*-linked sugar chains. J Neurochem 66:852–859
5. Kumagai T, Sato T, Natsuka S, Kobayashi Y, Zhou D, Shinkai T, Hayakawa S, Furukawa K (2010) Involvement of murine β-1,4-galactosyltransferase V in lactosylceramide biosynthesis. Glycoconj J 27:685–695
6. Sato T, Taka J, Aoki N, Matsuda T, Furukawa K (1997) Expression of β-N-acetylgalactosaminylated *N*-linked sugar chains is associated with functional differentiation of bovine mammary gland. J Biochem 122:1068–1073
7. Kitamura N, Guo S, Sato T, Hiraizumi S, Taka J, Ikekita M, Sawada S, Fujisawa H, Furukawa K (2003) Prognostic significance of reduced expression of β-*N*-acetylgalactosaminylated *N*-linked oligosaccharides in human breast cancer. Int J Cancer 105:533–541
8. Sato T, Takahashi M, Kawado T, Takayama E, Furukawa K (2005) Effect of staurosporine on *N*-glycosylation and cell adhesion to fibronectin of SW480 human colorectal adenocarcinoma cells. Eur J Pharm Sci 25:221–227
9. Tadokoro T, Ikekita M, Toda T, Ito H, Sato T, Nakatani R, Hamaguchi Y, Furukawa K (2009) Involvement of galectin-3 with vascular cell adhesion molecule-1 in growth regulation of mouse BALB/3T3 cells. J Biol Chem 284:35556–35563
10. Baenziger JU, Fiete D (1979) Structural determinants of *Ricinus communis* agglutinin and toxin specificity for oligosaccharides. J Biol Chem 254:9795–9799
11. Ogata S, Muramatsu T, Kobata A (1975) Fractionation of glycopeptides by affinity column chromatography on concanavalin A-Sepharose. J Biochem 78:687–696
12. Cummings RD, Kornfeld S (1982) Characterization of the structural determinants required for the high affinity interaction of asparagine-linked oligosaccharides with immobilized *Phaseolus vulgaris* leukoagglutinating and erythroagglutinating lectins. J Biol Chem 257:11230–11234
13. Cummings RD (1994) Use of lectins in analysis of glycoconjugates. In: Lennarz WJ, Hart GW (eds) Guide to techniques in glycobiology. Academic, San Diego, pp 66–86
14. Jacob GS, Scudder P (1994) Glycosidases in structural analysis. In: Lennarz WJ, Hart GW (eds) Guide to techniques in glycobiology. Academic, San Diego, pp 280–299

Chapter 9

Solid-Phase Assay of Lectin Activity Using HRP-Conjugated Glycoproteins

Kyoko Kojima-Aikawa

Abstract

Various enzyme-conjugated probes have been widely used for detection of specific interactions between biomolecules. In the case of glycan-protein interaction, horseradish peroxidase (HRP)-conjugated glycoproteins (HRP-GPs) are useful for the detection of carbohydrate-binding activity of plant and animal lectins. In this chapter, a typical solid-phase assay of the carbohydrate-binding activity of *Sophora japonica* agglutinin I, a Gal/GalNAc-specific lectin, using HRP-conjugated asialofetuin is described. HRP-GPs are versatile tools for probing lectin activities in crude extracts, screening many samples at one time, and applicable not only for solid-phase binding assays but also samples which are dot- or Western-blotted onto the membrane.

Key words Glycoprotein, Solid-phase assay, Horseradish peroxidase, 96-Well plate, Carbohydrate-binding activity

1 Introduction

The carbohydrate-binding properties of lectins have traditionally been determined by hemagglutination tests using red blood cells and precipitation reactions using glycoconjugates. The minimum concentrations of carbohydrates completely inhibiting two or four hemagglutinating doses and the concentrations required for a 50 % decrease in agglutination or precipitation are usually used to express the relative specificity of lectins. Although the hemagglutination inhibition assay is the most convenient and widely used method, it is not precisely quantitative and is affected by experimental conditions. Precipitin formation in the precipitation reaction requires a much longer incubation time than the hemagglutination reaction of a lectin and a glycoconjugate having polyvalent ligands. Furthermore, crude extracts often contain hemolysin and other interfering substances; hence, these two methods are not appropriate for wide-range detection of lectin activity in crude extracts. Instead of the hemagglutination

Jun Hirabayashi (ed.), *Lectins: Methods and Protocols*, Methods in Molecular Biology, vol. 1200,
DOI 10.1007/978-1-4939-1292-6_9, © Springer Science+Business Media New York 2014

test, various probes have been applied to lectin studies. Radiolabeled glycoproteins have been used for lectin characterization as a sensitive reagent [1–4]. Neoglycoproteins prepared by covalently attaching carbohydrate residues to a carrier protein [5, 6], synthetic cluster glycosides constructed using aspartic acid to which 6-aminohexyl glycosides were linked [7, 8], and neoglycolipids prepared by conjugation of lipid and carbohydrate chains from a glycoprotein [9] have also been utilized for lectin binding studies as custom-made reagents. Some applications of enzyme labeling have been reported for lectin detection. Neoglycoprotein-enzyme conjugates [10] and neoglycoenzymes [11] have been employed to quantitate plant and animal lectins. Carbohydrate-enzyme conjugates prepared by glycosylation of malate dehydrogenase were used to study the interaction of jacalin lectin [12].

In this study, we prepared horseradish peroxidase (HRP)-conjugated glycoproteins (GPs) by direct coupling of formyl HRP to natural GPs for use as a convenient probe assaying carbohydrate-binding activity. Their usefulness was demonstrated through detection and characterization of lectin activities in both purified lectins and crude bark extracts [13].

2 Materials

1. *Sophora japonica* agglutinin I [14].
2. Asialofetuin: Prepared from fetuin (Sigma) by desialylation with 25 mM H_2SO_4 at 80 °C for 1 h.
3. 1 mM acetate buffer (pH 4.0).
4. Tris-buffered saline (TBS; 10×): 1.5 M NaCl, 0.1 M Tris–HCl, pH 7.6.
5. Substrate solution (*see* **Note 1**): 100 mM citrate-phosphate buffer, pH 5.0, containing 0.04 % *o*-phenylenediamine and 0.007 % H_2O_2 (*see* **Note 2**).
6. 4 M H_2SO_4.
7. Blocking solution: 3 % bovine serum albumin in TBS.
8. Horseradish peroxidase (HRP I-C, Toyobo, Osaka, Japan) (*see* **Note 3**).
9. Acetate-buffered saline (ABS; 10×): 1.5 M NaCl, 0.1 M sodium acetate, pH 6.0.
10. 0.2 M sodium bicarbonate (pH 9.5).
11. Sodium borohydride (*see* **Note 4**).
12. Phosphate-buffered saline, pH 7.2 (PBS).
13. A 96-well plastic plate (ELISA plate #3801-096, Iwaki).

3 Methods

3.1 Preparation of HRP-GPs (See Notes 5 and 6)

1. Dissolve 5 mg of HRP in 1 mL of 1 mM sodium acetate buffer (pH 4); then add 0.8 mL freshly prepared 37 mM sodium periodate in water (*see* **Note 7**).
2. Incubate for 20 min at room temperature.
3. Add and dissolve 30 mg glucose in the HRP solution (*see* **Note 8**).
4. Dialyze the solution against 1 mM sodium acetate buffer (pH 4) at 4 °C with several changes overnight.
5. Prepare a 10 mg/mL glycoprotein solution (e.g., asialofetuin) in 0.2 M sodium bicarbonate (pH 9.5).
6. Remove the HRP solution from the dialysis tubing and add to 0.5 mL of the glycoprotein solution.
7. Incubate for 2 h at room temperature.
8. Add 60 μL of 0.4 % of sodium borohydride and incubate for 2 h at 4 °C (*see* **Note 9**).
9. Separate HRP-glycoprotein conjugates from uncoupled asialofetuin and HRP by gel filtration chromatography on a Toyopearl HW-50 (3 × 74 cm) column equilibrated with ABS.
10. Continuously collect the eluates (4 mL/tube) and monitor by measuring the absorbance at 280 nm. The first peak fractions are HRP-asialofetuin conjugates, and later peak fractions are uncoupled HRP and asialofetuin. The first peak fractions are combined and dialyzed against PBS.

3.2 Preparation of Bark Extract

1. Mill the bark of *S. japonica* in a pestle and mortar with 5 vol of 10 mM TBS.
2. Centrifuge the homogenate at 9,000 × *g* for 30 min, and use the supernatant as crude bark extract.

3.3 Binding and Inhibition Assay on 96-Well Plates with HRP-GPs

1. Prepare serial dilutions of B-SJA-I (62.5 μg/mL) and the crude bark extract of *S. japonica* (protein concentration, 10 mg/mL).
2. Add 100 μL each of sample solutions to each well of 96-well plastic plate to immobilize proteins on the bottom of the wells and incubate for 18 h at 4 °C.
3. Discard the solution by tapping the plate.
4. Add 300 μL of TBS to each well, leave for 3 min, and then discard the solution. Repeat this step three times.
5. Add 300 μL of blocking solution to each well and incubate for 30 min.
6. Discard the solution, and add 100 μL of blocking solution containing HRP-asialofetuin (10 μg/mL) to each well. Incubate for 1 h at 4 °C.

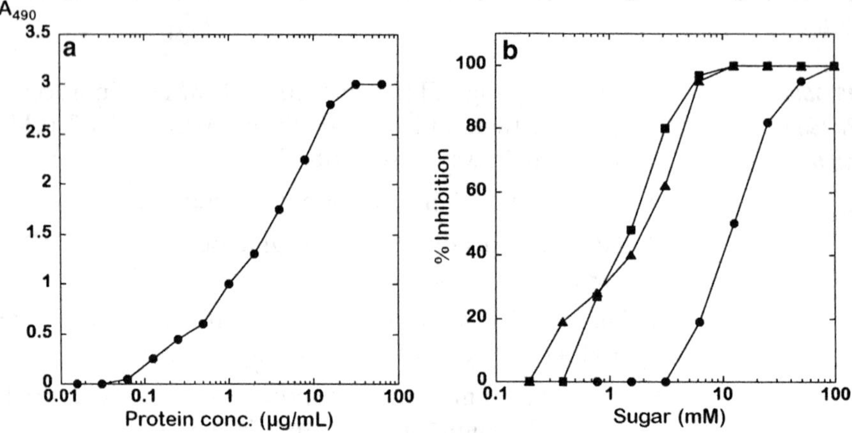

Fig. 1 Binding of HRP-asialofetuin to B-SJA-I (**a**) and inhibition assay with saccharides (**b**). The purified B-SJA-I solution was serially diluted and immobilized on 96-well plastic plates. The concentration of HRP-asialofetuin was 10 μg/mL, and the averages of triplicate determinations are plotted. Inhibitors: D-galactose (*filled circle*), *N*-acetyl-D-galactosamine (*filled square*), and lactose (*filled triangle*)

7. Discard the solution.
8. Add 300 μL of TBS to each well, leave for 3 min, and then discard the solution. Repeat this step three times.
9. Add 150 μL of substrate solution to each well and incubate for 7 min.
10. Stop the enzyme reaction by adding 4 M H_2SO_4.
11. Measure color development spectrophotometrically by absorbance at 490 nm using a plate reader (Model 3550, BIO-RAD) (Figs. 1 and 2).

4 Notes

1. This solution should be prepared freshly each time. Warm 100 mM citrate-phosphate buffer, pH 5.0, to 37 °C before use.
2. Hydrogen peroxide is usually available as a 30 % solution. Keep cold and use when fresh.
3. The quality of horseradish peroxidase can be determined by measuring the ratio of the HRP absorbance at 403 and 280 nm. OD 403 nm/280 nm should be at least 3.0. HRP is a glycoprotein having *N*-linked oligosaccharide chains, Manα1-3(Manα1-6)(Xylβ1-2)Manβ1-4GlcNAcβ1-4(Fucα1-3)GlcNAc. Therefore, intact HRP can be a convenient ready-made probe to detect mannose-specific lectins [15].

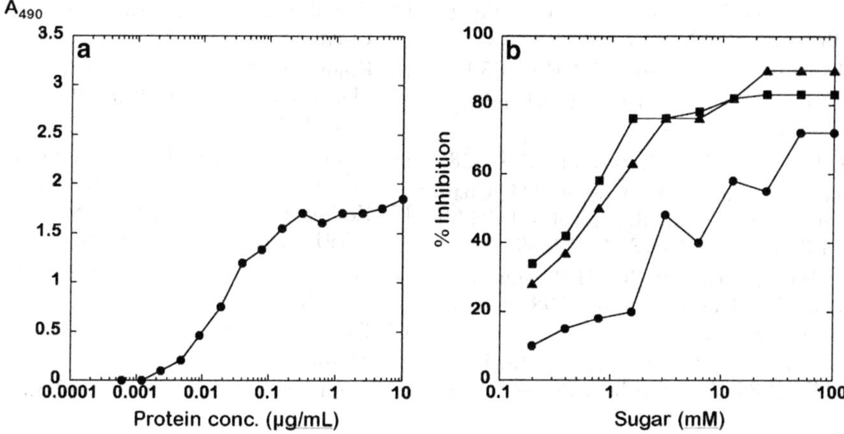

Fig. 2 Binding of HRP-asialofetuin to *Sophora japonica* bark extract (**a**) and inhibition assay with saccharides (**b**). The crude bark extract (protein concentration, ca. 10 mg/mL) was serially diluted and immobilized on a 96-well plastic plate. The averages of triplicate determinations are plotted. Inhibitors: D-galactose (*filled circle*), *N*-acetyl-D-galactosamine (*filled square*), and lactose (*filled triangle*)

4. Store in a small chamber with a desiccant.
5. A large number of enzymes have been used to label proteins, especially antibodies [16–18]. The most commonly used are HRP, alkaline phosphatase, and b-galactosidase. Urease and glucose oxidase are also in more limited use. The ideal enzyme is cheap, very stable, small, and easily conjugated.
6. Two methods commonly are used for the preparation of HRP conjugation, the glutaraldehyde method and the periodate method. The glutaraldehyde method has a relatively low coupling efficiency.
7. Periodate treatment of carbohydrates opens the ring structure and allows these moieties (aldehyde groups) to react with free amino groups in proteins.
8. This step is necessary to consume excess periodate and stop the reaction.
9. The Schiff's bases that have formed must be reduced by adding sodium borohydride.

References

1. Cooper DN, Lee SC, Barondes SH (1983) J Biol Chem 258:8745–8750
2. Lesniak AP, Liu EH (1984) Anal Biochem 142:140–147
3. Lehrman MA, Haltiwanger RS, Hill RI (1986) J Biol Chem 261:7426–7432
4. Kohuken RE, Berger EA (1987) Biochemistry 26:3949–3957

5. Stowell CP, Lee YC (1980) Adv Carbohydr Chem Biochem 37:225–281
6. Lee RT (1982) Biochemistry 21:1045–1050
7. Lee RT, Lin P, Lee YC (1984) Biochemistry 23:4255–4261
8. Lee RT, Lee YC (1987) Glycoconj J 4:317–328
9. Mizuochi T, Loveless RW, Lawson AM, Chai W, Lachmann PJ, Childs RA, Thiel S, Feizi T (1989) J Biol Chem 264:13834–13839
10. Gabius HJ, Engelhardt R, Hellmann KP, Hellmann T, Ochsenfahrt A (1987) Anal Biochem 165:349–355
11. Gabius S, Hellmann KP, Hellmann T, Brick U, Gabius HJ (1989) Anal Biochem 182:447–451
12. Kim B, Cha GS, Meyerhoff ME (1990) Anal Chem 62:2663–2668
13. Kojima K, Matsu-ura Y, Ogawa H, Seno N, Matsumoto I (1994) Plant Physiol Biochem 32:217–224
14. Ueno M, Ogawa H, Matsumoto I, Seno N (1991) J Biol Chem 266:3146–3153
15. Kojima K, Ogawa H, Seno N, Matsumoto I (1991) Carbohyr Res 275–282
16. Nakane PK, Kawaoi A (1974) J Histochem Cytochem 22:1084–1091
17. Farr AG, Nakane PK (1981) J Immunol Methods 47:129–144
18. Avrameas S (1972) Histochem J 4:321–330

Chapter 10

A Simple Viral Neuraminidase-Based Detection for High-Throughput Screening of Viral Hemagglutinin–Host Receptor Specificity

Nongluk Sriwilaijaroen and Yasuo Suzuki

Abstract

The correlation between precise interactions of influenza A virus hemagglutinins with host cell surface glycans having terminal sialic acids and host range specificity has provoked the development of a high-throughput viral-receptor specificity assay. Here, we describe the use of the virus itself as a specific antibody coupled to enzymes (virus with neuraminidase spikes) for determining its binding specificity to glycans, a strategy that reduces not only the cost but also the tedious steps of adding primary and secondary antibodies and washing between each step. All of the steps, including coating the glycopolymers onto microtiter plates, virus binding, and visual and quantitative detection of fluorescence products that correlate well with the amount of glycan-bound viruses, can be done within 3 h. This simple, rapid, sensitive, and reliable strategy is an ideal method for detection of high-throughput influenza virus receptor-binding preference not only for studies on viral evolution and transmission but also for viral surveillance in pandemic preparedness, leading to efficient prevention and control of the disease.

Key words Influenza virus, Receptor specificity, Sialylglycoconjugate, Evolution, Transmission, Pandemic, Surveillance, Hemagglutinin, Neuraminidase

1 Introduction

Influenza A virus, a member of the family of *Orthomyxoviridae*, has eight segments of single-stranded, negative-sense RNA in a genome that encodes structural proteins, including polymerase basic and acidic proteins (PB2, PB1, and PA), hemagglutinins (HAs), nucleoproteins, neuraminidases (NAs), matrix proteins (M1 and M2), and nuclear export proteins, and nonstructural proteins [1]. The virus subtypes are divided according to sequences of their surface antigenic HA and NA, which bind to sialic acids (most common being *N*-acetylneuraminic acid) on the host cell surface; HA binding promotes viral fusion to the host cell and NA binding cleaves sialic acid residues from carbohydrate side chains to facilitate the spread of virions [2]. The RNA polymerase of influenza A

Jun Hirabayashi (ed.), *Lectins: Methods and Protocols*, Methods in Molecular Biology, vol. 1200,
DOI 10.1007/978-1-4939-1292-6_10,

virus without proofreading and the segmented genome of the virus cause continuing point mutations and reassortment of viral gene segments, resulting in the generation of new influenza strains with epidemic or pandemic potentials [3–5]. Change in HA receptor binding preference from avian α2-3-linked sialic acid to human α2-6-linked sialic acid receptors plays a critical role in initiation of crossing the species barrier and emergence of influenza pandemic viruses. Receptor binding assays are useful not only for surveillance of change in the binding preference of an influenza virus from terminal α2-3- to terminal α2-6-linked sialic acid receptors but also for determination of virus binding preference in the substructures of receptors by using proteins/lipids conjugated with natural [6] or synthetic [7, 8] glycans with detailed fine structures. Studies on virus–receptor binding specificity in combination with cellular studies on the distribution of sugars on epithelial cells in influenza virus-target organs and viral genomic sequence analysis should lead to a better understanding of viral evolution and viral transmission that may enable potential prevention and control of the spread of new variants to human populations.

Several receptor-binding specificity assays including direct binding to sialylglycoconjugates coated on an ELISA plate [9] or on a chip (microglass slide) [2, 8] and competitive inhibition assay with α2-3 and α2-6 sialoconjugated protein fetuin [10–12] have been developed with the same principles of immobilization of specific receptors on a solid phase and the use of one or two specific antibodies to detect the degree of receptor–virus binding via enzymatic colorimetric reaction or fluorescence intensity according to conjugation of the antibody. The lack of a universal influenza virus antibody as well as a lack of antibodies against some influenza virus subtypes has limited virus–receptor binding analysis. We took advantage of the influenza virus having NA spikes acting as an antibody conjugated with enzymes by itself, and we could thus assess receptor-binding specificity of all influenza virus subtypes directly after viral HA binding to glycans on a solid phase. This assay with the use of an intact virus not only potentially reflects the natural binding of the viral HA but is also a fast, easy, low-cost, and reproducible assay and can be used for high-throughput screening of rapid surveillance of newly emerging influenza virus pandemic strains and for investigation and understanding of viral adaptation to the host.

2 Materials

2.1 Viruses

2.1.1 General Materials for Virus Preparation

1. Class II biological safety cabinet (*see* **Note 1**).
2. Phosphate-buffered saline (PBS), pH 7.2–7.4.
3. High-speed centrifuge.

4. –80 °C freezer.
5. Egg incubator automatically turning the eggs 45° from one direction to the other every hour (or manually turning the eggs at least once a day, if such an incubator is not available) set up at 40–60 % humidity, 39 °C for growing chicken embryos in fertilized eggs until 10 days old and 34 °C for growing influenza A viruses in inoculated eggs.
6. Humidified 37 °C, 5 % CO_2 incubator for culturing influenza A virus-uninfected/infected MDCK cells.
7. Vortex mixer.
8. 15-mL and 50-mL conical polypropylene tubes (such as TPP cultureware, Cat. No. 91015 and 91050, respectively).

2.1.2 Amplification of Avian Influenza Virus in Embryonated Chicken Eggs

1. 9- to 11-day-old specific pathogen-free embryonated chicken eggs.
2. Egg candler or light bulb in a dark room for checking and locating the air sac area and live embryo inside the egg.
3. 1-mL syringe with a 25-G×5/8″ needle for injection of influenza virus into the allantoic cavity.
4. Syringe (size, 10 mL) and needle (size, 18-G×11/2″) for harvesting allantoic fluid.
5. Sterile forceps.
6. Cotton wool.
7. 70 % (v/v) ethanol.
8. 1.2-mL cryogenic vials.

2.1.3 Amplification of Avian Influenza Virus in MDCK Cells

1. Madin-Darby canine kidney (MDCK) cells (ATCC CCL-34).
2. Minimum essential medium (MEM) with Earle's salt, L-glutamine, and nonessential amino acids (Nacalai Tesque Cat. No. 21443-15).
3. 10,000 U/mL penicillin and 10,000 μg/mL streptomycin (Gibco-BRL, Cat. No. 15140-122).
4. Acetylated trypsin, Type V-S from bovine pancreas (Sigma, Cat. No. T6763).
5. Cell culture flask, CellBind Surface area 75 cm^2 with vent cap, sterile, polystyrene (Corning, Cat. No. 3290).

2.1.4 Hemagglutination Assay

1. PBS, pH 7.2–7.4.
2. Guinea pig erythrocytes.
3. 96-well PVC (flexible) U-bottom plate, non-sterile (BD Biosciences, Cat. No. 353912 or Corning, Cat. No. 2797).
4. 4 °C refrigerator or icebox.

2.2 Receptor-Binding Specificity Assay

2.2.1 Receptor-Binding Assay

1. Influenza A viruses.
2. PBS, pH 7.2–7.4.
3. 0.1 % Tween-20 in PBS (PBST).
4. Either natural or synthetic sialylated glycans conjugated to a protein carrier or a lipid carrier can be used for the study (*see* Table 1).
5. Flat-bottom 96-well ELISA plates (*see* Table 2).
6. 96-well plate washer/dispenser.
7. Icebox.
8. Pipette tips.
9. Matrix™ Reagent reservoirs (such as Thermo Scientific, Cat. No. 8093-8096).
10. 8- or 12-channel pipettes (300 and 1,200 μL).
11. 1.5-mL Eppendorf tubes (Fisher Scientific).
12. 5-mL Eppendorf tubes (Bio-BIK).
13. 15-mL conical polypropylene tubes (such as TPP cultureware, Cat. No. 91015).

2.2.2 Neuraminidase-Based Detection

1. 2′-(4-methylumbelliferyl)-α-D-5-*N*-acetylneuraminic acid (MU-Neu5Ac) (Sigma, Cat. No. M8639 or Toronto Research Chemicals, Inc. (TRC), Cat. No. M334200).
2. −20 °C freezer.
3. 10 mM sodium acetate buffer with $CaCl_2$ and $MgCl_2$, pH 6.0, plus 0.5 % bovine serum albumin (BSA; Gibco-BRL, Cat. No. 15260-037).
4. 100 mM sodium carbonate-sodium bicarbonate buffer, pH 10.6.
5. 96-well solid black polystyrene flat-bottom plate (Corning, Cat. No. 3915).
6. Fluorometer microplate reader (such as Berthold Mithras LB940).
7. GraphPad Prism and Excel or other software for data analysis.

3 Methods

3.1 Reagent Preparation

3.1.1 Phosphate-Buffered Saline (PBS), pH 7.2–7.4

1. Prepare 10× PBS stock by dissolving 800 g NaCl, 20 g KCl, 115 g Na_2HPO_4 or 217 g $Na_2HPO_4 \cdot 7H_2O$, and 20 g KH_2PO_4 in 1 L distilled water. Make up the total volume to 10 L with distilled water. Store at room temperature (RT).
2. Prepare 1× PBS working solution by making a 1:10 (v/v) dilution of 10× PBS in distilled water and store at RT. If the solution is proposed to be used in cell/virus cultivation experiments, it must be autoclaved before storing at 4 °C.

Table 1
Example list of choices of available sialylglycoconjugates that can be used for the assay

Name	Structure	Manufacturer	Comment
Glycolipid (Ganglioside)			
3′SL-Cer (α2-3GM3) II^3-(Neu5Ac)LacCer	Neu5Acα2-3Galβ1-4Glcβ1-Cer	Calbiochem, Cat. No. 345733	Isolated from several kinds of animal tissues, such as human liver, and bovine milk and human erythrocytes [14–17]
α2-3GM1b IV^3(Neu5Ac)Gg4Cer	Neu5Acα2-3Galβ1-3GalNAcβ1-4 Galβ1-4Glcβ1-Cer	–	Isolated from bovine brain [18]
α2-3GD1a IV^3(Neu5Ac)II^3(Neu5Ac)Gg4Cer	Neu5Acα2-3Galβ1-3GalNAcβ1-4(Neu5Acα2-3)Galβ1-4Glcβ1-Cer	Calbiochem, Cat. No. 345736	
α2-3/α2-6sialylparagloboside IV^3/IV^6(Neu5Ac)nLc4Cer	Neu5Acα2-3/ α2-6Galβ1-4GlcNAcβ1-3Galβ1-4Glcβ1-Cer	–	Prepared from erythrocytes or feces of newborn baby (meconium) [15, 17]
Glycoprotein			
Neoglycoprotein 3′SLN-BSA	Neu5Acα2-3Galβ1-4GlcNAcβ1-BSA	Dextra Laboratories, Cat. No. NGP0301	[19]
6′SLN-BSA	Neu5Acα2-6Galβ1-4GlcNAcβ1-BSA	–	Prepared by conversion of 4-nitrophenyl-6′SLN (Tokyo Chemical Industry Co., Ltd., Cat. No. N0856) to 4-aminophenyl-6′SLN by catalytic reduction followed by glycoconjugation with BSA using *N*-(3-maleimidobenzoyloxy) succinimide (Nacalai Tesque, Cat. No. 58626-38-3) [19]
3′SLN/6′SLN- pAP-poly α-L-glutamic acid (α-PGA)	Neu5Acα2-3/ α2-6Galβ1-4GlcNAcβ1-*p*AP-α-PGA[a]	–	[7, 13]
3′SLN/6′SLN-polyacrylamide (PA)	Neu5Acα2-3/α2-6Galβl-4GlcNAc-PA	–	[20]
3′SL/6′SL-polystyrene (PS)	Neu5Acα2-3/α2-6Galβl-4Glc-PS	–	Isolated from bovine milk [21]

[a]Sialylglycopolymer (SGP) structures used as receptors for influenza virus binding specificity assay in this study. α2-3SGP with a molecular weight (MW) of 323,441 and α2-6SGP with MW of 470,885 bear 46.5 and 77.0 mol% of sialyloligosaccharide moieties, respectively, on their poly-α-L-glutamic acid backbones

Table 2
Example list of 96-well plates that should be selected according to carrier type

Carrier type	Plate type	Manufacturer	Comment
Protein/ polyglutamic acid carrier	Universal-BIND™ Surface, Stripwell 96-well flat-bottom plate	Corning Life Sciences, Cat. No. 2504	A blocking step is not required
	Nunc 96-well flat-bottom MaxiSorp™ plate, polystyrene	Thermo Fisher Scientific, Cat. No. 456529	After the coating step, block the plate with 250 μL of 2 % PVP (polyvinylpyrrolidone; TCI Tokyo Kasei, Cat. No. P0472) or 1 % BSA (Gibco-BRL, Cat. No. 15260-037) in PBS for 2 h at RT
Lipid (i.e., ceramide) carrier	Nunc 96-well flat-bottom PolySorp™ plate, polystyrene	Thermo Fisher Scientific, Cat. No. 456529	After the coating step, block the plate with 250 μL of 1 % defatted BSA (Sigma, Cat. No. A-6003) in PBS for 2 h at 37 °C

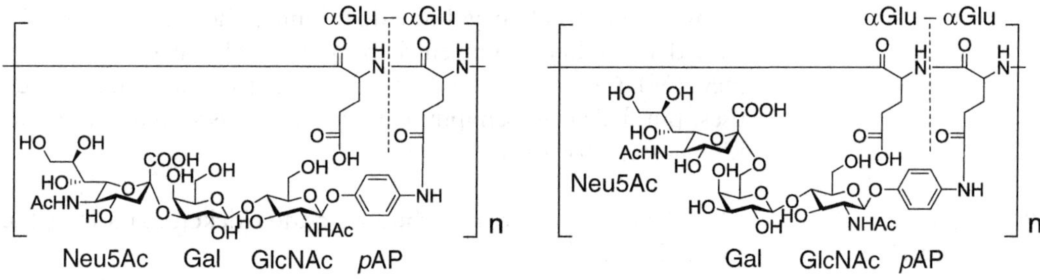

Fig. 1 Chemical structures of poly α-L-glutamic acid (PGA) backbone carrying Neu5Acα2-3Galβ1- (*left panel*) or Neu5Acα2-6Galβ1- (*right panel*) -4GlcNAcβ1-*p*-aminophenyl

3.1.2 0.1 % Tween-20 in PBS (PBST)

1. Add Tween-20 to PBS to give a final concentration of 0.1 % (v/v) Tween-20. Store at RT.

3.1.3 Sialylglyco-polymers

Neu5Acα2-3/α2-6Galβ1-4GlcNAcβ1-*p*-aminophenyl-poly(α-L-glutamic acid) polymers (3′SLN/6′SLN-*p*AP-α-PGA) (*see* Fig. 1) extensively used for direct binding assays for human and avian influenza viruses [9, 13] were used here for examples in assessment of receptor-binding specificity of influenza viruses detected by the neuraminidase-based assay.

1. Prepare 1 mg/mL stock solution of each polymer in distilled water and store in aliquots at –20 °C.
2. Prepare 20 ng/μL working solution by a 1:50 dilution of a stock solution in distilled water and store at 4 °C.

3.1.4 MU-Neu5Ac

1. Prepare 4 mM stock solution in distilled water and store in aliquots at –20 °C.

3.1.5 10 mM Sodium Acetate Buffer with $CaCl_2$ and $MgCl_2$, pH 6.0

1. Dissolve 0.82 g CH_3COONa, 5.844 g NaCl, 1.11 g $CaCl_2$, and 0.1 g $MgCl_2{\cdot}6H_2O$ in 100 mL of distilled water. Adjust pH to 6.0 by addition of about 24 μL of acetic acid. Adjust the volume to 1 L, aliquot, autoclave, and store at 4 °C. Add 0.5 % BSA to make a working solution and store at 4 °C.

3.1.6 100 mM Sodium Carbonate–Sodium Bicarbonate Buffer, pH 10.6

1. Prepare 100 mM Na_2CO_3 by dissolving 5.3 g Na_2CO_3 in 500 mL of distilled water.
2. Prepare 100 mM $NaHCO_3$ by dissolving 1.68 g $NaHCO_3$ in 200 mL of distilled water.
3. Mix 45 mL of 100 mM Na_2CO_3 with 5 mL of 100 mM $NaHCO_3$ to make 100 mM Na_2CO_3–$NaHCO_3$ buffer (pH 10.6).

3.2 Virus Preparation

All experiments with influenza viruses should be carried out in a class II biological safety cabinet for prevention of cross-contamination (*see* **Note 2**). If virus populations in the original specimens are lower than 8 hemagglutination units (HAU; *see* Subheading 3.2.3:

Determination of Hemagglutination Unit), the virus should be amplified in embryonated chicken eggs or Madin-Darby canine kidney (MDCK) cells (*see* **Note 3**). Supernatant/concentrated viruses/purified virus/semipurified virus can be used in the receptor-binding specificity assay.

3.2.1 Amplification of Avian Influenza Virus in Embryonated Chicken Eggs

1. Candle and mark areas of the air sac and chicken to indicate the injection site.
2. Swab above the line of the air sac with cotton wool soaked in 70 % (v/v) ethanol solution.
3. Pierce the egg shell above the line of the air sac with sterile forceps.
4. Inject 200 μL of the influenza virus diluted in PBS with 100 U/mL penicillin and 100 μg/mL streptomycin (PS) (1/2–1 HAU) using a 1-mL syringe with a 25-G×5/8″ needle into the allantoic cavity of a 10-day-old embryonated egg.
5. After sealing the holes, incubate inoculated eggs at 34 °C for 2 days.
6. Chill the eggs at 4 °C overnight.
7. Crack the eggs and remove the egg shell with sterile forceps.
8. Pierce the egg sac and draw as much allantoic fluid as possible using a 10-mL syringe with a 18-G×11/2″ needle.
9. Clarify the virus suspension at 1,750×*g* for 10 min at 4 °C.
10. Pellet the virus in supernatant at 18,270×*g* for 3 h and resuspend in PBS, aliquot, and store at −80 °C until use (*see* **Note 4**).

3.2.2 Amplification of Human Influenza Virus in MDCK Cells

1. Wash an MDCK cell confluent monolayer, prepared in an appropriate plate depending on the amount virus of needed, two times with PBS or MEM with PS.
2. Add the influenza A virus diluted in an infection medium, MEM with PS and 5 μg/mL acetylated trypsin (1 mL for a 75 cm^2 flask, varying based on the flask/plate being used) with a final concentration of approximately 1/2 to 1 HAU.
3. Allow the virus to be adsorbed onto the cell monolayer for 1 h at 37 °C with intermittent shaking every 15 min to prevent the cells from drying.
4. Add 14 mL infection medium to each flask. Gently rock the flask.
5. Place the flask with inoculated cells in a 5 % CO_2 incubator for 3–5 days at 37 °C or until a cytopathogenic effect has appeared.
6. Clarify the virus suspension at 1,750×*g* for 10 min at 4 °C.
7. Pellet the virus in supernatant at 18,270×*g* for 3 h and resuspend in PBS, aliquot, and store at −80 °C until use.

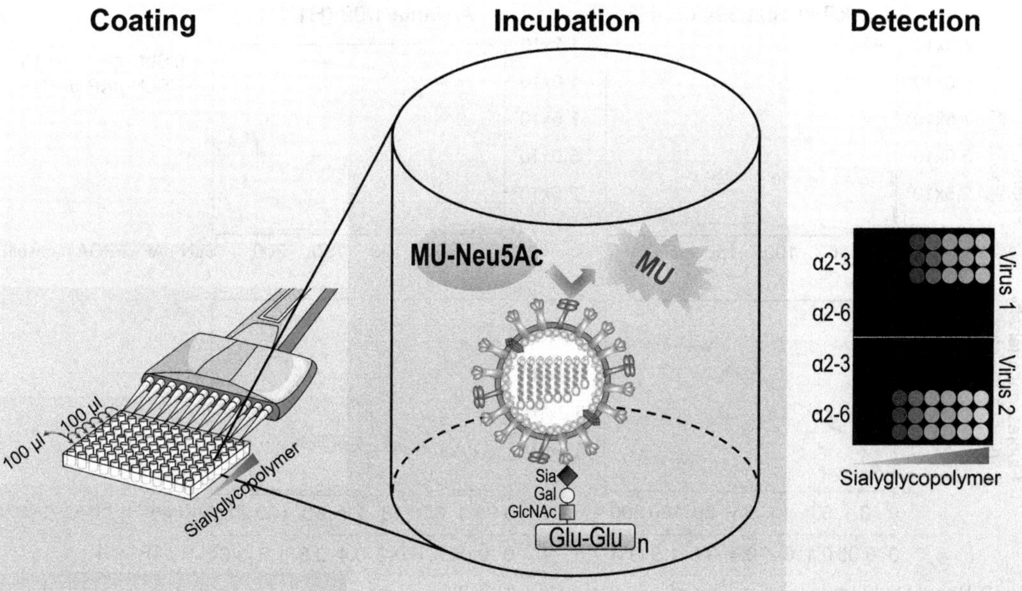

Fig. 2 Schematic diagram of hemagglutinin–receptor binding assay with detection by neuraminidase-based fluorescence. Coating step: serial dilutions of each glycopolymer are performed using a multi-channel pipette onto a 96-well flat-bottom plate and covalently immobilized by UV-irradiation at 254 nm for 5 min. Incubation step: after washes three times with PBS, intact viruses are added and incubated on ice for 2 h. Detection step: following removal of unbound virus by three washes with PBST, the virus binding is detected by addition of MU-Neu5Ac substrate cleaved by the viral neuraminidase to a fluorescent MU product detectable at Ex355/Em460. *See* color figure in the online version

3.2.3 Determination of Hemagglutination Unit (Relative Quantities of Virus Particles)

1. Twofold serially dilute the virus in PBS in a 96-well U-bottom plate.
2. Add guinea pig erythrocytes (0.5 % cell suspension in PBS) and mix by gently shaking.
3. Allow the erythrocytes to settle for 2 h at 4 °C and observe; unbound erythrocytes sink to the bottom of the well forming a halo or a circle, whereas virus-bound erythrocytes form a lattice coating the well (hemagglutination). A 0.75 % concentration of guinea pig erythrocytes may be used if wanted.

3.3 Sialylglycopolymer-Coated Plates

Serially dilute sialylglycopolymer in PBS twofold in a 96-well plate (*see* **Note 5**) to give final polymer concentrations ranging from 0 to 200 ng/well (0–2 ng/μL) (equivalent to SLN-*p*AP concentrations of 0–3 nM) as **steps 1–5** below (*see* Fig. 2).

1. Dilute each sialylglycopolymer working solution in PBS to be 4 ng/μL in an appropriate tube depending on the volume needed.
2. Dispense 100 μL of PBS to each well of a 96-well plate.

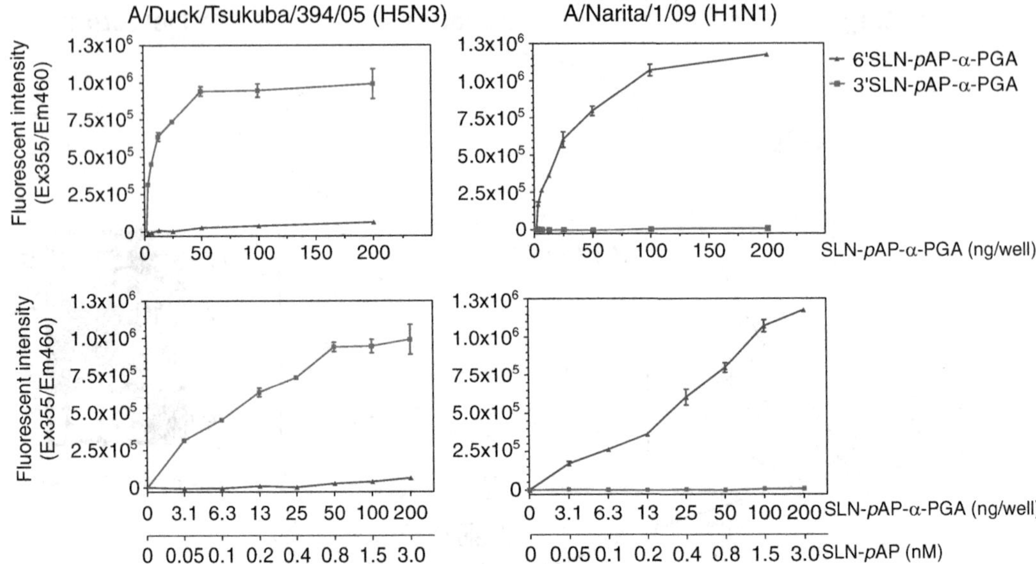

Fig. 3 Receptor-binding preference of avian isolates (H5N3) and clinical isolates from a human pandemic (H1N1). Thirty-two HAU of A/Duck/Tsukuba/394/05 (H5N3) (*left panel*) or A/Narita/1/2009 (H1N1) (*right panel*) in PBST was added to a plate coated with 0–200 ng/well of α2-3- (*red*) or α2-6- (*blue*) pAP-α-PGA and incubated at 4 °C for 2 h. The amount of bound viruses was detected in the presence of MU-Neu5Ac. The *upper panel graphs* show linear plots between fluorescence intensity of released MU versus concentration of the polymer (both axes on linear scales). The lower panel graphs show semi-log plots between fluorescence intensity of released MU versus log concentration of the polymer (concentration on the x-axis stretched to a log scale). Each point is the mean ± SEM of three replicates. The virus binding curves show that an avian A/Duck/Tsukuba/394/05 (H5N3) virus preferentially binds an α2-3 sialylated polymer, whereas a human A/Narita/1/2009 (H1N1) virus shows α2-6 specificity

3. Add 100 μL of 4 ng/μL polymer to each well in the first column of the plate and mix well by pipetting.
4. Transfer 100 μL from each well in the first column to each well in the next column. Mix well by pipetting.
5. Continue to transfer the solution and mix well until reaching the column before the last column. After mixing, discard 100 μL of the solution from this column. Thus, each well in the last column has no polymer representing as a negative control.
6. Expose the plate with the final well volume of 100 μL to a UV light at 254 nm for 5 min at RT (*see* **Notes 5** and **6**).
7. Wash the plate 3 times with PBS (*see* **Note 7**).

3.4 Receptor-Binding Specificity Assay

3.4.1 Incubation

1. Add 50 μL of virus diluted in PBST (*see* **Note 8**) to each well (*see* **Note 9**). In Fig. 3, 32 HAU of each virus are used.
2. Incubate the plate on ice or in a cold room (4 °C) for 2 h with gentle shaking (*see* **Note 10**).
3. Remove unbound virus particles by 3–5 washes with PBST (*see* **Note 11**).

3.4.2 Detection

1. Dilute 4 mM MU-Neu5Ac solution in 10 mM sodium acetate buffer with $CaCl_2$ and $MgCl_2$, pH 6.0, plus 0.5 % BSA to be 40 μM just before use.
2. Dispense 50 μL of 40 μM MU-Neu5Ac substrate solution to each well.
3. Incubate the plate at 37 °C for 1 h. Development of fluorescence intensity due to MU released from MU-Neu5Ac by sialidase activity of viral neuraminidase may be checked roughly on a UV box with a wavelength of about 355 nm.
4. Terminate the reaction by transferring 25 μL of the reaction solution to a 96-well black flat-bottom plate containing 100 μL of 100 mM sodium carbonate–sodium bicarbonate buffer, pH 10.6.
5. Fluorescence intensities of the MU products are read at Ex355/Em460 (*see* **Notes 12** and **13**).

3.4.3 Analysis

1. Use Microsoft Excel to subtract the values of fluorescence intensity of the test well from that of the negative control well without sialylglycopolymers in the same row of the 96-well plate.
2. Use GraphPad Prism or other graph software such as Microsoft Excel to make the curve.
3. Plot fluorescence intensity on the y-axis as a function of concentration of sialylglycopolymer on the x-axis.
4. Viral hemagglutinin–sialylglycopolymer binding data span a wide range of values as an exponential increase (*see* Fig. 3, *upper panel*). If equal spacing of data for seeing all data clearly is desired, change the x-axis scale from linear to log scale (*see* Fig. 3, *lower panel*).
5. Level of fluorescence intensity reflects level of virus binding; a point with higher fluorescence intensity indicates that there are more virus particles binding to the corresponding polymer at that point.

4 Notes

1. When working with a highly pathogenic virus, a Class III biological safety cabinet is required for safety of the user and the environment.
2. Every material exposed to an influenza virus, which is a biosafety hazard, should be cleaned or disposed of according to institution guidelines.
3. Since influenza A virus contains genomic RNA segments and has very high mutation rates, the virus undergoes very rapid alterations in its genomic materials, and the variant with optimal viability within the host environment becomes dominant

[22, 23]. For accurate data interpretation of influenza virus–receptor binding specificity, not only should appropriate cells be selected for virus amplification but also the virus should not be propagated more than 5 times to avoid newly formed viruses different from the original isolates.

4. Viruses and reagents that are stored at −20 °C or a lower temperature should be stored as aliquots to prevent excessive freeze-thawing in order to avoid damage to the viruses/reagents. When thawed for use but not used immediately, the viruses and reagents either in a stock solution or working solution should be held on ice.

5. Glycolipids should be completely dissolved in an organic solvent, such as ethanol, that has no effect on a plastic plate. Acetone and chloroform, which may melt the plate, should not be used. After making serial dilutions of glycolipids in ethanol in a 96-well Nunc PolySorp flat-bottom plate, the organic solvent in the well is removed by evaporation at 37 °C in an air chamber until dryness.

6. If a glycoprotein is coated on a MaxiSorp surface, serial dilutions of glycopolymers in PBS in a 96-well Nunc MaxiSorp flat-bottom plate should be left at RT for 2–3 h before washing three times with PBS.

7. After coating the plates with sialylglycoconjugates, the plates can be kept for long periods at 4 °C. It is not necessary to use the coated plates immediately.

8. Equal distribution of virus particles throughout the solution will provide accurate and precise results. To ensure viral distribution, dilute the virus in a small volume of PBST, vortex before addition of PBST to the final volume, and vigorously vortex again.

9. The sialylglycoconjugate-coated 96-well plate should be held on ice before and during distribution of the virus suspension into the wells to maintain enzyme activity of neuraminidase and prevent cleavage of sialic acid from saccharide chains, either bound or unbound with hemagglutinin, by sialidase activity of neuraminidase.

10. To prevent virus aggregation during the incubation period, the 96-well plate on ice should be gently rocked continuously on a laboratory shaker.

11. It is important to thoroughly wash out unbound materials. If they are not thoroughly washed out, they will produce a falsely elevated fluorescence level in neuraminidase-based detection.

12. In the case of high-fluorescence background (observed in negative control wells without sialylglycopolymer), unpurified materials in MU-Neu5Ac stock solution should be removed by

extraction with 2 volumes of water-saturated methyl-tert butyl ether (HPLC grade) (ether–water, 2:1 (v/v)) per 1 volume of MU-Neu5Ac solution for three times: each time, after shaking vigorously, remove the ether (upper) phase. To remove the remaining ether, bubble nitrogen gas through the MU-Neu5Ac solution until there is no ether smell. Divide Mu-Neu5Ac stock solution into aliquots and keep at −20 °C.

13. If MU-Neu5Ac solution alone does not give high fluorescence intensity, the problem of high background could be due to virus binding to the plate. To prevent/reduce nonspecific binding to the plate surface (not found when using the Costar Universal-BIND Surface plate), block the plate with a blocking solution (*see* Table 1) after coating with sialylglycoconjugates, and also a blocking reagent, such as 0.5 % BSA, 1 % Block-Ace, 1 % skim milk, 1 % polyvinyl pyrrolidone (PVP) + 1 % egg albumin or 3 % PVP, may be added to the virus solution.

References

1. Sriwilaijaroen N, Suzuki Y (2012) Molecular basis of the structure and function of H1 hemagglutinin of influenza virus. Proc Jpn Acad Ser B Phys Biol Sci 88:226–249
2. Stevens J, Blixt O, Paulson JC, Wilson IA (2006) Glycan microarray technologies: tools to survey host specificity of influenza viruses. Nat Rev Microbiol 4:857–864
3. Xu X, Smith CB, Mungall BA, Lindstrom SE, Hall HE, Subbarao K, Cox NJ, Klimov A (2002) Intercontinental circulation of human influenza A(H1N2) reassortant viruses during the 2001–2002 influenza season. J Infect Dis 186:1490–1493
4. Kilbourne ED (2006) Influenza pandemics of the 20th century. Emerg Infect Dis 12:9–14
5. Neumann G, Noda T, Kawaoka Y (2009) Emergence and pandemic potential of swine-origin H1N1 influenza virus. Nature 459: 931–939
6. Sriwilaijaroen N, Kondo S, Yagi H, Hiramatsu H, Nakakita S, Yamada A, Ito H, Hirabayashi J, Narimatsu H, Kato K, Suzuki Y (2012) Bovine milk whey for preparation of natural *N*-glycans: structural and quantitative analysis. Open Glycoscience 5:41–50
7. Suzuki Y, Asai A, Suzuki T, Hidari K, Murata T, Usui T, Takeda S, Yamada K, Noguchi T (2009) Method for determination of recognition specificity of virus for receptor sugar chain. US Patent 0181362 A1
8. Blixt O, Head S, Mondala T, Scanlan C, Huflejt ME, Alvarez R, Bryan MC, Fazio F, Calarese D, Stevens J, Razi N, Stevens DJ, Skehel JJ, van Die I, Burton DR, Wilson IA, Cummings R, Bovin N, Wong CH, Paulson JC (2004) Printed covalent glycan array for ligand profiling of diverse glycan binding proteins. Proc Natl Acad Sci U S A 101:17033–17038
9. Yamada S, Suzuki Y, Suzuki T, Le MQ, Nidom CA, Sakai-Tagawa Y, Muramoto Y, Ito M, Kiso M, Horimoto T, Shinya K, Sawada T, Usui T, Murata T, Lin Y, Hay A, Haire LF, Stevens DJ, Russell RJ, Gamblin SJ, Skehel JJ, Kawaoka Y (2006) Haemagglutinin mutations responsible for the binding of H5N1 influenza A viruses to human-type receptors. Nature 444:378–382
10. Cointe D, Leroy Y, Chirat F (1998) Determination of the sialylation level and of the ratio α-(2 → 3)/α-(2 → 6) sialyl linkages of *N*-glycans by methylation and GC/MS analysis. Carbohydr Res 311:51–59
11. Gambaryan AS, Matrosovich MN (1992) A solid-phase enzyme-linked assay for influenza virus receptor-binding activity. J Virol Methods 39:111–123
12. Kobasa D, Takada A, Shinya K, Hatta M, Halfmann P, Theriault S, Suzuki H, Nishimura H, Mitamura K, Sugaya N, Usui T, Murata T, Maeda Y, Watanabe S, Suresh M, Suzuki T, Suzuki Y, Feldmann H, Kawaoka Y (2004) Enhanced virulence of influenza A viruses with the haemagglutinin of the 1918 pandemic virus. Nature 431:703–707
13. Totani K, Kubota T, Kuroda T, Murata T, Hidari KI, Suzuki T, Suzuki Y, Kobayashi K, Ashida H, Yamamoto K, Usui T (2003)

Chemoenzymatic synthesis and application of glycopolymers containing multivalent sialyloligosaccharides with a poly(L-glutamic acid) backbone for inhibition of infection by influenza viruses. Glycobiology 13:315–326

14. Suzuki Y, Matsunaga M, Matsumoto M (1985) *N*-Acetylneuraminyllactosylceramide, GM3-NeuAc, a new influenza A virus receptor which mediates the adsorption-fusion process of viral infection. Binding specificity of influenza virus A/Aichi/2/68 (H3N2) to membrane-associated GM3 with different molecular species of sialic acid. J Biol Chem 260:1362–1365
15. Suzuki Y, Nagao Y, Kato H, Matsumoto M, Nerome K, Nakajima K, Nobusawa E (1986) Human influenza A virus hemagglutinin distinguishes sialyloligosaccharides in membrane-associated gangliosides as its receptor which mediates the adsorption and fusion processes of virus infection. Specificity for oligosaccharides and sialic acids and the sequence to which sialic acid is attached. J Biol Chem 261:17057–17061
16. Suzuki Y (1994) Gangliosides as influenza virus receptors. Variation of influenza viruses and their recognition of the receptor sialo-sugar chains. Prog Lipid Res 33:429–457
17. Suzuki Y, Nakao T, Ito T, Watanabe N, Toda Y, Xu G, Suzuki T, Kobayashi T, Kimura Y, Yamada A et al (1992) Structural determination of gangliosides that bind to influenza A, B, and C viruses by an improved binding assay: strain-specific receptor epitopes in sialo-sugar chains. Virology 189:121–131
18. Suzuki Y, Matsunaga M, Nagao Y, Taki T, Hirabayashi Y, Matsumoto M (1985) Ganglioside GM1b as an influenza virus receptor. Vaccine 3:201–203
19. Yamashita S, Yoshida H, Uchiyama N, Nakakita Y, Nakakita S, Tonozuka T, Oguma K, Nishikawa A, Kamitori S (2012) Carbohydrate recognition mechanism of HA70 from *Clostridium botulinum* deduced from X-ray structures in complexes with sialylated oligosaccharides. FEBS Lett 586:2404–2410
20. Suzuki T, Horiike G, Yamazaki Y, Kawabe K, Masuda H, Miyamoto D, Matsuda M, Nishimura SI, Yamagata T, Ito T, Kida H, Kawaoka Y, Suzuki Y (1997) Swine influenza virus strains recognize sialylsugar chains containing the molecular species of sialic acid predominantly present in the swine tracheal epithelium. FEBS Lett 404:192–196
21. Tsuchida A, Kobayashi K, Matsubara N, Muramatsu T, Suzuki T, Suzuki Y (1998) Simple synthesis of sialyllactose-carrying polystyrene and its binding with influenza virus. Glycoconj J 15:1047–1054
22. Sriwilaijaroen N, Kondo S, Yagi H, Wilairat P, Hiramatsu H, Ito M, Ito Y, Kato K, Suzuki Y (2009) Analysis of *N*-glycans in embryonated chicken egg chorioallantoic and amniotic cells responsible for binding and adaptation of human and avian influenza viruses. Glycoconj J 26:433–443
23. Takemae N, Ruttanapumma R, Parchariyanon S, Yoneyama S, Hayashi T, Hiramatsu H, Sriwilaijaroen N, Uchida Y, Kondo S, Yagi H, Kato K, Suzuki Y, Saito T (2010) Alterations in receptor-binding properties of swine influenza viruses of the H1 subtype after isolation in embryonated chicken eggs. J Gen Virol 91: 938–948

Chapter 11

Lectin Affinity Electrophoresis

Yuka Kobayashi

Abstract

An interaction or a binding event typically changes the electrophoretic properties of a molecule. Affinity electrophoresis methods detect changes in the electrophoretic pattern of molecules (mainly macromolecules) that occur as a result of biospecific interactions or complex formation. Lectin affinity electrophoresis is a very effective method for the detection and analysis of trace amounts of glycobiological substances. It is particularly useful for isolating and separating the glycoisomers of target molecules. Here, we describe a sensitive technique for the detection of glycoproteins separated by agarose gel-lectin affinity electrophoresis that uses antibody-affinity blotting. The technique is tested using α-fetoprotein with lectin (*Lens culinaris* agglutinin and *Phaseolus vulgaris* agglutinin)-agarose gels.

Key words Affinity electrophoresis, Lectin affinity electrophoresis (LAE), Sugar chains, Glycan, Glycoisomer, Affinophoresis, Capillary electrophoresis

1 Introduction

The affinity electrophoresis method utilizes the difference in strength-specific (affinity) binding of a substance with other substances. In lectin affinity chromatography, beads immobilized in agarose and acrylate function as ligands while a solution flows past the beads. Lectin affinity electrophoresis (LEA), on the other hand, uses a charged antigen in the presence of lectin. Various techniques and ligands can be used to perform affinity electrophoresis. Three kinds of lectin electrophoresis have been developed to date: LAE, affinophoresis, and capillary electrophoresis.

In LAE, the gel contains lectin as the ligand. Lectins have binding specificity for sugar chains. This property of lectins is utilized in LAE to separate glycoproteins based on their sugar chains. Some studies using LAE have shown that sugar chains on glycoproteins are different in cancer tissue than in normal tissue. The sugar chains may also differ depending on the type of cancer. In recent years, the separation of glycoproteins by LAE has also been used for diagnostic purposes [1–3].

Jun Hirabayashi (ed.), *Lectins: Methods and Protocols*, Methods in Molecular Biology, vol. 1200,
DOI 10.1007/978-1-4939-1292-6_11, © Springer Science+Business Media New York 2014

Affinophoresis, "an alternative method of affinity electrophoresis performed with the specific use of a soluble conjugate called affinophore," is performed using an affinophore, which is the conjugate of a polyionic polymer and an affinity ligand. Affinity ligands have an overall + or − charge because they are coupled to a charged water-soluble polymer. Because of these charges, molecules that bind the ligand move when an electric field is applied. The migration of a target protein changes during affinophoresis as a result of the difference between the mobility of the protein itself and the mobility of the protein-affinophore complex [4–6].

Capillary electrophoresis is different from slab gel electrophoresis in that a capillary that is negatively charged on the inside is used to conduct charge during electrophoresis. Although capillary electrophoresis can only analyze one sample at a time, there are advantages to the method: it yields quantitative results easily and requires little time. Moreover, even though a large number of samples can be loaded for migration in traditional electrophoresis, each sample is still analyzed one by one [7–10].

This chapter describes a method for LAE in agarose gels with antibody-affinity blotting.

2 Materials

2.1 Lectin

Lectins react with specific moieties within complicated sugar chain structures. The specificity and strength of binding depend on the type of lectin used. Table 1 lists the various lectins and the glycoproteins to which they bind, analyzed by lectin affinity electrophoresis in a gel.

2.2 Antigen

Glycoprotein, e.g., human α-fetoprotein (AFP) from cord blood or hepatocellular carcinoma (HCC) cells, alkaline phosphatase (ALP), α1-antitrypsin (A1AT), transferrin (Tf).

2.3 Antibody

1. Primary antibody, e.g., anti-human AFP antibody (rabbit), anti-alkaline phosphatase (ALP) antibody (rabbit), anti-α1-antitrypsin (A1AT) antibody (rabbit), anti-transferrin (Tf) antibody (rabbit).
2. Horseradish peroxidase-conjugated anti-rabbit immunoglobulin G antibody (goat).

2.4 Buffers

1. Tris-buffered saline (TBS): 20 mM Tris–HCl, pH 7.5, 500 mM NaCl.
2. Dilution buffer: TBS, 1 % gelatin.
3. Veronal buffer: 50 mM barbital/barbital-Na buffer, pH 8.6.

2.5 Other

Nitrocellulose (NC) membrane.

Table 1
Separation pattern of each lectin-agarose gel

Lectin	Glycoprotein			
	A1AT	ALP	AFP	Tf
Aleuria aurantia lectin (AAL)			AA1–AA4	
Canavalia ensiformis agglutinin (Con A)	C1–C3		C1–C2	C1–C2
Datura stramonium agglutinin (DSA)			D1–D5	
Lens culinaris agglutinin (LCA)	L1–L4		L1–L3	L1–L3
Phaseolus vulgaris agglutinin (PHA)-E4	P1–P4		P1–P5	P1–P5
Ricinus communis agglutinin (RCA120)			R1–R3	
Triticum vulgaris agglutinin (WGA)		W1–W		

A1AT α1-antitrypsin, *ALP* alkaline phosphatase, *AFP* α-fetoprotein, *Tf* transferrin

3 Methods

3.1 Preparation of Antibody-Coated Membranes

1. Prepare antibody-coated membranes by gently shaking mixed cellulose (MC) ester membranes (5 × 7.5 cm) in a glass tray (6 × 11 × 2 cm) with 5 mL of 50 μg/mL affinity-purified polyclonal horse antibodies (e.g., anti-human AFP) in Tris-buffered saline (TBS; 20 mM Tris–HC1, pH 7.5, containing 500 mM NaCl) for 30 min.
2. Blot the membranes with filter paper.
3. Fix the antibody with glutaraldehyde vapor for 30 min.
4. Neutralize the membranes with 0.02 % $NaBH_4$ in TBS for 2 min.
5. Wash the membranes with TBS.
6. Block the membranes with 0.5 % Tween 20 in TBS for 30 min.
7. Wash with TBS.
8. Dry in vacuo and store at 4 °C.

3.2 Preparation of Lectin-Agarose Gels

For electrophoresis, 1.0-mm thick agarose gel plates are used, prepared on GelBond films with 1 % agarose (Litex) in a gel buffer of barbital/barbital-Na (ionic strength 0.050, pH 8.6), with or without lectins.

1. Add 19 mg of $MgCl_2$, 19 mg of lactic acid, and 28 mg of polyvinyl to 52 mL of veronal buffer.
2. Dissolve completely on a hot plate with stirring.

3. Cool to room temperature.
4. Add distilled water to a total volume of 100 mL.
5. After setting the acrylate plate, incubate the GelBond at 50 °C.
6. Measure 40 mg of agarose in an Erlenmeyer flask.
7. Add 3.8 mL of magnesium- and calcium-containing veronal buffer, prepared in **step 1**.
8. Dissolve agarose completely on a hot plate with stirring.
9. Cool the dissolved agarose to 50 °C.
10. Add 0.2 mL (4 mg/mL) of lectin.
11. Pour the lectin-agarose solution without introducing bubbles. Allow the gel to solidify at room temperature.
12. Refrigerate the gel until use.

3.3 Lectin Affinity Electrophoresis Conditioning

1. Dissolve glycoprotein (e.g., human AFP) in water. Mix 6 μL of sample with 1 μL of 0.02 % bromophenol blue (BPB) to prepare a glycoprotein solution with a concentration >200 ng/mL.
2. Prepare a 1.0-mm-thick agarose gel with/without lectins, e.g., *Canavalia ensiformis* agglutinin (Con A), *Lens culinaris* agglutinin (LCA), and *Phaseolus vulgaris* agglutinin (PHA-E4).
3. Add an aliquot (2–5 μL) of the glycoprotein solution to the gel.
4. Use 50 mM barbital-NaOH buffer, pH 8.6, as the electrode buffer (*see* **Note 1**).
5. Electrophorese with a constant voltage of 200 V (15 V/cm), and maintain the gel temperature at 5–10 °C.
6. Stop electrophoresis when BPB has migrated >4.0 cm from the origin (approximately 50 min).

3.4 Antibody-Affinity Blotting

1. Use capillary blotting to transfer glycoproteins separated by LAE to nitrocellulose (NC) membranes pre-coated with purified antibody in order to visualize the glycoprotein antigens with high sensitivity. After electrophoresis, cover the lectin-agarose gels with antibody-pre-coated NC membranes soaked in TBS, followed by filter paper pads and an agarose plate weighing approximately 10 g/cm^2 (*see* **Note 2**).
2. Blot for 30 min.
3. Wash each membrane twice in 10 mL of washing buffer (0.05 % Tween 20 in TBS) for 5 min at room temperature.
4. Shake gently and slowly for 5 min to remove antigens.

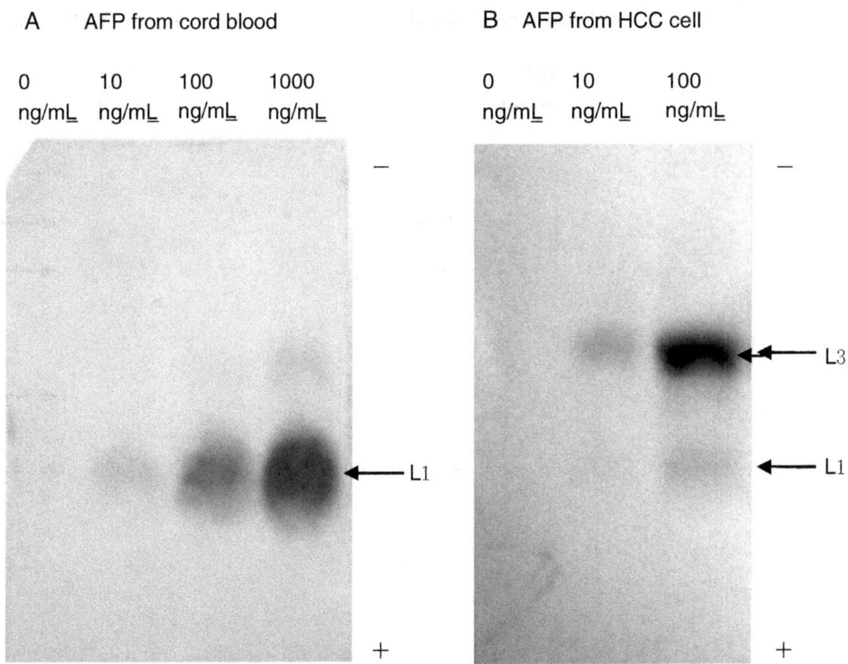

Fig. 1 AFP separated with an LCA agarose gel. LCA is a mannose- and core fucose-specific lectin. AFP-L1 and AFP-L3 are the major bands identified. AFP from cord blood does not separate (Refs. 11–14). AFP from HCC cells separates into discrete bands: AFP-3 corresponds to lectin-bound AFP, and AFP-L1 is not specific for lectin or its isoforms (glycosylation type). A previous study showed that AFP-L3 has core fucose in an *N*-linked sugar chain (Refs. 11–14)

5. Treat each membrane for 30 min at 37 °C with rabbit immunoglobulins to the antigen diluted 500-fold in 10 mL of dilution buffer (TBS containing 1 % gelatin).
6. Wash the membranes twice for 10 min with washing buffer.
7. Gently shake for 30 min in 10 mL of affinity-purified goat anti-rabbit IgG (H+L)-HRP conjugate diluted 1,000-fold with dilution buffer.
8. Wash twice more in a similar manner (*see* **Note 3**).
9. Develop color by treating the membranes with 0.025 % 3,3′-diaminobenzidine tetrahydrochloride and 0.005 % $H_2 0_2$ in 0.05 M Tris–HCl, pH 7.6, for 30 min.
10. For color development, the following solutions have also been tested: 0.05 % 4-chloro-1-naphthol dissolved in TBS with methanol (1/6 volume) and 0.015 % H_2O_2; 0.02 % 4-methoxy-1-naphthol dissolved in 0.05 M Tris–HCI, pH 7.6, with methanol (1/50 volume) and 0.005 % H_2O_2 (*see* **Note 4**).
11. Scan the colored membranes (Figs. 1 and 2) (*see* **Note 5**).

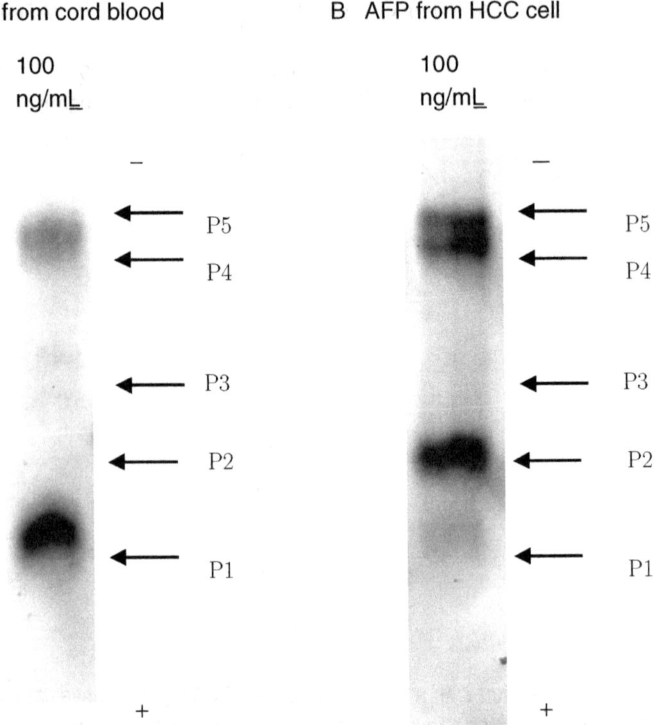

Fig. 2 AFP separated with a PHA-E4 agarose gel. PHA-E4 is a galactose-specific lectin. Five major bands are identified: AFP-P1, -P2, -P3, -P4, and -P5. Isomers in AFP bands with the lowest numbers have low or no affinity for their respective lectins (P1), and those in bands with higher numbers have higher affinities for their respective lectins (P5) (Refs. 11–14)

4 Notes

1. Glycoprotein and lectin bind strongly at low temperatures. Electrophoretic separation is better under low-temperature conditions.
2. The lectin agarose gel dries out easily: do not leave the gel exposed.
3. Wash the nitrocellulose membrane immediately: do not leave the membrane exposed.
4. There are many staining methods, such as the POD Immunostain Set (Wako Pure Chemical Industries, Ltd.)
5. The distribution of separated AFP bands can be expressed as a percentage of the sum of the intensities of all the AFP bands obtained for each of the lectins used (Figs. 3 and 4).

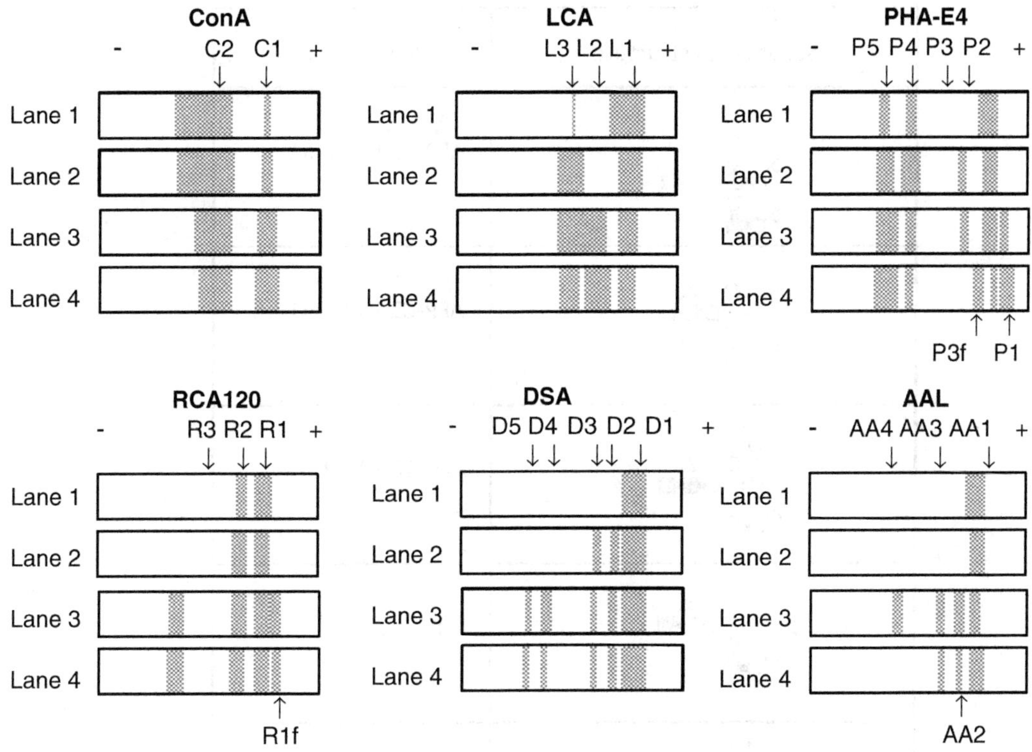

Lane 1, AFP from cord blood
Lane 2, AFP from hepatocarcinoma
Lane 3, AFP from gastrointestinal carcinoma
Lane 4, AFP from yolk sac tumor

Fig. 3 Separation images of lectin-agarose gels. AFP glycosylation isomers separate as four isomers (AA1–AA4) on the AAL-agarose gel, four isomers (C1–C2) on the Con A-agarose gel, five isomers (D1–D5) on the DSA-agarose gel, three isomers (L1–L3) on the LCA-agarose gel, five isomers (P1–P5) on the PHA-E4-agarose gel, and three isomers (R1–R3) on the RCA120-agarose gel (Refs. 15–17)

Sugar chain structure	AFP-isoforms
503 54	AFP-C2-L1-P2-R1
	AFP-C2-L3-P2-R1
608 59	AFP-C2-P4-R2
	AFP-C2-P2
405 42	AFP-C2-P4
	AFP-C2-P5-R1
	AFP-C1-P5-R3

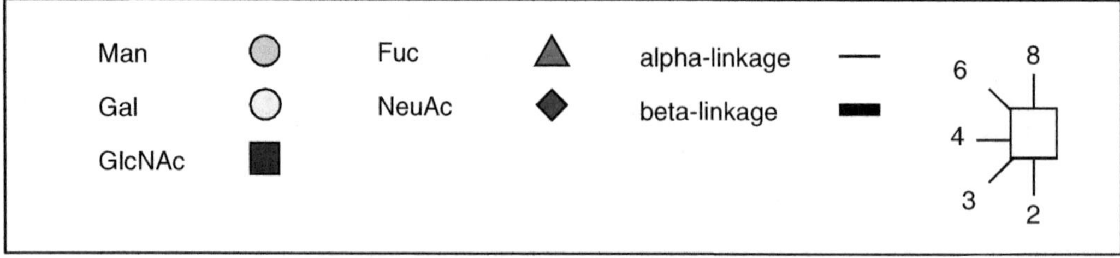

Fig. 4 Sugar chain structures of the AFP isoforms identified in Fig. 3

References

1. Taketa K, Ichikawa E, Tago H et al (1985) Antibody-affinity blotting, a sensitive technique for the detection of α-fetoprotein separated by lectin affinity electrophoresis in agarose gels. Electrophoresis 6:492–497
2. Taketa K, Hirai H (1989) Lectin affinity electrophoresis of α-fetoprotein in cancer diagnosis. Electrophoresis 10:562–567
3. Kagebayashia C, Yamaguchi I, Akinaga A et al (2009) Automated immunoassay system for AFP-L3 % using on-chip electrokinetic reaction and separation by affinity electrophoresis. Anal Biochem 388:306–311
4. Shimura K (1990) Progress in affinophoresis. J Chromatogr 510:251–270
5. Shimura K, Kasai K (1995) Determination of the affinity constants of concanavalin A for monosaccharides by fluorescence affinity probe capillary electrophoresis. Anal Biochem 227: 186–194
6. Shimura K, Kasai K (1996) Affinophoresis: selective electrophoretic separation of proteins using specific carriers. Methods Enzymol 271:203–218
7. El Rassi Z, Mechref Y (1996) Recent advances in capillary electrophoresis of carbohydrates. Electrophoresis 17:275–301
8. Kakehi K, Kinoshita M (2009) Capillary lectin-affinity electrophoresis for glycan analysis. Methods Mol Biol 534:93–105
9. Nakajima K, Kinoshita M, Oda Y et al (2004) Screening method of carbohydrate-binding proteins in biological sources by capillary affinity electrophoresis and its application to determination of *Tulipa gesneriana* agglutinin in tulip bulbs. Glycobiology 14:793–804
10. Nakajima K, Kinoshita M, Matsushita N et al (2006) Capillary affinity electrophoresis using lectins for the analysis of milk oligosaccharide structure and its application to bovine colostrum oligosaccharides. Anal Biochem 348: 105–114
11. Taketa K, Fujii Y, Taga H (1993) Characterization of E-PHA-reactive alpha-fetoprotein isoforms by two-dimensional lectin affinity electrophoresis. Electrophoresis 12:1333–1337
12. Taketa K, Fujii Y, Aoi T et al (1993) Identification of alpha-fetoprotein-P4 sugar chain as alpha 2→6 monosialylated biantennary complex-type oligosaccharides with an exposed galactose on the mannose alpha 1→6 arm by two-dimensional extended agarose gel-lectin affinity electrophoresis. Electrophoresis 14:798–804
13. Taketa K, Ichikawa E, Sato J et al (1989) Two-dimensional lectin affinity electrophoresis of alpha-fetoprotein: characterization of erythroagglutinating phytohemagglutinin-dependent microheterogeneity forms. Electrophoresis 10: 825–829
14. Taketa K, Sekiya C, Namiki M et al (1990) Lectin-reactive profiles of alpha-fetoprotein characterizing hepatocellular carcinoma and related conditions. Gastroenterology 99:508–518
15. Taketa K (1995) AFP. Rinsho Kensa 39:68–70
16. Yamashita K, Taketa K, Nishi S et al (1993) Sugar chains of human cord serum alpha-fetoprotein: characteristics of N-linked sugar chains of glycoproteins produced in human liver and hepatocellular carcinomas. Cancer Res 53:2970–2975
17. Kobayashi Y, Tateno H, Dohra H et al (2012) A novel core fucose-specific lectin from the mushroom *Pholiota squarrosa*. J Biol Chem 287:33973–33982

Chapter 12

Capillary-Based Lectin Affinity Electrophoresis for Interaction Analysis Between Lectins and Glycans

Mitsuhiro Kinoshita and Kazuaki Kakehi

Abstract

Capillary affinity electrophoresis (CAE) is a powerful technique for glycan analysis, and one of the analytical approaches for analyzing the interaction between lectins and glycans. The method is based on the high-resolution separation of fluorescently labeled glycans by capillary electrophoresis (CE) with laser-induced fluorescence detection (LIF) in the presence of lectins (or glycan binding proteins). CAE allows simultaneous determination of glycan structures in a complex mixture of glycans. In addition, we can calculate the binding kinetics on a specific glycan in the complex mixture of glycans with a lectin. Here, we show detailed procedures for capillary affinity electrophoresis of fluorescently labeled glycans with lectins using CE-LIF apparatus. Its application to screening a sialic acid binding protein in plant barks is also shown.

Key words Capillary electrophoresis, Lectins, Glycans, Capillary affinity electrophoresis

1 Introduction

Carbohydrate chains in glycoconjugates play important roles such as modulation of their functions and structures. In the extracellular environment, carbohydrate chains exert effects on cellular recognition in infection, cancer and immune responses, but details on the mechanisms still remain unsolved. Therefore, analysis of carbohydrate chains is of primary importance for understanding the role of carbohydrate chains on various biological phenomena.

Capillary electrophoresis (CE) allows high-resolution analysis of carbohydrate chains. A combination of CE and laser-induced fluorescence (LIF) detection is a powerful technique for ultrahigh sensitive detection of carbohydrate chains labeled with fluorescent tags such as 8-aminopyrene-1,3,6-trisulfonate (APTS) [1–4] and 2-aminobenzoic acid (2-AA) [5]. We applied the technique to the analysis of carbohydrate chains derived from various glycoprotein samples [6–8]. CE can be also employed to observe the interaction between carbohydrates and carbohydrate-binding proteins (i.e., lectins), and the technique is called "capillary affinity

Jun Hirabayashi (ed.), *Lectins: Methods and Protocols*, Methods in Molecular Biology, vol. 1200,
DOI 10.1007/978-1-4939-1292-6_12, © Springer Science+Business Media New York 2014

electrophoresis (CAE)." CAE is an analytical approach by which changes in the migration of the carbohydrate molecules are observed in the presence/absence of lectins. Shimura et al. determined dissociation constants between concanavalin A (Con A) and neutral monosaccharides using the affinity probe synthesized by coupling rhodamine B and *p*-amino-phenyl-α-D-mannopyranoside [9]. Taga et al. also reported the model study for the simultaneous determination of the interactions between a mixture of a fluorescently labeled *N*-linked glycans and *Lens culinaris* agglutinin (LCA), and demonstrated the efficacy of CAE method for simultaneous estimation of the association constants of a few common *N*-linked glycans [10]. The CAE method requires only small amount of samples (carbohydrates and lectins), and does not require separation of the free and bound molecules prior to the analysis. Another important point is that the interactions are observed in the solution state; in contrast, most methods other than CAE are based on the interactions between carbohydrates and immobilized lectins. Thus, CAE is one of the convenient and versatile methods for the determination of carbohydrate–lectin interaction.

The principle of CAE for the application of glycan profiling is shown in Fig. 1. At the first step, a mixture of negatively charged fluorescently labeled glycans (A, B, and C in this case) is analyzed by capillary electrophoresis in an electrolyte that does not contain

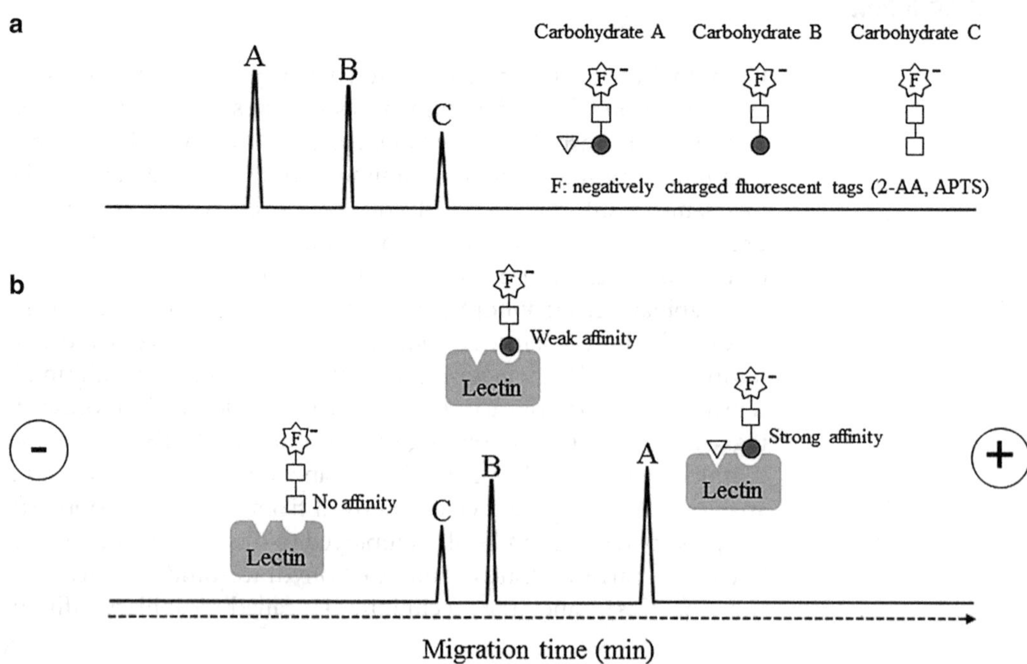

Fig. 1 Principle of capillary affinity electrophoresis. (**a**) In the absence of alectin. (**b**) In the presence of a lectin

a lectin (Fig. 1a). When the electroosmotic flow (EOF)-suppressed capillary is used, the negatively charged fluorescent labeled glycans (A, B, and C) are migrated in the order of mass to charge ratios. In the following step, the same sample is analyzed in the presence of a lectin of which specificity is well established. When the lectin recognizes glycan A (peak A in Fig. 1a), the glycan interacts with the lectin, and the peak is observed later. On the contrary, glycan C (peak C) does not show affinity to the lectin, and is observed at the same migration time as that observed in the absence of the lectin. Glycan B (peak B) shows weak affinity to the lectin, and the peak is observed slightly later than without the lectin. Thus, the migration order of the glycans changes as shown in Fig. 1b. By repeating the procedures described above using an appropriate set of lectins, all the glycans in the mixture can be categorized.

As mentioned above, the most important advantages of CAE is that the binding specificity of each glycan in the complex mixture can be simultaneously determined and subtle differences in the binding specificities of glycans having similar structural characteristics can be easily compared. In addition, it should be emphasized that the interaction by CAE is observed in the solution state, although many reported methods such as affinity chromatography or surface plasmon resonance (SPR) require immobilization of lectins or glycans. Here, we describe a method for determination of the interaction between fluorescently labeled carbohydrates and lectins, profiling of carbohydrates derived from glycoproteins, glycosaminoglycans, and animal milk samples, determination of affinity constants of the binding reaction between a lectin and a carbohydrate, and screening lectin activity in biological samples [11–15].

2 Materials

2.1 Equipments

1. A CE system equipped with a laser-induced fluorescence detection system (Beckman-Coulter). An Ar-laser (488 nm) or a He-Cd laser (325 nm) are available for this system.
2. An eCAP N-CHO coated capillary (Beckman-Coulter, 50 μm i.d., 375 μm o.d.).
3. A centrifugal vacuum evaporator (SpeedVac, Savant, Framingdale, NY).

2.2 Reagents

1. Peptide-N^4-(acetyl-β-D-glucosaminyl) asparagine amidase (*N*-glycoamidase F) from Roche Molecular Biochemicals.
2. Neuraminidase (*Arthrobacter ureafacience*) from Nakalai Tesque.
3. 2-Aminobenzoic acid (2-AA) (Tokyo Kasei) (*see* **Note 1**).
4. 8-Aminopyrene-l,3,6-trisulfonate (APTS) is from Beckman-Coulter (*see* **Note 2**).

5. Sodium cyanoborohydride ($NaBH_3CN$) (Sigma-Aldrich) (*see* **Note 3**). Caution: This compound is toxic; injurious to skin, eyes, and mucosa. Avoid inhalation.
6. Sephadex G-25 (fine grade) and Sephadex LH-20 (GE Healthcare BioSciences).
7. Mineral oil, n_D 1.4670, d 0.838 (Sigma-Aldrich).
8. 100 mM Tris-acetate buffer (pH 7.4) for CAE (*see* **Note 4**).
9. Polyethylene glycol (average molecular masses 70,000, PEG70000) (Wako Pure Chemicals) (*see* **Note 5**).

2.3 Preparation of N-Glycans from Glycoprotein Samples

1. A glycoprotein sample (1–200 μg) is suspended in 1 % SDS/1 % 2-mercaptoethanol (80 μL).
2. The mixture is kept in the boiling water bath for 10 min.
3. After cooling, an aqueous 10 % NP-40 solution (48 μL) and 0.3 M phosphate buffer (pH 7.5, 58 μL) are added.
4. After addition of *N*-glycoamidase F (1 unit/4 μL), the mixture is incubated at 37 °C for 24 h.
5. After keeping the mixture in the boiling water bath for 5 min, the mixture is mixed with 95 % EtOH (600 μL).
6. After centrifugation at 12,000 × *g* for 15 min, the supernatant containing the mixture of the released *N*-glycans is collected, and lyophilized to dryness by a centrifugal evaporator.

2.4 Preparation of Asialo-N-Glycans

1. A sialo-*N*-glycan sample (equivalent to 20 μg glycoprotein) is dissolved in 50 mM sodium acetate buffer (pH 5.0, 20 μL).
2. The mixture is incubated at 37 °C for 12 h.
3. After keeping the mixture in the boiling water bath for 5 min and centrifugation at 12,000 × *g* for 15 min, the supernatant containing the mixture of the asialo-*N*-glycans is collected, and lyophilized to dryness by a centrifugal evaporator.

2.5 Oligosaccharides for CAE

1. 2-AA-labeled milk oligosaccharide: 2-fucosyl-lactose (2-FL), Fucα1-2Galβ1-4Glc; Lewis X trisaccharide (Le^X), Galβ1-4(Fucα1-3)GlcNAc; difucosyl-lactose (LDFT), Fucα1-2Galβ1-4(Fucα1-3)Glc; Lacto-*N*-neotetraose (LNnT), Galβ1-4Glc NAcβ1-3Galβ1-4Glc; Lacto-*N*-fucopentaose (LNFPIII), Galβ1-4(Fucα1-3)GlcNAcβ1-3Galβ1-4Glc; Digalactosyl lacto-*N*-neohexaose (DGLNnH), Galα1-3Galβ1-4GlcNAcβ1-6[Galα1-3Galβ1-4GlcNAcβ1-3]Galβ1-4Glc) (*see* Subheading 3.1.3).
2. A mixture of APTS-labeled sialo-*N*-glycans from human α1-acid glycoprotein (*see* Subheading 3.2.5).

3. A mixture of APTS-labeled asialo-*N*-glycans from human α1-acid glycoprotein (*see* Subheading 3.2.3).
4. A mixture of oligomers of APTS-labeled GlcNAc (*see* Subheading 3.2.4).

2.6 Lectins for Capillary Affinity Electrophoresis

Concanavalin A (Con A), wheat germ agglutinin (WGA), *Datura stramonium* agglutinin (DSA), *Aleuria aurantia* lectin (AAL), *Ulex europaeus* agglutinin (UEA-I), *Ricinus communis* agglutinin (RCA120), soybean agglutinin (SBA), and *Maackia amurensis* lectin (MAM) are obtained from J-Oil Mills. *Pseudomonas aeruginosa* lectin (PA-I) is from Sigma Aldrich. *Aspergillus oryzae* lectin (AOL) is purchased from Gekkeikan. *Tulipa gesneriana* agglutinin (TGA), *Crocus sativus* lectin (CSL), and *Rhizopus stolonifer* lectin (RSL) are isolated and purified from the bulbs of tulip and crocus, and *Rhizopus stolonifer* (IFO 30816), respectively [11, 14, 16].

2.7 Preparation of the Solutions Containing Lectins for CAE

Con A. Con A requires calcium ion in the binding reaction. Calcium chloride dihydrate (0.7 g) is dissolved in 100 mM Tris-acetate buffer (pH 7.4, 100 mL). A portion (250 μL) is added to 100 mM Tris-acetate buffer (pH 7.4, 24.75 mL), and used for the electrolyte solution (100 mM Tris-acetate buffer containing 0.5 mM Ca^{2+}) of Con A. Con A (1.0 mg, 104 kDa) is dissolved in the buffer containing calcium ion (96 μL). The solution (100 μM Con A) is diluted with the same buffer and used for the electrolyte. *AAL.* A portion (250 μL) of the commercially available AAL (2 mg/mL, 27 μM, 72 kDa) solution is diluted with 100 mM Tris-acetate buffer (pH 7.4, 28 μL) to make up 24 μM solution. All other lectins are dissolved in 100 mM Tris-acetate buffer (pH 7.4) at the concentrations of 100 μM. *DSA* (1.0 mg, 86 kDa) in 116 μL: *RCA120* (1.0 mg, 120 kDa) in 83 μL: *SBA* (1.0 mg, 120 kDa) in 83 μL: *SSA* (1.0 mg, 160 kDa) in 62 μL: *MAM* (1.0 mg, 130 kDa) in 76 μL: *PA-I* (1.0 mg, 110 kDa) in 91 μL: *WGA* (1.0 mg, 43 kDa) in 232 μL: *UEA-I* (1.0 mg, 26.7 kDa) in 375 μL: *TGA* (1.0 mg, 43.2 kDa) in 232 μL: *RSL* (1.0 mg, 28 kDa) in 357 μL: *CSL* (1.0 mg, 48 kDa) in 208 μL: *AOL* (1.0 mg, 35 kDa) in 285 μL. Store the lectin solutions in aliquots at −20 °C until use.

3 Methods

It is important to select an appropriate set of lectins for the observation of clear interactions with various oligosaccharides. The interactions of each lectin with glycan/monosaccharide residues are illustrated in Fig. 2 using typical *N*-glycans and milk oligosaccharides as model structures. The lectins which show specificities toward mannose (Man), *N*-acetyl-D-glucosamine (GlcNAc),

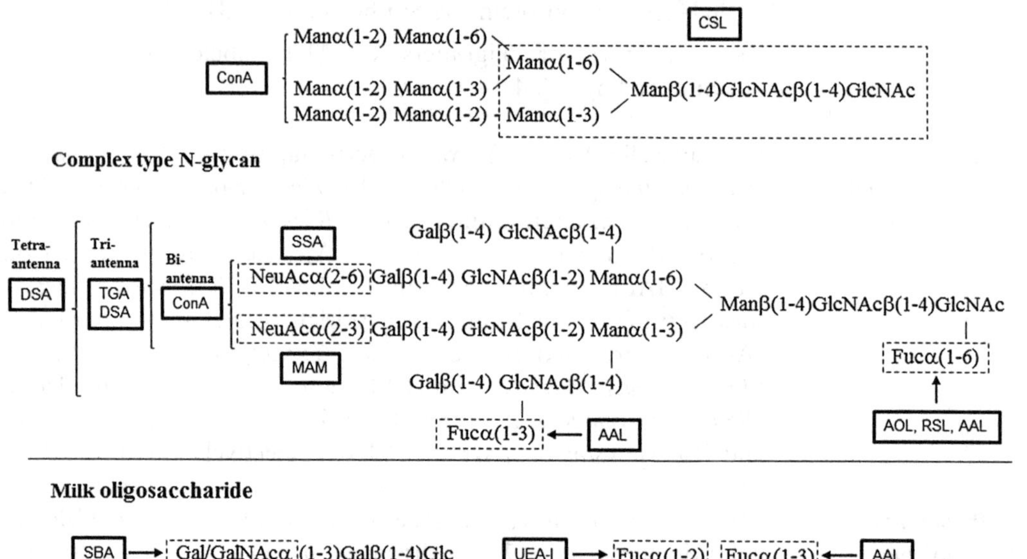

Fig. 2 Selection of lectins for capillary affinity electrophoresis. Abbreviations used for the structures: *GlcNAc* *N*-acetyl-D-glucosamine, *GalNAc* *N*-acetyl-D-galactosamine, *Man mannose*, *Gal* galactose, *Fuc* fucose, *NeuAc* *N*-acetyl-neuraminic acid

galactose (Gal), *N*-acetyl-D-galactosamine (GalNAc), fucose (Fuc), and sialic acids (NeuAc) are selected. Some of these lectins can recognize biatennary, triantennary, and tetraantennary structures of *N*-glycans. A set of lectins is also available for the analysis of other glycans such as those derived from mucin-type glycoproteins and glycolipids. All the solutions of lectins are diluted with 100 mM Tris-acetate buffer (pH 7.4), and a portion (5 μL) of the solution is mixed with the running buffer (5 μL) containing 1 % (for APTS-labeled glycans) or 10 % PEG70000 (for 2-AA-labeled glycans). The mixed solution is used as the electrolyte for CAE.

3.1 Capillary Affinity Electrophoresis of 2-AA-Labeled Glycans

3.1.1 Preparation of 2-AA-Labeled Glycans

1. To a sample of oligosaccharide (ca. 0.1–1 nmol) or a mixture of oligosaccharides (typically obtained from 1 to 100 μg of glycoprotein samples), is added a solution (200 μL) of 2-AA and $NaBH_3CN$, prepared by freshly dissolving both reagents (30 mg each) in methanol (1 mL) containing 4 % CH_3COONa and 2 % boric acid.
2. The mixture is kept at 80 °C for 60 min. After cooling, water (200 μL) is added to the reaction mixture.

3. The reaction mixture is applied to a small column (1 cm × 30 cm) of Sephadex LH-20 equilibrated with 50 % aqueous methanol. The earlier eluting fractions which contain fluorescently labeled oligosaccharides are collected and evaporated to dryness. The residue is dissolved in water (100 μL), and a portion (typically 10 μL) is used for CAE.

3.1.2 Capillary Affinity Electrophoresis of 2-AA-Labeled Glycans

Capillary affinity electrophoresis is performed with a P/ACE MDQ glycoprotein system (Beckman) equipped with a laser-induced fluorescence detection system. For the analysis of 2-AA-labeled carbohydrates, a He-Cd laser (325 nm) is used for detection with an emission band path filter at 405 nm. Separation is performed using a eCAP N-CHO capillary (30 or 50 cm total length, 50 μm i.d.) (*see* **Note 6**) after automatic injection using the pressure method (1.0 psi, 10 s). Separation is done with negative polarity at 25 °C. Data are collected and analyzed with a standard 32 Karat software (version 7.0, Beckman Coulter).

1. Prior to the analysis, the capillary is rinsed with 100 mM Tris-acetate buffer (pH 7.4) containing 0.5 % PEG70000 for 1 min at 20 psi.
2. The capillary is filled with the same buffer containing 5 % PEG70000 for 1 min at 20 psi.
3. A sample solution containing 2-AA-labeled oligosaccharides is introduced for 10 s at 1.0 psi.
4. Analysis is performed under constant voltage mode at an electric field of 600 V/cm.
5. The capillary is rinsed with the same buffer containing 0.5 % PEG70000 for 1 min at 20 psi.
6. The capillary is filled with a lectin solution at the specified concentration in the same buffer containing 5 % PEG70000.
7. Analysis is performed under constant voltage mode at an electric field of 600 V/cm. When necessary, the above procedures (**steps 5–7**) are repeated using different concentrations of the lectin or the different lectins.
8. The electropherogram observed in the presence of the lectin is compared with that observed in the absence of the lectin.

3.1.3 Profiling of Milk Oligosaccharides

Figure 3 shows the results on the analysis of a mixture of typical milk oligosaccharides using capillary affinity electrophoresis. In this example, milk oligosaccharides are labeled with 2-AA. The electropherogram at the bottom shows the separation of six milk oligosaccharides (2-FL, Le^X, LDFT, LNnT, LNFPIII, and DGLNnH) (*see* **Note 7**).

In the presence of RCA120 at 0.8 μM, the peak intensity of LnNT is dramatically decreased. This means that LNnT contains

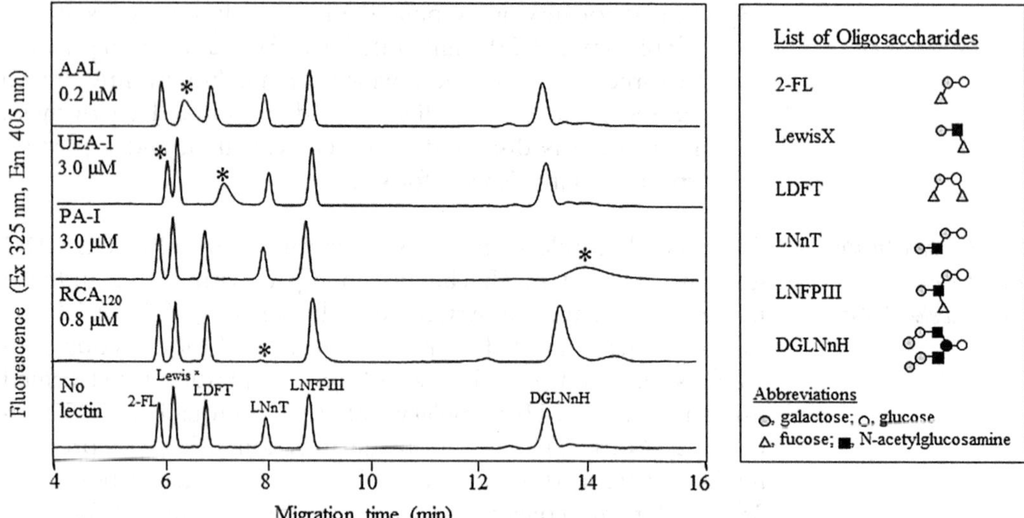

Fig. 3 Capillary affinity electrophoresis of 2-AA labeled milk oligosaccharides. Analytical conditions: capillary, eCAP N-CHO capillary (30 cm total length, 20 cm effective length, 50 μm i.d.); conditioning buffer, 100 mM Tris-acetate (pH 7.4) containing 0.5 % PEG70000; running buffer, 100 mM Tris-acetate (pH 7.4) containing 5 % PEG70000; applied voltage, 18 kV (reverse polarity); sample injection, pressure method (1.0 psi, 10 s). Fluorescence detection is performed with a 405 nm emission filter by irradiating with a He-Cd laser-induced 325 nm-light. The symbols are: Glc (*open circle*), GlcNAc (*filled square*), Gal (*filled circle*), and Fuc (*open triangle*)

LacNAc residue (*see* **Note 8**). PA-I at 3.0 μM shows retardation of DGLNnH. This indicates the presence of αGal residue at the non-reducing terminal of DGLNnH molecule. UEA-I which recognizes the Fuc residue linked through α1-2 linkage clearly interacts with 2-FL and LDFT. In contrast, AAL which also recognizes the Fuc residue interacts only with the Le^X glycan at 0.2 μM (*see* **Note 9**).

3.1.4 Analysis of Glycome of Human Cancer Cells

Glycans linked to proteins are synthesized by concerted actions of glycan-related enzymes (glycosyltransferases, glycosidases, sugar nucleotide synthases, etc.). Figure 4 shows an example for the analysis of *N*-glycans derived from human gastric adenocarcinoma cells (MKN45) using CAE. The electropherogram at the bottom shows the separation of total *N*-glycans in the absence of lectin. High-mannose type *N*-glycans (M6, M7, M8, and M9) are observed between 15 and 18 min. Disialo-biantennary *N*-glycan (A2F) is assigned using a standard sample. The peaks observed at 8–14 min are completely disappeared by digestion with sialidase and the peaks observed between 18 and 24 min are increased (data not shown).

In the presence of AAL at 0.75 μM, most peaks of sialo-N-glycans (8–14 min) and asialo-*N*-glycans (18–24 min) are completely disappeared, indicating that complex-type *N*-glycans in MKN45 cells have one or more Fuc residues (*see* **Note 10**). ConA which recognizes the high-mannose type and biantennary

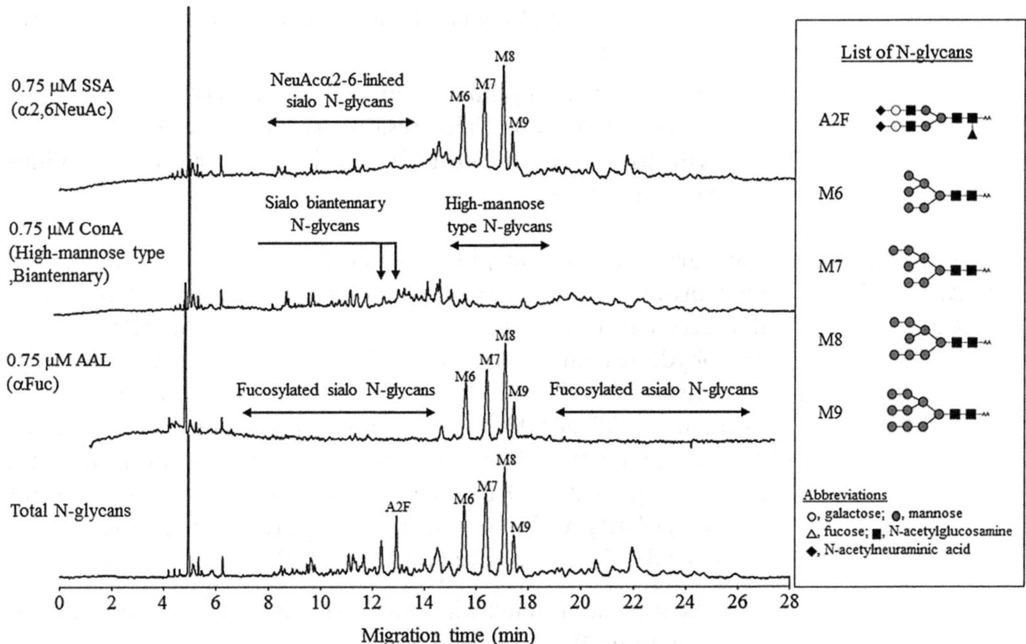

Fig. 4 Capillary affinity electrophoresis of 2-AA-labeled *N*-glycans derived from human gastric carcinoma (MKN45 cells). Analytical conditions: capillary, eCAP N-CHO capillary (50 cm total length, 40 cm effective length, 50 μm i.d.); running buffer, 100 mM Tris-acetate (pH 7.4) containing 5 % PEG70000; applied voltage, 30 kV (reverse polarity); sample injection, pressure method (1.0 psi, 5 s). Fluorescence detection is performed with a 405 nm emission filter by irradiating with a He-Cd laser-induced 325 nm-light

N-glycans interacts with A2F, M6, M7, M8, and M9. SSA which recognizes NeuAc residues linked through α2-6 linkage clearly interacts with A2F and most of sialo-*N*-glycans. But asialo-*N*-glycans (18–24 min) does not show interaction with SSA. From these results, we can easily obtain information on glycans of a complex mixture of *N*-glycans obtained from biological samples such as cancer cells.

3.2 Capillary Affinity Electrophoresis of APTS-Labeled Glycans

3.2.1 Preparation of APTS-Labeled Glycans

1. To a sample of oligosaccharide (ca. 0.1–1 nmol) or a mixture of oligosaccharides (typically obtained from 1 to 100 μg of glycoprotein samples), is added a solution (5 μL) of 100 mM APTS in 15 % aqueous acetic acid.
2. A freshly prepared solution of 1 M $NaBH_3CN$ in tetrahydrofuran (5 μL) is added to the mixture.
3. The mixture is overlaid with mineral oil (100 μL) to prevent evaporation of the reaction solvent and to obtain good reproducibility.
4. The mixture is kept for 90 min at 55 °C.
5. Water (200 μL) is added to the mixture. The fluorescent yellowish aqueous phase is collected, and applied to a column of

Sephadex G-25 (1 cm i.d., 50 cm length) previously equilibrated with water.

6. The earlier eluting fluorescent factions are collected and evaporated to dryness. The residue is then dissolved in water (100 μL), and a portion (20 μL) is used for capillary affinity electrophoresis.

3.2.2 Capillary Affinity Electrophoresis of APTS-Labeled Glycans

Capillary affinity electrophoresis is performed with a P/ACE MDQ glycoprotein system (Beckman) equipped with a laser-induced fluorescence detection system. For the analysis of APTS-labeled carbohydrates, an argon-laser (488 nm) is used for detection with an emission band path filter at 520 nm. Separation is performed using an eCAP N-CHO coated capillary (10 cm effective length, 30 cm total length, 50 μm i.d.) after automatic injections using the pressure method (1.0 psi, 10 s). Separation is performed using normal polarity at 25 °C. Data are collected and analyzed with a standard 32 Karat software (version 7.0, Beckman Coulter).

1. The capillary is filled with 100 mM Tris-acetate buffer (pH 7.4) containing 0.5 % PEG70000 for 1 min at 20 psi.
2. A sample solution containing APTS-labeled oligosaccharides is introduced from cathodic end for 10 s at 1.0 psi.
3. Analysis is performed under constant voltage mode at an electric field of 200 V/cm.
4. The capillary is rinsed with the same buffer containing 0.5 % PEG70000 for 1 min at 20 psi.
5. The capillary is filled with a lectin solution at the specified concentration in the same buffer containing 0.5 % PEG70000.
6. Analysis is performed under constant voltage mode at an electric field of 200 V/cm. When necessary, the above procedures (**steps 4–6**) are repeated using different concentrations of the lectin or the different lectins.
7. The electropherogram observed in the presence of the lectin is compared with that observed in the absence of the lectin.

3.2.3 Analysis of N-Linked Glycans Expressed on α1-Acid Glycoproteins from Human Serum

An example for CAE is shown in Fig. 5 using a mixture of APTS-labeled asialo-*N*-glycans derived from human α1-acid glycoprotein. The upper-left box shows the migrations of five asialo-N-glycans derived from α1-acid glycoprotein in the absence of lectin. All glycans are well resolved within 8 min, and good resolutions between AII and AIII and also between AIV and AV are achieved. AIII of triantennary glycan and AV of tetraantennary glycan contain a fucose residue as Le^X type residues. To confirm structure characteristics of these glycans, we can use the set of Con A, TGA, and AAL. Con A obviously decreases the peak of biantennary glycan (AI), and the peak of AI disappears at 3.0 μM concentration of Con A (*see* **Note 11**). Other triantennary and tetraantennary glycans

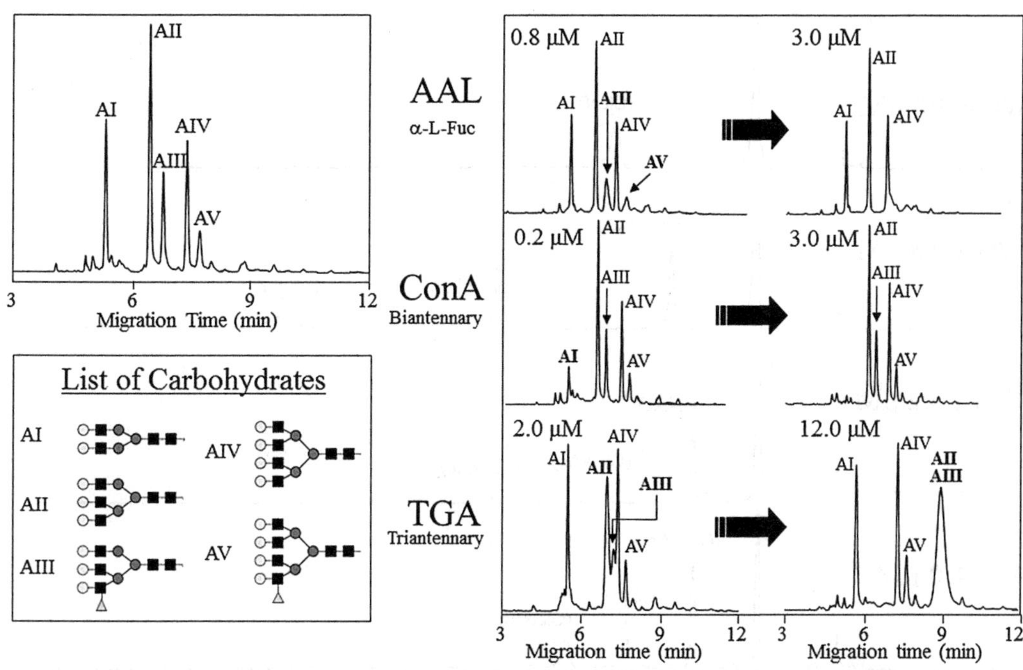

Fig. 5 Profiling of APTS-labeled asialo-*N*-linked glycans derived from human α1-acid glycoprotein by capillary affinity electrophoresis. Analytical conditions: capillary, eCAP N-CHO capillary (30 cm total length, 10 cm effective length, 50 μm i.d.); running buffer, 100 mM Tris-acetate (pH 7.4) containing 0.5 % PEG70000; applied voltage, 6 kV (normal polarity); sample injection, pressure method (1.0 psi, 10 s). Fluorescence detection was performed with a 520 nm emission filter by irradiating with an Ar laser-induced 488 nm-light. The symbols were: GlcNAc (*filled square*), Man (*open circle*), Gal (*filled circle*), Fuc (*open triangle*), and NeuAc (*open rhombus*)

do not show changes in migration times in the presence of Con A. Triantennary glycans are confirmed using the electrolyte which contains TGA. At 2.0 μM of TGA, both triantennary glycans of AII and AIII are observed slightly later. At 12.0 μM of TGA, AII and AIII are fused into a broad peak and observed after those of tetraantennary glycans (AIV and AV). Finally, the glycans having fucose residues attached to GlcNAc residue of lactosamine at the nonreducing positions are determined using the electrolyte containing AAL. At 0.8 μM of AAL, peak intensities of AIII and AV are obviously decreased, and these peaks are disappeared at 3.0 μM of AAL (*see* **Note 12**). As shown in Fig. 5, the set of Con A, TGA, and AAL can determine the structure characteristics of the glycans present in human α1-acid glycoprotein.

3.2.4 Kinetic Analysis of Lectin–Glycan Interactions

It is important to determine the stoichiometry of the binding reaction as well as categorization of carbohydrate chains for understanding the biological significance of carbohydrate chains attached to glycoproteins. In Fig. 6, we show the interactions between WGA and a mixture of APTS-labeled GlcNAc oligomers. The mixture of oligomers of GlcNAc shows interesting change in migrations

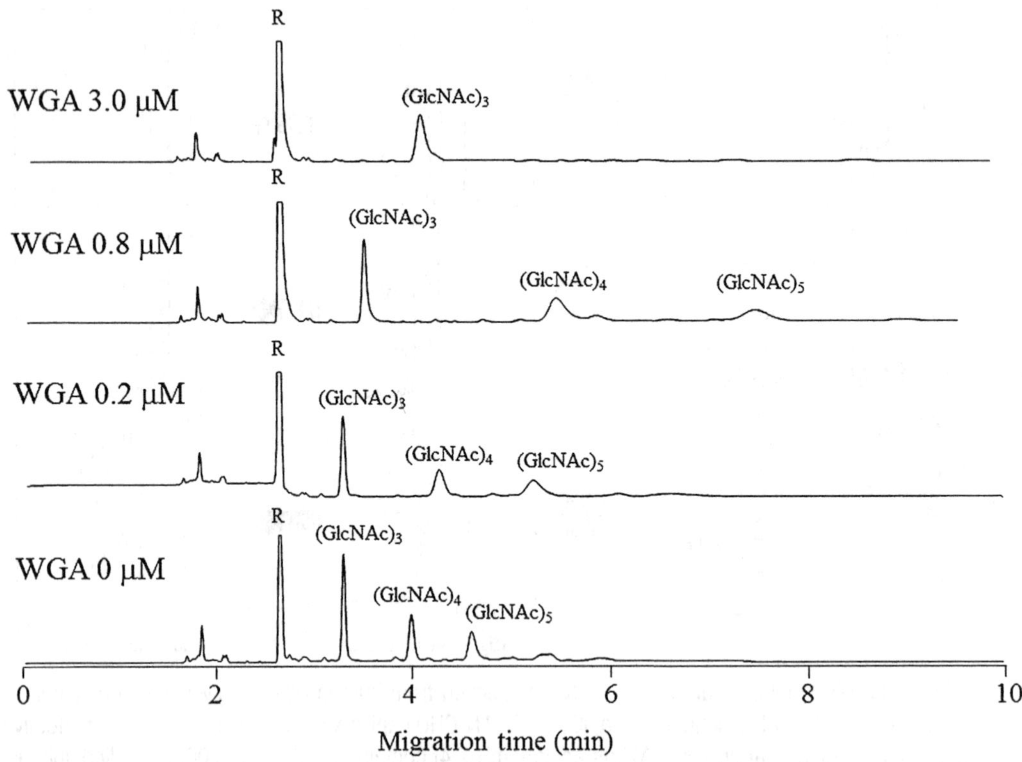

Fig. 6 Capillary affinity electrophoresis of a mixture of APTS-labeled *N*-acetylglucosamine oligomers in the presence of WGA in the electrolyte. R is due to the reagent (APTS). The numbers with the peaks indicate tri (3)-, tetra (4)-, and penta (5)-saccharide, respectively. An aqueous solution (20 μL) of the mixture (2 μg) of APTS-labeled *N*-acetylglucosamine oligomers was used as the sample solution. WGA was dissolved at the concentrations of (a) 0, (b) 0.20, (c) 0.80, and (d) 3.0 μM. Other analytical conditions were the same as in Fig. 5

in the presence of WGA at various concentrations. Trisaccharide ($GlcNAc_3$) shows weak affinity to WGA even at 3.0 μM of the lectin concentration. Tetrasaccharide ($GlcNAc_4$) begins to move at the slower velocity in the presence of 0.80 μM of WGA than that observed in the absence of WGA. Pentasaccharide ($GlcNAc_5$) begins to move at the slower velocity in the presence of 0.20 μM of WGA concentration. At higher concentrations of WGA, its migration velocities become obviously smaller than those of tetrasaccharide. Larger oligosaccharides than pentasaccharides show similar behaviors. These data indicate that the larger oligomers showed higher affinities to WGA. Taga et al. calculated association constants (K_a's) of oligosaccharides to a lectin using a few disaccharides as model and derived the equation as shown in Fig. 7. In the equation, t is the migration time of the ligand (APTS derivatives of GlcNAc oligomers in this case) in the presence of a protein at the concentration of [P] (WGA in this case), t_1 is the migration time of the ligand in the absence of the protein. We can easily obtain the K_a's by plotting the relationship between $(t-t_1)^{-1}$ and $[P]^{-1}$. It should be noticed that we do not have to determine the concentrations of the ligand (i.e., APTS-oligosaccharides).

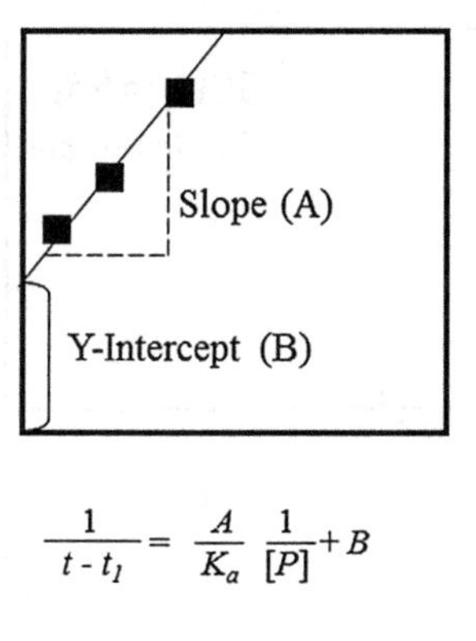

$$\frac{1}{t - t_1} = \frac{A}{K_a}\frac{1}{[P]} + B$$

Sugar	Ka
$(GlcNAc)_3$	$0.56 \times 10^6\ M^{-1}$
$(GlcNAc)_4$	$1.05 \times 10^6\ M^{-1}$
$(GlcNAc)_5$	$2.54 \times 10^6\ M^{-1}$

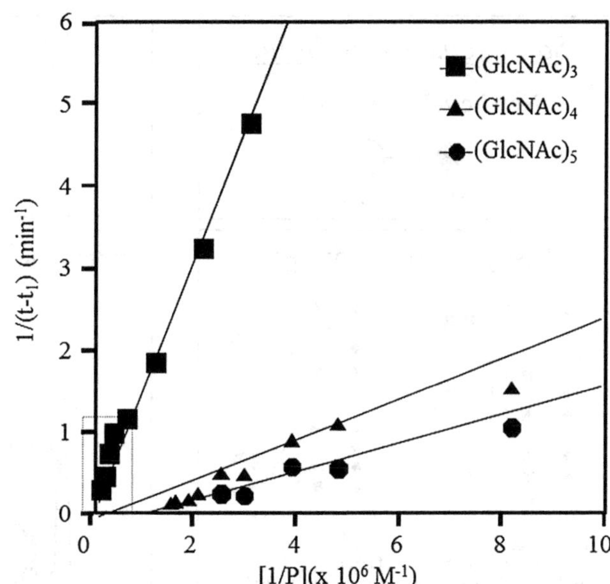

Fig. 7 Plots of $(t - t_1)^{-1}$ vs. $[P]^{-1}$ for trisaccharides, tetrasaccharides, and pentasaccharides of GlcNAc. *Closed square*, trisaccharide; *closed triangle*, tetrasaccharide; *closed circle*, pentasaccharide

This is quite important in the studies on the binding stoichiometry of a complex mixture of carbohydrate chains derived from biological samples, because it is difficult to determine the accurate concentration of each carbohydrate in a complex mixture of carbohydrates derived from biological samples. Using the data in Fig. 6, we plot the relationship between $(t - t_1)^{-1}$ and $[P]^{-1}$. Trisaccharides, tetrasaccharides, and pentasaccharides show good linear relationships with the concentrations of WGA (Fig. 7), and their binding constants were 0.56×10^6, 1.05×10^6, and $2.54 \times 10^6\ M^{-1}$ for trisaccharides, tetrasaccharides, and pentasaccharides, respectively. The results obtained by the present technique are well comparable to the reported values [12].

3.2.5 Screening of Lectin Activity in Biological Samples

Many screening methods such as agglutination assay for finding carbohydrate-binding proteins in crude extracts obtained from biological samples, essentially require a large amount of carbohydrates. In addition, it is often laborious to obtain carbohydrates for screening studies. CAE is available for detection of lectin activities in crude extracts from natural sources using an appropriate set of oligosaccharides (i.e., glycan library). At the initial step, a mixture of standard oligosaccharides of known compositions is analyzed in the absence of lectin. Then, the same mixture is analyzed in the running buffer containing crude extracts. If changes of migrations of one or more oligosaccharides are observed, such changes indicate that carbohydrate-binding proteins are present in the crude extract. It should be emphasized that quite small amount of

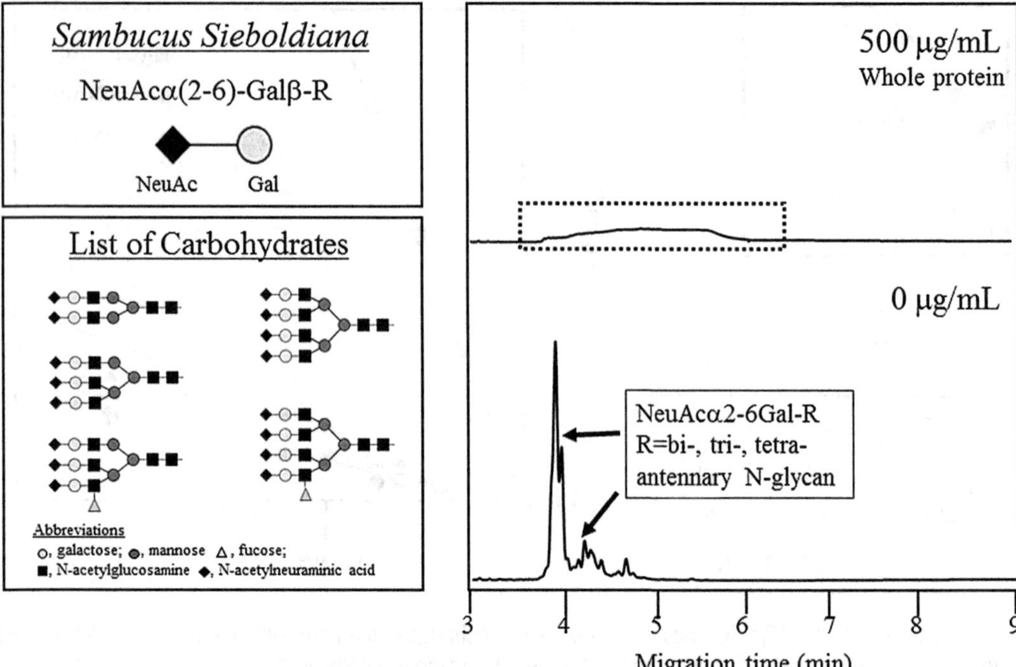

Fig. 8 Finding lectin activity in the crude extract of the barks of *S. sieboldiana*. An aqueous solution (10 μL) of the mixture of APTS-labeled sialo-*N*-glycans derived from human α1-acid glycoprotein (10 μg) was assayed in the presence of crude protein fractions from *S. sieboldiana* bark at the concentrations of 0 and 500 μg/mL as protein. Other analytical conditions were the same as in Fig. 5

carbohydrate samples are required for the assay. In addition, we can use a mixture of glycans derived from glycoprotein samples.

An example is shown in Fig. 8. The presence of sialic acid-specific *Sambucus sieboldiana* agglutinin (SSA) in barks of Japanese elderberry (*S. sieboldiana*) was reported [17, 18]. We examine the presence of the lectins by observing the change of migrations using a mixture of sialo-*N*-glycans derived from human α1-acid glycoprotein in the electrolyte containing the crude extract from the barks of *S. sieboldiana*. As shown in Fig. 8, the sialo-*N*-glycan peaks are disappeared at a concentration of 500 μg/mL crude protein fraction. In contrast, migrations of asialo-*N*-glycans are not affected (data not shown). These data clearly indicate that the crude extract contained a sialic acid-binding protein.

4 Notes

1. Reagents are of the highest grade commercially available, and used without further purification.
2. APTS reagent is also available from other commercial sources, but the reagent from other suppliers often contains

SO_3^--positional isomers which give minor peaks due to isomers in the analysis by CE.

3. The reagent is stable in the desiccator at room temperature for several months, but the fresh reagent should be used.
4. Running buffer for CAE is passed through an ultrafiltration membrane (0.40 K_a's m pore size) and degassed under reduced pressure before use.
5. Polyethylene glycol (PEG70000) is added in the electrolyte to minimize the electroosomotic flow. In addition, protein adsorption on the capillary surface is suppressed when polyethylene glycol is added in the running buffer at low concentrations.
6. The capillary of which inner surface is chemically modified with polyvinyl alcohol is employed for CAE, because lectins are often adsorbed on the capillary surface. A DB-1 capillary with dimethylpolysiloxane (DB-1) of the same size (GL Science Co. Ltd., Nishi-Shinjuku, Tokyo, Japan) is also available instead of eCAP N-CHO capillary. A DB-1 capillary shows weak EOF at neutral pH, and polyethylene glycol should be added to suppress EOF.
7. The smaller oligosaccharides are migrated faster than the larger ones based on their mass–charge ratios.
8. The interaction between RCA120 and Gal residues is inhibited when terminal Gal residues at nonreducing ends are substituted at C2 position by Fuc residues. Affinity of RCA120 toward Gal residue is decreased if Glc/GlcNAc residues on lactose/lactosamine are substituted with Fuc residues.
9. AAL recognizes Fucα1-3 residue more specifically than Fucα1-2 residue.
10. It is already reported that MKN45 cells express a large amount of fucosylated *N*-glycans [19].
11. High-affinity interaction with biantennary glycan is not prevented, even if outer Gal residues are substituted at C-3/6 position by sialic acids.
12. Complex type *N*-glycans with core α1-6 Fuc residues are observed later using the electrolyte containing RSL.

References

1. Chen FT, Evangelista RA (1995) Analysis of mono- and oligosaccharide isomers derivatized with 9-aminopyrene-1,4,6-trisulfonate by capillary electrophoresis with laser-induced fluorescence. Anal Biochem 230:273–280
2. Guttman A, Pritchett T (1995) Capillary gel electrophoresis separation of high-mannose typeoligosaccharidesderivatizedby1-aminopyrene-3,6,8-trisulfonic acid. Electrophoresis 16: 1906–1911
3. Chen F, Dobashi TS, Evangelista RA (1998) Quantitative analysis of sugar constituents of glycoproteins by capillary electrophoresis. Glycobiology 8:1045–1052
4. Chen F, Evangelista R (1998) Profiling glycoprotein n-linked oligosaccharide by

capillary electrophoresis. Electrophoresis 19: 2639–2644

5. Anumula KR, Dhume ST (1998) High resolution and high sensitivity methods for oligosaccharide mapping and characterization by normal phase high performance liquid chromatography following with highly derivatization fluorescent anthranilic acid. Glycobiology 8:685–694
6. Kamoda S, Nomura C, Kinoshita M, Nishiura S, Ishikawa R, Kakehi K, Kawasaki N, Hayakawa T (2004) Profiling analysis of oligosaccharides in antibody pharmaceuticals by capillary electrophoresis. J Chromatogr A 1050:211–216
7. Kakehi K, Kinoshita M, Kawakami D, Tanaka J, Sei K, Endo K, Oda Y, Iwaki M, Masuko T (2001) Capillary electrophoresis of sialic acid-containing glycoprotein. Effect of the heterogeneity of carbohydrate chains on glycoform separation using an alpha1-acid glycoprotein as a model. Anal Chem 73:2640–2647
8. Kakehi K, Funakubo T, Suzuki S, Oda Y, Kitada Y (1999) 3-Aminobenzamide and 3-aminobenzoic acid, tags for capillary electrophoresis of complex carbohydrates with laser-induced fluorescent detection. J Chromatogr A 863:205–218
9. Shimura K, Kasai K (1995) Determination of the affinity constants of concanavalin A for monosaccharides by fluorescence affinity probe capillary electrophoresis. Anal Biochem 227: 186–194
10. Uegaki K, Taga A, Akada Y, Suzuki S, Honda S (2002) Simultaneous determination of the association constants of oligosaccharides to a lectin by capillary electrophoresis. Anal Biochem 309:269–278
11. Oda Y, Senaha T, Matsuno Y, Nakajima K, Naka R, Kinoshita M, Honda E, Furuta I, Kakehi K (2003) A new fungal lectin recognizing alpha(1-6)-linked fucose in the N-glycan. J Biol Chem 278:32439–32447
12. Nakajima K, Oda Y, Kinoshita M, Kakehi K (2003) Capillary affinity electrophoresis for the screening of post-translational modification of proteins with carbohydrates. J Proteome Res 2:81–88
13. Nakajima K, Kinoshita M, Matsushita N, Urashima T, Suzuki M, Suzuki A, Kakehi K (2006) Capillary affinity electrophoresis using lectins for the analysis of milk oligosaccharide structure and its application to bovine colostrum oligosaccharides. Anal Biochem 348:105–114
14. Nakajima K, Kinoshita M, Oda Y, Masuko T, Kaku H, Shibuya N, Kakehi K (2004) Screening method of carbohydrate-binding proteins in biological sources by capillary affinity electrophoresis and its application to determination of Tulipa gesneriana agglutinin in tulip bulbs. Glycobiology 14:793–804
15. Kinoshita M, Kakehi K (2005) Analysis of the interaction between hyaluronan and hyaluronan-binding proteins by capillary affinity electrophoresis: significance of hyaluronan molecular size on binding reaction. J Chromatogr B Analyt Technol Biomed Life Sci 816:289–295
16. Oda Y, Nakayama K, Abdul-Rahman B, Kinoshita M, Hashimoto O, Kawasaki N, Hayakawa T, Kakehi K, Tomiya N, Lee YC (2000) *Crocus sativus* lectin recognizes Man3GlcNAc in the N-glycan core structure. J Biol Chem 275:26772–26779
17. Kaku H, Tanaka Y, Tazaki K, Minami E, Mizuno H, Shibuya N (1996) Sialylated oligosaccharide-specific plant lectin from Japanese elderberry (*Sambucus sieboldiana*) bark tissue has a homologous structure to type II ribosome-inactivating proteins, ricin and abrin. cDNA cloning and molecular modeling study. J Biol Chem 271:1480–1485
18. Rojo MA, Yato M, Ishii-Minami N, Minami E, Kaku H, Citores L, Girbés T, Shibuya N (1997) Isolation, cDNA cloning, biological properties, and carbohydrate binding specificity of sieboldin-b, a type II ribosome-inactivating protein from the bark of Japanese elderberry (*Sambucus sieboldiana*). Arch Biochem Biophys 340:185–194
19. Naka R, Kamoda S, Ishizuka A, Kinoshita M, Kakehi K (2006) Analysis of total *N*-glycans in cell membrane fractions of cancer cells using a combination of serotonin affinity chromatography and normal phase chromatography. J Proteome Res 5(1):88–97

Chapter 13

Basic Procedures for Lectin Flow Cytometry

Kenta Moriwaki and Eiji Miyoshi

Abstract

Glycans located on the cell surface regulate cell–cell interaction, cell homing, and signal transmission, which are particularly important for communication among cells. Certain cell types contain unique cell surface glycan structures, which have been utilized as markers for characterization. Flow cytometry is a powerful technology that enables the examination of multiple parameters of individual cells (e.g., cell size, internal complexity, and surface marker expression level). In this chapter, we describe a step-by-step procedure on how to detect glycans on the cell surface of live cells by flow cytometer, using lectins.

Key words Glycosylation, Glycan, Lectin, Flow cytometry, FACS

1 Introduction

Many proteins and lipids on the cell surface are modified by the attachment of glycans. Glycosylation plays various roles in a wide variety of biological processes including viral and bacterial pathogenesis, embryonic development, cancer, and inflammation [1, 2]. Although glycan structures are extremely diverse and susceptible to alterations during various cellular events and the progress of diseases, specific glycan structures are characteristic features of certain cell types, such as embryonic stem and/or cancer stem cells [3, 4]. To determine cellular glycan structures in detail, physicochemical analyses, such as mass spectrometry and lectin chromatography, have been developed [5]. However, these types of analyses require special equipment and expertise. Lectins, which are proteins that can bind to characteristic glycans, have been appreciated as a useful tool to easily determine total cellular oligosaccharide structures [6]. Nowadays, naturally purified or chemically modified lectins are commercially available and used in many laboratories to examine glycan structures and analyze glycan signals by various methods.

Flow cytometry is a technology that simultaneously measures and analyzes multiple physical characteristics of cells as they flow in a fluid stream through a beam of light. A fluorescent probe, such as

Jun Hirabayashi (ed.), *Lectins: Methods and Protocols*, Methods in Molecular Biology, vol. 1200, DOI 10.1007/978-1-4939-1292-6_13, © Springer Science+Business Media New York 2014

fluorochrome-labeled antibody, is used to detect molecules and biological properties of interest. This technology enables a quantitative analysis of single cells even if the sample contains heterogeneous cell populations and limited numbers of cells are available. Another major application of flow cytometry is to sort cells according to fluorescent signal for further biological studies. By combining this technology and lectins, we are able to characterize glycan structure on the surface of live cells of interest and sort the cells expressing particular glycans for further analyses. Here we provide a simple staining protocol for flow cytometry using chemically modified lectins. This staining protocol can be used to not only examine the expression level of specific glycans on the cell surface but also sort cells carrying glycans of interest. Single or multiple fluorescent colors will be applied in accordance with the intended use. In addition, glycan structures in a specific population of a heterogeneous cell population can be examined by simultaneously using antibodies against cell surface antigens.

2 Materials

2.1 Cell Culture

1. Tissue culture plates (multiple vendors).
2. RPMI 1640, Dulbecco's modified Eagle medium (DMEM), or any other media depending on the cell type used.
3. Fetal bovine serum (FBS, multiple vendors) and antibiotics (penicillin, streptomycin, and/or any other antibiotics, multiple vendors).

2.2 Staining for Flow Cytometry

1. Trypsin/EDTA (multiple vendors).
2. Phosphate-buffered saline (PBS): 2.5 mM KCl, 1.5 mM KH_2PO_4, 135 mM NaCl, 8 mM Na_2HPO_4.
3. FACS buffer: PBS supplemented with 2 % FBS or bovine serum albumin (BSA).
4. (*Optional*) Anti-Fc receptor antibody or any equivalent reagent (multiple vendors).
5. Fluorochrome or biotin-labeled lectins (multiple vendors).
6. Purified, fluorochrome- or biotin-labeled antibodies (multiple vendors).
7. (*Optional*) Fluorochrome-labeled streptavidin (multiple vendors) if biotin-labeled lectin or antibody is used.
8. (*Optional*) Fluorochrome-labeled secondary antibodies (multiple vendors) if purified antibody is used.
9. FACS tubes (BD Biosciences), 96-well U-bottom plate (multiple vendors), flow cytometer (e.g., FACSCalibur, FACSAria, BD Biosciences).

3 Methods

3.1 Cell Preparation

1. Allow cells to grow in culture media containing 10 % FCS and antibiotics in a 5 % CO_2 humidified incubator at 37 °C (*see* **Note 1**).
2. For adherent cells, aspirate culture media and wash cells with PBS. For suspension cells, transfer cells to a conical tube and spin down at 500 × *g* for 5 min. After washing with PBS twice, go to **step 5**.
3. For adherent cells, add trypsin/EDTA to each dish and place back in the incubator until cells begin to detach (*see* **Note 2**).
4. Detach cells by pipetting with PBS and transfer to a conical tube. Spin down at 500 × *g* for 5 min.
5. Resuspend cells with FACS buffer and count cell number (*see* **Note 3**).

3.2 Staining

1. Transfer aliquots of 0.1–1 × 10^6 cells to FACS tube (or 96-well U-bottom plate) and then spin down at 500 × *g* for 5 min (*see* **Notes 4** and **5**).
2. (*Optional*) Since cells express Fc receptors, if you plan to co-stain cells with antibodies, resuspend cell pellets in 100 μL of FACS buffer containing anti-Fc receptor antibody or any equivalent reagent and place the tubes on ice for 15 min (*see* **Note 6**).
3. Wash cells with 1 μL or 200 μL of FACS buffer for FACS tubes or 96-well plates, respectively.
4. Make a master mix by diluting lectins and antibodies that you want to use in FACS buffer (*see* **Note 7**). Resuspend cells with 100 μL of the master mix (*see* **Note 8**). If you plan to stain cells with multiple colors, do not forget to make a single staining control for each color (*see* **Notes 9–12**).
5. Incubate for 30 min in the dark on ice or at 4 °C (*see* **Note 13**).
6. Wash cells twice with 1 μL or 200 μL of FACS buffer for FACS tubes or 96-well plates, respectively.
7. If all of the lectins and antibodies used for staining are directly conjugated to a fluorochrome, resuspend cells in 500 or 200 μL of FACS buffer for FACS tubes or 96-well plates, respectively. Go to **step 12**.
8. If biotin-conjugated lectin or antibodies or purified antibodies are used, add the appropriate fluorochrome-labeled second-step reagents diluted in 100 μL of FACS buffer to cells.
9. Incubate for 20 min in the dark on ice or at 4 °C.
10. Wash cells twice with 1 μL or 200 μL of FACS buffer for FACS tubes or 96-well plates, respectively.

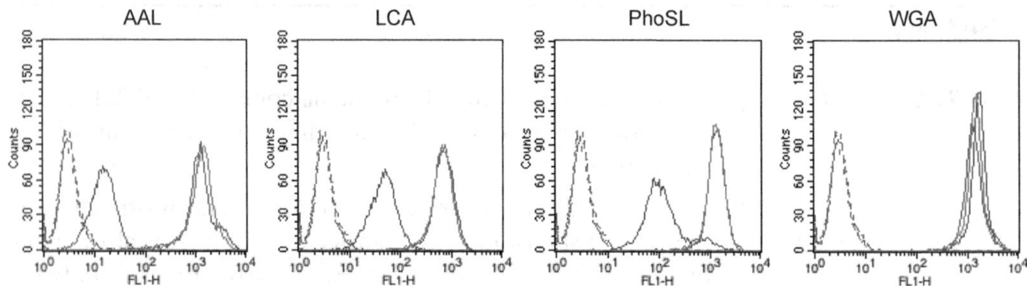

Fig. 1 Single staining analysis of various cells with FITC-labeled lectins in FACSCalibur. HCT116 cell line has a defect of fucosylation, which is caused by the mutation of GDP-mannose 4,6-dehydratase (GMDS) [9]. HCT116 cells were stably transfected with the vector carrying the *GMDS* gene (*red line*). *Blue line* indicates mock-transfected HCT116 cells. HCT116 cells were also cultured in the presence of 50 μM l-fucose to rescue fucosylation (*green line*). Strong staining with fluorescein isothiocyanate (FITC)-labeled AAL (*Aleuria aurantia* lectin), LCA (*Lens culinaris* agglutinin-A), and PhoSL (*Pholiota squarrosa* lectin), all of which recognize fucosylated glycan, was observed in fucosylation-rescued cells. Both LCA and PhoSL more specifically recognize core-fucose than Lewis-type fucose, compared to AAL. The residual staining in mock-transfected HCT116 cells could be due to nonspecific binding. All three cells showed similar staining intensity to FITC-labeled WGA (wheat germ agglutinin) lectin which preferentially recognizes the *N*-acetylgalactosamine residue. *Dotted line* shows non-staining control without FITC-labeled lectins. *y*-axis shows the intensity of fluorescence. When excited at 488 nm, FITC has a green emission which is mainly detected by the FL1 detector

11. Resuspend cells in 500 or 200 μL of FACS buffer for FACS tubes or 96-well plates, respectively.
12. Operate flow cytometer according to the manufacturer's protocol to acquire data and sort cells if needed (*see* **Notes 14** and **15**).

Typical examples of lectin flow cytometry to analyze cellular fucosylation in HCT116 cells are shown in Fig. 1. Other examples include E4-PHA lectin binding to hepatic progenitor cells (*see* Ref. 7) and application to characterize cancer stem cells (*see* Ref. 8).

4 Notes

1. Culture conditions, such as culture media, concentration of FCS and antibiotics, and the size of culture dishes vary depending on the experiment.
2. Trypsin/EDTA might degrade surface antigens and possibly intracellular antigens, and change the FSC (forward scatter) and SSC (side scatter) characteristics. Thus, before starting an experiment, examine whether trypsin/EDTA treatment affects the expression level of antigens of interest. If it does, try the following alternative way to detach cells from the dishes: EDTA alone and a cell scraper. For EDTA alone, incubate cells in 2 mM EDTA in PBS at 37 °C. It will take a longer time to detach cells than that for trypsin/EDTA treatment. You may

have difficulty getting a single-cell suspension. Scraping cells could potentially cause damage so that intracellular and extracellular antigens are compromised. Choose the best way for your experiment.

3. To distinguish between live and dead cells, it is recommended to use viability dyes such as trypan blue.
4. Cell number can be scaled up and down dependent on your experiment. If you are new to flow cytometry, use a higher number of cells to give yourself a margin of error. It is common to lose more cells than expected during the staining and washing procedures.
5. If you set up a new procedure or need to adjust instrument settings, it is helpful to prepare a few extra control samples with more cells in a larger volume. Running these cells before you start to collect data will give you time to adjust FSC and SSC parameters so that all the cells are on scale, and the background fluorescence is properly adjusted to a low position on the fluorescence scale.
6. Antibodies bind to many cell types, such as monocytes, by their nonspecific Fc ends. Thus, it is important to block the nonspecific binding by blocking the Fc receptors. Alternatively, F(ab) or F(ab') two fragments (antibodies without their Fc ends) can be used.
7. The optimum dilution which gives the best staining with minimum background and nonspecific binding must be determined experimentally for each assay. This is determined by using a series of dilutions in a titration experiment. High concentration of lectins may cause cell agglutination and death [10]. For purified lectins or antibodies, 1 μg/mL is suggested as the starting concentration.
8. To save staining reagents, the staining volume can be scaled down to 50 μL or less.
9. Every fluorochrome has characteristic excitation and emission spectra. The excitation spectrum is the range of light wavelengths that add energy to a fluorochrome, causing it to emit light in another range of lower wavelengths, which is called the emission spectrum. Flow cytometers use band-pass filters to separate the appropriate ranges of fluorescence emission. However, the emission spectra of several different fluorochromes overlap. This spectral overlap, if uncorrected, leads to a fluorochrome signal being detected by an inappropriate detector. To correct for this spectral overlap, a process of fluorescence compensation using single staining controls must be performed.
10. The choice of fluorochromes is critical to the success of flow cytometry experiments. To choose appropriate combination of fluorochromes, the brightness of fluorochromes, spectral overlap

between fluorochromes, the chemical and physical properties of fluorochromes, and lasers equipped in a flow cytometer must be considered.

11. Prepare proper controls depending on your experiment: (a) unstained control for each cell type used to determine auto-fluorescence that cells may have, (b) fluorochrome-conjugated second-step reagent control used to determine background binding, and (c) single staining control used as compensation controls to determine the extent of fluorochrome spectral overlap.
12. The viability dyes such as 7-amino actinomycin D and propidium iodide can be added to gate out dead cells in flow analysis after acquiring data.
13. Keep cells and reagents on ice or at 4 °C. The low temperature prevents the capping and internalization of antibodies binding surface antigens. Another way to prevent this is to add 0.05 % sodium azide in the FACS buffer.
14. Keep cells on ice and cover with foil until analysis on flow cytometer. For best results, analyze the cells as soon as possible. If you anticipate waiting longer, consider fixing cells, which can preserve them for at least several days.
15. Because each flow cytometer has different operating characteristics, each person or laboratory must determine its optimal operating procedure.

References

1. Ohtsubo K, Marth JD (2006) Glycosylation in cellular mechanisms of health and disease. Cell 126:855–867
2. Marth JD, Grewel PK (2008) Mammalian glycosylation in immunity. Nat Rev Immunol 8:874–887
3. Nairn AV, Aoki K, dela Rosa M et al (2012) Regulation of glycan structures in murine embryonic stem cells: combined transcript profiling of glycan-related genes and glycan structural analysis. J Biol Chem 287: 37835–37856
4. Adamczyk B, Tharmalingam T, Rudd PM (2012) Glycans as cancer biomarkers. Biochim Biophys Acta 1820:1347–1353
5. Rakus JF, Mahal LK (2011) New technologies for glycomic analysis: toward a systematic understanding of the glycome. Annu Rev Anal Chem 4:367–392
6. Cummings RD, Etzler ME (2009) Chapter 45: antibodies and lectins in glycan analysis. In: Varki A, Cummings RD, Esko JD et al (eds) Essentials of glycobiology, 2nd edn. Cold Spring Harbor, Cold Spring Harbor, NY
7. Sasaki N, Moriwaki K, Uozumi N et al (2009) High levels of E4-PHA-reactive oligosaccharides: potential as marker for cells with characteristics of hepatic progenitor cells. Glycoconj J 26:1213–1223
8. Moriwaki K, Okudo K, Haraguchi N et al (2011) Combination use of anti-CD133 antibody and SSA lactin can effectively enrich cells with high tumorigenicity. Cancer Sci 102:1164–1170
9. Moriwaki K, Noda K, Furukawa Y et al (2009) Deficiency of GMDS leads to escape from NK cell-mediated tumor surveillance through modulation of TRAIL signaling. Gastroenterology 137:188–198
10. Varki A, Etzler ME, Cummings RD et al (2009) Chapter 26: discovery and classification of glycan-binding proteins. In: Varki A, Cummings RD, Esko JD et al (eds) Essentials of glycobiology, 2nd edn. Cold Spring Harbor, Cold Spring Harbor, NY

Chapter 14

Histochemical Staining Using Lectin Probes

Yoshihiro Akimoto and Hayato Kawakami

Abstract

In histochemistry and cytochemistry, lectins are often used as probes for the localization of carbohydrates in cells and tissues. With lectins, cells and tissues can be identified as a particular type or a group in situ. Various lectins have been used for mapping of normal cells and tissues, pathological diagnosis such as malignant transformation, and identification of cell lineages during development. This chapter describes light and electron microscopic methods using lectin probes for determining carbohydrate localization in cells and tissues.

Key words Histochemistry, Direct and indirect detection methods, Light microscopy, Electron microscopy, Pre-embedding and post-embedding methods

1 Introduction

Glycans are examined histochemically by using carbohydrate-specific antibodies and lectins. Various lectins have been extracted from plants, animals, and microorganisms. Compared with antibodies, lectins have useful specificities for complex glycans and are less expensive. Changes in glycans during development or under pathological conditions can be observed histochemically by the use of lectins [1]. Lectins are used as markers for malignant transformation of tumors and differentiation of a specific kind of cell or tissue [2]. Histochemical research on lectins has been reviewed in several books [2, 3].

Most of the methods for conventional immunohistochemistry using antibody are also applicable for lectin histochemistry. As lectins are not species specific, lectins can be more readily used than antibodies. Certain groups of lectins have an affinity for *N*-linked, *O*-linked glycoproteins or both types. For example, to examine *N*-linked glycans, one should use lectins that recognize the mannose core or the complex type of *N*-glycans, i.e., lectins such as Con A, LCA, DSA, PVA etc.; whereas to examine *O*-linked glycans, lectins reacting with GalNAc, GlcNAc, Gal, Fuc, or sialic acid, such as ABA, BPA, PNA, SBA, etc., should be used.

Jun Hirabayashi (ed.), *Lectins: Methods and Protocols*, Methods in Molecular Biology, vol. 1200,
DOI 10.1007/978-1-4939-1292-6_14, © Springer Science+Business Media New York 2014

Table 1
A summarized guide to the selection of lectin-labeling methods

		Electron microscopy	
Label	Light microscopy	Pre-embedding method	Post-embedding method
Fluorescent dyes	R	NA	NA
Q dot	R	A	NR
HRP, ALP	R	R(HRP), A(ALP)	NR
Ferritin	NA	A	A
Colloidal gold	NR	NR	R

R recommended, *A* applicable, *NR* applicable but not recommended, *NA* not applicable, *ALP* alkaline phosphatase

Whereas in the case of immunohistochemistry ordinary proteins tend to lose their activity during fixation, glycans detected by lectin histochemical methods usually maintain their reactivity even after chemical fixation. For light microscopic studies, 4 % formaldehyde, Bouin's fluid, or ethanol fixation should be used. For electron microscopic studies, 2.5 % glutaraldehyde, 4 % formaldehyde, or a mixture of 2.5 % glutaraldehyde and 2 % formaldehyde are recommended. Lectin-binding sites can be visualized with different kinds of labels (e.g., fluorescent dyes, quantum dot (Q dot), horseradish peroxidase (HRP), ferritin, or colloidal gold). The most appropriate label for selection depends on whether the binding sites are to be examined at the light or electron microscopic level. A summarized guide to the selection of labeling methods is shown in Table 1.

For transmission electron microscopy (TEM), pre- and post-embedding methods are distinguished. In the case of the former, the cells or tissues are incubated with the desired lectin before embedding and sectioning, whereas in the latter, ultrathin sections are cut from resin-embedded specimens and then incubated with the lectin.

For light or electron microscopy, both direct and indirect methods are applicable. The simplest method is the direct method in which lectins directly conjugated to labels are used. In the indirect method, the cells or tissue sections are incubated with unconjugated lectin or biotinylated lectin; and then the bound lectin is detected by use of a labeled antibody against the lectin or by a streptavidin-conjugated label, respectively. Detection of the lectin-binding sites by direct method is sometimes less sensitive as compared with that by the indirect method. The indirect method employing biotinylated lectin is a useful technique with high sensitivity and with low background.

Several reviews and databases concerning lectin histochemistry are available and should be read before planning to use lectin histochemistry [1–4]. This chapter describes methods to localize glycans at both light and electron microscopic levels by use of lectins.

2 Materials

2.1 Tissue Fixation

1. Phosphate-buffered saline (PBS): Dissolve 8.00 g NaCl, 0.20 g KCl, 0.24 g KH_2PO_4, and 1.44 g Na_2HPO_4 in 1 L of distilled water.
2. 20 % (w/v) formaldehyde: Prepare formaldehyde solution by adding 20 g paraformaldehyde (EM grade) to 80 mL distilled water preheated to about 60 °C. Add a few drops of 10 N NaOH to clarify the solution and cool. This solution can be stored in the refrigerator for about 3 weeks.
3. 4 % (v/v) formaldehyde: Dilute 1 volume of 20 % formaldehyde solution with 4 volumes of PBS.
4. 2.5 % (v/v) glutaraldehyde–PBS: Dilute 1 volume of 25 % glutaraldehyde solution (EM grade, TAAB, Berkshire, UK) with 9 volumes of PBS.

2.2 Glycan Detection for Light Microscopy

2.2.1 Fluorescence-Labeling Method

1. Moisture chamber.
2. 5 % (w/v) Bovine serum albumin (BSA, Fraction V [Sigma, Saint Louis, MO, USA])—PBS.
3. Biotinylated lectins (Vector Lab, Cambridgeshire, UK).
4. 0.2 M phosphate buffer (pH 7.4): Prepare by using 0.2 M NaH_2PO_4 to adjust the pH of a 0.2 M Na_2HPO_4 solution to 7.4.
5. Fluorescent-conjugated streptavidin.
6. Anti-bleach reagent.
7. Nail polish.

2.2.2 HRP-Labeling Method

Steps 1–4, the same as Subheading 2.2.1.

5. HRP-conjugated avidin–biotin complex (Vector Lab).
6. 3, 3′-diaminobenzidine-4HCl (DAB [Sigma]).
7. 0.05 % (w/v) DAB-0.005 % (w/v) H_2O_2–PBS: Prepare a fresh solution by dissolving 50 mg DAB in 100 mL PBS. Add 16.7 μL of 30 % hydrogen peroxide just before starting the reaction.

2.3 Glycan Detection for Electron Microscopy

2.3.1 Post-embedding Method for TEM

Embedding with LR White Resin

1. Ethanol.
2. 50, 70, and 90 % aqueous solutions of ethanol.
3. LR-White (hard grade, London Resin Company, Berkshire, UK).
4. LR White-0.5 % (w/v) 2-ethoxy-2-phenylacetophenone, which accelerates the polymerization of LR-White.
5. A 1:1 mixture and 2:1 mixture of LR-White and ethanol.
6. 8-mm gelatin capsules (SPI Supplies/Structure Probe, West Chester, PA, USA).
7. Black light blue fluorescent lamp (FL15BLB 15 W, wavelength: 350 nm; Panasonic, Osaka Japan).

Sectioning and Lectin Reaction

1. 200-mesh nickel grids.
2. Light microscope.
3. Jewelers forceps (Vigor #5), anti capillary tweezer, platinum wire loop.
4. 12- or 18-mm colloidal gold-labeled streptavidin (Jackson Lab Research).
5. 2 % (w/v) uranyl acetate in distilled water.
6. 4 % (w/v) lead citrate in distilled water.

2.3.2 Pre-embedding Method for TEM

Lectin Reaction

1. HRP-conjugated lectins (EY Laboratory San Mateo OA, USA).
2. Others are the same as Subheading 2.2.2 except biotinylated lectins and HRP-conjugated avidin–biotin complex.

Post-fixation of Tissue

1. 1 % (v/v) glutaraldehyde–PBS: Dilute 1 volume of 25 % (v/v) glutaraldehyde solution (EM grade, TAAB, Berkshire, UK) with 24 volumes of PBS.
2. 1 % (w/v) OsO_4–0.1 M phosphate buffer (pH 7.4): Mix equal volumes of 2 % (w/v) OsO_4 in distilled water and 0.2 M phosphate buffer (pH 7.4).
3. 50, 70, 90, 100 % (v/v) Ethanol-distilled water.
4. 8-mm gelatin capsules (SPI Supplies/Structure Probe, West Chester, PA, USA).
5. Epon 812 (Plastic embedding resin [TAAB]).

Sectioning and Post-staining

1. Ultramicrotome (Ultracut UCT [Leica]).
2. 200-mesh copper grids.
3. 4 % (w/v) lead citrate in distilled water.
4. Single-edge razor blade.
5. Scotch tape.
6. Light microscope.

2.4 Controls

1. Inhibitory sugars.
   Mannose, fucose, galactose, *N*-acetylgalactosamine, *N*-acetylglucosamine, *N*-acetylneuraminic acid, etc. (Sigma).
2. Glycosidases.
   Exoglycosidases (neuraminidase, etc.), endoglycosidases (*O*-glycosidase, PNGase F, etc.) (New England Biolabs, Ipswich, MA, USA).

3 Methods

3.1 Tissue Fixation

1. Slice animal tissue into small pieces (size 5×5×2 mm) with razor blade as quickly as possible (*see* **Note 1**).
2. Fix tissue with cold 4 % formaldehyde–PBS or cold 2.5 % glutaraldehyde–PBS at 4 °C by gently shaking for 1 h for light or electron microscopy, respectively.

3.2 Glycan Detection for Light Microscopy

Fluorescence labeling method (Subheading 3.2.1) and HRP-labeling methods (Subheading 3.2.2) employing the avidin–biotin indirect method are described below. Specimens including paraffin section, cryostat section, semithin frozen section, and cell monolayers can be used (*see* **Note 2**).

3.2.1 Fluorescence-Labeling Method

1. Wash the specimens three times for 5 min each time with PBS.
2. Incubate the specimens with 5 % BSA–PBS (*see* **Note 3**) at room temperature for 30 min to block nonspecific binding of lectin (*see* **Note 4**).
3. Incubate the specimens with biotinylated lectin (5–50 μg/mL 0.1 % BSA–PBS, *see* **Note 5**) for 1 h in a moisture chamber at room temperature.
4. Wash the specimens three times for 5 min each time with cold PBS.
5. Incubate the specimens with fluorescence-labeled streptavidin (0.5 μg/mL PBS) and 4′,6-diamidino-2-phenylindole, dihydrochloride (DAPI, 0.5 μg/mL PBS, for nuclear staining) for 1 h at room temperature.
6. Wash the specimens three times for 5 min each time with cold PBS.
7. Mound with glycerol containing anti-bleach reagent.
8. Seal around the edges of the coverslip by using nail polish.
9. Observe specimen under a fluorescence microscope.

3.2.2 HRP-Labeling Method

1. Wash the specimens three times for 5 min each time with PBS.
2. Immerse specimens in 0.6 % H_2O_2–PBS at room temperature for 15 min to eliminate endogenous peroxidase activity (*see* **Note 6**).
3. Wash the specimens three times for 5 min each time with cold PBS.
4. Repeat **steps 1–4** above.
5. Incubate the specimens with HRP-conjugated avidin–biotin complex for 1 h at room temperature.

6. Wash the specimens three times for 5 min each time with cold PBS.
7. Incubate specimens with freshly prepared 0.05 % DAB-0.005 % H_2O_2–PBS at room temperature for 2 – 10 min while monitoring the reaction under a microscope. React specimens until they become dark (*see* **Note 7**).
8. When the specimens become dark, stop the reaction by removing the DAB solution and wash the specimens three times for 5 min each time with cold PBS.
9. Post-stain with hematoxylin or methyl green.
10. Observe under a light microscope. Figure 1a shows a light micrograph of 17-day-old chick embryonic skin stained with RCA-I lectin, which binds specifically to galactose residues. Strong staining is seen along the cell membrane of intermediate layer of the epidermis.

3.3 Glycan Detection for Electron Microscopy

The post-embedding method (Subheading 3.3.1) and pre-embedding method (Subheading 3.3.2) are described below.

3.3.1 Post-embedding Method for TEM

In most cases, hydrophilic resins such as LR-White or Lowicryl K4M can be used for embedding and ultrasectioning [4, 5]. Double labeling on both sides of ultrathin sections is possible with different kinds of lectins in combination with different sizes of colloidal gold particles by floating grids on solutions as shown in Fig. 2.

Embedding with LR White Resin

1. Fix tissue or cells with 2.5 % glutaraldehyde–PBS for 1 h at 4 °C.
2. Wash three times for 5 min each time with cold PBS.
3. Dehydrate the specimens in 50 % ethanol–distilled water (DW) for 15 min, 70 % ethanol–DW for 15 min, 90 % ethanol–DW for 20 min, and 100 % ethanol twice for 20 min each time.
4. Infiltrate the specimens with a 1:1 mixture of LR White and 100 % ethanol twice for 1 h each time and then with a 2:1 mixture of LR White and 100 % ethanol, again twice for 1 h each time (*see* **Note 8**).
5. Infiltrate the specimens with LR White-0.5 % (w/v) 2-ethoxy-2-phenylacetophenone three times for 1 h each time. Shake gently using a rocking plate.
6. Transfer the specimens into gelatin capsules filled with LR-White-0.5 % (w/v) 2-ethoxy-2-phenylacetophenone.
7. Polymerize LR White by use of near UV irradiation (wavelength: 350 nm) for 12 h at 4 °C (*see* **Note 9**).

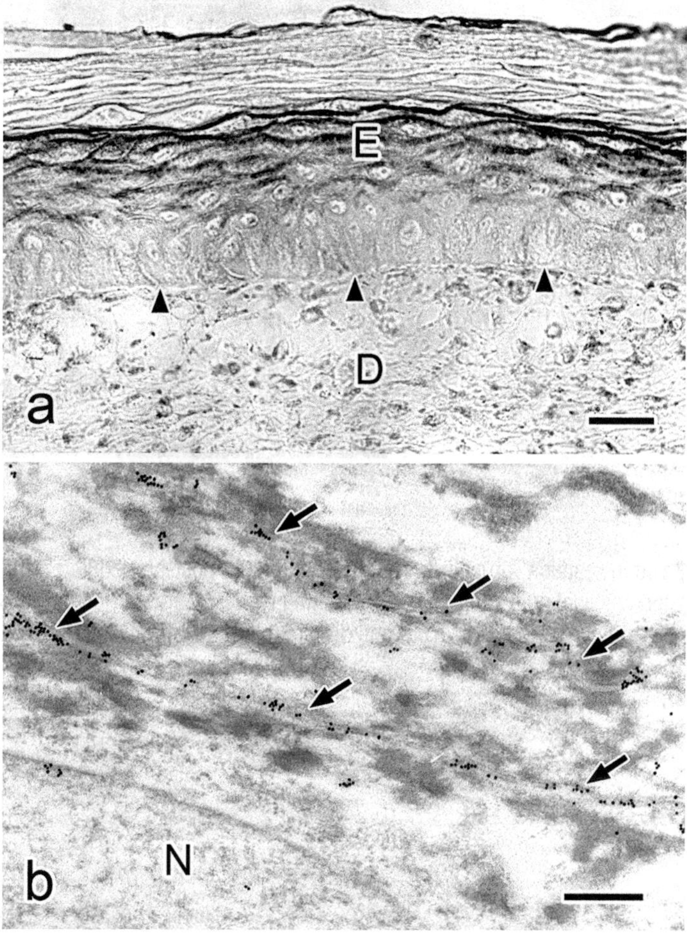

Fig. 1 Light (**a**) and electron (**b**) micrographs of skin from a 17-day-old chick embryo stained with RCA-I lectin, which binds specifically to galactose residues. The skin was fixed with 4 % formaldehyde. (**a**) This cryostat section was incubated with lectin and labeled with HRP. Strong staining is seen along the cell membrane in the intermediate layer of the epidermis. The *arrowheads* indicate the boundary between epidermis and dermis (*D*). Bar, 10 μm. (**b**) This LR White ultrathin section was incubated with lectin and labeled with colloidal gold. Strong labeling is seen along the cell membrane of the intermediate cells of the epidermis (*arrows*). *N* nucleus. Bar, 1 μm

Sectioning and Lectin Reaction

1. Cut ultrathin sections and mount them on nickel grids.
2. Incubate with 5 % BSA–PBS for 30 min.
3. Incubate with biotinylated lectins (5–50 μg/mL 0.1 % BSA–PBS) at room temperature for 1 h.
4. Wash three times for 5 min each time with PBS.
5. Incubate with colloidal gold (5–20 nm in diameter)-conjugated streptavidin at room temperature for 1 h.

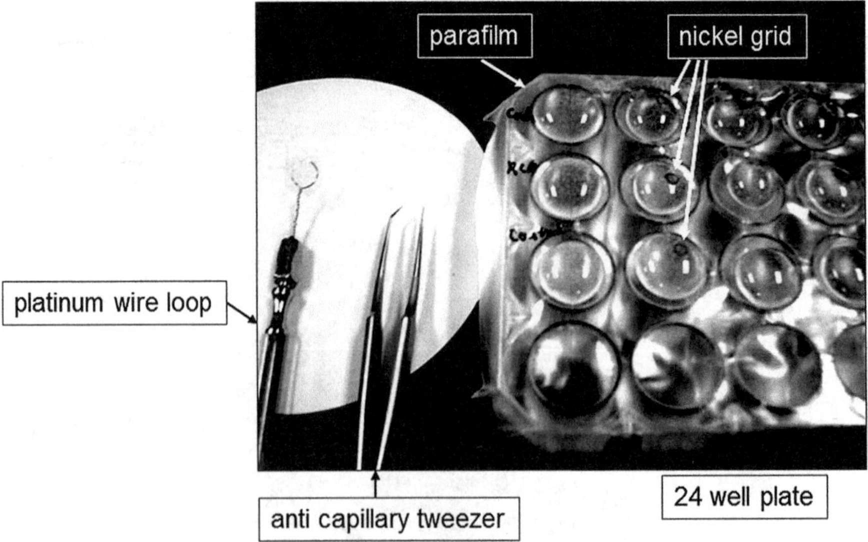

Fig. 2 Lectin reaction for electron microscopy by the post-embedding method. Incubation of specimen with lectin and colloidal gold conjugate and washing with PBS are carried out by floating grids on a droplet of each solution. Grids are transferred from one droplet to the next droplet by using a platinum wire loop or anti capillary tweezers

6. Wash three times for 5 min each time with PBS.
7. Post-fix with 2 % glutaraldehyde–PBS at room temperature for 10 min.
8. Wash three times for 5 min each time with distilled water.
9. Stain with 2 % uranyl acetate for 5 min.
10. Wash three times for 1 min each time with distilled water.
11. Stain with 4 % lead citrate solution for 1 min.
12. Wash three times for 1 min each time with distilled water.
13. Observe under a transmission electron microscope. Figure 1b is an electron micrograph of intermediate layer of 17-day-old chick embryonic skin stained with RCA-I lectin. Colloidal gold particles are seen along the cell membrane (arrows). It shows that glycoconjugates containing galactose are present along the cell membrane of intermediate cells.

3.3.2 Pre-embedding Method for TEM

For a detailed protocol of the pre-embedding method, *see* Refs. [6, 7].

Lectin Reaction

1. Wash cryostat sections three times for 5 min each time with PBS.
2. Immerse sections in 0.6 % H_2O_2–PBS at room temperature for 15 min to eliminate endogenous peroxidase activity (*see* **Note 6**).
3. Wash the sections three times for 5 min each time with cold PBS.

4. Incubate the specimens with 5 % BSA–PBS at room temperature for 30 min to block nonspecific binding of the lectin.
5. Wash the specimens three times for 5 min each time with PBS.
6. Incubate the specimens with HRP-conjugated lectin (5–50 μg/mL PBS) for 24 h in a moisture chamber at 4 °C.
7. Wash the specimens six times for 5 min with cold PBS.
8. Incubate the specimens in 0.05 % DAB solution for 10 min at room temperature.
9. Remove the specimens from the DAB solution. Initiate the HRP reaction with freshly prepared 0.05 % DAB-0.005 % H_2O_2–PBS and incubate the specimens at room temperature for 2–10 min while monitoring the reaction under a microscope. React specimens until they become dark.
10. When the specimens become dark, stop the reaction by removing the DAB solution and wash the specimens three times for 5 min each time with cold PBS.

Post-fixation of Tissues

1. Fix the labeled specimens by incubation for 10 min at 4 °C with 1 % glutaraldehyde–PBS.
2. Wash the specimens three times for 5 min each time with cold PBS.
3. Place the glass slides on ice in a fume hood and osmicate them by incubation for 1 h at 4 °C with 1 % OsO_4–0.1 M phosphate buffer (pH 7.4).
4. Wash the specimens three times for 5 min each time with cold distilled water.
5. Dehydrate the specimens by passage through a graded series of ethanol (50, 70, 90, 100 %), incubating them for 5 min in each concentration.
6. Fill a gelatin capsule with Epon 812; and before the specimen has dried, invert the resin-filled capsule over the tissue specimen on the glass slide.
7. Polymerize the resin at 60 °C for 24 h.
8. Detach the Epon-embedded specimens from the glass slide by heating the slide to 170 °C on a hot plate until the gelatin capsule containing resin can be snapped off the surface of the slide.

Sectioning and Post-staining

1. Observe the specimen under a binocular light microscope after attaching the Epon-embedded specimen to a glass slide by using Scotch tape and determine the area to be trimmed.
2. Trim the resin block by using a single-edge razor blade.
3. Cut ultrathin sections of the trimmed block with an ultramicrotome.

4. Place section on a copper grid and stain with 4 % lead citrate in distilled water for 4 min at room temperature.
5. Observe the stained sections under a transmission electron microscope.

3.4 Controls

Both negative and positive histochemical control experiments should be carried out to confirm the specificity of the lectin binding reaction. Negative controls should include (1) the addition of the appropriate inhibitory sugars (0.1–0.2 M) specific for each lectin to the HRP-conjugated lectin–PBS to compete with the target for lectin binding; (2) omission of the lectin; (3) destruction of target glycans by incubation with the appropriate glycosidases. The positive control should include incubation of the cells or tissue with a lectin whose binding to the cell or tissue has already been established experimentally.

3.4.1 Glycosidase Treatment

1. Incubate specimens with glycosidase in appropriate buffer for 2 h at room temperature.
2. Wash three times for 5 min each time with PBS.
3. Proceed to the lectin staining.

4 Notes

1. The tissue should be fixed by perfusion with cold fixative prior to removal from the animal. Perfusion fixation yields excellent morphology, as it provides rapid and uniform fixation.
2. When glycans of glycolipids are examined, paraffin sections should not be used, as organic solvent such as xylene, acetone, or ethanol will dissolve the lipids.
3. Do not use normal serum for blocking, as glycoconjugates present in serum competitively inhibit specific binding of the lectin. Instead of BSA, blocking reagent NB2025 (NOF Corporation), which is a synthetic polymer-based reagent that was developed for blocking nonspecific adsorption in immunoassay plates, should be used.
4. When the staining intensity is weak, antigen retrieval treatment sometimes improves it. For this procedure, immerse slides into 20 mM Tris–HCl buffer (pH 9.0). Autoclave slides at 105 °C for 10 min. Cool the slides to room temperature and wash with PBS, and then proceed to the blocking treatment.
5. As each lectin has a different affinity for a given sugar residue, the optimal lectin concentration to use should be determined by light microscopic observation before electron microscopic experiments are performed.

6. This step can be omitted if there is no endogenous peroxidase activity in the tissue.
7. When the DAB reaction product diffuses away from the primary site of reaction because of over-reaction for DAB deposition, samples should be reacted for a shorter period of time.
8. Dimethyl formamide can also be used instead of ethanol.
9. Instead of photopolymerization, polymerization can be conducted at 55 °C for 24 h.

References

1. Brooks SA, Leathem AJC, Schumacher U (1997) Lectin histochemistry: a concise practical handbook. BIOS Scientific Publishers in association with the Royal Microscopical Society—(Royal Microscopical Society microscopy handbooks, vol 36)
2. Sharon N, Lis H (2003) Lectins, 2nd edn. Kluwer Academic, Dordrecht, The Netherlands
3. Varki A, Cummings RD, Esko JD, Freeze HH, Hart GW, Etzler ME (2009) Essentials of glycobiology, 2nd edn. Cold Spring Harbor Laboratory Press, New York
4. Kawakami H, Hirano H (1993) Lectin histochemistry and cytochemistry. In: Ogawa K, Barka T (eds) Electron microscopic cytochemistry and immunocytochemistry in biomedicine. CRC, Boca Raton, FL, pp 319–332
5. Yamashita S (2010) The post-embedding method for immunoelectron microscopy of mammalian tissues: a standardized procedure based on heat-induced antigen retrieval. In: Schwartzbach SD, Osafune T (eds) Immunoelectron microscopy: methods and protocols, vol 657, Methods in molecular biology. Springer, New York, NY, pp 237–248
6. Oliver C (1994) Pre-embedding labeling methods. In: Javois LC (ed) Immunocytochemical methods and protocols, vol 34, Methods in molecular biology. Springer, New York, NY, pp 315–319
7. Akimoto Y, Kawakami H (2010) Pre-embedding electron microscopy methods for glycan localization in chemically fixed mammalian tissue using horseradish peroxidase-conjugated lectin. In: Schwartzbach SD, Osafune T (eds) Immunoelectron microscopy: methods and protocols, vol 657, Methods in molecular biology. Springer, New York, NY, pp 217–224

Chapter 15

Equilibrium Dialysis Using Chromophoric Sugar Derivatives

Tomomitsu Hatakeyama

Abstract

Equilibrium dialysis has been used to examine the binding affinity of ligands to proteins. It is a simple and reliable method, which requires only inexpensive equipment. For analysis of lectin–sugar interactions, the lectin and sugar are placed in the individual chambers separated by the membrane to allow the sugar to diffuse into the lectin chamber. After equilibrium has been reached, the concentrations of the sugar in both chambers are determined to evaluate the sugar-binding affinity of lectin. In this chapter, an example of the equilibrium dialysis experiment using the chromophoric derivatives of galactose and *N*-acetylgalactosamine is demonstrated, which reveals the difference in the affinity as well as specificities of two different carbohydrate-binding sites present in the B-chains of the plant lectin ricin.

Key words Equilibrium dialysis, Ricin, Galactose, *N*-acetylgalactosamine, Nonlinear regression analysis

1 Introduction

Lectin–carbohydrate interactions have been increasingly recognized as one of the most important molecular recognition systems in organisms to mediate information between molecules and cells. For the determination of lectin–carbohydrate interactions, various principles are used, as described in the other chapters of this book. Among the methods to determine static binding constants between proteins and ligands, equilibrium dialysis is one of the most simple and direct methods, and it can be performed using only inexpensive equipment, e.g., dialysis cells and membranes. There is no pretreatment required, such as immobilization of proteins or ligands, which may affect the binding efficiency. This approach provides straightforward results, because the binding of ligands is directly measured.

The principle of equilibrium dialysis is illustrated in Fig. 1. The experiment is performed using a dialysis apparatus containing two small chambers with semipermeable dialysis membranes between them. Suitable membranes with different pore sizes are selected and

Jun Hirabayashi (ed.), *Lectins: Methods and Protocols*, Methods in Molecular Biology, vol. 1200,
DOI 10.1007/978-1-4939-1292-6_15, © Springer Science+Business Media New York 2014

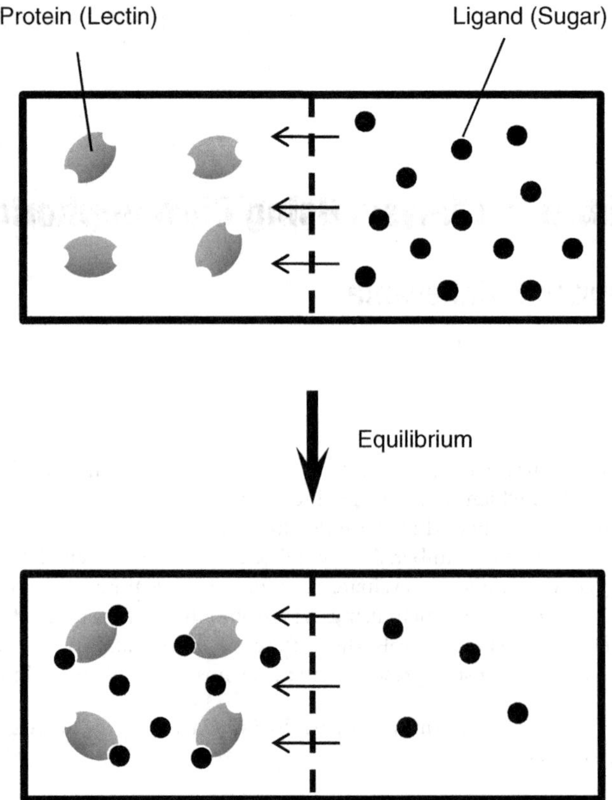

Fig. 1 Principle of equilibrium dialysis. The lectin and the sugar are placed in the chambers separated with a dialysis membrane. After equilibration has been reached, the concentration of the sugar in the lectin chamber (*left*) may be higher than in the sugar chamber (*right*) because of the binding of the sugar to the lectin. By measuring the sugar concentrations in both chambers, bound and free sugar concentrations are determined and used to calculate the association constant and the number of the binding sites of the lectin

the selection is based on molecular sizes of the proteins and ligands to permit the diffusion of ligands into the lectin chamber. After placing the lectin and sugar in the individual chambers separated by the membrane, sugar is allowed to diffuse into the lectin chamber. After equilibrium has been reached, the concentrations of the sugar in both chambers are determined to evaluate the amount of the sugar bound to lectin. Although the concentrations of sugars can be directly determined [1, 2], it is convenient to use those labeled with radioactive isotopes [3–6] or chromophoric compounds [7–10]. Although use of radioactive sugars enables highly sensitive determinations, they need to be handled carefully to prevent contamination. Chromophoric sugar derivatives are, on the other hand, much easier to handle, and can be easily determined by a spectrophotometer. However, potential effects of chromophoric moieties on the interaction with lectin should also be considered.

Here, an example of equilibrium dialysis using the plant lectin ricin is demonstrated. As its ligands, *p*-nitrophenyl β-D-galactopyranoside (NP-Gal) and *p*-nitrophenyl *N*-acetyl-β-D-galactosaminide (NP-GalNAc) are used. Such *p*-nitrophenyl derivatives of sugars are commercially available for use as substrates for glycosidases. Ricin is a highly toxic plant lectin, composed of A- and B-chains linked with a disulfide bond. While the B-chain has galactose-specific lectin activity, the A-chain shows strong cytotoxicity by attacking the eukaryotic 28S ribosomal RNA by *N*-glycosidase activity [11]. The A-chain is translocated into the cytosol with the help of the B-chain that binds to galactose-containing glycoconjugates on the target cell surface. Because binding to the specific complex carbohydrates on target cell surfaces is an obligatory step for the toxic action of ricin, its carbohydrate-binding properties have been investigated.

2 Materials

1. Dialysis membrane washing buffer: 1 mM ethylenediamine tetraacetate (EDTA).
2. Blocking solution for dialysis cells: 2 % aqueous bovine serum albumin (BSA) solution.
3. Dialysis buffer: 4.4 mM sodium phosphate buffer (pH 6.9) containing 90 mM NaCl.
4. Sugars: *p*-Nitrophenyl β-D-galactopyranoside (NP-Gal) and *p*-nitrophenyl *N*-acetyl-β-D-galactosaminide (NP-GalNAc) are from Sigma.
5. Lectin: Ricin is purified from the castor beans produced in the Philippines using ion-exchange chromatography and gel filtration by the method of Hara et al. [12].
6. Equilibrium dialyzer and dialysis membrane: The equilibrium dialyzer is from Technilab Instruments. Several other types of apparatuses for equilibrium dialysis are now commercially available (*see* **Note 1**). The dialysis membrane made of regenerated cellulose (molecular mass 10-kDa cutoff) is from Sanko Jun-yaku, Japan.

3 Methods

1. Wash the dialysis membrane by boiling in a 1 mM EDTA solution for 5 min, then rinse thoroughly in deionized water.
2. Fill dialysis cells of the equilibrium dialyzer with a 2 % BSA solution, and let stand for 1 h at room temperature to precoat the inner surface of the cells to prevent nonspecific interaction with the lectin, then wash the cells with deionized water.

3. Set up the equilibrium dialyzer, joining two chambers with a dialysis membrane placed between them.
4. Place the NP-Gal or NP-GalNAc solution (0.01–2 mM, 200 μL) in the dialysis buffer into one of the chambers separated by the dialysis membrane.
5. Place the same volume of the lectin solution (5 mg/mL) in the dialysis buffer into the other chamber (*see* **Note 2**).
6. Allow the sugar to diffuse across the membrane until equilibrium has been reached (~24 h) at a constant temperature (17 °C in this experiment) (*see* **Note 3**).
7. Withdraw aliquots (125 μL) of the solutions from the chambers with a microsyringe.
8. Determine the concentration of the sugar photometrically using a molar extinction coefficient of 1×10^4 M^{-1} cm^{-1} at 313 nm [9]. The binding ratios of the sugars to the lectin are plotted as shown in Fig. 2, and the Scatchard plot is demonstrated in Fig. 3 (*see* **Note 4**). The number of binding sites (n) per lectin molecule and the association constant (K_a) are calculated by nonlinear regression analysis (*see* **Notes 5** and **6**).

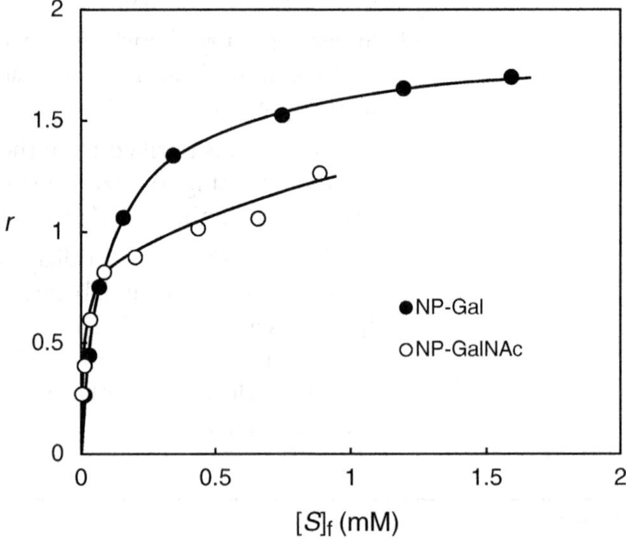

Fig. 2 The binding curves of NP-Gal and NP-GalNAc to ricin. The buffer for the equilibrium dialysis experiment is 4.4 mM phosphate buffer (pH 6.9) containing 90 mM NaCl and at 17 °C. The molar ratios of the bound sugar to lectin molecule (r) are plotted against the free sugar concentration ($[S]_f$)

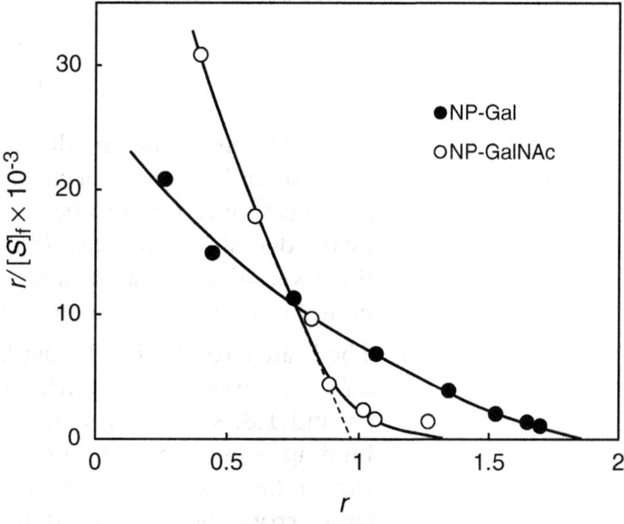

Fig. 3 Scatchard plot for the binding of NP-Gal and NP-GalNAc to ricin. The plot for the binding of NP-Gal indicates a biphasic curve with a maximum *r* of 1.8, suggesting that there are two binding sites for NP-Gal with different affinities. By analyzing the data for the binding of NP-Gal in Fig. 2 using nonlinear regression, two association constants, $K_{a1} = 2.3 \times 10^4$ M^{-1} and $K_{a2} = 3.8 \times 10^3$ M^{-1} for the high-affinity binding site and the low-affinity binding site [3, 13, 14], respectively, are obtained. In the case of NP-GalNAc, the maximum *r* is approximately 1.3 and K_{a2} for NP-GalNAc is too small to be calculated, suggesting that only one of the two binding sites in ricin actually has affinity for GalNAc ($K_{a1} = 6.5 \times 10^4$ M^{-1})

4 Notes

1. Various types of equilibrium dialyzers are available from manufacturers, such as the following:
   Rapid Equilibrium Dialysis (RED) Device (Thermo Scientific)
   Micro-Equilibrium Dialyzer (Harvard Apparatus)
   Reusable 96-well Micro-Equilibrium Dialysis Device (HTDialysis, LLC)
2. Addition of the same concentration of the sugar into the lectin chamber will shorten the time required for equilibration.
3. Constantly rotating the dialysis cells is often recommended to shorten the equilibration time.
4. The Scatchard plot is based on the following equations:

$$\frac{r}{[S]_f} = -K_a \cdot r + n \cdot K_a \quad (1)$$

$$r = \frac{[S]_b}{[P]} \quad (2)$$

where $[S]_f$ and $[S]_b$ are the concentrations of the free and lectin-bound sugar, respectively, which are determined from the sugar concentrations between the two chambers separated by the dialysis membrane, $[P]$ is the lectin concentration, K_a is the association constant, and n is the maximum number of bound sugar per protein molecule (Eqs. 1 and 2).

5. The Scatchard plot for the binding of NP-Gal to ricin shows a biphasic curve (Fig. 3) with a maximum binding ratio (r) of around 1.8, suggesting that there may be two carbohydrate-binding sites with different affinities. Because calculation of the binding constants from the Scatchard plot may give rise to large errors, the analysis is done by nonlinear regression using the data shown in Fig. 2 with the following equation [3, 13], assuming that there are two binding sites with different association constants:

$$r = \frac{K_{a1} \cdot [S]_f}{1 + K_{a1} \cdot [S]_f} + \frac{K_{a2} \cdot [S]_f}{1 + K_{a2} \cdot [S]_f} \quad (3)$$

where r and $[S]_f$ are the molar ratio of the bound sugar to the lectin molecule and the concentration of free sugar, respectively (Eq. 3). K_{a1} and K_{a2} are the two association constants for the different carbohydrate-binding sites of ricin.

6. Nonlinear regression analysis can be performed using commercially available software, such as GraphPad Prism (GraphPad Software).

References

1. Mo H, Van Damme EJ, Peumans WJ, Goldstein IJ (1993) Purification and characterization of a mannose-specific lectin from Shallot (Allium ascalonicum) bulbs. Arch Biochem Biophys 306:431–438
2. Ahmad S, Khan RH, Ahmad A (1999) Physicochemical characterization of *Cajanus cajan* lectin: effect of pH and metal ions on lectin carbohydrate interaction. Biochim Biophys Acta 1427:378–384
3. Zentz C, Frénoy JP, Bourrillon R (1978) Binding of galactose and lactose to ricin. Equilibrium studies. Biochim Biophys Acta 536:18–26
4. Cho M, Cummings RD (1996) Characterization of monomeric forms of galectin-1 generated by site-directed mutagenesis. Biochemistry 35: 13081–13088
5. Roberts DD, Goldstein IJ (1984) Reexamination of the carbohydrate binding stoichiometry of lima bean lectin. Arch Biochem Biophys 230:316–320
6. Kavan D, Kubícková M, Bílý J, Vanek O, Hofbauerová K, Mrázek H, Rozbeský D, Bojarová P, Kren V, Zídek L, Sklenár V, Bezouska K (2010) Cooperation between subunits is essential for high-affinity binding of *N*-acetyl-D-hexosamines to dimeric soluble and dimeric cellular forms of human CD69. Biochemistry 49:4060–4067
7. Hatakeyama T, Ohba H, Yamasaki N, Funatsu G (1989) Binding of saccharides to ricin E

isolated from small castor beans. J Biochem 105:444–448

8. Hatakeyama T, Matsuo N, Shiba K, Nishinohara S, Yamasaki N, Sugawara H, Aoyagi H (2002) Amino acid sequence and carbohydrate-binding analysis of the *N*-acetyl-D-galactosamine-specific C-type lectin, CEL-I, from the Holothuroidea, *Cucumaria echinata*. Biosci Biotechnol Biochem 66:157–163
9. Van Wauwe JP, Loontiens FG, De Bruyne CK (1973) The interaction of *Ricinus communis* hemagglutinin with polysaccharides and low molecular weight carbohydrates. Biochim Biophys Acta 313:99–105
10. Conte IL, Keith N, Gutiérrez-Gonzalez C, Parodi AJ, Caramelo JJ (2007) The interplay between calcium and the in vitro lectin and chaperone activities of calreticulin. Biochemistry 46:4671–4680
11. Olsnes S (2004) The history of ricin, abrin and related toxins. Toxicon 44:361–370
12. Hara K, Ishiguro M, Funatsu G, Funatsu M (1974) An improved method of the purification of Ricin D. Agric Biol Chem 38:65–70
13. Montfort W, Villafranca JE, Monzingo AF, Ernst SR, Katzin B, Rutenber E, Xuong NH, Hamlin R, Robertus JD (1987) The three-dimensional structure of ricin at 2.8 Å. J Biol Chem 262:5398–5403
14. Hatakeyama T, Yamasaki N, Funatsu G (1986) Evidence for involvement of tryptophan residue in the low-affinity saccharide binding site of ricin D. J Biochem 99: 1049–1056

Chapter 16

Centrifugal Ultrafiltration-HPLC Method for Interaction Analysis Between Lectins and Sugars

Kanji Hori and Makoto Hirayama

Abstract

The centrifugal ultrafiltration-HPLC method is a simple and rapid method for analyzing the binding interaction between lectins and sugars (oligosaccharides). In this method, a lectin is mixed with a fluorescent-labeled oligosaccharide in buffer and the unbound oligosaccharide recovered by centrifugal ultrafiltration is isolated and quantified by high-performance liquid chromatography. The binding activity is defined as a ratio (percentage) of the amount of bound oligosaccharide to that added, where the former is obtained by subtracting the amount of unbound oligosaccharide from the latter. The oligosaccharide-binding specificity of a lectin can be determined by comparing the binding activities with a variety of fluorescent-labeled oligosaccharides. The association constant and the optimum pH and temperature of the binding interaction between lectins and fluorescent-labeled oligosaccharides can be easily analyzed by this method.

Key words Lectin–sugar interaction, Oligosaccharide-binding specificity, Centrifugal ultrafiltration, HPLC (high-performance liquid chromatography), Fluorescent-labeled oligosaccharides, Kinetic analysis, Optimum pH/temperature

1 Introduction

There are several methods for analyzing the interaction between lectins and sugars, most of which require the immobilization of either lectins or sugars. In centrifugal ultrafiltration-HPLC (high-performance liquid chromatography) method, however, the interaction analyses are performed using the free state of both lectins and sugars. Thus, centrifugal ultrafiltration is an alternative to equilibrium dialysis and is now recognized as a simple and rapid analytical method to determine equilibrium binding between molecules [1, 2]. In 1992, Katoh et al. first applied this method to analyze the interaction between lectins and sugars [3, 4]. We further developed it as a centrifugal ultrafiltration-HPLC method to determine routinely the oligosaccharide-binding specificity and binding kinetics of newly isolated lectins [5–9]. In this method,

Jun Hirabayashi (ed.), *Lectins: Methods and Protocols*, Methods in Molecular Biology, vol. 1200,
DOI 10.1007/978-1-4939-1292-6_16, © Springer Science+Business Media New York 2014

the binding specificity can be determined by comparing the binding activities of a lectin with a variety of fluorescent-labeled oligosaccharides. The binding activity is defined as the ratio (%) of the amount of bound oligosaccharide to that added in the assay system. The amount of bound oligosaccharide is estimated by subtracting the amount of unbound oligosaccharide that is determined by HPLC of the filtrate recovered by centrifugal ultrafiltration of the reaction mixture of lectin and pyridylaminated (PA-) oligosaccharide. We confirmed the reliability of this method by examining the oligosaccharide-binding specificity of a known plant lectin, concanavalin A (Seikagaku Corporation, Japan) [5]. Moreover, the oligosaccharide-binding specificity of a newly isolated lectin was the same in both methods using the centrifugal ultrafiltration-HPLC method (free lectin) and the lectin-immobilized HPLC column (immobilized lectin) [5].

Thus, without immobilizing either lectins or sugars, the oligosaccharide-binding specificity of lectins is easily and rapidly determined by this method. Although it depends on the number of oligosaccharides examined, the binding property of a lectin can be analyzed in a few days for about 20 kinds of typical oligosaccharides, including preparation of the sample solution and reagents. The association constant and the number of carbohydrate-binding sites in a lectin molecule can be determined by this method [5, 6]. It is also possible to determine the optimum pH/temperature and time for achieving to equilibrium of lectin–carbohydrate interaction using this method [5]. The centrifugal ultrafiltration-HPLC method requires some equipment and reagents, including centrifugal ultrafiltration devices, a centrifuge, HPLC apparatus with a fluorometer and a chromatocorder, and fluorescent-labeled oligosaccharides.

2 Materials

Prepare all solutions using ultrapure water (with an electric resistivity of about 18 MΩ cm at 25 °C or for HPLC grade) and analytical grade reagents. Prepare and store all reagents at room temperature unless indicated otherwise.

2.1 Fluorescent-Labeled Oligosaccharides

1. PA-oligosaccharides: PA-derivatives of various kinds of oligosaccharides are commercially available (Takara, Japan) (*see* **Note 1**). PA-oligosaccharides, which are not commercially available, can be manually prepared from non-labeled oligosaccharides using a Pyridylamination Manual Kit (Takara) or a PALSTATION® Pyridylamination Reagent Kit and a semi-automated PA-derivatization apparatus (PLASTATION®, Takara) [5, 10] (*see* **Note 2**). Store the PA-oligosaccharides in a freezer (below −20 °C) until actual use.

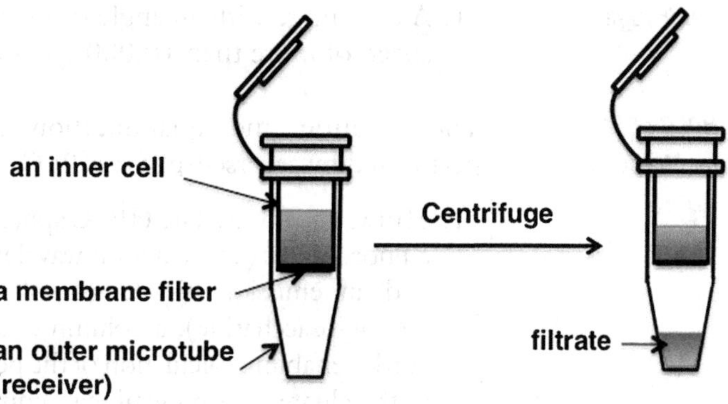

Fig. 1 A centrifugal ultrafiltration device. The device is composed of an inner cell with a membrane filter having a cut off value of 5, 10, or 30 kDa and an outer microtube (filtrate receiver). The reaction mixture in the inner cell is centrifuged at 10,000 × *g* for 30 s and the filtrate is recovered in the outer microtube (receiver). The filtrate is subjected to reversed-phase HPLC to quantify unbound PA-oligosaccharide

2. Working solutions of PA-oligosaccharides: Prepare PA-oligosaccharides each at 300 nM in 50 mM Tris–HCl buffer, pH 7.0 and store the working solutions in a freezer until actual use (*see* **Note 3**).

2.2 Lectin Solution

1. Lectin solution: Prepare a lectin solution at 500 nM in 50 mM Tris–HCl buffer, pH 7.0 (*see* **Note 4**).

2.3 Centrifugal Ultrafiltration Devices

1. Centrifugal ultrafiltration devices: Nanosep® Centrifugal Devices are obtained from Pall Life Sciences, USA. The device is composed of an inner cell with a membrane filter having a cut off value of 5, 10, or 30 kDa and an outer microtube (filtrate receiver) (Fig. 1). Unless there is a special reason, use the cells with a membrane filter having a cut off value of 10 kDa (*see* **Note 5**).
2. Washing of the inner cells of centrifugal ultrafiltration devices: Take off the inner cells from the centrifugal ultrafiltration devices, immerse them into ultrapure water in a beaker, and sonicate for about 5 min in a sonicator. After discarding the washing water, rinse them once more with ultrapure water. Thereafter, set the washed inner cells into microtubes and centrifuge them at 10,000 × *g* for 5 min to remove remaining water in the inner cells (*see* **Note 6**).
3. Setting of centrifugal ultrafiltration devices: Set back the washed inner cells in outer microtube receivers (counterpart for inner cell). These devices can be directly used for centrifugal ultrafiltration (*see* **Note 7**).

2.4 Centrifuge

1. A centrifuge with an angle rotor for microtubes and a rotation speed of more than 10,000 × *g* can be used.

2.5 HPLC of PA-Oligosaccharides

The isolation and quantification of PA-oligosaccharides are performed by reversed-phase HPLC.

1. HPLC apparatus: The HPLC apparatus should be equipped with a fluorometer (an excitation wavelength of 320 nm (Ex 320 nm) and an emission wavelength of 400 nm (Em 400 nm) for PA-oligosaccharide), a column oven (40 °C), and a chromatocorder enabling calculation of the peak area of PA-oligosaccharide in the eluate. It is preferable to equip an autosampler.
2. Column: TSKgel ODS-80TM (4.6 × 150 mm) (Tosoh Corporation, Japan) unless indicated otherwise (*see* **Note 8**).
3. Solvent: 10 % (v/v) methanol in 0.1 M ammonium acetate.
4. Elution: Isocratic elution with the above solvent at a flow rate of 1.0 mL/min (*see* **Note 8**).
5. Detection: Monitor the fluorescence of PA-oligosaccharide in eluate at Ex 320 nm and Em 400 nm (*see* **Note 9**).
6. Quantification of PA-oligosaccharide: Measure the peak area of PA-oligosaccharide in the eluate using a chromatocorder and estimate its amount from the peak area of a known amount of the same PA-oligosaccharide in the same HPLC (*see* **Note 10**).

3 Methods

3.1 Binding Reaction of Lectins with PA-Oligosaccharides

1. Mix 90 μL of a lectin solution (500 nM) and 10 μL of a PA-oligosaccharide solution (300 nM) in an inner cell with a membrane filter (a cut off value of 10 kDa) of a centrifugal ultrafiltration device (Fig. 2). Prepare the reaction mixture in duplicate.
2. Incubate the reaction mixture at room temperature (or in an ice bath) for 60 min (*see* **Note 11**).
3. Centrifuge the reaction mixture at 10,000 × *g* for 30 s on a centrifugal ultrafiltration device with a centrifuge.
4. Recover the filtrate including unbound PA-oligosaccharide (Sample) (*see* **Note 12**).
5. As a blank, mix 90 μL of 50 mM Tris–HCl buffer, pH 7.0 and 10 μL of a PA-oligosaccharide solution (300 nM). Prepare the blank solution in duplicate (Fig. 2).
6. Incubate the blank solution at room temperature for 60 min.
7. Centrifuge it at 10,000 × *g* for 30 s on a centrifugal ultrafiltration device with a centrifuge.
8. Recover the filtrate including unbound PA-oligosaccharide from the blank experiment (Blank).

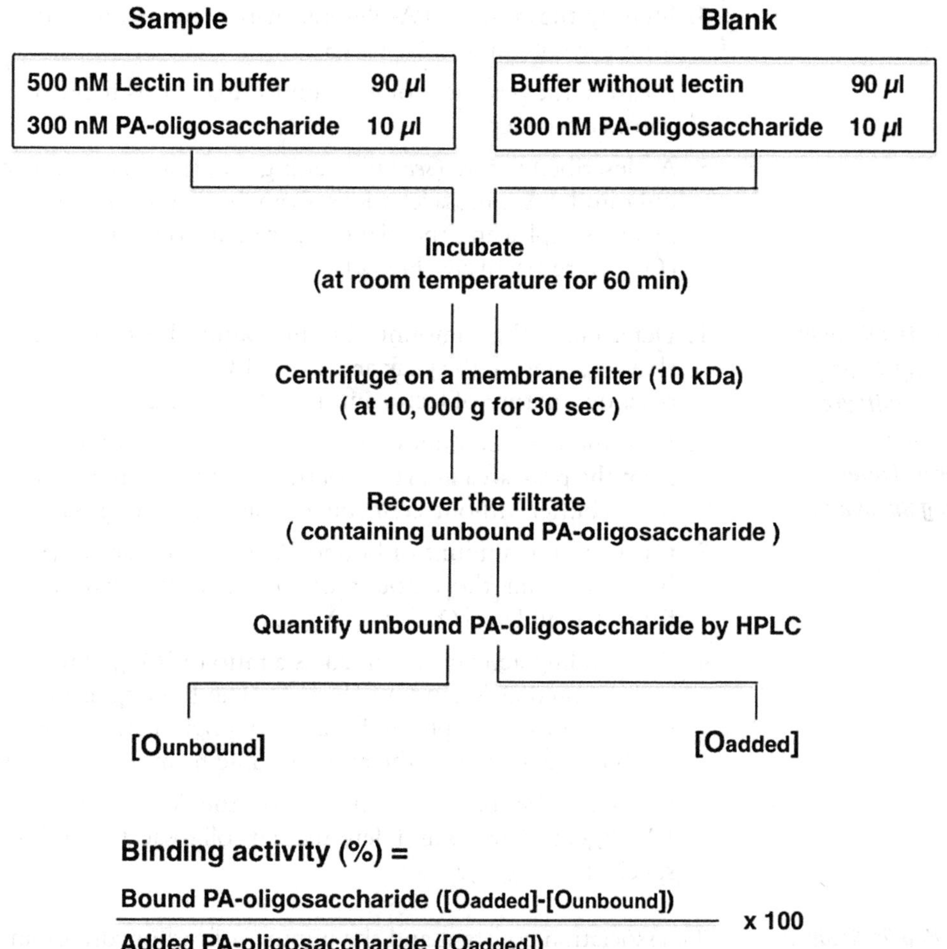

Fig. 2 The procedure of a centrifugal ultrafiltration-HPLC method to determine the binding activity of lectin with PA-oligosaccharide. The sample and blank solutions are incubated at room temperature for 60 min and then centrifuged at 10,000 × *g* for 30 s in centrifugal ultrafiltration devices, respectively. The recovered filtrates are subjected to reversed-phase HPLC with isocratic elution to quantify the unbound PA-oligosaccharide. The amount of unbound PA-oligosaccharide from sample solution is defined as $[O_{unbound}]$ whereas that from blank solution as $[O_{added}]$. The amount of bound PA-oligosaccharide $[O_{bound}]$ is obtained from a formula of $[O_{added}] - [O_{unbound}]$. Binding activity is expressed as a ratio (%) of the amount of bound PA-oligosaccharide ($[O_{added}] - [O_{unbound}]$) to that added ($[O_{added}]$). The oligosaccharide-binding specificity of a lectin is determined by comparing the binding activities with a variety of PA-oligosaccharides examined

3.2 Isolation and Quantification of PA-Oligosaccharides by HPLC

1. Apply an aliquot of the filtrate, which was obtained by centrifugal ultrafiltration of the reaction mixture (Sample) or the blank solution (Blank), to a TSKgel ODS 80TM column (4.6 × 150 mm) in a column oven (40 °C), which had been equilibrated with 10 % (v/v) methanol in 0.1 M ammonium acetate.
2. Elute the column at a flow rate of 1.0 mL/min with the same solvent and monitor the eluate at Ex 320 nm and Em 400 nm for PA-oligosaccharide.

3. Identify the peak of PA-oligosaccharide by its retention time from the column (*see* **Note 13**).
4. Measure the peak area of unbound PA-oligosaccharide using a chromatocorder.
5. As described before (*see* Subheading 2.5, **item 6**), quantify the unbound PA-oligosaccharides from both the sample experiment (Sample) and the blank experiment to get $[O_{unbound}]$ and $[O_{added}]$, respectively (Fig. 2).

3.3 Determination of Binding Activity (Semiquantitative Assay) and Oligosaccharide-Binding Specificity

1. Determine the amount of unbound PA-oligosaccharide ($[O_{unbound}]$) from the peak area in HPLC of the filtrate from the reaction mixture of lectin and PA-oligosaccharide.
2. Determine the amount of added PA-oligosaccharide ($[O_{added}]$) from the peak area in HPLC of the filtrate from the blank solution of buffer without lectin, but containing PA-oligosaccharide.
3. Calculate the amount of bound PA-oligosaccharide ($[O_{bound}]$) by subtracting the amount of unbound PA-oligosaccharide from that added ($[O_{added}] - [O_{unbound}]$).
4. The binding activity is defined as a ratio of $[O_{bound}]$ to $[O_{added}]$ and denoted as % binding (Fig. 2). The binding assay should be performed in duplicate for a PA-oligosaccharide to evaluate the binding activity as the average value from duplicate assays.
5. Compare the binding activities of the lectin with various PA-oligosaccharides and determine its oligosaccharide-binding specificity (Fig. 3).

3.4 Kinetic Analysis of Binding (Quantitative Assay)

The association constant and the number of carbohydrate-binding sites of a lectin molecule for PA-oligosaccharide can be determined using the centrifugal ultrafiltration-HPLC method [5, 6].

1. Prepare the reaction mixtures of 90 μL of a lectin at a constant concentration and 10 μL of PA-oligosaccharide at various concentrations in 50 mM Tris–HCl buffer, pH 7.0, and incubate them for 30 min at room temperature (or in an ice bath) (*see* **Note 14**).
2. Determine the concentration of unbound and bound PA-oligosaccharides in each reaction mixture according to the method described in Subheadings 3.1–3.3.
3. Divide the concentration of bound PA-oligosaccharide by the total concentration of lectin added, the value of which is defined as r. Next, divide this value (r) by the concentration of unbound PA-oligosaccharide, of which the value is defined as r/c (μM^{-1}).
4. To make the Scatchard plot, plot r/c (μM^{-1}) and r along the vertical and horizontal axes, respectively, and draw a straight

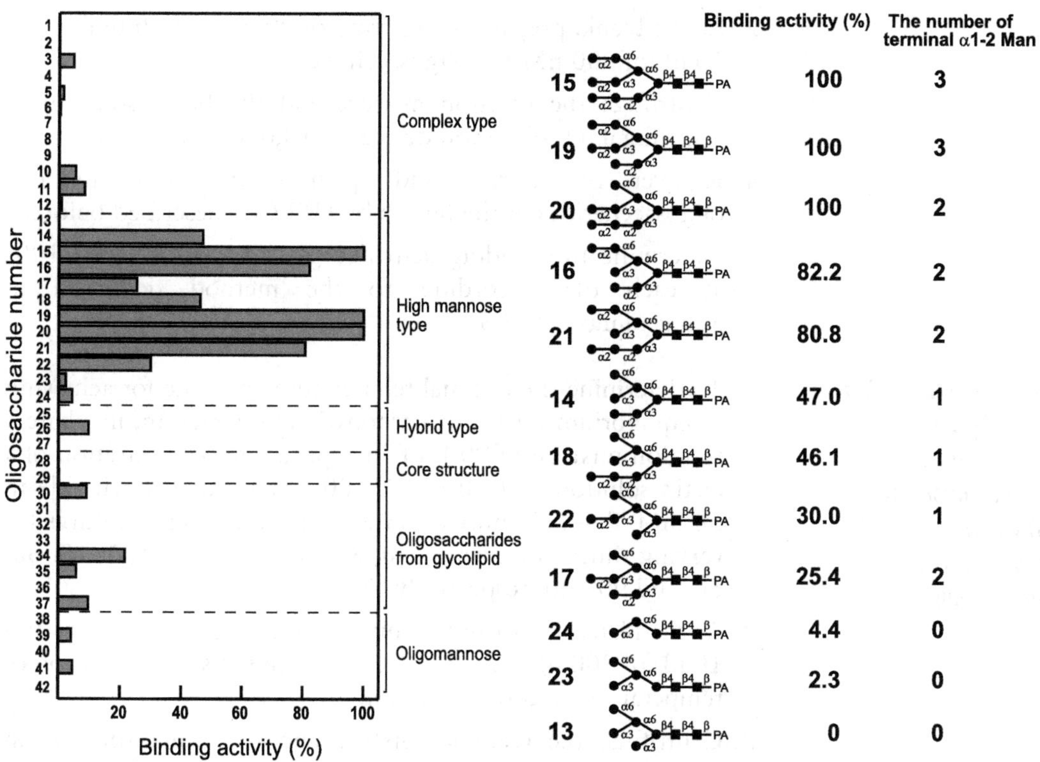

Fig. 3 Binding activities of a lectin from a green alga *Boodlea coacta* (BCA) to 42 PA-oligosaccharides by a centrifugal ultrafiltration-HPLC method. Binding activity was expressed as a ratio (%) of the amount of bound PA-oligosaccharide to that added. The assay was performed in duplicate for each PA-oligosaccharide, and the activity is expressed as the average value from duplicate assays. The assays were reproducible without any significant difference. The results suggest that BCA exclusively recognizes nonreducing terminal α1–2-linked mannose and the clustering of those residues might contribute to enhance BCA affinity. In the schematic representation of structures of PA-oligosaccharides (high-mannose types), *closed circle* and *closed square* indicate mannose and *N*-acetylglucosamine, respectively. The structures of all PA-oligosaccharides examined, which are indicated with "oligosaccharide number" in this figure, can be seen in a literature reproduced from Ref. 8

line fitting the plots. Estimate the association constant (K_a) from the slope of the line and the number of carbohydrate-binding sites in a molecule from the articulation point of the line at the horizontal axis.

3.5 Determination of Optimum pH for Lectin–Oligosaccharide Interaction

1. To determine the optimal pH for lectin–oligosaccharide interaction, prepare a reaction mixture of 90 μL of an appropriate concentration of a lectin solution and 10 μL of 300 nM PA-oligosaccharide at room temperature for 60 min in buffers at different pH; for example, 50 mM 2-(*N*-morpholino) ethanesulfonic acid (MES) buffer (pH 5.0, 6.0), 50 mM sodium phosphate buffer (pH 7.0), and 50 mM Tris–HCl buffer (pH 7.0, 8.0, 9.0) [5].

2. As the blank, prepare the mixture of 90 μL of each buffer and 10 μL of 300 nM PA-oligosaccharide.
3. Centrifuge the reaction mixture and the blank solution in centrifugal ultrafiltration devices at 10,000 × *g* for 30 s.
4. Recover the filtrate and quantify the unbound PA-oligosaccharides in the filtrate by HPLC as described before.
5. Determine the binding activities (%, $[O_{bound}]/[O_{added}] \times 100$) at each pH according to the method described in Subheadings 3.1–3.3.

3.6 Determination of Optimum Temperature and Equilibrium Time for Lectin–Carbohydrate Interaction

1. To determine the optimal temperature and time for achieving to equilibrium for lectin–carbohydrate interaction, incubate a reaction mixture of 90 L of an appropriate concentration of a lectin solution and 10 μL of 300 nM PA-oligosaccharide in 50 mM Tris–HCl buffer, pH 7.0, at varying temperatures for varying times; for example, 0, 37 and 50 °C for 1, 5, 10, 30, 60, and 90 min, respectively [5].
2. As the blank, incubate the mixture of 90 μL of the buffer and 10 μL of 300 nM PA-oligosaccharide in the same condition of temperatures and reaction times.
3. Centrifuge the reaction mixture and the blank solution in centrifugal ultrafiltration devices at 10,000 × *g* for 30 s.
4. Recover the filtrate and quantify the unbound PA-oligosaccharides in the filtrate by HPLC as described before.
5. Determine the binding activities (%, $[O_{bound}]/[O_{added}] \times 100$) at various temperatures for various reaction times, according to the method described in Subheadings 3.1–3.3.

4 Notes

1. As fluorescent-labeled oligosaccharides are detectable with high sensitivity with a fluorometer, they can often be used to examine the interaction between lectins and oligosaccharides. Fluorogenic reagents such as 2-aminopyridine (AP or PA-), fluorescein (FS-), and *P*-aminobenzoic acid ethyl ester (ABEE-) are often used for labeling. PA-oligosaccharides can be obtained from several makers (Takara, Glyence, Funakoshi, Sigma-Aldrich, Oxford GlycoSystems, etc.). The purity of the commercially available PA-oligosaccharides may vary among makers. Based on our experience, the PA-oligosaccharides from Takara can be used directly without purification after purchase. In routine assay, we often use 42 kinds of PA-oligosaccharides as shown in Fig. 3, including complex type *N*-glycans (bi-, tri-, tetra-antennary) with/without non-reducing terminal *N*-acetylneuraminic acid and core α1-6

fucose, high-mannose type *N*-glycans with M5 to M9, hybrid type *N*-glycans, core pentasaccharides with/without core α1-6 fucose, glycolipid-derived ones, and free oligomannoses [5–8]. Otherwise, the oligosaccharide-binding specificity of a newly isolated lectin may be characterized in advance using about 15 different types of PA-oligosaccharides, including bi- and tri-antennary complex type *N*-glycans, high-mannose type *N*-glycans with M5 and M9, core pentasaccharides with/without core α1-6 fucose, glycolipid-derived ones having nonreducing terminal α- or β-galactose and α- or β-*N*-acetylgalactosamine [9].

2. Add 20 μl of 2-aminopyridine in acetic acid to 50 nmol each of lyophilized oligosaccharides and heat the solution at 90 °C for 60 min. To this add 20 μL of borane dimethylamine in acetic acid and heat at 80 °C for 60 min. To remove excess reagents, subject the reaction solution to normal phase HPLC on a TSKgel NH_2-60 column (4.6×250 mm) using a linear gradient of acetonitrile in 50 mM acetic acid–triethylamine, pH 7.3, at a flow rate of 1.0 mL/min at 40 °C. Monitor the eluate for fluorescence of PA-oligosaccharide at Em 320 nm and Ex 400 nm, and collect the peak fractions of PA-oligosaccharides. Take a portion of the PA-oligosaccharide preparation, dry it with a Speed Vac, and subject to gas phase acid hydrolysis in 4 *N* HCl-4 *N* TFA (1:1, v/v) at 100 °C for 4 h. Apply an aliquot of the hydrolyzate to reversed-phase HPLC on a TSKgel ODS-80TM column (4.6×150 mm) equilibrated with 10 % (v/v) methanol in 0.1 M ammonium acetate and eluted with the same solvent at a flow rate of 1.0 mL/min at 40 °C. Measure the peak area of PA-monosaccharide, the reducing terminal residue which was liberated from PA-oligosaccharide by acid hydrolysis. At the same time, subject various amounts of a corresponding PA-monosaccharide (commercially available) to the same HPLC to make a standard curve for quantification. Quantify the free PA-monosaccharide liberated from PA-oligosaccharides using a standard curve of an authentic PA-monosaccharide, and estimate the amount of the PA-oligosaccharide from which the free PA-monosaccharide was liberated.

3. To prepare a 300 nM solution, add 1,616 μL of 50 mM Tris–HCl buffer, pH 7.0 to 50 μL each (500 pmol) of commercially available PA-oligosaccharides (Takara, Japan), which are provided at 10 μM in the frozen state. Divide a 300 nM solution of a PA-oligosaccharide into several portions in microtubes and stock them in a freezer until actual use. The working solutions as well as the original solutions of PA-oligosaccharides can be stored for at least 1 year with no significant loss of their fluorescence intensity.

4. The concentration of a lectin can be changed, depending on the strength of its carbohydrate-binding activity or hemagglutination activity. The buffer used can also be changed according to the optimum pH of carbohydrate-binding activity of the target lectin.
5. Centrifugal ultrafiltration devices can also be obtained from other makers.
6. Washing of the inner cells should be required to remove impurities which may be present in the cells and/or membrane filters and which may give impurity peaks with fluorescence in HPLC of the filtrate unless the cells are washed. The washing procedure described here is easy and sufficient to remove impurities that may interfere in HPLC, and it should be carried out just before use.
7. The outer microtubes of Nanosep® Centrifugal Devices can be used without washing based on our experience, although it is best to wash them before use.
8. The filtrate, which is obtained from the binding experiment with one kind of PA-oligosaccharide, gives only a fluorescence peak of unbound PA-oligosaccharide in reversed-phase HPLC on a TSKgel ODS-80TM combined with isocratic elution. When the mixture of PA-oligosaccharides is analyzed, the HPLC using the other column(s) and elution program with a gradient elution should be employed to separate them.
9. When the other fluorescent-labeled oligosaccharides such as FS- or ABEE-oligosaccharides are used, change to the excitation and emission wavelength accordingly to detect their fluorescence.
10. The standard curve for quantifying PA-oligosaccharide shows a straight line in a wide range of concentration. Therefore, it is possible to estimate the amount of PA-oligosaccharide from the peak area of a known amount of the same PA-oligosaccharide, without making the standard curve of PA-oligosaccharide. However, take note that the peak areas of PA-oligosaccharides may differ with each another, depending on the oligosaccharide structures.
11. The binding reaction of lectin with PA-oligosaccharide achieves equilibrium in about 30 min at room temperature or in an ice bath in the assay system described here [5].
12. About 30–40 μL of the filtrate is obtained by centrifugation at 10,000 × g for 30 s.
13. Identify the peak of unbound PA-oligosaccharide in the filtrate obtained from the reaction mixture (Sample) and from the retention time of PA-oligosaccharide in the filtrate from the blank solution (Blank). Any types of PA-oligosaccharides elute within

about 10 min in the HPLC with isocratic elution used here, enabling the rapid determination of many samples. They elute in this order of glycolipid-derived ones (2.9–4.1 min), oligomannoses (3.2–5.0), high-mannose types (3.6–5.4), complex type without core fucose (3.5–5.6), and complex type with core fucose (7.4–9.3) in a typical run.

14. The kinetic analysis can also be performed using the reaction mixtures of a PA-oligosaccharide solution at a constant concentration and lectin solutions at various concentrations.

Acknowledgement

This work was supported by a Grant-in-Aid for Scientific Research (B) from Japan Society for the Promotion of Science (JSPS).

References

1. Martin RL, Daley LA, Lovric Z et al (1989) The "regulatory" sulfhydryl group of *Penicillium chrysogenum* ATP sulfurylase. J Biol Chem 264:11768–11775
2. Gegg CV, Roberts DD, Segel IH et al (1992) Characterization of the adenine binding sites of two Dolichos biflorus lectin. Biochemistry 31:6938–6942
3. Katoh H, Satomura S, Matsuura S (1992) Analysis of lectin properties with membrane ultrafiltration and high-pressure liquid chromatography. J Biochem 111:623–626
4. Katoh H, Satomura S, Matsuura S (1993) Analytical method for sugar chain structures involving lectins and membrane ultrafiltration. J Biochem 113:118–122
5. Hori K, Sato Y, Ito K et al (2007) Strict specificity for high-mannose type *N*-gylcans and primary structure of a red alga *Eucheuma serra* lectin. Glycobiology 17:479–491
6. Sato Y, Okuyama S, Hori K (2007) Primary structure and carbohydrate binding specificity of a potent anti-HIV lectin isolated from the filamentous cyanobacterium *Oscillatoria agardhii*. J Biol Chem 282:11021–11029
7. Sato Y, Morimoto K, Hirayama M et al (2011) High mannose-specific lectin (KAA-2) from the red alga *Kappaphycus alvarezii* potently inhibits influenza virus infection in a strain-independent manner. Biochim Biophys Res Commun 405:291–296
8. Sato Y, Hirayama M, Morimoto K et al (2011) High mannose-binding lectins with preference for the cluster of α1-2-mannose from the green alga *Boodlea coacta* is a potent entry inhibitor of HIV-1 and influenza viruses. J Biol Chem 286:19446–19458
9. Hung LD, Sato Y, Hori K (2011) High-mannose N-glycan-specific lectin from the red alga *Kappaphycus striatum* (Carrageenophyte). Phytochemistry 72:855–861
10. Hase S, Koyama S, Daiyasu H et al (1986) Structure of a sugar chain of a protease Inhibitor isolated from barbados pride (*Caesalpinia pulcherrima* Sw.) seeds. J Biochem 100:1–10

Chapter 17

Surface Plasmon Resonance as a Tool to Characterize Lectin–Carbohydrate Interactions

Yasuro Shinohara and Jun-ichi Furukawa

Abstract

Biosensors based on surface plasmon resonance (SPR) monitor changes in refractive index in the vicinity of a surface in a real-time manner, which allows rapid, label-free characterization of the interactions of various types of molecules, from quantitative measurements of binding kinetics, thermodynamics, and concentrations in complex samples to epitope analysis. This method is usually capable of analyzing affinities in the range of millimolar to picomolar and is sensitive (typically, the concentration range of the analyte is $0.1–100 \times K_d$ and the typical volumes needed are in the range of 50–150 μL). There are two major applications of SPR biosensors for the analysis of lectin–carbohydrate interactions: detailed characterization of the interaction (e.g., specificity, affinity, kinetics, stoichiometry) and screening of lectin and carbohydrate/glycoconjugate interactions for diagnosis, identification of endogenous ligands, or binding properties of interest. Care should be taken, since the interaction of lectin and carbohydrate on the solid phase is complicated by the nonhomogeneous conditions under which binding occurs. However, this may in fact mimic some biological conditions, such as those occurring in cell–cell interactions.

Key words Avidity, Affinity, Biotinylation, Cluster effect, Glycoblotting, Immobilization, Kinetics, Mass transport limitation, Multivalency, Surface plasmon resonance

1 Introduction

1.1 Biosensors Based on Surface Plasmon Resonance (SPR)

The phenomenon of SPR, first observed by Wood in 1902 [1], was exploited by Otto [2] and Kretschmann and Raether [3] as an optical excitation of surface plasmons in the late 1960s, and was used as a chemical detection method by Nylander et al. in 1982 [4]. An automatic instrument (BIAcore) was commercialized in 1990 by Pharmacia Biosensor (now part of GE Healthcare) for the measurement of biomolecular interactions [5–8]. By using SPR to measure mass concentration-dependent changes in refractive index close to the sensor surface, BIAcore provided a unique approach for the analysis of molecular interactions, enabling label-free monitoring of biomolecular interactions in real time. The successful launch of BIAcore was also attributable to easy-to-use instrumentation

Jun Hirabayashi (ed.), *Lectins: Methods and Protocols*, Methods in Molecular Biology, vol. 1200,
DOI 10.1007/978-1-4939-1292-6_17, © Springer Science+Business Media New York 2014

based on the development of an integrated optical detection system, sensor surfaces with a biospecific coating, and a microfluidic cartridge for controlled sample delivery [9].

In Kretschmann's configuration constructed for the optical detector in the original BIAcore system, a microscope slide was coated with a thin metal film and the slide was coupled to a high-refractive-index prism as shown in Fig. 1 (*left*). When a beam of polarized light travels from a higher index medium to a lower index medium, it will be totally reflected at the interface if the incident angle is above a critical angle. However, when the interface is coated with a thin layer of metal (e.g., gold), energy from a specific angle is transferred to the free electrons in the metal surface, which leads to total attenuation of the light. The angle at which this happens depends on the refractive index of the interface between metal surface and a dielectric (in this case a solution). One of the parameters that influence the refractive index is any change of mass on the metal surface, in other words the presence of biomolecules on or close to the metal surface. If a biomolecule is immobilized on the surface and an appropriate analyte for that biomolecule is present in solution, binding will occur, causing the refractive index close to the surface to change [5, 6]. In general, adhesive reactions are detected as an increase in the SPR response, while dissociation reactions are detected as a decrease in the SPR response. The sensor device is composed of a sensor chip, which is a carboxymethylated dextran-coated thin film of gold covering a thin glass plate, a prism which is placed on the glass surface of the chip, and a microfluidic cartridge which provides the proper reaction fluid to the surface of the chip. Today, many other manufacturers have joined this market, and readers are advised to refer to recently published review [10].

A simple interaction experiment typically involves immobilizing one molecule of a binding pair on the sensor chip surface (ligand) and injecting a series of concentrations of its partner (analyte) across the surface (Fig. 1, *right*). Changes in the index of refraction at the surface where the binding interaction occurs are detected by the hardware and recorded as RUs (resonance units) in the control software, where 1,000 RU represents a shift in resonance angle of 0.1°. Curves are generated from the RU trace and are evaluated by fitting algorithms that compare the raw data to defined binding models. These fits allow determination of a variety of thermodynamic constants, including the apparent affinity of the binding interaction. SPR is one of many methods for the assessment of protein–protein and protein–ligand interactions. SPR biosensors are subsequently applied to studies of lectin–carbohydrate interactions [11, 12]. There are two major applications utilizing SPR biosensors for the analysis of lectin–carbohydrate interactions: detailed characterization of the interaction (e.g., specificity, affinity,

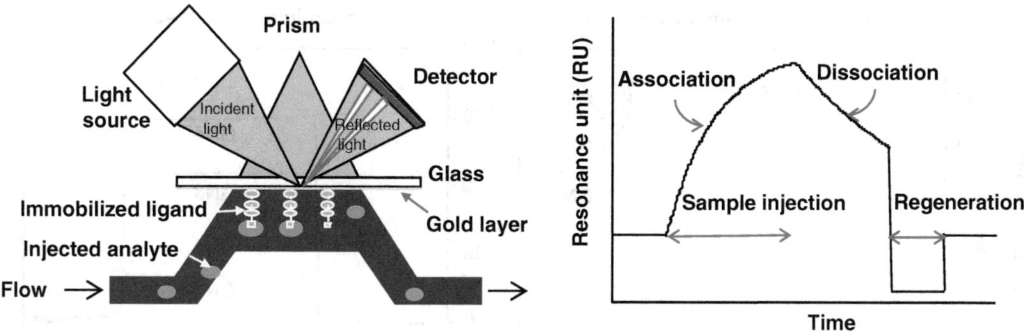

Fig. 1 Schematics of the optical detection system in Kretschmann configuration with light source, sensor chip, flow cells, and array detector. At *left*, below a prism, covered with a sensor chip with a gold layer on which a ligand is immobilized. The surface is irradiated with polarized light with a range of incident angles. Energy from a specific angle is transferred to the free electrons in the metal surface, which leads to total attenuation of the light. The angle at which this happens depends on the refractive index of the interface between metal surface and a dielectric (in this case a solution). At *right*, various phases of an SPR experiment are shown

kinetics, stoichiometry) and screening of lectin and carbohydrate/glycoconjugate interactions for diagnosis, identification of endogenous ligands, or binding properties of interest. Combination of a spatially resolved measuring device with SPR gave rise to a novel technological platform known as SPR imaging (SPRi) [13].

1.2 Lectin–Carbohydrate Interactions on Solid-Phase Surfaces

Unlike molecular interaction analysis techniques such as microdialysis, microcalorimetry, and NMR, where interactions are usually monitored in solution, for SPR either lectin or carbohydrate needs to be immobilized because SPR monitors binding events on solid-phase surfaces. This often makes the analysis complicated, as Rich and Myszka suggest that the quality and presentation of SPR work are often poor [14]. It is important to discriminate between intrinsic affinity (strength of a single bond) and avidity (apparent affinity synergistically achieved through the strength of multiple bond interactions), especially when interactions are monitored on a solid-phase surface. Avidity is a concept originally introduced to describe the binding behavior of antibodies with different valencies (e.g., Fab, IgG, IgM), where each binding interaction may be readily broken; however, presence of multiple binding sites may reinstate the interaction by preventing the molecule to diffuse away and enhancing the collision frequency [15–17].

In case of lectin–carbohydrate interaction analysis by SPR, it was found that the measured affinity can be quite different depending on the choice of the immobilized partner (i.e., lectin-immobilized or carbohydrate-immobilized assays) [18]. As an example, measurements were performed using WGA and *N*-acetylchitooligosaccharides. As shown in Fig. 2, observed sensorgrams

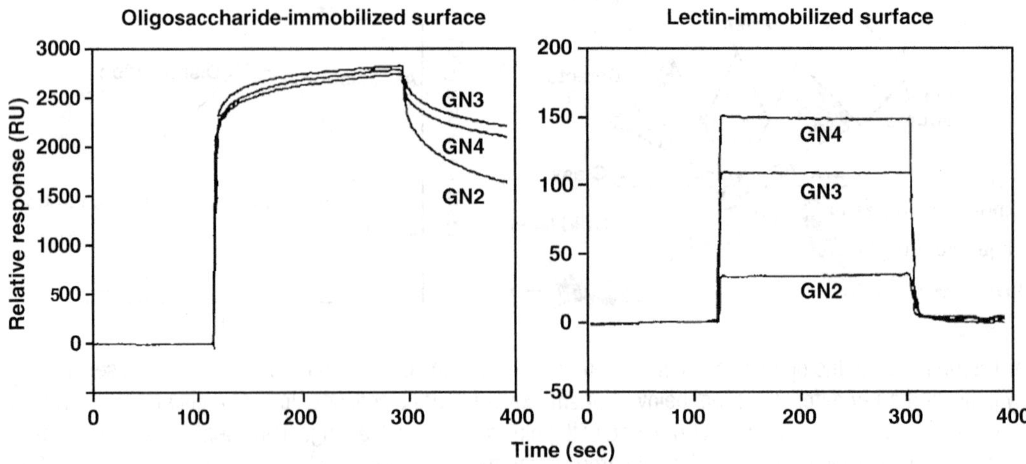

Fig. 2 Sensorgrams showing the interaction of WGA and chitooligosaccharides in the oligosaccharide (*left*)- or lectin (*right*)-immobilized assay. Reproduced with proper permission obtained from Oxford University press as published in Shinohara et al. 1997 [18]

are quite different depending on which binding partner is immobilized. Affinity constant (K_a) values obtained with immobilized lectin were in the range of 10^3–10^4 M^{-1}, in good agreement with previous data obtained by classical methods. Conversely, K_a values approached 10^8 M^{-1} in the carbohydrate-immobilized assay, and were >10,000 times higher than those observed in the lectin-immobilized assay. The increased K_a was mainly related to changes in the dissociation rate constant k_{diss}. The dissociation rate was much lower in the carbohydrate-immobilized assay than in the lectin-immobilized assay. By analyzing various other lectin–carbohydrate interactions by SPR, it was found that this effect was most remarkable for interactions where the association rate constant (k_{ass}) is extremely fast (*see* **Note 1**). Indeed, lectin–carbohydrate interactions can be characterized by their extremely fast k_{ass}. In addition, oligomerization of lectin enhanced the avidity due to a significant reduction in k_{diss}. These phenomena could be explained by considering the nonhomogeneous conditions under which binding occurred. The reaction in a nonhomogeneous state is limited by the mass transport effect, and by positively utilizing this effect, the apparent affinity between lectin and carbohydrate can be readily modulated using the same molecules. This appears to be an important factor during lectin–carbohydrate interactions. These observations shed light onto the cluster effect, the increase in affinity caused by the spatial density of specific sugars on a ligand, which has been established by Lee et al. [19, 20].

1.3 Detailed Characterization of Lectin–Carbohydrate Interactions (e.g., Specificity, Affinity, Kinetics, Stoichiometry)

Since many factors (e.g., immobilization density and multivalency) can affect binding parameters such as k_{ass}, k_{diss}, and K_a when lectin–carbohydrate interactions are assessed by SPR, it is important to adequately design the SPR experiment for the specific purpose of the study. Lectin-immobilized assay systems often provide the best and most reproducible method to obtain intrinsic affinity. Affinity constants can also be accurately measured, using competition assays, where an analyte is bound to an immobilized ligand at the surface. Upon binding of investigational ligands in solution, the analyte is displaced from the surface by competition, leading to a decrease in SPR signal (for reviews, *see* refs. 21, 22). On the other hand, carbohydrate-immobilized assays can provide useful information and insight into the nature of lectin–carbohydrate interactions, since lectin–carbohydrate interactions often occur on solid-phase surfaces such as cell surfaces coated with a variety of glycoconjugates (i.e., *N*- and *O*-linked glycans derived from glycoproteins, glycosaminoglycans, and glycosphingolipids).

1.4 Screening of Carbohydrate-Binding Protein and Carbohydrate/Glycoconjugate Interactions

In addition to the detailed characterization of lectin and carbohydrate/glycoconjugate interactions, SPR provides unique measures for diagnosis, identification of endogenous ligands, and screening of binding properties of lectins and carbohydrates. For example, Kazuno et al. reported multi-sequential SPR analysis of haptoglobin–lectin complex in sera of patients with malignant and benign prostate diseases. Following the injection of serum sample onto immobilized anti-haptoglobin antibody, the bound haptoglobin was then sequentially interacted with various lectins (Fig. 3). The response against SNA-1 of prostate cancer group was significantly higher than those of the control and benign prostate disease [23]. Detection and quantitation of autoantibodies by SPR may be also useful for diagnosis. Harrison et al. successfully characterized IgM antibody binding to asialo-GM1 glycolipids using SPR- and GM1-containing liposomes [24]. Metzger et al. reported specific detection of β2-glycoprotein I-reactive autoantibodies in sera from antiphospholipid syndrome patients upon covalent immobilization of β2-glycoprotein I onto planar carboxyl-terminated self-assembled monolayer (SAM) surfaces or carboxymethyldextran-coated hydrogel matrix surfaces [25]. de Boer et al. developed an SPR-based natural glycan microarray containing 144 glycans, which was used to detect anti-glycan antibodies in sera from *S. mansoni*-infected individuals [26]. Since 2000, the combination of SPR biosensors and mass spectrometry has been used for quantification and identification of binding partners. Such approaches were further evolved by the introduction of array formats and on-chip approaches [27]. Though their utilization for lectin–carbohydrate interactions is scarce, these techniques will be useful for a large number of applications including ligand fishing in wide-scale glycomics. Detailed characterization of these interactions can be multiplexed

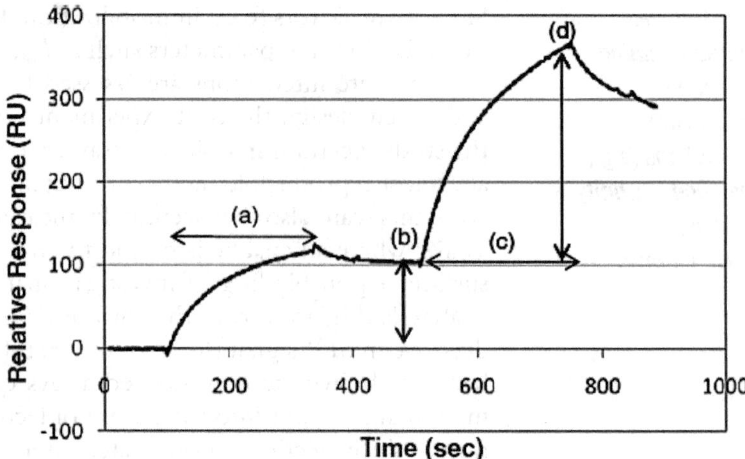

Fig. 3 Typical sensorgram of multi-sequential analysis on SPR. (*a*) Injection of serum sample on sensor chip C1-immobilized anti-haptoglobin antibody for 4 min (first step). (*b*) Binding amount of haptoglobin at 2 min after the end of sample injection. (*c*) Lectin injection followed by serum for 4 min (second step). (*d*) Binding amount of lectin shown as the difference between point 10 s before the end of lectin injection and point (*b*). Reproduced with proper permission obtained from Elsevier Inc. as published in Kazuno et al. 2011 [23]

by employing highly parallel formats such as glycan or lectin arrays. Examples include a glycan microarray screening assay for glycosyltransferase specificities [28] and for detection of influenza virus host specificity [29].

2 Materials

Interaction analysis protocols for SPR were established on a BIAcore 2000 (Uppsala, Sweden) with BIAcore Control 3.1 and BIA Evaluation 3.1 software.

2.1 Lectins and Oligosaccharides

1. Lectin: Wheat germ agglutinin (WGA), *Phaseolus vulgaris* erythroagglutinating (E-PHA or PHA-E4) and leukoagglutinating (L-PHA or L4-PHA) (J-Oil Mills, Tokyo, Japan).
2. Preparation of mannose-binding protein (MBP)-carbohydrate recognition domain (CRD) has been described [30].
3. Oligosaccharides: Chitooligosaccharides (di-*N*-acetylchitobiose [GN2], tri-*N*-acetylchitotriose [GN3], and tetra-*N*-acetylchitotetraose [GN4]) (J-Oil Mills, Tokyo, Japan).
4. *N*-glycans (M5, M6, M7D1, M7D3, M8, M9, NGA2, NA2, NA2B, NA3, and NA4) (Oxford GlycoSiences, which is acquired by Celltech, now part of UCB Group) (*see* **Note 2**).

2.2 Immobilization of Lectin

1. Amine coupling kit (1-ethyl-3-(3-dimethylaminopropyl) carbodiimide (EDC) and *N*-hydroxysuccinimide (NHS) and 1 M ethanolamine hydrochloride, pH 8.5) (GE Healthcare, UK).
2. Sensor Chip CM5 (GE Healthcare, UK).
3. 10 mM sodium acetate buffer (pH 3.5–5.5) for preconcentration assay.
4. 50 mM H_3PO_4 for regeneration.

2.3 Biotinylation of Carbohydrate

1. 4-biotinamidophenylacetylhydrazide (BPH) is synthesized as described in ref. 31 (*see* **Note 3**).
2. 30 % acetonitrile (v/v).
3. Formate buffer (50 mM, pH 3.5).
4. TSKgel ODS80 column 4.6×25 mm (Tosoh, Tokyo, Japan).
5. 70 mM phosphate buffer (pH 6.8) containing 14 % acetonitrile as HPLC eluent.

2.4 Immobilization of Carbohydrate

1. Sensor chip SA (GE Healthcare, UK).

2.5 Interaction Analysis by SPR

1. HBS buffer [10 mM HEPES (pH 7.4), 0.15 M NaCl, 1 mM $CaCl_2$, and 0.05 % Biacore surfactant P20].
2. 50 mM H_3PO_4 for regeneration.

2.6 Chemoselective Glycan Purification Technique (Glycoblotting) and Subsequent Tag Conversion

1. Anthraniloyl hydrazine (Ah) (Tokyo Chemical Industry Co., Ltd., Tokyo, Japan).
2. BlotGlyco H (Sumitomo Bakelite Co., Tokyo, Japan).
3. MassPREP HILIC μElution plate (Waters, MA, USA).
4. EZ-Link Biotin hydrazide (Thermo Scientific, CA, USA).
5. Microsorb-MV 100-5 Amino column, 250×4.6 (Varian, CA, USA).
6. HPLC-grade acetonitrile.
7. 0.2 % acetic acid in water (v/v).

3 Methods

3.1 Immobilization of Lectin

A standard amine coupling procedure is normally sufficient for the immobilization of proteins onto the sensor surface. For accurate quantitative analysis, the immobilization level needs to be low to minimize the possibility of the binding rate becoming limited by diffusion. The maximum binding capacity (R_{max}) of the immobilized ligand should be in the range of 50–150 RU (resonance unit). Flow cell 1 is left unmodified for use as a reference surface. When multiple lectins are immobilized onto different flow cells to compare the binding properties of each lectin, the immobilization levels of each lectin should be kept comparable. The following equation (Eq. 1) is used to determine an appropriate immobilization level that will generate an R_{max} of 50–150 RU:

$$\text{RL} = (\text{ligand MW} / \text{analyte MW}) \times R_{max} \times (1 / \text{Sm}), \qquad (1)$$

where MW is the molecular weight (of the ligand or the analyte), RL (ligand response) is the amount of immobilized ligand in RU, and Sm is the stoichiometry as defined by the number of binding sites on the ligand. Immobilization levels can be controlled by four factors: ligand concentration, pH, activation time by EDC/NHS, and injection time of ligand solution.

The electrostatic attraction of lectins onto the CM dextran surface is important for efficient immobilization (referred to as the preconcentration step). Since the CM dextran surface carries a net negative charge when the pH is above 3.5, the pH of the immobilization buffer should be between 3.5 and the lectin electrostatic point. Under these conditions, the ligand acquires a positive charge and is effectively preconcentrated into the negatively charged CM dextran matrix. Preconcentration assays are useful to determine the optimal pH of the immobilization buffer. Typically, the optimal pH for preconcentration will be 0.5–1 pH units below the *p*I of the protein. Since immobilization by covalent coupling cannot be repeated on the same sensor chip surface, preconcentration is an important strategy to control the immobilization level for experimental optimization and efficiency.

Once optimal pH is determined, the sensor surface is first activated with a 1:1 mixture of 0.4 M EDC and 0.1 M NHS to create reactive succinimide esters. The standard activation time is 7 min; this can be varied from 1 to 10 min to create fewer or more, respectively, reactive groups on the sensor chip surface depending on the immobilization level required. The ligand (lectin) is then injected in low-salt buffer lacking primary amines at an optimized pH. Lectin concentrations are typically in the range of 1–100 μg/mL. Finally, unreacted esters are blocked with ethanolamine. The volume or the concentration of ligand injected can be varied to adjust the immobilization level.

3.2 Immobilization of Carbohydrates

Glycoproteins/glycopeptides can be immobilized onto the sensor surface by amine coupling as described [12, 13]. Takimori et al. reported the interaction of different glycoforms of IgG (prepared by sequential exoglycosidase digestions) and bovine neonatal Fc receptor [32]. Glycosphingolipids can be directly immobilized via self-assembly chemistry onto a Sensor Chip HPA (hydrophobic association), where the surface is composed of long-chain alkanethiol molecules that form a flat, quasi-crystalline hydrophobic layer [33].

Several techniques have been reported to immobilize carbohydrates onto the sensor surface, which include covalent binding of

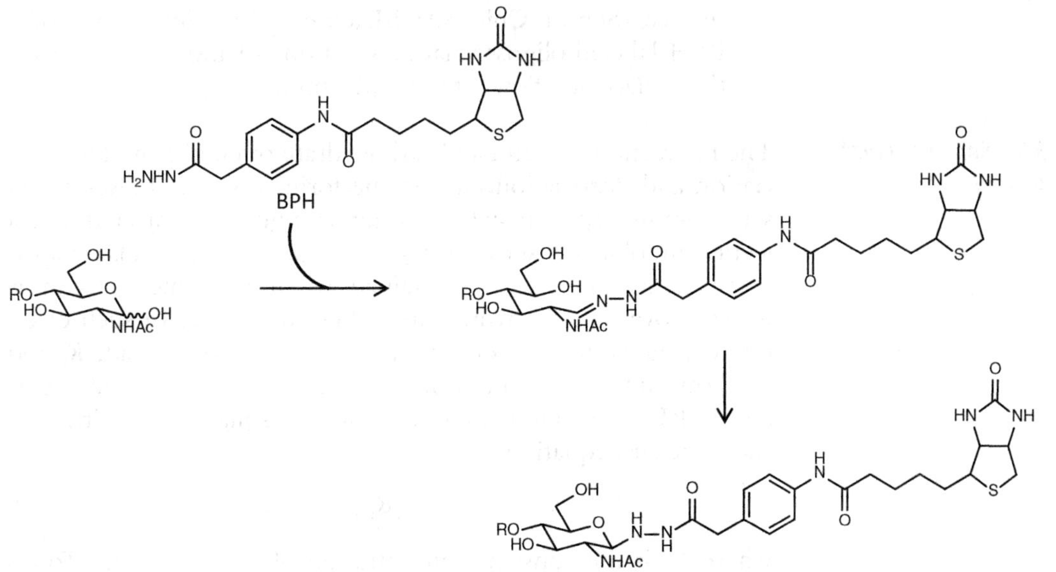

Fig. 4 Structure of BPH and the chemistry of adduct formation with oligosaccharide

fluorescent (2-aminobenzamide or anthranilic acid)-labeled glycans to an epoxide-activated chip [26] and the introduction of a thioctic acid moiety via reductive amination followed by direct sulfur–gold immobilization [34]. Biotinylation of the reducing end of the carbohydrate followed by immobilization to a streptavidin-preimmobilized surface may be the most frequently used method for the immobilization of carbohydrates. Though a number of methods for carbohydrate biotinylation are reported, we employ hydrazide chemistry since this strategy is relatively simple, and the majority of the product exists in the cyclic β-form, which mimics the naturally occurring *N*-glycan structure (Fig. 4) [31, 35].

1. The oligosaccharide (1–10 nmol, in 10 μL of water) is incubated with a fourfold molar excess of 4-biotinamidophenylacetylhydrazide (BPH) in 30 % acetonitrile (10 μL) at 90 °C for 1 h. After the reaction, 20 μL of formate buffer (50 mM, pH 3.5) is added and the mixture is stored at 4 °C for 12 h to promote tautomerization from the acyclic Schiff base-type hydrazone to a stable β-glycoside (*see* **Note 4**).
2. BPH-labeled oligosaccharides are purified by reversed-phase chromatography on a TSK gel ODS80 column (4.6 × 250 mm; Tosoh, Tokyo, Japan) with ultraviolet (UV) detection at 252 nm. Separation is performed at 40 °C at a flow rate of 0.5 mL/min. Elution is performed isocratically with 70 mM phosphate buffer (pH 6.8) containing 14 % acetonitrile.
3. The purified BPH-labeled oligosaccharide (~10 pmol) is injected onto a streptavidin-preimmobilized BIAcore sensor

surface (Sensor Chip SA; BIAcore). Typically, 10 μL of a BPH-labeled oligosaccharide solution (~1 μM) is passed over the surface at a flow rate of 2 μL/min.

3.3 Scatchard Plot Analysis

The interaction of lectin and carbohydrate often displays fast association and dissociation rates in the form of square-pulse-shaped sensorgrams, especially when monomeric glycan is injected over a lectin-immobilized surface (Fig. 2, *right*). Square-pulse-shaped sensorgrams can also be obtained when monomeric lectin is injected over a carbohydrate-immobilized surface. In such cases, rate constants may be too fast to be determined; instead, K_a and R_{max} (maximum concentration of the lectin-oligosaccharide complex in RUs) are determined by linear least-squares curve fitting of the Scatchard equation:

$$R_{eq} / C_0 = K_a R_{max} - K_a R_{eq}, \tag{2}$$

where C_0 is the constant concentration of injected lectin. To do this, the R_{eq} values collected at several carbohydrate (or lectin) concentrations are plotted against R_{eq}/C_0, and K_a and R_{max} are calculated from the slope and intercept, respectively. It is important to note that affinity measurements at equilibrium should always include a titration of the concentrations to determine the range corresponding to 0.1–10 times K_d ($1/K_a$).

1. BPH-labeled oligosaccharides (M5, M6, M7D1, M7D3, M8D1D3, and M9) are labeled by BPH as described.
2. For SPR analyses, HBS buffer (10 mM HEPES (pH 7.4), 0.15 M NaCl, 1 mM $CaCl_2$, and 0.05 % BIAcore surfactant P20) is used throughout.
3. Superose 12 (GE Healthcare, UK) is used to purify monomeric mannose-binding protein-A/C carbohydrate recognition domain (MBP-A/C CRD) (*see* **Note 5**).
4. Interaction analysis is performed by injecting MBP-CRD solution (0.46–3.7 μM) at a flow rate of 20 μL/min. After monitoring for 3 min, dissociation is initiated by introduction of the buffer. The obtained sensorgrams are shown in Fig. 5a. The chip surface can be regenerated by treatment with 50 mM H_3PO_4.
5. The equilibrium binding level (R_{eq}) is obtained from the apparent equilibrium binding level (in RUs) by correcting for the background resonance, which is obtained by injecting the same concentrations of MBP-C CRD onto an unmodified sensor surface. The R_{eq} values collected at several MBP-C CRD concentrations are plotted against R_{eq}/C_0, and K_a and R_{max} are calculated from the slope and intercept, respectively (Fig. 5b, Table 1).

The affinity and binding specificity of MBP-A and -C against various *N*-glycans were determined quantitatively. Both K_a and

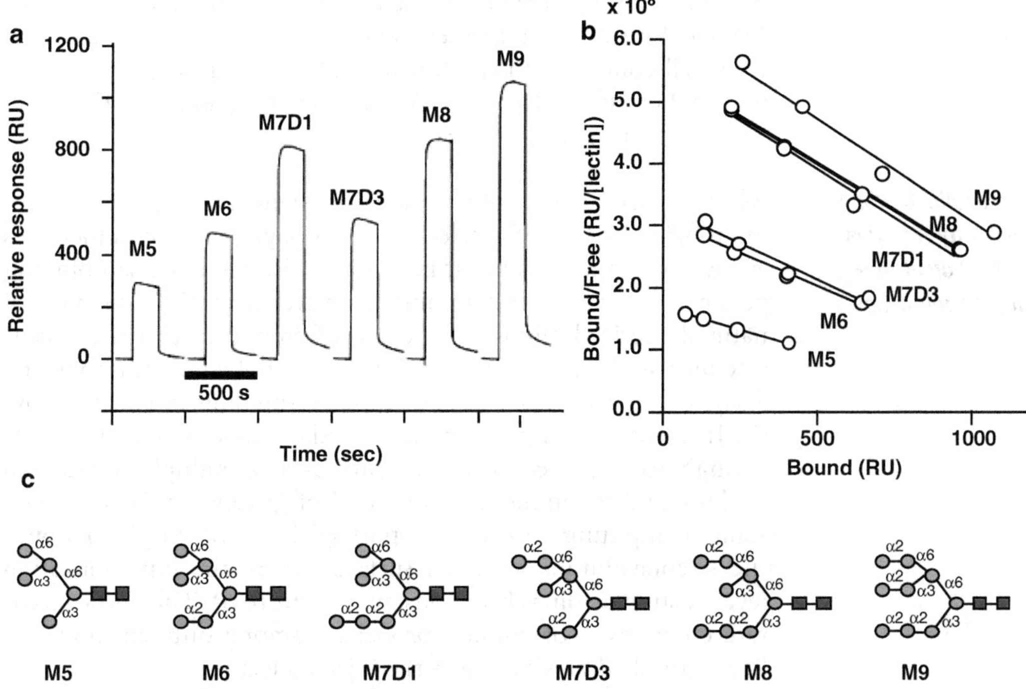

Fig. 5 Interaction between MBP-C CRD and immobilized high-mannose-type oligosaccharides. (**a**) Sensorgrams showing the interactions. Each lectin was introduced onto the surface at a concentration of 3.7 M. *RU* Response units. (**b**) Scatchard plot of the interactions. *x*-axis, R_{eq} in response units; *y*-axis, $R_{eq} = C_0$, where C_0 is the injected MBP-C concentration, which varied from 0.46 to 3.7 μM. (**c**) Structure of oligosaccharides used in this study. *Circle* Man, *square* GlcNAc. Reproduced with proper permission obtained from Portland Press Ltd. as published in Lee et al. 1999 [36]

Table 1
Calculated K_a and R_{max} by the Scatchard plot analysis

	MBP-A		MBP-C	
	$K_a(M^{-1}) \times 10^{-4}$	R_{max}(RU)	$K_a(M^{-1}) \times 10^{-4}$	R_{max} (RU)
M5	7.2	627	13.5	1,222
M6	6.2	564	21.0	1,457
M7D1	16.7	731	30.8	1,769
M7D3	5.6	720	23.6	1,403
M8D 1D3	12.6	694	30.4	1,806
M9	19.1	660	33.7	1,902

Reproduced with proper permission obtained from Portland Press Ltd. as published in Lee et al. 1999 [36]

R_{max} for MBP-C increased as the number of nonreducing terminal Manα1-2 residue(s) increased (Fig. 5b, c). M7D1, M8, and M9, which all contain Manα1-2Manα1-6Man branch, had about two-fold higher affinity for MBP-A than M5, M6, and M7D3, which lack this branch [36, 37].

3.4 Lectin-Binding Specificity Analysis in a Carbohydrate-Immobilized Assay

As described, since affinity measurement in a carbohydrate-immobilized assay with introduction of oligomeric lectin increases avidity (apparent affinity), it may not be feasible to obtain binding parameters such as rate constants. However, since the availability of naturally derived oligosaccharides is often limited and their accurate microscale quantitation is difficult, carbohydrate-immobilized assays provide a useful measure for lectin-binding specificity analysis. It is worth noting that immobilized oligosaccharides are stable enough for repeated measurements during multiple interaction analysis, and the immobilization level of glycans can be kept constant by injecting an excess of purified biotinylated glycans onto the streptavidin-immobilized surface, where the immobilization level of streptavidin is kept constant (*see* **Note 6**). This allows accurate comparison of binding properties among different immobilized carbohydrates as well as the injected lectins.

An example is shown using *Phaseolus vulgaris* erythroagglutinating and leukoagglutinating lectins (E- and L-PHA) [38]. They are homotetrameric legume lectins (isolectins). E- and L-PHA have a strong affinity toward complex-type *N*-glycans containing either bisecting GlcNAc or 1,6-linked LacNAc, respectively.

1. Purified BPH-labeled oligosaccharides (10 pmol) are immobilized onto a streptavidin Sensor Chip SA as described.
2. Lectins (10 μg/mL in HBS) are introduced onto the surface at a flow rate of 20 μL/min. Interactions between lectin and oligosaccharides are monitored as the change in SPR response at 25 °C. After 3 min, the flow is switched from sample to HBS buffer in order to initiate dissociation. Sensor surfaces are regenerated with 50 mM H_3PO_4. The obtained sensorgrams are shown in Fig. 6. Note that each sensorgram is directly comparable as the immobilized glycan levels and lectin concentrations are constant.
3. The apparent k_{diss} is analyzed by fitting the dissociation phase directly to the following equation:

$$R_t = R_0 \exp(-k_{diss} t) \quad (3)$$

using nonlinear least-squares analysis. R_0 represents the amplitude of the dissociation process. The k_{ass} is then analyzed by fitting the association phase directly to the following equation:

$$R_t = C k_{ass} R_{max} / (C k_{ass} + k_{diss})[1 - \exp\{-(C k_{ass} + k_{diss})t\}], \quad (4)$$

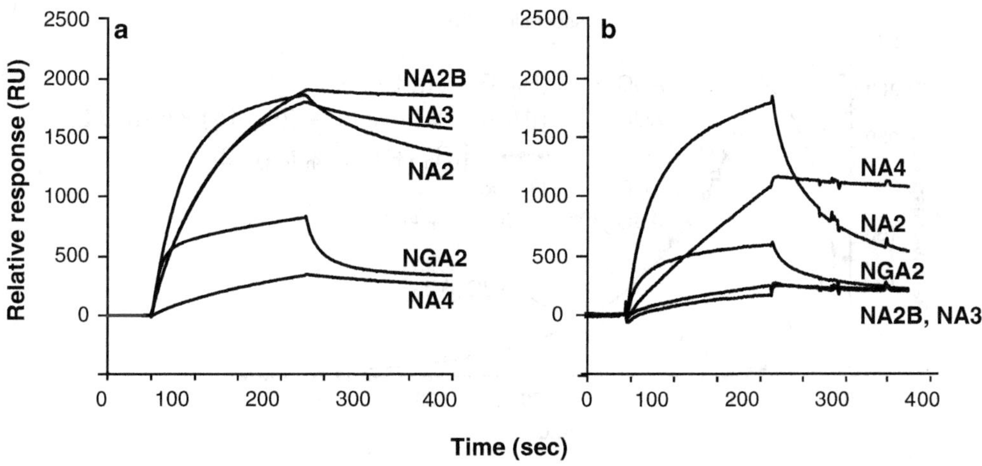

Fig. 6 Sensorgrams showing the interaction of E-PHA (**a**) and L-PHA (**b**) with immobilized NGA2, NA2, NA2B, NA3, and NA4. Each lectin was introduced onto the surface at a concentration of 10 μg/mL. Reproduced with proper permission obtained from American Society for Biochemistry and Molecular Biology as published in Kaneda et al. 2002 [38]

Table 2
Kinetic parameters obtained for the interaction analysis of E_4-PHA with immobilized NA2, NA2B, and NA3

	$K_{ass} \times 10^5\ M^{-1}S^{-1}$	$K_{diss} \times 10^{-4}\ S^{-1}$
NA2	3.1	11.5
NA2B	1.3	1.2
NA3	1.6	5.4

Reproduced with proper permission obtained from American Society for Biochemistry and Molecular Biology as published in Kaneda et al. 2002

where R_{max} represents the maximum concentration of the lectin–oligosaccharide complex (in RUs), and C is the constant concentration of injected lectin. The apparent rate constant values are calculated for NA2, NA2B, and NA3 by fitting the relevant sensorgrams using a simple one-to-one interaction model, and these are summarized in Table 2.

4. The area under the curve for the dissociation phase ($AUC_d^{0\rightarrow\infty}$, *see* **Note 7**) is calculated by the sum of $AUC_d^{0\rightarrow tn}$ and $AUC_d^{tn\rightarrow\infty}$. The former can be calculated using the trapezoidal rule in Excel software (Microsoft). The time interval used is 1 s. The latter can be calculated by fitting the dissociation phase of the last 30 s to the equation $R_t = R_0 \exp(-k_{diss}t)$, calculated by the equation $AUC_d^{tn\rightarrow\infty} = R_n/k_{diss}$ (Fig. 7). The calculated AUC for the interactions of E-PHA and L-PHA with immobilized NGA2, NA2, NA2B, NA3, and NA4 are shown in Fig. 8.

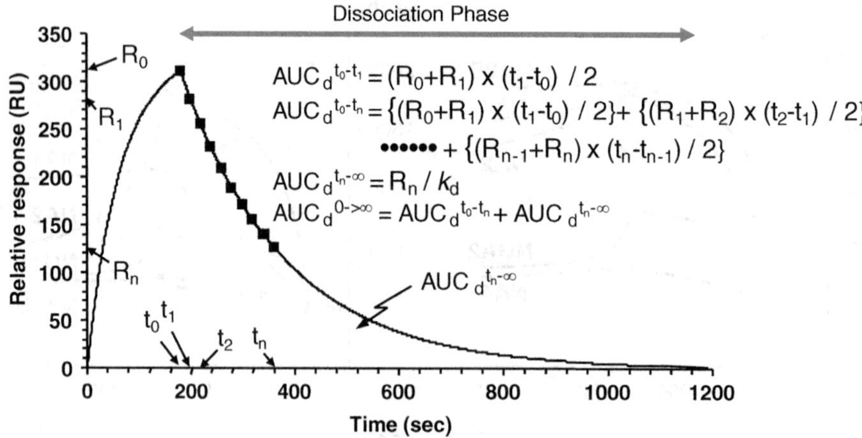

Fig. 7 Calculation of the area under the curve at the dissociation phase ($AUC_d^{0\rightarrow\infty}$) using the trapezoidal rule. The dissociation rate constant (k_{diss}) was calculated using the last 30 s of the dissociation phase and by directly fitting the dissociation curve to the equation $R_t = R_0 \exp(-k_{diss}t)$. Reproduced with proper permission obtained from Elsevier Inc. as published in Sota et al. 2003 [37]

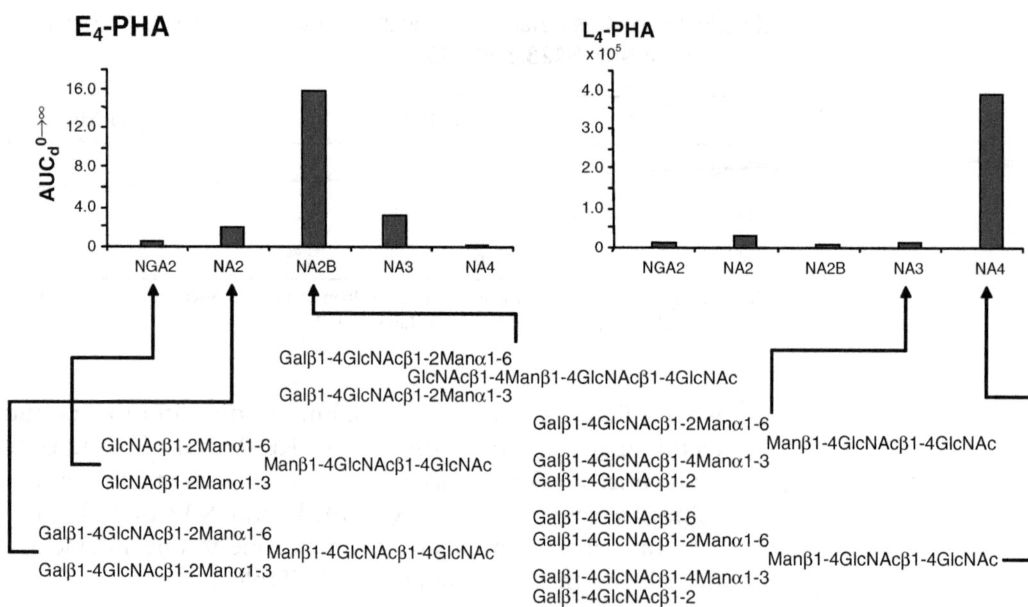

Fig. 8 Calculated $AUC_d^{0\rightarrow\infty}$ for the interactions of E-PHA (*left*) and L-PHA (*right*) with immobilized NGA2, NA2, NA2B, NA3, and NA4. After the injection of each lectin at a concentration of 10 μg/ml, association and dissociation were monitored for 3 min $AUC_d^{0\rightarrow\infty}$ was calculated using the trapezoidal rule. AUC_d was extrapolated over an infinite time interval using nonlinear regression. Reproduced with proper permission obtained from American Society for Biochemistry and Molecular Biology as published in Kaneda et al. 2002 [38]

3.5 A Strategy to Immobilize Glycans from Natural Sources by Chemoselective Glycan Purification (Glycoblotting) and Subsequent Tag Conversion

Posttranslational protein glycosylation changes the biological and physical properties of glycoconjugates, which include functions as signals or ligands to control their distribution, antigenicity, metabolic fate, stability, and solubility. One novel technique combining glycomics and SPR could be to determine lectin–carbohydrate interactions in a glycome-wide manner. To do this, it is important to immobilize glycans from natural sources. de Boer et al. reported chromatographic fractionation of fluorescent (2-aminobenzamide or anthranilic acid)-labeled glycans followed by printing onto an epoxide-activated chip [26]. Furukawa et al. reported a streamlined method of purification, chromatographic fractionation, and immobilization onto a solid support for SPR analysis based on a chemoselective glycan purification technique (glycoblotting) and subsequent tag conversion (Fig. 9) [39].

1. Oligosaccharides released from α_1-acid glycoprotein are blotted onto a hydrazide-functionalized polymer (BlotGlyco H) and recovered as fluorescent glycans labeled with anthraniloyl hydrazine (Ah).
2. Ah-labeled oligosaccharides are separated and fractionated by normal-phase HPLC. Separation of 2 Ah-labeled oligosaccharides is performed using an amino column. A gradient elution is applied at a flow rate of 200 μL/min using 0.1 % acetic acid in acetonitrile (solvent A) and 0.2 % acetic acid in water containing 0.2 % triethylamine (A/B = 90/10(0 min) → A/B = 90/10(30 min) → A/B = 77/23(30.1 min) → A/B = 77/23(35 min) → A/B = 56/44(110 min) → A/B = 5/95(110.1 min) → A/B = 5/95(125 min). Oligosaccharides are detected by in-line fluorescence (excitation at 330 nm and emission at 420 nm).
3. Each oligosaccharide's fraction is collected, added to 10 μL of 0.2 mM biotin hydrazide, and then incubated for 1 h at 90 °C to promote the imine exchange conversion of Ah derivatives to biotin derivatives.
4. Excess biotin hydrazide is removed using a simple solid-phase extraction (MassPREP HILIC μElution plate, Waters).
5. BPH-labeled oligosaccharides are used for SPR analyses as described.

4 Notes

1. Karlsson et al. numerically describe the kinetics between a soluble analyte, introduced in a flow system, and an immobilized ligand [40]. When molecule A is introduced to the surface where molecule B is immobilized, molecule A is transported to the surface by convection and diffusion in the thin-layer flow cell, and the concentration of complex AB is measured.

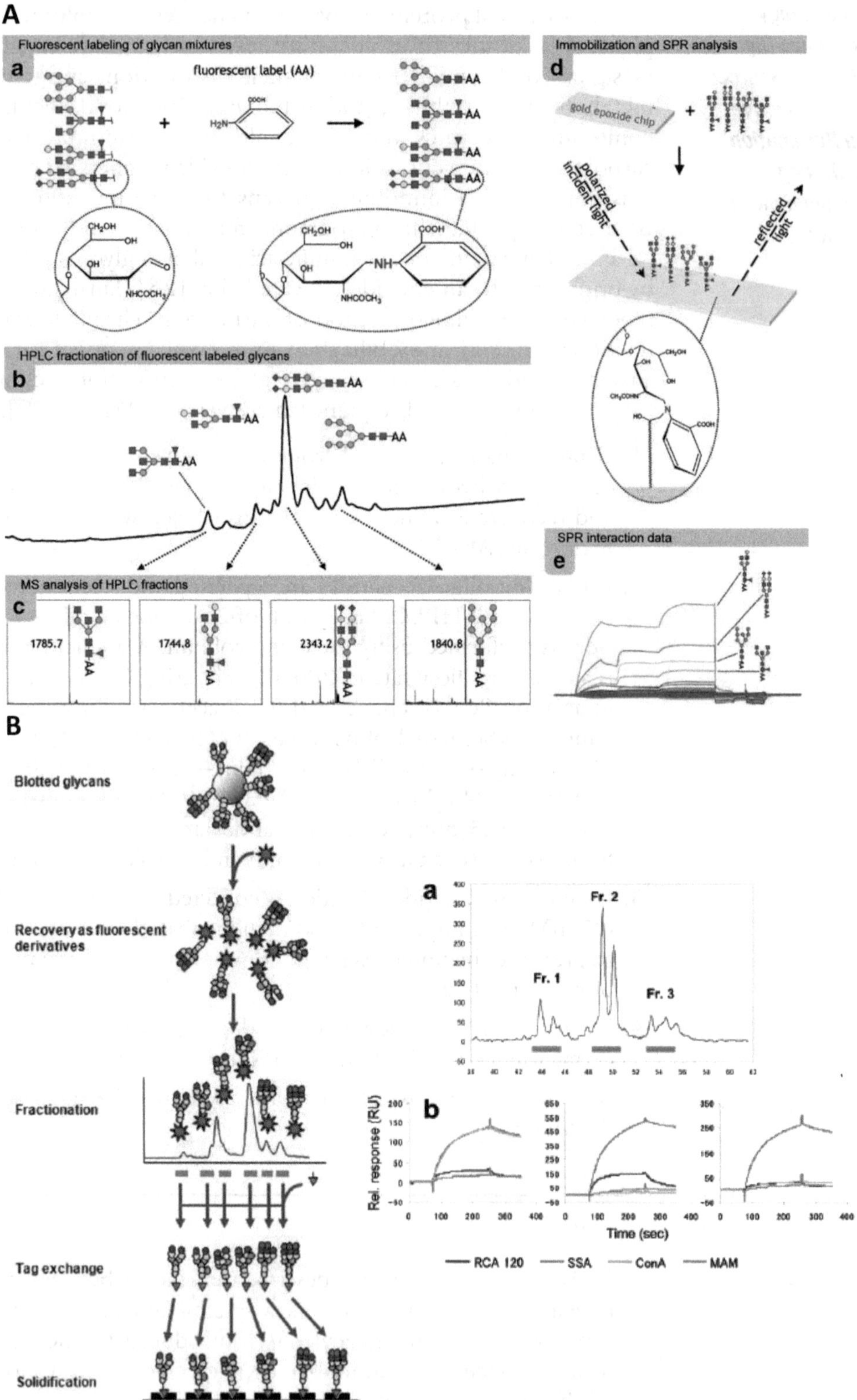

Fig. 9 Strategies to immobilize glycans from natural sources. (**A**) Chromatographic fractionation of fluorescent (2-aminobenzamide or anthranilic acid)-labeled glycans followed by printing onto an epoxide-activated chip.

The situation is described in Fig. 10a. A_0 and A_s are molecules in the eluent stream and at the sensor surface, respectively. The k_m and k_{-m} are mass transport constants that describe the flux of molecules to the sensor surface and out from the sensor surface. The k_{ass} and k_{diss} are kinetic constants, and k_f and k_r are the resulting forward and reverse rate constants that determine the overall rate of AB complex formation at the surface. k_f and k_r are described as

$$k_f = k_{ass} / \left(1 + k_{ass}[B] / k_M\right). \quad (5)$$

$$k_r = k_{diss} / \left(1 + k_a[B] / k_M\right), \quad (6)$$

where k_M is the mass transport coefficient. $[B]$ is the concentration of free sites at the surface.

When the mass transport flux is much higher than the heterogeneous rate constants (i.e., when $k_M >> k_{ass}[B]$), k_f and k_r are quite similar to k_{ass} and k_{diss}, respectively. Thus, $k_{ass}[B]/k_M$ is the limiting coefficient which can describe how the apparent kinetic constants differ from the intrinsic kinetic constants. When $k_{ass}[B]/k_M \gg 0$, the apparent and intrinsic kinetic values will not agree. This situation occurs when the immobilized density is high (high $[B]$), when large molecules are used as an analyte and/or the flow rate is low (low k_M) and, most importantly, when the interaction has a fast k_{ass} (Fig. 10b). The interaction of carbohydrate and various lectins is characterized by

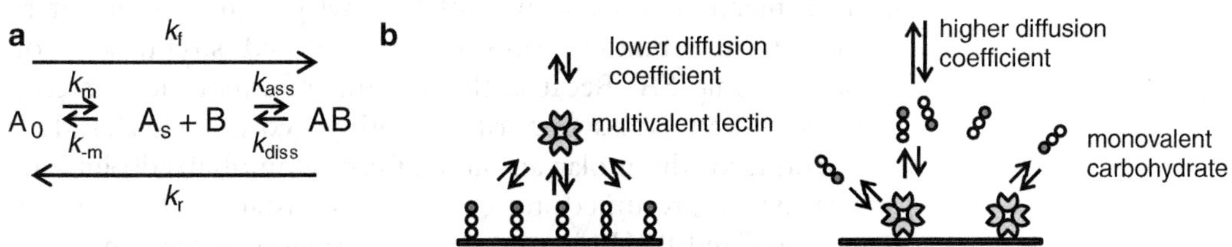

Fig. 10 Interaction at the surface. (**a**) Immobilized ligand and injected analyte present in the flow system are shown as B and A, respectively. $[A_0]$ is the concentration of the injected sample, and $[A_s]$ is the concentration of A at the surface. k_m and k_{-m} are mass transport rate constants, k_{ass} and k_{diss} are the rate constants for the reaction between A and B, and k_f and k_r are the overall rate constants. (**b**) Possible explanation of the enhanced avidity observed when the carbohydrate is immobilized and multivalent lectin is injected

Fig. 9 (continued) Reproduced with proper permission obtained from Springer as published in deBoer et al. 2008 [26]. (**B**) A streamlined method of purification, chromatographic fractionation, and immobilization onto a solid support for SPR analysis based on chemoselective glycan purification technique (glycoblotting) and subsequent tag conversion. Reproduced with proper permission obtained from American Chemical Society as published in Furukawa et al. 2008 [39]

intrinsic but fast association; for example, an extremely fast k_{ass} ($>10^7$ M^{-1} s^{-1}) was reported for P-selectin [41]. Readers are advised to refer to reviews discussing about carbohydrate clusters, lectin multivalency, and avidity [42–44].

2. Various *N*-glycans are commercially available from other providers such as Carbosynth Limited (Berkshire, UK).
3. Various biotin hydrazide reagents that differ in spacer arm length and solubility are commercially available (e.g., EZ-Link Biotin Hydrazide, and EZ-Link Biotin-LC-Hydrazide from Thermo Scientific). These reagents should also work well for the immobilization of carbohydrates.
4. The key to efficient adduct formation is to carry out the reaction in a small volume (~20 μL) in a tightly sealed reaction tube, in which almost all solvent is vaporized on heating at 90 °C [45]. Formate buffer is added after cooling the reaction tube on ice to liquefy the vaporized gas. This method can be applied to sialylated oligosaccharides without any loss of sialyl residues.
5. We experienced that the interaction of monomeric MBP-A and -C and galectin-1 with immobilized glycans gave square-pulse-shaped sensorgrams. The contamination of trace amounts of dimer or higher oligomers will affect the shape of the sensorgram, e.g., a biphasic interaction consisting of fast k_{ass}/slow k_{ass} and fast k_{diss}/slow k_{diss}. Therefore, gel chromatography purification should be performed before SPR analysis to remove heterogeneous oligomeric component(s).
6. The injection of ~10 pmol of biotinyl glycan is sufficient to saturate the binding sites of immobilized streptavidin on Sensor Chip SA. Because the amount of streptavidin is constant and an excess of purified biotinylated oligosaccharide is introduced, the molar amount of each immobilized oligosaccharide is presumed to be nearly constant. The surface-immobilized BPH oligosaccharides are fairly stable: almost no decrease in response is detected during 20 repeated injections of lectin or regeneration solution (H_3PO_4). The immobilized BPH-labeled oligosaccharide is also suitable for in situ enzymatic digestion on the sensor surface [31].
7. We use $AUC_d^{0\to\infty}$ as a quantitative index for evaluating sensorgrams obtained from the interactions of lectins with immobilized oligosaccharides. Because this method does not require model fitting, except for extrapolation using nonlinear regression, the obtained values are free from miscalculations due to inadequate models. We observed a good correlation between $AUC_d^{0\to\infty}$ and the elution order derived by immobilized lectin affinity chromatography. This quantitative comparison allowed clarification of the role of loop B and C in the recognition

of structural motifs. Note that $AUC_d^{0\to\infty}$ is a meaningful parameter, resulting from both the concentration of bound lectin and avidity, which is described by

$$AUC_d^{0\to\infty} = R_0 / k_{diss}, \tag{7}$$

where R_0 represents the amplitude of the dissociation process and k_{diss} is the dissociation rate constant.

AUC_d was recently used to assess serial samples obtained from pregnant women at high risk for fetal and neonatal alloimmune thrombocytopenia. Maternal anti-HPA-1a alloantibodies in sera, responsible for most cases of severe FNAIT, were analyzed using an HPA-1aa-immobilized surface and a consistent correlation between AUC_d and the fetal and neonatal platelet count was found [46].

References

1. Wood RW (1902) On a remarkable case of uneven distribution of light in a diffraction grating spectrum. Phil Mag 4:396–402
2. Otto A (1968) Excitation of nonradiative surface plasma waves in silver by the method of frustrated total reflection. Z Phys 216: 398–410
3. Kretschmann E, Raether H (1968) Radiative decay of nonradiative surface plasmon excited by light. Z Naturforschung 23:2135
4. Nylander C, Liedberg B, Lind T (1982) Gas detection by means of surface plasmon resonance. Sens Actuators 3:79–88
5. Fagerstam LG, Frostell A, Karlsson R, Kullman M, Larsson A, Malmqvist M, Butt H (1990) Detection of antigen-antibody interactions by surface plasmon resonance. J Mol Recogn 3:208–214
6. Granzow R, Reed R (1992) Interactions in the fourth dimension. Bio/Technology 10: 390–393
7. Fagerstam LG, Karlsson AF, Karlsson R, Perssou B, Riinnberg I (1992) Biospecific interaction analysis using surface plasmon resonance detection applied to lunetic, binding site and concentration analysis. J Chromatogr 597:397–410
8. Schuster SC, Swanson RV, Alex LA, Bourret RB, Simon MI (1993) Assembly and function of a quaternary signal transduction complex monitored by surface plasmon resonance. Nature 365:343–347
9. Löfås S (2007) Biacore – creating the business of label-free protein-interaction analysis. In: Marks RS, Cullen DC, Karube I, Lowe CR, Weetall HH (eds) Handbook of biosensors and biochips. Wiley, New York, NY
10. de Mol NJ, Fischer MJ (2010) Surface plasmon resonance: a general introduction. Methods Mol Biol 627:1–14
11. Shinohara Y, Kim F, Shimizu M, Goto M, Tosu M, Hasegawa Y (1994) Kinetic measurement of the interaction between an oligosaccharide and lectins by a biosensor based on surface plasmon resonance. Eur J Biochem 223: 189–194
12. Hutchinson AM (1994) Characterization of glycoprotein oligosaccharides using surface plasmon resonance. Anal Biochem 220: 303–307
13. Kodoyianni V (2011) Label-free analysis of biomolecular interactions using SPR imaging. Biotechniques 50:32–40
14. Rich RL, Myszka DG (2006) Survey of the year 2005 commercial optical biosensor literature. J Mol Recognit 19:478–534
15. Greenbury CL, Moore DH, Nunn LA (1965) The reaction with red cells of 7S rabbit antibody, its sub-units and their recombinants. Immunology 8:420–431
16. Crothers DM, Metzger H (1972) The influence of polyvalency on the binding properties of antibodies. Immunochemistry 9:341–457
17. Hornick CL, Karuch F (1972) Antibody affinity. 3. The role of multivalence. Immunochemistry 9:325–340
18. Shinohara Y, Hasegawa Y, Kaku H, Shibuya N (1997) Elucidation of the mechanism enhancing the avidity of lectin with oligosaccharides on the solid phase surface. Glycobiology 7:1201–1208
19. Lee YC, Townsend RR, Hardy MR, Lonngren J, Arnarp J, Haraldsson M, Lonn H (1983) Binding of synthetic oligosaccharides to the

hepatic Gal/GalNAc lectin. Dependence on fine structural features. J Biol Chem 258: 199–202

20. Lee RT, Lin P, Lee YC (1984) New synthetic cluster ligands for galactose/Nacetylgalactosamine-specific lectin of mammalian liver. Biochemistry 23:4255–4261
21. Duverger E, Lamerant-Fayel N, Frison N, Monsigny M (2010) Carbohydrate-lectin interactions assayed by SPR. Methods Mol Biol 627:157–178
22. de Mol NJ (2010) Affinity constants for small molecules from SPR competition experiments. Methods Mol Biol 627:101–111
23. Kazuno S, Fujimura T, Arai T, Ueno T, Nagao K, Fujime M, Murayama K (2011) Multi-sequential surface plasmon resonance analysis of haptoglobin-lectin complex in sera of patients with malignant and benign prostate diseases. Anal Biochem 419:241–249
24. Harrison BA, MacKenzie R, Hirama T, Lee KK, Altman E (1998) A kinetics approach to the characterization of an IgM specific for the glycolipid asialo-GM1. J Immunol Methods 212:29–39
25. Metzger J, von Landenberg P, Kehrel M, Buhl A, Lackner KJ, Luppa PB (2007) Biosensor analysis of beta2-glycoprotein I-reactive autoantibodies: evidence for isotype-specific binding and differentiation of pathogenic from infection-induced antibodies. Clin Chem 53: 1137–1143
26. de Boer AR, Hokke CH, Deelder AM, Wuhrer M (2008) Serum antibody screening by surface plasmon resonance using a natural glycan microarray. Glycoconj J 25:75–84
27. Remy-Martin F, El Osta M, Lucchi G, Zeggari R, Leblois T, Bellon S, Ducoroy P, Boireau W (2012) Surface plasmon resonance imaging in arrays coupled with mass spectrometry (SUPRA-MS): proof of concept of on-chip characterization of a potential breast cancer marker in human plasma. Anal Bioanal Chem 404:423–432
28. Peng W, Nycholat CM, Razi N (2013) Glycan microarray screening assay for glycosyltransferase specificities. Methods Mol Biol 1022:1–14
29. Stevens J, Blixt O, Paulson JC, Wilson IA (2006) Glycan microarray technologies: tools to survey host specificity of influenza viruses. Nat Rev Microbiol 4:857–864
30. Drickamer K (1989) Demonstration of carbohydrate-recognition activity in diverse proteins which share a common primary structure motif. Biochem Soc Trans 17:13–15
31. Shinohara Y, Sota H, Gotoh M, Hasebe M, Tosu M, Nakao J, Hasegawa Y, Shiga M (1996) Bifunctional labeling reagent for oligosaccharides to incorporate both chromophore and biotin groups. Anal Chem 68:2573–2579
32. Takimori S, Shimaoka H, Furukawa J, Yamashita T, Amano M, Fujitani N, Takegawa Y, Hammarström L, Kacskovics I, Shinohara Y, Nishimura S (2011) Alteration of the N-glycome of bovine milk glycoproteins during early lactation. FEBS J 278:3769–3781
33. Nakajima H, Kiyokawa N, Katagiri YU, Taguchi T, Suzuki T, Sekino T, Mimori K, Ebata T, Saito M, Nakao H, Takeda T, Fujimoto J (2001) Kinetic analysis of binding between Shiga toxin and receptor glycolipid Gb3Cer by surface plasmon resonance. J Biol Chem 276:42915–42922
34. Suda Y, Arano A, Fukui Y, Koshida S, Wakao M, Nishimura T, Kusumoto S, Sobel M (2006) Immobilization and clustering of structurally defined oligosaccharides for sugar chips: an improved method for surface plasmon resonance analysis of protein-carbohydrate interactions. Bioconjug Chem 17:1125–1135
35. Vila-Perelló M, Gutiérrez Gallego R, Andreu D (2005) A simple approach to well-defined sugar-coated surfaces for interaction studies. Chembiochem 6:1831–1838
36. Lee RT, Shinohara Y, Hasegawa Y, Lee YC (1999) Lectin-carbohydrate interactions: fine specificity difference between two mannose-binding proteins. Biosci Rep 19:283–292
37. Sota H, Lee RT, Lee YC, Shinohara Y (2003) Quantitative lectin-carbohydrate interaction analysis on solid-phase surfaces using biosensor based on surface plasmon resonance. Methods Enzymol 362:330–340
38. Kaneda Y, Whittier RF, Yamanaka H, Carredano E, Gotoh M, Sota H, Hasegawa Y, Shinohara Y (2002) The high specificities of Phaseolus vulgaris erythro- and leukoagglutinating lectins for bisecting GlcNAc or beta 1-6-linked branch structures, respectively, are attributable to loop B. J Biol Chem 27: 16928–16935
39. Furukawa J, Shinohara Y, Kuramoto H, Miura Y, Shimaoka H, Kurogochi M, Nakano M, Nishimura S (2008) Comprehensive approach to structural and functional glycomics based on chemoselective glycoblotting and sequential tag conversion. Anal Chem 80:1094–1101
40. Karlsson R, Roos H, Fägerstam L, Persson B (1994) Kinetic and concentration analysis using BIA technology. Methods 6:99–110
41. Alon R, Hammer DA, Springer TA (1995) Lifetime of the P-selectin-carbohydrate bond and its response to tensile force in hydrodynamic flow. Nature 374:539–542

42. Monsigny M, Mayer R, Roche AC (2000) Sugar-lectin interactions: sugar clusters, lectin multivalency and avidity. Carbohydr Lett 4: 35–52
43. Duverger E, Frison N, Roche AC, Monsigny M (2003) Carbohydrate-lectin interactions assessed by surface plasmon resonance. Biochimie 85:167–179
44. Brewer CF, Miceli MC, Baum LG (2002) Clusters, bundles, arrays and lattices: novel mechanisms for lectin-saccharide-mediated cellular interactions. Curr Opin Struct Biol 12: 616–623
45. Shinohara Y, Sota H, Kim F, Shimizu M, Gotoh M, Tosu M, Hasegawa Y (1995) Use of a biosensor based on surface plasmon resonance and biotinyl glycans for analysis of sugar binding specificities of lectins. J Biochem 117: 1076–1082
46. Bakchoul T, Bertrand G, Krautwurst A, Kroll H, Bein G, Sachs UJ, Santoso S, Kaplan C (2013) The implementation of surface plasmon resonance technique in monitoring pregnancies with expected fetal and neonatal alloimmune thrombocytopenia. Transfusion 53: 2078–2085

Chapter 18

Isothermal Calorimetric Analysis of Lectin–Sugar Interaction

Yoichi Takeda and Ichiro Matsuo

Abstract

Isothermal titration calorimetry (ITC) is a powerful tool for analyzing lectin–glycan interactions because it can measure the binding affinity and thermodynamic properties such as ΔH and ΔS in a single experiment without any chemical modification or immobilization. Here we describe a method for preparing glycan and lectin solution to minimize the buffer mismatch, setting parameters, and performing experiments.

Key words Isothermal titration calorimetry, Binding constant, Enthalpy, Intracellular lectin, Calreticulin, Malectin, ER-related *N*-glycan

1 Introduction

Lectins associate with free oligosaccharides or glycan motifs on various glycoconjugates by discriminating small differences in the glycan structures. The affinity of lectin–glycan interactions is sometimes controlled by the multivalency or the population of glycans [1]. However, quantitatively understanding the molecular basis of individual interactions between them can be useful to gain knowledge of biological phenomena through glycan–lectin interactions. In order for analysis of lower affinity interactions between animal lectins and glycans ($K_d = 10^{-3}$–10^{-6} M), analysis technique for higher affinity interactions such as conventional plate assays could not be applied directly due to washing out of ligand. Moreover, care should be taken to maintain the density of glycan on plates in order to avoid the experimental error of K_a values. Isothermal titration calorimetry (ITC) does not require any chemical modification or immobilization. Moreover, it directly measures the enthalpy change for a reaction. Also, the thermodynamic parameters such as K_a ΔH, ΔG, and ΔS can be estimated by data fitting (Fig. 1). Therefore, ITC is a powerful tool for analyzing lectin–glycan interactions [1]. Here we describe, as an example, a method for

Jun Hirabayashi (ed.), *Lectins: Methods and Protocols*, Methods in Molecular Biology, vol. 1200,
DOI 10.1007/978-1-4939-1292-6_18, © Springer Science+Business Media New York 2014

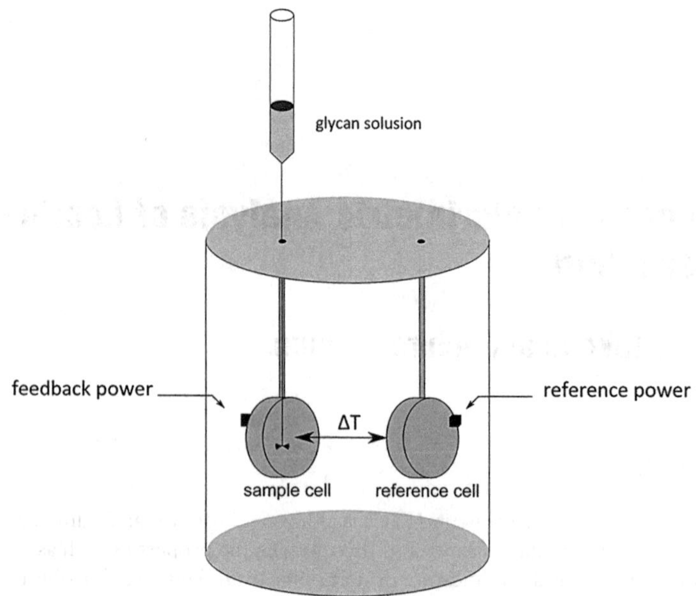

Fig. 1 Schematic of ITC. The syringe is filled with glycan solution and the sample cell is filled with lectin solution. The reference cell is filled with water or buffer used in the experiment. The feedback power to keep $\Delta T = 0$ is measured

measuring the interactions between synthetic high-mannose-type glycans [2, 3] and endoplasmic reticulum lectins (calreticulin and malectin) [4, 5], recently developed commercial instruments (Figs. 2 and 3).

2 Materials

2.1 Special Equipment

1. VP-ITC: MicroCal VP-ITC calorimeter from GE Healthcare (Northampton, MA).

   Auto-pipette (injection syringe).

   ThermoVac (for degassing sample).

   Hamilton 2.5 mL filling syringe.

   Plastic syringe (1 mL).
2. iTC200: iTC200 from GE Healthcare (Northampton, MA).

   Titration syringe.

   Hamilton 1.0 mL filling syringe.
3. UV-visible microplate reader.
4. Refrigerated centrifuge.

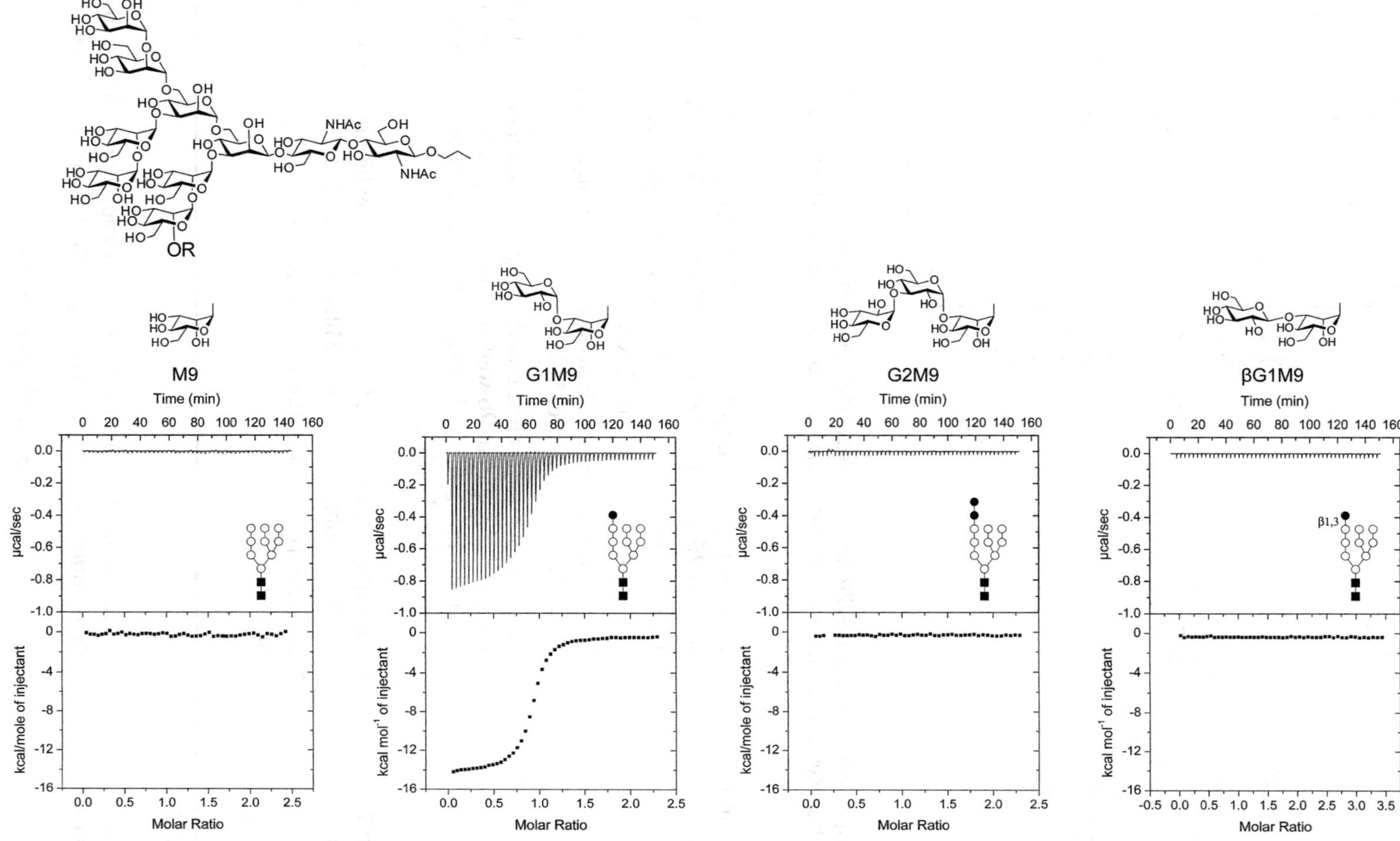

Fig. 2 Binding analyses of M9-*O*-propyl glycoside (Pr), G1M9-*O*-Pr, G2M9-*O*-Pr, and βG1M9-*O*-Pr (stereoisomer) to ER lectin chaperone calreticulin (Refs. 9–11). The thermogram was recorded by VP-ITC, [lectin] = 30 µM and [glycan] = 300 µM (c = 40). Molar heat values are plotted as a function of the molar ratio (*bottom*). The *solid line* represents the best-fit binding isotherm obtained from a single-site model

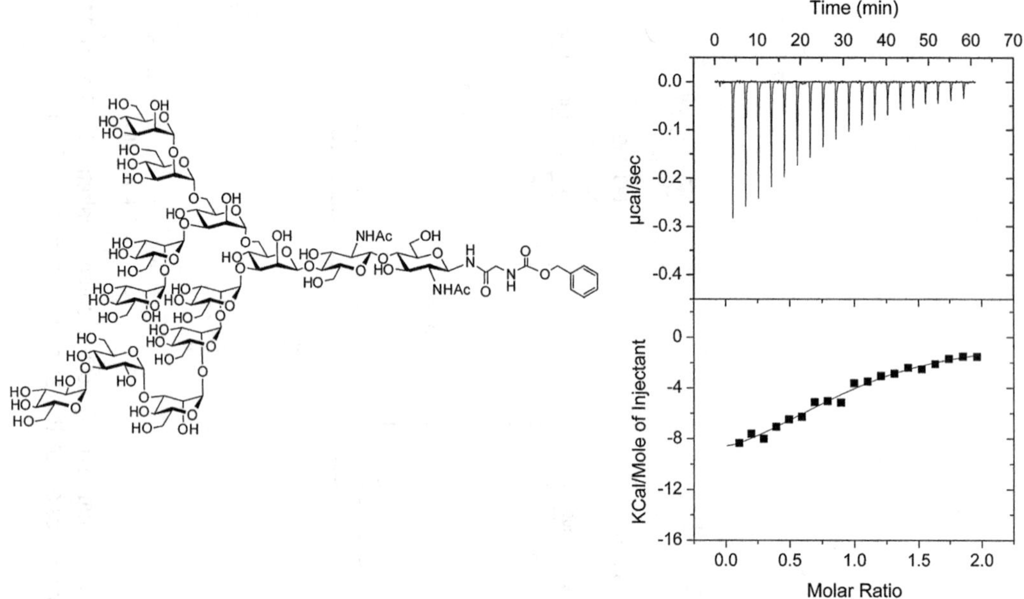

Fig. 3 Binding analyses of G2M9-Gly-Cbz (*left*) to malectin [12], which associates with the diglucosylated glycans. The thermogram was recorded by iTC200, [lectin] = 20 μM and [glycan] = 200 μM (c = 2.9)

2.2 Lectin Solution Preparation

1. Dialysis tubing: Slide-A-Lyzer G2, MWCO 2 K (Thermo, MA), 0.25–0.75 mL (for iTC200) or 1–3 mL (for VP-ITC).
2. Dialysis buffer: 2 L of 10 mM MOPS buffer (pH 7.4), including 5 mM $CaCl_2$ and 150 mM NaCl.
3. Protein concentrators: Amicon Ultra centrifugal filter devices, MWCO 2 K (Millipore, MA).
4. Bradford assay kit: Protein assay dye reagent concentrate (Bio-Rad, CA), UV transparent 96-well plate (Corning, NY), and lyophilized bovine serum albumin >96 % as a protein standard (Sigma-Aldrich, MO).

2.3 Glycan Solution Preparation

1. Centrifugal filter device: Ultrafree-MC, Low-binding Durapore PVDF membrane 0.22 μm (Millipore, MA).
2. Solvent: 10 mL of buffer which was used in Subheading 3.1, **step 3**.

3 Methods

3.1 Preparing Lectin Solution

1. Estimate the K_a value and determine the appropriate lectin concentration (*see* **Note 1**) according to the following equation:

$$5 < n \times [P] \times K_a < 250$$

n: estimated number of binding sites.
[*P*]: lectin concentration.
K_a: estimated association constant.

2. Concentrate the protein, if necessary, to the desired concentrations using an Amicon Ultra centrifugal filter device.
3. Dialysis at 4 °C against the buffer with three buffer changes (*see* **Notes 2** and **3**).
4. Determine the protein concentration with the protein assay dye reagent (*see* **Note 4**).
5. Degas the protein solution under vacuum using the ThermoVac (for VP-ITC) (*see* **Note 5**).

3.2 Preparing Glycan

1. Determine the glycan concentration. It should be 10–20 times higher than the lectin concentration in the cell.
2. Weigh glycan using a balance (*see* **Note 6**).
3. Dissolve glycan with the dialysis buffer used in Subheading 3.1, **step 3**, to 100 μL (for iTC200) or 300 μL (for VP-ITC).
4. Filter the glycan solution using the centrifugal filter device (*see* **Note 7**).
5. Degas the glycan solution under vacuum using the ThermoVac (for VP-ITC) (*see* **Note 5**).

3.3 Loading Lectin into the Cell

1. Remove the water in the cell and wash the cell three times with the dialysis buffer used in Subheading 3.1, **step 3**.
2. Insert the Hamilton gastight syringe with 300 μL (for iTC200) or 2.4 mL (for VP-ITC) of lectin solution into the cell, and touch the bottom of the cell with the tip of the syringe needle.
3. Inject the lectin solution into the cell while holding the needle tip 1 mm from the bottom. Do not raise the syringe during loading.
4. After ca. 50 μL (for iTC200) or 250 μL (for VP-ITC) of solution spills out of the top of the cell stem, repeat an abrupt spurt of solution one or two times to remove bubbles in the cell.
5. Remove excess solution by touching the tip of the syringe to the point where the cell stem meets the cell port (sample volume: 1.4181 mL for VP-ITC, 0.1977 mL for iTC200).

3.4 Loading the Titration Syringe

3.4.1 VP-ITC

1. Place a plastic tube containing 300 μL of glycan solution in the tube holder.
2. Insert the titration syringe into the pipette stand.
3. Click on the "Open Fill Port" button and attach the tube of the plastic loading syringe to the fill port of the titration syringe.
4. Slowly withdraw the plunger of the plastic loading syringe to extract the glycan solution.

Table 1
Method parameters

Parameter	Value	
	VP-ITC	iTC200
Total injections	50	19
Cell temp. (°C)	20	20
Reference power	10	5
Initial delay (s)	60	60
Stirring speed (RPM)	300	1,000

Table 2
Injection parameters

Parameter	Value	
	VP-ITC	iTC200
Injection volume (μL)	6 (2.0)	2 (0.4)
Duration (s)	14.4 (2.0)	4 (0.8)
Spacing (s)	180	150
Filter period (s)	2	5

The injection volume and duration time of the first injection should be changed to the values in parentheses

5. After the solution exits the top fill port, immediately click the "Close Fill Port" button.
6. Remove the plastic loading tube to the fill port of the titration syringe.
7. Click on the "Purge → Refill" button to dislodge air bubbles that may rest in the titration syringe (twice, sequentially).
8. Carefully remove the pipette from its stand by picking it straight up, avoiding bending the long needle. Wash the pipette by the ultrapure water and carefully wipe the water using a cotton bud.

3.4.2 iTC200

1. Place a microcentrifuge tube containing ca. 100 μL of glycan solution in the tube holder.
2. The titrant solution is automatically filled according to the software instructions for controlling the device.

3.5 Setting Parameters

Set method parameters and injection parameters for VP-ITC or iTC200 as follows (*see* Tables 1 and 2).

3.6 Performing an Experiment

1. Carefully insert the titration syringe into the cell and ensure that the syringe is firmly seated.
2. Enter the file name.
3. Click the "Start" button to begin the experiment.

3.7 Data Analysis

1. When all injections of the titrant solution are completed, the software automatically calculates the peak areas and plots the normalized area data in kcal/mol of glycan vs. the molar ratio of glycan/lectin (*see* **Note 8**).
2. The stoichiometric and thermodynamic parameters, n, K_a, ΔH, ΔG, and ΔS, can be calculated by a nonlinear, least-squares analysis of the data using the accessory software Origin (Origin Lab, MA) (*see* **Notes 9** and **10**).

4 Notes

1. A favorable concentration of lectin is dependent on K_a, n, and ΔH (heat change in cal/mol). The unitless constant, c, which is derived by Wiseman et al. [6], is useful for experimental design:

$$c = n \times [P] \times K_a$$

When c ranges from 5 to 250, a sigmoidal thermogram, which provides accurate estimates of K_a, is obtained.

2. If the glycan–lectin interaction involves the release or the uptake of protons, the observed enthalpy change involves the enthalpy of protonation or ionization for each proton absorbed or released by the buffer. Therefore, use of buffers with ΔH_{ion} ~0 (phosphate, acetate, formate, citrate, etc.) is recommended [7]. However, the K_a (association constant) or the n (number of binding sites) values will not be affected by using Good's buffers.
3. The buffer solution used in the experiment should be chosen on the basis of the stability of the lectin. If required, up to 5 mM 2-mercaptoethanol or 2 mM tris(2-carboxyethyl)phosphine can be used. Dithiothreitol should not be used.
4. Protein quantification can be performed by other alternative methods such as the measurement of UV absorbance at 280 nm, BCA, and Lowry assays.
5. Degassed lectin and glycan solutions should be used for experiments with VP-ITC.
6. If glycan sample which cannot be weighed on a balance has to be used, it should be quantified by the phenol-sulfuric acid method [8] using concentrated sulfuric acid (Sigma-Aldrich) and 5 % (w/w) aqueous phenol solution, prepared by adding 5 g of redistilled reagent grade phenol (Sigma-Aldrich) to 95 g of ultrapure water.

7. Dust, other particulates, and air bubbles can cause artifacts in the baseline of the ITC thermogram. The sample should be centrifuged in microcentrifuge tubes and removed to pellet particles.
8. Control experiments should be conducted by injecting the glycan solution into the buffer.
9. When the interaction of a limited amount of glycan against lectin is investigated, experiments should be performed under lower *c* value conditions, and consequently, the data would not provide a sigmoidal thermogram. In these cases, care should be taken to minimize the mismatch between the glycan and lectin solutions to increase the signal-to-noise ratio.
10. Even if *c* is beyond the recommended range, the K_a value can still be estimated. In this case, the curve fitting might have to be performed by fixing the *n* value. The experiment depicted in Fig. 3 was conducted when $c=2.9$.

References

1. Dam TK, Brewer CF (2002) Thermodynamic studies of lectin–carbohydrate interactions by isothermal titration calorimetry. Chem Rev 102:387–429
2. Matsuo I, Totani K, Tatami A, Ito Y (2006) Comprehensive synthesis of ER related high-mannose-type sugar chains by convergent strategy. Tetrahedron 62:8262–8277
3. Matsuo I, Wada M, Manabe S, Yamaguchi Y, Otake K, Kato K, Ito Y (2003) Synthesis of monoglucosylated high-mannose-type dodecasaccharide, a putative ligand for molecular chaperone, calnexin, and calreticulin. J Am Chem Soc 125:3402–3403
4. Schallus T, Jaeckh C, Feher K, Palma AS, Liu Y, Simpson JC, Mackeen M, Stier G, Gibson TJ, Feizi T, Pieler T, Muhle-Goll C (2008) Malectin: a novel carbohydrate-binding protein of the endoplasmic reticulum and a candidate player in the early steps of protein *N*-glycosylation. Mol Biol Cell 19:3404–3414
5. Williams DB (2006) Beyond lectins: the calnexin/calreticulin chaperone system of the endoplasmic reticulum. J Cell Sci 119: 615–623
6. Wiseman T, Williston S, Brandts JF, Lin LN (1989) Rapid measurement of binding constants and heats of binding using a new titration calorimeter. Anal Biochem 179: 131–137
7. Cooper A, Johnson C (1994) Introduction to microcalorimetry and biomolecular energetics. In: Jones C, Mulloy B, Thomas A (eds) Microscopy, optical spectroscopy, and macroscopic techniques. Humana, New York, NY, pp 109–124
8. Saha AK, Brewer CF (1994) Determination of the concentrations of oligosaccharides, complex type carbohydrates, and glycoproteins using the phenol-sulfuric acid method. Carbohydr Res 254:157–167
9. Arai MA, Matsuo I, Hagihara S, Totani K, Maruyama J, Kitamoto K, Ito Y (2005) Design and synthesis of oligosaccharides that interfere with glycoprotein quality-control systems. Chembiochem 6:2281–2289
10. Ito Y, Hagihara S, Arai MA, Matsuo I, Takatani M (2004) Synthesis of fluorine substituted oligosaccharide analogues of monoglucosylated glycan chain, a proposed ligand of lectin-chaperone calreticulin and calnexin. Glycoconj J 21:257–266
11. Ito Y, Hagihara S, Matsuo I, Totani K (2005) Structural approaches to the study of oligosaccharides in glycoprotein quality control. Curr Opin Struct Biol 15:481–489
12. Takeda Y, Seko A, Sakono M, Hachisu M, Koizumi A, Fujikawa K, Ito Y (2013) Parallel quantification of lectin–glycan interaction using ultrafiltration. Carbohydr Res 375:112–117

Chapter 19

Carbohydrate–Lectin Interaction Assay by Fluorescence Correlation Spectroscopy Using Fluorescence-Labeled Glycosylasparagines

Mamoru Mizuno

Abstract

Fluorescence correlation spectroscopy (FCS) is a high-throughput system for the assay of interactions in solution and can be used to measure the numbers of molecules and molecular size in micro-regions. FCS can be used to measure interactions in environments that are close to those in vivo. It is a useful technique for measuring bioactive substances, screening inhibitors, and detecting the binding of materials, as well as for determining K_d and IC_{50} values. Glycosyl amino acids with natural oligosaccharides are useful for the interaction assay of oligosaccharides. Fluorescence probes can be introduced into the glycosyl amino acid while the whole structure of the oligosaccharide is maintained. Carbohydrate–lectin interaction in a solution assay system can be analyzed easily by FCS using fluorescence-labeled glycosylasparagine.

Key words Fluorescence correlation spectroscopy (FCS), Glycosylasparagine, Solution assay system, Brownian motion, Fluctuation, Single-molecule detection

1 Introduction

The study of carbohydrates (glycomics) is very interesting as a post-genome and post-proteome. Interaction assay under conditions similar to those in vivo is very important for understanding the biological interactions between various species and oligosaccharides [1, 2]. Recently, fluorescence correlation spectroscopy (FCS) has been studied as a system for the assay of interactions in solution [3, 4].

FCS is an experimental technique used to measure the numbers of molecules and molecular size from the fluctuation of molecules in micro-regions. The Brownian motion of molecules passing through a micro-region is detected as fluctuations in the fluorescence signal. Interaction is determined by analyzing the resulting fluorescence signal.

Jun Hirabayashi (ed.), *Lectins: Methods and Protocols*, Methods in Molecular Biology, vol. 1200,
DOI 10.1007/978-1-4939-1292-6_19, © Springer Science+Business Media New York 2014

FCS has some advantages over conventional methods such as ELISA and surface plasmon resonance. For example, (1) troublesome pretreatment steps such as immobilization of samples on the microplate and the sensor chip are unnecessary, (2) nonspecific adhesion to plates or vessels has very little influence on the interaction assay, (3) highly sensitive and low-noise analyses can be performed, (4) high-throughput assays are performed in solutions in which the molecules can move freely, and (5) behavior at the molecular level can be observed in a micro-region.

Of course, there are some caveats with this method. For example, (1) FCS needs a fluorescence modification step, unlike isothermal titration calorimetry; (2) a fluorescence dye that is unlikely to enter the triplet state has to be selected, because the triplet state is a noise component in FCS analysis; and (3) interaction assays between substrates that have similar molecular weights are difficult. At least four times the molecular weight is required.

Here, we describe an interaction assay between oligosaccharide and lectin using FCS. Glycosylasparagines were used as carbohydrate substrates. Using glycosylasparagines has some advantages: (1) they can be obtained easily in large quantities from eggs, (2) the fluorescence labeling is simple and uses the amino group of an asparagine moiety, and (3) the sugar structure of the reduced terminal is not changed, because reductive amination is not used in the labeling step [5].

2 Materials

Prepare all solutions by using ultrapure water (prepared by purifying deionized water to achieve a sensitivity of 18 MΩ-cm at 25 °C) and analytical grade reagents. Prepare all reagents at room temperature.

2.1 Fluorescence-Labeled Glycosylasparagine

1. Preparation of glycosylasparagine: Glycosylasparagine with the disialo complex-type *H-Asn(Sialo)-OH* and glycosylasparagine with the asialo complex-type *H-Asn(Asialo)-OH*, derived from egg yolk, are prepared as previously described [6]. High-mannose-type asparagine *H-Asn[M6]-OH* and hybrid-type asparagine *H-Asn[M5GN4]-OH* are obtained from ovalbumin [7] (for structures, *see* Fig. 1).
2. Fluorescence modification of glycosylasparagine: Glycosylasparagine (1.1 μmol) is dissolved in 40 μl of *N,N*-dimethylformamide (DMF) (*see* **Note 1**). A 2 M solution of *N,N*-diisopropylethylamine (DIEA) in DMF (1.6 μL, 3.0 equiv.), 1 % aqueous $NaHCO_3$ (27 μL, 3.0 equiv.), and 1.5–2.0 equiv. of 5-(and-6)-carboxytetramethylrhodamine succinimidyl ester (5(6)-TAMRA-X, SE) are added (*see* **Note 2**), and the

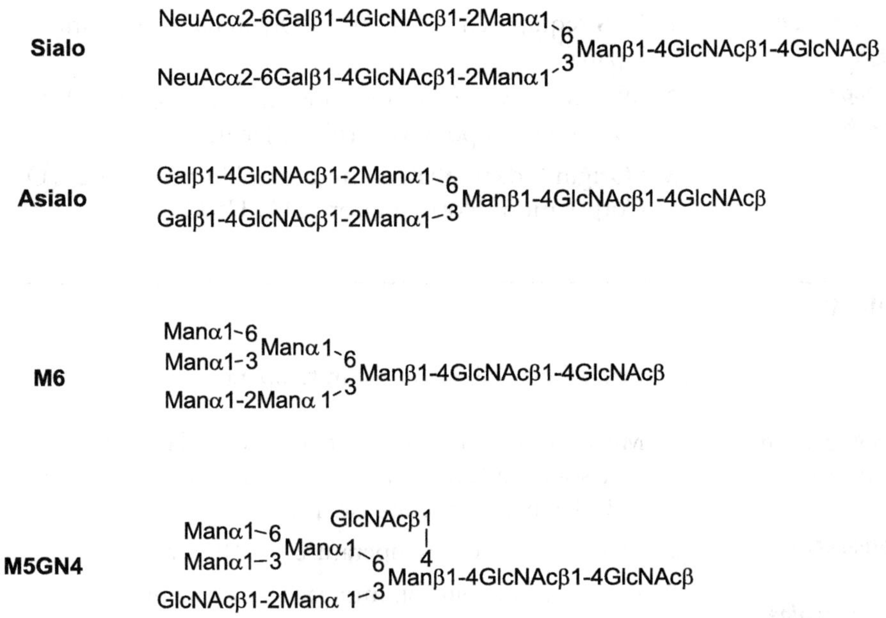

Sialo

NeuAcα2-6Galβ1-4GlcNAcβ1-2Manα1-6
Manβ1-4GlcNAcβ1-4GlcNAcβ
NeuAcα2-6Galβ1-4GlcNAcβ1-2Manα1-3

Asialo

Galβ1-4GlcNAcβ1-2Manα1-6
Manβ1-4GlcNAcβ1-4GlcNAcβ
Galβ1-4GlcNAcβ1-2Manα1-3

M6

Manα1-6
Manα1-3 Manα1-6
Manβ1-4GlcNAcβ1-4GlcNAcβ
Manα1-2Manα 1-3

M5GN4

Manα1-6 GlcNAcβ1
Manα1-3 Manα1-6 4
Manβ1-4GlcNAcβ1-4GlcNAcβ
GlcNAcβ1-2Manα 1-3

Fig. 1 Structures of glycosylasparagines prepared from hen's egg

mixture is stirred at room temperature in the dark for 3 h (*see* **Note 3**). After HPLC purification, fluorescence-labeled glycosylasparagine is obtained (64–97 %) (*see* **Note 4**).

3. 5-(and-6)-Carboxytetramethylrhodamine, succinimidyl ester (5(6)-TAMRA-X, SE): (Funakoshi Co. Ltd., Tokyo, Japan).
4. Fluorescence-labeled glycosylasparagine: Store at −20 °C wrapped in aluminum foil.
5. Fluorescence-labeled glycosylasparagine solution: Fluorescence-labeled glycosylasparagine is dissolved in PBS buffer and 10- and 2 nM solutions are prepared. Store at 4 °C wrapped in aluminum foil.
6. PBS buffer: 10 mM phosphate-buffered saline, pH 7.4.

2.2 Lectins and Inhibitors

1. Lectin: Concanavalin A (Con A), wheat germ agglutinin (WGA), *Ricinus communis* agglutinin (RCA120). Lectin is dissolved in PBS and a 100 μM lection solution is prepared. Store at 4 °C. Prepare ten different concentrations (0.2, 1.0, 10, 20, 100, 200, 1, 2, 10, and 20 μM) of lectin solution just before FCS measurement.
2. Inhibitor: α-Methyl mannoside, *N*-acetylglucosamine. Inhibitor is dissolved in PBS and a 100 mM inhibitor solution is prepared. Store at 4 °C. Prepare eight different concentrations (2, 4, 20, 40, 200, 400, 2, and 20 mM) of inhibitor solution just before FCS measurement.

2.3 Fluorescence Correlation Spectroscopy Components

1. FCS equipment: MF20™ (Olympus Corporation, Tokyo, Japan).
2. 384-well plate: Glass bottom microplate MP0384120 (Olympus Corporation, Tokyo, Japan).
3. "Origin" data analysis and graphing software (OriginLab Corporation, Northampton, MA, USA).

3 Methods

Perform all procedures at room temperature.

3.1 FCS Interaction Assay Between Lectins and Fluorescence-Labeled Glycosylasparagine

1. Mix each concentration of fluorescence-labeled glycosylasparagine solution (28 μL) and each concentration of lectin solution (28 μL) in a microtube (*see* **Note 5**).
2. Incubate at room temperature for 10 min.
3. Put 50 μL of solution into each well of a plate.
4. Perform FCS measurement (10 s × 5 times/sample).
5. Measure the diffusion times at each lectin concentration (*see* **Note 6**).
6. Calculate the degree of binding [fluorescence-labeled glycosylasparagine + lectin] complex by using the diffusion time (*see* **Note 6**) (Fig. 2).

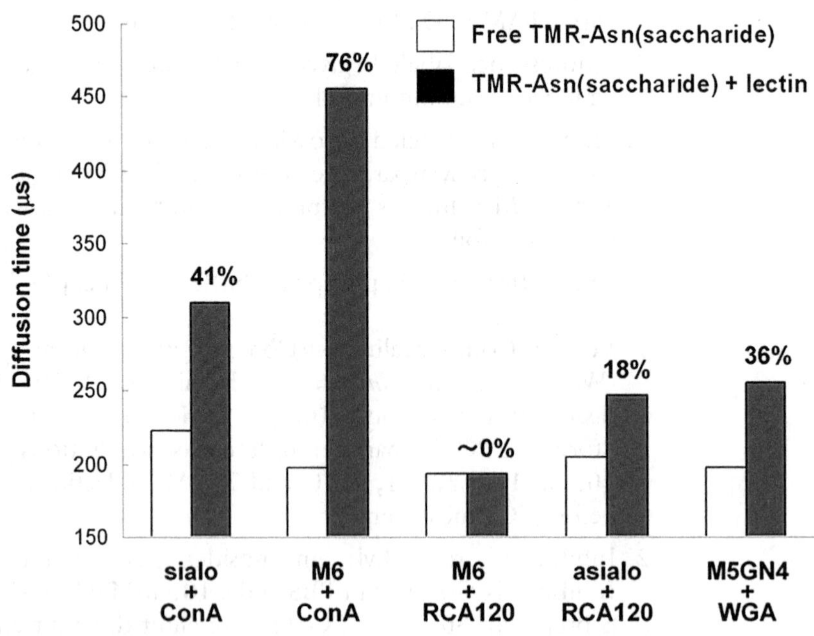

Fig. 2 Interaction of TMR-labeled glycosylasparagines with lectins measured by FCS. The concentration of *TMR-Asn(M6)-OH* is 5 nM, and Con A is 5 μM

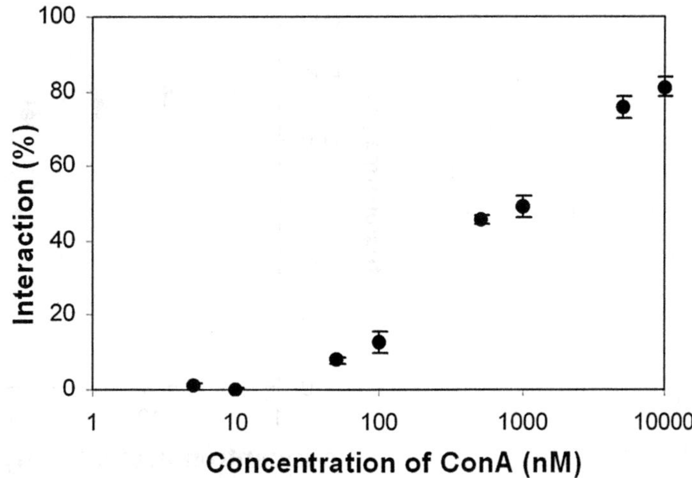

Fig. 3 Binding of *TMR-Asn(M6)-OH* by Con A by FCS. The concentration of *TMR-Asn(M6)-OH* is 5 nM

7. Fit the data by using the following formula and "Origin" software. From the results of the fitting the K_d value is calculated (Fig. 3):

$$\Upsilon = \frac{Kd + [A] + [B] - \sqrt{(Kd + [A] + [B])^2 - 4[A][B]}}{2[A]}$$

Υ = degree of binding of [fluorescence-labeled glycosylasparagine + lectin] complex.

$[A]$ = concentration of fluorescence-labeled glycosylasparagine [M].

$[B]$ = concentration of lectin [M].

3.2 Inhibition Assay

1. Mix each concentration inhibitor solution (20 μL) and 20 μM lectin solution (20 μL) in a microtube.
2. Incubate at room temperature for 10 min.
3. Mix an aliquot of the solution (28 μL) and 10 nM fluorescence-labeled glycosylasparagine solution (28 μL) in another microtube.
4. Place 50 μL of the solution into the well of a plate.
5. Perform FCS measurement (10 s × 5/sample).
6. The inhibition assay is performed by observing the diffusion time (*see* **Note 6**) (Fig. 4).

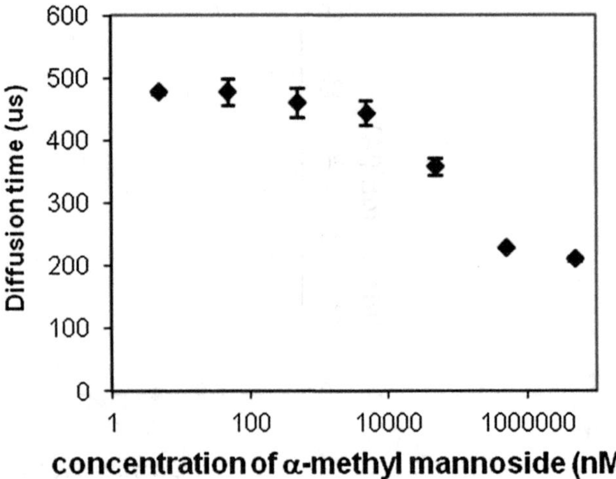

Fig. 4 Inhibition of Con A binding to *TMR-Asn(M6)-OH* by FCS. The concentration of *TMR-Asn(M6)-OH* was 5 nM, and the inhibitor was α-methyl mannoside

4 Notes

1. Glycosylasparagine is not soluble in DMF. After water has been added, all of the substrates and reagents are soluble.
2. The triplet state is a noise component in FCS analysis. To perform a reliable analysis it is best that the ratio of the triplet state is low. Choose a dye that is unlikely to enter the triplet state. FITC is not recommended because it can easily enter the triplet state [3, 4].
3. The reaction vessel is wrapped in aluminum foil.
4. Preparative HPLC is performed on an Inertsil ODS three column (20×250 mm) (GL Sciences Inc., Tokyo, Japan). Solvent system A is 0.1 % TFA in water, and solvent system B is 0.1 % TFA in acetonitrile. Elution is performed with buffer A to buffer B at 20:80–40:60 (40 min). Detection is performed at a wavelength of 254 nm. Flow rate is 10 mL/min. Mass numbers of fluorescence-labeled glycosylasparagines are determined by matrix-assisted laser desorption ionization-time-of-flight (MALDI-TOF) mass spectrometry with a Voyager™ RP (PerSeptive Biosystems Inc., Framingham, MA). Mass numbers are calculated as averages. MALDI-TOF mass spectrometry is performed in the positive ion mode by using α-cyano-4-hydroxycinnamic acid as a matrix.
5. Fluorescence-labeled glycosylasparagine is used at 10 and 2 nM and lectin at 0.2, 1.0, 10, 20, 100, 200, 1, 2, 10, and 20 μM; a total of 20 samples are used.
6. Data are processed by using MF20 software.

Acknowledgement

We thank Dr. Noriko Kato, Dr. Naoaki Okamoto, and Mr. Masayoshi Kusano (OLYMPUS CORPORATION) for operation of the MF20.

References

1. Hirabayashi J (2003) Oligosaccharide microarrays for glycomics. Trends Biotechnol 21:141–143
2. Mellet CO, Fernández JMG (2002) Carbohydrate microarrays. Chem BioChem 3: 819–822
3. Eggeling C, Fries JR, Brand L et al (1998) Monitoring conformational dynamics of a single molecule by selective fluorescence spectroscopy. Proc Natl Acad Sci U S A 95:1556–1561
4. Kask P, Palo K, Dirk Ullmann D et al (1999) Fluorescence-intensity distribution analysis and its application in biomolecular detection technology. Proc Natl Acad Sci U S A 96:13756–13761
5. Mizuno M, Noguchi M, Imai M et al (2004) Interaction assay of oligosaccharide with lectins using glycosylasparagine. Bioorg Med Chem Lett 14:485–490
6. Seko A, Koketsu M, Nishizono M et al (1997) Occurrence of a sialylglycopeptide and free sialylglycans in hen's egg yolk. Biochim Biophys Acta 1335:23–32
7. Tai T, Yamashita K, Ogata-Arakawa M et al (1975) Structural studies of two ovalbumin glycopeptides in relation to the endo-beta-*N*-acetylglucosaminidase specificity. J Biol Chem 250:8569–8575

Part II

Emerging Techniques for Lectin-Based Glycomics

Chapter 20

Lectin-Based Glycomics: How and When Was the Technology Born?

Jun Hirabayashi

Abstract

Lectin-based glycomics is an emerging, comprehensive technology in the post-genome sciences. The technique utilizes a panel of lectins, which is a group of biomolecules capable of deciphering "glycocodes," with a novel platform represented by a lectin microarray. The method enables multiple glycan–lectin interaction analyses to be made so that differential glycan profiling can be performed in a rapid and sensitive manner. This approach is in clear contrast to another advanced technology, mass spectrometry, which requires prior glycan liberation. Although the lectin microarray cannot provide definitive structures of carbohydrates and their attachment sites, it gives useful clues concerning the characteristic features of glycoconjugates. These include differences not only in terminal modifications (e.g., sialic acid (Sia) linkage, types of fucosylation) but also in higher ordered structures in terms of glycan density, depth, and direction composed for both *N*- and *O*-glycans. However, before this technique began to be implemented in earnest, many other low-throughput methods were utilized in the late twentieth century. In this chapter, the author describes how the current lectin microarray technique has developed based on his personal experience.

Key words Lectin-based glycomics, Frontal affinity chromatography, Lectin microarray, Dissociation constant, Evanescent-field-activated fluorescence detection, Proteomics

1 The Dawn of Lectin Glycomics: A Personal View

Glycomics is defined as the field of studies that "profile" the glycome, i.e., the complete set of glycans and glycoconjugates that cells produce under specific conditions of time, space, and environment [1]. Although the first paper describing "glycome/glycomics" was published in PubMed in 2001 [2], a crude idea of glycome/glycomics had already been discussed by several glycoscientists toward the end of the twentieth century, i.e., shortly after the concept of the proteome had been provided by V.C. Washinger et al. [3]. V.C. Reinfold (University of New Hampshire) was one of the pioneering scientists of the field, with his specialty being mass spectrometry (MS). To the best of my knowledge, the first mention of glycome/glycomics in public was made by this author,

Jun Hirabayashi (ed.), *Lectins: Methods and Protocols*, Methods in Molecular Biology, vol. 1200, DOI 10.1007/978-1-4939-1292-6_20, © Springer Science+Business Media New York 2014

J. Hirabayashi (Teikyo University at that time), in 1999 in an oral presentation entitled "Invitation to the glycopeptide glycome project of *Caenorhabditis elegans*" [4]. In that symposium session, however, relatively few attendees showed any apparent interest in the idea of glycome/glycomics or in the proposed approach. However, soon after a special issue of the newly founded journal "Proteomics" was dedicated to glycobiology; this issue was edited by N.H. Packer (Sydney), who selected this author's talk from the symposium as one of the topics [5]. N. Tanigichi (Osaka University at the time) was the only other contributor to that special issue who also used the new glycome/glycomics terminology [6]. In my paper, I presented a primitive but practical approach about how to proceed with the glycome project in relation to *C. elegans*, which in 1998 became the first multicellular organism to have its genome fully sequenced [7]. I also contributed to a millennium issue of Trends in Glycoscience and Glycotechnology, describing the *C. elegans* glycome project, in which I addressed the issue of the glycoprotein and defined the following three criteria as core strategies [8]:

1. The whole set of glycans in a single organism needs to be identified.
2. Glycopeptides are targeted.
3. As attributes to specify glycopeptides, the cosmid ID, molecular weight (M_r), and dissociation constants (K_ds) for a set of lectins are adopted.

The first criterion is none other than the concept of glycome. The second one is essential to link the glycome information to a genome database in terms of glycosylation sites. The third criterion addresses a major issue that accompanies any glycome project. Though it is best to determine all the covalent structures of all types of glycans, this was obviously impractical. Instead, this author proposed the adoption of K_ds for a set of lectins to characterize each glycan structure. One reason for this is that lectins are known to behave as in vivo "decipherers" of complex glycoconjugates [9–13]. In fact, plant lectins have proven to be useful tools to study animal cells, discriminating subtle differences between glycans as described by E.V. Damme (*see* Chapter 1). The other reason for selecting lectins is more practical, and based on the fact that I had just succeeded in constructing a glycan profiling system by reinforcing a conventional technique, i.e., frontal affinity chromatography (FAC). FAC was developed by K. Kasai as a most reliable method to determine biomolecular interactions based on a clear principle and by using a simple procedure (*see* Chapter 21). Thus, I considered that obtaining K_ds for a set of lectins would provide useful clues to the structure of each glycan if combined with m/z values obtained by MS (mass spectrometry).

Unfortunately, the proposed *C. elegans* glycome project was not successful in achieving its objectives, firstly because the separation of glycopeptides (criterion 2) was found to be much more difficult than expected compared with non-glycosylated peptides, and secondly because glycan structures of *C. elegans* are rather unusual; in fact, the structures were totally unknown at that time except for the conserved high-mannose-type *N*-glycans [14, 15]. To this extent, it was only recently that complex-type *N*-glycans of this model organism were elucidated [16–18]. However, various attempts at elucidating the *C. elegans* glycome led to the generation of several important technologies: these included reinforced FAC [19–21] and the glyco-catch method [16, 22]. As described by C. Sato (*see* Chapter 22), the former method was automated with a highly sensitive fluorescence detection system in collaboration with Shimadzu Co. Ltd. (Kyoto) in a Japanese Government-funded high-priority project [23]. The glyco-catch method was drastically improved by combination with a high-throughput proteomics liquid chromatography (LC)-MS strategy [24]. This served as a large-scale glycoprotein identification system with *N*-glycosylation sites that was named isotope-coded glycosylation-site-specific tagging (IGOT) [25, 26].

2 World-Wide Consortium for Functional Glycomics

It should be mentioned that in 2001 the worldwide Consortium for Functional Glycomics (CFG), which was funded by a US National Institute of General Medical Sciences (NIGMS) glue grant, commenced its activities. The CFG is a large research initiative to "define paradigms by which protein–carbohydrate interactions mediate cell communication" (for details, *see* NIGMS press release: http://www.nigms.nih.gov/News/Results/Paulson.htm). To achieve this goal, the CFG studies the functions of (1) the three major classes of mammalian glycan-binding proteins (GBPs, or lectins), i.e., C-type lectin, galectin, and SIGLEC; (2) immune receptors that bind carbohydrates, i.e., CD1, T cell receptors, and anti-carbohydrate antibodies; and (3) GBPs of microorganisms that bind to host cell glycans as receptors. The most powerful research tool, which the CFG originally developed, is the glycan microarray [27–29]. This tool enables multiple interaction analyses between a panel of >400 immobilized, structure-defined synthetic glycans and a variety of GBPs. However, the main purpose of the CFG is to elucidate the function of glycans along with "-omics" strategies, where a diverse range of native lectins is targeted with a view to possible uses in the field of human health (e.g., development, immunology, infection). In this context, the purpose of our lectin-based glycomics is significantly different, where the aim is to develop new technologies for glycomics (described in this Subheading) with the aid of both natural (*see* Part I) and engineered lectins (*see* Part IV). The functional analysis of endogenous lectins is yet another challenging issue (*see* Part III).

3 Trends in Lectin-Based Glycomics

Glycans are considered the third group of bio-informative molecules, acknowledged after nucleic acids and proteins. All of these biomolecules demonstrate huge structural diversity, which is attributed to the same diversification mechanism of "dehydration condensation" of relatively few components (i.e., 4 nucleotides, 20 amino acids, ~10 monosaccharides). Nevertheless, structural complexity is more evident in glycans because these compounds are often branched with multiple glycosidic bonds for a single monosaccharide [30]. The occurrence of anomeric isomers further increases their complexity. However, the most complex issue surrounding glycans concerns their biosynthetic features. In short, they are not directly coded by genes as occurs for proteins. In fact, they are the product of the activities of a series of enzymes (e.g., glycosyltransferases, sulfotransferases, glycosidases) and sugar nucleotide transporters, whose expressions are significantly altered by the types and states of cells. Glycans act on proteins and lipids in the same manner on some occasions, and yet in a completely different way on others. This has led to a certain aversion to glycans by many researchers. From a practical viewpoint of "biologics" (pharmaceutical protein-based drugs) development, the fact that glycan structures are highly sensitive to cell culture conditions (pH, temperature, nutrients, etc.) is a critical issue. Moreover, glycan structures sometimes differ from one site to another depending on peptide sequences, probably because a significant number of the enzymes involved in glycosylation processes favor (or not) particular local peptide structures and geometries to which glycans are attached. Given these differences, it seems natural that Francis Crick did not refer to glycans when he mentioned his "central dogma" in relation to DNA, RNA, and protein [31].

As described above, soon after the concept of the "proteome" was presented by V.C. Washinger et al. [3], an analogous idea of the glycome was generated around 1999. A search of annual publication trends for glycome/glycomics in the PubMed database (http://www.ncbi.nlm.nih.gov/pubmed/) was carried out to analyze the uptake of this terminology. Based on the following queries, proteome/proteomics (queried by "proteome OR proteomics"), glycome/glycomics (queried by "glycome OR glycomics"), and "lectin" (Fig. 1), it is clear that the number of papers addressing "proteome/proteomics" (total number 56,700, as of 18 Aug. 2013) accounted for as many as 40 times more than those addressing "glycome/glycomics" (1,314). However, both categories of papers show a significant increase from around the year 2000: proteome/proteomics papers first appear in 1995 while those of glycome/glycomics do so after 2000, with particularly large increases having taken place over the last few years.

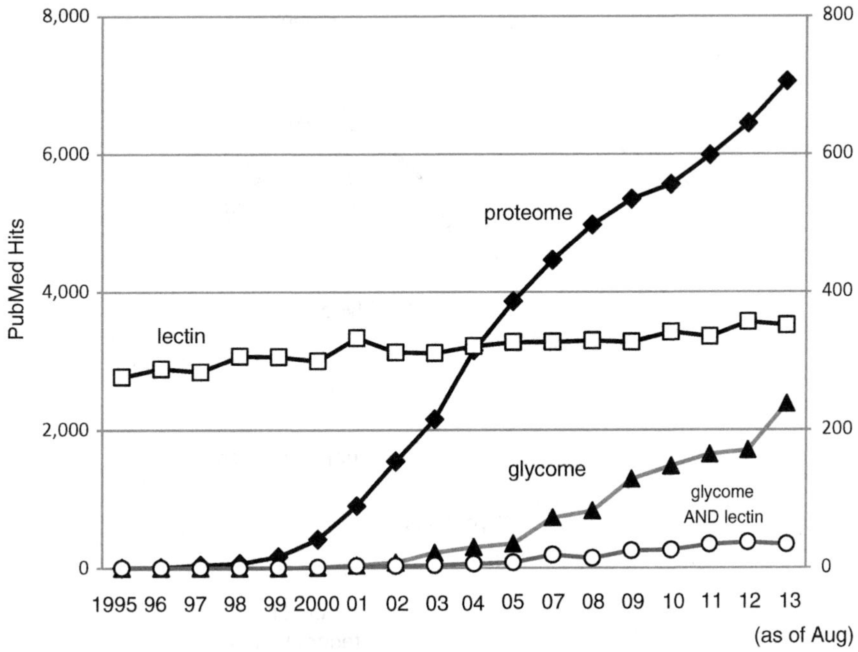

Fig. 1 *Line graph* representation showing PubMed hit trends for queries on proteome, glycome, and lectin. Annual hits with queries of "lectin" (*open square*) and "proteome" (*closed diamond*) are shown relative to the *left vertical axis*, while those of "glycome" (*closed triangle*) and "glycome" AND "lectin" (*open circle*) are shown with respect to the *right vertical axis*

On the other hand, the annual number of papers regarding "lectin" has remained mostly constant (~2,000); the total number of lectin papers to the present is over 106,000, with lectins first appearing in PubMed back in 1946. Though the total number of papers regarding both "glycome/glycomics" AND "lectin" is relatively small (240 as of 18 Aug. 2013), this is also on the increase. This observation indicates that papers on glycomics are highly dependent on the co-presence of lectins as well. However, considering the total number of glycome/glycomics papers (1,314), the number of "lectin-based glycomics" papers is relatively small, accounting for just 18 %. This probably means that lectin-based glycomics is still an emerging topic in the field of glycomics, while the application of some of the leading technologies has been attempted successfully. For example, MS, one of the most used techniques in glycomics, is offered as a basic infrastructure by many companies, whereas the lectin microarray has been commercialized by only a few start-up companies. Also, it is important to note that just 322 hits were recorded when the query "proteome/proteomics" AND "glycome/glycomics" was searched for, accounting for just 0.63 % of all proteome/proteomics papers.

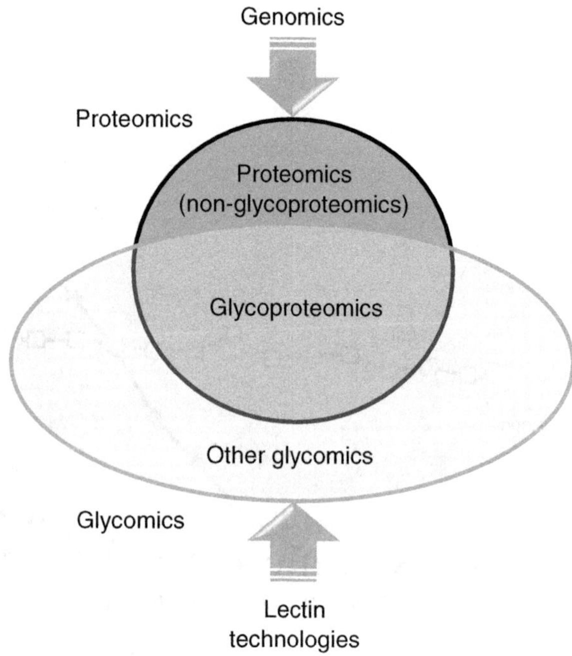

Fig. 2 Relation between proteomics and glycomics

What does this result mean? Considering the fact that most proteins in the eukaryotic cell are glycosylated, substantial features of the proteome should also belong to "glycoproteome" [32]. The result is similar when the query "glycoproteome/glycoproteomics" is searched for (369 hits), accounting for 0.65 % (i.e., 369/56,700) of proteome/proteomics papers. On the other hand, the number of hits increases when "proteome/proteomics" AND "lectin" is queried (792 hits), but the result still only accounts for 1.4 % of this group of papers. This discrepancy may imply that a significant number of proteome/proteomics papers actually deal with lectins as tools to analyze a range of targets of glycoproteins; however many authors do not yet count their studies as belonging to the glycome/glycoproteome field. In this context, the proteome comprises both non-gycoproteome and glycoproteome components (Fig. 2), the former representing cytoplasmic and nucleoplasmic proteins that are not glycosylated (except for *O*-GlcNAc proteins) [33], while the latter represent secreted and membrane-integrated (e.g., receptor) proteins, which are largely subjected to glycosylation in eukaryotes. On the other hand, glycomics also deals with other groups of glycomics, which include glycolipid (often termed glycolipidomics [34, 35]), glycosaminoglycan (glycosaminoglycomics [36–38]), and free (liberated) glycans. Polysaccharides are also a target of glycomics.

Taken together, both the underlying concept and the actual means of carrying out glycomics research have not been widely accepted in the field of proteomics. This is despite the fact that leading technologies developed to address lectin-based glycomics have proven to be effective in various fields of life science, sometimes with great impact. However, this is an inevitable consequence of the development of the field, because without these technologies nobody would be able to broach these essential but difficult issues of biology.

4 Technologies That Came Before the Lectin Microarray

The first four papers in the literature to report on the lectin microarray were published in 2005 [39–42]. However, the conceptual idea and platform for the lectin microarray were developed earlier in the decade by an Israel company, Procognia (http://procogniail.com/). Two related techniques had actually been developed before the lectin microarray; these were serial lectin affinity chromatography and FAC. The former is a semiquantitative procedure to profile glycan structures targeting *N*-linked glycans in the form of either radiolabeled *N*-glycans or glycopeptides as described by K. Yamashita and T. Okura (*see* Chapter 7). The latter (FAC) was initially developed as a quantitative affinity method to determine relatively weak biomolecular interactions between immobilized ligands (e.g., lectins) and labeled analytes as described by K. Kasai (*see* Chapter 21). FAC employs a range of detection methods, i.e., radioisotopes [43], MS [44], and fluorescence detection (FD) [45]. Of these, FAC-FD enables precise and reproducible determination in a systematic manner of interactions between lectins and glycans, expressed in terms of K_d values.

The first study using FAC-FD was made on evolutionarily diverse members of galectins, in which detailed sugar-binding specificity was elucidated for the first time through a systematic analysis using 45 standard pyridylaminated (PA) oligosaccharides [46]. In that study, a consensus recognition rule applied to all galectins thus far investigated was presented [46]. Later, while the FAC-FD system was fully automated through a Japanese Government-funded project focused on structural glycomics (2003–2005), manual systems are also in use in many other laboratories [47–50]. The number of FAC-FD papers published thus far stands at 52 (as of 22 Aug. 2013): these include studies of diverse families of lectins involving anti-carbohydrate antibodies derived from animals, plants, and microorganisms. K_d values reported for 18 representative lectins are summarized in Table 1: i.e., PSA (*Pisum sativum*), LCA (*Lens culinaris*), AOL (*Aspergillus oryzae*), AAL (*Aleuria aurantia*), MAL (*Maackia amurensis*), SSA (*Sambucus sieboldiana*), ECA (*Erythrina cristagalli*), RCA120

Table 1
Dissociation constants (K_ds) determined by frontal affinity chromatography (FAC)

Sugar	Lectin abbreviations[a]																	
ID[b]	PSA	LCA	AOL	AAL	MAL	SSA	ECA	RCA	GSL-II	ACG	BPL	ABA	Jacalin	PNA	WFA	SBA	Cal	GSL-IB
1	–	–	–	–	–	–	–	–	–	–	–	–	221	–	–	–	13	–
2	128	139	–	–	–	–	–	–	–	–	–	–	98	–	–	–	19	–
3	–	–	–	–	–	–	–	–	–	–	–	–	–	–	–	–	250	443
4	135	–	137	–	–	–	–	–	–	–	–	–	170	–	–	23		441
5	141	72	–	–	–	–	–	–	–	–	–	–	–	–	–	–	143	497
6	128	–	–	–	–	–	–	–	–	–	–	–	300	–	–	–	16	–
7	96	57	146	–	–	–	–	–	–	–	–	–	–	–	–	–	125	751
8	127	54	–	–	–	–	–	–	–	–	–	–	–	–	–	–	–	422
9	–	76	–	–	–	–	–	–	–	–	–	–	882	–	–	–	83	–
10	–	73	–	–	–	–	–	–	–	–	–	–	783	–	–	–	111	–
11	59	40	–	–	–	–	–	–	–	–	–	–	–	–	–	–	–	–
12	–	59	–	–	–	–	–	–	–	–	–	–	833	–	–	–	–	301
13	117	58	–	–	–	–	–	–	–	–	–	–	865	–	–	–	–	–
14	8	18	8	11	–	–	–	–	–	–	–	–	157	–	–	–	40	182
15	–	106	–	–	–	–	–	–	–	–	–	543	88	–	–	–	19	514
16	–	–	45	–	–	ND	–	ND	–	–	–	369	540	–	–	–	59	ND
17	–	69	–	–	–	–	–	–	–	–	–	182	379	–	–	–	53	–
18	–	91	–	–	–	ND	–	ND	–	–	–	521	–	–	–	–	9	ND
19	–	–	–	–	–	–	–	–	55	–	–	433	–	–	–	–	–	425

20	–	–	–	–	–	ND	ND	ND	57	–	ND	870	ND	ND	ND	ND	ND	–
21	–	–	–	–	–	–	–	–	49	–	–	–	–	–	–	–	–	–
22	–	–	–	–	–	–	–	–	57	–	–	–	–	–	292	–	–	ND
23	10	20	8	19	–	–	–	–	–	–	–	488	136	–	–	–	32	441
24	23	9	8	19	–	–	–	–	–	–	–	195	773	–	–	–	91	–
25	40	14	8	15	–	–	–	ND	200	–	–	654	618	–	–	–	19	ND
26	–	–	13	24	ND	ND	–	ND	58	ND	–	–	763	–	–	–	–	ND
27	–	–	56	–	–	–	31	9	–	–	709	–	101	–	208	–	21	655
28	–	–	50	–	–	ND	40	5	–	–	412	–	–	–	181	507	ND	657
29	–	–	–	–	–	–	33	7	–	–	–	352	645	ND	268	–	27	–
30	–	95	–	–	–	–	42	4	–	–	–	1538	792	–	–	–	48	–
31	–	107	–	–	–	ND	ND	–	ND	–	ND	571	ND	ND	ND	ND	7	ND
32	–	–	–	–	417	–	17	2	–	28	199	–	638	–	212	527	21	502
33	–	–	0.1	–	–	ND	22	2	–	–	245	–	–	–	175	462	4	ND
34	–	–	–	–	ND	ND	381	ND	–	ND	108	382	788	–	171	–	ND	ND
35	–	–	–	–	166	–	13	2	–	8	95	–	–	–	114	407	–	–
36	–	–	–	–	–	–	19	2	–	–	37	–	–	–	100	363	–	–
37	–	–	–	–	163	–	15	1	–	7	74	–	848	–	111	331	–	–
38	11	24	7	18	–	–	35	9	–	–	–	–	153	–	272	–	32	–
39	21	17	6	9	–	ND	34	5	–	–	–	–	799	–	211	654	48	763
40	26	10	8	19	–	ND	35	8	–	–	–	347	–	–	184	–	91	–
41	22	17	7	15	–	ND	35	4	–	–	676	532	–	–	207	682	43	–

(continued)

Table 1
(continued)

Sugar	Lectin abbreviations[a]																	
ID[b]	PSA	LCA	AOL	AAL	MAL	SSA	ECA	RCA	GSL-II	ACG	BPL	ABA	Jacalin	PNA	WFA	SBA	Cal	GSL-IB
42	28	21	7	18	546	–	20	3	–	26	269	–	551	–	208	506	42	–
43	74	57	9	21	–	–	24	2	–	–	358	–	–	–	146	629	9	–
44	–	–	9	22	ND	ND	39	ND	79	ND	–	3030	752	–	183	559	ND	ND
45	–	–	10	27	196	–	17	2	–	8	147	–	–	–	125	416	–	–
46	–	–	13	31	ND	ND	20	ND	270	ND	307	–	776	–	156	–	ND	ND
47	–	–	17	–	ND	ND	15	ND	54		ND	–	–	680	–	196	622	ND
48	–	–	13	20	ND	ND	12	ND	70	ND	235	–	782	–	138	414	ND	ND
49	–	–	16	–	167	–	14	2	–	6	138	–	–	–	99	364	–	–
50	–	–	–	–	–	–	19	2	–	–	69	–	–	–	161	–	–	–
51	–	–	–	–	–	–	17	2	–	–	49	–	644	–	166	439	–	–
52	–	–	–	–	724	3.4	39	6	–	70	704	–	803	–	–	–	18	–
53	–	–	–	–	–	3.2	37	4		–	58	–	–	497	–	269	–	32
54	–	–	–	–	–	2.1	–	11	–	–	–	–	543	–	–	–	26	–
55	–	–	–	–	–	1.8	–	8	–	–	–	–	–	–	–	783	–	–
56	–	–	–	–	–	2.5	–	15	–	–	–	–	–	–	–	792	–	–
57	32	33	7	16		ND	–	41	9	–	ND	ND	–	ND	–	ND	ND	ND
58	35	41	7	19	12	–	–	–	–	2	ND	–	ND	–	ND	ND	ND	–
59	33	30	–	–	ND	ND	ND	ND	ND	ND	ND	ND	ND	ND	ND	ND	ND	ND

60	–	–	–	–	–	–	498	30	–	–	543	ND	–	509	277	690	–	–
61	–	–	–	–	–	–	411	42	–	–	142	ND	–	–	12	60	–	–
62	–	–	–	–	–	–	–	–	–	–	53	ND	–	50	–	–	–	–
63	–	–	–	–	–	11.1	–	–	–	–	–	ND	–	–	–	–	–	–
64	–	–	–	–	–	–	–	–	–	–	–	ND	–	–	20	265	–	–
65	–	–	–	–	–	–	–	–	–	–	175	ND	–	101	–	–	–	–
66	–	–	–	–	–	–	–	–	–	5	–	ND	–	–	–	–	–	–
67	–	–	–	–	–	–	–	–	–	–	99	ND	–	141	–	–	–	–
68	–	–	–	–	–	–	–	–	–	4	–	ND	–	–	–	–	–	–
69	–	–	–	–	–	–	–	–	–	3	–	ND	–	–	–	–	–	–
70	–	–	–	–	–	–	–	–	–	–	–	ND	–	–	–	52		–
71	–	–	–	–	–	–	319	–	–	–	87	ND	–	–	57	76	–	–
72	–	–	–	–	–	–	364	–	–	–	–	ND	–	–	47	36	–	–
73	–	–	6	15	–	–	375	–	–	–	–	ND	–	–	–	–	–	ND
74	–	–	–	–	–	–	–	–	–	–	113	ND	–	–	–	735	–	–
75	–	–	18	–	–	–	–	–	–	–	–	ND	–	–	–	–	–	97
76	–	–	–	28	–	–	–	–	–	–	–	ND	–	–	–	–	–	118
77	–	–	–	–	–	–	37	6	–	–	–	ND	–	–	–	–	–	–
78	–	–	–	–	–	–	–	69	–	–	283	ND	–	–	–	–	–	150
79	–	–	–	–	–	–	–	–	–	–	107	ND	–	–	–	–	–	–
80	–	–	13	8	–	–	–	–	–	–	548	ND	–	–	–	–	–	–
81	–	–	80	–	–	–	299	–	–	–	182	ND	–	–	285	–	–	–

(continued)

Table 1 (continued)

Sugar	Lectin abbreviations[a]																	
ID[b]	PSA	LCA	AOL	AAL	MAL	SSA	ECA	RCA	GSL-II	ACG	BPL	ABA	Jacalin	PNA	WFA	SBA	Cal	GSL-IB
82	–	–	–	–	–	–	241	–	–	–	135	ND	–	–	239	679	–	–
83	–	–	–	–	–	–	40	2	–	–	64	ND	–	–	141	–	–	–
84	–	–	–	–	–	–	29	2	–	–	158	ND	–	–	294	–	–	–
85	–	–	–	–	–	–	–	21	–	–	–	ND	–	528	–	–	–	682
86	–	–	–	–	–	–	–	25	–	–	–	ND	–	409	–	607	–	–
87	–	–	–	–	–	–	–	31	–	6	–	ND	–	432	–	–	–	–
88	–	–	–	–	–	–	–	28	–	3	–	ND	–	433	–	–	–	892
89	–	–	–	–	–	–	–	62	–	–	–	ND	ND	–	–	–	–	71
90	–	–	–	–	–	–	514	29	–	–	–	ND	–	–	–	–	603	–
91	–	–	–	–	–	–	41	6	–	–	690	ND	–	–	–	819	–	–
92	–	–	–	–	–	–	44	7	–	–	–	ND	–	–	–	–	–	–
93	–	–	–	–	–	–	54	7	–	–	–	ND	–	–	–	–	–	–
94	–	–	–	–	–	–	–	–	30	–	–	ND	–	–	–	–	–	–
95	113	–	–	–	–	–	–	–	36	–	–	ND	–	–	–	–	–	–
96	–	–	–	–	–	–	443	–	–	–	262	ND	–	–	–	–	–	–
97	–	–	23	3	–	–	–	–	–	–	–	ND	–	–	–	–	–	–
98	–	–	18	11	–	–	–	–	–	–	–	ND	–	–	–	–	–	–

ND not determined

[a]For lectin abbreviations, *see* text (also *see* Table 45.1)

[b]For sugar IDs and structures, *see* Fig. 45.1

(*Ricinus communis*), GSL-II (*Griffonia simplicifolia*), ACG (*Agrocybe cylindracea*), BPL (*Bauhinia purpurea*), ABA (*Agaricus bisporus*), Jacalin (*Artocarpas integliforia*), WFA (*Wisteria floribunda*), SBA (soy bean), Calsepa (*Calystegia sepium*), and GSL-IB4 (*G. simplicifolia*).

FAC is a chromatographic procedure that uses a single column or a pair of columns and a series of purified (i.e., standard) glycans. As such, it is difficult to carry out simultaneous analyses allowing multiple (>20) glycan–lectin interactions to be studied. Moreover, FAC deals primarily with authentic standard glycans which have UV-activated fluorescent tags.

5 Crude Image of Lectin Microarray

Despite FAC's shortcomings, it is important to emphasize that the results of FAC analyses with >100 lectins and >100 standard glycans (designated the "hect-by-hect" project; see ref. 23) provided us with a crude image of the "glycan signature" (Fig. 3). Lectin profiling data (here 18 lectins from Table 1) provide us with two points of note: one is that each lectin has a unique glycan-binding specificity as was understood previously, and the other is that each glycan also has a unique lectin-binding pattern (shown by the bolded rectangle in Fig. 3). For the sake of simplicity, binding affinities between lectins and PA-glycans are grouped into six classes as shown in Fig. 3; i.e., $K_a = 0$ (*white*), 0–1 (*pale gray*), 1–10 (*modest gray*), 10–100 (*medium gray*), 100–1,000 (*dark gray*), and >1,000 (*darkest gray*). Despite such a generalized classification, the differentiation between the 80 standard glycans is evident (glycans where data are missing for two or more lectins are excluded). To improve the level of differentiation among glycans, it would obviously be advantageous if either a higher resolution method existed for lectin-binding detection or if more lectins are used. Under such conditions, the performance of the lectin microarray would be able to discriminate even complex mixtures of glycans and glycoconjugates under the concept of higher orders of lectin recognition, which include density, depth, and direction [51].

6 Development of the Lectin Microarray

As described above, earlier technologies (e.g., serial lectin affinity chromatography, frontal affinity chromatography) did not meet with the standards required for lectin-based glycomics. An overriding need for a new technology enabling multiple glycan–lectin interactions to be analyzed in a comprehensive and high-throughput manner was thus generated [52]. As shown in Fig. 4, the general procedure for the lectin microarray is quite simple. A panel of lectins (>20) is immobilized on a glass slide using appropriate surface

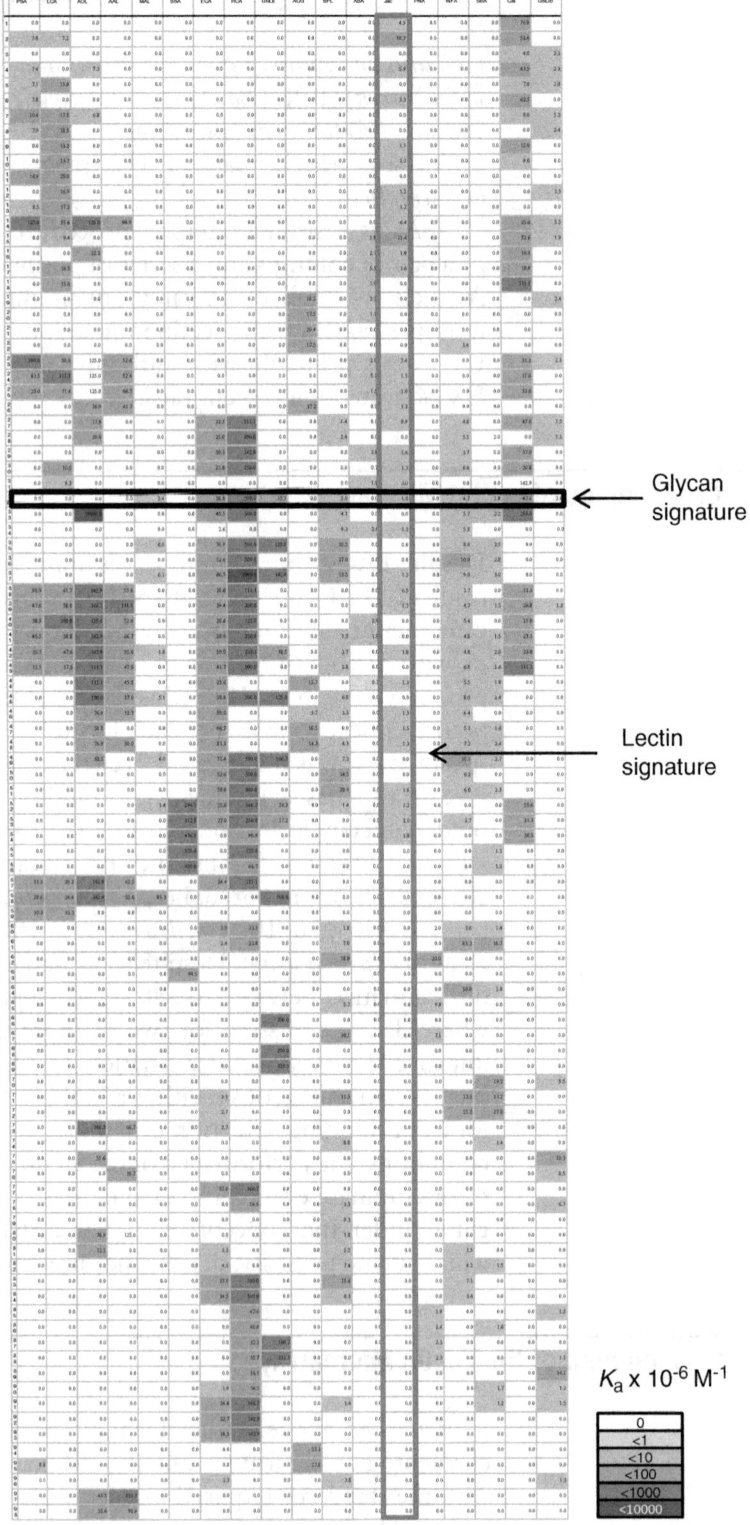

Fig. 3 FAC data provide a faithful representation of glycan profiling by means of lectins. While a conventional approach to lectin specificity analysis results in elucidation of a lectin signature, here a glycan signature can be obtained from such a comprehensive analysis in terms of a set of affinity data to lectins

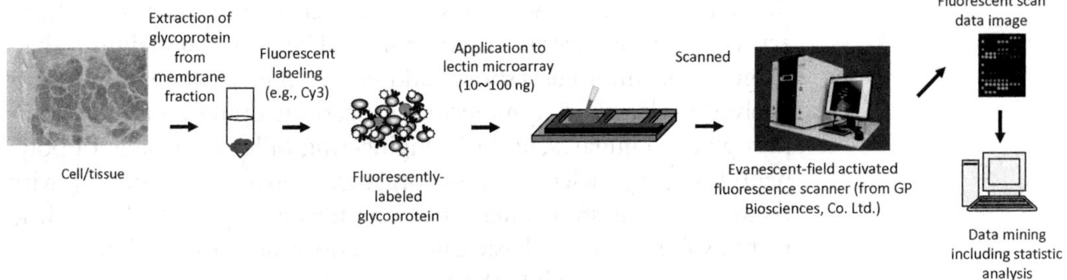

Fig. 4 General scheme of glycan profiling analysis with lectin microarray

chemistry methods (e.g., *N*-hydroxysuccinimide or epoxy activated), to which fluorescently labeled glycoprotein is applied to allow sufficient interactions in solution. Usually, extensive washing is performed to remove any unbound probe, and the resulting fluorescence intensity of each lectin spot is measured by a confocal-type fluorescence scanner. Most lectin microarray systems employ this type of scanner [39–41].

In contrast to the above, the authors' group adopted a unique detection approach, i.e., evanescent-field-activated fluorescence detection. In this method, excitation light is introduced from both sides of the glass slide at the critical angle required to achieve total internal reflection at the interfaces between the glass slide and the liquid phase. An evanescent wave is generated within a limited space ("near-field optic") from the surface. When Cy3-labeled glycoprotein is used as a probe (excitation wavelength of 530 nm), the near-field optic should be within 120 nm [52]. The evanescent field fluorescence detection system thus enables observation of the liquid phase, and consequently, in situ interaction between lectin and glycoprotein under equilibrium conditions. This feature provides substantial advantages over earlier methods in terms of both sensitivity and reproducibility.

7 Remaining Challenges

As pointed out by E.V. Damme (*see* Chapter 1), a conventional definition of lectin seems no longer applicable in today's era of lectin engineering. Lectins are now free from "nonenzyme" and "non-immunoglobulin origin" restrictions, and the presence of many carbohydrate-binding domains/modules is evident in numerous hydrolytic enzymes as pointed by Z. Fujimoto et al. (*see* Chapter 39). Further to this, siglecs, a family of sialic acid-binding lectins from higher vertebrates, are members of the immunoglobulin superfamily. As described by P. Gemeiner et al. (*see* Chapter 37), boronate exhibits intrinsic affinity to carbohydrates, and DNA/RNA aptamers can be engineered as "lectins." As described by D. Hu et al. (*see* Chapter 43) and T. Ogawa and T. Shirai (*see* Chapter 44), desired lectins can now be designed under the molecular evolutionary concept of "lectin engineering." Therefore, remaining challenges for lectinologists should address (1) elucidation of how lectins evolved as deciphering molecules for carbohydrates and more complex glycoconjugates, and (2) engineering of lectins (made of polypeptides, polynucleotides, or organic/inorganic chemicals) with control of their stereochemistry. The former approach should lead to the solution of the basic question concerning the "origin of carbohydrates" [53], while the latter should lead to the generation of new lectin-based medicines.

References

1. Bertozzi CR, Sasisekharan R (2009) Glycomics. In: Varki A, Cummings RD, Esko JD, Freeze HH, Stanley P, Bertozzi CR, Hart GW, Etzler ME (eds) Essentials of glycobiology, 2nd edn. Cold Spring Harbor Laboratory Press, Cold Spring Harbor, NY, Chapter 48
2. Feizi T (2000) Progress in deciphering the information content of the 'glycome' – a crescendo in the closing years of the millennium. Glycoconj J 17:553–565
3. Wasinger VC, Cordwell SJ, Cerpa-Poljak A, Yan JX, Gooley AA, Wilkins MR, Duncan MW, Harris R, Williams KL, Humphery-Smith I (1995) Progress with gene-product mapping of the Mollicutes: *Mycoplasma genitalium*. Electrophoresis 16:1090–1094
4. Hirabayashi J, Kasai K (1999) *C. elegans* glycome project. Glycoconj J 16:S33. In XV international symposium on glycoconjugates: abstracts, 22–27 Aug 1999, Tokyo, Japan. Chapman & Hall, London
5. Hirabayashi J, Arata Y, Kasai K (2001) Glycome project: concept, strategy and preliminary application to *Caenorhabditis elegans*. Proteomics 1:295–303
6. Taniguchi N, Ekuni A, Ko JH, Miyoshi E, Ikeda Y, Ihara Y, Nishikawa A, Honke K, Takahashi M (2001) Proteomics 1:239–247
7. The C. elegans Sequencing Consortium (1998) Genome sequence of the nematode *C. elegans*: a platform for investigating biology. Science 282:2012–2018
8. Hirabayashi J, Kasai K (2000) Glycomics, coming of age! Trends Glycosci Glycotechnol 12:1–5
9. Kasai K, Hirabayashi J (1996) Galectins: a family of animal lectins that decipher glycocodes. J Biochem 119:1–8
10. Kasai K (1997) Galectin: intelligent glue, non-bureaucratic bureaucrat or almighty supporting actor. Trends Glycosci Glycotechnol 9: 167–170
11. Gabius H-J (2000) Biological information transfer beyond the genetic code: the sugar code. Naturwissenschaften 87:108–121
12. Rüdiger H, Siebert HC, Solís D, Jiménez-Barbero J, Romero A, von der Lieth CW, Diaz-Mariño T, Gabius H-J (2000) Medicinal chemistry based on the sugar code: fundamentals of lectinology and experimental strategies with lectins as targets. Curr Med Chem 7: 389–416
13. Lis H, Sharon N (2007) Lectins, 2nd edn. Springer, Dordrecht
14. Natsuka S, Adachi J, Kawaguchi M, Nakakita S, Hase S, Ichikawa A, Ikura K (2002) Structural analysis of *N*-linked glycans in *Caenorhabditis elegans*. J Biochem 131:807–813
15. Cipollo JF, Costello CE, Hirschberg CB (2002) The fine structure of *Caenorhabditis elegans* N-glycans. J Biol Chem 277: 49143–49157
16. Hirabayashi J, Hayama K, Kaji H, Isobe T, Kasai K (2002) Affinity capturing and gene assignment of soluble glycoproteins produced by the nematode *Caenorhabditis elegans*. J Biochem 132:103–114
17. Takeuchi T, Hayama K, Hirabayashi J, Kasai K (2008) *Caenorhabditis elegans N*-glycans containing a Gal-Fuc disaccharide unit linked to the innermost GlcNAc residue are recognized by *C. elegans* galectin LEC-6. Glycobiology 18:882–890
18. Titz A, Butschi A, Henrissat B, Fan YY, Hennet T, Razzazi-Fazeli E, Hengartner MO, Wilson IB, Künzler M, Aebi M (2009) Molecular basis for galactosylation of core fucose residues in invertebrates: identification of *Caenorhabditis elegans N*-glycan core alpha1,6-fucoside beta1,-4-galactosyltransferase GALT-1 as a member of a novel glycosyltransferase family. J Biol Chem 284:36223–36233
19. Hirabayashi J, Arata Y, Kasai K (2000) Reinforcement of frontal affinity chromatography

for effective analysis of lectin–oligosaccharide interactions. J Chromatogr A 890:261–271

20. Arata Y, Hirabayashi J, Kasai K (2001) Application of reinforced frontal affinity chromatography and advanced processing procedure to the study of the binding property of a *Caenorhabditis elegans* galectin. J Chromatogr A 905:337–343
21. Hirabayashi J, Arata Y, Kasai K (2003) Frontal affinity chromatography as a tool for elucidation of sugar recognition properties of lectins. Methods Enzymol 362:353–368
22. Hirabayashi J, Hashidate T, Kasai K (2002) Glyco-catch method: a lectin affinity technique for glycoproteomics. J Biomol Tech 13: 205–218
23. Hirabayashi J (2004) Lectin-based structural glycomics: glycoproteomics and glycan profiling. Glycoconj J 21:35–40
24. Mawuenyega KG, Kaji H, Yamuchi Y, Shinkawa T, Saito H, Taoka M, Takahashi N, Isobe T (2003) Large-scale identification of *Caenorhabditis elegans* proteins by multidimensional liquid chromatography-tandem mass spectrometry. J Proteome Res 2:23–35
25. Kaji H, Saito H, Yamauchi Y, Shinkawa T, Taoka M, Hirabayashi J, Kasai K, Takahashi N, Isobe T (2003) Lectin affinity capture, isotope-coded tagging and mass spectrometry to identify N-linked glycoproteins. Nat Biotechnol 21:667–672
26. Kaji H, Isobe T (2013) Stable isotope labeling of N-glycosylated peptides by enzymatic deglycosylation for mass spectrometry-based glycoproteomics. Methods Mol Biol 951:217–227
27. Blixt O, Head S, Mondala T, Scanlan C, Huflejt ME, Alvarez R, Bryan MC, Fazio F, Calarese D, Stevens J, Razi N, Stevens DJ, Skehel JJ, van Die I, Burton DR, Wilson IA, Cummings R, Bovin N, Wong CH, Paulson JC (2004) Printed covalent glycan array for ligand profiling of diverse glycan binding proteins. Proc Natl Acad Sci U S A 101:17033–17038
28. Paulson JC, Blixt O, Collins BE (2006) Sweet spots in functional glycomics. Nat Chem Biol 2:238–248
29. Blixt O, Razi N (2006) Chemoenzymatic synthesis of glycan libraries. Methods Enzymol 415:137–153
30. Laine RA (1994) A calculation of all possible oligosaccharide isomers both branched and linear yields 1.05 x 10(12) structures for a reducing hexasaccharide: the isomer barrier to development of single-method saccharide sequencing or synthesis systems. Glycobiology 4:759–767
31. Crick F (1970) Central dogma of molecular biology. Nature 227:561–563
32. Narimatsu H, Sawaki H, Kuno A, Kaji H, Ito H, Ikehara Y (2010) A strategy for discovery of cancer glyco-biomarkers in serum using newly developed technologies for glycoproteomics. FEBS J 277:95–105
33. Wells L, Vosseller K, Hart GW (2001) Glycosylation of nucleocytoplasmic proteins: signal transduction and O-GlcNAc. Science 291:2376–2378
34. Zareim M, Müthingm J, Peter-Katalinić J, Bindila L (2010) Separation and identification of GM1b pathway Neu5Ac- and Neu5Gc gangliosides by on-line nanoHPLC-QToF MS and tandem MS: toward glycolipidomics screening of animal cell lines. Glycobiology 20:118–126
35. Taki T (2012) An approach to glycobiology from glycolipidomics: ganglioside molecular scanning in the brains of patients with Alzheimer's disease by TLC-blot/matrix assisted laser desorption/ionization-time of flight MS. Biol Pharm Bull 35:1642–1647
36. Zamfir A, Seidler DG, Schönherr E, Kresse H, Peter-Katalinić J (2004) On-line sheathless capillary electrophoresis/nanoelectrospray ionization-tandem mass spectrometry for the analysis of glycosaminoglycan oligosaccharides. Electrophoresis 25:2010–2016
37. Minamisawa T, Suzuki K, Hirabayashi J (2006) Multistage mass spectrometric sequencing of keratan sulfate-related oligosaccharides. Anal Chem 78:891–900
38. Takegawa Y, Araki K, Fujitani N, Furukawa J, Sugiyama H, Sakai H, Shinohara Y (2011) Simultaneous analysis of heparan sulfate, chondroitin/dermatan sulfates, and hyaluronan disaccharides by glycoblotting-assisted sample preparation followed by single-step zwitter-ionic-hydrophilic interaction chromatography. Anal Chem 83:9443–9449
39. Angeloni S, Ridet JL, Kusy N, Gao H, Crevoisier F, Guinchard S, Kochhar S, Sigrist H, Sprenger N (2005) Glycoprofiling with micro-arrays of glycoconjugates and lectins. Glycobiology 15:31–41
40. Pilobello KT, Krishnamoorthy L, Slawek D, Mahal LK (2005) Development of a lectin microarray for the rapid analysis of protein glycopatterns. Chembiochem 6:985–989
41. Zheng T, Peelen D, Smith LM (2005) Lectin arrays for profiling cell surface carbohydrate expression. J Am Chem Soc 127:9982–9983
42. Kuno A, Uchiyama N, Koseki-Kuno S, Ebe Y, Takashima S, Yamada M, Hirabayashi J (2005) Evanescent-field fluorescence-assisted lectin microarray: a new strategy for glycan profiling. Nat Methods 2: 851–856

43. Ohyama Y, Kasai K, Nomoto H, Inoue Y (1985) Frontal affinity chromatography of ovalbumin glycoasparagines on a concanavalin A-sepharose column. A quantitative study of the binding specificity of the lectin. J Biol Chem 260:6882–6887
44. Ng ES, Chan NW, Lewis DF, Hindsgaul O, Schriemer DC (2007) Frontal affinity chromatography-mass spectrometry. Nat Protoc 2: 1907–1917
45. Tateno H, Nakamura-Tsuruta S, Hirabayashi J (2007) Frontal affinity chromatography: sugar–protein interactions. Nat Protoc 2:2529–2537
46. Hirabayashi J, Hashidate T, Arata Y, Nishi N, Nakamura T, Hirashima M, Urashima T, Oka T, Futai M, Muller WE, Yagi F, Kasai K (2002) Oligosaccharide specificity of galectins: a search by frontal affinity chromatography. Biochim Biophys Acta 1572:232–254
47. Kamiya Y, Kamiya D, Yamamoto K, Nyfeler B, Hauri HP, Kato K (2008) Molecular basis of sugar recognition by the human L-type lectins ERGIC-53, VIPL, and VIP36. J Biol Chem 283:1857–1861
48. Fujii Y, Kawsar SM, Matsumoto R, Yasumitsu H, Ishizaki N, Dogasaki C, Hosono M, Nitta K, Hamako J, Taei M, Ozeki Y (2011) A D-galactose-binding lectin purified from coronate moon turban, Turbo (Lunella) coreensis, with a unique amino acid sequence and the ability to recognize lacto-series glycosphingolipids. Comp Biochem Physiol B Biochem Mol Biol 158:30–37
49. Isomura R, Kitajima K, Sato C (2011) Structural and functional impairments of polysialic acid by a mutated polysialyltransferase found in schizophrenia. J Biol Chem 286: 21535–21545
50. Watanabe M, Nakamura O, Muramoto K, Ogawa T (2012) Allosteric regulation of the carbohydrate-binding ability of a novel conger eel galectin by D-mannoside. J Biol Chem 287:31061–31072
51. Smith DF, Cummings RD (2013) Application of microarrays for deciphering the structure and function of the human glycome. Mol Cell Proteomics 12:902–912
52. Hirabayashi J, Yamada M, Kuno A, Tateno H (2013) Lectin microarrays: concept, principle and applications. Chem Soc Rev 42:4443–4458
53. Hirabayashi J (1996) On the origin of elementary hexoses. Q Rev Biol 71:365–380

Chapter 21

Frontal Affinity Chromatography (FAC): Theory and Basic Aspects

Ken-ichi Kasai

Abstract

Frontal affinity chromatography (FAC) is a versatile analytical tool for determining specific interactions between biomolecules and is particularly useful in the field of glycobiology. This article presents its basic aspects, merits, and theory.

Key words Frontal affinity chromatography, FAC, Biorecognition, Lectin profiling, Glycobiology, Weak interaction, Molecular interaction, Theory

1 Introduction

Since every biological phenomenon is supported and controlled by molecular recognition performed by biomolecules, versatile research tools that provide accurate information on specific interaction between biomolecules are essential for better understanding of life. Although a variety of methodologies have appeared, every currently available approach has its own weaknesses and limitations. Moreover, procedures suitable for analysis of weak interactions are small in number, in spite of having a large choice for those applicable to strong interactions. This is due to inevitable difficulties associated with studies of weak interactions. Prompt improvement of this situation is desirable because the necessity of understanding weak interactions in biological systems is growing rapidly. Dynamic aspects of cellular activities are principally under control of weak interactions, e.g., recognition of a signaling molecule by its receptor, instead of strong ones, because rapid switch-on and -off are essential for cells to respond accordingly to rapidly changing circumstances. Therefore, not only rapid formation of a complex between interacting molecules but also rapid dissociation of the formed complex is critical. From such a viewpoint, strong interactions are not preferable because the dissociation rate

Jun Hirabayashi (ed.), *Lectins: Methods and Protocols*, Methods in Molecular Biology, vol. 1200,
DOI 10.1007/978-1-4939-1292-6_21,

constant of the complex is usually very small and this results in slow switch-off. For rapid switch-off, the complex should dissociate rapidly. This is the reason why most dynamic phenomena in living organisms are considered to be under control of weak interactions.

Researchers wishing to study a certain biological system controlled by weak interactions, however, always encounter a number of difficulties. For example, accurate quantification of the complex is always challenging because the fraction of the complex formed between interacting molecules is extremely low. However, the amount of experimental materials available is usually very limited. Quantification of the complex after isolation from a reaction mixture (e.g., precipitation by using antibody), one of the most popular and reliable techniques routinely employed for studies on strongly interacting systems, does not work, because the separation process itself disturbs the equilibrium state and will result in dissociation of the complex and make its accurate quantification impossible. For investigation of weak interactions, development of procedures that allow quantification of an extremely minute amount of the complex without disturbing the equilibrium state has been desired. Glycobiology is still a typical field where researchers continue to struggle against these difficulties, because phenomena to be clarified are principally supported by weak interactions between various binding proteins and complex glycan molecules, biomaterials of extremely heterogeneous and of low abundance. Acquisition of an amount of purified glycan enough for research from natural sources is usually extremely difficult. Nevertheless, efficient and practical techniques that allow multiplication (e.g., polymerase chain reaction) and/or synthesis of complex glycans (e.g., automated solid-phase synthesizer) have not yet appeared. We needed to develop practical research tools that would overcome such disadvantages, because molecular-based understanding of interactions involving complex glycans is one of the most important frontiers in biological sciences. Frontal affinity chromatography (FAC) that we have developed [1–5] is at present one of the most capable procedures that overcomes the disadvantages mentioned above and provides valuable information on biorecognition of complex glycans.

Affinity chromatography was first invented as an efficient method for purification of biomolecules, but we expected that its utility was not limited only for preparative purposes but had the potential to become a powerful analytical tool for biospecific interaction. This is because affinity chromatography is a combination of an efficient separation principle (chromatography) and the ability of biomolecules to recognize their counterparts (bioaffinity). We intended to make the best use of this advantage and tried to create a new analytical tool for biorecognition. To realize this possibility, it was necessary to determine how to relate chromatographic results with parameters of molecular interaction, e.g., an equilibrium constant. We finally found that the adoption of a frontal chromatographic mode instead of a zonal chromatographic mode is extremely advantageous for the purpose, and we established FAC. Application of FAC successfully provided us with valuable information on various weakly interacting systems including enzymes and sugar-binding proteins (lectins). It is noteworthy to mention that accurate and detailed binding properties of many lectins would not have been obtained without FAC. The ability of FAC to overcome the difficulties mentioned above comes from its unique properties listed below.

1. Theoretical basis is simple and straightforward because FAC deals with only an equilibrium state (more precisely, dynamic equilibrium state) established between an immobilized ligand and a soluble analyte.
2. Quantification of the complex formed between interacting molecules can be made without disruption of the equilibrium state.
3. FAC is principally suited for analysis of weak interactions in contrast to almost all currently available methods which work only for strongly interacting systems, because it is based on measurements of the extent of leakage of analyte molecules from the adsorbent in the column.
4. From an operational viewpoint, neither special equipment nor sophisticated skills are required because the simplest elution mode, isocratic elution, is applied throughout the chromatographic run. This contributes greatly to the robustness of the method.
5. An accurate elution volume can be determined easily by calculating integrated multiple data, thus minimizing the influence of noise in the measurement of signals and resulting in acquisition of reliable equilibrium constants (K_d values).
6. Once a minimum set of basic parameters for a given column is collected, K_d values for other analytes can be determined without knowing their exact concentrations. This has a significant advantage from an operational viewpoint.
7. Even for a weak interaction, it is unnecessary to raise the concentration of the analyte in order to enhance complex formation and signal intensity. Regardless of binding strength, the analyte concentration can be kept at the lowest level that allows drawing of its elution profile.
8. It is economical because only an ordinary high-performance liquid chromatography (HPLC) system present in most laboratories is enough to conduct the experiments.

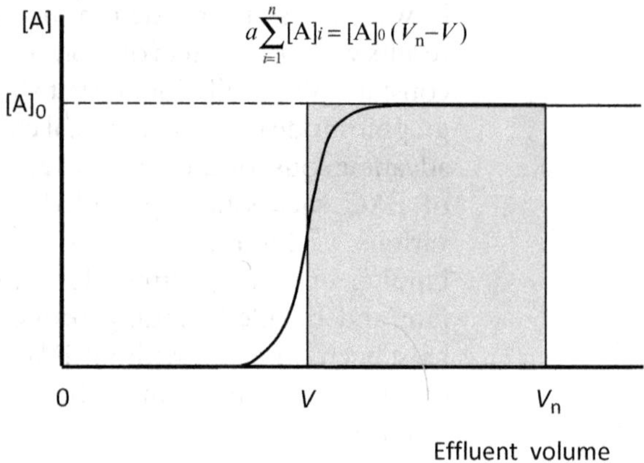

Fig. 1 Determination of elution volume (V) in frontal chromatography

It is necessary to keep in mind that FAC should be carried out under a constant temperature because an equilibrium state is susceptible to temperature change. In the case of trypsin, we found a ten times increase in binding strength when temperature was lowered from 25 to 5 °C. It seems to be advantageous to do experiment at lower temperatures if the extent of retardation of an analyte is found too small.

1.1 Measurement of Elution Volume

In FAC, an analyte solution of a relatively larger volume than that of the column is applied, and the effluent volume at which the analyte begins to leak (elution volume, V) is measured. Accurate estimation of the elution volume is made possible by this chromatographic mode (Fig. 1). V can be considered as the effluent volume at which a hypothetical boundary of the analyte solution would appear if the boundary was undisturbed in any way, though the analyte appears actually from the column as a shape of a forehead (front). If the shape of the front is symmetric around the midpoint, V can be estimated from the effluent volume corresponding to [A]/2, where [A] is the concentration of the analyte, although it is not often the case. However, even if the shape of the elution curve is not ideal, an accurate V value can be deduced as follows. If the concentration of the analyte is measured constantly by collecting an equal volume of fractions (or monitoring at constant intervals), V can be determined by using the following equation (Eq. 1):

$$V = na - a\frac{\sum_{i=1}^{n}[A]_i}{[A]_0} \qquad (1)$$

where a is the volume of one fraction, n is the number of an optionally selected fraction in the plateau region of the elution curve, $[A]_0$ is the initial concentration of A, and $[A]_i$ is the concentration of fraction i. As shown in Fig. 1, the area under the elution curve up to fraction n, $a\sum_{i=1}^{n}[A]_i$, is equal to that of the rectangle, $[A]_0(V_n - V)$. In other words, the calculated V value is the position of the hypothetical elution front if the boundary of the analyte was undisturbed during passage through the column. Although V includes the volume of tubing from the outlet of the column to a fraction collector or a detector, they can be neglected because we only consider the difference from the elution volume of a reference substance having no affinity for the column. In ordinary chromatography, accurate determination of a peak position is not easy especially for a low, broad, and asymmetrical peak. Frontal chromatography, however, allows accurate determination of the elution volume even if the shape of an actual elution front is not ideal due to microscopic nonideal effects. Calculations are straightforward by using one of the commercially available table calculation software.

2 Principle and Theory

2.1 Correlation Between Chromatographic Data and Parameters of Interaction

In FAC, a relatively large volume of dilute analyte solution is continuously applied to a small column packed with an affinity adsorbent, which contains immobilized ligands (Fig. 2). An important point for success in experiments is that a relatively weak affinity adsorbent that does not tightly bind the target analyte but allows its leakage should be used. An affinity adsorbent unusable for preparative purposes due to its weakness is better for the present analytical purpose. Unlike ordinary zonal chromatography, an elution curve composed of a front area and a plateau region is obtained. Retardation of the analyte (A) having affinity for the immobilized ligand (B) occurs due to interaction during passage through the column. The amount of retarded A is equal to the area surrounded by the two elution curves in Fig. 3: the right curve (II) being that of the analyte and the left curve (I) being that of a reference substance having no affinity for the adsorbent. This area corresponds to the amount of A forming complex with the immobilized ligand in the column, and is equal to the rectangle, $[A]_0(V - V_0)$, where $[A]_0$ is the initial concentration of A, V is the elution volume of A, and V_0 is that of the reference substance. The larger the amount of A bound to B, the larger will be the value of $(V - V_0)$. $[A]_0(V - V_0)$ is equal to "specifically" adsorbed A, from which degree of saturation of B can be deduced. Therefore, it is a function of the dissociation constant K_d, the amount of B, and $[A]_0$.

The essential point of the present consideration is how to relate the dissociation constant with the data obtained from chromatographic experiments. Parameters for chromatography are defined as follows: $[B]_0$, amount of immobilized ligand per unit volume (expressed as concentration) of the affinity adsorbent; V, bed volume of the column; B_t, total amount of the immobilized ligand;

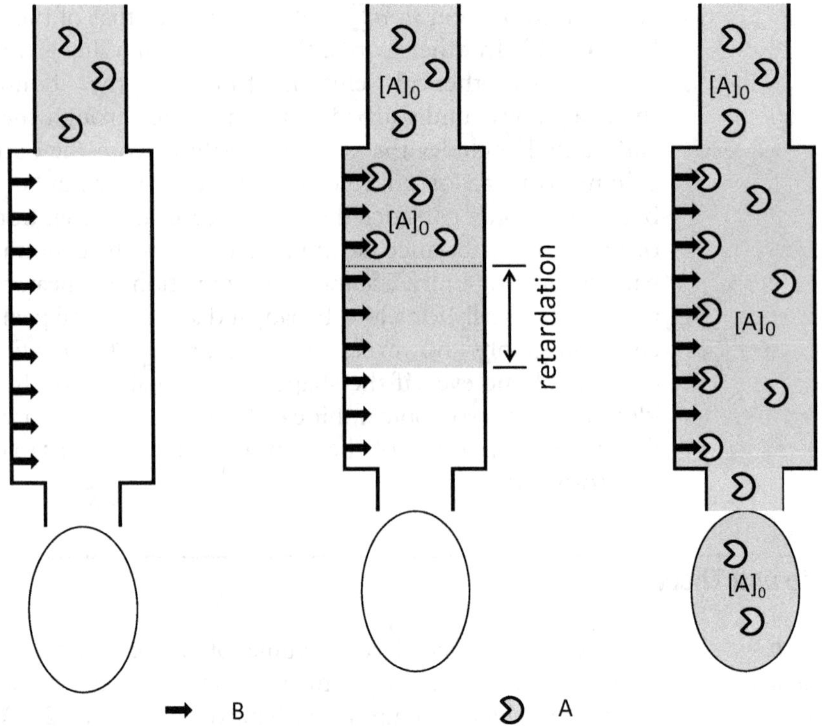

Fig. 2 Schematic presentation of frontal affinity chromatography. A and B represent the analyte and immobilized ligand, respectively

i.e., $V[\mathrm{B}]_0$. Relation between the dissociation constant and chromatographic parameters is as follows (Eq. 2):

$$K_\mathrm{d} = \frac{[\mathrm{A}][\mathrm{B}]}{[\mathrm{AB}]} = \frac{[\mathrm{A}]_0\{[\mathrm{B}]_0 - [\mathrm{A}]_0(V - V_0)/v\}}{[\mathrm{A}]_0(V - V_0)/v} = \frac{B_\mathrm{t}}{V - V_0} - [\mathrm{A}]_0 \quad (2)$$

This equation can be rearranged to the following form (Eq. 3):

$$[\mathrm{A}]_0(V - V_0) = \frac{B_\mathrm{t}[\mathrm{A}]_0}{[\mathrm{A}]_0 + K_\mathrm{d}} \quad (3)$$

This equation is equivalent to the Michaelis-Menten equation of enzyme kinetics and in principle to the Langmuir's adsorption isotherm. Equation 3 gives a hyperbolic curve shown in Fig. 4, indicating that the column becomes saturated at the infinite concentration of A. B_t and K_d correspond to the maximum velocity and the Michaelis constant, respectively, and can be determined by using one of linear plots derived from Eq. 3:

$$\frac{1}{[\mathrm{A}]_0(V - V_0)} = \frac{K_\mathrm{d}}{B_\mathrm{t}}\frac{1}{[\mathrm{A}]_0} + \frac{1}{B_\mathrm{t}} \quad (4)$$

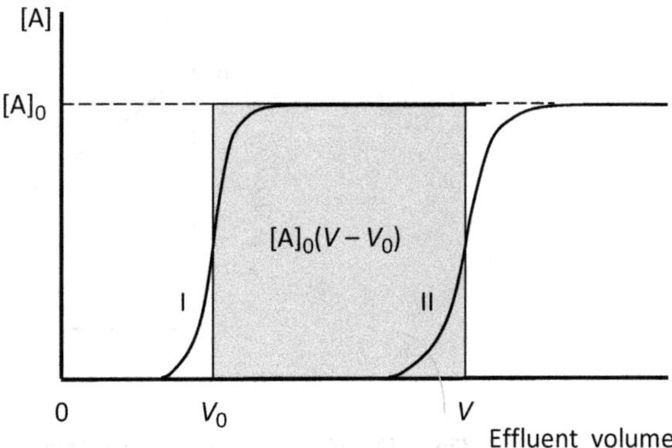

Fig. 3 Elution profiles in frontal affinity chromatography. Curve II is the elution pattern of the analyte, curve I is that of a reference substance which has no affinity for the adsorbent

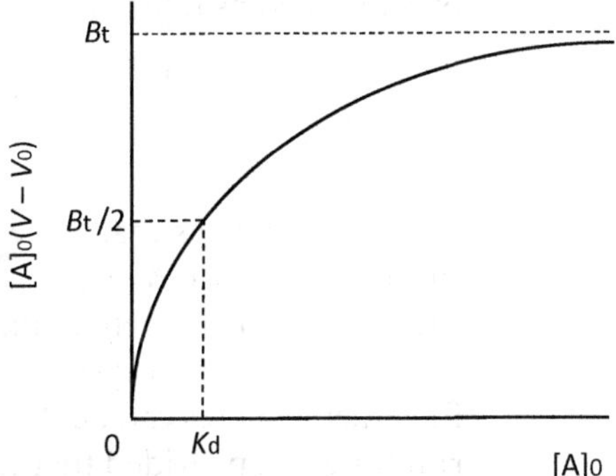

Fig. 4 $[A]_0(V - V_0)$ vs. $[A]_0$ plot. This plot is analogous to the Michaelis-Menten plot. Values of B_t and K_d can be calculated from the coordinates of the two asymptotes of the hyperbola

This equation (Eq. 4) corresponds to the Lineweaver-Burk plot of enzyme kinetics. However, the following Woolf-Hofstee-type equation is more practical and straightforward in the case of FAC:

$$[A]_0(V - V_0) = B_t - K_d(V - V_0) \quad (5)$$

If we apply the analyte changing its concentration, measure V values and make a $[A]_0(V - V_0)$ vs. $(V - V_0)$ plot, the slope and the intercept on the ordinate give $-K_d$ and B_t, respectively (Fig. 5). B_t indicates the amount of immobilized ligand molecules actually retaining binding ability.

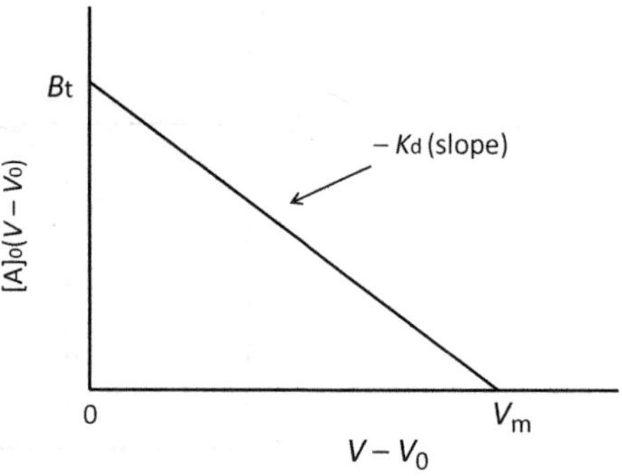

Fig. 5 $[A]_0(V - V_0)$ vs. $(V - V_0)$ plot. This plot is analogous to the Woolf-Hofstee plot. The slope gives $-K_d$, and the intercept on the ordinate is B_t

If we apply an extremely dilute analyte solution, in other words, $[A]_0$ is set negligibly low in comparison with K_d, Eq. 3 can be simplified as follows:

$$(V - V_0) = \frac{B_t}{K_d} \tag{6}$$

This indicates that the extent of retardation is proportional to the reciprocal of K_d (i.e., association constant). Therefore, once the B_t value of a given affinity column is obtained by a concentration dependence analysis by using an appropriate analyte, K_d values for other analytes can be determined by only one chromatographic run for each, provided that their concentrations are adequately low (e.g., lower than 1 % of K_d). This feature is analogous to enzyme kinetics, i.e.; at low concentrations of the substrate, the reaction rate becomes proportional to the substrate concentration. This provides a great advantage from an experimental viewpoint. It is not necessary to know the correct concentration of the analyte, as in the case of enzyme kinetics in which we need not know the enzyme concentration. Moreover, even for a weakly interacting analyte (having a large K_d), it is unnecessary to raise the concentration of the analyte. This is a great help for researchers because extremely valuable experimental materials will not be wasted.

2.2 Dependence of Elution Volume on $[A]_0$

Now, let us consider the aspects concerning chromatography. Equation 7 can be derived from Eq. 5:

$$V = V_0 + \frac{B_t}{[A]_0 + K_d} \tag{7}$$

This equation is useful to grasp the fundamental feature of frontal affinity chromatography. A plot of V vs. $[A]_0$ is also a hyperbola and the two asymptotes correspond to $-K_d$ and V_0, respectively (Fig. 6). As concentration of the analyte is decreased,

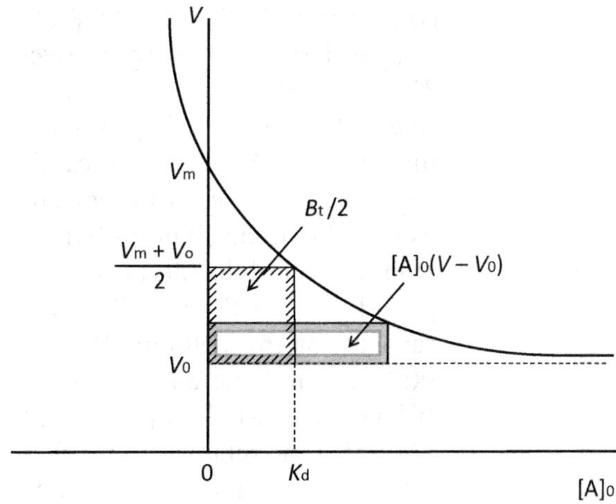

Fig. 6 V vs. $[A]_0$ plot. Dependency of elution volume on concentration of A. Both elution volume and adsorbed amount can be predicted for a given $[A]_0$, provided that the values of K_d and B_t are known

extent of retardation increases. If $[A]_0$ is negligibly small compared to K_d, V approaches the maximum value, V_m (Eq.8):

$$V_m = V_0 + \frac{B_t}{K_d} \tag{8}$$

V_m becomes apparently independent from $[A]_0$. When $[A]_0$ increases, V becomes smaller. However, V cannot be smaller than V_0. Therefore, V_0 is considered as the limit of V when $[A]_0$ approaches infinity, i.e., the immobilized ligand becomes saturated. At a certain concentration of $[A]_0$, the amount of adsorbed A corresponds to the area of the dotted rectangle in Fig. 6. It is easily seen that when $[A]_0 = K_d$, adsorbed A becomes equal to $B_t/2$, i.e., half of the maximum capacity, and V becomes $(V_m + V_0)/2$. Therefore, the elution volume varies from V_m to V_0 depending on $[A]_0$, and the amount of adsorbed A varies from $[A]_0(V_m - V_0)$ to B_t.

The reason why V becomes equal to V_m (which is independent of $[A]_0$) is explained as follows. In a region where $[A]_0$ is very small, $[A]_0(V_m - V_0)$ becomes approximately proportional to $[A]_0$. Therefore, when $[A]_0$ doubles, the amount of adsorbed A also doubles and does not result in any change in V. This is analogous to enzyme kinetics in which velocity is approximately proportional to the substrate concentration if the latter is negligibly small in comparison to the Michaelis constant, K_m.

2.3 Consideration for Designing Affinity Adsorbent

Equation 8 is helpful for designing affinity adsorbents. This equation can predict the results when we intend to concentrate a certain substance from a dilute solution using an affinity adsorbent. Let us assume that we wish to prepare an affinity adsorbent in order to concentrate a lectin (e.g., MW. 50,000) and succeed in immobilizing a sugar derivative having relatively high affinity for the lectin (e.g., $K_d = 10^{-7}$ M) at $[B]_0 = 10^{-4}$ M. If we apply a dilute lectin solution (e.g., $[A]_0 = 10^{-7}$ M, 5 µg/mL) continuously to a column (bed volume V = 10 mL), V will be ca. 5,000 mL. This means that we can apply as much as ca. 4,000 mL to the column before leakage of the lectin will occur (binding strength is assumed to be unchanged after an immobilization process). The amount of adsorbed lectin, $[A]_0(V - V_0)$, will be ca. 25 mg, which corresponds to half of the capacity of the column. If $[A]_0 = 10^{-6}$ M (50 µg/mL), V is ca. 900 mL, and the amount of adsorbed lectin will be ca. 45 mg. If $[A]_0 = 10^{-5}$ M (500 µg/mL), the lectin will begin to leak at ca. 100 mL. The retained lectin then is ca. 50 mg, which corresponds to the maximum capacity of the adsorbent.

Let us consider also the case of a weak affinity adsorbent. Suppose we have a column of the same size containing another sugar derivative of $K_d = 10^{-4}$ M at $[B]_0 = 10^{-4}$ M. Even if $[A]_0 = 10^{-6}$ M, the lectin will begin to leak at ca. <20 mL. Only 0.5 mg of the lectin will be retained.

This method is not directly applicable to ordinary zonal chromatography because [A] is subject to change during passage through the column. However, in limited cases, i.e., when [A] is negligible in comparison to K_d, the system can be treated similarly. In this case, the value $[B]_0/K_d$ can be considered as ratio of elution volume to bed volume. Therefore, for example, if we prepare an affinity adsorbent for which $[B]_0$ is ten times K_d, A will appear at a volume of ten times the bed volume.

2.4 Simplified System in Which $[A]_0$ Can Be Neglected

Equation 8 can be rearranged as follows:

$$K_d = \frac{B_t}{V_m - V_0} \tag{9}$$

This means that we can determine K_d without considering the term $[A]_0$. In other words, [B] can be considered as $[B]_0$ because [AB] is negligibly small compared to $[B]_0$. This relation is very useful because we can determine K_d even if the exact concentration of A is unknown (e.g., unpurified protein, material of unknown molecular weight, etc.), provided that the elution profile can be drawn by an appropriate procedure (e.g., measurement of enzyme activity, immunochemical quantification, isotope label, and mass spectrometry).

This relation can also be used to compare K_d values of a system under various conditions (e.g., pH dependence of binding). K_d is always inversely proportional to the extent of retardation.

2.5 Presence of Two Analytes

If a mixture of two analytes having different affinities for the adsorbent is applied to the column, two-step elution will be observed. If the concentrations of both analytes are adequately low, it can be considered that chromatography is carried out by using two columns simultaneously, and each K_d can be obtained from each elution volume.

2.6 Effect of Counter-Ligand (I)

Now, let us consider a more complicated situation where a soluble counterpart molecule (I) which binds to A, like a competitive inhibitor for an enzyme, is present. The effect of I on elution of A provides information on the interaction between A and I. To simplify, we consider only limited cases where $[A]_0$ is extremely low, i.e., $[A]_0 \ll K_d$. In the presence of I, binding of A to B is inhibited and the elution volume of A decreases to V_I. After the column is equilibrated with a solution of I (to $[I]_0$), A dissolved in the same solution is applied. To simplify, $[I]_0$ is also assumed to be adequately large in comparison to $[A]_0$. Under these conditions, the amount of adsorbed A decreases from $[A]_0(V_m - V_0)$ to $[A]_0(V_I - V_0)$, and therefore (Eq. 10):

$$\frac{V_I - V_0}{V_m - V_0} = \frac{1}{1 + [1]/K_I} = \frac{1}{1 + [I]_0/K_I} \tag{10}$$

where K_I is dissociation constant of AI complex, and can be calculated by using the following equation (Eq. 11):

$$K_I = \frac{V_I - V_0}{V_m - V_I}[I]_0 \tag{11}$$

If $[I]_0 = K_I$, V_I will be the middle of V_0 and V_m. For more accurate determination of K_I, V_I values in the presence of various concentrations of I are measured and analyzed by means of the following equation (Eq. 12):

$$V_I = V_0 + K_I \frac{V_m - V_I}{[I]_0} \tag{12}$$

A plot of V_I vs. $(V_m - V_I)/[I]_0$ gives a straight line (Fig. 7). The intercept on the ordinate corresponds to V_0 and the slope to K_I. It is apparent that V_0 is reached when $[I]_0$ approaches infinity. This means that A becomes saturated with I. This procedure is analogous to the analysis of competitive inhibition in enzyme kinetics. The interaction between A and I is measured indirectly in terms of the decrease in the elution volume, like the decrease in velocity in the case of enzyme kinetics. One of the advantages of this indirect method is that a variety of counter ligands having wide range of K_I values can be determined by using a single column.

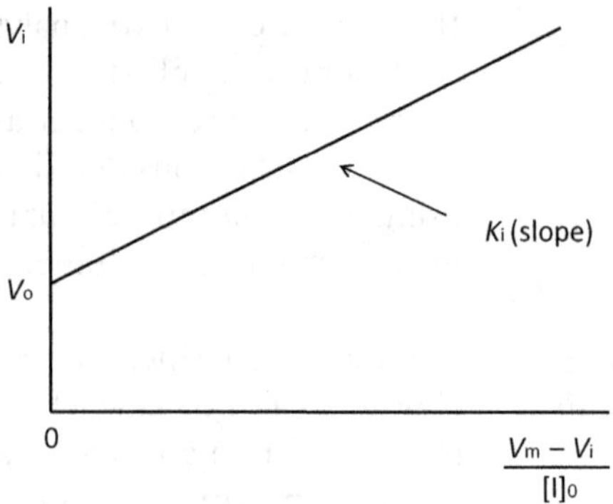

Fig. 7 V_i vs. $(V_m - V_i)/[I]_0$ plot. Effect of a counter ligand (I) which binds to A. The dissociation constant of I can be calculated from the slope. This procedure resembles the analysis of competitive inhibition in enzyme kinetics

Even if the interaction between A and I is weak, it is not necessary to raise the concentration of A. Moreover, this indirect method is not susceptible to nonspecific interaction, which often interferes with direct methods.

2.7 Extension to Analytes Having Multiple Binding Sites

We have considered analyte molecules having only one binding site. Now, we extend the application to a multivalent system, for example, a dimer of homologous binding protein. The binding site of each monomer is assumed to be independent and unaffected by the situation of the other. It is also assumed that the dimer protein binds to an immobilized ligand with only one site at a time. Under these conditions, the concentration of the analyte can be considered doubled, and the following equation (Eq. 13) is obtained (detail of derivation is omitted):

$$V = V_0 + \frac{B_t}{[A]_0 + K_d / 2} \tag{13}$$

This equation means that we shall have an apparent K_d of half the intrinsic K_d. In other words, the apparent K_d is half of the true K_d. Under the conditions where $[A]_0 \ll K_d$, the following equation (Eq. 14) analogous to Eq. 8 can be obtained:

$$V_m = V_0 + 2\frac{B_t}{K_d} \tag{14}$$

Therefore, the extent of retardation, $(V_m - V_0)$, is doubled. The intrinsic K_d value can be obtained as follows (Eq. 15):

$$K_d = \frac{2B_t}{V_m - V_0} \tag{15}$$

For a multivalent protein that has n identical and independent binding sites, we can similarly derive the following equation:

$$K_d = \frac{nB_t}{V_m - V_0} \qquad (16)$$

Therefore, the extent of retardation will be n times that of monovalent protein, and the intrinsic K_d is n times the apparent K_d.

3 Conclusion

At its primitive stage, when HPLC was not yet available, FAC was rather time-consuming and required a relatively large amount of analyte. In spite of these demerits, it proved to be useful for providing us with valuable information on biorecognition, e.g., contribution of subsite interaction on binding properties of proteolytic enzymes [3], binding properties of inactivated enzymes [6], first quantitative evaluation of recognition of various oligosaccharides by concanavalin A [7, 8], and serum albumin–drug interaction [9]. However, significant improvements in both instruments and materials have now made FAC a highly sensitive, accurate, rapid, stable, and convenient procedure for research on weak interactions [10–17]. In its application to the field of glycobiology, the use of a small column of immobilized lectin and a variety of fluorescent oligosaccharides (e.g., pyridylaminated-sugars) have made FAC an extremely effective and high-throughput analytical tool for profiling lectins in terms of their binding properties (both binding specificity and binding strength) as described in other chapters in this book. Although FAC gives only equilibrium constants, this can be adequately compensated for, because of the extremely high sensitivity in determining K_d values, which none of the other modern powerful procedures have yet to achieve.

References

1. Kasai K, Ishii S (1975) Quantitative analysis of affinity chromatography of trypsin: a new technique for investigation of protein-ligand interaction. J Biochem 77:262–264
2. Kasai K, Oda Y, Nishikata M et al (1986) Frontal affinity chromatography: theory for its application to studies on specific interaction of biomolecules. J Chromatogr 376:33–47
3. Nishikata M, Kasai K, Ishii S (1977) Affinity chromatography of trypsin and related enzymes. I: quantitative comparison of affinity adsorbents containing various arginine peptides. J Biochem 82:1475–1484
4. Kasai K, Ishii S (1978) Affinity chromatography of trypsin and related enzymes. V. Basic studies of quantitative affinity chromatography. J Biochem 84:1051–1060
5. Kasai K, Ishii S (1978) Studies on the interaction of immobilized trypsin and specific ligands by using quantitative affinity chromatography. J Biochem 84:1061–1069
6. Yokosawa H, Ishii S (1977) Anhydrotrypsin: new features in ligand interactions revealed by affinity chromatography and thionine replacement. J Biochem 81:647–656
7. Oda Y, Kasai K, Ishii S (1981) Studies on the specific interaction of concanavalin A and saccharides by affinity chromatography: application of quantitative affinity chromatography to a multivalent system. J Biochem 89:285–296

8. Ohyama Y, Kasai K, Nomoto H et al (1985) Frontal affinity chromatography of ovalbumin glycoasparagines on a concanavalin A-Sepharose column: a quantitative study of the binding specificity of the lectin. J Biol Chem 260: 6882–6887
9. Nakano NI, Oshio T, Fujimoto Y et al (1978) Study of drug-protein binding by affinity chromatography: interaction of bovine serum albumin and salicylic acid. J Pharm Sci 67: 1005–1008
10. Zhang B, Palcic MM, Schriemer DC et al (2001) Frontal affinity chromatography coupled to mass spectrometry for screening mixtures of enzyme inhibitors. Anal Biochem 299:173–182
11. Hirabayashi J, Arata Y, Kasai K (2000) Reinforcement of frontal affinity chromatography for effective analysis of lectin-oligosaccharide interactions. J Chromatogr 890:262–272
12. Arata Y, Hirabayashi J, Kasai K (2001) Sugar-binding properties of the two lectin domains of the tandem repeat-type galectin LEC-1 (N32) of *Caenorhabditis elegans*: detailed analysis by an improved frontal affinity chromatography method. J Biol Chem 276: 3068–3077
13. Hirabayashi J, Hashidate T, Arata Y et al (2002) Oligosaccharide specificity of galectins: a search by frontal affinity chromatography. Biochim Biophys Acta 1572:232–254
14. Kamiya Y, Yamaguchi Y, Takahashi N et al (2005) Sugar-binding properties of VIP36, an intracellular animal lectin operating as a cargo receptor. J Biol Chem 280:37178–37182
15. Matsubara H, Nakamura-Tsuruta S, Hirabayashi J et al (2007) Diverse sugar-binding specificities of marine invertebrate C-type lectins. Biosci Biotechnol Biochem 71:513–519
16. Kawsar SM, Fujii Y, Matsumoto R et al (2008) Isolation, purification, characterization and glycan-binding profile of a D-galactoside specific lectin from the marine sponge, *Halichondria okadai*. Comp Biochem Physiol B Biochem Mol Biol 150:349–357
17. Sato C, Yamakawa N, Kitajima K (2010) Measurement of glycan-based interactions by frontal affinity chromatography and surface plasmon resonance. Methods Enzymol 478: 219–232

Chapter 22

Frontal Affinity Chromatography: Practice of Weak Interaction Analysis Between Lectins and Fluorescently Labeled Oligosaccharides

Chihiro Sato

Abstract

Frontal affinity chromatography (FAC) is a simple and effective method that is applicable to the analysis of interactions between glycans and glycan-recognition proteins, including lectins, with weak affinity ranging from 10^{-4} to 10^{-6} (M) in terms of dissociation constant (K_d). Using conventional instruments, such as a high-performance liquid chromatography (HPLC) system equipped with pump, injector, (fluorescent) detector, and data recorder, the dissociation constants for weak glycan-based interactions can be easily determined with high throughput and accuracy. Notably, if the glycans are labeled with fluorescent dyes, only a small amount of glycans is required for the analysis. Fluorescent labeling of glycans is a common technique, and an increasing number of fluorescent-labeled glycans are commercially available. In this chapter, an advanced FAC method using fluorescent-labeled glycans is described.

Key words Frontal affinity chromatography, Fluorescent-labeling, Lectins, Glycan, Dissociation constant, High-performance liquid chromatography (HPLC)

1 Introduction

Cell surfaces are covered by dense layers of glycoconjugates that are frequently involved in extracellular communications through their glycan moieties. Therefore, analyses of glycan-based interactions, including those involving glycan-recognizing proteins, particularly lectins, and other glycans, are expected to provide a greater understanding of cellular biological activities [1]. Lectins purified from plants and invertebrates have been often used for the detection of specific glycan epitopes. Recently, several vertebrate lectins were shown to be important functional proteins for essential cellular activities, including the immune response and developmental processes [2, 3]. To gain an insight into the regulation of these activities, it is important to characterize the types and strength of interactions between lectins and their target molecules in solution.

Jun Hirabayashi (ed.), *Lectins: Methods and Protocols*, Methods in Molecular Biology, vol. 1200,
DOI 10.1007/978-1-4939-1292-6_22, © Springer Science+Business Media New York 2014

Molecular interactions are typically investigated using methods based on surface plasmon resonance (SPR) and isothermal titration calorimetry (ITC). However, these methods are often not suitable for the analysis of glycan-based interactions, which are relatively weak and rapid compared with those between proteins. In addition, because large quantities of specific glycan structures can be difficult to obtain chemically due to the selectivity of hydroxyl groups and existence of anomers, it is often necessary to analyze target interactions with only a small amount of sample. For these reasons, sensitive approaches for detecting glycan-based interactions, such as those involving clustering ligands and lectins, are needed [4]. An ideal method would also be more assessable than SPR- and ITC-based instruments, which are not found in most laboratories due to their high cost.

Frontal affinity chromatography (FAC) is a simple, effective, and efficient method to measure weak molecular interactions, particularly those involving carbohydrates [5–7]. The principles of this approach are described in Chapter 21 and in several recent reviews [5, 8, 9]. The merits of this method are threefold. *First*, weak interactions can be analyzed using conventional and relatively inexpensive machines. *Second*, this approach requires only small amounts of glycan samples if the glycans are labeled with fluorescent dyes. *Third*, the system components can be arranged for the specific molecular interactions of interest. Recently an automated FAC machine was established and a high-throughput analysis has been achieved with this machine [9].

Here, a method for establishing a FAC system based on conventional HPLC that is capable of analyzing glycan-based interactions is described.

2 Materials

All solutions should be prepared using ultrapure water (18 Ωm^{-1}) and analytical grade reagents.

2.1 Equipment

1. HPLC system: Any HPLC system equipped with an isocratic pump, injector, appropriate detector, and recording instrument can be used (*see* **Note 1**). A PC-based instrument capable of processing data using software such as Microsoft Excel is preferable. Our system consists of an isocratic pump (PU-980i, Jasco, Tokyo Japan), injector with a 2-mL sample loop (PEEK), affinity resin column (4.0 mm × 10 mm, 126 μL, GL Science), and fluorescent detector (FP2025, Jasco) connected to a Chromato-PRO integrator (Run Time Corp., Kanagawa, Japan), as schematically illustrated in Fig. 1. For high-throughput analysis, an automated FAC-FD (fluorescence detection) system developed by Hirabayashi et al. [9] can be used for analyzing up to 100 interactions within 10 h.

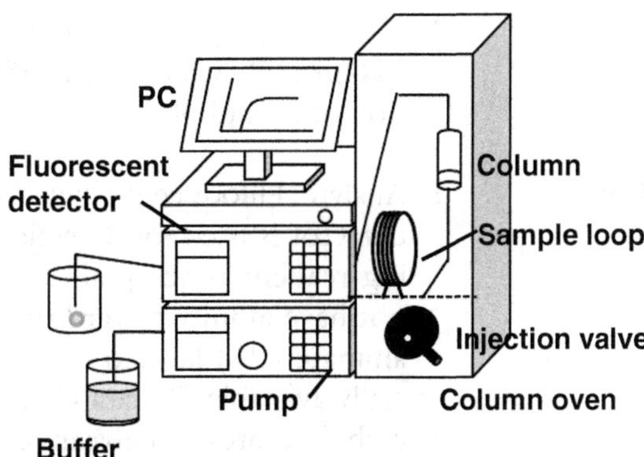

Fig. 1 Schematic illustration of a FAC system. An HPLC pump is connected to an injector with a 2 mL sample loop and is equipped with an affinity resin column. The elution of analyte is detected by a fluorometric detector and the data is recorded using PC-based software

2. Column oven or incubator: Depending on the conditions, a column oven, such as CTO-6A (Shimadzu Corp., Kyoto, Japan), incubator, or water bath is needed for maintaining appropriate column and sample loop temperatures, typically 25 or 37 °C (*see* **Note 2**).
3. Solvent: Depending on the conditions, an appropriate solvent is necessary. Degassed phosphate-buffered saline (PBS) (10 mM, pH 7.2; 137 mM NaCl, and 2.7 mM KCl) may be suitable for equilibrating the affinity resin and dissolving analytes. Solvents must be free of fluorescent contaminants if fluorescent detection system is used.

2.2 Preparation of Affinity Resin

1. Lectins: Prepare the target glycan-binding proteins to be analyzed (*see* **Note 3**). Many plant lectins and several invertebrate lectins are commercially available as described by Y. Kobayashi et al. (*see* Chapter 45). Recombinant lectins can be easily obtained if their nucleotide sequences are known.
2. Immobilization of lectins onto resin/gels: Native proteins can be immobilized onto affinity resin as well as commercially available *N*-hydroxy succinimide (NHS)-activated-Sepharose (GE Healthcare Life Science) or NHS-activated agarose (Thermo Scientific) (*see* **Note 4**). Tags such as $(His)_6$, Fc, protein A, GST, and biotin can be used to immobilize lectins onto the appropriate resin such as Ni-Sepharose, Protein G/A-Sepharose, Ig-Sepharose, Glutathione-Sepharose, and Streptavidin-Sepharose (GE Healthcare Life Science), respectively, without destroying glycan-binding sites. In addition, an affinity adsorbent with adequate affinity for the target molecule should be prepared (volume ~0.5 mL).

3. Column preparation: Fill the column (4.0 mm × 10 mm, 126 μL, GL Science) with affinity resin and avoid the introduction of air bubbles.

2.3 Preparation of Analyte

1. Analyte: Fluorescent-labeled analytes are required if a fluorescent detector is used for detection. Although any fluorescent labeling reagent is acceptable, the selected fluorescent label must not have affinity toward the affinity resin. Libraries of pyridylaminate (PA)-labeled carbohydrates and glycans are commercially available (Takara Co., Shiga, Japan). In addition, specific carbohydrates and glycans can be easily labeled with a PA labeling kit (Takara Co.). Rhodamine green (RG)-labeled glycan chains and 1,2-diamino-4,5-methylenedioxybenzene (DMB)-labeled sialic acid can also be used as analytes. To evaluate correct V_0, the prepared carbohydrate, glycan, or other target molecules must not bind to the affinity resin (*see* **Note 5**).
2. Concentration: Carbohydrates and glycans should be dissolved in PBS or an appropriate buffer at concentrations ranging from 0 to 30 nM.

3 Methods

3.1 Operation of the FAC System

1. Start the HPLC system at 25 °C and set the excitation (Ex) and emission (Em) wavelengths of the fluorescence detector for the target label (e.g., Ex. 320 nm and Em. 400 nm for PA-glycans, Ex. 373 nm and Em. 488 nm for DMB-glycans, and Ex. 503 nm and Em. 530 nm for RG-glycans).
2. Equilibrate the column with PBS at a flow rate of 0.125 mL/min until a flat baseline is achieved.
3. Turn the injector to the "load position" and inject 20 mL air using a syringe to completely empty the sample loop.
4. Inject approximately 2 mL analyte solution dissolved in PBS to completely fill the 2 mL sample loop (50–100 μL of excess analyte solution is required to completely fill the 2 mL sample loop).
5. Turn the injector to the "inject position" and flow the analyte into the column at a rate of 0.125 mL/min.
6. The eluted analyte is monitored for 8 min using the fluorescent detector (1 data/2 s) to generate an elution curve (Fig. 2).
7. The next sample can be analyzed from **step 3** when the baseline fluorescence returns to baseline.

3.2 Data Analysis

1. Import the recorded data into the selected software program.
2. Calculate the elution volume of the analyte (V) (Fig. 2). V represents the volume of analyte at which half of the plateau concentration is attained (*see* Chapter 21).

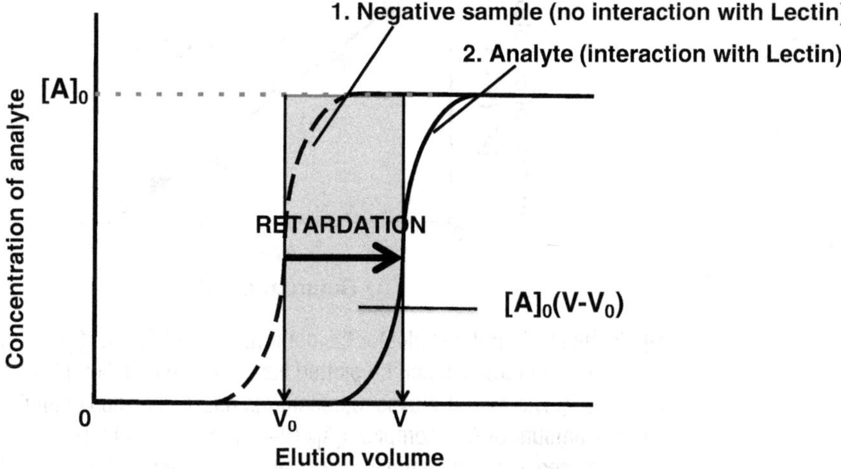

Fig. 2 Typical FAC profiles. The immobilized lectin (**b**) is equipped with the column and a volume in excess of the total column volume of analyte (**a**) is then eluted. Line 1 is the curve for the negative sample that does not interact with **b**. The elution volume (V_0) (*see* Chapter 21) represents the void volume of the column. Line 2 is the curve for the analyte that interacts with **b**. Retardation of analyte ($V - V_0$) indicates a specific interaction between A (analyte) and B (lectin). The shaded area, $[A]_0(V - V_0)$, represents the amount of the complex [AB]

3. Calculate the elution volume of the non-binding molecule (negative sample) (V_0) (Fig. 2). V_0 represents the volume of non-binding molecule at which half of the plateau concentration is attained (see Chapter 21).
4. Plot and describe the graph (X-axis: retardation volume of analyte A ($V - V_0$), Y-axis: amount of analyte A and lectin B complex $\{[A]_0(V - V_0)\}$) based on the principles of FAC (*see* Chapter 21).
5. Calculate the B_t (*see* **Note 6**) and K_d with data obtained using different concentrations of the same analyte. The intercept of the Y-axis represents the B_t value and the slope represents K_d based on the equation $\{[A]_0(V - V_0) = B_t - K_d(V - V_0)\}$ (*see* Chapter 21) (Fig. 3). After calculating the B_t value of the column, the K_d of a different analyte can be determined using only a single concentration of the analyte based on the equation $K_d = B_t(V - V_0)$, if the concentration of the analyte (nM) is markedly less than the K_d (μM ~ mM).

4 Notes

1. FAC is a flexible method to analyze weak molecular interactions. HPLC systems can be used for FAC because the flow rate of the column can be easily adjusted. An open column may also be used to determine the dissociation constant [6, 10]. Any type of detector, such as fluorescent and UV detectors [10],

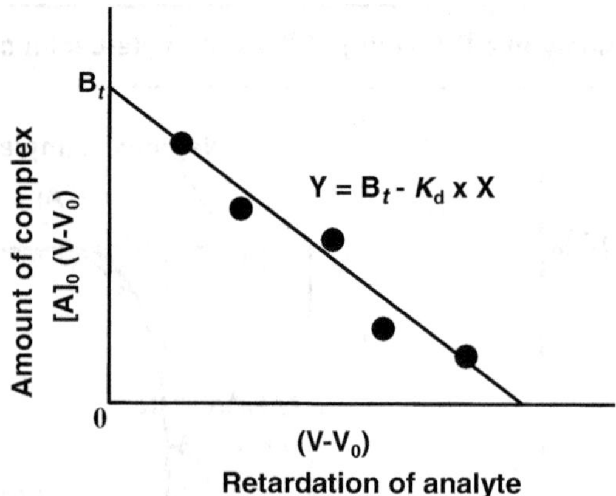

Fig. 3 The Woolf–Hofstee plot for the determination of B_t and K_d. Data obtained for different concentrations can be plotted based on the equation $[A]_0 (V - V_0) = -K_d (V - V_0) + B_t$. The *X*- and *Y*-axes represent the retarded elution of analyte A $(V - V_0)$ and the amount of A-B complex ($[A]_0 (V - V_0)$), respectively. The intercept of the *Y*-axis is used to calculate the B_t (effective lectin content). The K_d value is obtained from the slope (=$-K_d$).

and mass spectrometer [11] is suitable for FAC if analytes can be detected. A pulsed amperometric detector might be useful for the detection of poly-ol materials.

2. The operation temperature of the column oven or incubator is dependent on the target molecules and conditions to be analyzed. Stability of the lectins and analytes should be considered when deciding the operation temperature. As molecular interactions are occasionally stronger at lower temperatures, K_d may be determined when a difference in retardation cannot be detected at high temperature.
3. This FAC method is suitable for analyzing not only lectin-carbohydrate interactions (Fig. 4) but also other weak molecular interactions, such as glycan-small molecule [12, 13] (Fig. 5) and glycan-glycan interactions.
4. For immobilization, NHS functional groups bind amino groups of basic or N-terminal amino acids of lectins. Sometimes such basic amino acids are important for the binding of glycans. If a binding feature disappears after immobilization, a different cross-linking method (SH- or COOH-) may be used. Conjugation of a lectin with a target ligand can improve immobilization.
5. It is critical to know the V_0 of the column. Therefore, the measurement of a molecule with no binding affinity toward the resin is required. Although any non-binding molecule is theoretically acceptable, it is preferable to use labeled carbohydrates or glycans with similar molecular size and features to the analyte to be used.
6. B_t represents the effective immobilized lectin content and is experimentally obtained from Woolf-Hofstee-type plots (Fig. 3).

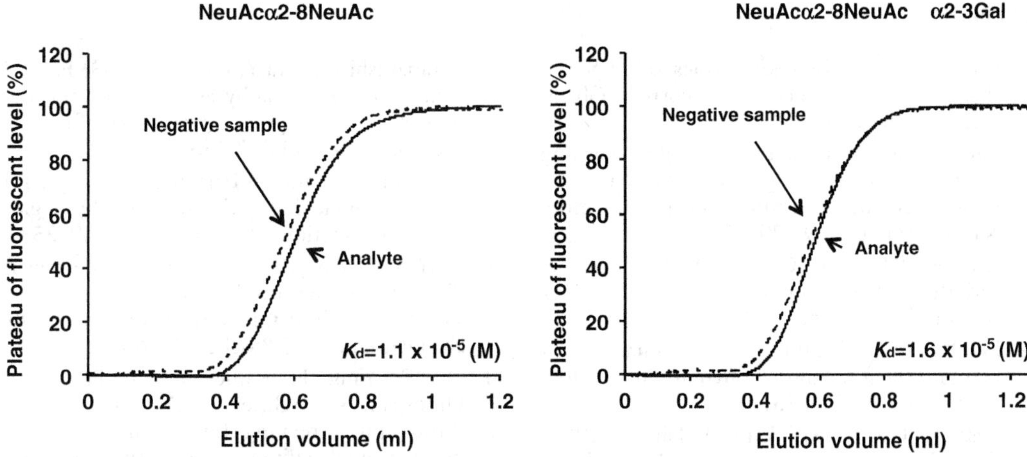

Fig. 4 Profiles of glycan and glycan-binding protein interactions. Sialic acid-binding immunoglobulin-like lectin-7 (Siglec-7) and disialyl-ligand interactions were observed by FAC. Extracellular Siglec-7 immunoglobulin domains conjugated with Fc protein were adsorbed with protein G-Sepharose and filled into the column. Two disialylated analytes, Neu5Acα2-8Neu5Acα2-RG (*left* panel) and Neu5Acα2-8Neu5Acα2-3-Gal-RG (*right* panel) [14], were subjected to Siglec7-immobilized column. Elutions were monitored using fluorescent detector. *Dotted* line represents the elution profile of negative sample (PA-labeled glucose). K_ds were shown in the panels

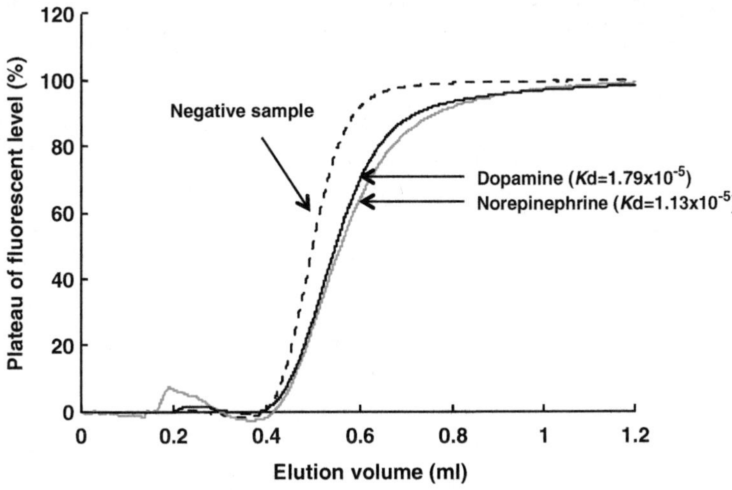

Fig. 5 A FAC profile of glycan and small molecule interactions. Polysialic acid (polymer of sialic acid) was immobilized onto the resin [12] and filled into the column. Neurotransmitters such as dopamine (*black*), norepinephrine (*grey*), and acetylcholine (negative sample, *dotted* line) were subjected to the polysialic acid-immobilized resin. Elutions were monitored using UV detector. The K_ds were shown in the panels

References

1. Varki A (1993) Biological roles of oligosaccharides: all of the theories are correct. Glycobiology 3:97–130
2. Rabinovich GA, Toscano MA (2009) Turning 'sweet' on immunity: galectin-glycan interactions in immune tolerance and inflammation. Nat Rev Immunol 9:339–352
3. Crocker P, Paulson JC, Varki A (2007) Siglecs and their roles in immune system. Nat Rev Immunol 7:255–266
4. Lee RT, Lee YC (2000) Affinity enhancement by multivalent lectin-carbohydrate interaction. Glycoconj J 17:543–551
5. Kasai K, Oda Y, Nishikata M, Ishii S (1986) Frontal affinity chromatography: theory for its application to studies on specific interactions of biomolecules. J Chromatogr 376:33–47
6. Oda Y, Kasai K, Ishii S (1981) Studies on the specific interaction of concanavalin A and saccharides by affinity chromatography. Application of quantitative affinity chromatography to a multivalent system. J Biochem 89: 285–296
7. Arata Y, Hirabayashi J, Kasai K (1997) The two lectin domains of the tandem-repeat 32-kDa galectin of the nematode Caenorhabditis elegans have different binding properties. Studies with recombinant protein. J Biochem 121: 1002–1009
8. Hirabayashi J, Arata Y, Kasai K (2003) Frontal affinity chromatography as a tool for elucidation of sugar recognition properties of lectins. Methods Enzymol 362:353–368
9. Tateno H, Nakamura-Tsuruta S, Hirabayashi J (2007) Frontal affinity chromatography: sugar-protein interactions. Nat Protoc 2:2529–2537
10. Kasai K, Ishii S (1973) Unimportance of histidine and serine residues of trypsin in the substrate binding function proved by affinity chromatography. J Biochem 74:631–633
11. Ng E, Yang F, Kameyama A, Palcic M, Hindsgaul O, Schriemer D (2005) High-throughput screening for enzyme inhibitors using frontal affinity chromatography with liquid chromatography and mass spectrometry. Anal Chem 77:6125–6133
12. Isomura R, Kitajima K, Sato C (2011) Structural and functional impairments of polysialic acid by a mutated polysialyltransferase found in schizophrenia. J Biol Chem 286: 21535–21545
13. Sato C, Yamakawa N, Kitajima K (2010) Analysis of glycan-protein interaction by frontal affinity chromatography and Biacore. Methods Enzymol 478:219–232
14. Tanaka H, Nishiura Y, Takahashi T (2006) Stereoselective synthesis of oligo-alpha-(2,8)-sialic acids. J Am Chem Soc 128:7124–7125

Chapter 23

Differential Glycan Analysis of an Endogenous Glycoprotein: Toward Clinical Implementation—From Sample Pretreatment to Data Standardization

Atsushi Kuno, Atsushi Matsuda, Sachiko Unno, Binbin Tan, Jun Hirabayashi, and Hisashi Narimatsu

Abstract

There are huge numbers of clinical specimens being stored that contain potential diagnostic marker molecules buried by the coexistence of high-abundance proteins. To utilize such valuable stocks efficiently, we must develop appropriate techniques to verify the molecules. Glycoproteins with disease-related glycosylation changes are a group of useful molecules that have long been recognized, but their application is not fully implemented. The technology for comparative analysis of such glycoproteins in biological specimens has tended to be left behind, which often leads to loss of useful information without it being recognized. In this chapter, we feature antibody-assisted lectin profiling employing antibody-overlay lectin microarray, the most suitable technology for comparative glycoanalysis of a trace amount of glycoproteins contained in biological specimens. We believe that sharing this detailed protocol will accelerate the glycoproteomics-based discovery of glyco-biomarkers that has attracted recent attention; simultaneously, it will increase the value of clinical specimens as a gold mine of information that has yet to be exploited.

Key words Glycoprotein, Glycan analysis, Lectin microarray, Clinical specimen, Biomarker

1 Introduction

It is evident that almost all secreted proteins are glycosylated via the glycosynthetic pathway in the endoplasmic reticulum and the Golgi apparatus. Because this glycosylation is characteristic of the extent of cell differentiation and the state of the cell, i.e., the origin of the tissue, its developmental stage, and the presence of malignancy, blood glycoproteins consist of a mixture of heterogeneous molecules derived from many origins [1, 2]. Thus, glycoproteins that are present in serum and that exhibit cancer-associated changes in glycosylation (glyco-alteration) have potential as biomarkers (glyco-biomarkers) for cancer diagnosis. Owing to the rapid advances in glycomics/glycoproteomics technologies, numerous glycoproteins

Jun Hirabayashi (ed.), *Lectins: Methods and Protocols*, Methods in Molecular Biology, vol. 1200,
DOI 10.1007/978-1-4939-1292-6_23, © Springer Science+Business Media New York 2014

have now been identified as candidate glyco-biomarkers. These glyco-biomarkers have been attracting a great deal of attention in the "discovery phase" [3, 4], and are expected to move toward clinical implementation by a process similar to that followed for alpha-fetoprotein (AFP). In the early 1990s, increased fucosylation of complex-type *N*-glycans was detected in some glycoproteins from hepatocellular carcinoma (HCC) patients [5]. More than 30 % of all AFP glycoforms were found to react to a fucose-binding lectin, *Lens culinaris* agglutinin (LCA). This fraction, designated as AFP-L3, was subsequently approved by the US FDA in 2005 as the first glycoprotein biomarker. Such a scenario is the ideal for identification of subsequent candidate molecules; however, it requires a systematic verification procedure for selection of the most appropriate candidate from hundreds identified in the preceding discovery phase. The lack of such a system has been a major obstacle in the mass spectrometry-based development of glyco-biomarkers.

Lectin microarray is a twenty-first century technology for glycan analysis of proteins [6]. Although the early developments of the methodology focused on high-sensitivity analysis of the microheterogeneity of glycan structures on a target glycoprotein [7–9], the majority of its applications seem to have shifted to the comparative glycome analysis of crude samples, e.g., cultured cells [10–13], bacteria [14, 15], and viruses [16]. We note that direct labeling is mostly used as the detection principle, in which more than 100 ng of the glycoprotein is usually needed prior to Cy3 labeling to assure highly practical and reproducible analysis (although the analysis itself can be satisfactorily performed with even 1 ng of the analyte sample). This is clearly a serious disadvantage for the analysis of less available endogenous glycoproteins, e.g., those contained in clinical samples. To overcome this problem, antibody-overlay lectin microarray was developed as an alternative approach to detecting specific interactions between a target glycoprotein (analyte) and multiple lectins immobilized on a microarray. This is achieved with the aid of antibodies raised against the "core protein" moiety [17]. In addition, the only pretreatment or prior processing required with the use of a specific antibody (either polyclonal or monoclonal) is immunoprecipitation. This extreme simplicity allows us to analyze sub-picomole (nanogram) amounts of a target glycoprotein preparation by means of lectin microarray, with a much greater throughput. In fact, this technical modification of the lectin microarray was made specifically for glyco-biomarker verification [18], and enables high-throughput glycan analysis of over a hundred clinical samples targeting a particular candidate glycoprotein [19, 20]. However, despite the attention it has attracted, the use of this verification method has not been sufficiently popularized in the field of glyco-biomarker development. This can be demonstrated by the fact that there are few studies concerning glyco-biomarkers except those published by a small

Table 1
Examples of focused glycan profiling by antibody-overlay lectin microarray analysis

Target glycoprotein	Clinical specimen		Detection method	Reference
Tissue				
Podoplanin	FFPE testis tissue from seminoma patient	3.6 mm^3	ALP	Kuno et al. [17]
PSA, MME	Frozen OCT-embedded prostate tissue from prostate cancer	200 μg total protein	ALP	Li et al. [21]
Serum				
Fetuin-A	Sera of patients with pancreatic cancer	100 μL	DL	Kuwamoto et al. [22]
AGP	Sera of chronic hepatitis C patients	0.5 μL	ALP	Kuno et al. [19]
Mac-2 binding protein	Sera of chronic hepatitis C patients and healthy volunteers	2.0 μL	ALP	Kuno et al. [20]
CPN2, CSF 1R, SPARCL1, ICOSLG, PIGR,	Sera of HCC patients and healthy volunteers	10 μL	ALP	Kaji et al. [23]
Other body fluid				
Transferrin	Cerebrospinal fluid		ALP	Futakawa et al. [24]
L1CAM	Bile of patients with CC and hepatolithiasis	100 μL	ALP	Matsuda et al. [25]

ALP anitbody-assisted lectin profiling method, *DL* direct labeling method

number of groups focusing on glyco-biomarker development (*see* Table 1). In this chapter, we would like to introduce the simple and versatile methodology of lectin microarray by example, with detailed protocols including sample pretreatment for two endogenous glycoproteins, Mac-2 binding protein (M2BP) and MUC1.

2 Materials

Prepare all solutions using ultrapure water (e.g., Milli-Q water). Prepare and store all reagents at room temperature (RT) unless otherwise indicated.

2.1 M2BP Immunoprecipitation Components

1. Phosphate buffered saline (PBS).
2. Tris buffered saline (TBS): 10 mM Tris–HCl, 150 mM NaCl, pH 7.6.

3. TBS containing 1.0 % Triton X-100 (TBSTx).
4. Magnetic beads (SA-MB): Dynabeads® MyOne Streptavidin T1 (Life Technologies, Carlsbad, CA).
5. Antibody solution (bio-Ab): biotinylated goat anti-human M2BP polyclonal antibody (R&D Systems, Inc., Minneapolis, MN) in PBS (50 ng/μL).
6. (Optional) Biotinylation reagent: biotin labeling kit-NH_2 (Dojindo Laboratories, Kumamoto, Japan).
7. (Optional) Affinity column for antibody purification: HiTrap Protein G HP, 1 mL (GE Healthcare UK Ltd., Little Chalfont, UK).
8. Elution buffer (EB): TBS containing 0.2 % SDS.
9. Low-retention tubes: Eppendorf® Protein LoBind Microcentrifuge Tubes, 1.5 mL (Eppendorf Co., Hamburg, Germany).
10. Low-retention tips: epT.I.P.S. LoRetention, 2–200 μL (Eppendorf Co.).
11. Magnet stand (MgS): DynaMag-2 (Life Technologies).
12. Reaction mixer: Thermomixer Comfort (Eppendorf Co.).
13. Multipurpose spin down mixer: Bug Crasher GM-01 (Taitec Co., Nishikata, Japan).
14. Heat block: Dry Thermo Unit (Taitec Co.).

2.2 M2BP Western Blot Components

1. Electrophoresis gel: 5–20 % Gradient gel (DRC, Tokyo, Japan).
2. PVDF membrane: Immun-Blot® PVDF Membrane for protein blotting (0.2 μm) (Bio-Rad Laboratories, Hercules, CA).
3. Blotter: Trans-Blot® Turbo™ Transfer System (Bio-Rad Laboratories).
4. Blocking solution (BS): 4.0 % Block Ace powder (DS Pharma Biomedical Co., Ltd., Osaka, Japan) in Milli-Q water.
5. TBS containing 0.05 % Tween-20 (TBST).
6. Alkaline phosphatase–streptavidin (Jackson ImmunoResearch Laboratories Inc., Philadelphia, PA).
7. WB substrate: Western Blue® stabilized substrate for alkaline phosphatase (Promega, Madison, WI).

2.3 M2BP Antibody-Overlay Lectin Microarray Components

1. Lectin microarray slides: LecChip™ (GlycoTechnica Ltd., Yokohama, Japan, *see* Fig. 1).
2. Array scanner: GlycoStation™ Reader 1200 (GlycoTechnica Ltd.).
3. Imaging software: Array Pro Analyzer v. 4.5 (Media Cybernetics, Inc., Bethesda, MD) or GlycoStation ToolsPro Suite v. 2.0 (GlycoTechnica Ltd.).

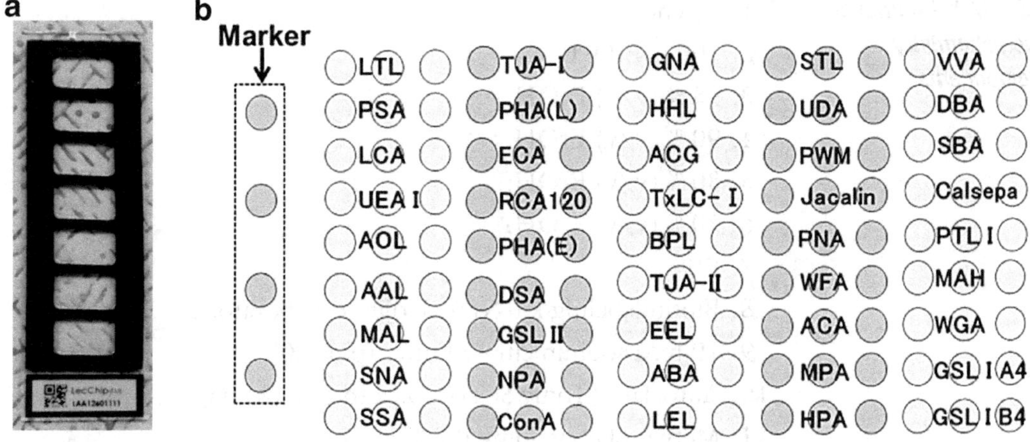

Fig. 1 Overview of LecChip instrument (**a**) and array format (**b**)

4. Chip incubator: to keep mixing the reaction solution on the chip under constant conditions at 20 °C with high humidity, set Bio shaker SHM-311 (SCINICS) in the incubator MIR-153 (SANYO Electric Co., Ltd., Moriguchi, Japan).
5. Humidified incubation chamber: incubation chamber, dark orange (Cosmo Bio Co., Ltd., Tokyo, Japan).
6. Probing buffer (PB): TBS containing 1 % (v/v) Triton X-100, 1.0 mM $CaCl_2$, and 1.0 mM $MnCl_2$.
7. Chip cleaner wipes: Kimtex white 115 mm × 230 mm (Nippon Paper Crecia Co., Ltd., Tokyo, Japan).
8. PBS containing 1.0 % Triton X-100 (PBSTx).
9. Antibody solution (bio-Ab): biotinylated goat anti-human M2BP polyclonal antibody in PBS (50 ng/μL).
10. Blocking reagent (BR): IgG from human serum (about 10 mg/mL, Sigma-Aldrich Co., St. Louis, MO).
11. Cy3-labeled streptavidin solution (Cy3-SA): Cy3-streptavidin (GE Healthcare) in PBS (50 ng/μL).

2.4 MUC1 Cell Culture Components

1. Media: RPMI1640 medium containing 5 % fetal bovine serum (FBS), serum-free RPMI1640 medium (Life Technologies).
2. Antibiotics: 100 units/mL penicillin, 100 mg/mL streptomycin (Life Technologies).
3. 150 cm^2 cell culture flasks (Becton Dickinson Co., Franklin Lakes, NJ).
4. CO_2 incubator (Thermo Fisher Scientific Inc., Fremont, CA).
5. Centrifugal filters: Amicon Ultra 10 kDa (Merck, Darmstadt, Germany).

2.5 MUC1 Immunohistochemistry Components

1. Xylene.
2. 100 % EtOH.
3. 95 % (v/v) EtOH.
4. 90 % (v/v) EtOH.
5. 80 % (v/v) EtOH.
6. 70 % (v/v) EtOH.
7. PBS.
8. Biotin labeling kit-NH_2 (Dojindo Laboratories).
9. 10 mM sodium citrate buffer (pH 6.0).
10. Autoclave (Tomy Seiko Co., Ltd., Tokyo, Japan).
11. Methanol containing 0.3 % H_2O_2.
12. VECTASTAIN ABC Kit (Vector Laboratories, Ltd., Burlingame, CA).
13. 2 % normal horse serum.
14. ImmPACT™ DAB Peroxidase Substrate (Vector Laboratories, Ltd).
15. Vector hematoxylin (Vector Laboratories, Ltd.).
16. VECTASHIELD mounting medium (Vector Laboratories, Ltd.).
17. Incubation chamber: a "Chip incubator" for lectin microarray analysis is recommended (*see* Subheading 2.3).
18. 1 % (w/v) eosin solution (Wako Pure Chemical Industries, Ltd.).
19. 0.2 % HCl–EtOH solution.

2.6 Dissection of Tissue Fragments and Protein Extraction Components

1. Surgical formalin-fixed paraffin-embedded tissue (FFPT) sections (5 μm thickness).
2. Disposable scalpels No. 12 (As One Corp., Osaka, Japan).
3. Low-retention tubes: Eppendorf® Protein LoBind microcentrifuge tubes, 1.5 mL (Eppendorf Co.).
4. 10 mM sodium citrate buffer (pH 6.0).
5. Coprecipitant: 50 % slurry of Avicel PH-101 (Sigma-Aldrich, Co.) in PBS.
6. PBS containing 0.5 % NP-40.
7. PBS.
8. Centrifuge (Tomy Seiko Co., Ltd.).
9. Sonicator (Tamagawa Seiki Co., Ltd., Iida, Japan).
10. Microscopy: Olympus CKX41 (Olympus Co., Tokyo, Japan).

2.7 MUC1 Immunoprecipitation Components

1. Anti-sialyl-MUC1 antibody (MY.1E12).
2. SA-MB (*see* Subheading 2.1, **item 4**).
3. DynaMag™-2 magnetic particle concentrator (Life Technologies).
4. PBS.
5. PBSTx.
6. PBS containing 0.2 % SDS.
7. Reaction mixer: Thermomixer Comfort (Eppendorf Co.).

2.8 MUC1 Antibody-Overlay Lectin Microarray Components

1. Lectin microarray slides.
2. GlycoStation™ Reader 1200 (GlycoTechnica Ltd.).
3. PBSTx.
4. Blocking reagent (BR): IgG from human serum (about 10 mg/mL, Sigma-Aldrich Co.).
5. Cy3-labeled streptavidin solution (Cy3-SA): Cy3-streptavidin (GE Healthcare) in PBS (50 ng/μL).
6. Imaging software: Array Pro Analyzer v. 4.5 (Media Cybernetics, Inc.) or GlycoStation ToolsPro Suite v. 2.0 (GlycoTechnica Ltd.).
7. Chip incubator.

3 Methods

Lectin microarray enables high-sensitivity glycan analysis of glycoproteins in clinical specimens. In this situation, quite a small amount of the analyte glycoproteins will be handled during the sample preparation and lectin microarray analysis. Therefore, low-retention tips and tubes should be used to prevent unintended loss of proteins. In addition, all buffer solutions used here contain surfactants such as SDS and Triton X-100 (*see* **Note 1**). Handle all reagents and perform all processes at RT unless otherwise specified.

3.1 Differential Glycan Analysis of M2BP from Culture Supernatants

M2BP is ubiquitously expressed in several kinds of cancer cells at different expression levels with various glycosylation modifications (A. Kuno, unpublished observation). In this protocol, we determined the conditions for the sample preparation and lectin microarray analysis based on the experimental results for six hepatocellular carcinoma cell lines (Huh7, HepG2, HAK1A, HAK1B, KYN-1, and KYN-2), which were cultivated in serum-free culture media.

3.1.1 Pretreatment of Magnetic Beads

1. Put the required amount of 10 mg/mL SA-MB (e.g., [number of samples + one for reserve] × 10 μL) in a microtube (*see* **Note 2**).
2. Place the tube on an MgS until the solution becomes clear.

3. After discarding the supernatant and removing bubbles (if any), take the tube from the MgS and add TBSTx (5× the volume of SA-MB).
4. After mixing thoroughly and flash-spinning (F-S) the SA-MB solution with a Bug Crasher, place the tube on the MgS until the solution becomes clear.
5. Repeat the washing process twice.
6. After discarding the supernatant, resuspend the SA-MB with TBSTx (1/2 volume of the beads).
7. Store the twofold-concentrated SA-MB (2× SA-MB) at 4 °C until use for immunoprecipitation.

3.1.2 Immuno-precipitation of M2BP

1. Put the culture supernatant containing 1 μg of the total protein, previously determined by the Bradford protein assay kit, into a 1.5 mL microtube.
2. Adjust the reaction solution to 93 μL with TBSTx.
3. Add 2 μL of bio-Ab and incubate at 4 °C for 30 min with gentle mixing with the reaction mixer.
4. F-S and add 5 μL of the pretreated 2× SA-MB, and then further incubate for 30 min.
5. F-S and keep the tube on the MgS until the reaction solution becomes clear.
6. Separate the supernatant and keep it as the "pass-through fraction (TF)."
7. Take the tube from the MgS and add TBSTx (200 μL).
8. Resuspend the SA-MB conjugated with Ab and M2BP with thorough mixing using the Bug Crasher.
9. F-S and keep the tube on the MgS until the reaction solution becomes clear.
10. Repeat the washing process twice.
11. After discarding the supernatant, resuspend the SA-MB in EB (10 μL) with gentle tapping.
12. Heat-treat at 70 °C for 5 min and then immediately place on ice (1 min).
13. Next keep at RT for 5 min, F-S, and add 10 μL of TBSTx.
14. F-S and keep the tube on the MgS until the reaction solution becomes clear.
15. Separate the supernatant and keep it as the "elution fraction (EF)."
16. Store both the TF and EF at –30 °C until used for subsequent quantitative and qualitative analysis.

3.1.3 Quantitation and Characterization of Enriched M2BP by Western Blot Analysis

1. Electrophorese half the volume of EF (10 μL) on a 5–20 % gradient gel under the following conditions: current 20 mA/gel, running time 40–50 min (until the front line migrates about 4 cm).
2. Transfer the electrophoresed proteins to a PVDF membrane by blotting.
3. Wash the membrane twice with TBS for 5 min.
4. Immerse the membrane in TBS at 37 °C for 30 min.
5. Wash the membrane three times with TBST for 5 min.
6. Immerse the membrane in 0.15 ng/μL bio-Ab diluted in TBST at 37 °C for 50 min.
7. Wash the membrane three times with TBST for 5 min.
8. Immerse the membrane in the alkaline phosphatase–streptavidin solution diluted 10,000-fold with TBST at 37 °C for 40 min.
9. Wash the membrane three times with TBST for 5 min.
10. Place the membrane in TBS and incubate for 5 min.
11. Immerse the membrane in the WB substrate.
12. Stop the chromogenic reaction by placing the membrane in Milli-Q water (*see* the expression pattern for each cell line in Fig. 2a).

3.1.4 Antibody-Overlay Lectin Microarray

1. Bring the LecChip, which is prepacked in a lightproof zipper bag, from the freezer (−20 °C) and leave on a bench without opening until it reaches RT.
2. Open the package and wash each well on the chip three times with 100 μL of PB.
3. Place the chip on a chip cleaner to remove excess solution (*see* **Note 3**).
4. Add PB (100 μL) into each well and store the chip in a humidified incubation chamber at 4 °C.
5. Remove the PB completely, and add 52 μL of PBSTx and 8 μL of EF to each well.
6. Incubate the chip in the chip incubator overnight.
7. Add 2 μL/well of BR and incubate the chip for 30 min.
8. Wash each well three times with 60 μL of PBSTx, and add 56 μL of PBSTx and 2 μL of BR.
9. Mix the solution gently, add 2 μL of bio-Ab, and incubate the chip for 1 h.
10. Remove the antibody solution completely and wash each well three times with 60 μL of PBSTx.
11. Add 56 μL of PBSTx and 4 μL of Cy3-SA and incubate the chip for 30 min.

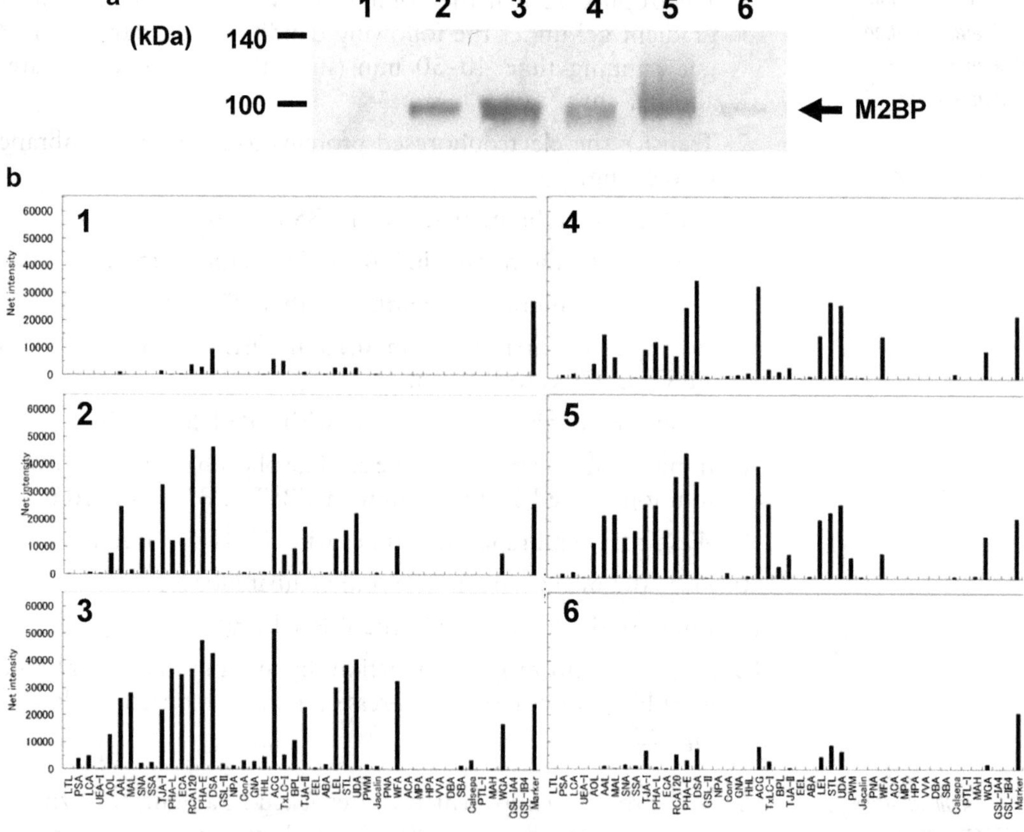

Fig. 2 Quantitation and characterization of endogenous human M2BP from culture supernatants of six hepatocellular carcinoma cell lines (*1*, Huh7; *2*, HepG2; *3*, HAK1A, *4*, HAK1B; *5*, KYN-1; *6*, KYN-2) by western blot (**a**) and lectin microarray (**b**)

12. Remove the solution completely and wash each well three times with 60 μL of PBSTx.
13. Wipe the sides and back of the chip with the chip cleaner while keeping the buffer in each well of the chip.
14. Scan the chip with the scanner (*see* **Note 4**).
15. Save the resulting fluorescent images as TIFF files and convert them to signal counts with the imaging software.
16. Calculate the net intensity value for each spot by subtracting the background value from the mean of the signal intensity values of three spots.
17. Characterize or compare the obtained glycan profiles using MS Excel (*see* the signal pattern obtained for each cell line in Fig. 2b).

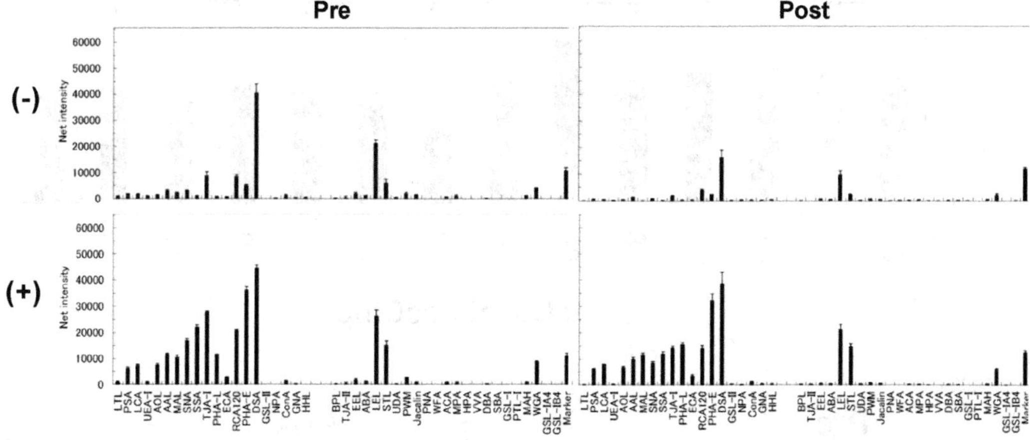

Fig. 3 Effect of the contaminant glycoproteins in the detection antibody solution prior to biotinylation on glycan profiles of human M2BP. Glycan profiles were obtained in the presence (+) or absence (–) of serum M2BP by overlaying the detection antibody (pre: commercial anti-human M2BP antibody was used without purification; post: the same antibody was used after repurification)

3.1.5 (Optional) Antibody Biotinylation

If a biotinylated antibody is not commercially available, free antibodies are routinely labeled with a biotin labeling kit-NH_2. The biotinylated antibody should be characterized by both western blot and lectin microarray before its use in immunoprecipitation and the differential glycan analysis.

3.1.6 (Optional) Antibody Purification

If a background signal for the negative control (without M2BP) is obtained on some lectins, repurification of the detection antibody using a protein G HP column may be helpful to reduce signal noise (*see* Fig. 3).

3.2 Differential Glycan Analysis of Human Serum M2BP

3.2.1 Serum Pretreatment

1. Dilute 10 μL of serum with PBS (88 μL) and then add 10 % SDS solution (2 μL) in a tube.
2. Heat-treat the tube at 95 °C for 20 min and put on ice for 1 min.
3. Mix gently and F-S with the Bug Crasher.
4. Store at –80 °C until use for immunoprecipitation (*see* **Note 5**).

3.2.2 Immuno-precipitation of Serum M2BP

1. Perform this using 20 μL of the heat-treated serum solution.
2. Refer to the protocol for culture supernatants (*see* Subheading 3.1, **step 2**).

3.2.3 Antibody-Overlay Lectin Microarray

1. Perform this using 8 μL of the EF.
2. Refer to the protocol for culture supernatants (*see* Subheading 3.1, **step 4**).

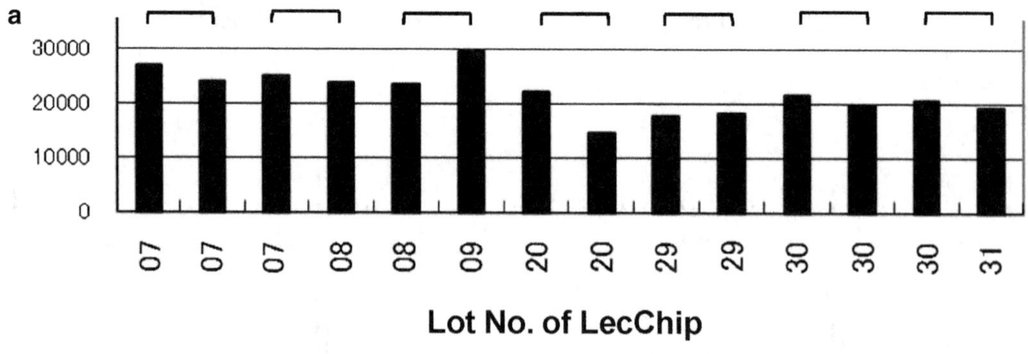

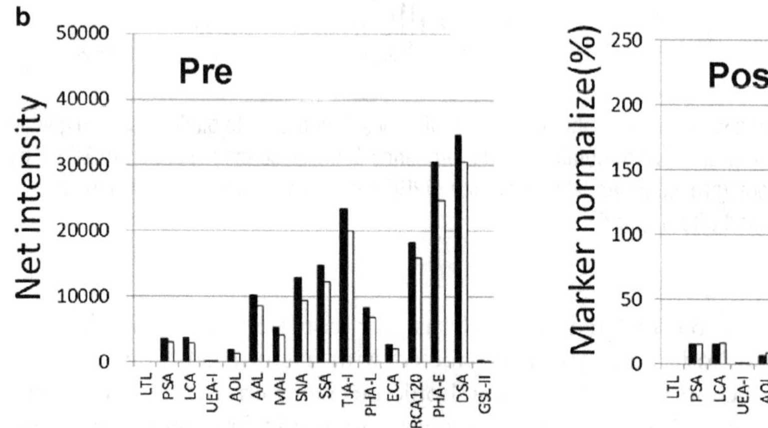

Fig. 4 Importance of an internal standard for data acquisition. (**a**) Effect of the time after powering up the system on inter- or intraday variation of the array signal intensity. When we performed a large-scale validation (about 500 patient sera) of serum M2BP using the lectin microarray, over 70 slides were used with some product lot variation. The figure shows the signal intensity of the marker in the first well of the last chip in each antibody-overlay lot. The *brackets* indicate the measurements made in the morning and afternoon of the same day. The results clearly indicate that the time from powering up until use affects the signals, as does lot variation. (**b**) Signal intensity of the positive controls (IP elution fraction from type-A serum) before and after standardization. The same IP elution fraction was added to the first well of the chips from different lots and scanned on different days

3.2.4 (Optional) Data Standardization

The relative intensity of the lectin-positive samples can be determined from the ratio of their fluorescent intensity to that of the internal standard spots designated as "markers" (*see* Fig. 1). Also *see* **Note 6** for statistical analysis using the array data and the results in Fig. 4.

The following three protocols are optimized for the differential glycan analysis of sialylated MUC1 derived from cell culture supernatants, tissue sections, and body fluids. Quantification of proteins in these types of samples is quite difficult. Although the glycan profiling of sialyl-MUC1 described in this paragraph is not a direct quantitative measurement of the alterations, the signals obtained are dose dependent and enable us to analyze the glycosylation alteration in a quantitative manner using statistics.

3.3 Differential Glycan Analysis Targeting Sialyl-MUC1 from Culture Supernatants of Biliary Tract Cancer Cell Lines

3.3.1 Preparation of Culture Supernatants

1. Prepare RPMI1640 medium supplemented with 5 % FBS, 100 units/mL penicillin, and 100 mg/mL streptomycin for cell culture.
2. Culture 100,000 cells/cell line in 40 mL of the culture medium in a 150 cm^2 flask in 5 % CO_2 at 37 °C.
3. After cells are confluent, wash cells five times with serum-free RPMI1640 medium to completely remove FBS.
4. Culture cells in serum-free RPMI1640 medium for 2 days.
5. Collect the supernatant and filter it through a 0.22 μm filter.
6. Concentrate the supernatant tenfold by ultrafiltration.

3.3.2 Immuno-precipitation of Sialyl-MUC1 with Monoclonal Antibody, MY.1E12

1. Purify MY.1E12 from the culture supernatant of hybridoma cells (*see* **Note 7**) and label using a biotin labeling kit-NH_2.
2. React biotin-labeled MY.1E12 (500 ng) with SA-MB in 20 μL of PBSTx in a reaction mixer set at 4 °C and 1,400 rpm for 1 h.
3. Wash the beads conjugated with biotin-labeled MY.1E12 three times with 200 μL of PBSTx.
4. Dilute 20 μL of the culture supernatant to 40 μL with PBSTx and add to the beads.
5. Incubate the mixture overnight at 4 °C for the reaction to occur.
6. Wash the beads three times with 200 μL of PBSTx and add 10 μL of PBS containing 0.2 % SDS to the beads.
7. Elute the bound material by heat denaturing at 95 °C for 10 min.
8. Collect the supernatant for complete depletion of the contaminating biotinylated MY.1E12 antibody.
9. Add 40 μL of SA-MB to the collected solution.
10. Incubate for 1 h at 4 °C.
11. Collect the supernatant as the immunoprecipitated sample.

3.3.3 Antibody-Overlay Lectin Microarray Analysis Targeting Sialyl-MUC1

1. Dilute the immunoprecipitated sialyl-MUC1 solutions (2.5–20 μL) obtained from KMC cells to 60 μL with PBSTx and apply to the lectin microarray slides (*see* **Note 8**).
2. Place the slides in a humidified incubation chamber and incubate at 20 °C overnight.
3. After incubation, add 2 μL/well of BR and incubate at 20 °C for 30 min.
4. Wash each slide three times with 60 μL of PBSTx, add 100 ng of Cy3-SA in PBSTx and incubate at 20 °C for 25 min.
5. Wash the slide three times with 60 μL of PBSTx and scan with the GlycoStation™ Reader (*see* **Note 9**).
6. Analyze the data with the Array Pro Analyzer.

The dilution curves of sialyl-MUC1 obtained from KMC cell culture supernatants confirm the detection of lectin signals for sialyl-MUC1 in the linear response range using 5 μL supernatants (*see* Fig. 5b).

3.3.4 Differential Glycan Profiling of Sialyl-MUC1 from Biliary Tract Cancer Cell Lines

1. Apply 5 μL of immunoprecipitated sialyl-MUC1 solutions obtained from 20 μL of eight biliary tract cancer cell lines to the lectin microarray slides for antibody-overlay lectin microarray.

Glycan profiling data are obtained in four cell lines as sialyl-MUC1 positive (KMC, TGBC-TKB-1, TGBC-TKB-44, and SZchA-1) (*see* Fig. 5c) and not obtained in four cell lines as sialyl-MUC1 negative (KMCH, TGBC-TKB-2, SkchA-1, and MZchA-2).

3.4 Antibody-Overlay Lectin Microarray Analysis of Sialyl-MUC1 Derived from Surgical Cholangiocarcinoma FFPT Sections

3.4.1 Immunohistochemistry of Surgical Tissue Sections with MY.1E12

1. Deparaffinize the FFPT sections by soaking twice in xylene for 10 min and then soak twice in 100 % EtOH for 10 min.
2. Soak the sections in 95, 90, 80, and 70 % EtOH for 10 min each and wash with PBS.
3. Immerse the sections in 10 mM sodium citric acid buffer (pH 6.0) and autoclave at 110 °C for 10 min for antigen retrieval.
4. Wash the sections with PBS and incubate with MeOH containing 0.3 % H_2O_2 for 15 min to block the endogenous peroxidase.
5. Wash the sections twice with PBS and incubate with PBS containing 2 % normal horse serum at RT for 15 min to block nonspecific binding.
6. Remove the excess blocking solution (*see* **Note 10**) and incubate the sections with the primary antibody solution (biotin-labeled MY.1E12, 0.5 μg/mL in PBS containing 2 % normal horse serum) in a humidified incubation chamber at RT for 1 h.
7. Wash the sections twice with PBS for 3 min and incubate with the secondary reagents (VECTASTAIN ABC reagents) at RT for 30 min in the humidified incubation chamber.
8. Wash the sections three times with PBS and stain with ImmPACT DAB reagents.
9. Wash the sections under tap water flow and incubate with hematoxylin for 20 s to counterstain.
10. Wash the section under warm tap water flow for 10 min and mount with VECTASHIELD mounting medium. A typical immunohistologic image is shown in Fig. 6a.

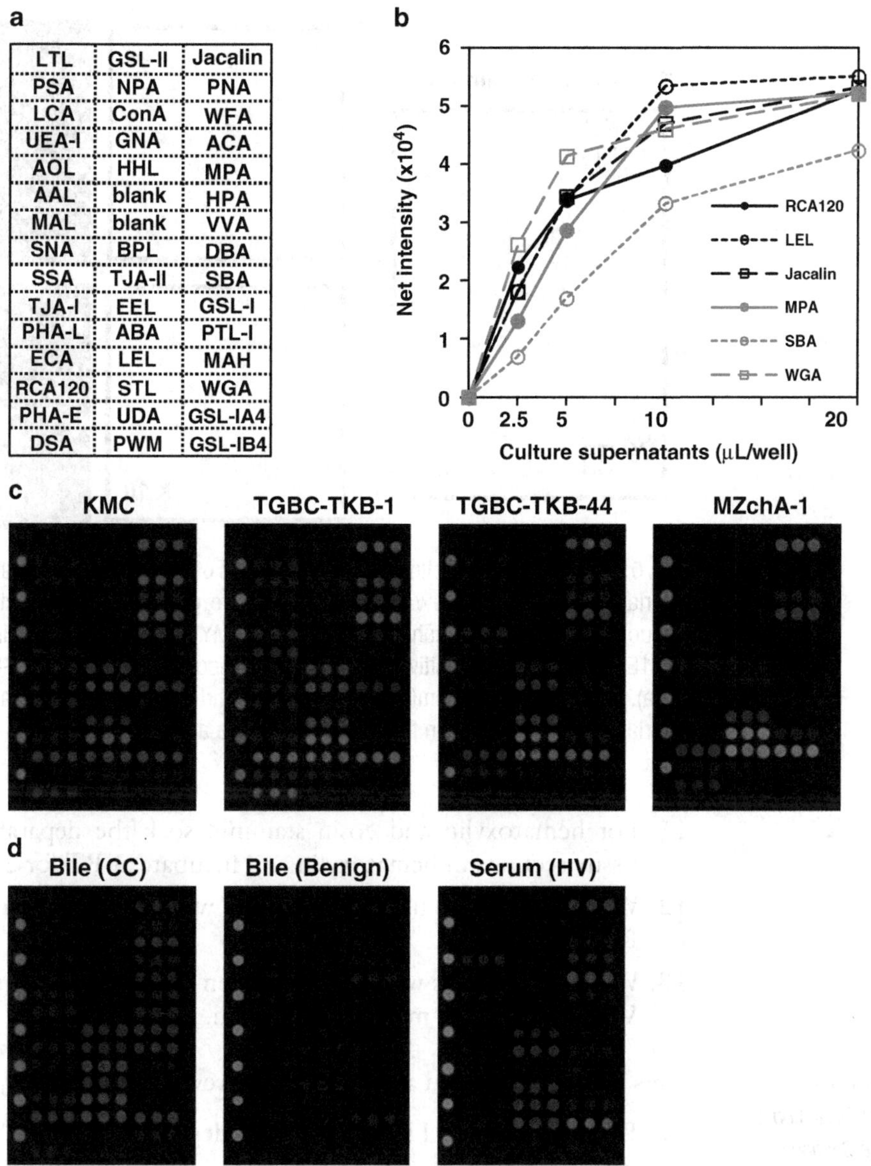

Fig. 5 Antibody-overlay lectin microarray analysis of sialyl-MUC1 obtained from culture supernatants, bile, and serum. (**a**) Array format of in-house lectin array. (**b**) Dose dependency of the antibody-overlay lectin microarray in cell culture supernatants of KMC. KMC cell culture supernatants were applied to the lectin microarray at various volumes (2.5–20 μL/well) to determine the appropriate amount required for a reliable analysis. (**c**) Glycan profiles of the sialyl-MUC1-producing biliary tract cancer cell lines. Immunoprecipitation of sialyl-MUC1 was performed using the same volume of samples. (**d**) Typical profiles of human body fluid specimens (bile and serum). Bile (CC), pooled bile from CC patients; Bile (Benign), pooled bile from patients with hepatolithiasis; serum (HV), serum from a healthy volunteer without any hepatic disease

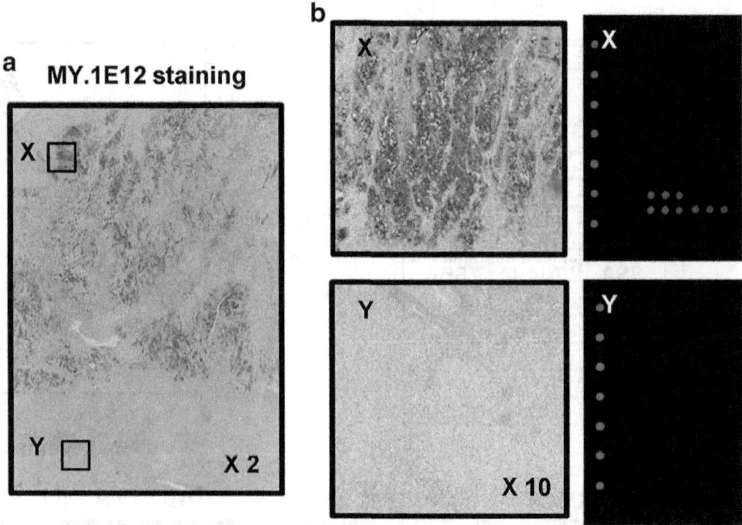

Fig. 6 Antibody-overlay lectin microarray analysis of tissue sections. An antibody-overlay lectin microarray was performed on the glycoproteins extracted from tissue sections. (**a**) Immunohistochemistry with MY.1E12. The lesion stained with MY.1E12 (*X*: MY.1E12-positive area) and unstained lesion (*Y*: MY.1E12-negative area). (**b**) Immunohistochemistry with MY.1E12 and scanned images of antibody-overlay lectin microarray in the MY.1E12-positive and -negative areas

11. For hematoxylin and eosin staining, soak the deparaffinized tissue sections in hematoxylin and incubate at RT for 20 s.
12. Wash the sections under flowing tap water and soak in eosin at RT for 5 min.
13. Wash the sections with tap water (ten passes) and mount with VECTASHIELD mounting medium.

3.4.2 Dissection and Protein Extraction from Tissue Sections

This step is performed as described in previous studies [26, 27].

1. Deparaffinize the FFPT sections as described above.
2. Wash the sections with PBS and visualize under a microscope after hematoxylin staining.
3. Wash the sections with PBS and scratch the tissue fragments from the sections off the slides with a scalpel under a microscope (*see* **Note 11**).
4. Typical staining for the comparative analysis between MY.1E12-positive and -negative tissue areas is shown in Fig. 6a, b.
5. Collect the fragments into a 1.5 mL microtube containing 200 μL of 10 mM sodium citrate buffer acid.
6. Add 4 μL of a cellulose solution (50 % slurry of Avicel) to each suspension as a coprecipitant and incubate at 95 °C for 1 h for antigen retrieval.

7. Centrifuge the tube at 20,000×*g* at 4 °C for 5 min, remove the supernatant and add 200 μL of PBS to the pellet.
8. Centrifuge the tube at 20,000×*g* at 4 °C for 5 min, remove the supernatant and add 20 μL of PBS containing 0.5 % NP-40 to the tube for solubilization.
9. Sonicate the solution gently three times for 10 s and leave the tube on ice for 1 h.
10. Centrifuge the tube at 20,000×*g* at 4 °C for 5 min. The supernatants are used as the protein extraction samples.

3.4.3 Immunoprecipitation of Sialyl-MUC1 from Tissue Lysates with Monoclonal Antibody, MY.1E12

1. Perform this procedure using 20 μL of the protein extraction samples.
2. Refer to the protocol for immunoprecipitation using culture supernatants (*see* Subheading 3.3, **step 2**).
3. The resulting supernatants are designated as the immunoprecipitated samples from tissue lysates.

3.4.4 Glycan Profiling of Sialyl-MUC1 by Antibody-Overlay Lectin Microarray

1. Perform this procedure using the immunoprecipitated samples.
2. Refer to the protocol for immunoprecipitation using culture supernatants (*see* Subheading 3.3, **step 3**).
3. As the amount of the target protein obtained is quite small, the total volume of EF (20 μL) should be applied to each well.

Typical scan data from lectin microarray for the comparative analysis between MY.1E12-positive (X) and -negative (Y) tissue areas in Fig. 6b show specific detection of the glycan profile of sialyl-MUC1.

3.5 Differential Glycan Analysis of Sialyl-MUC1 Derived from Bile and Serum of Normal Controls and Patients with Cholangiocarcinoma and Benign Diseases

1. Immunoprecipitation of sialyl-MUC1 from bile and serum with monoclonal antibody, MY.1E12:
   Based on our previous studies, the outcome will be best with 20 μL of bile or serum specimens. Refer to the protocol for immunoprecipitation using culture supernatants (*see* Subheading 3.3, **step 2**). The resulting supernatants are designated as immunoprecipitated samples from bile and serum.
2. Glycan profiling of sialyl-MUC1 with antibody-overlay lectin microarray:
   Perform it using the immunoprecipitated samples. Refer to the protocol for antibody-overlay lectin microarray using culture supernatants (*see* Subheading 3.3, **step 3**). Dilute 5 μL of immunoprecipitated sialyl-MUC1 with PBSTx to 60 μL and apply to the lectin microarray slides (*see* **Note 12**).

Typical scan data of bile or serum sialyl-MUC1s are shown in Fig. 5d. Unlike the results in serum, although some of the bile samples from benign disease patients show a high level of sialyl-MUC1, biliary sialyl-MUC1 increases significantly in CC compared with benign bile duct disease [28]. The results for the bile and serum samples represent quantitative alterations and qualitative alterations, respectively.

4 Notes

1. The procedures from pretreatment to lectin array analysis involve very small amounts of target proteins. Use of buffer solutions containing minimal levels of surfactants is effective to reduce physical adsorption into tubes and tips, which results in stable recovery of the target proteins. Particularly when handling heat-denatured samples, it is necessary to keep surfactants in buffers at a certain level to avoid aggregation of proteins.
2. For pretreatment of the SA-MB, when five times the required volume of the beads exceeds 1.5 mL, use a 2.0 mL low-retention tube. When more beads are necessary, divide the beads into multiple tubes, wash, and follow the procedure until twofold-concentrated beads are obtained. Combine the obtained 2× SA-MB before storing. When a Bug Crasher is not available, mix the solutions thoroughly by tapping and inverting the tube and then F-S.
3. Handle the LecChip on a sheet of Kimtex throughout the washing process. As shown in Fig. 7, after removing the solutions from the wells, place the LecChip facedown to absorb excess solution from the wells. Turn the chip face up and immediately add the solution of the following step to prevent wells from drying. A handheld electric pipettor that can dispense a desired amount of solution is useful for efficient processing.

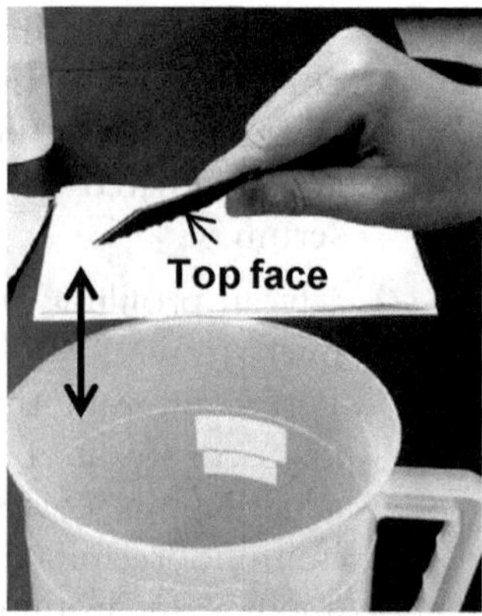

Fig. 7 Efficient procedure for removing solutions from the chip without contamination. Turn the chip facedown and shake three times to remove the excess solution

4. Check the scanned images for undesired stain, etc., on the bottom or sides of LecChips. If any is present, remove it with a Kimwipe soaked with ultrapure water, then dry off the area with a dry Kimwipe. If the stain remains, use 70 % ethanol, followed by ultrapure water, then dry off with a Kimwipe. Do not wipe roughly as the rubber is only lightly covering the chip.
5. When handling serum from HBV- or HCV-infected patients, we inactivate the virus by heat treatment in a P2 room prior to the experiment. For complete inactivation, we dilute the serum tenfold with PBS containing 0.2 % SDS, then heat at 95 °C for 20 min. This treatment may affect the yield of the target protein in the following immunoprecipitation depending on the type of antibodies used.
6. When handling a large number of samples, be aware of the following:
   (a) Prepare the required amounts of biotinylated antibodies, TBSTx, and magnetic beads for the immunoprecipitation of "number of samples + one reserve" prior to the experiment.
   (b) Wash the magnetic beads prior to every use. Do not store the beads after washing or use stored washed beads.
   (c) The number of samples used for one immunoprecipitation should be no more than 22, as handling many samples leads to time gaps within the batch because of washing procedures, etc. The whole process is relatively short, so perform several batches of immunoprecipitation in one day when handling more than 22 samples.
   (d) Use a total of 22 samples including the negative (TBSTx instead of serum) and positive (commercial serum derived from a healthy volunteer) control for each immunoprecipitation.
   (e) Note that signal intensities obtained can vary depending on the scanner conditions (e.g., time from turning on the lamp until use) and even when scanning the same chip at the same gain setting (Fig. 4a). Therefore, data standardization is necessary when comparing a large number of samples (>50) (Fig. 4b).

In addition, as ten LecChip are manufactured and supplied as one lot, inter-lot variation of chips is not negligible when handling more than 70 samples. Therefore, we use aliquots of the same immunoprecipitation EF as the positive control in one well per lot to confirm there is no large deviation within the lot. In our previous studies performed in accordance with this protocol, the positive control was prepared by a single immunoprecipitation of the required amount of the immunoprecipitation EF of a commercial type-A serum. Antibody-overlay lectin microarray was conducted on up to eight chips per day allowing one well per chip for the positive control.

7. Anti-sialyl-MUC1 monoclonal antibody, MY.1E12 was generated by the group of Dr. Irimura (University of Tokyo). This antibody recognizes the epitope involving sialyl alpha 2-3galactosyl beta 1–3 *N*-acetylgalactosaminide linked to a distinct threonine residue in the MUC1 tandem repeat [29].
8. All lectin microarray slides used in this study were produced in-house (Fig. 5a).
9. All scan data were obtained at a camera gain of 110 and an exposure time of 199 ms.
10. Tissue slides are not washed after removing the excess blocking solution in this step.
11. Tissue fragments from an area of 3 mm^2 and with 5 μm thickness are needed for reliable analysis.
12. The amounts of serum sialyl-MUC1 differ between cholangiocarcinoma, benign disease patients, and normal controls. However, in our experience, 5 μL of the enriched sialyl-MUC1 is sufficient for differential glycan profiling in all groups.

Acknowledgments

This work was supported in part by a grant from New Energy and Industrial Technology Development Organization of Japan. The authors would like to acknowledge Prof. J. Shoda (University of Tsukuba) for kindly providing surgical specimens. We thank A. Togayachi and T. Sato (AIST) for cell cultivation. We also thank H. Tateno, J. Murakami, and K. Suzuki (AIST) for in-house production of the lectin microarrays.

References

1. Dennis JW, Laferté S, Waghorne C et al (1987) Beta1-6 branching of Asn-linked oligosaccharides is directly associated with metastasis. Science 236:582–585
2. Granovsky M, Fata J, Pawling J et al (2000) Suppression of tumor growth and metastasis in Mgat5-dificient mice. Nat Med 6:306–312
3. Taniguchi N, Hancock W, Lubman DM et al (2009) The second golden age of glycomics: from functional glycomics to clinical applications. J Proteome Res 8:425–426
4. Pan S, Chen R, Aebersold R et al (2011) Mass spectrometry based glycoproteomics – from a proteomics perspective. Mol Cell Proteomics 10:1–14
5. Sato Y, Nakata K, Kato Y et al (1993) Early recognition of hepatocellular carcinoma based on altered profiles of alpha-fetoprotein. N Eng J Med 328:1802–1806
6. Hirabayashi J, Yamada M, Kuno A et al (2013) Lectin microarrays: concept, principle and applications. Chem Soc Rev 42:4443–4458
7. Angeloni S, Ridet JL, Kusy N et al (2005) Glycoprofiling with micro-arrays of glycoconjugates and lectins. Glycobiology 15: 31–41
8. Pilobello KT, Krishnamoorthy L, Slawek D et al (2005) Development of a lectin microarray for the rapid analysis of protein glycopatterns. ChemBiochem 6:985–989
9. Kuno A, Uchiyama N, Koseki-Kuno S et al (2005) Evanescent-field fluorescence-assisted lectin microarray: a new strategy for glycan profiling. Nat Methods 2:851–856
10. Ebe Y, Kuno A, Uchiyama N et al (2006) Application of lectin microarray to crude samples: differential glycan profiling of lec mutants. J Biochem 139:323–327

11. Pilobello KT, Slawek DE, Mahal LK (2007) A ratiometric lectin microarray approach to analysis of the dynamic mammalian glycome. Proc Natl Acad Sci U S A 104:11534–11539
12. Tateno H, Uchiyama N, Kuno A et al (2007) A novel strategy for mammalian cell surface glycome profiling using lectin microarray. Glycobiology 17:1138–1146
13. Tao SC, Li Y, Zhou J et al (2008) Lectin microarrays identify cell-specific and functionally significant cell surface glycan markers. Glycobiology 18:761–769
14. Hsu KL, Pilobello KT, Mahal LK (2006) Analyzing the dynamic bacterial glycome with a lectin microarray approach. Nat Chem Biol 2:153–157
15. Yasuda E, Tateno H, Hirabayashi J et al (2011) Lectin microarray reveals binding profiles of *Lactobacillus casei* strains in a comprehensive analysis of bacterial cell wall polysaccharides. Appl Environ Microbiol 77:4539–4546
16. Krishnamoorthy L, Bess JW Jr, Preston AB et al (2009) HIV-1 and microvesicles from T cells share a common glycome, arguing for a common origin. Nat Chem Biol 5:244–250
17. Kuno A, Kato Y, Matsuda A et al (2009) Focused differential glycan analysis with the platform antibody-assisted lectin profiling for glycan-related biomarker verification. Mol Cell Proteomics 8:99–108
18. Narimatsu H, Sawaki H, Kuno A et al (2010) A strategy for discovery of cancer glyco-biomarkers in serum using newly developed technologies for glycoproteomics. FEBS J 277:95–105
19. Kuno A, Ikehara Y, Tanaka Y et al (2011) Multilectin assay for detecting fibrosis-specific glyco-alteration by means of lectin microarray. Clin Chem 57:48–56
20. Kuno A, Ikehara Y, Tanaka Y et al (2013) A serum "sweet-doughnut" protein facilitates fibrosis evaluation and therapy assessment in patients with viral hepatitis. Sci Rep 3:1065
21. Li Y, Tao SC, Bova GS et al (2011) Detection and verification of glycosylation patterns of glycoproteins from clinical specimens using lectin microarrays and lectin-based immunosorbent assays. Anal Chem 83:8509–8516
22. Kuwamoto K, Takeda Y, Shirai A et al (2010) Identification of various types of alpha2-HS glycoprotein in sera of patients with pancreatic cancer: Possible implication in resistance to protease treatment. Mol Med Rep 3:651–656
23. Kaji H, Ocho M, Togayachi A et al (2013) Glycoproteomic discovery of serological biomarker candidates for HCV/HBV infection-associated liver fibrosis and hepatocellular carcinoma. J Proteome Res 12:2630–2640
24. Futakawa S, Nara K, Miyajima M et al (2012) A unique N-glycan on human transferrin in CSF: a possible biomarker for iNPH. Neurobiol Aging 33:1807–1815
25. Matsuda A, Kuno A, Matsuzaki H et al (2013) Glycoproteomics-based cancer marker discovery adopting dual enrichment with *Wisteria floribunda* agglutinin for high specific glycodiagnosis of cholangiocarcinoma. J Proteomics 85:1–11
26. Matsuda A, Kuno A, Ishida H et al (2008) Development of an all-in-one technology for glycan profiling targeting formalin-embedded tissue sections. Biochem Biophys Res Commun 370:259–263
27. Kuno A, Matsuda A, Ikehara Y et al (2010) Differential glycan profiling by lectin microarray targeting tissue specimens. Methods Enzymol 478:165–179
28. Matsuda A, Kuno A, Kawamoto T et al (2010) *Wisteria floribunda* agglutinin-positive mucin 1 is a sensitive biliary marker for human cholangiocarcinoma. Hepatology 52:174–182
29. Takeuchi H, Kato K, Denda-Nagai K et al (2002) The epitope recognized by the unique anti-MUC1 monoclonal antibody MY.1E12 involves sialyl alpha 2-3galactosyl beta 1-3N-acetylgalactosaminide linked to a distinct threonine residue in the MUC1 tandem repeat. J Immunol Methods 270:199–209

Chapter 24

Lectin-Microarray Technique for Glycomic Profiling of Fungal Cell Surfaces

Azusa Shibazaki and Tohru Gonoi

Abstract

Lectin microarrays are rows of lectins with different carbohydrate-binding specificities spotted on surfaces of glass slides. Lectin microarray technique enables glycomic analyses of carbohydrate composition of fungal cell walls. We will describe an application of the technique in analyzing cell surface glycome of yeast-form fungal cells in the living state. The analysis reveals genus- and species-dependent complex cell surface carbohydrate structures of fungi, and enabled us, therefore, to suggest that cell walls of yeast cells, which have been considered to have relatively simple structures, actually have a more complex structure containing galactose and fucose. This shows that the technique can be used to find new insights into the study of phylogenetic relations and into the classification of cells in the fungal kingdom based on cell wall glycome.

Key words Lectin microarray, Fungi, Yeast, Cell surface, Carbohydrate, Glycan, Glycome, Phylogeny, Cell tracker, Evanescent-field fluorescence

1 Introduction

The fungal kingdom, which diverged from the animal and plant kingdoms more than a billion years ago [1], includes 1.5 million species [2]. Fungi are heterotrophic organisms that use organic matter in the environment. Many fungi are symbiotic with other organisms, but some fungi cause agricultural problems as plant pathogens, while some cause mild-to-severe health problems in humans and animals. Fungal cell surface structures have diversified significantly during their evolution and may be related to their pathogenicity [3].

Currently, fungi are classified based on morphological, physiological, and biochemical properties, and DNA sequences of ribosomal and/or multiple other genes. Consistencies among these properties and sequences are important in the classification. Carbohydrate compositions of fungal cell walls are also an important index in classification; for example, the presence of xylose in

Jun Hirabayashi (ed.), *Lectins: Methods and Protocols*, Methods in Molecular Biology, vol. 1200, DOI 10.1007/978-1-4939-1292-6_24, © Springer Science+Business Media New York 2014

fungal cell walls is a useful marker for identification of fungi belonging to Basidiomycota [4].

We applied lectin microarray technique [5, 6] to analyze cell surface glycome of phylogenetically diverse yeast-form fungal cells, including model organisms like baker's yeast, *Saccharomyces cerevisiae*, and fission yeast, *Schizosaccharomyces pombe*, and human pathogenic yeasts, including several *Candida* species, *Malassezia furfur*, and *Rhodotorula mucilaginosa*. Although the number of fungal species analyzed so far is still limited, the relationship between phylogeny [7, 8] and fungal cell surface carbohydrates has been progressively emerging, indicating that carbohydrate compositions of fungal cell surfaces are a good index in the classification of fungi. The lectin microarray technique is also easy to use and useful for performing various types of carbohydrate analyses of fungal cell surfaces.

2 Materials

We use Milli-Q water for making all aqueous solutions. All chemicals are analytical grade and used without further purification.

2.1 Lectin Microarray Slides

An example list of lectins spotted on the slide, their concentrations, a spotting format, and a construction method of microarray slides used in the present experiments are described by Kuno et al. (*see* Chapter 23 of this book; also *see* refs. 5, 6) Briefly,

1. Dissolve lectins in a spotting solution (Matsunami, Japan) at a concentration of 0.5 mg/mL.
2. Spot the lectin solutions on epoxy-coated glass slides (Schott AG, Mainz, Germany) using a noncontact microarray printing robot (Microsys 4000 System; Genomic Solutions Inc., Ann Arbor, MI, or an equivalent machine) with a spot diameter size of 500 μm spaced at 650 μm intervals. Each lectin is typically spotted in triplicate.
3. After spotting, remove excess non-immobilized lectins by washing with Tris-buffered saline (10 mM Tris, 137 mM NaCl, 2.7 mM KCl, 8 mM Na_2HPO_4, 2 mM KH_2PO_4, pH 7.4) containing 1 % (v/v) Triton-X.
4. Incubate the glass slides in a chamber with more than 80 % humidity at 25 °C for 18 h to further immobilize the lectins.
5. After incubation, carefully attach a silicone rubber sheet with eight wells to the glass slide and block with 100 μL of blocking solution (Stabiguard Choice; SurModics Inc., Eden Prairie, MN) at 20 °C for 1 h.

2.2 Cell Culture

1. Culture yeast-form fungal cells originated from a single strain in appropriate liquid culture medium, e.g., (1) YPD medium: 2 % polypeptone, 1 % yeast extract, and 2 % glucose; (2) YM medium: 0.5 % polypeptone, 0.3 % yeast extract, 0.3 % malt extract, and 1 % glucose; or other media, at the appropriate temperatures, periods, and conditions, depending on the strains and purpose of experiments.

2.3 Collection and Washing of Cultured Cells

1. Washing and suspension medium: Phosphate-buffered saline (PBS; 137 mM NaCl, 2.7 mM KCl, 8 mM Na_2HPO_4, 2 mM KH_2PO_4, pH 7.0).
2. Bovine serum albumin (BSA), final concentration at 10 mg/ mL in PBS (BSA/PBS, *see* **Note 1**).

2.4 Fluorescent Staining of Fungal Cells

1. For preparation of a stock solution of Cell-Tracker Orange CMRA® (Invitrogen), see below (*see* **Notes 2** and **3**).
2. Dilute Cell-Tracker Orange CMRA® stock solution using PBS at a final concentration of 25 μM in PBS (*see* **Note 4**).

2.5 Fluorescent Scanner

For scanning, we use an evanescent-field fluorescence scanner (SC-Profiler; Moritex Corp., Tokyo), which enables us to detect fluorescent signals from cells bound within 200 nm from the surface of lectin-microarray slide.

2.6 Other Machinery and Tools

1. Cell counter: Burker-Turk hemacytometer or an equivalent tool.
2. A conventional low-speed centrifuge with a swing rotor is required for collecting and washing fungal cells, and for washing microarray slides.
3. A conventional water bath incubator at temperature range covering 30–40 °C.
4. An ice bucket.
5. A dry bath incubator to keep at 20 °C.

2.7 Computer Software

1. For fluorescence scanning analysis: Array Pro analyzer ver. 4.5 (Media Cybernetics Inc., Bethesda, MD).
2. For cluster analysis: Cluster 3.0 [9].

3 Methods

3.1 Cell Culture

Ascomycetous (e.g., *Saccharomyces* sp., *Schizosaccharomyces* sp., *Candida* sp., *Pichia* sp.) and Basidiomycetous yeast cells (e.g., *Rhodotorula* sp., *Ustilago* sp., *Malassezia* sp.) are typically cultured in YPD and YM media, respectively. However, culture conditions, including medium, temperature, and period, can be arbitrarily changed depending on fungal strains and purpose of the experiment.

It is recommended that the genes and/or species of the strains be validated by sequencing of either the D1/D2 region of the 26S rRNA gene, the Internal Transcribed Spacer (ITS) region of rRNA, and/or other appropriate genes, either before or after the culture.

3.2 Harvesting and Fluorescent Staining of Fungal Cells

1. Transfer fungal cell suspension to a 1.5-mL plastic microtube and microcentrifuge at 3,000 × *g* for 1 min. Remove the culture medium by aspiration or pipetting. Add 1 mL PBS to the microtube and resuspend the cells. Repeat the cell rinse once.
2. Resuspend the cells in PBS and count the cell number using a hemacytometer or any cell-counting tool. Adjust the cell concentration to 1×10^7 cells per 1 mL PBS (*see* **Note 5**).
3. Centrifuge the microtube, remove the PBS in the tube, and then resuspend the cells in 1 mL 25 μM Cell-Tracker solution as prepared in Subheading 2.4 (*see* **Notes 4** and **5**).
4. Incubate the cells at 37 °C for 45 min while protecting the solution from light (*see* **Notes 6** and **7**).
5. After the incubation, remove the solution containing Cell-Tracker® by centrifugation at 150 × *g* for 1 min, followed by aspiration or pipetting.
6. Resuspend the cells in 1 mL of ice-chilled PBS containing 10 mg/mL BSA (BSA/PBS, *see* **Note 1**). Repeat the cell rinse once and then resuspend the cells in 1 mL BSA/PBS on ice.

3.3 Hybridization to the Lectin Microarray

1. Overlay the lectin microarray slide in each well with BSA/PBS suspension of Cell-Tracker®-labeled cells at a volume of 100 μL/well. Incubate the slide at 20 °C for 1 h (*see* **Note 8**).
2. Put the microarray slide in a 50-mL plastic tube (Falcon) in a swing rotor of a centrifuge. Remove unbound cells by centrifugation at 1 × *g* for 5 min.
3. Remove the microarray slide from the plastic tube (Falcon) and put on a flat table, and then remove the BSA/PBS on the surface of the microarray slides. Add 100 μL PBS to each well of the slide using a micropipette (*see* **Note 9**).

3.4 Scanning of Fluorescent Signals

1. The evanescent-field fluorescence scanner should be used under Cy3 mode. Analyze data using analyzer software (Array Pro analyzer ver. 4.5; Media Cybernetics Inc., Bethesda, MD) (*see* **Note 10**). The net intensity of each spot should be calculated by subtracting the background from the signal intensity. Present data as relative ratios of the fluorescence intensities for each spot of lectin probes. Data may be typically presented as fluorescence intensity vs. lectin names (Fig. 1), or the intensity vs. cell numbers (Fig. 2), and so on (*see* **Note 11**).

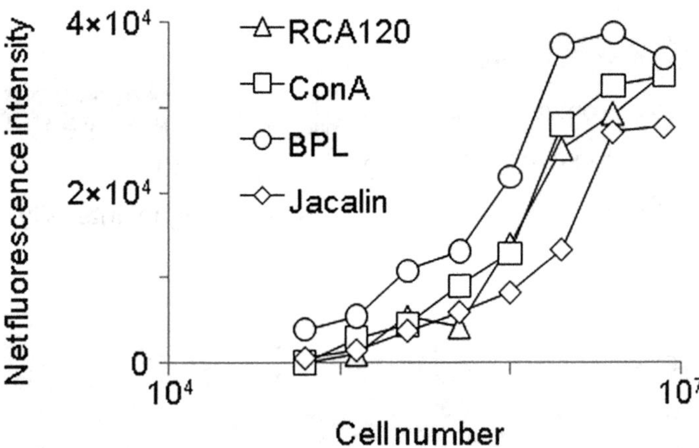

Fig. 1 Lectin microarray analysis of cell surface glycans in *Schizosaccharomyces pombe*. Yeast cells loaded with Cell-Tracker® were applied on the arrays at cell concentrations between 6×10^4 and 8×10^6 per well. Net fluorescence intensities (ordinate) are shown against the cell number in logarithmic scale (abscissa)

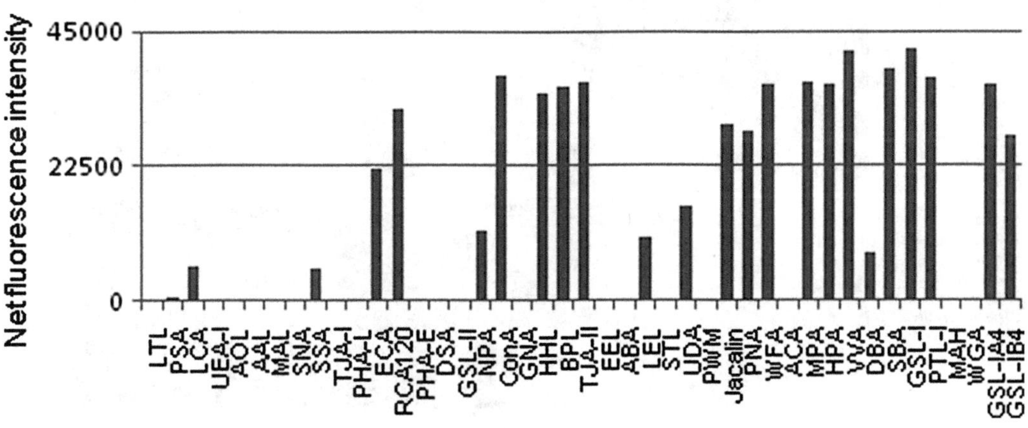

Fig. 2 An example of fluorescence output. *Schizosaccharomyces pombe* cells were stained with Cell-Tracker®, and analyzed by the lectin microarray method. The evanescent-field fluorescence scanner, SC-Profiler (Moritex Corp., Tokyo), and the analysis software, Array Pro analyzer ver. 4.5 (Media Cybernetics Inc., Bethesda, MD) were used in the analysis. Ordinate: net fluorescence intensities; horizontal axes: abbreviated lectin names

3.5 Cluster Analysis of Fungal Cell Surface Glycome

Glycomic studies of fungal cell surface structure have been performed in only a limited number of species and strains, but these studies have suggested that fungal cell surface glycans are closely related to fungal cell phylogeny [6]. An example of a dendrogram analysis using the software Cluster 3.0 [9] is shown in Fig. 3 (*see* **Note 12**).

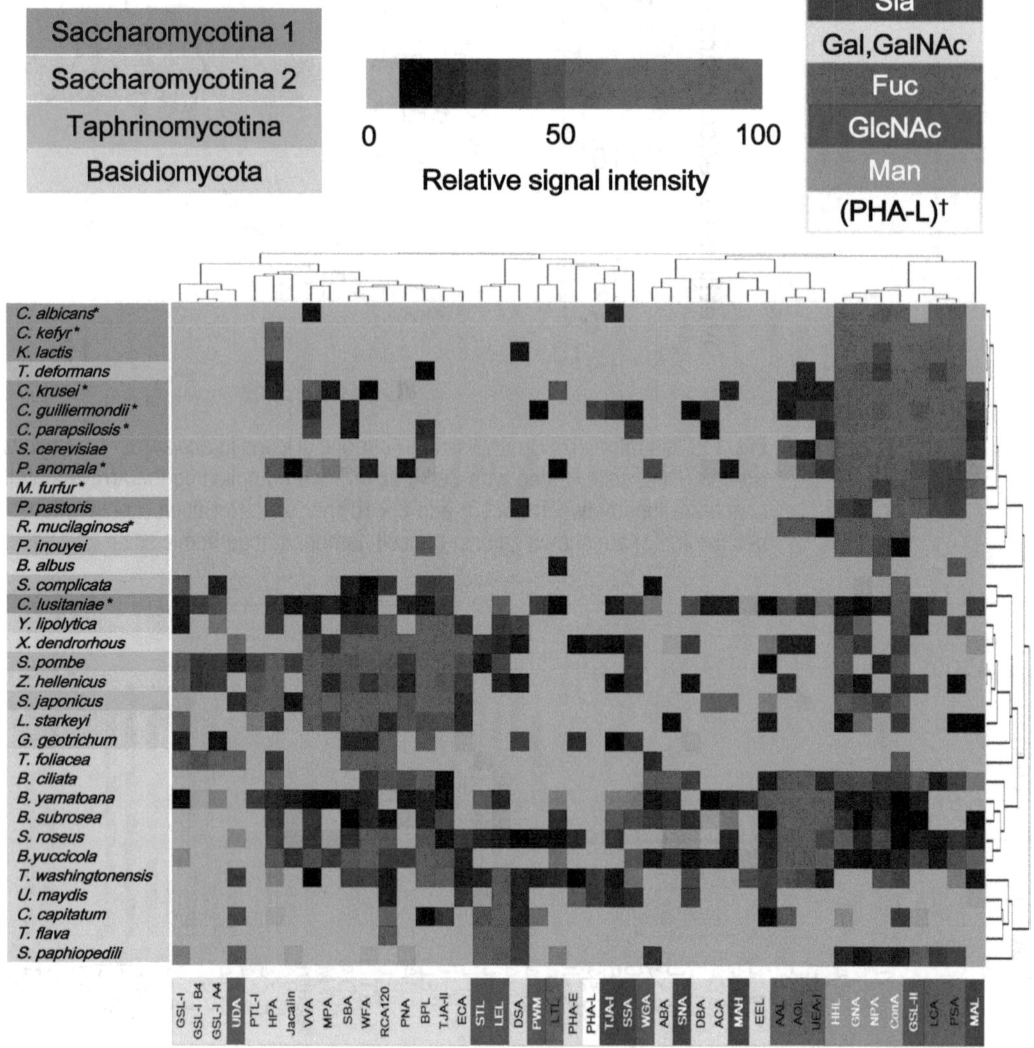

Fig. 3 Cluster analysis of cell surface glycans of fungi. Names of fungi are colored *pink*, *orange*, *blue*, and *light green*, respectively, for Saccharomycotina group 1, group 2, Taphrinomycotina, and Basidiomycota. Possible binding specificities of lectins are shown as different colors below the main panel. Relative signal intensities are digitized and shown as different colors. Cluster analyses were performed using computer software (Cluster3.0) [9]. Abbreviated lectin names are shown on the *horizontal axis* of the main panel, and yeast strains are shown on the *vertical axis*. Human pathogenic fungi are marked with an *asterisk*. Binding specificities of lectins are shown in different colors (*dagger*) The lectin, PHA-L, binds *N*-linked tri/tetra-antennary oligosaccharide containing galactose, *N*-acetylglucosamine, and mannose. PSA and LCA bind mannose as well as fucose [10, 11]. The illustration is adapted from ref. 12 with permission

4 Notes

1. BSA prevents cells from aggregating with each other.
2. Cell Tracker® (M.W. 550) should be kept frozen and protected from light. Before opening a container of Cell Tracker®, it should be brought back to room temperature.

3. For making stock solution, dissolve 50 μg powder of Cell Tracker® in 10 μL DMSO, resulting in a final concentration of 10 mM. The solution may be stable for 6 months at −20 °C.
4. Prepare only required amount of Cell Tracker® PBS solution every time. The PBS solution cannot be stored stably.
5. Usually 1 mL of cell suspension is enough for use in an experiment with one microarray slide.
6. Aluminum foil may be used to cover the tube and incubator to protect them from light.
7. Appropriate incubation conditions, including culture medium, temperature, and time length, may vary depending on strains. For example, *Candida albicans* cells, a prevalent human pathogen, may be incubated at 30 °C instead of 37 °C to avoid filamentous growth [13]. Culture conditions for other fungi should be determined experimentally or based on literature and also depending on the purpose of the experiment.
8. The microarray slides with fluorescently labeled cells should be protected from light.
9. Perform quickly to avoid drying the surface of the microarray slide.
10. The fluorescent lamp of the scanner should be turned on for at least 30 min before starting measurement to stabilize the lamp.
11. The experimental results may vary depending on the surface conditions of array slides and the states of fungal cells. It is strongly recommended, therefore, that the experiment be repeated more than three times using different array slides under the same culture and experimental conditions.
12. Similar and other types of analyses can be performed using other software.

Acknowledgement

This work was supported by research grants of the Ministry of Education, Culture, Sports, Science, and Technology in Japan [grants #21406003], by the National BioResource Project (http://www.nbrp.jp/), and by the Cooperative Research Grant of NEKKEN, 2012.

References

1. Bruns T (2006) Evolutionary biology: a kingdom revised. Nature 443:758–761
2. Hawksworth D (1991) The fungal dimension of biodiversity: magnitude, significance, and conservation. Mycol Res 95:641–655
3. Bartnicki-Garcia S (1968) Cell wall chemistry, morphogenesis, and taxonomy of fungi. Annu Rev Microbiol 22:87–108
4. Suzuki M, Nakase T (1988) The distribution of xylose in the cells of Ballistosporous yeasts—

application of high performance liquid chromatography without derivatization to the analysis of xylose in whole cell hydrolysates. Gen Appl Microbiol 34:95–103

5. Kuno A, Uchiyama N, Koseki-Kuno S, Ebe Y, Takashima S, Yamada M, Hirabayashi J (2005) Evanescent-field fluorescence-assisted lectin microarray: a new strategy for glycan profiling. Nat Methods 2:851–856
6. Tateno H, Uchiyama N, Kuno A, Togayachi A, Sato T, Narimatsu H, Hirabayashi J (2007) A novel strategy for mammalian cell surface glycome profiling using lectin microarray. Glycobiology 17:1138–1146
7. James TY, Kauff F, Schoch CL, Matheny PB, Hofstetter V, Cox CJ, Celio G, Gueidan C, Fraker E, Miadlikowska J et al (2006) Reconstructing the early evolution of Fungi using a six-gene phylogeny. Nature 443: 818–822
8. Hibbett DS, Binder M, Bischoff JF, Blackwell M, Cannon PF, Eriksson OE, Huhndorf S, James T, Kirk PM, Lucking R et al (2007) A higher-level phylogecation of the Fungi. Mycol Res 111:509–547
9. de Hoon MJ, Imoto S, Nolan J, Miyano S (2004) Open source clustering software. Bioinformatics 20:1453–1454
10. Kornfeld K, Reitman ML, Kornfeld R (1981) The carbohydrate-binding specificity of pea and lentil lectins. Fucose is an important determinant. J Biol Chem 256:6633–6640
11. Schwarz FP, Puri KD, Bhat RG, Surolia A (1993) Thermodynamics of monosaccharide binding to concanavalin A, pea (*Pisum sativum*) lectin, and lentil (*Lens culinaris*) lectin. J Biol Chem 268:7668–7677
12. Shibazaki A, Tateno H, Akikazu A, Hirabayashi J, Gonoi T (2011) Profiling the cell surface glycome of five fungi using lectin microarray. J Carbohydr Chem 30:147–164
13. Bastidas RJ, Heitman J (2009) Trimorphic stepping stones pave the way to fungal virulence. Proc Natl Acad Sci U S A 106:351–352

Chapter 25

Application of Lectin Microarray to Bacteria Including *Lactobacillus casei/paracasei* Strains

Emi Yasuda, Tomoyuki Sako, Hiroaki Tateno, and Jun Hirabayashi

Abstract

Since 2005, lectin microarray technology has emerged as a simple and powerful technique for comprehensive glycan analysis. By using evanescent-field fluorescence detection technique, it has been applied for analysis of not only glycoproteins and glycolipids secreted by eukaryotic cells but also glycoconjugates on the cell surface of live eukaryotic cells. Bacterial cells are known to be decorated with polysaccharides, teichoic acids, and proteins in the peptide glycans of their cell wall and lipoteichoic acids in their phospholipid bilayer. Specific glycan structures are characteristic of many highly pathogenic bacteria, while polysaccharides moiety of lactic acid bacteria are known to play a role as probiotics to modulate the host immune response. However, the method of analysis and knowledge of glycosylation structure of bacteria are limited. Here, we describe the development of a simple and sensitive method based on lectin microarray technology for direct analysis of intact bacterial cell surface glycomes. The method involves labeling bacterial cells with SYTOX Orange before incubation with the lectin microarray. After washing, bound cells are directly detected using an evanescent-field fluorescence scanner in a liquid phase. The entire procedure takes 3 h from putting labeled bacteria on the microarray to profiling its lectin binding affinity. Using this method, we compared the cell surface glycomes from 16 different strains of *L. casei/paracasei*. The lectin binding profile of most strains was found to be unique. Our technique provides a novel strategy for rapid profiling of bacteria and enables us to differentiate numerous bacterial strains with relevance to the biological functions of surface glycosylation.

Key words *L. casei*, Polysaccharide, Lectin, Lactic acid bacteria, Glycosylation, CSA, CSL, SYTOX Orange

1 Introduction

It is well documented that bacterial cell surface components and structures are critical factors for pathogenesis, host–microbe interaction, immune modulation, and symbiosis. Toll-like receptors (TLRs) expressed on mammalian epithelial and immune cells act as pattern recognition receptors, which are individually responsible for a variety of different bacterial components such as lipopolysaccharides (LPS) from gram-negative bacteria, peptidoglycan (PG), lipoteichoic acid (LTA), and wall teichoic acid (WTA) from

Jun Hirabayashi (ed.), *Lectins: Methods and Protocols*, Methods in Molecular Biology, vol. 1200,
DOI 10.1007/978-1-4939-1292-6_25,

gram-positive bacteria, flagella, lipoproteins, and nucleic acids. TLRs transfer signals to the innate as well as the acquired immune system [1]. Bacterial cell surface components are also recognized by other mammalian signaling molecules such as NOD1 and NOD2, which are intracellular proteins functioning as cytosolic sensors in the regulation of inflammatory responses [2].

In general, non-pathogenic as well as pathogenic bacteria often change their cell surface by having mutations in its genome or plasmids during a comparatively short period. Even getting the perfect bacterial sequence information, it is rather difficult to analyze their cell surface glycome in relation to them. In the light of recent research on the role of bacterial cell surface structures/components in probiotic or symbiotic action [3], polysaccharides (PSs) from each bacterial cell wall could have unique roles that are crucially important [4–7]. As described in reports concerning *Bacteroides fragilis* PSs [8–10], bacterial cells have different impacts on host cells, even when various mutants having defects in different predictive PS biosynthesis genes or having their deletions showed different immune modulating effects towards cultured splenocytes and T cells. However, it is still not known what kind of PS structure is important for their activity and how these molecules exert their activities on host cells. In addition, it has been shown that extracellular PSs of *Streptococcus thermophilus* [11] and group B *Streptococcus* capsular PSs are highly diverse [12]. It is well known that pathogenic bacteria, such as hemolytic streptococci, often change their outer surface glycan profile to escape the host immune defense mechanism [13].

Lactic acid bacteria (LAB) are industrially important microorganisms for fermented food production. Many strains of LAB are used as "probiotics," which are defined as microbes that exert a beneficial effect on the host. In addition to their role in immune modulation, various factors produced by LAB have been proposed as active interaction factors with mammalian host cells [14]. For example, soluble proteins produced by probiotic bacteria are known to regulate survival and growth of intestinal epithelial cells [15–17]. Although the cell surface components of LAB such as S-layer proteins, LTA, WTA, and PSs are proposed as immune modulators [3], the active components directly involved in immune modulation are largely unknown.

Probiotic activity of the *L. casei* strain Shirota (YIT 9029) has been extensively analyzed. We have focused on the role of cell wall PSs in the immune modulation activities of YIT 9029 and identified a cluster of genes essential for the biosynthesis of high molecular mass PS-1 moiety [7]. To investigate the dynamism and diversity of bacterial outer surface structures, especially bacterial glycomes, in relation to their functional characterization, a powerful methodology with high throughput and versatility is needed.

Lectin microarray was developed to profile the complex features of glycans expressed in various forms [18–24]. We adopted a unique evanescent-field activated fluorescence detection principle to detect highly sensitive and reproducible lectin–glycoconjugate interactions on a glass slide [21, 24]. Using this detection principle, Tateno et al. (2007) developed an application method enabling detection of direct interaction between lectins and whole mammalian cells [22]. Though Ku-Lung Hsu et al. [19, 20, 23] similarly applied lectin microarray technology to targeting bacterial cells, they utilized a confocal detection method. Since the weak binding between the bacterial PSs and lectins come away in process of drying a glass slide, these interactions are not detected by their system.

Here, we describe the development of a practical methodology for profiling bacterial cell surface glycomes to use SYTOX Orange [25]. Using this technique, we have shown substantial differences in glycan profiles among the 16 strains belonging to the same *Lactobacillus* species [26].

Before using SYTOX Orange [27, 28], we examined the applicability of Cy3 labeling, which is usually used in the fluorescent labeling of glycoproteins and cell surface proteins for microarray analysis. Cells were labeled by employing a standard fluorescein isothiocyanate methodology [29]. Using 10 μg/mL Cy3, we successfully labeled $1–2 \times 10^9$ cells. However, nonspecific binding between Cy3-labeled cells and the glass microarray slides was detected. This observed binding was probably due to increased hydrophobicity of the cells. We also tested Cell Tracker, a labeling reagent for whole mammalian cell staining [22]. However, the fluorescence intensity of *L. casei/paracasei* cells ($1–2 \times 10^{10}$) labeled with Cell Tracker (50 μg/mL) was too low to be analyzed by this technology.

Our technique for characterizing cell wall structures is both simple to perform and highly reproducible. This assay can be easily extended to other bacterial species. In addition, it is most effective to analyze a novel strain to use some genetically different strains of same genus and/or species or reported strains.

2 Materials

2.1 Bacterial Strains and Culture Conditions

1. Bacterial strains used in this study are listed in Table 1. *L. casei* YIT 9029 is a commercial strain used in the production of fermented milk. *ΔcpslC* is a mutant of *L. casei* YIT 9029, in which the *cpslC* gene was knocked out [7]. YIT 0180 is the neotype strain of *L. casei* [26].
2. Cells are cultured in MRS medium (Becton, Dickinson and Company, Franklin Lakes, NJ) for 22 h at 37 °C under aerobic conditions.

Table 1
***Lactobacillus casei/paracasei* strains used in this study**

	Strains			
	YIT No.[a]	Identification No.[b]	Source/designation	References
1	YIT 0001	ATCC 27139	S-1 (A Murata)	[39]
2	YIT 0003	IAM 1045 =JCM 20024 (2007-)	Cheese	
3	YIT 0005	ATCC 25302	Saliva	[32]
4	YIT 0006	ATCC 25303	Saliva	[32]
5	YIT 0007	JCM 1109	Human intestine	
6	YIT 0009	NIRD C-9	Human	[33]
7	YIT 0015	JCM 1053	T. Mitsuoka S2-5	
8	YIT 0047	NIRD A-121	Human	[33, 40]
9	YIT 0091	IPOD 1766	YPS-1 (collection in Yakult Central Institute for Microbiological Research)	
10	YIT 0123	ATCC 27216	Saliva of child, Type of *L. casei* s.s. *alactosus*	
11	YIT 0128	ATCC 4646	Dental caries	
12	YIT 0226	PHLS A357/84	Human blood	
13	YIT 0289	PHLS A22/73	Endocarditis	
14	YIT 0290	PHLS A198/89	Endocarditis	
15	YIT 9029	Strain Shirota	Original collection of Yakult	[7, 39, 40]
16	YIT 0180[T]	ATCC 334	Emmental cheese, Neotype strain	[26]
17	*ΔcpslC*		The mutant of YIT 9029	[7]

[a]Registration number of the culture collections preserved in Yakult Central Institute for Microbiological Research, Tokyo, Japan. Other strains were purchased from the American Type Culture Collection (Manassas, VA), Institute of Molecular and Cellular Biosciences (The University of Tokyo, Japan), Japan Collection of Microorganisms (Wako, Japan), National Institute for Research in Dairying collection (Karnal, India), International Patent Organism Depositary (Tsukuba, Japan) and Public Health Laboratory Service (London, United Kingdom)
[b]Identification number for each strain is given in parentheses, ATCC, American Type Culture Collection (USA); IAM, Institute of Molecular and Cellular Biosciences (Japan); JCM, Japan Collection of Microorganisms (Japan); NIRD, National Institute for Research in Dairying collection (India); IPOD, International Patent Organism Depositary (Japan); PHLS, Public Health Laboratory Service (UK)

2.2 Lectin Microarray Preparation and Lectin Specificities

1. Prepare the lectin microarray. We use commercial lectin microarray or self-made one described in Chapter 24 [21, 22, 24]. One of the most important lectins is a rhamnose (Rha)-biding lectin [30] to analyze the bacterial cells. Though Rha is not commonly found in the cell surface PSs of mammalian and yeast, it is often found in many bacterial cells [31–33]. Therefore, we use CSL as a Rha-binding lectin isolated from

chum salmon (*Oncorhynchus keta*) eggs [34, 35] in this study (Fig. 1a). Similarly as four sugars: glucose (Glc), galactose (Gal), *N*-acetylglucose (GlcNAc), and *N*-acetylgalactose (GalNAc) are frequently found in the bacterial cell wall [31–33], lectins to bind these sugars are also essential. Glycan-binding specificities of the lectins used in this article are listed in Table 2 (also *see* Table 1). Each lectin concentration was 0.5 mg/mL.

2. Wash the lectin-immobilized glass slides with TBS containing 0.02 % NaN_3 and store at 4 °C until use.

3 Methods

3.1 Fluorescent Staining of Lactobacillus casei/paracasei Cells

1. Harvest the cells cultured in 4 ml MRS (1×10^9–2×10^{10}) by centrifugation (4,000 ×*g* for 5 min at 4 °C) and wash three times with PBS (10 mM phosphate-buffered saline, pH 7.0).
2. Suspend the cells in 4 mL of 70 % ethanol and then agitate (100 rpm for 30 min at RT) using a Personal-11 Shaker (Taitec Co., Ltd., Saitama, Japan) (*see* **Note 1**).
3. Harvest the cells by centrifugation (4,000 ×*g* for 5 min at 4 °C) and wash three times with PBS.
4. Suspend the cells in 4 mL of PBS before incubation with 10 μM SYTOX Orange Nucleic Acid Stain (Molecular Probes Co., Ltd., Eugene, OR) for 5 min at RT [27, 28] (*see* **Note 1**).
5. Wash the labeled cells three times with PBS and suspend in 360 μL of PBS containing 1 % BSA (PBS/BSA) finally. Measure the fluorescence intensity of 2×10^8 labeled cells using ALVO™X3 (PerkinElmer, Shelton, CT) within 1 h of labeling.
6. Make a microscopic examination of the cells labeled with SYTOX Orange. They are embedded in VECTASHIELD with 4′,6-diamidino-2-phenylindole (DAPI) (Vector Laboratories, Burlingame, CA) [36]. Slide glasses are examined using a Leica Q550FW system and the fluorescent images were subsequently analyzed using Image-Pro Plus software (Media Cybernetics, Silver Spring, MD) [37] (*see* **Note 2**).

3.2 Lectin Microarray Analysis

1. Suspend *L. casei/paracasei* cells labeled with SYTOX Orange in PBS/BSA, and add to each well of a glass slide containing immobilized lectins (100 μL/0.5–4 × 10^9 cells/well) followed by incubation at 4 °C for 1 h (*see* **Note 1**).
2. Remove unbound cells by immersing the inverted lectin microarray in at least 1 L of cold PBS at 4 °C for 1 h (Fig. 2).
3. Detect bound cells with lectins immobilized on a glass slide with an evanescent-field fluorescence scanner (Fig. 2, *see* **Notes 1** and **3**).

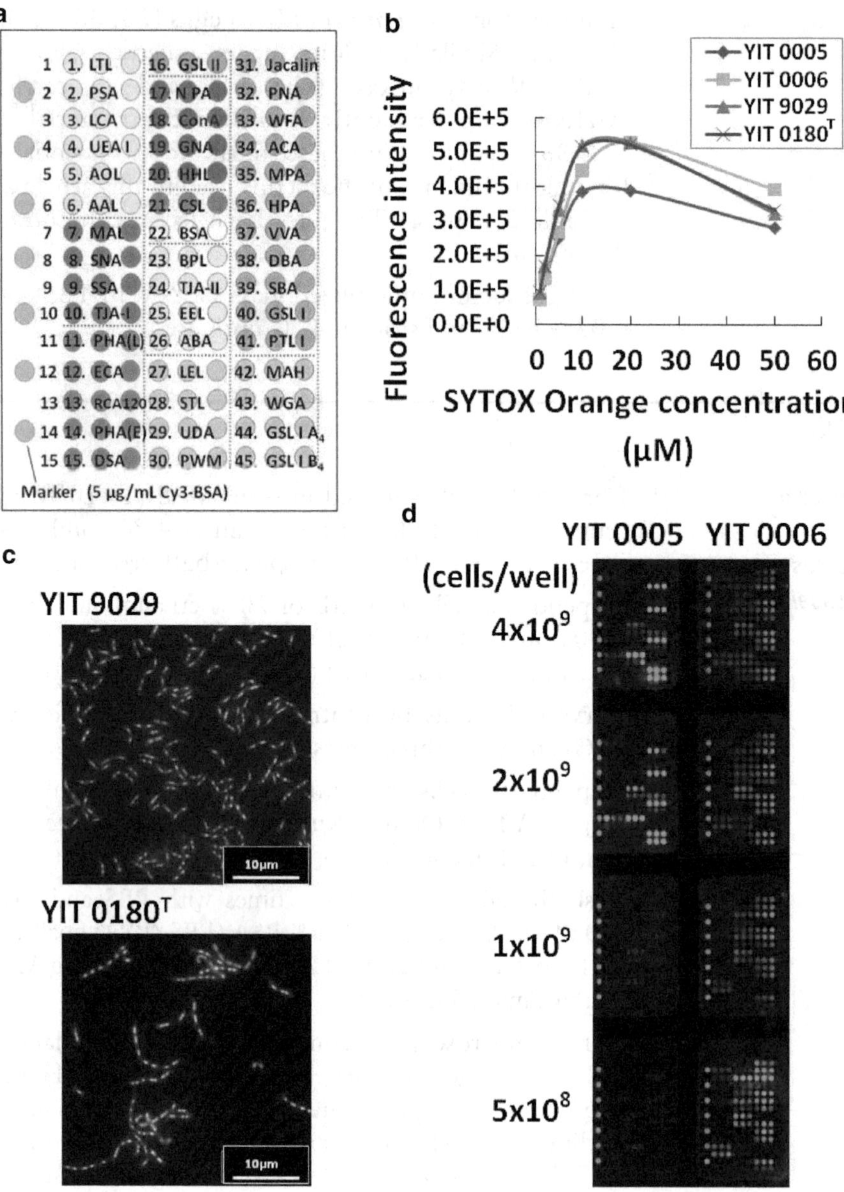

Fig. 1 The optimization of lectin microarray analysis for bacterial cells. (**a**) Lectin microarray format. Glycan-binding specificities of the lectins used in this article are listed in Table 2. Each lectin concentration was 0.5 mg/mL. (**b**) The effect of SYTOX Orange concentration on cell fluorescence intensity. *Lactobacillus casei* YIT 0005, YIT 0006, YIT 9029 and YIT 0180^T cells were grown to stationary phase (as described in Subheading 2), before labeling with SYTOX Orange (1–50 μg/ml, as described in Subheading 3). The fluorescence intensity of 1×10^8 cells was measured by ARVO X3 (PerkinElmer) using a Cy3 filter. The highest fluorescence intensity was 10 μM for each strain. (**c**) The fluorescent images of *L. casei* cells labeled with SYTOX Orange and DAPI. *L. casei* YIT 9029 (*above*) and *L. casei* YIT 0180^T (*below*) cells were labeled with 10 μM SYTOX Orange, and re-labeled with DAPI on a glass slide. *Red images* were obtained with SYTOX Orange and *green images* were obtained with SYTOX Orange and DAPI. The cells were observed to be thin and long using SYTOX Orange, and round using DAPI. (**d**) Dose-dependent fluorescent signals of *L. casei* cells in the lectin microarray. Varying numbers of cells labeled with SYTOX Orange were allowed to bind with the lectin array (0.5–4 × 10^9 cells/well). Bound cells were detected using an evanescent-field fluorescent scanner

Table 2
Sugar-binding specificities of the lectins

	Abbreviation	Origin	Sugar binding specificity
1	LTL	*Lotus tetragonolobus*	Fucα1-3GlcNAc, Sia-Le^x and Le^x
2	PSA	*Pisum sativum*	Fucα1-6GlcNAc (core fucose) and α-Man
3	LCA	*Lens culinaris*	Fucα1-6GlcNAc (core fucose) and α-Man, α-Glc
4	UEA-I	*Ulex europaeus*	Fucα1-2Galβ1-4Glc(NAc)
5	AOL	*Aspergillus oryzae*	Fucose
6	AAL	*Aleuria aurantia*	Fucose
7	MAL	*Maackia amurensis*	Siaα2-3Galβ1-4Glc(NAc)
8	SNA	*Sambucus nigra*	Siaα2-6Galβ1-4Glc(NAc)
9	SSA	*Sambucus sieboldiana*	Siaα2-6Gal β1-4Glc(NAc)
10	TJA-I	*Trichosanthes japonica*	Siaα2-6Gal β1-4Glc(NAc) and Galβ1-4Glc(NAc)
11	PHA(L)	*Phaseolus vulgaris*	Tetraantennary complex-type N-glycans
12	ECA	*Erythrina cristagalli*	Galβ1-4Glc(NAc)
13	RCA120	*Ricinus communis*	Galβ1-4Glc(NAc)
14	PHA(E)	*Phaseolus vulgaris*	Bisecting GlcNAc and biantennary N-glycans
15	DSA	*Datura stramonium*	polyLacNAc and branched LacNAc
16	GSL-II	*Griffonia simplicifolia*	GlcNAc and agalactosylated N-glycans
17	NPA	*Narcissus pseudonarcissus*	Non-substituted α1-6Man
18	ConA	*Canavalia ensiformis*	α-Man (no binding in the presence of bisecting GlcNAc)
19	GNA	*Galanthus nivalis*	Non-substituted α1-6Man
20	HHL	*Hippeastrum hybridum*	Non-substituted α1-6Man
21	CSL	Chum salmon eggs	Rha, Galα1-4Galβ1-4Glc
22	BSA	Bovine serum albumin	Negative control
23	BPL	*Bauhinia purpurea alba*	Galβ1-3GalNAc
24	TJA-II	*Trichosanthes japonica*	β-GalNAc and Fuc α1-2Gal
25	EEL	*Euonymus europaeus*	Galα1-3(Fucα1-2)Gal
26	ABA	*Agaricus bisporus*	Gal and Galβ1-3GalNac
27	LEL	*Lycopersicon esculentum*	Poly-LacNAc and (GlcNAc)n
28	STL	*Solanum tuberosum*	Poly-LacNAc and (GlcNAc)n
29	UDA	*Urtica dioica*	Poly-LacNAc and (GlcNAc)n
30	PWM	*Phytolacca americana*	Poly-LacNAc and (GlcNAc)n
31	Jacalin	*Artocarpus integrifolia*	Galβ1-3GalNAcα-Ser/Thr (T) and GalNAcα-Ser/Thr (Tn)

(continued)

Table 2 (continued)

	Abbreviation	Origin	Sugar binding specificity
32	PNA	*Arachis hypogaea*	Galβ1-3GalNAcα-Ser/Thr (T)
33	WFA	*Wisteria floribunda*	Terminal GalNAc (e.g., GalNAcb1-4GlcNAc)
34	ACA	*Amaranthus caudatus*	Galβ1-3GalNAcα-Ser/Thr (T)
35	MPA	*Maclura pomifera*	Galβ1-3GalNAcα-Ser/Thr (T) and GalNAcα-Ser/Thr (Tn)
36	HPA	*Helix pomatia*	GalNAc (Tn)
37	VVA	*Vicia villosa*	GalNAc and GalNAcα-Ser/Thr (Tn)
38	DBA	*Dolichos biflorus*	GalNAcα-Ser/Thr (Tn) and GalNAcα1-3GalNAc
39	SBA	*Glycine max*	Terminal GalNAc (especially GalNAcα1-3Gal)
40	GSL-I	*Griffonia simplicifolia*	αGalNAc, GalNAcα-Ser/Thr (Tn) and α-Gal
41	PTL-I	*Psophocarpus tetragonolobus*	αGalNAc and Gal
42	MAH	*Maackia amurensis*	Siaα2-3Gal
43	WGA	*Triticum vulgaris*	Multivalent Sia and (GlcNAc)n
44	GSL-I A_4	*Griffonia simplicifolia*	GalNAc
45	GSL-I B_4	*Griffonia simplicifolia*	αGal

GSL-I is a mixture of five isolectins

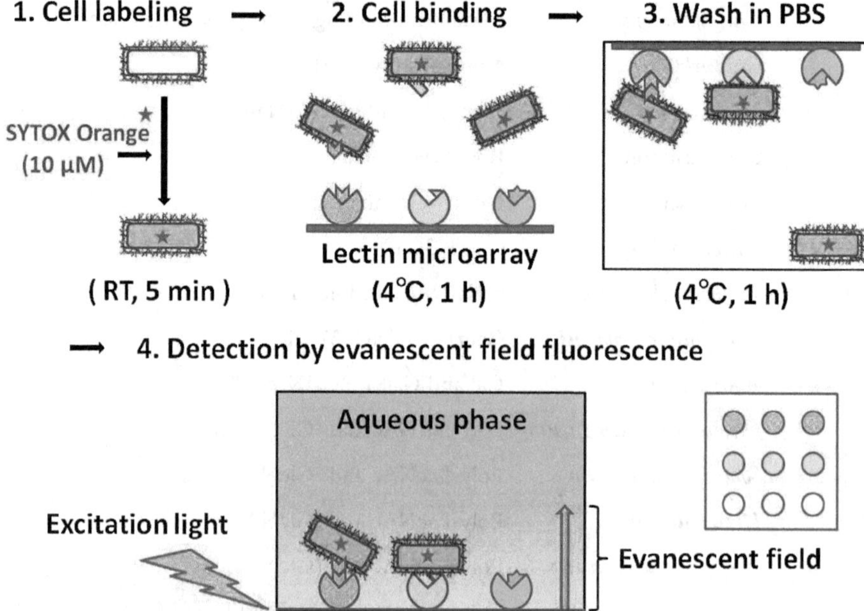

Fig. 2 The profiling system of bacterial cell wall polysaccharides using lectin microarray

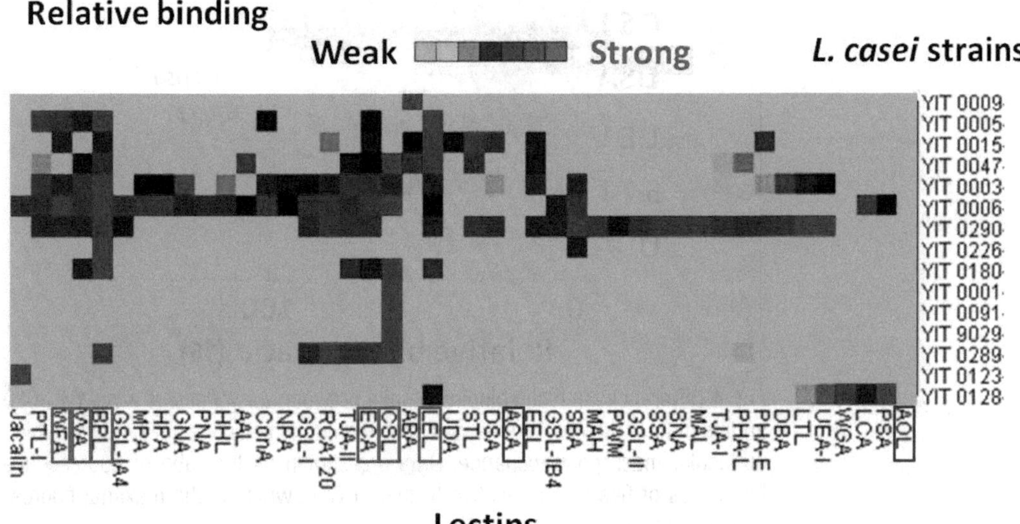

Fig. 3 Relative binding of 16 *Lactobacillus casei* strains with respect to lectin binding. The lectin-binding signals for each strain were normalized with the highest signal. Unsupervised hierarchical clusters were generated for the *L. casei* strains and their glycan-profiles obtained by lectin microarray. The levels of lectin binding signals are indicated by the color change from *green* (low binding levels) to *red* (high binding levels). The *red box* and *blue box* indicate the lectins that bound to more than five *L. casei* strains, and that did not bind to any *L. casei* strains, respectively

4. Normalize the lectin binding signals for each strain against the most intense signal. Data are shown as the ratio of fluorescence intensities of the 44 lectins relative to the maximal fluorescence intensity on the lectin microarray.
5. Analyze the unsupervised hierarchical clusters of *L. casei/paracasei* strains and their glycan-profiles obtained by lectin microarray. The levels of lectin binding signals are indicated by the color change from green (low binding levels) to red (high binding levels) (Fig. 3, also *see* **Note 9**).

3.3 Carbohydrate Inhibition Assay

It is possible to confirm that the signals as described above result from the binding affinity between lectins and bacterial strain by carbohydrate inhibition assay.

1. Add 0.1 M D-Gal, D-Glc, D-galactosylpyranosyl-(1 → 4)-D-Glc (Lac), D-mannose (Man), L-Rha, D-fructofuranosyl-(2 → 1)-D-glucopyranoside (Suc), or L-fucose to 2×10^9 cells labeled with SYTOX Orange in PBS/BSA.
2. Calculate the ratio of the effect of carbohydrate inhibition as a percentage of control signals with no added carbohydrate. For example, the results of CSL and YIT 9029 and/or YIT 0047 are shown in Fig. 4 (also *see* **Note 7**).

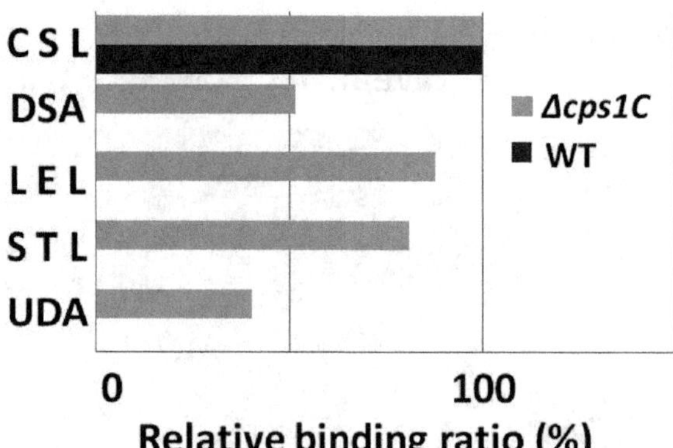

Fig. 4 Differences in lectin binding affinity between *Lactobacillus casei* YIT 9029 and its mutant the *cps1C* gene was knocked out, which does not synthesize high molecular-mass polysaccharide. Data are shown as the ratio of fluorescence intensities of five lectins relative to that of CSL, which is the maximal fluorescence intensity

4 Notes

1. Prepare the bacterial cells ($2–4 \times 10^9$ cells) labeled with SYTOX Orange [27, 28], whose maximum excitation wavelength is similar to Cy3 [29]. The higher fluorescence intensity of labeled cells is more suitable for lectin microarray analysis. Measure the fluorescence intensity of 1×10^8 cells labeled with different concentrations of SYTOX Orange (Fig. 1b). The highest fluorescence intensity is obtained at 10 μM for all the four strains, which is a 100-fold higher than the maximum concentration recommended in the standard protocol for labeling bacterial cells. It is essential to treat bacterial cells with 70 % ethanol for 30 min before staining with SYTOX Orange in order to obtain full fluorescence intensity.
2. Observe the labeled cells with fluorescence microscope not to effect on the cells treated with 70 % ethanol. As shown in Fig. 1c, *L. casei* YIT 9029 (above) and *L. casei* YIT 0180^T (below) cells are labeled with 10 μM SYTOX Orange, and relabeled with DAPI on a glass slide. Red images are obtained with SYTOX Orange and green images are obtained with SYTOX Orange and DAPI. The cells are observed to be thin and long using SYTOX Orange, and round using DAPI. DAPI stained only the center of the cells because the dye binds to AT-specific chromosomal DNA [36]. By contrast, SYTOX Orange stains the whole area of the cell two or three times more strongly than DAPI. The observed general staining of SYTOX Orange is explicable because the dye binds not only to double-stranded DNA, as described in the supplier's manual, but also to RNAs that are distributed throughout the cytoplasm.

3. Determine the most suitable concentration of SYTOX Orange to label bacterial cells in consideration of its stability. Labeled cells with 10 μM SYTOX Orange are stable for 2 days at 4 °C and for 6 weeks at −20 °C not to reduce the fluorescence intensity. As described above, it is concluded that 10 μM is the most suitable concentration of SYTOX Orange to label *L. casei/paracasei*.
4. Determine the required number of SYTOX Orange-labeled cells by using a few strains. We used two different strains of *L. casei*, YIT 0005 and YIT 0006, at 0.5, 1, 2, or 4×10^9 cells/well. As shown in Fig. 1d, only 2.0 and 4.0×10^9 cells/well gave the full fluorescence intensity and reproducible results in both strains. It is concluded that $2–4 \times 10^9$ cells/well is suitable for *L. casei/paracasei*.
5. Explore the lectin binding affinity of some genetically different strains of same genus and/or species or reported strains. For example, the lectin binding affinity of four *L. casei* strains (YIT 0005, YIT 0006, YIT 9029, and YIT 0180^T) was shown in Fig. 5. Bacterial cells (4×10^9 cells/well) were subjected to glycan profiling by lectin microarray. Data are the average ± S.D. of triplicate determinations as described in Subheading 3. Each strain shows a strain specific lectin binding affinity. Red box indicates the signals binds to Rha-binding lectin (CSL) [22, 34, 35] and YIT 9029 binds only to CSL. Meanwhile, YIT 0005, YIT 0006, and YIT 0180^T have similar lectin binding profile. Thus, strong signals are commonly detected in these three strains for ECA (asialo complex-type *N*-glycan binder), BPL and TJA-II (Gal binders), LEL (chitin binder), and *O*-glycan binders (WFA, MPA, HPA, VVA, SBA, GSL-I, and PTL-I).
6. Characterize the lectin binding activity of a bacterial strain in consideration of its genetic information and/or constituting sugars in its cell wall. We confirmed the sugar composition of YIT 9029 PS using the ABEE labeling kit [38]. Rha, Glc, Gal, and other sugars (data not shown) are detected, which is consistent with the results of Nagaoka et al. [31]. The binding pattern of YIT 9029 does not directly reflect the sugar composition of its cell wall PS. It is known that PS from YIT 9018, a parental strain of YIT 9029 produced by removing the bacteriophage φFSW from YIT 9018 genome [39], contains Rha, Glc, Gal, GlcNAc, and GalNAc [35]. Similarly, to analyze the relationship between lectin-binding profiles and sugar composition, we confirmed the sugar composition of PSs from YIT 0005, YIT 0006, and YIT 0180^T. While the presence of Glc, GlcNAc, and GalNAc is confirmed in all the strains, Rha is not detected in YIT 0005, and Gal is only confirmed in YIT 9029. The sugar compositions of YIT 0005 and YIT 0006 in our study are consistent with those reported by Simelyte et al. [32]

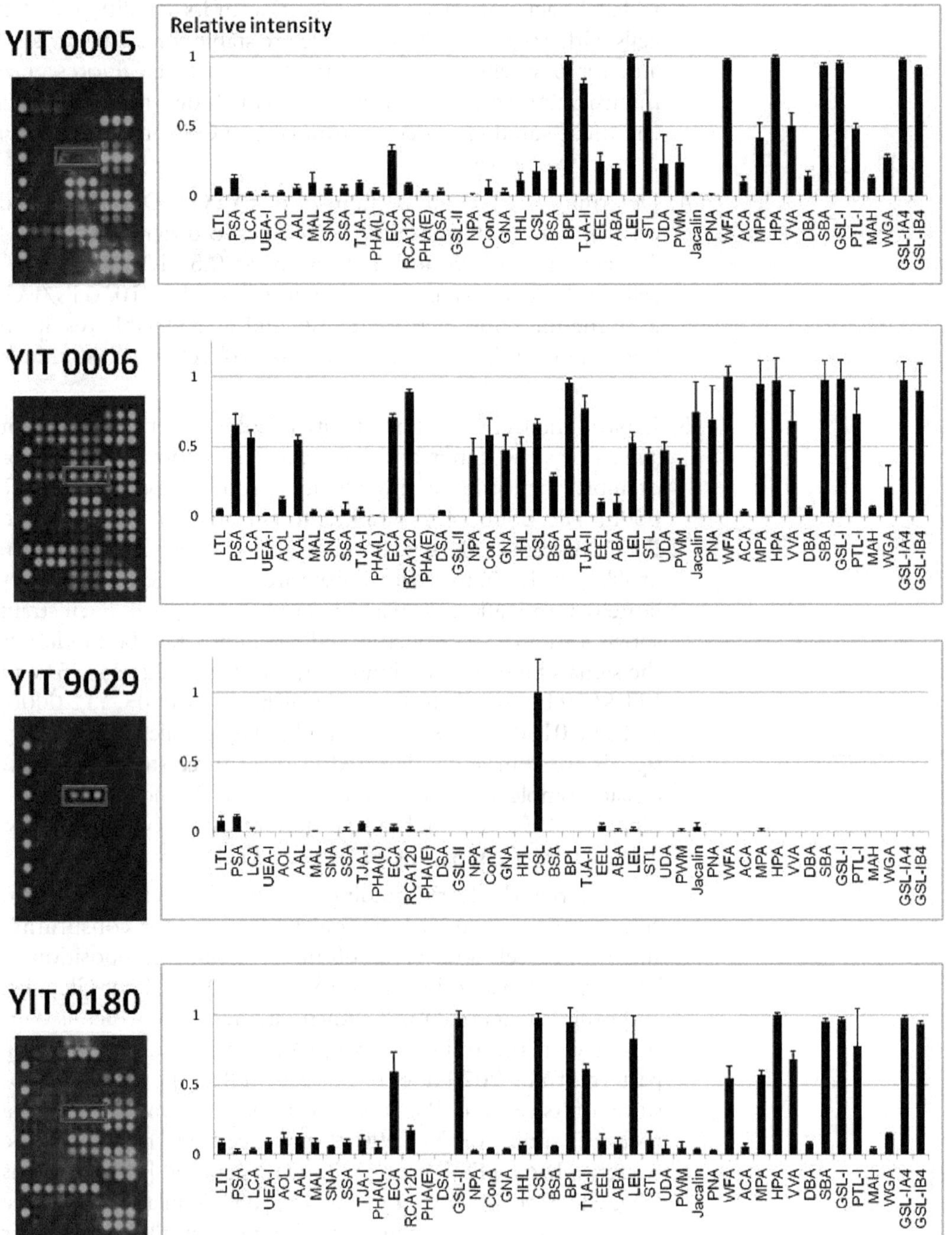

Fig. 5 Differential profiling of four *Lactobacillus casei* strains. Bacterial cells (4×10^9 cells/well) were subjected to glycan profiling by lectin microarray. Each strain shows a strain specific lectin binding affinity. *Red box* indicates the signals bound to Rha-binding lectin (CSL). Though YIT 0005, YIT 0006 and YIT 0180^T cells bound to a variety of lectins, YIT 9029 cells bound to CSL only. Data are the average ±S.D. of triplicate determinations

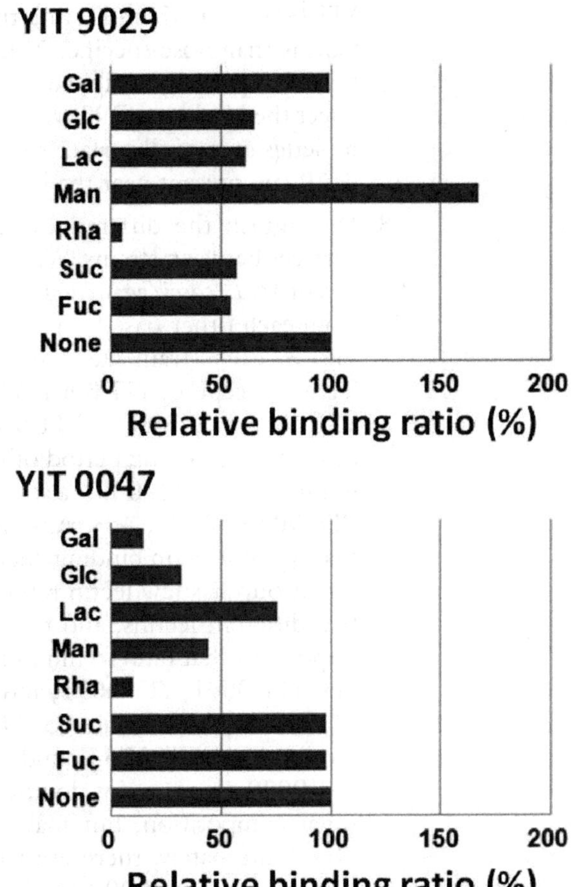

Fig. 6 Carbohydrate inhibition assay. Relative binding ratio of *Lactobacillus casei* YIT 9029 (*above*) and YIT 0047 (*below*) are shown. Gal, Glc, Lac, Man, Rha, Suc and Fuc were added at 0.1 M to 2×10^9 YIT 9029 or YIT 0047 cells labeled with SYTOX Orange. None, control assay, without sugars. Rha competes to inhibit CSL in both strains. The other sugars were not as effective in inhibiting CSL

(YIT 0005 and YIT 0006 correspond to *L. casei* ATCC 25302 and *L. casei* ATCC 25303, respectively). Based on our results, the binding profiles of the strains analyzed are not just determined by the sugar composition of the cell wall PSs, they are partly determined by the specificity of lectin binding.

7. Measure the binding specificity of bacterial strain and lectins by using monosaccharides and/or disaccharides as competitors in consideration of constituting PS in its cell wall. We performed the binding specificity of YIT 9029 and YIT 0047 (NIRD A-121) to CSL using monosaccharides and disaccharides as competitors (Gal, Glc, Man, Rha, Suc, Fuc, and Lac) to measure. The binding ratio of YIT 9029 and YIT 0047 are shown in Fig. 6. Rha completely inhibits the binding of YIT 9029 and

YIT 0047 to CSL, indicating the binding of both the strains to CSL is rhamnose-specific. Although the binding of YIT 0047 to CSL is partially inhibited by Gal and Glc, these sugars do not affect the binding of YIT 9029 to CSL, suggesting Gal and Glc moieties in the cell surface structure of YIT 0047, but not YIT 9029, are present near the binding site of CSL.

8. Distinguish the differences in the glycome of bacterial cell surfaces between strains. We carried out lectin microarray analysis of 16 *L. casei/paracasei* strains, which are indistinguishable from each other based on 16S-rRNA sequences. As shown in Fig. 3, each strain has a unique binding profile to various lectins, except for YIT 0001, YIT 0091, and YIT 9029 (Fig. 3). YIT 0091 is a clone of YIT 9029 that has been stored in our institute over a long period of time (M. Onoue, personal communication). YIT 0001 also had the same binding profile as YIT 9029 [7, 39, 40]. Among 16 *L. casei/paracasei* strains, two types of lectin binding characteristics could be recognized; one group has few lectin responders that bind to only one or two different lectins, and the other group has multiple lectin responders that bind to multiple lectins with different specificities. YIT 0001, YIT 0091, and YIT 9029 bound only to CSL, YIT 0009 bound only to ABA, YIT 0123 bound only to Jacalin, and YIT 0226 bound only to BPL and SBA. Similar to YIT 9029, these strains do not necessarily have PSs with simple sugar composition, but may contain various sugar compositions. Alternately, there are strains that bind to a number of lectins, e.g., YIT 0003 binds 23 lectins, YIT 0006 binds 24 lectins, and YIT 0290 binds 27 lectins. From all the binding profiles we could not draw a specific binding profile for *L. casei/paracasei* species, but rather recognized strain-specific profiles for individual strains. None of the strains bound ACA and AOL, which are specific for Gal-β1-3GalNAc-Thr/Ser and Fuc, respectively.

9. Agglutination itself does not affect the binding to lectins for our system (Fig. 5). In fact, some strains in this study, for instance YIT 0005 and YIT 0180^{T} are confirmed to undergo auto-agglutination in MRS medium. However, none of the *L. acidophilus* strains binds to any lectins because of auto-agglutination [41].

10. Lectin microarray is more useful to analyze the cell wall glycome changes between a mutant and wild type. We confirmed the *ΔcpsIC* mutant of YIT 9029 with YIT 9029 as a typical example of a strain with few lectin binders. As described in **Note 6**, the cell wall PS of YIT 9029 contains several sugar molecules, including Glc, Gal, Rha, GlcNAc, and GalNAc with different linkages. The *ΔcpsIC* mutant of YIT 9029 was constructed by site-specific deletion mutagenesis within the *cpsIC*

gene encoding a putative rhamnosyltransferase, which is essential for the synthesis of PS-1, and most probably lacks a certain glycosyltransferase activity [7]. The glycan profile of *ΔcpslC* was clearly different from YIT 9029. In addition to binding CSL, the *ΔcpslC* mutant cells bound to an asialo complex-type *N*-glycan binder, DSA, and chitin binders, LEL, STL, and UDA (Fig. 4). Interestingly all these lectins have the binding specificity of the *N*-acetylglucosamine polymer. No other lectins bound to the mutant. Therefore, the deficiency of the *cpslC* gene resulted in a partial release of PS structures consisting of at least one GlcNAc moiety in the cell wall PS, in addition to the disappearance of the high molecular mass PS-1 [7].

Acknowledgments

We thank Yoshiko Kubo and Jinko Murakami of the Research Center for Medical Glycoscience at the National Institute of Advanced Industrial Science and Technology for help in preparation of the lectin microarray, Toshihiko Takada of the Yakult Central Institute for Microbiological Research for help with bacterial labeling methods and preparation of the electron microscopic images, and Dr. Koichi Watanabe for advice on choosing *L. casei/paracasei* strains. We deeply thank Mayumi Kiwaki and Tohru Iino of the Yakult Central Institute for Microbiological Research, Dr. Teruo Yokokura and the late Dr. Toshiaki Osawa, who always encouraged us and incited helpful discussions.

References

1. Akira S, Takeda K (2004) Toll-like receptor signaling. Nat Rev Immunol 4:499–511
2. Inohara N, Nuñes G (2003) NODs: intracellular proteins involved in inflammation and apoptosis. Nat Rev Immunol 3:371–382
3. Lebeer S, Vanderleyden SJ, De Keersmaecker SCJ (2008) Genes and molecules of L. supporting probiotic action. Microbiol Mol Biol Rev 72:728–764
4. Baik YS, Cheong WJ (2007) Development of SPE for recovery of polysaccharides and its application to the determination of monosaccharide composition of the polysaccharide sample of a lactobacillus KLB 58. J Sep Sci 30:1509–1515
5. Kullberg MC (2008) Soothing intestinal sugars. Nature 453:602–604
6. Liu CH, Lee SM, VanLare JM et al (2008) Regulation of surface architecture by symbiotic bacteria mediates host colonization. Proc Natl Acad Sci U S A 105:3951–3956
7. Yasuda E, Serata M, Sako T (2008) Suppressive effect on activation of macrophages by *Lactobacillus casei* Strain Shirota genes determining the synthesis of cell wall-associated polysaccharides. Appl Environ Microbiol 74: 4746–4755
8. Coyne MJ, Tzianabos AO, Mallory BC et al (2001) Polysaccharide biosynthesis locus required for virulence of *Bacteroides fragilis*. Infect Immun 69(7):4342–4350
9. Cobb BA, Wang Q, Tzianabos AO et al (2004) Polysaccharide processing and presentation by the MHCII pathway. Cell 117:677–687
10. Mazmanian SK, Liu CH, Tzianabos AO et al (2005) An immunomodulatory molecule of symbiotic bacteria directs maturation of the host immune system. Cell 122:107–118
11. Vaningelgem F, Zamfir M, Mozzi F et al (2004) Biodiversity of exopolysaccharides produced by *Streptococcus thermophilus* strains is reflected in their production and their molecular

and functional characteristics. Appl Environ Microbiol 70:900–912

12. Cieslewicz MJ, Chaffin D, Glusman G et al (2005) Structural and genetic diversity of group B *Streptococcus* capsular polysaccharides. Infect Immun 73:3096–3103
13. Bentley SD, Aanensen DM, Mavroidi A et al (2006) Genetic analysis of the capsular biosynthetic locus from all 90 Pneumococcal serotypes. PLoS Genet 2:e31
14. Feng Y, Xiao-Min W (2007) Difference in gene expression of macrophage between normal spleen and portal hypertensive spleen indentified by cDNA microarray. World J Gastroenterol 13:3369–3373
15. Pretzer G, Snel J, Molenaar D et al (2005) Biodiversity-based identification and functional characterization of the mannose-specific adhesion of *Lactobacillus plantarum*. J Bacteriol 187:6128–6136
16. Fang Y, Cao H, Cover TL et al (2007) Soluble proteins produced by probiotic bacteria regulate intestinal epithelial cell survival and growth. Gastroenterology 132:562–575
17. Kankainen M, Paulin L, Tynkkynwn S et al (2009) Comparative genomic analysis of *Lactobacillus rhamnosus* GG reveals pili containing a human-mucus binding protein. Proc Natl Acad Sci U S A 106:17193–17198
18. Angeloni S, Ridet JL, Kusy N et al (2005) Glycoprofiling with micro-arrays of glycoconjugates and lectins. Glycobiology 15:31–41
19. Hsu K-L, Mahal LK (2006) A lectin microarray approach for the rapid analysis of bacterial glycans. Nat Protoc 1:543–549
20. Hsu K-L, Pilobello KT, Mahal LK (2006) Analyzing the dynamic bacterial glycome with a lectin microarray approach. Nat Chem Biol 2:153–157
21. Uchiyama N, Kuno A, Koseki-Kuno S et al (2006) Development of a lectin microarray based on an Evanescent-field fluorescence principle: a new strategy for glycan profiling. Method Enzymol 415:341–351
22. Tateno H, Uchiyama N, Kuno A et al (2007) A novel strategy for mammalian cell surface glycome profiling using lectin microarray. Glycobiology 17:1138–1146
23. Hsu K-L, Gildersleeve JC, Mahal LK (2008) A simple strategy for the creation of a recombinant lectin microarray. Mol BioSyst 4:654–662
24. Hirabayashi J, Yamada M, Kuno A et al (2013) Lectin microarrays: concept, principle and applications. Chem Soc Rev 42:4443–4458
25. Yasuda E, Tateno H, Hirabayashi J et al (2011) Lectin microarray reveals binding profiles of *Lactobacillus casei* strains in a comprehensive analysis of bacterial cell wall polysaccharides. Appl Environ Microbiol 77:4539–4546
26. Dicks LM, Plessis EM, Dellaglio F et al (1996) Reclassification of *Lactobacillus casei* subsp. *casei* ATCC 393 and *Lactobacillus rhamnosus* ATCC 15820 as *Lactobacillus zeae* nom. rev., designation of ATCC 334 as the neotype of *L. casei* subsp. *casei*, and rejection of the name *Lactobacilllus paracasei*. Int J Syst Bacteriol 46:337–340
27. Yan X, Habbersett RC, Cordek JM et al (2000) Development of a mechanism-based, DNA staining protocol using SYTOX Orange nucleic acid stain and DNA fragment sizing flow cytometry. Anal Biochem 286:138–148
28. Yan X, Habbersett RC, Yoshida TM et al (2005) Probing the kinetics of SYTOX Orange stain binding to double-stranded DNA with implications for DNA analysis. Anal Chem 77: 3554–3562
29. Shida K, Kiyoshima-Shibata J, Nagaoka M et al (2006) Induction of interleukin-12 by *Lactobacillus* strains having a rigid cell wall resistant to intracellular digestion. J Dairy Sci 89:3306–3317
30. Tateno H (2010) SUEL-related lectins, a lectin family widely distributed throughout organisms. Biosci Biotechnol Biochem 74(6): 1141–1144
31. Nagaoka M, Muto M, Nomoto K et al (1990) Structure of polysaccharide-peptidoglycan complex from the cell wall of *Lactobacillus casei* YIT 9018. J Biochem 108:568–571
32. Šimelyte E, Rimpiläinen M, Lehtonen L et al (2000) Bacterial cell wall-induced arthritis: chemical composition and tissue distribution of four Lactobacillus strains. Infect Immun 68: 3535–3540
33. Yokokura T, Kodaira S, Ishiwa H et al (1974) Lysogeny in Lactobacilli. J Gen Mirobiol 84: 277–284
34. Shirai T, Watanabe Y, Lee M et al (2009) Structure of Rha-binding lectin CSL3: unique pseudo-tetrameric architecture of a pattern recognition protein. J Mol Biol 391:390–403
35. Watanabe Y, Abolhassani M, Tojo Y et al (2009) The function of Rha-binding lectin in innate immunity by restricted binding Gb3. Dev Comp Immunol 33:187–197
36. Schweizer D (1981) Counterstain-enhanced chromosome banding. Hum Genet 57:1–14
37. Takada T, Matsumoto K, Nomoto K (2004) Development of multi-color FISH method for analysis of seven *Bifidobacterium* species in human feces. J Microbiol Methods 58:413–421

38. Yasuno S, Kokubo K, Kamei M (1999) New method for determining the sugar composition of glycoproteins, glycolipids, and oligosaccharides by high-performance liquid chromatography. Biosci Biotechnol Biochem 63:1353–1359
39. Shimizu-Kadota M, Kiwaki M, Sawaki S et al (2000) Insertion of bacteriophage phiFSW into the chromosome of *Lactobacillus casei* Shirota (S-1): characterization of attachment sites and integrase gene. Gene 249:127–134
40. Yuki N, Watanabe K, Mike A et al (1999) Survival of a probiotic, *Lactobacillus casei* strain Shirota, in the gastrointestinal tract: Selective isolation from feces and identification using monoclonal antibodies. Int J Food Microbiol 48:51–57
41. Annuk H, Hynes SO, Hirmo S et al (2001) Characterisation and differentiation of lactobacilli by lectin typing. J Med Microbiol 50: 1069–1074

Chapter 26

Live-Cell Imaging of Human Pluripotent Stem Cells by a Novel Lectin Probe rBC2LCN

Hiroaki Tateno, Yasuko Onuma, and Yuzuru Ito

Abstract

We performed comprehensive glycome analysis of a large set of human pluripotent stem cells (hPSCs) using a high-density lectin microarray. We found that a recombinant lectin, rBC2LCN, binds exclusively to all of the undifferentiated hPSCs tested, but not to differentiated somatic cells. rBC2LCN can be used for both the staining and sorting of fixed and live hPSCs. rBC2LCN could serve as a novel detection reagent for hPSCs, particularly given that rBC2LCN is cost effective and, unlike conventional antibodies which require mammalian cells for their production, is easy to produce in a large amount (0.1 g/L) in an *Escherichia coli* expression system. Here we describe protocols for the fluorescence staining of fixed and live hPSCs and their detection by flow cytometry.

Key words Pluripotent stem cells, iPS, ES, Staining, rBC2LCN, Lectin

1 Introduction

hPSCs with properties of self-renewal and pluripotency are attractive cell sources for cell replacement therapies [1, 2]. Extensive research has been conducted with these cells to produce a range of cell types. Several pluripotent stem cell-based therapeutics have entered the clinical trial stage. In 2012, preliminary outcomes were published concerning clinical trials conducted with retinal pigment epithelial (RPE) cells derived from human embryonic stem cells (hESCs) to treat patients with dry age-related macular degeneration and Stargardt's macular dystrophy [3].

Glycans are located on the outermost cell surface and reflect numerous cell properties, including the degree of differentiation and malignancy. Indeed, well-known pluripotency markers such as SSEA3, SSEA4, Tra-1-60, and Tra-1-81 are glycans. Therefore glycans serve as appropriate cell surface marker targets for the evaluation of cells. To develop a glycan-targeting probe, which can be used to identify, isolate, and separate hPSCs, we performed comprehensive glycome analysis of a large set of human induced

Jun Hirabayashi (ed.), *Lectins: Methods and Protocols*, Methods in Molecular Biology, vol. 1200,
DOI 10.1007/978-1-4939-1292-6_26, © Springer Science+Business Media New York 2014

pluripotent stem cells (hiPSCs) and human embryonic stem cells (hESCs) using a high-density lectin microarray [4]. We found that a lectin designated rBC2LCN (recombinant N-terminal domain of BC2L-C), identified from *Burkholderia cenocepacia*, bound exclusively to all of the undifferentiated hPSCs tested, but not to differentiated somatic cells [4]. Fluorescently labeled rBC2LCN can thus be used as an effective staining reagent for fixed hPSCs [5]. Furthermore, it has been possible to develop a protocol to stain live hPSCs. In this protocol, fluorescently labeled rBC2LCN is supplemented into the cell culture medium. It should be noted that rBC2LCN has little or no toxicity to hPSCs even at a high concentration (~100 μg/mL), as confirmed by DNA microarray [5]. Recently, we searched for glycoprotein ligands of rBC2LCN expressed on hPSCs [6] and found that podocalyxin, a hyperglycosylated type 1 transmembrane protein highly expressed on hPSCs, was found to be a predominant cell surface ligand of rBC2LCN [6]. In frontal affinity chromatography analyses, rBC2LCN exhibited significant affinity to a mucin-type *O*-glycan comprising an H type 3 structure isolated from human 201B7 iPSCs, indicating that H type 3 is a novel pluripotency marker recognized by rBC2LCN [6, 7]. rBC2LCN is a small single-chain protein (16 kDa), which can be easily purified to homogeneity by one-step sugar-immobilized affinity chromatography from soluble fractions of *Escherichia coli* (>0.1 g/L). In contrast, antibodies are large and complex proteins (>140 kDa) that require mammalian cells for their production. Thus, rBC2LCN has high potential to serve as a novel type of detection reagent for the targeting of hPSCs. Here we present two staining protocols of hPSCs using rBC2LCN:

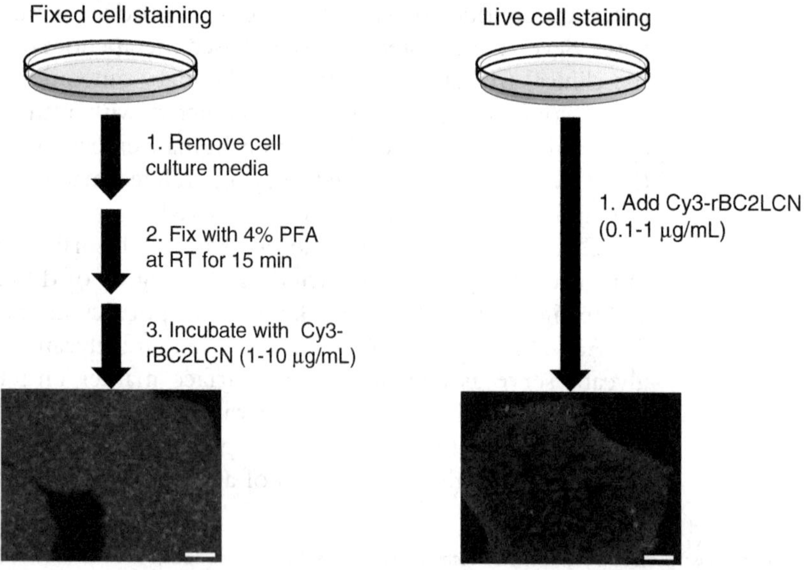

Fig. 1 Fluorescence staining of fixed (*left*) and live (*right*) 201B7 hiPSCs. Scale bars: 50 μm

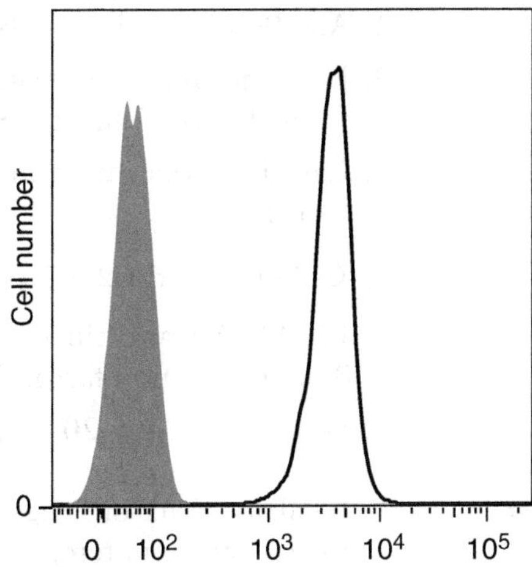

Fig. 2 Flow cytometry analysis of 201B7 hiPSCs using rBC2LCN

"fixed staining" and "live staining" (Fig. 1). In addition, the application of rBC2LCN to flow cytometry is described (Fig. 2).

2 Materials

1. rBC2LCN (Wako Pure Chemicals, Osaka).
2. Cy3 Mono-reactive Dye, Protein Array Grade (GE Healthcare).
3. R-Phycoerythrin Labeling Kit—NH2 (Dojindo, Tokyo).
4. FITC-I (Dojindo, Tokyo).
5. 0.8 mL volume miniature column.
6. PBS.
7. PBS/BSA: PBS containing 1 % BSA.
8. 4 % paraformaldehyde.
9. 24-Well plates.
10. 201B7 hiPSCs.
11. Versene (EDTA) 0.02 % (Lonza).
12. Accutase (Funakoshi).
13. 40 μm Cell Strainer (Stemcell Technologies).
14. 5 mL polystyrene round bottom tubes (BD).
15. TBS.

3 Methods

3.1 Cy3 Labeling

1. Incubate 10 μL of 1 mg/mL rBC2LCN in PBS with Cy3-NHS (for 10 μg protein-labeling equivalent) at room temperature in the dark for 1 h.

2. Add 90 μL of TBS (*see* **Note 1**).
3. Pass through a miniature column containing 100 μL of Sephadex G-25 fine to remove unreacted Cy3 (*see* **Note 2**).
4. Calculate concentration according to the following equation (Eq. 1):

$$[\text{Cy3}-\text{labeled rBC2LCN}]=[\text{A280}-(0.08\times\text{A552})]/13{,}688 \quad (1)$$

13,688: Molar extinction coefficient of rBC2LCN.
0.08: Correction factor for Cy3.

5. Store at 4 °C or −20 °C in the dark.

3.2 FITC Labeling

1. Incubate 10 μL of 1 mg/mL rBC2LCN in PBS with 10 μg/mL FITC-I at room temperature in the dark for 1 h.
2. Add 90 μL of TBS.
3. Pass through a miniature column containing 100 μL of Sephadex G-25 fine to remove unreacted FITC.
4. Calculate concentration according to the following equation (Eq. 2):

$$[\text{FITC}-\text{labeled rBC2LCN}]=[\text{A280}-(0.22\times\text{A500})]/13{,}688 \quad (2)$$

13,688: Molar extinction coefficient of rBC2LCN.
0.22: Correction factor for FITC.

5. Store at 4 °C or −20 °C in the dark.

3.3 R-Phycoerythrin (PE) Labeling

1. Label rBC2LCN according to the manufacturer's instructions and store at 4 °C until use.
2. Use the same protein concentration as the input protein concentration (*see* **Note 3**).
3. Store at 4 °C or −20 °C in the dark.

3.4 Fixed Staining

1. Culture hPSCs in 24-well plates (*see* **Note 4**).
2. Remove cell culture media.
3. Wash with PBS, 500 μL/well.
4. Fix with 200 μL/well of 4 % paraformaldehyde at room temperature for 15 min (*see* **Note 5**).
5. Incubate with 200 μL/well of Cy3-labeled rBC2LCN (1–10 μg/mL) at room temperature for 1 h.
6. Remove the solution.
7. Wash with PBS, 500 μL/well.
8. Observe under fluorescence microscope.

For an example of this staining method, *see* Fig. 1 (*left*).

3.5 Live Staining

1. Add Cy3-labeled rBC2LCN (0.1–1 μg/mL) into cell culture media.
2. Incubate for 1–2 h in CO_2 incubator.
3. Observe under fluorescence microscope (*see* **Note 6**).

For an example of this staining method, *see* Fig. 1 (*right*).

3.6 Flow Cytometry

1. Culture hPSCs on 6-well plates.
2. Wash once with PBS. Recover hPSCs with 1 mL of Versene (EDTA) 0.02 % or Accutase by incubating for 8 min in CO_2 incubator (*see* **Note 7**).
3. Remove the solution.
4. Add 2 mL of hPSC cell culture media.
5. Pipette with a 1 mL micropipette.
6. Recover cells.
7. Centrifuge at 2,300 × *g* for 1 min.
8. Resuspend with PBS/BSA (1×10^5 cells/100 μL).
9. Add 0.1–10 μg/mL of FITC- or PE-labeled rBC2LCN.
10. Incubate on ice for 1 h.
11. Centrifuge at 2,300 × *g* for 1 min.
12. Remove supernatants and resuspend with 1 mL of PBS/BSA.
13. Centrifuge at 2,300 × *g* for 1 min.
14. Remove supernatants and resuspend with 500 μL PBS/BSA.
15. Transfer to 5 μL polystyrene round-bottom tube and pass through 40 μm cell strainer.
16. Analyze by flow cytometry.

For an example of this staining method, *see* Fig. 2.

4 Notes

1. Use TBS to block unreacted fluorescence dye.
2. Commercially available desalting columns could also be used.
3. In this protocol, no desalting step is required. Thus the protein concentration should be the same as the input protein concentration.
4. Other types of plates such as 6- and 12-well plates and cover slips can also be used.
5. Cells can be fixed at 4 °C for 10–60 min. Chose fixation conditions depending on the antibodies to be used, co-staining, and cell types.

6. Long exposure time greatly affects cell viability due to phototoxicity.
7. Cells tend to die during this step, so gentle treatment is necessary.

Acknowledgement

Human iPS cell line 201B7 (HPS0063) was obtained from the RIKEN Bioresource Center.

References

1. Takahashi K, Tanabe K, Ohnuki M, Narita M, Ichisaka T, Tomoda K, Yamanaka S (2007) Induction of pluripotent stem cells from adult human fibroblasts by defined factors. Cell 131:861–872
2. Nakagawa M, Koyanagi M, Tanabe K, Takahashi K, Ichisaka T, Aoi T, Okita K, Mochiduki Y, Takizawa N, Yamanaka S (2008) Generation of induced pluripotent stem cells without Myc from mouse and human fibroblasts. Nat Biotechnol 26:101–106
3. Schwartz SD, Hubschman JP, Heilwell G, Franco-Cardenas V, Pan CK, Ostrick RM, Mickunas E, Gay R, Klimanskaya I, Lanza R (2012) Embryonic stem cell trials for macular degeneration: a preliminary report. Lancet 379:713–720
4. Tateno H, Toyoda M, Saito S, Onuma Y, Ito Y, Hiemori K, Fukumura M, Nakasu A, Nakanishi M, Ohnuma K, Akutsu H, Umezawa A, Horimoto K, Hirabayashi J, Asashima M (2011) Glycome diagnosis of human induced pluripotent stem cells using lectin microarray. J Biol Chem 286:20345–20353
5. Onuma Y, Tateno H, Hirabayashi J, Ito Y, Asashima M (2013) rBC2LCN, a new probe for live cell imaging of human pluripotent stem cells. Biochem Biophys Res Commun 431:524–529
6. Tateno H, Matsushima A, Hiemori K, Onuma Y, Ito Y, Hasehira K, Nishimura K, Ohtaka M, Takayasu S, Nakanishi M, Ikehara Y, Ohnuma K, Chan T, Toyoda M, Akutsu H, Umezawa A, Asashima M, Hirabayashi J (2013) Podocalyxin is a glycoprotein ligand of the human pluripotent stem cell-specific probe rBC2LCN. Stem Cells Transl Med 2:265–273
7. Hasehira K, Tateno H, Onuma Y, Ito Y, Asashima M, Hirabayashi J (2012) Structural and quantitative evidence for dynamic glycome shift on production of induced pluripotent stem cells. Mol Cell Proteomics 11:1913–1923

Chapter 27

Carbohydrate-Binding Specificity of Lectins Using Multiplexed Glyco-Bead Array

Kazuo Yamamoto

Abstract

Multiplexed bead array is an application that allows us to quantify multiple ligands simultaneously by using flow cytometry. Glycopeptides are immobilized on multiplexed beads, and the glycan-binding specificities of several lectins are determined. This strategy is easy, rapid, and suitable for small amount of samples, and allows the reliable elucidation under the identical condition. Such a technology is useful for analyzing characteristics and functions of lectins.

Key words Lectin, Specificity, Multiplexed bead, Flow cytometry, Fluorescent microsphere, Glycopeptide

1 Introduction

Glycans on the cell surface play important roles in cellular recognition including development, embryogenesis, immune recognition, and infection. Intracellular function of *N*-glycans attached to proteins is also reported in quality control of glycoproteins such as folding, degradation, sorting, and transport of glycoproteins. In every case, recognition of glycans by sugar-binding proteins, lectins, is essential for the regulation of these biological events. For understanding these glycobiological processes precisely, it is necessary to develop a large-scale analysis of glycome, especially elucidation of detailed interaction between glycans and lectin. Luminex microspheres are polystyrene microspheres with two spectrally distinct fluorochromes. Using ratios of these two fluorochromes, a spectral array is created encompassing 100 different microsphere sets. A third fluorochrome coupled to a reporter molecule makes it possible to measure the binding of the molecule to each microsphere surface [1]. In this chapter, we prepare arrays of multiplexed beads that bear *N*- and *O*-linked glycopeptides derived from several glycoproteins and demonstrate their binding to several lectins

Jun Hirabayashi (ed.), *Lectins: Methods and Protocols*, Methods in Molecular Biology, vol. 1200,
DOI 10.1007/978-1-4939-1292-6_27, © Springer Science+Business Media New York 2014

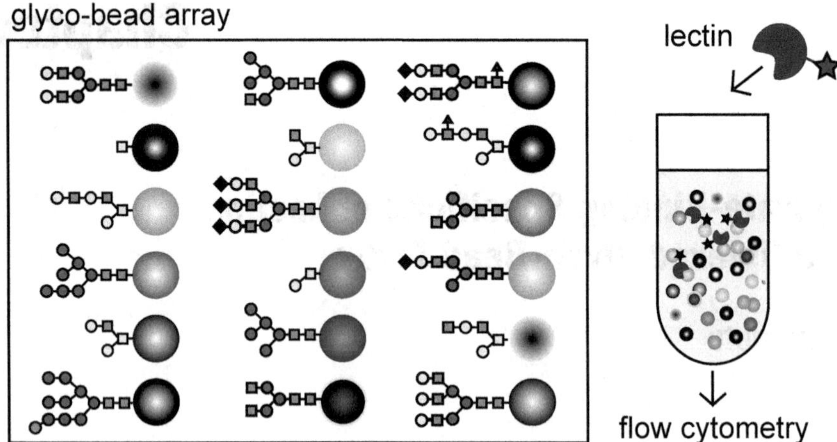

Fig. 1 Schematic illustration of sugar-binding assay of lectins using multiplexed bead-based suspension arrays

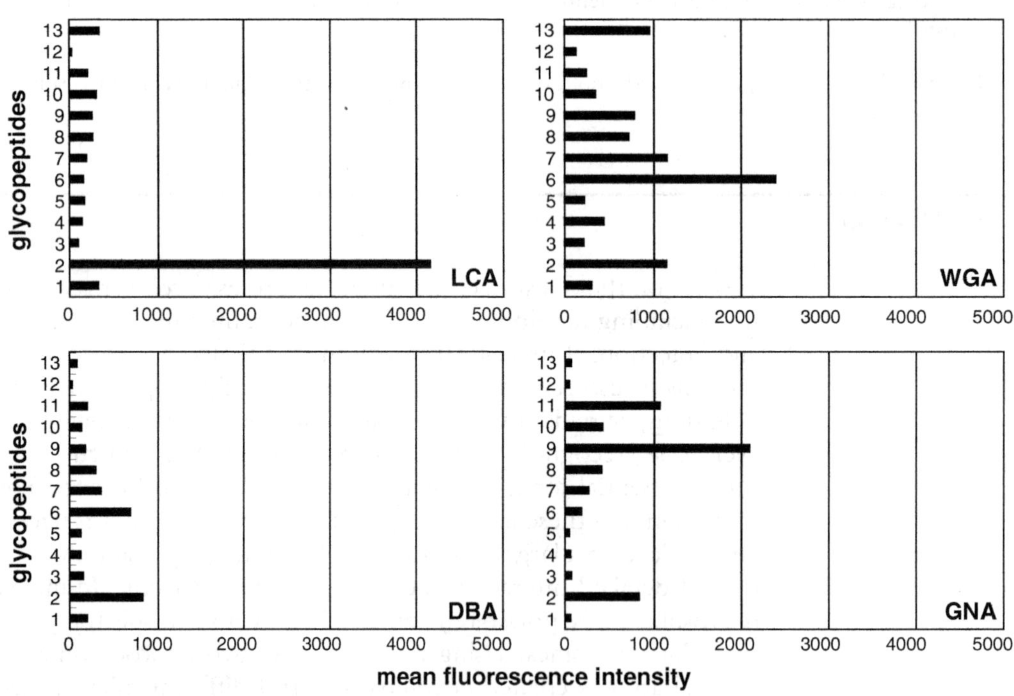

Fig. 2 Reactivities of several lectins with 13 *N*- and *O*-linked glycopeptides. *LCA Lens culinaris* agglutinin, *WGA* wheat germ agglutinin, *DBA Dolichos biflorus* agglutinin, *GNA Galanthus nivalis* agglutinin. Glycopeptides used are monosialyl biantennary (*1*) and triantennary (*2*) complex-type glycopeptides from bovine thyroglobulin, $Man_9GlcNAc_2$ (*3*), $Man_8GlcNAc_2$ (*4*), and $Man_7GlcNAc_2$ (*5*) high-mannose-type glycopeptides from porcine thyroglobulin, GP-I (*6*), GP-II (*7*), and GP-III (*8*) hybrid-type glycopeptides from chicken ovalbumin, $Man_6GlcNAc_2$ (*9*), $Man_5GlcNAc_2$ (*10*), and $Man_4GlcNAc_2$ (*11*) high-mannose-type glycopeptides from chicken ovalbumin, monosialyl *O*-glycan from bovine submaxillary mucin (*12*), and disialyl *O*-glycan from porcine submaxillary mucin (*13*)

to identify the wide range of specificity of these lectins (Figs. 1 and 2; also *see* ref. 2). This method provides easy, rapid, sensitive, and reproducible assays compared with conventional systems [3–7].

2 Materials

2.1 Instrumentation

1. Luminex 100/200 cytometer (Luminex, Austin, TX).
2. Bath sonicator: For example, Branson, ED-40 (Danbury, CT) (*see* **Note 1**).
3. Microcentrifuge for 1.5 mL microcentrifuge tubes (*see* **Note 2**).
4. Vacuum manifold for a 96-well filter plate (Millipore, Bedford, MA) (*see* **Note 3**).
5. Plate shaker (*see* **Note 4**).
6. Vortex mixer.
7. Rotator.
8. Data analysis and graphing software program (e.g., Excel (Microsoft), xPONENT (Luminex)).

2.2 Coupling of Glycopeptides to Carboxylated Microspheres

1. Glycopeptides or glycoproteins (*see* **Note 5**).
2. MicroPlex microspheres with carboxyl groups on surface (polystyrene beads, 5.5 μm in diameter) (Luminex). MicroPlex microspheres coded into 100 spectrally distinct color sets are available. Fluorescent microspheres are light sensitive and should be protected from light during storage and usage (*see* **Notes 6** and 7).
3. Coupling buffer I: 50 mM 2-(*N*-morpholino)ethanesulfonic acid (MES), pH 4.5.
4. EDC: 1-ethyl-3-(3-dimethylaminopropyl)carbodiimide hydrochloride (Thermo Scientific, Sunnyvale, CA). Make fresh as required.
5. Wash buffer: 0.15 M Tris–HCl, pH 8.0 containing 0.02 % (v/v) Tween 20.
6. Activation buffer: 0.1 M phosphate buffer, pH 6.2.
7. Coupling buffer II: 10 mM phosphate buffer, pH 7.4 containing 0.15 M NaCl.
8. Sulfo-NHS: *N*-hydroxysulfosuccinimide (Thermo Scientific). Make fresh as required.
9. Binding buffer: 10 mM phosphate buffer, pH 7.4, 0.15 M NaCl, and 10 mg/mL bovine serum albumin.
10. PBST: 10 mM phosphate buffer, pH 7.4, 0.15 M NaCl, and 0.05 % (v/v) Tween 20.
11. Biotin-labeled lectins (Seikagaku Kogyo, Tokyo, Japan).

12. *R*-phycoerythrin (PE)-conjugated streptavidin (Molecular Probes) (*see* **Note 8**).
13. 96-Well PVDF filter plate with pore size 1.2 μm (MultiScreen-BV, etc., Millipore).

3 Methods

3.1 Coupling of Glycopeptides to Carboxylated Microspheres

3.1.1 One-Step Coupling of Glycopeptides to Carboxylated Microspheres

1. Dispense 2.5×10^6 of carboxylated microspheres into a 1.5 mL microcentrifuge tube.
2. Centrifuge the microspheres at 20,000 × *g* for 1 min.
3. Remove the supernatant carefully without disturbing the pellet.
4. Add 100 μL of coupling buffer I. Vortex and sonicate.
5. Add 50 nmol of glycopeptides (in 100 μL of coupling buffer I) (*see* **Notes 9–11**).
6. Before use, add 1.0 mL of water to 20 mg of EDC powder (*see* **Note 12**).
7. Add 10 μL aliquots of the fresh EDC solution to the microspheres. Vortex immediately.
8. Incubate for 30 min at room temperature with occasional sonication in the dark.
9. Repeat **steps 7** and **8** three times.
10. Centrifuge the microspheres at 20,000 × *g* for 1 min.
11. Remove the supernatant carefully without disturbing the pellet.
12. Add 200 μL of wash buffer. Vortex.
13. Centrifuge the microspheres at 20,000 × *g* for 1 min.
14. Repeat **steps 11–13**.
15. Remove the supernatant carefully without disturbing the pellet.
16. Resuspend the microspheres in 100 μL of wash buffer and store in the dark at 4 °C (*see* **Note 13**).
17. Count the number of microspheres.

3.1.2 Two-Step Coupling of Glycopeptides to Carboxylated Microspheres

EDC coupling agent helps link amino groups of glycopeptides with carboxyl groups on the microspheres. If the glycopeptides have the sialic acid's carboxyl group, the use of two-step coupling method is recommended to avoid unintended reaction.

Activation of Microspheres

1. Dispense 2.5×10^6 microspheres into a 1.5 mL microcentrifuge tube.
2. Centrifuge the microspheres at 20,000 × *g* for 1 min.

3. Remove the supernatant carefully without disturbing the pellet.
4. Wash twice with 200 μL of activation buffer.
5. Resuspend the microspheres in 80 μL of activation buffer.
6. Before use, make a 50 mg/mL sulfo-NHS solution in activation buffer.
7. Add 10 μL of the sulfo-NHS solution to the microsphere suspension and vortex gently.
8. Make a 100 mg/mL EDC solution in activation buffer.
9. Add 10 μL of the EDC solution to the microsphere suspension. Vortex gently.
10. Incubate the suspension for 1 h at room temperature with occasional sonication in the dark.
11. Centrifuge the microspheres at 20,000 ×*g* for 1 min (*see* **Note 14**).
12. Remove the supernatant carefully without disturbing the pellet.
13. Add 200 μL of coupling buffer II to the activated microspheres and vortex.
14. Centrifuge the microspheres at 20,000 ×*g* for 1 min.
15. Remove the supernatant carefully without disturbing the pellet.
16. Resuspend the microspheres in 100 μL of coupling buffer II and move to the following coupling steps.

Coupling, Blocking, and Storage

1. Add 50 nmol of glycopeptides (in 100 μL of water) to the activated microspheres and vortex.
2. Incubate the suspension for 2 h at room temperature with occasional sonication in the dark.
3. Centrifuge the microspheres at 20,000 ×*g* for 1 min.
4. Remove the supernatant carefully without disturbing the pellet.
5. Add 200 μL of wash buffer and vortex.
6. Centrifuge the microspheres at 20,000 ×*g* for 1 min.
7. Repeat **steps 4–6**.
8. Remove the supernatant carefully without disturbing the pellet.
9. Resuspend the microspheres in 100 μL of wash buffer and store in the dark at 4 °C (*see* **Note 13**).
10. Count the number of microspheres.

3.2 Lectin and Multiplexed Bead Binding Assay

1. Prepare a microsphere mixture by diluting the glycopeptide-coupled microsphere stocks at a concentration of 10^5/mL.
2. Dispense 20 μL of microsphere mixture into wells of 96-well filter plate.
3. Add 100 μL of biotin-labeled lectin solution (5–10 μg/mL) in binding buffer to each well (*see* **Note 15**).
4. Mix the reactions gently by pipetting.
5. Incubate for 1 h at room temperature in the dark on a plate shaker with shaking at 550 rpm.
6. Aspirate the supernatant by vacuum manifold (*see* **Note 16**).
7. Wash each well three times with 100 μL of PBST and aspirate by vacuum manifold (*see* **Note 17**).
8. Dilute the *R*-phycoerythrin (PE)-labeled streptavidin to 2.5 μg/mL in binding buffer.
9. Add 100 μL of the diluted PE-streptavidin to each well.
10. Mix the reactions gently by pipetting.
11. Incubate for 30 min at room temperature in the dark on a plate shaker with shaking at 550 rpm.
12. Aspirate the supernatant by vacuum manifold.
13. Wash each well twice with 100 μL of PBST and aspirate by vacuum manifold.
14. Add 100 μL of PBST each well and mix the reactions gently by pipetting.
15. Analyze 50–75 μL on the Luminex 100/200 analyzer according to the manufacturer's protocol (flow rate 60 μL/min, bead count 100 each).

3.3 Flow Cytometric Analysis

1. Turn on the instrument and warm up for 30 min.
2. Dispense classification calibration microspheres and reporter calibration microspheres from storage. Vortex the calibration microspheres into each labeled sample tube.
3. Wrap the sample tubes or plates in aluminum foil to protect the microspheres from light. Allow the samples to warm to room temperature.
4. Flush the line using 1 mL of 70 % isopropanol or 70 % ethanol for the alcohol flush. The alcohol flush removes air bubbles.
5. Wash the line using sheath fluid.
6. Load the classification calibration microsphere tube into the sample port and start calibration according to the manufacturer's protocol.
7. Wash the line using sheath fluid.

8. Load the reporter calibration microsphere tube into the sample port and start calibration according to the manufacturer's protocol (*see* **Note 18**).
9. Make sure that the instrument has been warmed up and calibrated.
10. Set the condition for data collection according to the manufacturer's protocol (*see* **Note 19**).
11. Load the sample tube into the sample port and start data collection.
12. Progress of data collection can be seen on the histogram.

4 Notes

1. Microspheres should be distributed before use. Gentle vortexing or sonication is the effective method of separating aggregated microspheres.
2. In the coupling protocols, microspheres are separated from reaction mixtures or wash buffer by centrifugation. Microcentrifuge can pellet microspheres at 20,000 × *g* in 1 min.
3. Instead of centrifugation, a vacuum separation method is useful for binding and washing steps. Binding and washing are performed in filter-bottomed 96-well plates and resuspension is accomplished by repetitive pipetting.
4. Agitation with a plate shaker during incubation steps prevents microspheres from settling.
5. Preparation of glycopeptides is usually performed by Pronase digestion of glycoproteins followed by gel filtration and ion-exchange chromatography. Free amino groups are essential for coupling.
6. Coupling efficiency of glycopeptides to microspheres is measured by released oligosaccharides from glycopeptide-coupled microspheres, which have been treated with endoglycosidases or *N*-glycanases. PE-labeled lectins, which can bind to the glycopeptides, are also used to monitor the coupling efficiency.
7. To avoid photobleaching, wrap microsphere containers in aluminum foil. Once photobleached, the beads are no longer usable.
8. Alexa532 and Cy3 are also used as reporter fluorochrome.
9. For preparing control microspheres, triethanolamine hydrochloride is used instead of glycopeptides to block carboxyl groups.
10. When oligosaccharides are used as ligands, biotin-hydrazide (Dojindo Laboratories, Kumamoto, Japan) is sometimes introduced at reducing termini by reductive amination [8]. Biotin-labeled oligosaccharides are then incubated with

avidin-coupled microspheres (LumAvidin microspheres; Luminex).

11. Free amino groups are essential for coupling. If glycopeptides were previously solubilized in an amine-containing buffer (e.g., Tris–HCl, sodium azide), then desalting is required by gel filtration on a Sephadex G-10 column with water. Coupling efficiency of glycopeptides depends on the concentration of glycopeptides.
12. Use aliquots immediately and discard containers after use. Make a fresh 20 mg/mL EDC solution before each addition.
13. Microspheres should be protected from light during storage and usage. Freezing conditions and organic solvents should be avoided.
14. During activation, centrifuged microspheres form a sheetlike layer on the side of the microfuge tube. This formation is normal.
15. It is recommended that samples are filtered prior to mixing with microspheres.
16. Less than 5 psi is suitable when vacuum aspiration is performed.
17. **Steps** 7, **12**, and **13** can be omitted for a no-wash binding assay. This procedure is suitable when the binding reaction is performed in a tube.
18. Calibration microspheres are very concentrated, so you should wash the line with three wash cycles using sheath fluid after calibration.
19. Flow rate, sample volume, and bead concentration should be used according to the manufacturer's recommendation.

References

1. Nolan JP, Sklar LA (2002) Suspension array technology: evolution of the flat-array paradigm. Trends Biotechnol 20:9–12
2. Yamamoto K, Ito S, Yasukawa F, Konami Y, Matsumoto N (2005) Measurement of the carbohydrate-binding specificity of lectins by a multiplexed bead-based flow cytometric assay. Anal Biochem 336:28–38
3. Crowley JF, Goldstein IJ, Arnarp J, Lonngren J (1984) Carbohydrate binding studies on the lectin from *Datura stramonium* seeds. Arch Biochem Biophys 231:524–533
4. Yamamoto K, Tsuji T, Osawa T (2002) Affinity chromatography of oligosaccharides and glycopeptides with immobilized lectins. In: Walker JM (ed) Protein protocols handbook, 2nd edn. Humana, Totowa, NJ, pp 917–931
5. Kasai K, Oda Y, Nishikawa M, Ishii S (1986) Frontal affinity chromatography: theory for its application to studies on specific interactions of biomolecules. J Chromatogr 376: 11–32
6. Gupta D, Cho M, Cummings RD, Brewer CF (1996) Thermodynamics of carbohydrate binding to galectin-1 from Chinese hamster ovary cells and two mutants. Biochemistry 35: 15236–15243
7. Shinohara Y, Hasegawa Y, Kaku H, Shibuya N (1997) Elucidation of the mechanism enhancing the avidity of lectin with oligosaccharides on the solid phase surface. Glycobiology 7: 1201–1208
8. Shinohara Y, Sota H, Kim F, Shimizu M, Gotoh M, Tosu M, Hasegawa Y (1995) Use of a biosensor based on surface plasmon resonance and biotinyl glycans for analysis of sugar binding specificities of lectins. J Biochem 117: 1076–1082

Chapter 28

Supported Molecular Matrix Electrophoresis: A New Membrane Electrophoresis for Characterizing Glycoproteins

Yu-ki Matsuno and Akihiko Kameyama

Abstract

Protein blotting is often used for identification and characterization of proteins on a membrane to which proteins separated by gel electrophoresis are transferred. The transferring process is sometimes problematic, in particular, for mucins and proteoglycans. Here, we describe a novel membrane electrophoresis technique, termed supported molecular matrix electrophoresis (SMME), in which a porous polyvinylidene difluoride (PVDF) membrane filter is used as the separation support. Proteins separated by this method can be immunoblotted without any transferring procedures.

Key words Immunoblots, Polyvinylidene difluoride membrane, Mucins, Proteoglycans, Polyvinylalcohol, Polyvinylpyrrolidone

1 Introduction

Mucins have great potential as novel clinical biomarkers for various malignant tumors, and they are typically characterized by large molecular mass (~2 MDa) and high carbohydrate content (50–90 % by weight), which can be attributed to a large number of *O*-linked glycans. In tumors, regulation of mucin expression is disrupted [1], and structural alterations of *O*-linked glycans have been reported [2]. Although several techniques have recently been developed for protein characterization, most of these cannot be used for characterization of mucins owing to their large size and heterogeneity. In particular, typical gel electrophoresis of mucins entails a variety of problems, such as mucins barely entering the resolving gel of SDS-PAGE and the transferring procedure that is required to characterize the separated proteins [3, 4]. Moreover, recovery of the transferred mucins for quantitative evaluations is often problematic. Recently, we reported a novel membrane electrophoresis method, termed supported molecular matrix

Jun Hirabayashi (ed.), *Lectins: Methods and Protocols*, Methods in Molecular Biology, vol. 1200,
DOI 10.1007/978-1-4939-1292-6_28,

electrophoresis (SMME) [5], in which a hydrophilic polymer such as polyvinylalcohol (PVA) serves as a separation medium within a porous polyvinylidene difluoride (PVDF) membrane filter. SMME enables the separation of large, heavily glycosylated proteins such as mucins and proteoglycans. After SMME, mucins on the membrane can be stained with Alcian blue [6] or labeled with monoclonal antibodies [7, 8] against a particular mucin in a manner analogous to western blotting. Lectins are also useful probes for the detection and characterization of SMME-separated proteins.

Here, we describe the separation and immunostaining of mucins by using SMME. Mucins are easily washed away from the PVA matrix embedded in the PVDF membrane by buffers containing a detergent such as Tween-20. SMME using polyvinylpyrrolidone (PVP) or polyethylene glycol (PEG) as the hydrophilic polymer sacrifice electrophoretic performance to immobilize mucins on the membrane. Thus, we have optimized the composition of the hydrophilic polymer matrix and a fixation process. In preparation of the SMME membrane, we added some PVP or PEG into a PVA solution to achieve a favorable balance between electrophoretic performance and immobilization efficiency. For fixation after electrophoresis, we investigated different physicochemical procedures. Organic solvents are often used as fixatives for immunostaining of tissues or cells [9]. Heat treatment has also been used for fixation [10]. We found that the best results were obtained by performing a sequential acetone and heat treatment. Using this protocol, proteins separated by SMME can be successfully stained with antibodies in a similar manner to western blotting.

2 Materials

Prepare all solutions using ultrapure water and analytical grade reagents.

2.1 SMME Components

1. Polyvinylidene difluoride (PVDF) filter membrane (*see* **Note 1**).
2. Methanol.
3. Hydrophilic polymer solution: Polyvinylalcohol (PVA), polyethyleneglycol (PEG), or polyvinylpyrrolidone (PVP) are dissolved in a running buffer to 0.25 % (w/v) concentration (*see* **Note 2**). Store at room temperature or at 4 °C.
4. Running buffer: 0.1 M pyridine-formic acid buffer (pH 4.0). This buffer is prepared by dissolving pyridine with water to make 0.1 M, followed by a pH adjustment with formic acid. Store at room temperature or at 4 °C (*see* **Note 3**).
5. Filter paper.

2.2 Blotting Components

1. Acetone.
2. Glass fiber filter sheets (ATTO, Tokyo, Japan, AC-5972) (*see* **Note 4**).
3. Methanol.
4. Phosphate-buffered saline (PBS).
5. Washing solution: 0.05 % Tween 20 in PBS (PBST).
6. Blocking solution: 5 % bovine serum albumin (BSA) in PBS. Store at 4 °C.

2.3 Immunostaining Components

1. Antibodies: Anti-CA19-9 antibody (TFB, Tokyo, Japan, use at 1:300 dilution), anti-sialyl MUC1 antibody (MY.1E12, gift from Professor T. Irimura, use at 1:1,000 dilution), horseradish peroxidase-conjugated anti-mouse IgG (GE Healthcare, use at 1:2,000 dilution).
2. Enhanced Chemiluminescence reagents (Western Lightning Plus-ECL, PerkinElmer) (*see* **Note 5**).

3 Methods

3.1 Preparation of SMME Membrane

1. Cut a PVDF membrane to an appropriate size (typical length, 6 cm and width, 1 cm for one sample).
2. Immerse the membrane in methanol for a few minutes.
3. Transfer the membrane into a solution of 0.25 % hydrophilic polymers dissolved in the running buffer (*see* **Note 6**).
4. Incubate for 30 min with gentle shaking.
5. Pick up the membrane and put it on filter paper to remove excess solution remaining on the surface of the membrane (*see* **Note 7**).

3.2 Membrane Electrophoresis

1. Set the membrane in an apparatus for cellulose acetate membrane electrophoresis (*see* **Note 8**) (Fig. 1).

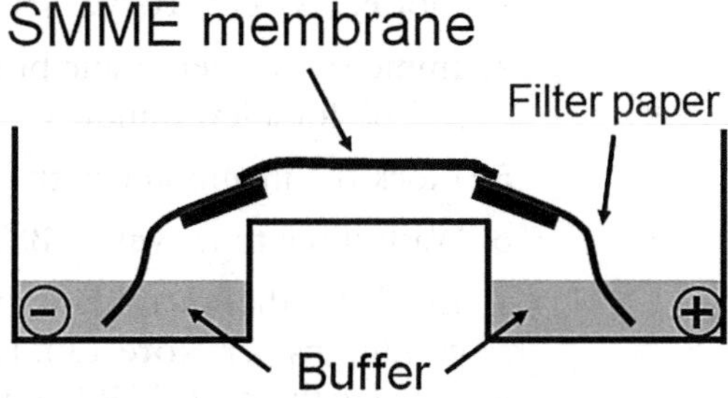

Fig. 1 Membrane electrophoresis assembly for SMME utilizing an apparatus for cellulose acetate membrane electrophoresis

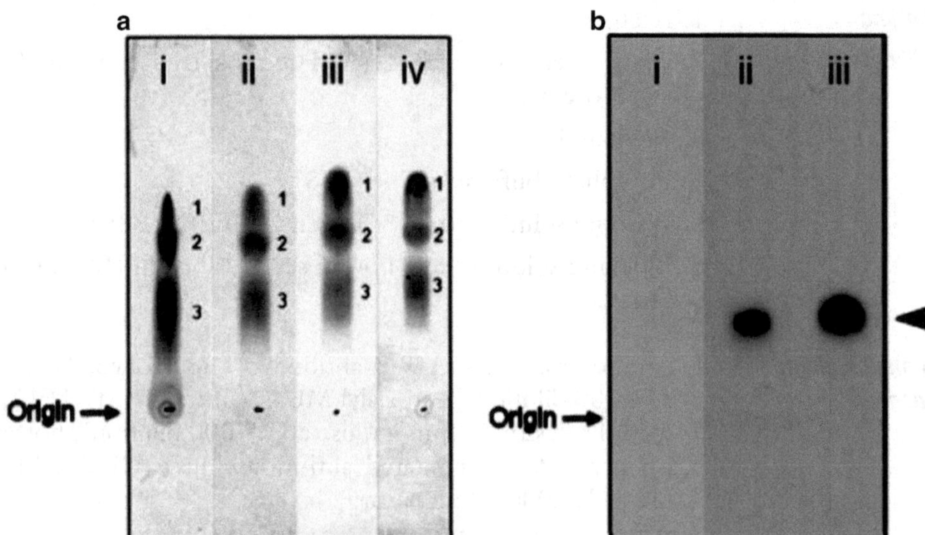

Fig. 2 Optimization of balance between resolving power and fixation efficiency. (**a**) Optimization of matrix for separation of porcine stomach mucin (PSM). PSM was separated using only PEG (*i*), mixed matrix (PEG:PVA = 4:1) (*ii*), mixed matrix (PEG:PVA = 3:2) (*iii*), and only PVA (*iv*). PSM was visualized by Alcian blue staining. (**b**) Immunostaining of mucins using mixed matrix. Mucin fractions from human bile were analyzed using PVA (*i*) and mixed matrix (PEG:PVA = 3:2 (*ii*), PEG:PVA = 4:1 (*iii*)), and then detected with anti-CA19-9 antibody (reproduced from ref. 7 with permission)

2. Apply samples (typically 1 μL) on the membrane (*see* **Note 9**). At most, 10 μg of protein can be applied.
3. Perform electrophoresis in constant current mode, typically at 1.0 mA/cm for 30 min.

3.3 Immunostaining

1. Incubate the membrane at room temperature in acetone for 30 min to fix the separated proteins.
2. Remove acetone from the membrane by air-drying at room temperature.
3. Place the membrane between two glass fiber filter sheets and then heat-press at 150 °C for 5 min using a TLC Thermal Blotter (ATTO, Tokyo, Japan) (*see* **Note 10**).
4. Immerse the membrane briefly in methanol and then incubate in PBS for a few minutes.
5. Block the membrane with blocking solution for 1 h.
6. Wash three times with PBST, 5 min each time.
7. Incubate the membrane with an antibody in appropriate conditions (*see* **Note 11**). For example, the membrane is incubated with anti-CA19-9 antibody (1:300 dilution in PBS-T) at 4 °C overnight (Fig. 2b). For the detection of sialyl-MUC1, MY.1E12 is used at a 1:1,000 dilution (Fig. 3).

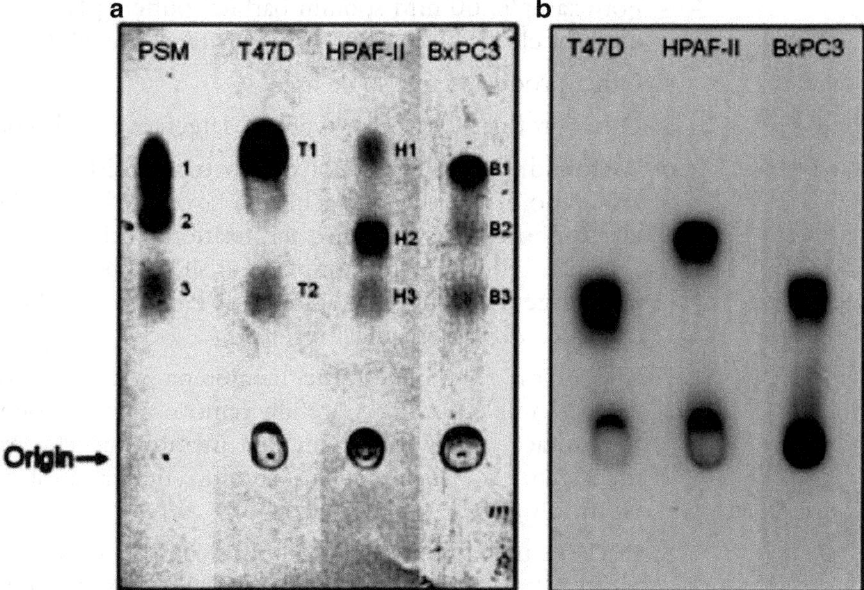

Fig. 3 SMME analysis of mucins from cancer cell lines. Mucins were visualized by Alcian blue staining (**a**) and immunostaining with MY.1E12 (**b**). Visualized spots are labeled by Arabic numerals with or without a capital letter, in which T, H, and B indicate cell lines T47D, HPAF-II, and BxPC3, respectively (reproduced from ref. 7) with permission)

8. Wash three times with PBST, 5 min each time.
9. Incubate the membrane with a second antibody conjugated to HRP in appropriate conditions (*see* **Note 12**). For example, the membrane is incubated with horseradish peroxidase-conjugated anti-mouse IgG diluted with PBS-T (1:2,000) for 1 h at room temperature.
10. Wash three times with PBST and three times with PBS, 5 min each time.
11. Visualize the target molecules using an appropriate method, such as ECL reagents followed by exposure to Amersham Hyperfilm ECL (GE Healthcare).

4 Notes

1. A commercial PVDF filter membrane for western blotting (e.g., Immobilon-P, Hybond-P, and BioTrace) can be used (typical pore size 0.45 μm).
2. Heating is effective to dissolve rapidly and may be required for some polymers. The polymer solution is for single use only.
3. This buffer is for electrophoresis of mucins or proteoglycans. Other buffers for membrane electrophoresis can also be used.

For example, 60 mM sodium barbital buffer (pH 8.6) can be used for electrophoresis of serum proteins (*see* ref. 6).

4. Other products can also be used.
5. Other products and other types of detection can also be used.
6. We found that the mixed polymer matrix consisting of a PEG/PVA ratio of 3:2 is best to achieve a favorable balance between electrophoretic performance and fixation efficiency (*see* ref. 7) (Fig. 2). However, the composition should be optimized for each target molecule. In addition, PVP is also available instead of PEG and affords similar results.
7. The excess solution on the membrane sometimes disturbs correct sample loading. Carefully remove only the solution on the surface and avoid drying the membrane. The prepared membrane can be kept in running buffer until use for electrophoresis.
8. Place the membrane to bridge both electrode sides as shown in Fig. 1. The membrane is kept wet with buffer supplied via filter papers from both electrode sides.
9. Apply samples as spots or lines. A line is convenient for densitometry. Mucins should be reduced and alkylated before applying to the membrane because mucins often form a huge network structure via intermolecular disulfide bonds. Otherwise, mucins are not introduced into the membrane. Furthermore, trypsin digestion followed by removing the digested peptides by filtration is useful for enrichment of mucins from crude samples (*see* refs. 5, 7).
10. This process is sometimes unnecessary for lower glycosylated proteins. For example, serum proteins are well fixed by incubation with only acetone.
11. Instead of antibody, lectins are also useful for the staining of mucins or other glycoproteins. For example, lectins labeled with biotin or a fluorescent tag can be used.
12. If biotin-labeled lectins are used as a first probe, horseradish peroxidase-conjugated avidin is used as a second probe.

Acknowledgements

This work was performed as a part of the R&D project of the Industrial Science and Technology Frontier Program supported by the New Energy and Industrial Technology Development Organization (NEDO). This work was also supported by a Grant-in-Aid for Scientific Research (KAKENHI 23310154) from the Japan Society for the Promotion of Science (JSPS).

References

1. Andrianifahanana M, Moniaux N, Batra SK (2006) Regulation of mucin expression: mechanistic aspects and implications for cancer and inflammatory diseases. Biochim Biophys Acta 1765:189–222
2. Brockhausen I (2006) Mucin-type *O*-glycans in human colon and breast cancer: glycodynamics and functions. EMBO Rep 7:599–604
3. Schulz BL, Packer NH, Karlsson NG (2002) Small-scale analysis of *O*-linked oligosaccharides from glycoproteins and mucins separated by gel electrophoresis. Anal Chem 74: 6088–6097
4. Andersch-Björkman Y, Thomsson KA, Holmén Larsson JM, Ekerhovd E, Hansson GC (2007) Large scale identification of proteins, mucins, and their *O*-glycosylation in the endocervical mucus during the menstrual cycle. Mol Cell Proteomics 6:708–716
5. Matsuno Y-K, Saito T, Gotoh M, Narimatsu H, Kameyama A (2009) Supported molecular matrix electrophoresis: a new tool for characterization of glycoproteins. Anal Chem 81: 3816–3823
6. Dong W, Matsuno Y-K, Kameyama A (2013) Serum protein fractionation using supported molecular matrix electrophoresis. Electrophoresis 34:2432–2439
7. Matsuno Y-K, Dong W, Yokoyamam S, Yonezawa S, Saito T, Gotoh M, Narimatsu H, Kameyama A (2011) Improved method for immunostaining of mucin separated by supported molecular matrix electrophoresis by optimizing the matrix composition and fixation procedure. Electrophoresis 32:1829–1836
8. Matsuno Y-K, Dong W, Yokoyama S, Yonezawa S, Narimatsu H, Kameyama A (2013) Identification of mucins by using a method involving a combination of on-membrane chemical deglycosylation and immunostaining. J Immunol Methods 394:125–130
9. Jamur MC, Oliver C (2010) Cell fixatives for immunostaining. Methods Mol Biol 588: 55–61
10. Li KW, Geraerts WP, van Elk R, Joosse J (1988) Fixation increases sensitivity of India ink staining of proteins and peptides on nitrocellulose paper. Anal Biochem 174:97–100

Part III

Techniques for Elucidating Functions of Endogenous Animal Lectins

Chapter 29

Overall Strategy for Functional Analysis of Animal Lectins

Norihito Kawasaki

Abstract

Animal lectins elicit biological functions through the interaction with glycan ligands. To clarify the functions of the lectins, both identification of their glycan ligand structure and assessment of impact of lectin–glycan interaction on the biological event are essential strategies. This chapter focuses on two of key useful methodologies for planning experiments based on the strategies. One is the detection of lectin–glycan interaction by the multivalent display of lectins and glycans. This methodology is a powerful means for identification of the glycan ligand structure and proteins and/or lipids carrying the glycan ligands for lectins. The other is the intervention of lectin–glycan interaction to assess the biological roles of lectins. Bioinformatics especially useful for animal lectins will be also described in this chapter. The concepts described in this chapter are versatile and applicable to a wide range of animal lectin research.

Key words Lectin–glycan interaction, Multivalency, Neoglycoconjugates, Gene-modified mouse, Glycan array, Bioinformatics

1 Introduction

This chapter gives an overview of strategies to elucidate functions of animal lectins. These strategies will be widely applicable to study diverse biological events such as immune system, nervous system, and cancer, which animal lectins are frequently associated with. For readers who seek to learn a particular method, other chapters of this book will be helpful.

Biological function of animal lectins is predominantly triggered by recognizing their specific glycan ligands [1, 2], though some animal lectins can elicit the function in a glycan-binding independent manner [3, 4]. In response to the glycan ligand binding, lectins induce a wide range of biological functions such as cell–cell adhesion, endocytosis, cellular activation, induction of bacteria killing [1, 2]. Therefore, identification of the glycan ligands and assessment of the role of lectin–glycan interactions is primary requisite to understand function of the lectins. Firstly, this chapter describes methodologies to identify glycan ligands. Because the affinity of animal lectins towards glycans is relatively weak

Jun Hirabayashi (ed.), *Lectins: Methods and Protocols*, Methods in Molecular Biology, vol. 1200,
DOI 10.1007/978-1-4939-1292-6_29, © Springer Science+Business Media New York 2014

($K_a \sim 10^4$ M^{-1}) compared to that of plant lectins [1, 2], the multivalent display of either glycans or lectins in the experimental systems is an appropriate method with a wide applicability. The second section describes the methods to identify proteins and/or lipids carrying the glycan ligands for lectins. Since binding of some animal lectins to the glycan ligands influences function of the glycan carriers, analysis of the interaction between lectins and the glycoproteins and/or glycolipids becomes of interest [5–7]. The third section describes the methods to modulate lectin–glycan interactions, which is an effective approach to assess impact of lectin–glycan interactions on the biological system. Briefly, intervention of lectin–glycan interaction can be achieved by alteration of glycan ligands, ablation of lectins, and inhibition of the interaction by sugar inhibitors. The fourth section describes the use of bioinformatics for animal lectin research. Such databases are useful to predict the function of lectins and identify novel animal lectins.

2 Identification of Glycan Ligands for Animal Lectins

This section summarizes the multivalent display of glycans and lectins for identification of glycan ligands of lectins. When multiple glycan ligands are located closely enough to interact with more than one lectins simultaneously, the interaction between lectins and glycans increases dramatically, resulting in thousands fold increase in the apparent affinity [1]. For example, trimeric asialoglycoprotein receptor binds to triantennary glycans more than a thousand-fold better than monoantennary glycans [8].

2.1 Glycan Array

Glycan arrays achieve the multivalent display of the glycan ligands by immobilizing the glycan ligands on the solid surface [9–11]. For a deeper understanding of basic principle and the development of glycan arrays, there are excellent reviews available [9–11]. The protocol is similar to the ELISA assay. Briefly, fluorescent lectins or cells expressing lectins of interest are applied on the glycan array carrying various glycans in small spots on the solid surfaces [12]. After quick washing, the remaining fluorescence in each spots of the array is measured by fluorescence detectors.

Several glycan arrays have been developed by different groups. There are mainly two types of methods to immobilize glycans onto the array. One is covalent-linking of glycans to the chemically functionalized array surface. For instance, the glycan array developed by the Consortium of Functional Glycomics (CFG) employs *N*-hydroxysuccinimide (NHS)-activated glass slides to immobilize glycans derivatized with a terminal amine linker attached at the reducing end [13], providing over 600 different mammalian glycan targets (http://www.functionalglycomics.org/static/consortium/resources/resourcecoreh.shtml). The other is non-covalent immobilization via hydrophobic interaction between lipid-derivatized glycans to the hydrophobic array surface. Fukui et al. have developed neoglycolipid-based glycan array where the glycans are lipid-derivatized with a 1,2-dihexadecyl-*sn*-glycero-3-phosphoethanolamine (DHPE) at the reducing end and immobilized onto the hydrophobic nitrocellulose membrane [14, 15]. This approach is also used to develop the natural glycolipid array to identify lectins specific to glycolipids [16].

Custom glycan arrays carrying a particular type of glycans are contributing to a deeper understanding on the function of lectins. The glycan arrays focused on sialic acid containing glycans have successfully used to determine the glycan-binding specificity of influenza virus hemagglutinins [17–20]. The glycan arrays with glycosaminoglycans or plant polysaccharides also have been developed successfully [21–24]. The glycan array focused on polysaccharides from pathogenic bacteria has been established by CFG (http://www.functionalglycomics.org/static/consortium/resources/timelined2p.shtml). To further improve the sensitivity of the array system, a "wash free" glycan array based on the evanescent-field fluorescence-assisted detection principle is established [25] (*see* Chapter 30 described by Tateno et al.).

Generation of efficient glycan arrays relies on structurally defined glycans synthesized enzymatically and/or chemically, or purified from native sources. With rapidly improving synthetic and purification methodologies [23, 26–34], the repertoire of glycans on the array has been increasing. A recent mathematical analysis provides an algorithm to evaluate diversity in repertoire of glycan ligands on the glycan array [35]. These supporting infrastructures in glycomics will create more diverse and customized glycan arrays available to investigators.

2.2 Multivalent Neoglycoconjugates

While glycan arrays achieve the multivalent presentation of glycan ligands on the solid surface, neoglycoconjugates can achieve such multivalent display in solution, increasing the compatibility to many experimental settings. Such multivalent neoglycoconjugates include sugar-derivatized BSA and polyacrylamide (PAA), nanoparticles decorated with glycans, and glycan dendrimers. These neoglycoconjugates can be labeled with fluorescent dyes, expanding their compatibility to fluorescence-based applications including flow cytometry and fluorescence microscopy. Moreover, these neoglycoconjugates have been used successfully to analyze in vivo function of animal lectins.

Sugar-derivatized BSA and PAA might be the most familiar neoglycoconjugates for lectin research [8, 36, 37]. Multiple glycans are chemically attached to a single molecule of BSA or PAA, forming multivalent glycan ligands. These probes can be obtained from several companies (Vector Laboratories and Lectinity) and CFG. Biotinylated PAA probes are also commercially available

(Glycotech). Biotinylated PAA probes can be fluorescently labeled by conjugating with fluorescent streptavidin. Zhou et al. used mannosylated BSA to analyze glycan-binding activity of a C-type lectin SIGNR1 in vitro and assess the effect of the interaction between mannosylated BSA and SIGNR1 on the immune response in vivo [38]. Hudson et al. reported the use of soluble PAA probes decorated with 6′-*O*-sulfated sialyl Lewis X to induce apoptosis of eosinophils through interaction with a siglec receptor, Siglec-F [39]. Fluorescent sugar-PAA has been used successfully to analyze endocytosis of lectins by flow cytometry and fluorescence microscopy [40, 41].

Nanoparticles ranging the molecular size up to 1,000 nm diameter decorated with glycans have also been developed. The glycan ligands are chemically linked to the virus-like particle, liposomes, and the beads [42–44]. Generally, the conjugation of chemically derivatized glycans to the particles and the lipids for liposomes employs simple and controllable chemistry [45–47]. The glycans with the 2-azidoethyl group can be used directly for the click reaction, or reduced to aminoethyl group for NHS-coupling [45, 47]. Biotinylated glycans can be directly conjugated to streptavidin [46]. Such facile protocols allow investigators to develop their own glycan-bearing particles with controlled multivalency of the glycans. The functionalized glycans can be made chemically [48, 49] and obtained from CFG or a commercial source (Elicityl). The beads and lipids for the conjugation are also commercially available from many sources (Life Technologies, Bangs Laboratories, Avanti Polar Lipids, and NOF Corporation). In addition to glycans, investigators can also modify the particles with other molecules. Glycan-bearing liposomes have been used as a carrier of therapeutic reagents in vivo. Briefly, proteins and drugs are encapsulated inside the liposomes and delivered to the lectin-expressing cells [43, 50–52]. In other studies, glycans and immunogenic proteins are displayed on the surface of the liposomes to analyze impact of lectin–glycan interaction on the immune response to the proteins [53, 54]. For other format of bead-based glycan probes, *see* Chapters 27 and 31 described by K. Yamamoto and M. Amano et al., respectively.

Chemical probes consisting of glycan ligands attached to dendrimers and linear polymers have been developed [55–58]. Such probes have been used to analyze the effect of the neoglycoconjugate architecture on lectin functions, which is important to develop high affinity inhibitors for lectins. Gestwicki et al. have synthesized several neoglycoconjugates with the same valency of glycans, but with different core structure (i.e., linear polymer or dendrimer) and demonstrated that the shape of the neoglycoconjugate is a factor for effective inhibition of lectins [59]. Based on the pentagon-like structure of Shiga-like toxin B subunit specific for Gb_3, the starfish-shaped Gb_3 decamer has been synthesized and shown a million fold higher affinity than monovalent Gb_3 [60].

2.3 Multivalent Display of Lectins

Multivalent display of lectins is another effective methodology to detect weak lectin–glycan interactions. This mimics the nature of animal lectins which can be found as multimer, increasing their apparent affinity to the glycan ligands [1, 6, 61, 62]. Monomeric lectins are likely to bind to glycan ligands in conditions both lectins and glycans become multivalent, which can be seen in cell–cell and cell–pathogen interactions [63]. Of note, the glycan-binding specificity obtained by different multivalent lectin probes shows consistent results [64, 65] (http://www.functionalglycomics.org/glycomics/publicdata/primaryscreen.jsp). Various methods which are similar to the one for multivalent presentation of glycans have been used. This chapter describes immobilization of lectins onto beads and cell surface membranes and multimerization of lectins in a solution.

Lectin-immobilized beads and columns have been used successfully to detect lectin–glycan interactions (*see* Chapters 7, 21, and 22 described by K. Yamashita and T. Oukura, K. Kasai, and C. Sato, respectively). The conjugation of lectins to the beads and columns often employs facile procedures such as NHS-coupling and biotin–streptavidin coupling.

Anchoring lectins onto the cell surface membrane is another way to increase avidity of lectins. Transmembrane lectins are expressed on irreverent cell lines such as CHO cells and tested for the binding to multivalent glycan probes [66, 67]. Interestingly, Otto et al. have reported an expression system which allows soluble proteins to be expressed as membrane bound proteins [68]. These cell lines can be used to analyze lectin–glycan interaction in the ELISA-based assay, flow cytometry, and fluorescence microscopy.

Another format utilizing membrane-bound lectins is the reporter assay converting the binding of lectins with glycans to reporter gene induction. As a reporter gene, β-galactosidase or green fluorescent protein (GFP) is used to monitor lectin–glycan interactions in the reporter cell lines BWZ.36 and 2B4, respectively [69, 70]. Several C-type lectins bearing immunoreceptor tyrosine-based activation motif (ITAM) are expressed on the surface of the reporter cells. The binding of lectins to the multivalent glycans induce crosslinking of the ITAM, initiating intracellular activation signals that lead to the expression of the reporter products [71, 72]. Since this assay does not require any washing step, it is suitable for detecting weak lectin–glycan interactions.

Soluble recombinant lectins can be multimerized in various ways. Generally, two common approaches have been used. The first approach employs the Fc region of immunoglobulin [73]. The lectins are expressed as a chimeric protein with immunoglobulin Fc region in mammalian cells. The Fc fusion lectins form a dimer via the disulfide bonds in the Fc regions. Subsequent purification can be performed using the Protein G or Protein A column. The purified Fc-fusion lectins can be further complexed with an antibody toward the Fc region, generating tetravalent lectin probes [24, 67].

The second approach is to use the streptavidin–biotin coupling. Lectins are expressed in bacteria with a biotinylation peptide sequence [74]. The biotinylation sequence is recognized by the biotin ligase BirA, which attaches a biotin molecule to the sequence. Subsequently, biotinylated lectins can be tetramerized with streptavidin [75, 76].

3 Identification of Proteins and Lipids Carrying the Glycan Ligands for Lectins

Identification of glycoprotein and glycolipid ligands of lectins has become of interest in animal lectin research. Indeed, some animal lectins including galectins are known to regulate functions of cell surface receptors by binding to the glycan ligands on the receptors [5–7]. As tools to identify specific binding of the glycoprotein ligands to lectins, lectin-affinity chromatography and the co-precipitation assay with the lectin-immobilized beads are described in this section. Recent advance in in situ detection of the interaction between a lectin and a glycoprotein or a glycolipid is also discussed. This includes chemical cross-linkers to covalently capture lectin-bound ligands, mammalian two-hybrid systems utilizing GFP complementation, fluorescence microscopic analysis to visualize lectin–glycolipid interactions. These approaches are versatile and robust methods for identification of ligands for lectins.

Affinity columns with immobilized animal lectins have been used successfully to identify glycoprotein ligands for galectins and C-type lectins [77–81]. Briefly the mixture of the glycoproteins from cell lysates and plasma is passed through the lectin-immobilized column. The column is washed and the glycoproteins captured in the column are eluted. The eluted fraction is resolved by SDS-PAGE and analyzed by mass spectrometry to determine the ligands for lectins. Similar to the lectin-immobilized column, co-precipitation using lectin-immobilized beads has been used for siglecs, galectins, and C-type lectins [82–87]. In these methods, mixtures of glycoproteins are incubated with lectin-immobilized beads. The beads are washed and the captured glycoproteins are determined by mass spectrometry. However, these methods tend to identify heavily glycosylated proteins as ligands for lectins possibly because the glycoproteins with a few glycans would not be tolerated in the experimental systems.

To overcome this potential problem, methods using chemical cross-linkers have been developed to identify the ligands for lectins in situ. Chemically modified sugars with photoreactive cross-linkers such as aryl azide have been introduced to the cells [88, 89]. The incorporated aryl-azide sugars are metabolically converted to substrates for endogenous glycan synthesis in the cell, thereby generating cross-linkable glycan ligands [90]. Han et al. synthesized 9-aryl-azide NeuAc and generated the B cells with modified sialic acids [62]. To identify the glycoprotein ligands for Siglec-2, soluble Siglec-2 proteins were incubated with the cells, followed by UV exposure. Subsequently the complex of Siglec-2 and the glycoproteins was precipitated from cell lysates, and the cross-linked glycoproteins were identified by mass spectrometry [91]. Another method to detect weak interactions between lectins and glycans in situ is the fluorescence reporter assay [92–94]. Briefly lectins are tagged with a fragment of fluorescent protein such as yellow fluorescent protein (YFP) and GFP. Glycoproteins are also tagged with the complementary fragment of YFP or GFP. If the lectin binds to the glycoprotein, the fluorescent protein is reconstituted and becomes fluorescent. Using this method, the weak lectin–glycan interactions in the secretory pathway inside the cell have been successfully monitored [92–94].

The detection of the binding between lectins and glycolipids in situ is also challenging. Fluorescently labeled glycolipids become a facile and reliable tool [95–97]. Briefly, cells were incubated with the fluorescent glycolipids to prepare the cells expressing fluorescently labeled glycolipids. Within 10 min of incubation, fluorescent glycolipids were inserted into the cell membrane [95]. These labeled cells are directly used for microscopic analysis to monitor interactions between glycolipids and lectins [95–97].

4 Intervention of Lectin–Glycan Interaction for Functional Analysis of Lectins

Once investigators have determined the glycan-binding specificity and glycoprotein/glycolipid ligands of lectins, next step is to analyze the role of the lectin–glycan interaction in the experimental system. For the purpose, the intervention of lectin–glycan interaction becomes a powerful strategy. Briefly, the methods discussed in this section and the following chapters are to assess the impact of (1) alteration of glycan ligand structure, (2) ablation of lectin expression, and (3) inhibition of the lectin–glycan interaction by sugar inhibitors on the biological readout.

4.1 Alternation of Glycan Ligand Structure

This section describes several methods to alter glycan ligand structure in mammalian cells. This includes the use of inhibitors for glycosylation, overexpression, knockdown, deletion of genes for glycosylation and glycoproteins (glycogenes), and the use of glycosylation-deficient cell lines.

Several small molecule inhibitors for glycosylation are available from commercial sources (Sigma-Aldrich and EMD Millipore). These membrane permeable compounds inhibit glycan processing in the cell, resulting in alteration of cell surface glycan structure [2]. The glycan-modified cells can be used to analyze the binding of lectins to the altered cell surface glycans [65, 98]. For the protocol, *see* Chapter 32 described by N. Kawasaki et al.

Cell surface glycosylation can be also controlled by overexpression and knockdown of glycogenes. The key is to find the appropriate target genes in the glycosylation pathway. For this regard, several databases for glycosylation will be helpful (http://www.genome.jp/kegg/glycan/) (http://www.functionalglycomics.org/glycomics/molecule/jsp/glycoEnzyme/geMolecule.jsp) (http://www.cazy.org/).

Mice lacking a glycogene are critical to analyze the function of lectins in vivo [99–103]. Investigators can check the availability of gene-modified mice through the Mouse Genome Informatics (http://www.informatics.jax.org/). Recent advance in genetics enables to control the gene deletion in a spatiotemporal manner, which provides a deeper understanding on the functions of lectins. Using a drug-inducible Cre recombinase system, several glycosyltransferase-deficient mice have been established. In these mice, the tissue specific Cre recombinase is produced only in response to the drug administration, enabling to ablate the targeted glycosyltransferases in a specific tissue and organ at any time point of interest [104, 105].

The lectin-resistant mutant cell lines derived from CHO cells have been used for decades as reliable tools to assess the impact of alteration of glycans [1, 2, 106]. It has been known that the binding of plant lectins to the cells results in cell death [1]. Accordingly, CHO cells were cultured in the presence of cytotoxic plant lectins, and a few colonies resistant to the plant lectin-induced cell death have been cloned. Subsequent glycan analysis as well as gene mapping has revealed that these survivors exhibit altered glycosylation caused by the mutation in glycogenes [106, 107]. To date, several CHO cell lines with defect in *N*-glycan, *O*-glycan, and glycosaminoglycan synthesis are established and available from American Type Culture Collection [106–108].

4.2 Ablation of Lectins

This section describes methods to abrogate lectins from the experimental system using cells and animals. This includes deletion and knockdown of lectin genes and introduction of glycan-binding deficient mutant lectins to the system.

Several lectin-deficient mice have been established and used to analyze the effect of ablation of lectins from the system [109–112] (http://www.informatics.jax.org/). The knockdown of lectin genes has been successfully used in cell culture. For the detail of protocol, *see* Chapter 34 described by M. Nonaka and T. Kawasaki. To gain further insights into the role of lectin–glycan interactions, the mutant lectins become a key experimental model. The mutant lectins possess mutations in the glycan-binding domains, which abrogates binding to the glycan ligands. Klaas et al. have established the mutant Siglec-1 knock-in mice in which the mutant Siglec-1 cannot bind to sialic acids [113]. This approach can also be applied to other formats of lectins including recombinant soluble lectins and cell lines expressing mutant lectins. An approach based on databases described in Subheading 5 will be helpful to identify the essential amino acid residues in lectins for glycan-binding activity.

4.3 Inhibition of the Lectin–Glycan Interaction by Sugar Inhibitors

Glycan probes can be used to inhibit lectin–glycan interaction. This includes the multivalent glycan probes described in Subheading 2.2. In some cases, inhibition of lectin–glycan interaction needs to be achieved by monomeric glycan probes. This is because clustering of animal lectins by the multivalent glycan probes may result in cellular signaling. For instance, a C-type lectin DC-SIGN induces cellular activation in response to binding to the mannosylated lipoarabinomannan, a soluble multivalent glycan ligand from *Mycobacterium tuberculosis* [114]. To provide affinity high enough to inhibit the lectin–glycan interaction, yet avoid clustering of the lectin, the synthetic glycan ligands have been developed [115, 116]. Such high affinity monomeric probes have been successfully used to elucidate the role of Siglec-2-sialic acid interaction in the B cell function [117]. Alternatively, antibodies against the lectins have been successfully used to inhibit lectin–glycan interactions [114].

5 Database for Functional Analysis of Animal Lectins

Computational database analysis is helpful to predict lectin functions [118, 119]. This includes the amino acid homology search of lectins to identify signaling motifs and carbohydrate recognition motifs. Since animal lectins can interact with other signaling molecules such as intracellular phosphatases and kinases which activate signaling events, this prediction may help to identify such signaling molecules associated with lectins. To find such signaling motifs, there are several databases available (http://www.genome.jp/tools/motif/) (http://www.motifsearch.com/) (http://www.ebi.ac.uk/Tools/pfa/iprscan/). Amino acid sequence search for glycan recognition motif have been used to determine the essential amino acids for glycan binding. To date, for most of lectin classes, these essential amino acids have been identified [1, 65, 120–124]. Moreover, investigators can screen proteins to identify putative glycan binding proteins based on the essential amino acid motifs for glycan binding activity of lectins [125, 126]. A recent study by Rademacher et al. provides an interesting approach to predict glycan-binding activity. The study employs tertiary structural motif search based on a crystal structure data from the protein data bank (http://www.rcsb.org/pdb/home/home.do) to find proteins bearing the structural similarity to a sialic acid binding motif [127].

6 Conclusion

Approaches described in this chapter build on the state-of-the-art techniques in glycobiology and carbohydrate chemistry. Considering the rapid advance of the field in the recent years, techniques will be further improved for generating more multiplex glycan structures, providing more versatile methods to prepare glycan ligands, and establishing more sensitive detection methods for weak lectin–glycan interactions. As a result, more glycan ligands and biological tools for animal lectin research will become available.

Acknowledgements

I would like to thank Dr. Kazuo Yamamoto for critical reading of the manuscript and Dr. Fumiko Kawasaki for her technical help in manuscript preparation.

References

1. Lis H, Sharon N (2003) Lectins, 2nd edn. Kluwer Academic Publishers, Dordrecht
2. Varki A, Cummings R, Esko J et al (1999) Essentials of glycobiology. Cold Spring Harbor Laboratory Press, New York
3. Carlin AF, Chang YC, Areschoug T et al (2009) Group B Streptococcus suppression of phagocyte functions by protein-mediated engagement of human Siglec-5. J Exp Med 206:1691–1699
4. Mitsuki M, Nara K, Yamaji T et al (2010) Siglec-7 mediates nonapoptotic cell death independently of its immunoreceptor tyrosine-based inhibitory motifs in monocytic cell line U937. Glycobiology 20:395–402
5. Ohtsubo K, Marth JD (2006) Glycosylation in cellular mechanisms of health and disease. Cell 126:855–867
6. Boscher C, Dennis JW, Nabi IR (2011) Glycosylation, galectins and cellular signaling. Curr Opin Cell Biol 23:383–392
7. Zhao Y, Sato Y, Isaji T et al (2008) Branched *N*-glycans regulate the biological functions of integrins and cadherins. FEBS J 275: 1939–1948
8. Lee RT, Lee YC (2000) Affinity enhancement by multivalent lectin-carbohydrate interaction. Glycoconj J 17:543–551
9. Rillahan CD, Paulson JC (2011) Glycan microarrays for decoding the glycome. Annu Rev Biochem 80:797–823
10. Oyelaran O, Gildersleeve JC (2009) Glycan arrays: recent advances and future challenges. Curr Opin Chem Biol 13:406–413
11. Liang PH, Wu CY, Greenberg WA et al (2008) Glycan arrays: biological and medical applications. Curr Opin Chem Biol 12: 86–92
12. Rillahan CD, Schwartz E, McBride R et al (2012) Click and pick: identification of sialoside analogues for siglec-based cell targeting. Angew Chem Int Ed Engl 51:11014–11018
13. Blixt O, Head S, Mondala T et al (2004) Printed covalent glycan array for ligand profiling of diverse glycan binding proteins. Proc Natl Acad Sci U S A 101:17033–17038
14. Palma AS, Zhang Y, Childs RA et al (2012) Neoglycolipid-based "designer" oligosaccharide microarrays to define beta-glucan ligands for Dectin-1. Methods Mol Biol 808: 337–359
15. Fukui S, Feizi T, Galustian C et al (2002) Oligosaccharide microarrays for high-throughput detection and specificity assignments of carbohydrate-protein interactions. Nat Biotechnol 20:1011–1017
16. Rinaldi S, Brennan KM, Goodyear CS et al (2009) Analysis of lectin binding to glycolipid complexes using combinatorial glycoarrays. Glycobiology 19:789–796
17. Nycholat CM, McBride R, Ekiert DC et al (2012) Recognition of sialylated poly-N-acetyllactosamine chains on *N*- and *O*-linked glycans by human and avian influenza A virus hemagglutinins. Angew Chem Int Ed Engl 51:4860–4863
18. Liao HY, Hsu CH, Wang SC et al (2010) Differential receptor binding affinities of

influenza hemagglutinins on glycan arrays. J Am Chem Soc 132:14849–14856
19. Childs RA, Palma AS, Wharton S et al (2009) Receptor-binding specificity of pandemic influenza A (H1N1) 2009 virus determined by carbohydrate microarray. Nat Biotechnol 27:797–799
20. Deng L, Chen X, Varki A (2013) Exploration of sia diversity and biology using sialoglycan microarrays. Biopolymers 99:650–665
21. Rogers CJ, Clark PM, Tully SE et al (2011) Elucidating glycosaminoglycan-protein-protein interactions using carbohydrate microarray and computational approaches. Proc Natl Acad Sci U S A 108:9747–9752
22. Takada W, Fukushima M, Pothacharoen P et al (2013) A sulfated glycosaminoglycan array for molecular interactions between glycosaminoglycans and growth factors or anti-glycosaminoglycan antibodies. Anal Biochem 435:123–130
23. Pedersen HL, Fangel JU, McCleary B et al (2012) Versatile high resolution oligosaccharide microarrays for plant glycobiology and cell wall research. J Biol Chem 287: 39429–39438
24. Palma AS, Feizi T, Zhang Y et al (2006) Ligands for the beta-glucan receptor, Dectin-1, assigned using "designer" microarrays of oligosaccharide probes (neoglycolipids) generated from glucan polysaccharides. J Biol Chem 281:5771–5779
25. Tateno H, Mori A, Uchiyama N et al (2008) Glycoconjugate microarray based on an evanescent-field fluorescence-assisted detection principle for investigation of glycan-binding proteins. Glycobiology 18:789–798
26. Koizumi A, Matsuo I, Takatani M et al (2013) Top-down chemoenzymatic approach to high-mannose-type glycan library: synthesis of a common precursor and its enzymatic trimming. Angew Chem Int Ed Engl 52:7426–7431
27. Wang N, Huang CY, Hasegawa M et al (2013) Glycan sequence-dependent Nod2 activation investigated by using a chemically synthesized bacterial peptidoglycan fragment library. Chembiochem 14:482–488
28. Murase T, Tsuji T, Kajihara Y (2009) Efficient and systematic synthesis of a small glycoconjugate library having human complex type oligosaccharides. Carbohydr Res 344:762–770
29. Martin CE, Broecker F, Eller S et al (2013) Glycan arrays containing synthetic Clostridium difficile lipoteichoic acid oligomers as tools toward a carbohydrate vaccine. Chem Commun (Camb) 49:7159–7161
30. Tam PH, Lowary TL (2009) Recent advances in mycobacterial cell wall glycan biosynthesis. Curr Opin Chem Biol 13:618–625
31. Nguyen TK, Arungundram S, Tran VM et al (2012) A synthetic heparan sulfate oligosaccharide library reveals the novel enzymatic action of D-glucosaminyl 3-*O*-sulfotransferase-3a. Mol Biosyst 8:609–614
32. Goodfellow JJ, Baruah K, Yamamoto K et al (2012) An endoglycosidase with alternative glycan specificity allows broadened glycoprotein remodelling. J Am Chem Soc 134: 8030–8033
33. Sugiarto G, Lau K, Qu J et al (2012) A sialyltransferase mutant with decreased donor hydrolysis and reduced sialidase activities for directly sialylating LewisX. ACS Chem Biol 7:1232–1240
34. Ruas-Madiedo P, de los Reyes-Gavilan CG (2005) Invited review: methods for the screening, isolation, and characterization of exopolysaccharides produced by lactic acid bacteria. J Dairy Sci 88:843–856
35. Rademacher C, Paulson JC (2012) Glycan fingerprints: calculating diversity in glycan libraries. ACS Chem Biol 7:829–834
36. Galanina OE, Chinarev AA, Shilova NV et al (2012) Immobilization of polyacrylamide-based glycoconjugates on solid phase in immunosorbent assays. Methods Mol Biol 808:167–182
37. Coombs PJ, Harrison R, Pemberton S et al (2010) Identification of novel contributions to high-affinity glycoprotein-receptor interactions using engineered ligands. J Mol Biol 396:685–696
38. Zhou Y, Kawasaki H, Hsu SC et al (2010) Oral tolerance to food-induced systemic anaphylaxis mediated by the C-type lectin SIGNR1. Nat Med 16:1128–1133
39. Hudson SA, Bovin NV, Schnaar RL et al (2009) Eosinophil-selective binding and pro-apoptotic effect in vitro of a synthetic Siglec-8 ligand, polymeric 6′-sulfated sialyl Lewis x. J Pharmacol Exp Ther 330:608–612
40. Tateno H, Li H, Schur MJ et al (2007) Distinct endocytic mechanisms of CD22 (Siglec-2) and Siglec-F reflect roles in cell signaling and innate immunity. Mol Cell Biol 27:5699–5710
41. Denda-Nagai K, Kubota N, Tsuiji M et al (2002) Macrophage C-type lectin on bone marrow-derived immature dendritic cells is involved in the internalization of glycosylated antigens. Glycobiology 12:443–450
42. Ribeiro-Viana R, Sanchez-Navarro M, Luczkowiak J et al (2012) Virus-like glycodendrinanoparticles displaying quasi-equivalent nested polyvalency upon glycoprotein platforms potently block viral infection. Nat Commun 3:1303
43. Unger WW, van Beelen AJ, Bruijns SC et al (2012) Glycan-modified liposomes boost

CD4+ and CD8+ T-cell responses by targeting DC-SIGN on dendritic cells. J Control Release 160:88–95
44. Totani K, Miyazawa H, Kurata S et al (2011) Magnetic beads-assisted mild enrichment procedure for weak-binding lectins. Anal Biochem 411:50–57
45. Kaltgrad E, O'Reilly MK, Liao L et al (2008) On-virus construction of polyvalent glycan ligands for cell-surface receptors. J Am Chem Soc 130:4578–4579
46. Barral P, Eckl-Dorna J, Harwood NE et al (2008) B cell receptor-mediated uptake of CD1d-restricted antigen augments antibody responses by recruiting invariant NKT cell help in vivo. Proc Natl Acad Sci U S A 105:8345–8350
47. Nycholat CM, Rademacher C, Kawasaki N et al (2012) In silico-aided design of a glycan ligand of sialoadhesin for in vivo targeting of macrophages. J Am Chem Soc 134: 15696–15699
48. Chernyak AY, Sharma GVM, Kononov LO et al (1992) 2-Azidoethyl glycosides: glycosides potentially useful for the preparation of neoglycoconjugates. Carbohydr Res 223: 303–309
49. Grun CH, van Vliet SJ, Schiphorst WE et al (2006) One-step biotinylation procedure for carbohydrates to study carbohydrate-protein interactions. Anal Biochem 354:54–63
50. Ikehara Y, Shiuchi N, Kabata-Ikehara S et al (2008) Effective induction of anti-tumor immune responses with oligomannose-coated liposome targeting to intraperitoneal phagocytic cells. Cancer Lett 260:137–145
51. Chen WC, Kawasaki N, Nycholat CM et al (2012) Antigen delivery to macrophages using liposomal nanoparticles targeting sialoadhesin/CD169. PLoS One 7:e39039
52. Kawasaki N, Vela JL, Nycholat CM et al (2013) Targeted delivery of lipid antigen to macrophages via the CD169/sialoadhesin endocytic pathway induces robust invariant natural killer T cell activation. Proc Natl Acad Sci U S A 110:7826–7831
53. Pfrengle F, Macauley MS, Kawasaki N et al (2013) Copresentation of antigen and ligands of siglec-G induces B cell tolerance independent of CD22. J Immunol 191:1724–1731
54. Macauley MS, Pfrengle F, Rademacher C et al (2013) Antigenic liposomes displaying CD22 ligands induce antigen-specific B cell apoptosis. J Clin Invest 123:3074–3083
55. Jimenez Blanco JL, Ortiz Mellet C, Garcia Fernandez JM (2013) Multivalency in heterogeneous glycoenvironments: hetero-glycoclusters, -glycopolymers and -glycoassemblies. Chem Soc Rev 42:4518–4531
56. Chabre YM, Roy R (2013) Multivalent glycoconjugate syntheses and applications using aromatic scaffolds. Chem Soc Rev 42: 4657–4708
57. Deniaud D, Julienne K, Gouin SG (2011) Insights in the rational design of synthetic multivalent glycoconjugates as lectin ligands. Org Biomol Chem 9:966–979
58. Kitov PI, Bundle DR (2003) On the nature of the multivalency effect: a thermodynamic model. J Am Chem Soc 125:16271–16284
59. Gestwicki JE, Cairo CW, Strong LE et al (2002) Influencing receptor-ligand binding mechanisms with multivalent ligand architecture. J Am Chem Soc 124:14922–14933
60. Kitov PI, Sadowska JM, Mulvey G et al (2000) Shiga-like toxins are neutralized by tailored multivalent carbohydrate ligands. Nature 403:669–672
61. Feinberg H, Guo Y, Mitchell DA et al (2005) Extended neck regions stabilize tetramers of the receptors DC-SIGN and DC-SIGNR. J Biol Chem 280:1327–1335
62. Han S, Collins BE, Bengtson P et al (2005) Homomultimeric complexes of CD22 in B cells revealed by protein-glycan cross-linking. Nat Chem Biol 1:93–97
63. Klaas M, Crocker PR (2012) Sialoadhesin in recognition of self and non-self. Semin Immunopathol 34:353–364
64. Kamiya Y, Yamaguchi Y, Takahashi N et al (2005) Sugar-binding properties of VIP36, an intracellular animal lectin operating as a cargo receptor. J Biol Chem 280:37178–37182
65. Kawasaki N, Matsuo I, Totani K et al (2007) Detection of weak sugar binding activity of VIP36 using VIP36-streptavidin complex and membrane-based sugar chains. J Biochem 141:221–229
66. Yamaji T, Teranishi T, Alphey MS et al (2002) A small region of the natural killer cell receptor, Siglec-7, is responsible for its preferred binding to alpha 2,8-disialyl and branched alpha 2,6-sialyl residues. A comparison with Siglec-9. J Biol Chem 277:6324–6332
67. Angata T, Varki A (2000) Cloning, characterization, and phylogenetic analysis of siglec-9, a new member of the CD33-related group of siglecs. Evidence for co-evolution with sialic acid synthesis pathways. J Biol Chem 275: 22127–22135
68. Otto DM, Campanero-Rhodes MA, Karamanska R et al (2011) An expression system for screening of proteins for glycan and protein interactions. Anal Biochem 411: 261–270
69. Sanderson S, Shastri N (1994) LacZ inducible, antigen/MHC-specific T cell hybrids. Int Immunol 6:369–376

70. Ohtsuka M, Arase H, Takeuchi A et al (2004) NFAM1, an immunoreceptor tyrosine-based activation motif-bearing molecule that regulates B cell development and signaling. Proc Natl Acad Sci U S A 101: 8126–8131
71. Yamasaki S, Matsumoto M, Takeuchi O et al (2009) C-type lectin Mincle is an activating receptor for pathogenic fungus, Malassezia. Proc Natl Acad Sci U S A 106:1897–1902
72. Robinson MJ, Osorio F, Rosas M et al (2009) Dectin-2 is a Syk-coupled pattern recognition receptor crucial for Th17 responses to fungal infection. J Exp Med 206:2037–2051
73. Flanagan ML, Arias RS, Hu P et al (2007) Soluble Fc fusion proteins for biomedical research. Methods Mol Biol 378:33–52
74. Schatz PJ (1993) Use of peptide libraries to map the substrate specificity of a peptide-modifying enzyme: a 13 residue consensus peptide specifies biotinylation in *Escherichia* coli. Biotechnology (N Y) 11:1138–1143
75. Yamamoto K, Kawasaki N (2010) Detection of weak-binding sugar activity using membrane-based carbohydrates. Methods Enzymol 478:233–240
76. Powlesland AS, Fisch T, Taylor ME et al (2008) A novel mechanism for LSECtin binding to Ebola virus surface glycoprotein through truncated glycans. J Biol Chem 283:593–602
77. Ohannesian DW, Lotan D, Thomas P et al (1995) Carcinoembryonic antigen and other glycoconjugates act as ligands for galectin-3 in human colon carcinoma cells. Cancer Res 55:2191–2199
78. Ozeki Y, Matsui T, Yamamoto Y et al (1995) Tissue fibronectin is an endogenous ligand for galectin-1. Glycobiology 5:255–261
79. Graham SA, Antonopoulos A, Hitchen PG et al (2011) Identification of neutrophil granule glycoproteins as Lewis(x)-containing ligands cleared by the scavenger receptor C-type lectin. J Biol Chem 286: 24336–24349
80. Kawasaki N, Lin CW, Inoue R et al (2009) Highly fucosylated N-glycan ligands for mannan-binding protein expressed specifically on CD26 (DPPVI) isolated from a human colorectal carcinoma cell line, SW1116. Glycobiology 19:437–450
81. Kumamoto Y, Higashi N, Denda-Nagai K et al (2004) Identification of sialoadhesin as a dominant lymph node counter-receptor for mouse macrophage galactose-type C-type lectin 1. J Biol Chem 279:49274–49280
82. Zhang M, Varki A (2004) Cell surface sialic acids do not affect primary CD22 interactions with CD45 and surface IgM nor the rate of constitutive CD22 endocytosis. Glycobiology 14:939–949
83. van den Berg TK, Nath D, Ziltener HJ et al (2001) Cutting edge: CD43 functions as a T cell counterreceptor for the macrophage adhesion receptor sialoadhesin (Siglec-1). J Immunol 166:3637–3640
84. Chen GY, Tang J, Zheng P et al (2009) CD24 and Siglec-10 selectively repress tissue damage-induced immune responses. Science 323:1722–1725
85. van Vliet SJ, Gringhuis SI, Geijtenbeek TB et al (2006) Regulation of effector T cells by antigen-presenting cells via interaction of the C-type lectin MGL with CD45. Nat Immunol 7:1200–1208
86. Pace KE, Lee C, Stewart PL et al (1999) Restricted receptor segregation into membrane microdomains occurs on human T cells during apoptosis induced by galectin-1. J Immunol 163:3801–3811
87. Walzel H, Schulz U, Neels P et al (1999) Galectin-1, a natural ligand for the receptor-type protein tyrosine phosphatase CD45. Immunol Lett 67:193–202
88. Bond MR, Zhang H, Vu PD et al (2009) Photocrosslinking of glycoconjugates using metabolically incorporated diazirine-containing sugars. Nat Protoc 4:1044–1063
89. Brunner J (1993) New photolabeling and crosslinking methods. Annu Rev Biochem 62:483–514
90. Laughlin ST, Agard NJ, Baskin JM et al (2006) Metabolic labeling of glycans with azido sugars for visualization and glycoproteomics. Methods Enzymol 415:230–250
91. Ramya TN, Weerapana E, Liao L et al (2010) In situ trans ligands of CD22 identified by glycan-protein photocross-linking-enabled proteomics. Mol Cell Proteomics 9: 1339–1351
92. Nyfeler B, Michnick SW, Hauri HP (2005) Capturing protein interactions in the secretory pathway of living cells. Proc Natl Acad Sci U S A 102:6350–6355
93. Nyfeler B, Reiterer V, Wendeler MW et al (2008) Identification of ERGIC-53 as an intracellular transport receptor of alpha1-antitrypsin. J Cell Biol 180:705–712
94. Chen Y, Hojo S, Matsumoto N et al (2013) Regulation of Mac-2BP secretion is mediated by its *N*-glycan binding to ERGIC-53. Glycobiology 23:904–916
95. Rocheleau JV, Petersen NO (2001) The Sendai virus membrane fusion mechanism studied using image correlation spectroscopy. Eur J Biochem 268:2924–2930
96. Schwarzmann G, Wendeler M, Sandhoff K (2005) Synthesis of novel NBD-GM1 and NBD-GM2 for the transfer activity of GM2-activator protein by a FRET-based assay system. Glycobiology 15:1302–1311

97. Lauterbach T, Manna M, Ruhnow M et al (2012) Weak glycolipid binding of a microdomain-tracer peptide correlates with aggregation and slow diffusion on cell membranes. PLoS One 7:e51222
98. Lagana A, Goetz JG, Cheung P et al (2006) Galectin binding to Mgat5-modified *N*-glycans regulates fibronectin matrix remodeling in tumor cells. Mol Cell Biol 26:3181–3193
99. Lowe JB, Marth JD (2003) A genetic approach to Mammalian glycan function. Annu Rev Biochem 72:643–691
100. Dennis JW, Pawling J, Cheung P et al (2002) UDP-N-acetylglucosamine:alpha-6-D-mannoside beta1,6 N-acetylglucosaminyltransferase V (Mgat5) deficient mice. Biochim Biophys Acta 1573:414–422
101. Furukawa K, Ohmi Y, Ohkawa Y et al (2011) Regulatory mechanisms of nervous systems with glycosphingolipids. Neurochem Res 36:1578–1586
102. Kawashima H, Fukuda M (2012) Sulfated glycans control lymphocyte homing. Ann N Y Acad Sci 1253:112–121
103. Takahashi M, Kuroki Y, Ohtsubo K et al (2009) Core fucose and bisecting GlcNAc, the direct modifiers of the N-glycan core: their functions and target proteins. Carbohydr Res 344:1387–1390
104. Bao X, Moseman EA, Saito H et al (2010) Endothelial heparan sulfate controls chemokine presentation in recruitment of lymphocytes and dendritic cells to lymph nodes. Immunity 33:817–829
105. Fu J, Wei B, Wen T et al (2011) Loss of intestinal core 1-derived O-glycans causes spontaneous colitis in mice. J Clin Invest 121:1657–1666
106. Patnaik SK, Stanley P (2006) Lectin-resistant CHO glycosylation mutants. Methods Enzymol 416:159–182
107. Bai X, Wei G, Sinha A et al (1999) Chinese hamster ovary cell mutants defective in glycosaminoglycan assembly and glucuronosyltransferase I. J Biol Chem 274: 13017–13024
108. Kaneko M, Kato Y, Kunita A et al (2004) Functional sialylated *O*-glycan to platelet aggregation on Aggrus (T1alpha/Podoplanin) molecules expressed in Chinese hamster ovary cells. J Biol Chem 279:38838–38843
109. Nitschke L, Carsetti R, Ocker B et al (1997) CD22 is a negative regulator of B-cell receptor signalling. Curr Biol 7:133–143
110. Saijo S, Fujikado N, Furuta T et al (2007) Dectin-1 is required for host defense against *Pneumocystis carinii* but not against Candida albicans. Nat Immunol 8:39–46
111. Hsu DK, Yang RY, Pan Z et al (2000) Targeted disruption of the galectin-3 gene results in attenuated peritoneal inflammatory responses. Am J Pathol 156:1073–1083
112. Zhang B, Zheng C, Zhu M et al (2011) Mice deficient in LMAN1 exhibit FV and FVIII deficiencies and liver accumulation of alpha1-antitrypsin. Blood 118:3384–3391
113. Klaas M, Oetke C, Lewis LE et al (2012) Sialoadhesin promotes rapid proinflammatory and type I IFN responses to a sialylated pathogen, *Campylobacter jejuni*. J Immunol 189:2414–2422
114. Geijtenbeek TB, Van Vliet SJ, Koppel EA et al (2003) Mycobacteria target DC-SIGN to suppress dendritic cell function. J Exp Med 197:7–17
115. Kelm S, Madge P, Islam T et al (2013) C-4 modified sialosides enhance binding to Siglec-2 (CD22): towards potent Siglec inhibitors for immunoglycotherapy. Angew Chem Int Ed Engl 52:3616–3620
116. Abdu-Allah HH, Watanabe K, Completo GC et al (2011) CD22-antagonists with nanomolar potency: the synergistic effect of hydrophobic groups at C-2 and C-9 of sialic acid scaffold. Bioorg Med Chem 19: 1966–1971
117. Kelm S, Gerlach J, Brossmer R et al (2002) The ligand-binding domain of CD22 is needed for inhibition of the B cell receptor signal, as demonstrated by a novel human CD22-specific inhibitor compound. J Exp Med 195:1207–1213
118. Damodaran D, Jeyakani J, Chauhan A et al (2008) CancerLectinDB: a database of lectins relevant to cancer. Glycoconj J 25: 191–198
119. Kumar D, Mittal Y (2011) AnimalLectinDb: an integrated animal lectin database. Bioinformation 6:134–136
120. Crocker PR, Paulson JC, Varki A (2007) Siglecs and their roles in the immune system. Nat Rev Immunol 7:255–266
121. Drickamer K (1992) Engineering galactose-binding activity into a C-type mannose-binding protein. Nature 360:183–186
122. Salomonsson E, Carlsson MC, Osla V et al (2010) Mutational tuning of galectin-3 specificity and biological function. J Biol Chem 285:35079–35091
123. Hirabayashi J, Kasai K (1994) Further evidence by site-directed mutagenesis that conserved hydrophilic residues form a carbohydrate-binding site of human galectin-1. Glycoconj J 11:437–442

124. Satoh T, Chen Y, Hu D et al (2010) Structural basis for oligosaccharide recognition of misfolded glycoproteins by OS-9 in ER-associated degradation. Mol Cell 40:905–916
125. Zelensky AN, Gready JE (2005) The C-type lectin-like domain superfamily. FEBS J 272:6179–6217
126. Drickamer K, Fadden AJ (2002) Genomic analysis of C-type lectins. Biochem Soc Symp 59–72
127. Rademacher C, Bru T, McBride R et al (2012) A Siglec-like sialic-acid-binding motif revealed in an adenovirus capsid protein. Glycobiology 22:1086–1091

Chapter 30

Evaluation of Glycan-Binding Specificity by Glycoconjugate Microarray with an Evanescent-Field Fluorescence Detection System

Hiroaki Tateno

Abstract

The glycan microarray is now an essential tool used to study lectins. With this technique, glycan-binding specificity can be easily assessed by incubation with an array immobilizing a series of glycans. Glycan microarrays have been developed by numerous research groups around the world. Among the available microarrays, our glycan microarray has two unique characteristics: one is the incorporation of an evanescent-field fluorescence detection system and the other is the use of multivalent glycopolymers. These two unique properties allow the highly sensitive detection of only nanogram quantities of lectins even in crude samples such as cell lysates and cell culture media. Thus, this system is suitable for the initial screening of lectins, lectin-like molecules, lectin candidates, and lectin mutants. Here I describe the protocols employed to analyze the glycan-binding specificity of lectins using our glycan microarray system.

Key words Glycoconjugate, Microarray, Evanescent-field, Glycan-binding specificity, Lectins

1 Introduction

Historically, lectins have been identified through hemagglutination, and the binding and precipitation of glycoproteins [1]. In addition, a variety of methods have been devised to analyze glycan-binding specificity, including enzyme-linked immunosorbent assay (ELISA) [2], equilibrium dialysis (ED) [3], surface plasmon resonance (SPR) [4], isothermal titration calorimetry (ITC) [5], frontal affinity chromatography (FAC) [6], and glycan microarray [7]. Glycan microarray is an emerging technology for lectin studies that allows the high-throughput investigation of glycan-binding activity and specificity of lectins [8]. Enormous amounts of information about the glycan-binding activity of lectins can be obtained by simply arranging hundreds of glycans immobilized on glass slides. Researchers around the world have been competing to expand the size of the glycan microarray library. The microarray developed by

Jun Hirabayashi (ed.), *Lectins: Methods and Protocols*, Methods in Molecular Biology, vol. 1200, DOI 10.1007/978-1-4939-1292-6_30, © Springer Science+Business Media New York 2014

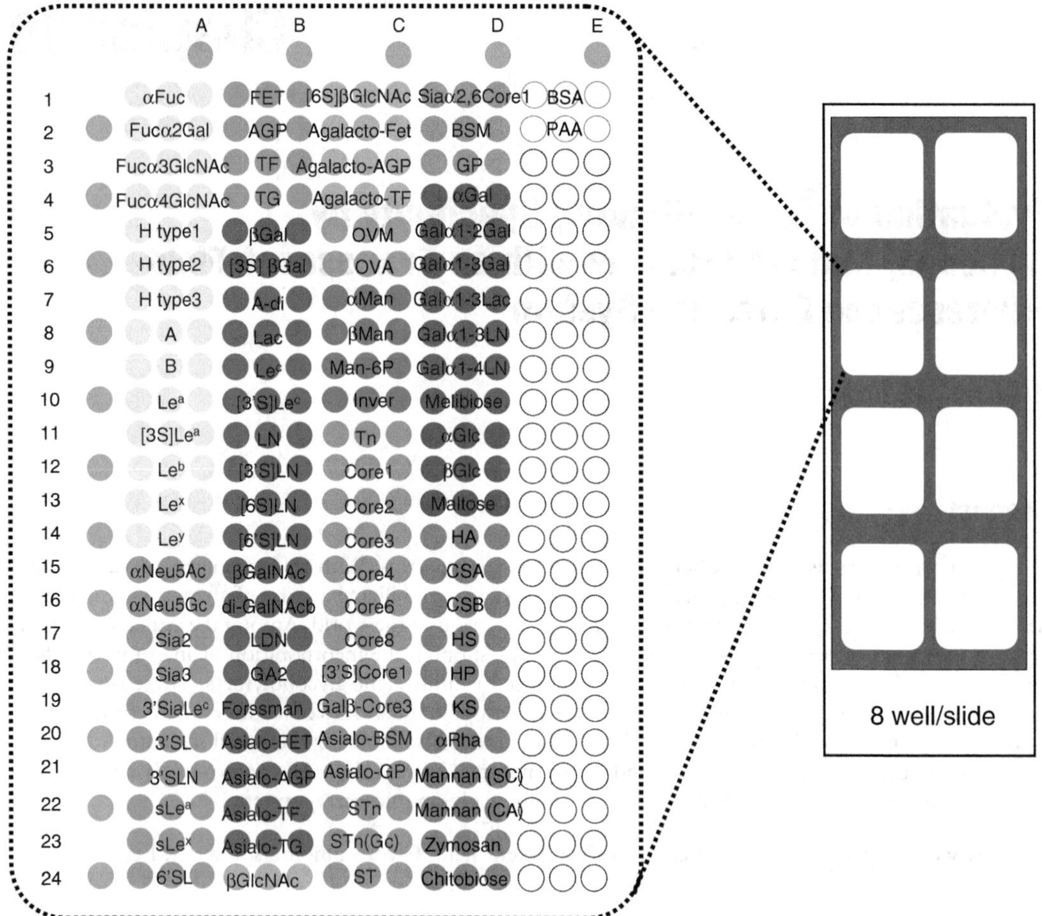

Fig. 1 Spot pattern of glycoconjugate microarray Ver4.3

the Consortium for Functional Glycomics (CFG) currently contains 610 mammalian-type glycans (http://www.functionalglycomics.org/). Our glycan microarray, named "glycoconjugate microarray with an evanescent-field fluorescence detection system", contains minimum essential glycans (98 glycans, Fig. 1) [9]. Unlike antibodies, since lectins in general have broad specificity to glycans, 98 glycans are enough to detect most lectins. The purpose of our system is "the first screening" of lectin activity. Detailed secondary analysis can be performed by other quantitative techniques such as frontal affinity chromatography, which provides affinity constants of lectins for more than a hundred glycans [6]. Our microarray has two unique features. (1) An evanescent-type detection system, but not a confocal-type, is incorporated into this system [9]. The detection system allows equilibrium analysis of the interaction between lectins and glycans. (2) The multivalent polyacrylamide glycopolymers, which are high affinity ligands of lectins, are immobilized on glass slides. These two characteristics provide high

Table 1
Representative examples of the application of glycoconjugate microarray with an evanescent-field fluorescence detection system

	Publication	Purpose	Ref
1	Hu et al. (2013) Biochem J 453, 261–270	Screening of glycan-binding specificity of lectin mutants	[10]
2	Shimokawa et al. (2012) Glycoconj J 29, 457–465	Analysis of glycan-binding specificity of GNA-like lectin	[11]
3	Hu et al. (2012) J Biol Chem 287, 20313–20320	Screening of lectin mutants with specificity to 6-sulfo-galactose created by ribosome display	[12]
4	Takahara et al. (2012) Infect Immun 80, 1699–1706	Analysis of glycan-binding specificity of mouse SIGNR1 and human DC-SIGN	[13]
5	Tateno et al. (2012) Glycobiology 22, 210–220	Screening of novel endogenous lectins	[9]
6	Tateno et al. (2011) J Biol Chem 286, 20345–20253	Analysis of glycan-binding specificity of recombinant lectin library	[14]
7	Yabe et al. (2010) FEBS J 277, 4010–4026	Analysis of glycan-binding specificity of DC-SIGN, DC-SIGNR, and LSECtin extend evidence for affinity to agalactosylated *N*-glycans	[15]
8	Tateno et al. (2010) J Biol Chem 285, 6390–6400	Analysis of glycan-binding specificity of human Langerin	[16]
9	Mitsunaga et al. (2009) J Biochem 146, 369–373	Analysis of glycan-binding specificity of human C21orf63	[17]
10	Yamasaki et al. (2009) Proc Natl Acad Sci U S A 106, 1897–1902	Analysis of glycan-binding specificity of mouse Mincle	[18]
11	Takeuchi et al. (2008) Biochem Biophys Res Commun 377, 303–306	Analysis of glycan-binding specificity of *C. elegans* C-type lectin-like molecules	[19]

sensitivity and high reproducibility to our glycan microarray system. Lectin quantities in the order of nanograms, even in crude samples such as *Escherichia coli* cell lysates [10] and cell culture media [9], can be directly analyzed without purification steps with the aid of an antibody. Furthermore, even live bacteria and viruses can be directly analyzed for their glycan-binding specificity. In contrast, quantitative methods such as ED, SPR, ITC, and FAC require purified lectins for the analysis that are not suitable for the initial screening, but are adequate for secondary analysis. Representative examples of the application of our glycan microarray are shown in Table 1. Here I describe three protocols to analyze lectins using the glycoconjugate microarray: (1) direct labeling method, (2) antibody precomplex method, (3) antibody overlay method (Fig. 2).

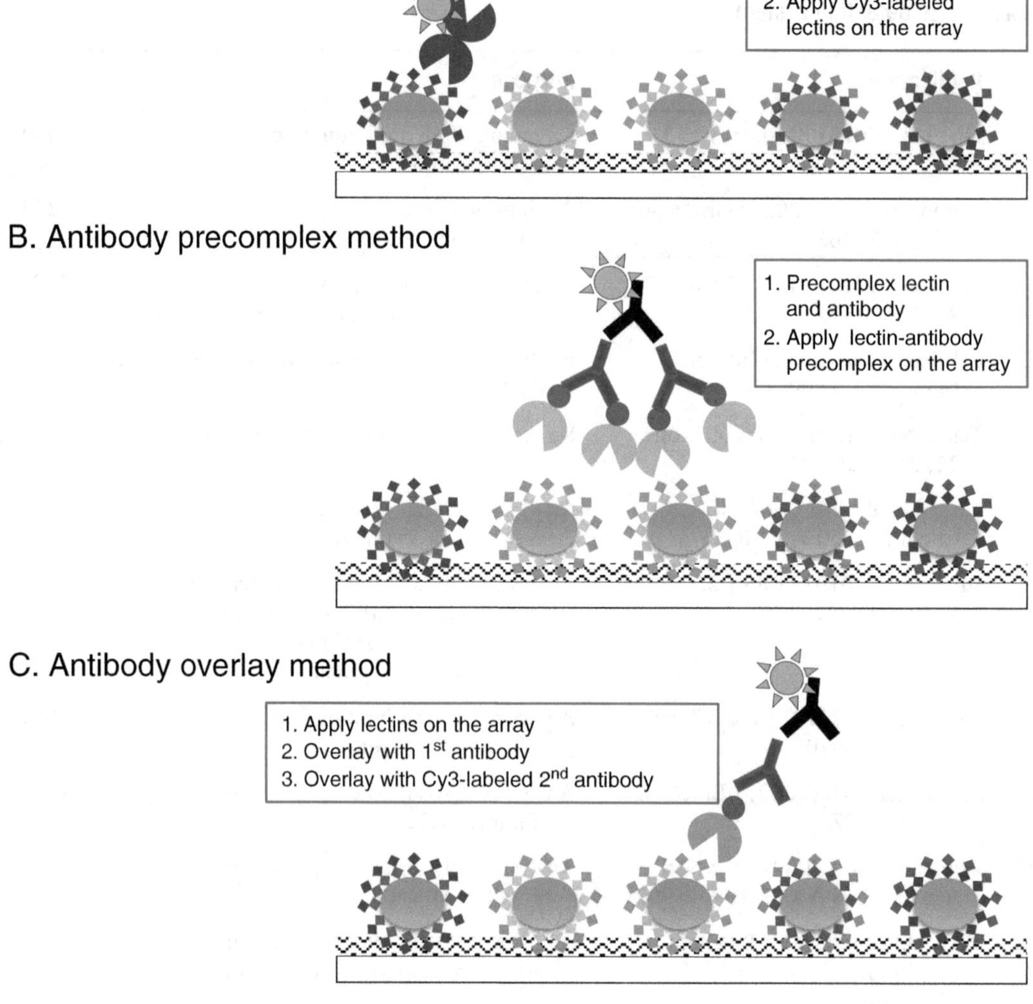

Fig. 2 Analytical protocols using glycoconjugate microarray

2 Materials

1. Cy3 Mono-reactive Dye, Protein Array Grade.
2. Probing solution: 25 mM Tris–HCl, pH 7.5, 140 mM NaCl (TBS), 2.7 mM KCl, 1 mM $CaCl_2$, 1 mM $MnCl_2$, and 1 % Triton X-100.
3. 0.8 mL volume miniature column.
4. Sephadex G-25.
5. PBS: 10 mM Na-phosphate, pH 7.2, 150 mM NaCl.
6. TBS: 25 mM Tris–HCl, pH 7.5, 140 mM NaCl.
7. GlycoStation™ Reader 1200 (GlycoTechnica).
8. ArrayPro Analyzer ver.4.5 (Media Cybernetics).

3 Methods

3.1 Direct Labeling Method

For a total image of the procedure, *see* Fig. 1 (also *see* **Note 1**).

1. Incubate 10 μg of lectins in PBS with Cy3-NHS (for 10 μg protein labeling equivalent) at room temperature in the dark for 1 h.
2. Add 90 μL of probing solution.
3. Pass through a miniature column containing 100 μL of Sephadex G-25 fine to remove unreacted Cy3.
4. Dilute Cy3-labeled lectins to 0.1–10 μg/mL with probing solution.
5. Apply 60–100 μL onto glycan microarray and incubate at 20 °C for 1 h.
6. Acquire fluorescence images using the evanescent-field activated fluorescence scanner (GlycoStation™ Reader 1200).
7. Quantify the fluorescence signal of each spot using an ArrayPro Analyzer ver.4.5 (Media Cybernetics, Bethesda, MD).

3.2 Antibody Precomplex Method

For a total image of the procedure, *see* Fig. 1 (also *see* **Note 2**).

1. Dilute lectin-containing solutions to 0.1, 1, 10 μg/mL with probing solution.
2. Add 1 μg/mL primary antibody and 1 μg/mL Cy3-labeled secondary antibody and incubate at room temperature for 15 min.
3. Apply 60 μL to a well of glycoconjugate microarray.
4. Incubate at 20 °C overnight.
5. Remove solutions from the array and apply 100 μL of fresh probing solution.
6. Acquire fluorescence images using the evanescent-field activated fluorescence scanner (GlycoStation™ Reader 1200).
7. Quantify the fluorescence signal of each spot using the Array Pro Analyzer ver.4.5 (Media Cybernetics, Bethesda, MD).

3.3 Antibody Overlay Method

For a total image of the procedure, *see* Fig. 1 (also *see* **Note 3**).

1. Dilute lectin-containing solutions to 10–100 μg/mL with probing solutions.
2. Apply 60 μL to a well of glycoconjugate microarray.
3. Incubate at 20 °C overnight.
4. Remove solutions from the array and apply 100 μL of fresh probing solution.
5. Apply 60 μL of probing solution containing 1 μg/mL primary antibody and 1 μg/mL Cy3-labeled secondary antibody.

6. Incubate at 20 °C for 1 h.
7. Remove solutions from the array and apply 100 μL of fresh probing solution.
8. Acquire fluorescence images using the evanescent-field activated fluorescence scanner (GlycoStation™ Reader 1200).
9. Quantify the fluorescence signal of each spot using Array Pro Analyzer ver.4.5 (Media Cybernetics, Bethesda, MD).

4 Notes

1. Try this method first if you have enough purified lectins with relatively higher affinity ($K_d < 100$ μM).
2. If the lectins of interest are predicted to have weak affinity (>100 μM), they should be cross-linked with antibody prior to incubation with glycoconjugate microarray to detect their weak glycan-binding activity by cluster effect.
3. Try this method if the lectins are expected to contain inactive forms.

References

1. Boyd WC, Shapleigh E (1954) Specific precipitating activity of plant agglutinins (lectins). Science 119:419
2. Blixt O, Collins BE, van den Nieuwenhof IM, Crocker PR, Paulson JC (2003) Sialoside specificity of the siglec family assessed using novel multivalent probes: identification of potent inhibitors of myelin-associated glycoprotein. J Biol Chem 278:31007–31019
3. Mega T, Hase S (1991) Determination of lectin-sugar binding constants by microequilibrium dialysis coupled with high performance liquid chromatography. J Biochem 109: 600–603
4. Shinohara Y, Kim F, Shimizu M, Goto M, Tosu M, Hasegawa Y (1994) Kinetic measurement of the interaction between an oligosaccharide and lectins by a biosensor based on surface plasmon resonance. Eur J Biochem 223:189–194
5. Dam TK, Gerken TA, Cavada BS, Nascimento KS, Moura TR, Brewer CF (2007) Binding studies of alpha-GalNAc-specific lectins to the alpha-GalNAc (Tn-antigen) form of porcine submaxillary mucin and its smaller fragments. J Biol Chem 282:28256–28263
6. Tateno H, Nakamura-Tsuruta S, Hirabayashi J (2007) Frontal affinity chromatography: sugar-protein interactions. Nat Protoc 2: 2529–2537
7. Blixt O, Head S, Mondala T, Scanlan C, Huflejt ME, Alvarez R, Bryan MC, Fazio F, Calarese D, Stevens J, Razi N, Stevens DJ, Skehel JJ, van Die I, Burton DR, Wilson IA, Cummings R, Bovin N, Wong CH, Paulson JC (2004) Printed covalent glycan array for ligand profiling of diverse glycan binding proteins. Proc Natl Acad Sci U S A 101: 17033–17038
8. Paulson JC, Blixt O, Collins BE (2006) Sweet spots in functional glycomics. Nat Chem Biol 2:238–248
9. Tateno H, Mori A, Uchiyama N, Yabe R, Iwaki J, Shikanai T, Angata T, Narimatsu H, Hirabayashi J (2008) Glycoconjugate microarray based on an evanescent-field fluorescence-assisted detection principle for investigation of glycan-binding proteins. Glycobiology 18:789–798
10. Hu D, Tateno H, Sato T, Narimatsu H, Hirabayashi J (2013) Tailoring GalNAcalpha1-3Galbeta-specific lectins from a multi-specific fungal galectin: dramatic change of carbohydrate specificity by a single amino-acid substitution. Biochem J 453:261–270
11. Shimokawa M, Fukudome A, Yamashita R, Minami Y, Yagi F, Tateno H, Hirabayashi J (2012) Characterization and cloning of GNA-like lectin from the mushroom *Marasmius oreades*. Glycoconj J 29:457–465

12. Hu D, Tateno H, Kuno A, Yabe R, Hirabayashi J (2012) Directed evolution of lectins with sugar-binding specificity for 6-sulfo-galactose. J Biol Chem 287:20313–20320
13. Takahara K, Arita T, Tokieda S, Shibata N, Okawa Y, Tateno H, Hirabayashi J, Inaba K (2012) Difference in fine specificity to polysaccharides of *C. albicans* mannoprotein between mouse SIGNR1 and human DC-SIGN. Infect Immun 80:1699–1706
14. Tateno H, Toyoda M, Saito S, Onuma Y, Ito Y, Hiemori K, Fukumura M, Nakasu A, Nakanishi M, Ohnuma K, Akutsu H, Umezawa A, Horimoto K, Hirabayashi J, Asashima M (2011) Glycome diagnosis of human induced pluripotent stem cells using lectin microarray. J Biol Chem 286:20345–20353
15. Yabe R, Tateno H, Hirabayashi J (2010) Frontal affinity chromatography analysis of constructs of DC-SIGN, DC-SIGNR and LSECtin extend evidence for affinity to agalactosylated N-glycans. FEBS J 277:4010–4026
16. Tateno H, Ohnishi K, Yabe R, Hayatsu N, Sato T, Takeya M, Narimatsu H, Hirabayashi J (2010) Dual specificity of Langerin to sulfated and mannosylated glycans via a single C-type carbohydrate recognition domain. J Biol Chem 285:6390–6400
17. Mitsunaga K, Harada-Itadani J, Shikanai T, Tateno H, Ikehara Y, Hirabayashi J, Narimatsu H, Angata T (2009) Human C21orf63 is a heparin-binding protein. J Biochem 146:369–373
18. Yamasaki S, Matsumoto M, Takeuchi O, Matsuzawa T, Ishikawa E, Sakuma M, Tateno H, Uno J, Hirabayashi J, Mikami Y, Takeda K, Akira S, Saito T (2009) C-type lectin Mincle is an activating receptor for pathogenic fungus, Malassezia. Proc Natl Acad Sci U S A 106:1897–1902
19. Takeuchi T, Sennari R, Sugiura K, Tateno H, Hirabayashi J, Kasai K (2008) A C-type lectin of *Caenorhabditis elegans*: its sugar-binding property revealed by glycoconjugate microarray analysis. Glycobiology 377:303–306

Chapter 31

Potential Usage for In Vivo Lectin Screening in Live Animals Utilizing Cell Surface Mimetic Glyco-nanoparticles, Phosphorylcholine-Coated Quantum Dots (PC-QDs)

Maho Amano, Hiroshi Hinou, Risho Miyoshi, and Shin-Ichiro Nishimura

Abstract

Utilizing glycosylated derivatives as a tag, we are able to explore novel counter-receptor of endogenous lectins or lectin-like molecules in vivo. We have established the standardized methodology including preparation of glycosylated derivatives and construction of a platform for tracing the molecules in vivo at first. Combined use of an aminooxy-terminated thiol derivative and a phosphorylcholine (PC) derivative provides quantum dots (QDs) with novel functions for the chemical ligation of ketone-functionalized compounds and the prevention of nonspecific protein adsorption concurrently. In order to track the derivatives in vivo, near-infrared (NIR) fluorescence imaging of QDs displaying various simple sugars (glyco-PC-QDs) after administration into the tail vein of the mouse can be performed. It has revealed that distinct long-term delocalization over 2 h can be observed depending on the species of glycans ligated to PC-QDs at least in the liver. Until today we have performed live animal imaging utilizing various kinds of sialyl glyco-PC-QDs. They are still retained stably in whole body after 2 h while they showed significantly different in vivo dynamics in the tissue distribution, suggesting that structure/sequence of the neighboring sugar residues in the individual sialyl oligosaccharides might influence the final organ-specific distribution, which should be equivalent to the distribution of sialic acid-recognizing lectins. Here we describe a standardized protocol using ligand-displayed PC-QDs for live cell/animal imaging by versatile NIR fluorescence photometry without influence of size-dependent accumulation/excretion pathway for nanoparticles (e.g., viruses) > 10 nm in hydrodynamic diameter by the liver.

Key words Quantum dot, In vivo screening, Lectin, Glyco-nanoparticle, Phosphorylcholine

1 Introduction

Recent successful approaches of noninvasive imaging for probing in vivo dynamics of glycoconjugates have provided valuable information on tissue distribution of individual glycosylated compounds including natural and neoglycoproteins, sugar-modified liposomes, and magnetite-based nanoparticles presenting glycans [1–6].

Jun Hirabayashi (ed.), *Lectins: Methods and Protocols*, Methods in Molecular Biology, vol. 1200, DOI 10.1007/978-1-4939-1292-6_31, © Springer Science+Business Media New York 2014

For example, Davis et al. developed a protocol for the construction of magnetic resonance imaging (MRI)-visible high Fe-content glyco-nanoparticles which carry E-/P-selectin ligand glycan that allows pre-symptomatic in vivo imaging of rat brain disease models [3]. Fukase et al. also reported an efficient method for the synthesis of glycoclusters labeled by ^{68}Ga-1,4,7,10-tetraazacyclododecane-1,4,7,10-tetraacetic acid (^{68}Ga-DOTA) and showed that the use of ^{68}Ga-DOTA labeling of dendrimer-type *N*-glycoclusters made noninvasive imaging of the in vivo dynamics and organ-specific accumulation of various *N*-glycans by positron emission tomography (PET) possible [6]. It is believed that these datasets of the dynamic bio-distribution profiles for individual glycosylated derivatives would become beneficial for further drug discovery research by means of the recognition of lectins or lectin-like molecules, as well as the development of highly sensitive diagnostic methodology based on MRI and PET.

It should be noted that evaluating the effect of carbohydrate moiety of various natural/non-natural glycoconjugates on bio-distribution and its lifetime appears to be difficult independently from the influence by the structure/property of aglycon (non-glycan) moieties, notably artificial scaffold materials and hydrophobic photosensitive probes as well as protein core structures in native glycoproteins. It is well known that glycans usually exhibit relatively weak affinity with their partner molecules such as cell surface receptors and lectins, even though specificity of the interaction may be very high [7–10]. To investigate the functional role of carbohydrate itself for controlling glycoprotein circulation and distribution in vivo, it seems likely that advent of a novel class of simple glycoprotein model, in which aglycon-scaffold should entail general globular protein-like structure/property and the potential to prevent nonspecific interaction with other biomolecules, cells, tissues, or artificial materials surfaces.

Gold nanoparticles presenting glycans are one of the simplest and most versatile glycoprotein models for probing and investigating functional roles of glycans in vitro using atomic force microscopy (AFM), transmission electron microscopy (TEM), or surface plasmon resonance (SPR) [11–13]. On the other hand, quantum dots (QDs) are fluorescence semiconductor nanoparticles that have unique optical properties, including narrow band and size-dependent luminescence with broad absorption, long-term photostability, and resistance to photobleaching [14–17]. As a scaffold to display carbohydrates and investigate their functions, advantages of QDs are summarized as follows: (1) QDs can be detected and monitored in vivo by simple fluorescent photometric analysis without any special and expensive equipment. (2) Multiple carbohydrates can be displayed on a single QD (100–1,000 molecules/particle), and the carbohydrate density on the single QD can be readily controlled. Thus enhanced affinity with target molecules is expected as a result of the glycoside cluster effect. (3) The QDs range in size between several nanometers and dozens of nanometers, the same level as typical folded proteins (~10 nm in diameter), is suitable for designing a new class glycoprotein models. It is therefore expected that the QDs will exhibit behavior and dynamics similar to common globular proteins distributing in living cells and animals. Because inorganic, metal-containing QDs are obvious that surface modification of QDs is required to be improved in solubility and stability in common physiological condition. For that purpose, many water-soluble QDs have been developed by coating with a variety of thiols and copolymers based on poly(ethylene glycols) (PEG)-like structures [18]. Thus far, Seeberger et al. developed a convenient method to prepare different sugar-capped PEGylated QDs that can be used for in vitro imaging and in vivo liver targeting, in which paraffin sections of mouse livers were prepared for the visualization of QDs by fluorescence microscopy [5]. However, it is noteworthy that nanoparticle hydrodynamic diameter is a crucial design parameter in the development of potential diagnostic and therapeutic agents, as well as an elaborate tool for exploring cell surface lectins existing in various tissues in the body. In fact, QDs > 10 nm in hydrodynamic diameter were proved to be accumulated by the liver designed specifically to capture and eliminate nanoparticles while smaller QDs < 5.5 nm resulted in rapid renal excretion. At present, there is no versatile QDs platform mimicking typical folded proteins (~10 nm in diameter) to have satisfactory functions both for displaying glycans and reducing nonspecific interactions with abundant serum and cellular proteins without loss of quantum yield of the original QDs.

For the purpose of purification of the explored lectin-like molecules, QDs is able to be replaced to magnetic beads, which is powerful for purification steps.

2 Materials

2.1 Preparation of Glyco-PC-QDs

1. $(Boc\text{-}ao\text{-}S)_2$ (11,11′-dithiobis{undec-11-yl 12-[*N*-(tert-butoxycarbonyl)aminooxyacetyl]amino hexa(ethyleneglycol)}) [19] and PC-SH (11-mercaptoundecyl phosphorylcholine) can be purchased from Medicinal Chemistry Pharmaceuticals (MCP) (Sapporo, Japan) via Funakoshi (*see* **Note 1**) (Tokyo, Japan) or APRO Life Science Institute, Inc. (Tokushima, Japan). Stock solutions of these in methanol (10 mM for $(Boc\text{-}ao\text{-}SH)_2$ and 100 mM for PC-SH) can be stored in well-sealed sample tubes at −20 °C for a month.
2. All organic solvents are of special grade. Milli-Q water is used as H_2O. Composition of mixed solvents is represented by v/v, except where otherwise noted.

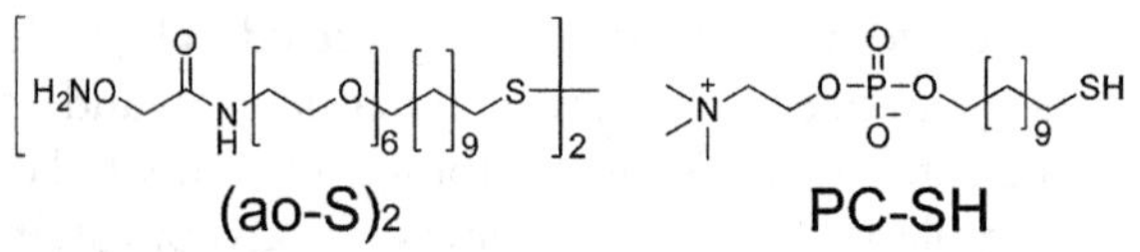

(ao-S)2 PC-SH

Fig. 1 Chemical structure of $(ao\text{-}S)_2$ and PC-SH

3. 4 M HCl in dioxane and 12 % (w/v) sodium borohydride in 14 M NaOH solution are purchased from Sigma-Aldrich (St. Louis, USA).
4. TOPO-coated QDs (CdSe/ZnS, $\lambda_{em} = 545$, 565, 585, 605, 655, 705, and 800 nm) can be purchased from Life Technologies™ (Carlsbad, USA). The solvent of each QD is replaced by *n*-hexane just before use as following: the purchased QD solvent (100 μL, 1 μM in *n*-decane) was diluted with methanol–isopropanol (75:25, v/v, 400 μL). The mixture was centrifuged at 15,000 × *g* to precipitate TOPO-QDs. The supernatant was discarded and remaining TOPO-QD pellet was re-dispersed in *n*-hexane (50 μL) to give a 2 μM solution (*see* **Note 2**).
5. 10 mM of *p*-ketonized phenyl glycosides [20] solution can be stored in well-sealed sample vials at −20 °C for several years (Fig. 1).

3 Methods

3.1 Preparation of Glyco-PC-QDs

The scheme of the reaction is shown in Fig. 2.

3.1.1 Preparation of $(ao\text{-}S)_2$ Solution

1. Add 20 μL of methanol and 10 μL of 4 M HCl in dioxane to 10 μL of 10 mM $(Boc\text{-}ao\text{-}S)_2$ in methanol.
2. Incubate the reaction mixture at 40 °C for 1 h, then neutralize the reaction with 40 μL of 1 M ammonium bicarbonate solution (*see* **Note 3**).
3. Concentrate the mixture to dryness by centrifugal evaporator, then dissolve the residue in 10 μL of methanol to obtain a 10 mM $(ao\text{-}S)_2$ solution (Fig. 3; also *see* **Note 4**).

3.1.2 Preparation of Aminooxy-Functionalized Quantum Dots (ao-PC-QDs)

1. Add 8 μL of 100 mM PC-SH in methanol, 10 μL of 10 mM $(ao\text{-}S)_2$ in methanol, 1 μL of 12 % (w/v) sodium borohydride in 14 M NaOH, 31 μL of H_2O to the 50 μL of 2 μM TOPO-QDs in *n*-hexane freshly prepared in a 1.5 mL Eppendorf tube, then shake intensely for 30 min at room temperature (Fig. 4; also *see* **Note 5**).

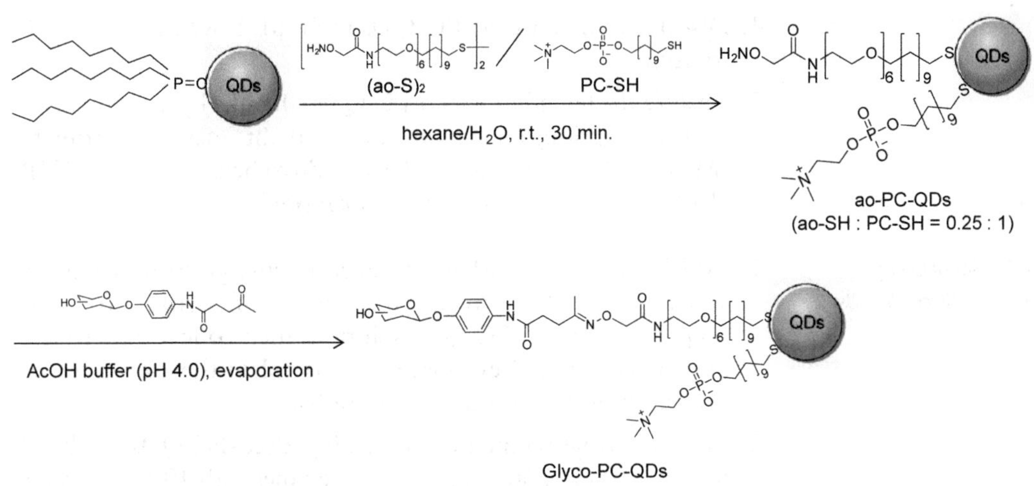

Fig. 2 Preparation of Glyco-PC-QDs

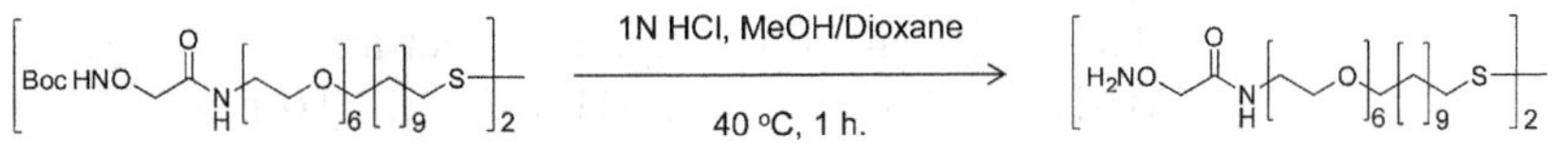

Fig. 3 Deprotection of $(Boc\text{-}ao\text{-}S)_2$

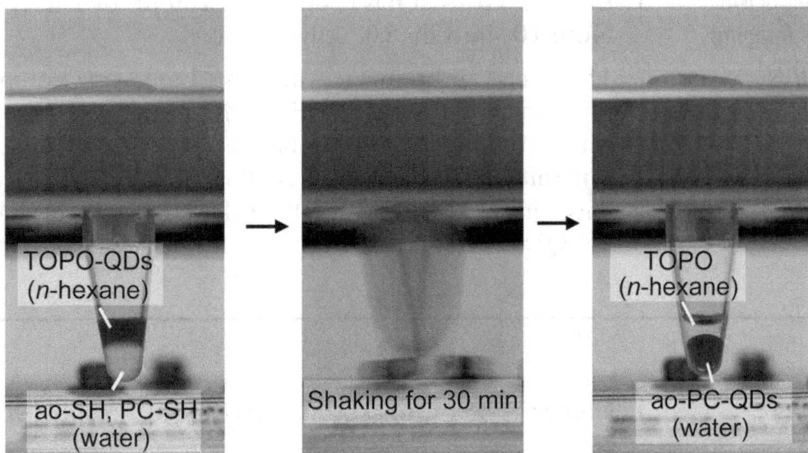

Fig. 4 Monitoring of the ligand exchange reaction. The progress of ligand exchange reaction on QDs can be monitored by the color shift from *n*-hexane layer to water layer. The TOPO ligand remaining in the *n*-hexane layer can be easily removed from the reaction mixture

2. Dilute the *n*-hexane layer before discarding the transparent *n*-hexane layer containing TOPO and repeat several times the dilution and separation process (*see* **Note 6**).
3. Move the water layer containing ao-PC-QDs to ultra-filtration tube (YM 50) and washed five times with 400 μL of H_2O.

4. Dissolve remaining ao-PC-QDs in 50 μL H_2O to give a 2 μM solution.
5. Analyze the ligand of the ao-PC-QDs (0.1 μL) by matrix-assisted laser desorption/ionization time-of-flight mass spectrometry (MALDI-TOFMS) using 2,5-dihydroxybenzoic acid (DHB) (1 μL, 10 mg/mL) as a matrix (*see* **Note 7**).

3.1.3 Glycoblotting to Prepare Glyco-PC-QDs

1. Add 10 μL of 10 mM *p*-ketonized phenyl glycosides solution and 10 μL of 100 mM sodium acetate buffer (pH 4.0) to the 50 μL of 2 μM ao-PC-QDs solution, then concentrate to dryness by centrifugal evaporator to complete the oxime formation reaction for 60 min (*see* **Note 8**).
2. Move the reaction mixture containing glyco-PC-QDs to ultrafiltration tube (YM 50) and washed five times with 400 μL of water.
3. Dissolve the glyco-PC-QDs in 50 μL of water to obtain 2 μM glyco-PC-QDs solution.
4. Apply the glyco-PC-QDs solution to MALDI-TOFMS target plate and analyze the ligand structure on the QD directly by MALDI-TOFMS using DHB (1 μL, 10 mg/mL) as a matrix (*see* **Note 9**).
5. Store the glyco-PC-QDs at 4 °C.

3.2 Live Animal NIR Fluorescence Imaging of Glyco-PC-QDs

1. Inject glyco-PC-QDs (100 pmol, 100 μL of 1 μM/saline, *see* **Note 10**) into the tail veins of mice.
2. The mice can be imaged utilizing Lumazone equipped with Photometrics Cascade II EMCCD camera. QDs are excited with 710 nm and emission filter is 800/12 nm bandpass filter. The suitable exposure time is 50 ms. Figure 5 is an example of mice injected with Lac-PC-QDs, Neu5Ac-PC-QDs, and PC-QDs as a control.

4 Notes

1. Contact www.funakoshi.co.jp/export (export@funakoshi.co.jp).
2. Eppendorf tube (polypropylene tube) is not suitable to store *n*-hexane. If you want to store this QD solution in *n*-hexane, glass tube is recommended.
3. Use well-sealed sample tube to avoid evaporation of the solvents and HCl.
4. Immediate use of this solution for the next ligand exchange process is recommended, although this solution can be stored in well-sealed sample vials at −20 °C for several days.
5. Use well-sealed sample tube. If the color of QD remaining in hexane layer, another 30 min shaking or more vigorous shaking frequency is required to complete the ligand exchange.

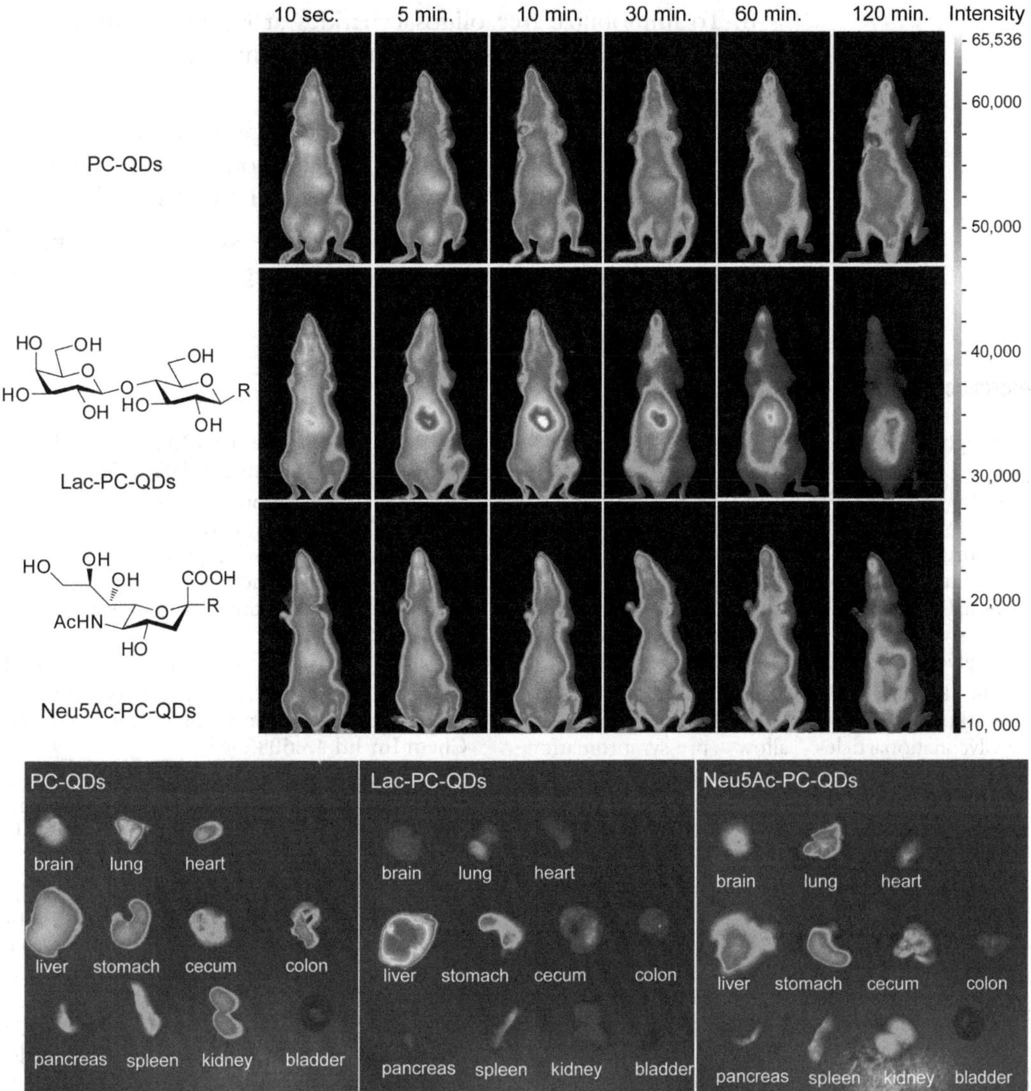

Fig. 5 Live animal imaging of Lac-PC-QDs, Neu5Ac-PC-QDs, and PC-QDs. Imaging of Lac-PC-QDs, Neu5Ac-PC-QDs, and PC-QDs as a control, after injection of 100 pmole glyco-PC-QDs (*top*). The mice were imaged using Lumazone equipped with Photometrics Cascade II EM CCD camera. QDs were excited with 710 nm and emission filter was 800/12 nm bandpass filter. Photographs of major organs isolated from three tested mice (PC-QDs, Lac-PC-QDs, and Nue5Ac-PC-QDs) at 2 h after administration (*bottom*). Lectins specific for each glycan are clearly shown in the tissues

6. If it is difficult to discard the *n*-hexane layer by pipetting completely, using a nitrogen or air stream directed into the tube evaporates remaining *n*-hexane.
7. You can find the ions of PC-SH and ao-SH from the MALDI-TOFMS spectrum, which indicates that QDs are covered with PC-SH and ao-SH. In addition, mixed disulfides generated from PC-SH and ao-SH are observed.

8. To immobilize free oligosaccharides at reducing end, longer incubation time (1–12 h) under higher temperature (50–90 °C) is required [21].
9. You can find the ions of PC-SH and glyco-ao-SH from the MALDI-TOFMS spectrum. In addition, mixed disulfides generated from PC-SH and glyco-ao-SH are observed.
10. QDs having both λ_{ex} and λ_{em} within 700–900 nm are required for the live animal NIR fluorescence imaging. You can use any type of QDs for live cell imaging.

References

1. Laughlin ST, Baskin JM, Amacher SL, Bertozzi CR (2008) In vivo imaging of membrane-associated glycans in developing zebrafish. Science 320:664–667
2. Andre S, Kozar T, Kojima S, Unverzagt C, Gabius H-J (2009) From structural to functional glycomics: core substitutions as molecular switches for shape and lectin affinity of *N*-glycans. Biol Chem 390:557–565
3. van Kasteren SI, Campbell SJ, Serres S, Anthony DC, Sibson NR, Davis BG (2009) Glyconanoparticles allow pre-symptomatic in vivo imaging of brain disease. Proc Natl Acad Sci U S A 106:18–23
4. Marradi M, Alcantara D, de la Fuente JM, Garcia-Martin ML, Cerdan S, Penades S (2009) Paramagnetic Gd-based gold glyconanoparticles as probes for MRI: tuning relaxivities with sugars. Chem Commun 26:3922–3924
5. Kikkeri R, Lepenies B, Adibekian A, Laurino P, Seeberger PH (2009) In vitro imaging and in vivo liver targeting with carbohydrate capped quantum dots. J Am Chem Soc 131:2110–2112
6. Tanaka K, Siwu ER, Minami K, Hasegawa K, Nozaki S, Kanayama Y, Koyama K, Chen WC, Paulson JC, Watanabe Y, Fukase K (2010) Noninvasive imaging of dendrimer-type *N*-glycan clusters: in vivo dynamics dependence on oligosaccharide structure. Angew Chem Int Ed 49:8195–8200
7. Lee YC (1992) Biochemistry of carbohydrate-protein interaction. FASEB J 6:3193–3200
8. Lee YC, Lee RT (1995) Carbohydrate-protein interactions: Basis of glycobiology. Acc Chem Res 28:321–327
9. Mammen M, Choi SK, Whitesides GM (1998) Polyvalent interactions in biological systems: implications for design and use of multivalent ligands and inhibitors. Angew Chem Int Ed 37:2754–2794
10. Lundquist JJ, Toone EJ (2002) The cluster glycoside effect. Chem Rev 102:555–578
11. de la Fuente JM, Eaton P, Barrientos AG, Rojas TC, Rojo J, Canada J, Fernandez A, Penades S (2001) Gold glyconanoparticles as water-soluble polyvalent models to study carbohydrate interactions. Angew Chem Int Ed 40: 2257–2261
12. Katz E, Willner I (2004) Integrated nanoparticle-biomolecule hybrid systems: synthesis, properties, and applications. Angew Chem Int Ed 43:6042–6108
13. Marradi M, Martin-Lomas M, Penades S (2010) Glyconanoparticles polyvalent tools to study carbohydrate-based interactions. Adv Carbohydr Chem Biochem 64:211–290
14. Reed MA, Randall JN, Aggarwal RJ, Matyi RJ, Moore TM, Wetsel AE (1988) Observation of discrete electronic states in a zero-dimensional semiconductor nanostructure. Phys Rev Lett 60:535–537
15. Rossetti R, Nakahara S, Brus LE (1983) Quantum size effects in the redox potentials, resonance Raman spectra, and electronic spectra of CdS crystallites in aqueous solution. J Chem Phys 79:1086–1088
16. Gleiter H (1992) Nanostructured materials. Adv Mater 4:474–481
17. Alivisatos AP (1996) Semiconductor clusters, nanocrystals, and quantum dots. Science 271:933–937
18. Medintz IL, Uyeda HT, Goldman ER, Mattoussi H (2005) Quantum dot bioconjugates for imaging, labelling and sensing. Nat Mater 4:435–446
19. Nagahori N, Abe M, Nishimura S-I (2009) Structural and functional glycosphingolipidomics by glycoblotting with an aminooxy-functionalized gold nanoparticle. Biochemistry 48:583–594

20. Ohyanagi T, Nagahori N, Shimawaki K, Hinou H, Yamashita T, Sasaki A, Jin T, Iwanaga T, Kinjo M, Nishimura S-I (2011) Importance of sialic acid residues illuminated by live animal imaging using phosphorylcholine self-assembled monolayer-coated quantum dots. J Am Chem Soc 133:12507–12517

21. Shimaoka H, Kuramoto H, Furukawa J-I, Miura Y, Kurogochi M, Kita Y, Hinou H, Shinohara Y, Nishimura S-I (2007) One-pot solid-phase glycoblotting and probing by transoximization for high-throughput glycomics and glycoproteomics. Chem Eur J 13:1664–1673

Chapter 32

Remodeling Cell Surface Glycans Using Glycosylation Inhibitors

Norihito Kawasaki

Abstract

Cell surface glycan remodeling is a useful method to modulate glycan–lectin interactions. In this chapter, a facile and reliable method to remodel mammalian cell surface *N*-glycans using inhibitors for *N*-glycan-processing enzymes is described. The method is widely applicable to many mammalian systems because those inhibitors work for the conserved glycosylation pathways among species.

Key words Glycan remodeling, Deoxymannojirimycin, Kifunensine, Swainsonine, Flow cytometry

1 Introduction

Modulation of lectin–glycan interactions is a useful strategy to investigate function of animal lectins. To modify the lectin–glycan interactions, alteration of cell surface glycans has been achieved successfully in many different ways. Overexpression of cellular glycosyltransferases has been utilized successfully for the purpose (*see* Chapter 33 described by Y. Naito-Matsui and H. Takematsu). Gene deletion for glycosyltransferases and glycoproteins of interest is a powerful strategy to elucidate functions of animal lectins in vivo. Cell surface glycan remodeling by using membrane permeable small compounds that inhibit cellular glycan processing is one of the most convenient methods, and is described in the chapter.

Inhibitors for enzymes involved in glycan synthesis can be used to alter glycan synthesis in the cell, resulting in remodeling of cell surface glycans. Plant alkaloids have been shown to inhibit glycosidases involved in *N*-glycan processing (Fig. 1) [1]. Deoxymannojirimycin (DMJ) and kifunensine (KIF) inhibit α-mannosidases in the ER and the Golgi, resulting in accumulation of high-mannose-type glycans on the cell surface. Swainsonine (SW) is an inhibitor for Golgi mannosidase II, thereby causing accumulation of the hybrid-type glycans on the cell surface. For *O*-glycans, benzyl-α-GalNAc serves as a decoy

Jun Hirabayashi (ed.), *Lectins: Methods and Protocols*, Methods in Molecular Biology, vol. 1200,
DOI 10.1007/978-1-4939-1292-6_32,

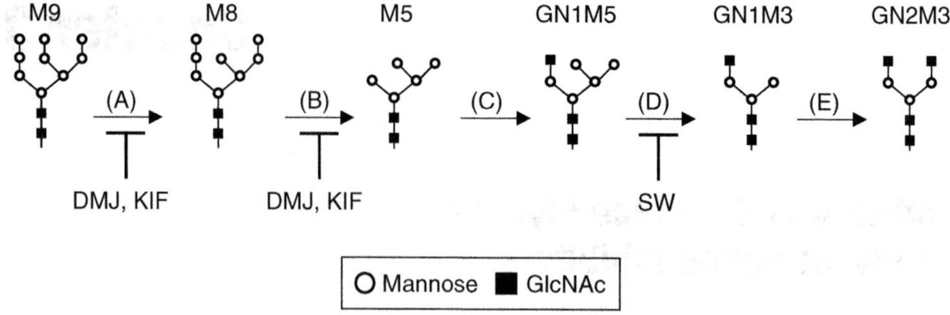

Fig. 1 Inhibition of mannose trimming on *N*-glycans by mannose analogs. Initial trimming of M9 to M5 is mediated by two enzymes, ER-mannosidase I (*A*) and Golgi-mannosidase I (*B*). These mannosidases are inhibited by DMJ and KIF, which results in accumulation of M9 and M8 on the cell surface [4]. The truncated high-mannose-type glycan, M5, is the following substrate for GlcNAc transferase I (*C*), generating GN1M5 structure. Further mannose trimming on GN1M5 to GN1M3 occurs through Golgi-mannosidase II (*D*), which is inhibited by SW. Thus SW treatment of cells results in accumulation of hybrid-type glycans on the cell surface. Finally, GlcNAc transferase II (*E*) transfers another GlcNAc on GN1M3, which initiates formation of complex-type glycans

Table 1
Representative plant lectins for cell surface glycan analysis

Lectin	Glycan-binding specificity	References
Canavalia ensiformis agglutinin (ConA)	Mannose, glucose	[1, 6]
Galanthus nivalis agglutinin (GNA)	α1,2 and 1,6-linked mannose	
Datura stramonium agglutinin (DSA)	GlcNAc and LacNAc	
Erythrina cristagalli agglutinin (ECA)	Asialylated galactose	
Ulex europaeus agglutinin I (UEA-I)	α1,2-Linked fucose	
Sambucus nigra agglutinin (SNA)	α2,6-Linked sialic acid	
Maackia amurensis Agglutinin (MAA I and II)	α2,3-Linked sialic acid and sulfated asialylated galactose	[7]

substrate for glycosyltranferases involved in *O*-glycan synthesis, resulting in GalNAc-terminated *O*-glycans (Tn antigen) [2]. Recently, synthetic fluorinated sugar analogs for sialyl- and fucosyltransferases have become available [3].

In this chapter, I describe a method to modulate cell surface *N*-glycans of HeLaS3 cells with DMJ, KIF, and SW. The protocol includes the culture of HeLaS3 cells in the presence of the inhibitors and the cell surface staining by plant lectins to assess the effect of the inhibitors. Representative plant lectins useful for cell surface glycan analysis are listed in Table 1 (*see* **Note 1**). In successful cases, significant changes of the cellular reactivity to the plant lectins will be observed. In the previous study, the binding of *Galanthus nivalis* agglutinin (GNA), a mannose specific plant lectin, to the HeLaS3 cells was enhanced by three- to four fold by the DMJ and KIF treatment (Fig. 2). SW treatment also increased the binding of GNA to the HeLaS3 cells to a lesser extent, presumably

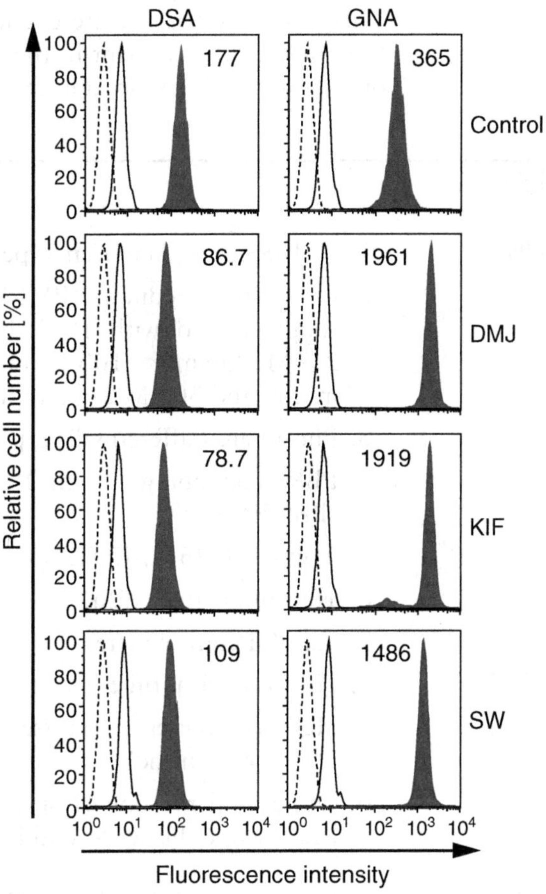

Fig. 2 Validation of cell surface glycan remodeling by lectin staining. HeLaS3 cells were cultured for 24 h in the presence of 1 mM DMJ, 8.6 μM, KIF, 58 μM SW, or vehicle (Control). Cells were harvested and stained with PE-labeled GNA or DSA (*filled*), or PE-streptavidin alone (*thin line*). Stained cells were analyzed by flow cytometry (reproduced from [4] with permission from Oxford University Press)

because GNA binds to the terminal mannose residues in hybrid-type glycans, whose expression was enhanced by SW treatment (Fig. 2). In contrast, the binding of *Datura stramonium* agglutinin (DSA), a LacNAc-specific plant lectin, to the cells was reduced by the treatment with DMJ or KIF by more than half, demonstrating the reduced expression of complex and hybrid-type glycans on the cell surface (Fig. 2). The reduction of complex-type glycans was also demonstrated by the reduced binding of DSA to SW-treated cells (Fig. 2).

Although the protocol described in this chapter is for DMJ, KIF, and SW, it is applicable for glycan remodeling using other inhibitors. Inhibitors may have different efficacy depend on their cell permeability, turnover of cell surface glycoconjugates, and expression level of the target enzymes in the cell of interest.

Therefore investigators are encouraged to optimize the dose and time period of treatment in their own experimental design based on the protocol described in this chapter.

2 Materials

2.1 Cell Culture

1. HeLaS3 cells (American Type Culture Collection) (*see* **Note 2**).
2. Cell culture medium: RPMI 1640 medium (Life Technologies) supplemented with 10 % heat-inactivated fetal calf serum, 2 mM glutamine, 100 U/mL penicillin, 100 μg/mL streptomycin, and 50 μM 2-mercaptoethanol.
3. Phosphate buffered saline, pH 7.4 (PBS).
4. PBS containing 1 mM ethylenediaminetetraacetic acid (PBS-EDTA).
5. Trypsin (0.25 %)-EDTA (Life Technologies).
6. 100-mm culture dish.
7. Six-well culture plate.
8. 15-mL conical tube.
9. Hemocytometer (Hausser Bright-Line hemocytometer, Hausser scientific).
10. Optical microscope (Fisher Scientific Micromaster Inverted Microscope, Fisher Scientific).

2.2 Glycosylation Inhibitors

1. Deoxymannojirimycin (DMJ) stock solution: 25 mM DMJ in H_2O. Add 1 mL of sterile H_2O into the vial of 5 mg DMJ (Sigma-Aldrich). Vortex the vial gently to dissolve DMJ powder completely. Transfer the solution into a 1 mL syringe (BD Biosciences) and pass through a sterile 0.22 μm syringe filter (PALL). Make aliquots of sterile DMJ solution in 1.5 mL screw tubes and store at −20 °C. Avoid multiple freeze/thaw cycles.
2. Kifunensine (KIF) stock solution: 4.3 mM KIF in H_2O. Add 1 mL of sterile H_2O into the vial of 1 mg KIF from Merck Millipore. Make sterile aliquots and store at −20 °C as described above.
3. Swainsonine (SW) stock solution: 5.8 mM SW in H_2O. Add 0.5 mL of sterile H_2O into the vial of 0.5 mg SW (Merck Millipore). Make sterile aliquots and store at −20 °C as described above.

2.3 Staining of Cell Surface Glycans by Fluorescent Plant Lectins for Flow Cytometry

1. Flow cytometer (FACS Calibur, BD Biosciences).
2. Centrifuge equipped with swing-bucket rotor for 96-well plates.
3. HEPES-buffered saline pH 7.4, containing 0.1 % bovine serum albumin, 1 mM $CaCl_2$, and 0.1 % NaN_3 (FACS buffer).

4. PE-labeled plant lectin DSA and GNA. Dilute biotin-labeled DSA or GNA (EY Laboratories) in FACS buffer in 1.5 mL tubes at final concentration of 10 μg/mL. Add phycoerythrin (PE)-conjugated streptavidin (BD Biosciences) at final concentration of 3 μg/mL (*see* **Note 3**). Incubate the tubes at 4 °C for 30 min in the dark to form the PE-conjugated DSA or GNA.
5. Propidium iodide (PI) working solution: 3 μg/mL PI in FACS buffer. Dissolve 5 mg of PI (Sigma-Aldrich) in 5 mL of FACS buffer. This PI stock solution is stored at 4 °C in the dark. Dilute this stock solution in FACS buffer to make PI working solution (*see* **Note 4**).
6. 96-well round-bottom tissue culture plate (BD Biosciences).
7. 5 mL round-bottom tube for flow cytometry (FACS tube, BD Biosciences).

3 Methods

3.1 Treatment of HeLaS3 Cells with the Glycosylation Inhibitors

On the day before the inhibitor treatment, cultured HeLaS3 cells in a 100-mm culture dish will be harvested and seeded in a 6-well culture plate.

1. Aspirate the culture supernatant.
2. Wash the culture dish with 5 mL of PBS (*see* **Note 5**).
3. Add 1 mL of trypsin-EDTA solution into the dish.
4. Incubate the dish at 37 °C in the 5 % CO_2 incubator for 1–3 min (*see* **Note 6**).
5. Gently tap the dish to detach the cells and harvest the cells with 9 mL of culture medium into a 15 mL conical tube.
6. Spin down the cells by centrifugation at 230 × *g* for 5 min.
7. Discard the supernatant and suspend the cells in 10 mL of the culture medium.
8. Count the cells by hemocytometer.
9. Seed the cells at 50 % confluency in a 6-well culture plate (1.0×10^6/well, 2 mL cell culture medium) so that they will be almost confluent on the day of treatment (*see* **Note 7**).
10. After 18 h-culture, add DMJ, KIF, or SW at 1 mM, 8.6 μM, or 58 μM, respectively into the culture. As a negative control, add the same amount of sterile water into the culture.
11. Incubate the cells for another 24 h at 37 °C in the 5 % CO_2 incubator.

3.2 Staining of Cells with Fluorescent Plant Lectins for Flow Cytometry

After 24 h incubation (*see* **Note 8**), reactivity of HeLaS3 cells to DSA and GNA will be assessed by flow cytometry.

1. Aspirate the cell culture supernatant.
2. Add 0.5 mL of PBS-EDTA into the 6-well culture plate (*see* **Note 9**).
3. Incubate the plate at 37 °C in the 5 % CO_2 incubator for 5–10 min.
4. Gently tap the dish to detach the cells and harvest the cells with 4.5 mL of FACS buffer into a 15 mL conical tube.
5. Spin down the cells by centrifugation at 230 × *g* for 5 min.
6. Discard the supernatant and suspend the cells in 5 mL of FACS buffer.
7. Count the cells by hemocytometer.
8. Suspend the cells at a concentration of 2×10^7/mL in FACS buffer.
9. Transfer cells at 2×10^5/well into a 96-well round bottom plate (*see* **Note 10**).
10. Spin down the cells by centrifugation at 510 × *g* for 3 min.
11. Discard the supernatant completely by decantation (*see* **Note 11**).
12. Vortex the plate well to break down cell pellet.
13. Add 40 μL of the PE-labeled DSA, GNA, or PE-streptavidin alone into the well.
14. Gently tap the plate to mix the cells (*see* **Note 12**).
15. Incubate the plate at 25 °C for 30 min in the dark.
16. Add 200 μL of FACS buffer and spin down the cells at 510 × *g* for 3 min.
17. Discard the supernatant completely by decantation and vortex well (*see* **Note 11**).
18. Repeat **steps 15** and **16**.
19. Suspend the cells in 200 μL of FACS buffer.
20. Transfer the cells into the 5 mL FACS tube.
21. Add 100 μL of PI working solution into the tube (*see* **Note 13**).
22. Analyze the binding of DSA and GNA to the cells with FACS Calibur (*see* **Note 14**).

4 Notes

1. To select the appropriate plant lectins, Table 1 as well as the web site for vendors is useful (EY Laboratories, VECTOR LABORATORIES).

2. Since the *N*-glycosylation pathway is conserved in all mammalian species, the use of glycosylation inhibitor is applicable to other cell lines. CHO, BW5147, and HEK 293 cells could be successfully treated under the same condition.
3. Lectin–streptavidin complex at a molar ratio of lectin to streptavidin, 4:1, enhances the detection sensitivity by increasing the avidity of the lectin probe [4].
4. PI is used to exclude dead cells in flow cytometric analysis. Dead cells affect assessment of lectin staining because of their nonspecific binding to plant lectins and streptavidin.
5. The residual medium frequently inhibits the trypsin digestion.
6. Cell death by the longer incubation with trypsin should be avoided.
7. Cell surface glycosylation can be affected by the cell density and metabolites in the cell culture [5]. To obtain reproducible effect of glycosylation inhibitors, it is important to keep the cell density constant in experiments.
8. The incubation time needs to be optimized when other inhibitors are used. DMJ, KIF, and SW show a maximum effect on the cell surface glycans after 20–26 h incubation.
9. It is important to harvest the cells with PBS-EDTA instead of trypsin-EDTA. Trypsin cleaves the cell surface glycoproteins, which may affect subsequent analysis.
10. Make four wells for each treatment condition. Three wells are used for staining by DSA, GNA, and streptavidin alone, respectively, and one well is left unstained for negative control. Staining the cells in a 96-well plate is convenient.
11. All residual liquid must be removed from the wells by tapping the plate once gently on absorbent paper.
12. Do not use the vortex. This causes contamination of samples with each other. Alternatively pipetting may be suitable to mix the cells.
13. Extra incubation is not required for PI staining.
14. Alternatively, you can use your own assay to verify the effect of inhibitors such as agglutination assay and western blotting with plant lectins, HPLC, and mass spectroscopy.

Acknowledgements

I would like to thank Dr. Kazuo Yamamoto for critical reading of the manuscript, Oxford University Press for the permission to use the Fig. 2 with modification, Dr. Fumiko Kawasaki for her technical help in manuscript preparation.

References

1. Varki A, Cummings R, Esko J et al (1999) Essentials of glycobiology. Cold Spring Harbor Laboratory, New York
2. Sakamaki T, Imai Y, Irimura T (1995) Enhancement in accessibility to macrophages by modification of mucin-type carbohydrate chains on a tumor cell line: role of a C-type lectin of macrophages. J Leukoc Biol 57:407–414
3. Rillahan CD, Antonopoulos A, Lefort CT et al (2012) Global metabolic inhibitors of sialyl- and fucosyltransferases remodel the glycome. Nat Chem Biol 8:661–668
4. Kawasaki N, Matsuo I, Totani K et al (2007) Detection of weak sugar binding activity of VIP36 using VIP36-streptavidin complex and membrane-based sugar chains. J Biochem 141: 221–229
5. Yang M, Butler M (2000) Effects of ammonia on CHO cell growth, erythropoietin production, and glycosylation. Biotechnol Bioeng 68: 370–380
6. Sharon N, Lis H (2003) Lectins, 2nd edn. Kluwer Academic Publishers, Boston
7. Geisler C, Jarvis DL (2011) Effective glycoanalysis with *Maackia amurensis* lectins requires a clear understanding of their binding specificities. Glycobiology 21:988–993

Chapter 33

Remodeling of Glycans Using Glycosyltransferase Genes

Yuko Naito-Matsui and Hiromu Takematsu

Abstract

Remodeling of glycans on the cell surface is an essential technique to analyze cellular function of lectin-glycan ligand interaction. Here we describe the methods to identify the responsible enzyme (glycosyltransferase) regulating the expression of the glycan of interest and to modulate the glycan expression by overexpressing the glycosyltransferase gene. For the identification of the responsible enzyme, we introduce a new method, CIRES (correlation index-based responsible-enzyme gene screening), that consists of statistical comparison of glycan expression profile obtained by flow cytometry and gene expression profile obtained by DNA microarray.

Key words Cellular glycan function, Lectin ligand, Biosynthesis pathway, Flow cytometry, DNA microarray, Statistical correlation, Transfection, Retrovirus infection

1 Introduction

Lectins usually function by interacting with their glycan ligands. Therefore, modulation of glycan ligand expression on the cells is useful for the functional analysis of lectins. However, sometimes it is difficult to efficiently regulate the targeted glycan expression by single-gene overexpression because several different enzymes share common molecules as substrates and targets in the multi-step and highly branched glycan biosynthesis pathway. In this chapter, we introduce the method to find responsible (e.g., rate-limiting) enzyme in the biosynthesis pathway for effective glycan expression modulation before describing transfection of glycosyltransferase gene to mammalian cells for glycan remodeling.

To find the responsible gene for the biosynthesis of a certain glycan, we established new methodology called CIRES (correlation index-based responsible-enzyme gene screening) [1]. CIRES is novel quantitative phenotype-genotype correlation analysis, utilizing cell surface phenotypes (glycan expression) and gene expression profiles obtained from multiple cells (Fig. 1). Relative glycan expression profile among multiple cell lines is obtained using flow

Jun Hirabayashi (ed.), *Lectins: Methods and Protocols*, Methods in Molecular Biology, vol. 1200,
DOI 10.1007/978-1-4939-1292-6_33,

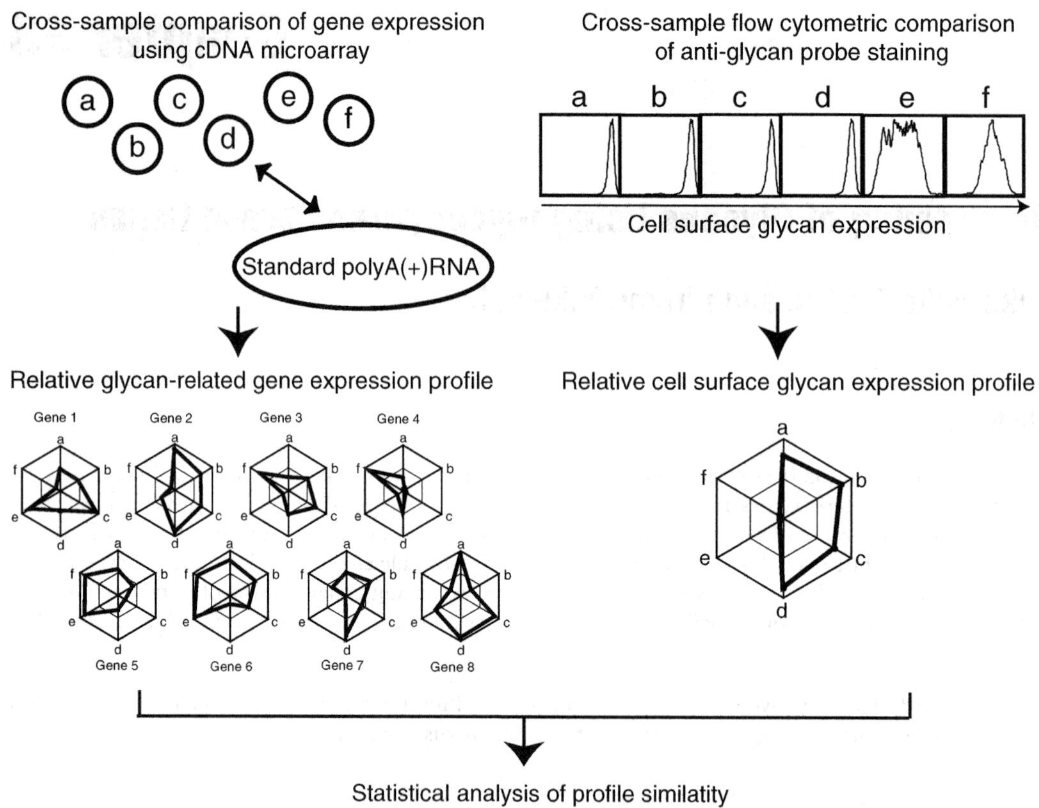

Fig. 1 Concept of CIRES: Example of comparison among six cell lines (**a**–**f**). Relative gene expression profile obtained by DNA microarray and relative glycan expression profile obtained by flow cytometry are compared to find responsible enzyme for the glycan biosynthesis. In the web graphs, the edge of the polygon corresponds to the strongest expression. The polygons in the left web graphs represent the relative gene expression profiles of eight different genes. The same set of cell lines is examined for cell surface glycan expression by flow cytometry (here, the relative mean fluorescence intensity is also expressed as web graph). Similarities and dissimilarities between the profiles are assessed using Pearson's correlation coefficient which has values ranging from −1 (inverse correlation) to 1 (perfect correlation) (figure was modified from ref. 1)

cytometry by staining the cells with glycan-specific probes/antibodies and calculating relative mean fluorescence intensity. The gene expression profile is concomitantly obtained by microarray [2]. Then, these glycan and gene expression profiles are assessed by calculating Pearson's correlation coefficient, and candidate genes are listed [1, 3, 4]. For the cross-comparison of multiple samples by microarray, each sample is compared with universal reference RNA (Clontech) to create a relative gene expression profile. Pearson's correlation coefficient ranges from −1 (negative correlation) to 1 (positive correlation) and 0 means no correlation; thus, CIRES can be used to identify both positively and negatively regulating enzymes. Overexpression of positively correlated gene or knockdown of negatively correlated gene is expected to induce the

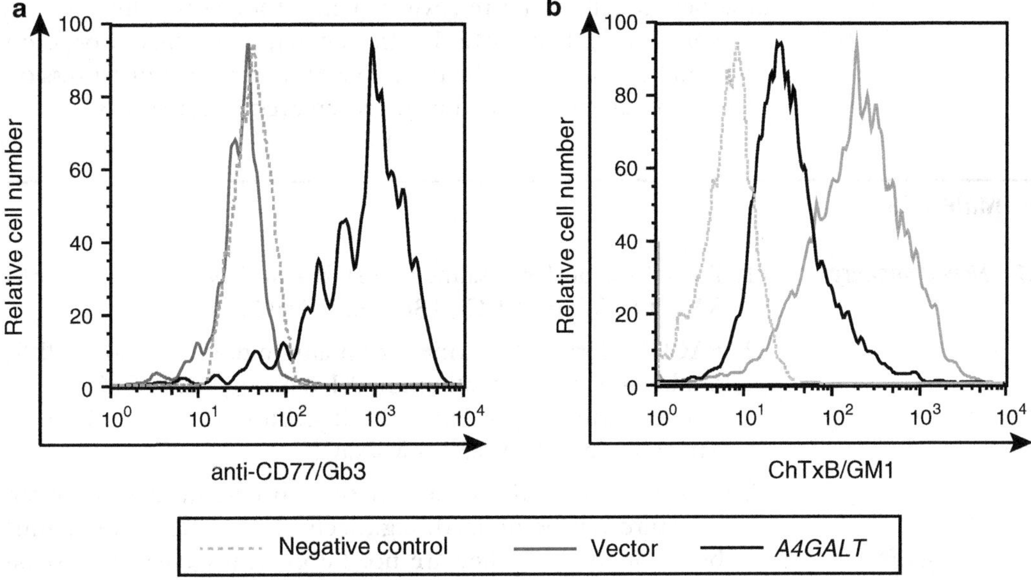

Fig. 2 Remodeling of the cell surface glycan expression by α1,4-galactosyltransferase (*A4GALT*) overexpression. (**a**) Cell surface expression of neutral glycosphingolipid Gb3 (CD77) was positively correlated with the gene expression of *A4GALT*. In order to have appropriate control, retrovirus vector used here expresses glycosyltransferase cDNA and EGFP via an internal ribosome entry site (IRES); thus, virus-infected cells were distinguished from non-infected cells by EGFP signal. Gb3-negative Namalwa cells were infected with retroviral vector encoding *A4GALT* and stained with anti-Gb3 monoclonal antibody. Positively correlated A4GALT expression efficiently remodeled cell surface Gb3 expression with single-gene manipulation. (**b**) Cell surface expression of ganglioside GM1 was negatively correlated with the gene expression of *A4GALT*. GM1-positive Namalwa cells were infected with retroviral vector encoding *A4GALT* and stained with cholera toxin B subunit (ChTxB) that binds to GM1. Although A4GALT is not directly involved in the GM1 biosynthesis, induction of A4GALT suppressed GM1. Further analysis revealed that A4GALT controls LacCer synthase to dominantly compete out GM3 synthase in the glycosphingolipid biosynthesis in human B cells (*see* ref. 4). Focusing on the negative correlation could result in the buildup of hypothesis, which is otherwise impossible to reach. The strength of CIRES is to rationally evaluate biological systems such as biosynthetic pathway under the unbiased and robust statistical calculation, which sometimes goes beyond researcher's gut feelings

expression of interested glycan (Fig. 2). The microarray data can be repeatedly used making it unnecessary to carry out microarray experiment as long as same set of cells is used for comparison. We have used six human B cell lines, Daudi, KMS-12BM, KMS-12PE, Namalwa, Raji, and Ramos, for cross-comparison and deposited in gene expression database as GEO Series GSE 4407 [1, 3], thus publically available for subsequent informatics analyses. This gene expression profiling was performed using a glycan-focused cDNA microarray (RIKEN human glycogene microarray, version 1) and a Gene Expression Omnibus (GEO) Platform (GPL#3465) [2, 5].

Transfection of a glycosyltransferase gene can be done using standard transfection methods. Here we describe two different methods: a conventional method using liposome for transfection of adherent cells and a retrovirus method for both adherent and suspension cells (*see* **Note 1**). In general, a stronger expression level can be achieved by the use of plasmid vector-based expression and a more efficient gene transfer is achieved by retrovirus.

2 Materials

2.1 Flow Cytometry

1. Phosphate-buffered saline (PBS): 8.1 mM Na_2HPO_4, 1.5 mM KH_2PO_4, 2.7 mM KCl, 136.8 mM NaCl.
2. FACS buffer: [1 % bovine serum albumin, 0.1 % NaN_3, PBS] or [1 % bovine serum albumin, 0.1 % NaN_3, 1 mM $CaCl_2$, Tris-buffered saline] (*see* **Note 2**). Tris-buffered saline: 10 mM Tris–HCl (pH 7.6), 154 mM NaCl.
3. Probes: Any kinds of probes that specifically detect glycan structure can be utilized, e.g., lectin-Fc chimera probe, antibody, or toxin. If they are not fluorescently labeled, fluorescence-labeled (e.g., FITC, R-phycoerythrin-conjugated) secondary probe/antibody is needed for detection by flow cytometry.
4. 12×75 mm FACS tube (BD Biosciences), and 40 μm Cell Strainer (BD Biosciences) or nylon mesh, e.g., Polystyrene Round-Bottom Tube with Cell-Strainer Cap (BD Biosciences).

2.2 Transfection of Glycosyltransferase Gene: Lipofectamine Method

1. Plasmid DNA.
2. Transfection reagent, e.g., Lipofectamine (Life Technologies) (*see* **Note 3**).
3. Opti-MEM (Life Technologies).
4. Target cells and appropriate growth medium containing serum.

2.3 Transfection of Glycosyltransferase Gene: Retrovirus Method

1. Plasmid DNA: MSCV (mouse stem cell virus) vector.
2. Virus packaging cells: Plat-A for human cells, Plat-E for rodent cells [6].
3. Cell culture medium for virus packaging cells (DMEM/FBS): DMEM containing 4.5 g/L glucose, 10 % fetal bovine serum, 2 mM L-glutamine.
4. BD CalPhos Mammalian Transfection Kit (BD Biosciences): 2 M Calcium Solution and 2× HEPES-buffered saline (HBS) are included.
5. 0.45 μm filter: Acrodisc 0.45 μm HT Tuffryn membrane (NIPPON Genetics).
6. Polybrene solution: 10 mg/mL polybrene in H_2O.
7. Target cells and appropriate growth medium containing serum.

3 Methods

3.1 Identification of Responsible Enzyme

1. Harvest cells, suspend them in FACS buffer, and incubate for 15–20 min at room temperature (*see* **Note 4**).
2. Aliquot the cells in 1.5 mL tube as 2×10^5 cells/sample.
3. Spin down the cells and remove the supernatant.
4. Resuspend the cells in 100 μL of probe/antibody-containing FACS buffer for the staining (*see* **Note 5**).
5. Incubate for 30 min–1 h on ice.
6. Wash with 1 mL of PBS or FACS buffer (*see* **Note 6**).
7. If the primary probe/antibody is not labeled with fluorescence, repeat **steps 4–6** with fluorescence-labeled secondary probe/antibody (incubate for 30 min on ice) (*see* **Note** 7).
8. Resuspend the cells in 500 μL of FACS buffer and transfer to 12×75 mm FACS tube through 40 μm Cell Strainer or nylon mesh.
9. Obtain fluorescence intensity data by flow cytometry (*see* **Note 8**).
10. Calculate the relative staining signal as the ratio of sample mean fluorescence intensity (MFI) divided by the control MFI.
11. Calculate Pearson's correlation coefficient between glycan expression profiles (obtained by flow cytometry) and gene expression profiles (obtained by microarray, *see* **Note 9**) using the correlation coefficient test.

3.2 Transfection of Transferase Gene

3.2.1 Lipofectamine Method (for Adherent Cells)

1. Subculture target cells in 60 mm dish the day before transfection (*see* **Note 10**).
2. Dilute 4 μg of plasmid DNA (glycosyltransferase cDNA-containing plasmid) into 250 μL of Opti-MEM (*see* **Note 11**).
3. Dilute 10–20 μL of Lipofectamine reagent into 250 μL of Opti-MEM.
4. Combine diluted plasmid DNA and Lipofectamine reagent, and incubate for 15 min at room temperature.
5. Replace culture medium with 2 mL of Opti-MEM.
6. Add combined DNA-Lipofectamine complex (500 μL) to the cells. Gently move dishes back and forth to distribute transfection solution evenly (*see* **Note 12**).
7. Incubate the cells at 37 °C for 1–5 h.
8. Add 2 mL of pre-warmed growth medium (appropriate growth medium used for the target cell culture) dropwise.
9. Incubate at 37 °C overnight.

10. Replace the culture medium with fresh growth medium and culture at 37 °C.
11. Harvest the cells 48 h after transfection for analysis (*see* **Note 13**).

3.2.2 Retrovirus Method (for Both Suspension and Adherent Cells)

1. Subculture packaging cells (e.g., Plat-A for human cells, Plat-E for rodent cells) in 10 cm dish the day before transfection (*see* **Note 14**).
2. 0.5 h prior to transfection, replace culture medium with 10 mL of pre-warmed fresh medium (DMEM/FBS).
3. Dilute 30 μg of plasmid DNA (MSCV vector plasmid containing glycosyltransferase cDNA) to 876 μL with sterile water and add 124 μL of 2 M Calcium Solution (*see* **Note 15**).
4. Add DNA–calcium solution dropwise to 1 mL of 2× HBS with constant bubbling (*see* **Note 16**).
5. Incubate the transfection solution (DNA-calcium mixture in HBS) at room temperature for 20 min to form calcium crystal conjugated with DNA.
6. Gently vortex the transfection solution.
7. Add the transfection solution dropwise to the packaging cells. Gently move dishes back and forth to distribute transfection solution evenly (*see* **Note 12**).
8. Incubate at 37 °C.
9. Remove medium 6 h after transfection and add 4 mL of fresh medium (DMEM/FBS).
10. Incubate at 37 °C for 24 h.
11. Collect and filter (Acrodisc 0.45 μm filter) culture supernatant which contains retrovirus. Aliquot the virus and store at −80 °C until infection (*see* **Note 17**).
12. Add 4 mL of fresh medium (DMEM/FBS) to the cells and incubate for another 24 h (*see* **Note 18**).
13. Collect and filter culture supernatant as in **step 11**. Directly proceed to infection step (**step 14**) or freeze collected virus at −80 °C.
14. If the virus is frozen, thaw it in 37 °C water bath and immediately place it on ice once it is thawed. Add polybrene (6 μg/mL) to the retrovirus-containing culture supernatant and incubate for 10 min on ice (*see* **Note 19**).
15. If target cells are suspension, prepare the target cells during this incubation time (*see* **step 15a**). If target cells are adherent, go to **step 15b**.
    (a) In the case of suspension cells, harvest the cells, count the number of cells, and aliquot into a 1.5 mL tube per infection sample (e.g., 4×10^5 cells/sample for infection in

24-well plate; *see* **Note 20**). Spin down the cells and remove the supernatant. Then, resuspend the cells in 500 μL of retrovirus-polybrene mixture (from **step 14**), transfer to 24-well culture plate, and spin-infect the cells with virus (1,120 × *g*, 90 min, 32 °C) (*see* **Note 21**). After centrifugation, add 500 μL of fresh growth medium (appropriate growth medium used for the target cell culture) to the cells (no pipetting) (*see* **Note 22**).

(b) In the case of adherent cells, add virus-polybrene mixture directly to the culture.

16. Incubate at 32 °C overnight.
17. Add fresh growth medium (scale up to larger well or dish, if needed) and incubate at 37 °C.
18. Harvest the cells >48 h after infection for analysis (*see* **Note 13**).

4 Notes

1. Here we introduce the method for gene overexpression; however, if the negatively correlated gene is selected to increase glycan expression, knock down that gene by RNAi.
2. Serum is often used as blocking reagent; however, they include an array of glycoproteins. Therefore, it is recommended to use non-glycosylated bovine serum albumin instead in the case of glycan staining. Some lectins such as selectins require calcium cation for their binding to glycan ligand. If this is the case, FACS buffer should contain calcium cation. Higher concentration of calcium could alter FSC/SSC value in flow cytometry analysis. Thus, calcium concentration should be suppressed to minimal. We use 1 mM $CaCl_2$ in such a case. Since calcium could cause salt precipitation, use Tris-buffered saline instead of PBS. Filter FACS buffer using 0.22 μm filter such as Stericup (Millipore) to remove precipitate.
3. Other reagents can also be used, although here we describe the method using Lipofectamine. Please refer to the manufacturer's protocol.
4. Choice of cell lines for CIRES is important. Choose cells showing different expression levels of focused glycan. It is not necessary to use the cells whose glycan expression is planned to be modified after identification of the responsible enzyme. Use of suspension cells is recommended because the fluorescent peak of staining is usually sharper in suspension cells than adherent cells, resulting in more reliable profiling. If it is required to use adherent cells, harvest the cells with 1–5 mM EDTA/PBS or cell scraper instead of trypsin to prevent trimming of cell surface proteins carrying glycans, and then pass-through 40 μm

Cell Strainer. Since cellular status affects the expression of glycans, use cells that are in log phase.

5. The probe/cells ratio is important. Before obtaining comparison data among various cell lines, perform titration using strongly positive cells.
6. If staining is performed in $CaCl_2$-containing FACS buffer, wash with FACS buffer.
7. Lectins often bind weakly to their ligands and therefore are easy to be washed out. It is recommended to pre-complex the primary and secondary probes if possible.
8. To cross-compare staining signals among various cell lines, adjust the mean fluorescence intensity (MFI) of the background (negative control sample) to around 10 for each cell line. Fluorescent signal could be affected by small difference in the setup of experiments. Therefore, it is important to normalize the signal.
9. For the comparison among multiple samples, we use universal reference RNA (Clontech) for comparison to each sample. Then, relative expression level against reference RNA is compared among various cell lines. For a sample size of six, a correlation coefficient of 0.81 indicates a statistical significance level of 5 %.
10. Prepare the cells to reach 70–90 % confluence on the day of transfection.
11. Filter plasmid DNA using 0.22 μm filter to prevent contamination of bacteria into the culture. To obtain control cells, prepare empty vector plasmid and use this to transfect cells. In the case of stable transfection, use linearized DNA for transfection (DNA amount can be reduced to 0.5–1 μg).
12. Do not rotate the dish as this will concentrate transfection precipitate in the center of the dish.
13. If stable clones should be established, start selection with selection reagent. If cells are sorted using cell sorter, culture the cells for 2 weeks before the sorting to make sure that the gene is stably integrated into the genome.
14. Cell density affects viral titer. Prepare the cells to reach 70 % confluence at transfection.
15. Filter plasmid DNA using 0.22 μm filter to prevent contamination of bacteria into the culture. To obtain control virus, prepare empty MSCV vector plasmid and use this to transfect cells.
16. Keep blowing bubbles into 2× HBS with a 1 mL pipette and an autopipettor while adding DNA-calcium solution dropwise. Constant motion of bubbles keeps overgrowth of calcium

phosphate crystal. Alternatively, slowly vortex 2× HBS while adding DNA-calcium solution.

17. Use of fresh virus (just after collection, not frozen) shows the highest titer, although the virus can be stored at −80 °C for a long period. The virus-containing culture medium often colors yellow to orange.
18. The titer of virus collected at 48 h is usually higher than that at 24 h. Usually, both of them show enough titer for infection.
19. We use DOTAP Liposomal Transfection Reagent (10 μL/1 mL of virus, Roche) instead of polybrene for infection of mouse primary B cells stimulated with lipopolysaccharide.
20. Use cells covering ~95 % of the area of the well, depending on the size of cells. During the centrifugation, the cells are pressed against the bottom of the wells and viruses are precipitated on them. To get stable transfectants, use 96- or 48-well plate. In the case of transient expression, use 24-well plate (if needed, prepare multiple wells per sample).
21. If high-temperature centrifuge is not available, use normal centrifuge by setting temperature at 32 °C (during centrifugation, the temperature will rise).
22. No need to resuspend the cells by pipetting. After the centrifugation, viruses are attached to the cells.

References

1. Yamamoto H, Takematsu H, Fujinawa R et al (2007) Correlation index-based responsible-enzyme gene screening (CIRES), a novel DNA microarray-based method for enzyme gene involved in glycan biosynthesis. PLoS One 2:e1232
2. Takematsu H, Kozutsumi, Y (2007) DNA microarrays in glycobiology. In: Comprehensive glycoscience from chemistry to systems biology, vol 2. Elsevier, Amsterdam, p 428–448
3. Naito Y, Takematsu H, Koyama S et al (2007) Germinal center marker GL7 probes activation-dependent repression of N-glycolylneuraminic acid, a sialic acid species involved in the negative modulation of B-cell activation. Mol Cell Biol 27:3008–3022
4. Takematsu H, Yamamoto H, Naito-Matsui Y et al (2011) Quantitative transcriptomic profiling of branching in a glycosphingolipid biosynthetic pathway. J Biol Chem 286:27214–27224
5. Gene Expression Omnibus (GEO) database. Website: http://www.ncbi.nlm.nih.gov/geo/
6. Morita S, Kojima T, Kitamura T (2000) Plat-E: an efficient and stable system for transient packaging of retroviruses. Gene Ther 7:1063–1066

Chapter 34

Functional Assay Using Lectin Gene Targeting Technologies (Over-Expression)

Motohiro Nonaka and Toshisuke Kawasaki

Abstract

Function of lectin depends on its amino acid sequence of carbohydrate-recognition domain (CRD), conformation, and extracellular/intracellular localization. Altering lectin gene expression by over-expression or knockdown is a powerful tool for analyzing its cellular function. Here, we describe a method of lectin gene over-expression, taking a C-type lectin, mannan-binding protein (MBP), as an example. Carbohydrate-binding ability of MBP, its subcellular localization, and functional co-localization with ligand glycoprotein are assayed comparing with an inactive mutant MBP.

Key words Tagged protein, Over-expression, Stable transformation, Carbohydrate-binding assay, Confocal microscopy

1 Introduction

Gene over-expression system is a relatively simple approach to probe the function of protein. Strategy to create N- or C-terminally tagged fusion proteins allows functional characterization as well as purification of the targeted proteins. Dominant negative mutant is designed not only to abolish its molecular function but also to inhibit the function of endogenous wild-type protein in the cell [1, 2]. The induction of loss of function can be achieved by introducing small interfering RNA (siRNA) [3]. Recently, new "genome editing" technologies have emerged, enabling to disrupt any genes in diverse range of cells (e.g., ZFN, TALEN, and CRISPR/Cas systems) [4–7]. Regardless, targeted gene knockdown by siRNA is still providing investigators with an easy, high-throughput alternative to study gene functions. As a whole, we now have many choices for gene targeting technologies. On the other hand, it is sometimes required to eliminate the possibility of artifact arising from unusual environment for cells. It is good to consider the robust experimental controls. Of note, some lectins require self-assembly for interaction with their carbohydrate ligands under physiological condition.

Jun Hirabayashi (ed.), *Lectins: Methods and Protocols*, Methods in Molecular Biology, vol. 1200, DOI 10.1007/978-1-4939-1292-6_34, © Springer Science+Business Media New York 2014

It is known that an exponential affinity enhancement between a clustered carbohydrate recognition domain (CRD) and a multivalent carbohydrate ligand, which is much greater than expected from the sum of the each CRD-carbohydrate interaction, can be achieved by allocating CRDs with appropriate orientation and spacing. Proper processing of the lectin molecule thus should be validated beforehand.

Mannan-binding protein (MBP), also called as mannan-binding lectin, is a Ca^{2+}-dependent (C-type) animal lectin [8]. MBP exhibits binding specificity to mannose, fucose, and *N*-acetylglucosamine. The molecular structure of MBP is composed of homo-oligomers of 32-kDa subunit [9]. The subunit has an N-terminal cysteine-rich region, a collagen-like domain, and a C-terminal CRD. The CRD structure is essential for its carbohydrate binding, and cysteine-rich region and collagen-like domain are critical for molecular multimerization. Human MBP occurs naturally in two forms, secretory serum MBP (S-MBP) and intracellular MBP (I-MBP), though they are translated from a single form of mRNA [10, 11]. S-MBP is well known as an important constituent of the innate immunity [12, 13]. S-MBP binds to HIV envelope glyprotein gp120 and inhibits the virus infection [14]. In contrast, the characterization of I-MBP had not been reported until our recent study [15]. We showed that I-MBP is mainly localized in the ER and exhibits glycan-binding affinity to intracellular gp120, suggesting the role of I-MBP during virus assembly. The point mutant I-MBP that lacks carbohydrate-binding ability did not bind and co-localize with gp120. Here I-MBP as an example, we describe methods for functional assays of lectin by gene overexpression. Wild-type I-MBP and its point mutant genes are tagged with green fluorescent protein (GFP) and stably transfected in human hepatoma HLF cells. Carbohydrate-binding abilities of I-MBPs are determined using mannan beads. I-MBP subcellular localization is analyzed by organelle staining, and functional co-localization between I-MBPs and gp120 is evaluated by double transfection of MBP and gp120 followed by confocal microscopic observation.

2 Materials

2.1 Transfection of Plasmids Encoding MBP Genes

1. Plasmids of interest: Plasmids encoding wild-type MBP-GFP and CS-mutant MBP-GFP (*see* **Note 1**).
2. Human hepatoma cell line HLF (JCRB 0405) (*see* **Note 2**).
3. Culture medium: Eagle's minimal essential medium (MEM) containing 10 % of fetal bovine serum (FBS) without antibiotics.
4. Lipofectamine® 2000 (Invitrogen).

5. Opti-MEM® (Invitrogen).
6. G418 disulfate salt (Sigma-Aldrich).
7. Fluorescent microscope (Zeiss Axioplan Universal microscope).

2.2 Carbohydrate-Binding Assay of Recombinant MBPs

1. HLF cells stably expressing wild-type or mutant MBP (HLF-WT MBP and HLF-CS-mutant MBP).
2. PBS, pH 7.4.
3. Cell scraper.
4. Lysis buffer: 50 mM HEPES, pH 7.5, 300 mM NaCl, 5 mM $CaCl_2$, 1 % NP-40, protease inhibitor cocktail (*see* **Note 3**).
5. Rotator (TAITEC Rotator RT-5).
6. Mannan-Sepharose 4B beads (*see* **Note 4**).
7. Wash buffer: 50 mM HEPES, pH 7.5, 300 mM NaCl, 5 mM $CaCl_2$, 0.1 % NP-40.
8. Dilution buffer: 60 mM imidazole, pH 7.8, 3.75 M NaCl, 60 mM $CaCl_2$.
9. Loading buffer: 20 mM imidazole, pH 7.8, 1.25 M NaCl, 20 mM $CaCl_2$.
10. Elution buffer: 20 mM imidazole, pH 7.8, 1.25 M NaCl, 4 mM EDTA.
11. 6× SDS Sample buffer: 375 mM Tris, pH 6.8, 12 % SDS, 60 % glycerol, 0.06 % bromophenol blue.

2.3 Subcellular Localization of Recombinant MBPs

1. HLF cells stably expressing wild-type or mutant MBP (HLF-WT MBP and HLF-CS-mutant MBP).
2. Nunc™ Lab-Tek™ II Chamber Slide, 2-well (Thermo Scientific).
3. 4 % paraformaldehyde (PFA)-PBS, pH 7.4.
4. Ice-cold 100 % methanol.
5. Blocking buffer: 1 % BSA-PBS, pH 7.4.
6. Mouse anti-ERp57 monoclonal antibody (Enzo Life Sciences).
7. Alexa Fluor 568 Goat anti-mouse IgG antibody (Invitrogen).
8. Vectashield mounting medium (Vector Laboratories).
9. Confocal microscope (OLYMPUS FluoView 1000 confocal microscope).

2.4 Functional Co-localization of Recombinant MBPs and Gp120 Glycoprotein

1. HLF cells stably expressing wild-type or mutant MBP (HLF-WT MBP and HLF-CS-mutant MBP).
2. Plasmid encoding gp120-RFP (*see* **Note 5**).
3. 4 % PFA-PBS, pH 7.4.
4. Confocal microscope.

3 Methods

Perform all procedures at room temperature unless specified.

3.1 Transfection of Plasmids Encoding MBP Genes

1. On the previous day of transfection, plate 2×10^5 cells of HLF cells in 2 mL of MEM-10 % FBS without antibiotics in 6-well plate. This will give 90–95 % confluence of the cells on the day of transfection.
2. Dilute 4.0 μg plasmid DNA in 250 μL of Opti-MEM and mix gently.
3. Mix 10 μL of Lipofectamine 2000 Reagent in 250 μL of Opti-MEM and pipette it gently. Incubate for 5 min at room temperature.
4. Within 20 min from **step 3**, gently mix DNA solution with the diluted Lipofectamine 2000 reagent. Incubate for 20 min (*see* **Note 6**).
5. During incubation of DNA-Lipofectamine solution, you may replace the medium in **step 1** with 2 mL of fresh MEM-10 % FBS without antibiotics.
6. Add approximately 500 μL of the complexes to each well. Mix the complexes and medium gently by rocking the plate back and forth.
7. Incubate cells at 37 °C in a CO_2 incubator for 24 h prior to subculture. You may also change the medium 4–6 h after the transfection.
8. Twenty-four hours after the transfection, cells are passaged at a 1:10 using fresh MEM-10 % FBS.
9. On the following day, start selection of transfected cells by incubating the cells with MEM-10 % FBS containing 1 mg/mL G418.
10. Change the media containing G418 every 2–3 days until most of the cells die (*see* **Note 7**).
11. Observe the cells by inverted fluorescent microscope with the excitation wavelength set at 470–490 nm and the detection wavelength set at 510–550 nm. You will see cells producing green fluorescence (Fig. 1).

3.2 Carbohydrate-Binding Assay of Recombinant MBPs

1. Prepare HLF-WT MBP and HLF-CS-mutant MBP cells in 10-cm culture dish (*see* **Note 8**).
2. Remove the culture medium and add 5 mL of ice-cold PBS.
3. Harvest cells by cell scraper, transfer to 15-mL tube, and centrifuge at $800 \times g$ for 2 min at 4 °C.
4. Remove the supernatant and resuspend the cell pellet with 1 mL of ice-cold PBS.

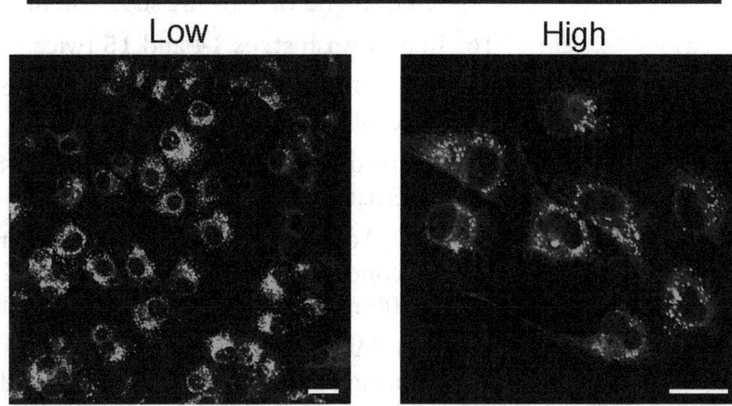

Fig. 1 Microscopic images of wild-type I-MBP. Human hepatoma HLF cells were stably transfected with wild-type MBP tagged with GFP. The images show that a number of HLF cells were producing green fluorescence and wild-type I-MBP was distinctly accumulated in cytoplasmic granules. This research was originally published in The Journal of Biological Chemistry. Nonaka *et al.*, Subcellular Localization and Physiological Significance of Intracellular Mannan-binding Protein. *The Journal of Biological Chemistry.* 2007; 282: 17908–17920. © The American Society for Biochemistry and Molecular Biology

5. Transfer the cell suspension into 1.5 mL tube and centrifuge at 800 ×*g* for 2 min.
6. After removing the supernatant, add 500 μL of ice-cold lysis buffer and the cells are completely dispersed by pipetting.
7. Incubate the lysate by rotating for 30 min at 4 °C (*see* **Note 9**).
8. Centrifuge the cell suspension at 15,000 ×*g* at 4 °C for 20 min, and collect the supernatant as cell lysate (*see* **Note 10**).
9. The resulting cell lysate is mixed with a half volume (~250 μL) of dilution buffer.
10. Equilibrate mannan-Sepharose 4B beads (~100 μL of 50 % bead slurry) by centrifugation and resuspension with 1 mL of loading buffer. Repeat this **step 10** twice.
11. Add 20 μL of mannan-Sepharose 4B beads (50 % bead slurry) to the diluted cell lysate.
12. Incubate the cell lysate-bead mixture with rotation for 1 h at 4 °C.
13. Centrifuge the tube at 2,500 ×*g* for 15 s at 4 °C and then remove the supernatant carefully.
14. Add 1 mL of wash buffer and resuspend the beads by gently pipetting.

15. Centrifuge the tube at 2,500 × *g* for 15 s at 4 °C, carefully remove, and discard the supernatant.
16. Repeat wash **steps 14** and **15** twice.
17. Add 80 μL of elution buffer and gently pipette the bound materials (*see* **Note 11**).
18. Centrifuge the tube at 2,500 × *g* for 15 s at 4 °C and save the supernatant.
19. Add ~16 μL of 6× SDS sample buffer to the eluate. For reducing conditions, add dithiothreitol (DTT) at final concentration of 100 mM and then boil the samples at 100 °C for 5 min.
20. SDS-PAGE on a 5–20 % gradient gel under nonreducing and reducing conditions. Recombinant MBPs are detected by immunoblotting with anti-GFP monoclonal antibody (*see* Fig. 2).

3.3 Subcellular Localization of Recombinant MBPs

1. Culture HLF cells stably expressing recombinant wild-type or mutant MBPs until ~80 % confluency in a glass slide chamber.
2. Remove the medium and gently wash the cells by adding 1 mL of PBS pH 7.4.
3. Add 1 mL of 4 % PFA-PBS and fix the cells by incubating for 15 min.
4. Discard PFA solutions and wash the wells with 1 mL of PBS. Repeat this **step 4** twice.
5. Place the chamber slide on ice and remove PBS. Add 1 mL of ice-cold 100 % methanol and incubate for 30–60 s (*see* **Note 12**). Discard methanol and wash the wells with 1 mL of PBS three times.
6. Incubate cells in 1 mL of blocking buffer for 30 min, and discard the solution.
7. Add 500 μL of primary antibody solution (1:200 dilution of anti-ERp57 antibody in blocking buffer). Incubate for 30 min.
8. Discard primary antibody solutions and wash the wells with 1 mL of PBS three times.
9. Add 500 μL of secondary antibody solution (1:200 dilution of Alexa Fluor 568 anti-mouse IgG in blocking buffer) and incubate for 30 min.
10. Discard secondary antibody solutions and wash the wells with 1 mL of PBS three times.
11. Remove the chamber from the slide and add one drop of Vectashield mounting medium.
12. Analyze the cells with a confocal microscope with the excitation wavelength set at 488 and 561 nm, and the detection wavelength set at 510–550 nm and 580–610 nm, respectively (*see* Fig. 3).

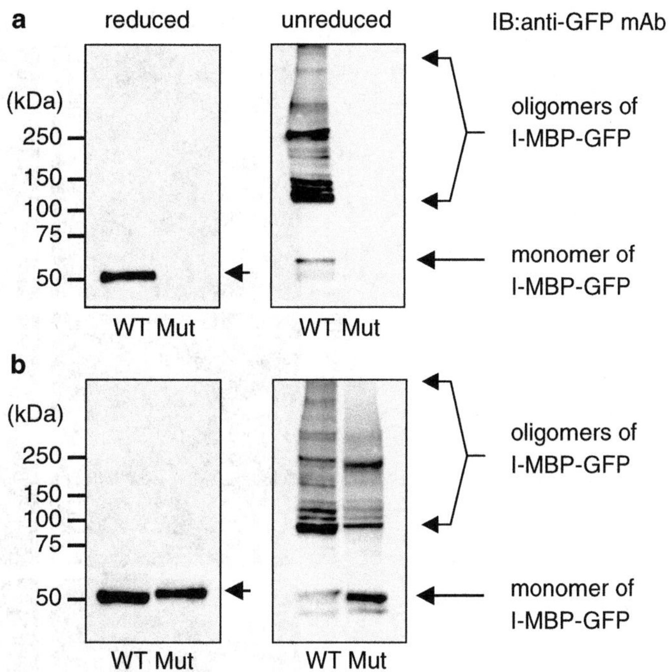

Fig. 2 Carbohydrate-binding assay of wild-type and CS-mutant I-MBPs. Cell lysates from stably transfected HLF cells were incubated with mannan-Sepharose 4B beads in the presence of Ca^{2+}. The bound materials were then eluted with a buffer containing EDTA. The eluted proteins were separated by SDS-PAGE under reducing (*left panel*) and nonreducing (*right panel*) conditions, and the immunoblots were probed with anti-GFP antibody (**a**). As a control experiment, cell lysates were subjected to SDS-PAGE without incubation with mannan-Sepharose 4B and I-MBPs were detected by anti-GFP antibody (**b**). The *arrows* on the *right* indicate the positions of the monomer and oligomers of I-MBP. Note that the CS-mutant I-MBP did not bind to Sepharose 4B-mannan beads while there was no big difference in the degree of oligomerization between the intracellular wild-type and CS-mutant I-MBPs. *WT* wild-type MBP, *Mut* CS-mutant I-MBP. This research was originally published in The Journal of Biological Chemistry. Nonaka *et al.*, Subcellular Localization and Physiological Significance of Intracellular Mannan-binding Protein. *The Journal of Biological Chemistry.* 2007; 282: 17908–17920. © The American Society for Biochemistry and Molecular Biology

3.4 Functional Co-localization of Recombinant MBPs and Gp120 Glycoprotein

1. According to protocol in Subheading 3.1, transfect HLF-WT MBP or HLF-CS-mutant MBP cells with gp120-RFP plasmid (*see* **Note 13**).
2. Culture the HLF cells doubly transfected with wild-type/ CS-mutant MBP and gp120 until 80 % confluency in a glass slide chamber.
3. Remove the medium and gently wash with 1 mL of PBS once.
4. Fix the cells with 1 mL of 4 % PFA-PBS at room temperature for 15 min.

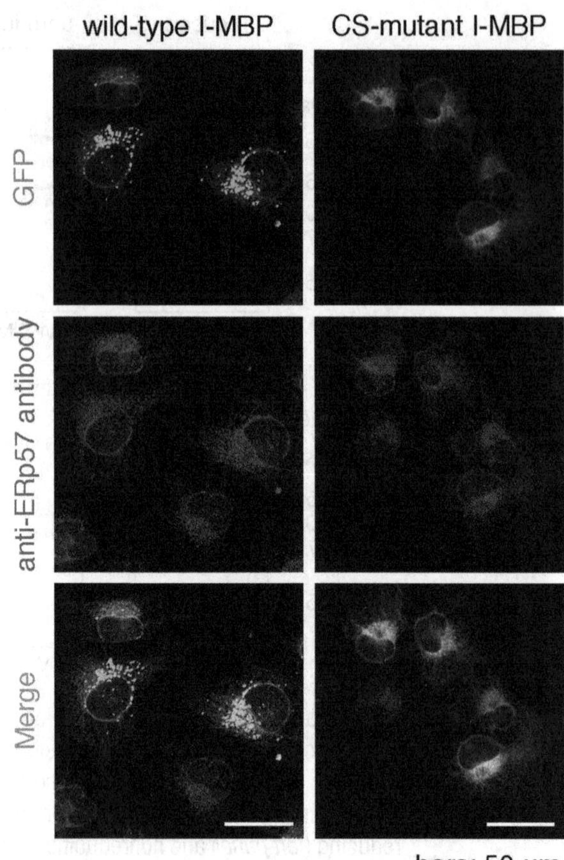

Fig. 3 Confocal microscopic images of subcellular localization of wild-type and CS-mutant l-MBPs in the ER. HLF cells stably transfected with wild-type or CS-mutant MBP tagged with GFP was incubated with anti-ERp57 antibody, followed by the staining with Alexa Fluor 568-conjugated secondary antibody. The images suggest that both l-MBPs are localized mainly in the ER. This research was originally published in The Journal of Biological Chemistry. Nonaka *et al.*, Subcellular Localization and Physiological Significance of Intracellular Mannan-binding Protein. *The Journal of Biological Chemistry*. 2007; 282: 17908–17920. © The American Society for Biochemistry and Molecular Biology

5. Discard the PFA solutions and wash with 1 mL of PBS three times.
6. Remove the chamber from slide and add one drop of Vectashield mounting medium.
7. Analyze the cells with a confocal microscope with the excitation wavelength set at 488 and 561 nm, and the detection wavelength set at 510–550 and 580–610 nm, respectively (*see* Fig. 4).

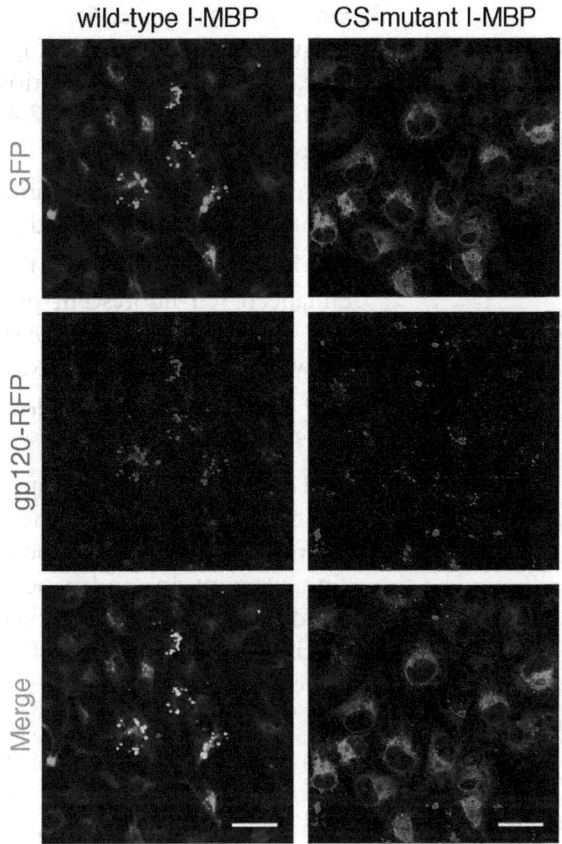

Fig. 4 Functional co-localization of l-MBP with gp120. The doubly transfected HLF cells expressing both gp120-RFP and wild-type or CS-mutant l-MBP-GFP were observed by confocal microscopy. The images show that wild-type but not mutant l-MBP-GFP was overlapped with gp120-RFP, suggesting that carbohydrate-binding activity of l-MBP is involved in the MBP-gp120 co-localization. This research was originally published in The Journal of Biological Chemistry. Nonaka *et al.*, Subcellular Localization and Physiological Significance of Intracellular Mannan-binding Protein. *The Journal of Biological Chemistry.* 2007; 282: 17908–17920. © The American Society for Biochemistry and Molecular Biology

4 Notes

1. In the mutant MBP construct, cysteine 236 and 244 at CRD have been replaced with serine residues by site-directed mutagenesis using PCR. Wild-type MBP and CS-mutant MBP are tagged at C-terminal with green fluorescent protein (pEGFP).
2. Recombinant human MBP produced in HLF cells shows higher oligomers of the MBP subunit and has the ability to activate complement via lectin pathway in a dose-dependent manner.

3. Since MBP requires Ca^{2+} to bind carbohydrates, the buffer must contain 5 mM $CaCl_2$. Buffers containing phosphate (e.g., PBS) form precipitation in the presence of Ca^{2+}. Tris-buffered saline (TBS) pH 7.4 can be substituted for HEPES-based buffer.
4. Mannan-Agarose beads (Sigma-Aldrich) can be substituted for mannan-Sepharose 4B beads.
5. HIV envelope protein gp120 is tagged at C-terminal with enhanced red fluorescent protein (pDsRed2-N1). This plasmid is available in NIH AIDS Reagent Program (https://www.aidsreagent.org/index.cfm).
6. The solution may appear cloudy. Complexes are stable for 6 h at room temperature.
7. G418 (neomycin) takes approximately 1 week to kill most of the cells. Depending on the antibiotic-resistant gene in your plasmid, you can choose selection markers such as puromycin, hygromycin, and zeocin. Puromycin induces relatively rapid cell death within several days at low concentrations. Optimal concentrations of the antibiotics should be determined individually.
8. For this experiment, you can also use transient transfectant cells. Transfect HLF cells with wild-type or mutant MBP in 10-cm dish and culture for 24–48 h.
9. This step is recommended to increase the yield of recombinant MBP. Instead of using a rotator, you can place the tube on ice and vortex every 5 min for 30 min.
10. It is recommended not to freeze cell lysates once you add lysis buffer since it may form a precipitate after thawing.
11. Since MBP requires Ca^{2+} to bind to carbohydrates, we usually use a Ca^{2+} chelator EDTA for elution. You can also use monosaccharide such as 100 mM mannose for elution.
12. Ice-cold methanol is used for dehydration, fixation, and permeabilization of cell membrane. You should choose the permeabilization protocol depending on the antigen of your interest. Detergents such as 0.1 % Triton can also be used for permeabilization after PFA fixation.
13. Instead of preparing gp120 transient transfectant cells, you can also make MBP-gp120 double-stable transfectant cells using different selection markers for MBP and gp120 plasmid. It is also possible to manually clone the double-stable transfectant cells even if two plasmids contain the same antibiotic-resistant gene. For this, the cells producing both green and red fluorescence are gently detached by scraper using inverted fluorescent microscope, and then subcultured into new culture dishes.

References

1. Prelich G (2012) Gene overexpression: uses, mechanisms, and interpretation. Genetics 190:841–854
2. Herskowitz I (1987) Functional inactivation of genes by dominant negative mutations. Nature 329:219–222
3. McManus MT, Sharp PA (2002) Gene silencing in mammals by small interfering RNAs. Nat Rev Genet 3:737–747
4. Urnov FD, Rebar EJ, Holmes MC et al (2010) Genome editing with engineered zinc finger nucleases. Nat Rev Genet 11:636–646
5. Gaj T, Gersbach CA, Barbas CF III (2013) ZFN, TALEN, and CRISPR/Cas-based methods for genome engineering. Trends Biotechnol 31:397–405
6. Cho SW, Kim S, Kim JM et al (2013) Targeted genome engineering in human cells with the Cas9 RNA-guided endonuclease. Nat Biotechnol 31:230–232
7. Mali P, Yang L, Esvelt KM et al (2013) RNA-guided human genome engineering via Cas9. Science 339:823–826
8. Ma Y, Shida H, Kawasaki T (1997) Functional expression of human mannan-binding proteins (MBPs) in human hepatoma cell lines infected by recombinant vaccinia virus: post-translational modification, molecular assembly, and differentiation of serum and liver MBP. J Biochem 122:810–818
9. Kawasaki N, Kawasaki T, Yamashina I (1983) Isolation and characterization of a mannan-binding protein from human serum. J Biochem 94:937–947
10. Mori K, Kawasaki T, Yamashina I (1984) Subcellular distribution of the mannan-binding protein and its endogenous inhibitors in rat liver. Arch Biochem Biophys 232:223–233
11. Mori K, Kawasaki T, Yamashina I (1988) Isolation and characterization of endogenous ligands for liver mannan-binding protein. Arch Biochem Biophys 264:647–656
12. Turner MW (2003) The role of mannose-binding lectin in health and disease. Mol Immunol 40:423–429
13. Ma Y, Uemura K, Oka S et al (1999) Antitumor activity of mannan-binding protein in vivo as revealed by a virus expression system: mannan-binding protein-dependent cell-mediated cytotoxicity. Proc Natl Acad Sci U S A 96: 371–375
14. Ezekowitz RA, Kuhlman M, Groopman JE et al (1989) A human serum mannose-binding protein inhibits in vitro infection by the human immunodeficiency virus. J Exp Med 169: 185–196
15. Nonaka M, Ma BY, Ohtani M et al (2007) Subcellular localization and physiological significance of intracellular mannan-binding protein. J Biol Chem 282:17908–17920

Chapter 35

Analysis of L-Selectin-Mediated Cellular Interactions Under Flow Conditions

Hiroto Kawashima

Abstract

Lymphocyte homing is mediated by a specific interaction between L-selectin expressed on lymphocytes and its ligands expressed on high endothelial venules (HEVs) in lymph nodes under physiological flow conditions. In this chapter, two methods for detecting L-selectin-mediated cellular interactions under shear stress mimicking physiological flow conditions are described. First, a modified Stamper–Woodruff cell-binding assay using leukocytes labeled with a fluorescent orange dye, CMTMR, is introduced. In this method, leukocytes are allowed to bind to frozen lymph node sections under shear stress and their binding to HEVs can be clearly visualized by fluorescence microscopy. Second, a parallel flow chamber assay is described. In this assay, leukocytes are allowed to roll on L-selectin ligand-expressing cells under various levels of shear stress and their adhesive interactions are recorded by a video camera equipped with an inverted microscope. These methods can be applied to determine the effects of various agents that might affect L-selectin-mediated lymphocyte homing and recruitment.

Key words L-Selectin, Lymphocyte homing, High endothelial venule, Rolling

1 Introduction

The selectins are a family of three C-type lectins that mediate rapid and reversible adhesive interactions between leukocytes and vascular endothelial cells under physiological flow [1–3]. L-Selectin is expressed on most leukocytes, whereas E- and P-selectin are expressed on activated endothelial cells. It is also known that P-selectin is expressed on activated platelets.

Lymphocytes migrate to the peripheral lymph nodes (PLNs), where foreign antigens accumulate and immune responses occur. L-Selectin plays an essential role in this process, as revealed by L-selectin-deficient mice [4]. The lymphocytes that migrate to PLNs through specialized blood vessels, called high endothelial venules (HEVs) [5], emigrate through the efferent lymphatics to the lymph unless they encounter their cognate antigens. The lymphocytes that emigrate to the lymph return to the bloodstream

Jun Hirabayashi (ed.), *Lectins: Methods and Protocols*, Methods in Molecular Biology, vol. 1200,
DOI 10.1007/978-1-4939-1292-6_35, © Springer Science+Business Media New York 2014

through the thoracic duct and migrate again into the PLNs. This circulatory process is called lymphocyte recirculation or lymphocyte homing, which increases the chance that the lymphocytes will encounter antigens. Thus, lymphocyte homing is important for the immune system to recognize foreign antigens efficiently.

HEVs have a characteristic cuboidal morphology and synthesize a unique sulfated glycan known as 6-sulfo sialyl Lewis X (Siaα2-3Galβ1-4[Fucα1-3(sulfo-6)]GlcNAcβ1-R; 6-sulfo sLe^x), which serves as a major L-selectin ligand [6–9]. Lymphocyte homing is achieved by a series of interactions between lymphocytes and HEVs [10]: (1) lymphocyte rolling mediated by the interaction between L-selectin on lymphocytes and its ligands on HEVs; (2) activation of lymphocytes by chemokines presented on the surface of HEVs by heparan sulfate [11, 12]; (3) firm attachment of lymphocytes to HEVs mediated by integrins; and (4) lymphocyte transmigration across HEVs. Notably, L-selectin is localized to the tips of microvilli [13], which is advantageous for the initial tethering and rolling on HEVs under physiological blood flow.

L-Selectin is expressed not only on lymphocytes but also on other types of leukocytes, such as neutrophils and monocytes. L-Selectin is involved in the infiltration of these leukocytes to sites of inflammation, which can be inhibited by anti-L-selectin mAb MEL-14 [14]. This effect has been verified by the study of L-selectin-deficient mice, which showed significant reduction of leukocyte infiltration to the inflammatory sites [4].

Here, I describe two methods to detect L-selectin-mediated cellular interactions under physiological shear stress. First, a modified Stamper–Woodruff cell-binding assay to detect L-selectin-mediated binding of fluorescently labeled leukocytes to HEVs in lymph node tissue sections under shear stress is described [15]. This assay was originally described by Stamper and Woodruff using unlabeled leukocytes [16] and has made an enormous contribution to the elucidation of the molecular mechanisms underlying lymphocyte homing. Second, a parallel flow chamber assay in which leukocytes are allowed to roll on L-selectin ligand-expressing cells under various levels of shear stress is described [15]. These two methods mimic physiological flow conditions and should be useful for determining the effects of various agents that might affect L-selectin-mediated lymphocyte homing and leukocyte infiltration to sites of inflammation.

2 Materials

2.1 Mice

1. Eight to 10-week-old C57BL/6 female mice (Japan SLC, Hamamatsu, Japan).
2. The mice are treated in accordance with the guidelines of the Animal Research Committee of the University of Shizuoka.

2.2 Cells

1. CHO cells stably expressing human CD34, human fucosyltransferase-VII (F7), human core 1β1,3-N-acetylglucosaminyltransferase (C1), human core 2 β1,6-N-acetylglucosaminyltransferase-I (C2), and mouse *N*-acetylglucosamine-6-*O*-sulfotransferase-2 (GlcNAc6ST-2) (CHO/CD34/F7/C1/C2/GlcNAc6ST-2) [17] (*see* **Note 1**) are cultured in DME/F-12 medium (Sigma-Aldrich, St. Louis, MO, USA) supplemented with 10 % fetal bovine serum (FBS) (HyClone, Logan, UT, USA) and penicillin-streptomycin (Invitrogen, Carlsbad, CA, USA) and are maintained in a humidified incubator (37 °C, 5 % CO_2) (*see* **Note 2**).

2.3 Buffers

1. PBS (10 mM Na_2HPO_4, 1.8 mM KH_2PO_4, 137 mM NaCl, 2.7 mM KCl, pH 7.4).
2. Buffer A (20 mM HEPES-NaOH, 0.15 M NaCl, 1 mM $CaCl_2$, 1 mM $MgCl_2$, pH 7.4).
3. ACK solution (150 mM NH_4Cl, 1 mM $KHCO_3$, 0.1 mM EDTA 2Na).

2.4 Microscopes and Related Instruments

1. BX-51 microscope (Olympus, Center Valley, PA, USA).
2. CKX-41 inverted microscope (Olympus).
3. Penguin 600CL digital camera (Pixera Co., Santa Clara, CA, USA).
4. ADT-40S CCD camera (Flovel Co., Ltd., Tokyo, Japan).

2.5 Other Instruments

1. Shaker platform (Double Shaker NR-3; TAITEC Co., Ltd., Nagoya, Japan).
2. Microm HM 525 Cryostat (Thermo Fisher Scientific, Waltham, MA, USA).
3. Moisture chamber (Multi-Purpose Incubation Chamber 10DO; Cosmo Bio Co., Ltd., Tokyo, Japan).
4. Hemocytometer (Neubauer-improved counting chambers, Hirschmann, Inc., Louisville, KY, USA).
5. Parallel plate flow chamber (GlycoTech Co., Gaithersburg, MD, USA).
6. Syringe pump Model 11 Plus (Harvard Apparatus Co., Holliston, MA, USA).

2.6 Other Materials

1. Tissue-Tek O.C.T. compound (Sakura Finetek, Tokyo, Japan).
2. Tissue-Tek Cryomolds (Sakura Finetek).
3. Aminosilane (APS)-coated glass slide (Matsunami Glass Ind., Ltd., Osaka, Japan).
4. Glutaraldehyde solution, 50 %.
5. Wax pencil (ImmEdge hydrophobic barrier pen, Vector Laboratories, Burlingame, CA, USA).

6. Bovine serum albumin.
7. Fluoromount (Diagnostic Biosystems, Pleasanton, CA, USA).
8. RPMI 1640 medium (Sigma-Aldrich).
9. Nylon mesh (100 μm).
10. CellTracker™ Orange CMTMR (5-(and-6)-(((4-chloromethyl) benzoyl)amino)tetramethylrhodamine; Lonza, Walkersville, MD, USA).
11. MEL-14 (Beckman Coulter Inc., Brea, CA, USA).
12. Rat IgG.
13. Mouse IgM.
14. 35-mm culture dishes (Corning, Inc., Corning, NY, USA).

3 Methods

3.1 Modified Stamper–Woodruff Cell-Binding Assay

3.1.1 Preparation of Tissue Sections

1. Place 4–5 PLNs from C57BL/6 female mice in a Tissue-Tek Cryomold.
2. Fill the Cryomold with Tissue-Tek O.C.T. compound, place it on dry ice until it is frozen, and store it at –80 °C until sectioning.
3. Cut cryostat sections (7 μm) using a Microm HM 525 Cryostat (*see* **Note 3**) and mount them on APS-coated glass slides.
4. Air-dry the sections for 10 min and fix them for 10 min in 0.5 % glutaraldehyde in PBS.
5. Dip the sections three times in tap water for 5 min each.
6. Remove excess fluid by gently tapping the glass slides against tissue paper.
7. Carefully wipe around the tissue sections.
8. Circle the tissue sections with a wax pencil (*see* **Note 4**).
9. Apply 300 μL of 3 % BSA in buffer A for 45 min (*see* **Note 5**).
10. Incubate sections with or without various agents to be tested (*see* **Note 6**) in 0.1 % BSA in buffer A for at least 30 min.
11. Keep the glass slides in a moisture chamber at 4 °C until use.

3.1.2 Preparation of Fluorescently Labeled Leukocyte Suspensions

1. Mince mouse spleens from three C57BL/6 female mice with a scalpel blade.
2. Gently press the tissues between frosted glass slides to squeeze out leukocytes into 5 mL RPMI 1640 containing 10 % FBS.
3. Allow the cell suspension to settle in a 15-mL centrifuge tube for 5 min to remove debris.
4. Pass the supernatant containing leukocytes through a nylon mesh (100 μm).

5. Transfer the pass-through fraction containing leukocytes into a new 15-mL centrifuge tube.
6. Centrifuge at 1,500 rpm (420×*g*) for 5 min at 4 °C.
7. Remove supernatant. Add 1 mL of ACK solution and incubate for 5 min at room temperature (*see* **Note 7**).
8. Add 3 mL of RPMI 1640 containing 10 % FBS.
9. Centrifuge at 1,500 rpm (420×*g*) for 5 min at 4 °C.
10. Remove the supernatant.
11. Incubate the cells with 2 mL of 3.5 μM CellTracker Orange CMTMR in RPMI 1640 without FBS for 10 min at 37 °C (*see* **Note 8**).
12. Add 8 mL of 0.1 % BSA in buffer A.
13. Centrifuge at 1,500 rpm (420×*g*) for 5 min at 4 °C.
14. Remove the supernatant.
15. Add 10 mL of 0.1 % BSA in buffer A and count the cell number using a hemocytometer.
16. Centrifuge at 1,500 rpm (420×*g*) for 5 min at 4 °C.
17. Remove the supernatant.
18. Incubate the cells in 0.1 % BSA in buffer A at a cell density of 1×10^7 cells/ml in the presence or absence of 10 μg/mL mAb MEL-14 or control rat IgG in 0.1 % BSA in buffer A for at least 10 min on ice before applying them to the glass slides in Subheading 3.1.3.

3.1.3 Adhesion Assay

1. Remove fluid from the glass slides prepared in Subheading 3.1.1 by gently tapping them against tissue paper.
2. Place the glass slides horizontally on a shaker platform operating at 60 rpm at 4 °C in a cold room.
3. Apply 100 μL (1×10^6 cells/section) of the leukocyte suspension (prepared in Subheading 3.1.2) onto each glass slide while the shaker platform is operating.
4. Continue the rotation for 30 min at 60 rpm.
5. Remove the slides one by one from the shaker platform (*see* **Note 9**).
6. Remove excess fluid from the glass slides by gently tapping them against tissue paper.
7. Gently dip the slides in buffer A in a 50-mL centrifuge tube to remove non-adherent leukocytes for 10 s (*see* **Note 10**).
8. Gently dip the slides in a 50-mL centrifuge tube containing 0.5 % glutaraldehyde in buffer A for 10 min (*see* **Note 10**).
9. Gently dip the sections twice in tap water (*see* **Note 10**).
10. Mount the sections with Fluoromount.

11. Observe the section by a BX-51 microscope using a 10× eyepiece and 20× objective lens.
12. Take pictures using a Penguin 600CL digital camera (*see* **Note 11**).

3.2 Leukocyte Rolling Assay

1. Culture CHO/CD34/F7/C1/C2/GlcNAc6ST-2 cells as monolayers in 35-mm culture dishes.
2. Wash the cell monolayers twice with DME/F-12 medium supplemented with 1 % FBS.
3. Incubate the cell monolayers with or without the various agents to be tested (*see* **Note 12**) in DME/F-12 medium supplemented with 1 % FBS for 10 min at RT.
4. Remove the lid of the cell culture dish and put it on the microscope stage.
5. Connect the vacuum silicone tube to the parallel plate flow chamber apparatus.
6. Attach the vacuum silicone tube to the vacuum and open the air valve to allow the parallel plate flow chamber apparatus to attach to the gasket with a thickness of 0.0254 cm and flow path width of 0.5 cm placed on the lid of the cell culture dish.
7. Suspend freshly prepared leukocytes from the spleens of wild-type (WT) mice in 0.1 % BSA in buffer A at 1×10^6 cells/mL in a 50-mL centrifuge tube (*see* **Notes 13** and **14**).
8. Place the 50-mL centrifuge tube in a holder and connect it to the input silicone tube that has been connected to the parallel plate flow chamber apparatus (*see* **Note 15**).
9. Connect the output silicone tube to the flow chamber apparatus on one end and to a 20-mL syringe on the other end.
10. Pull the syringe to get rid of air bubbles in the input and output silicone tubes (*see* **Note 16**).
11. Clamp the silicone tubes using paperclips to avoid air bubbles entering into the silicone tubes.
12. Disconnect the output silicone tube from the 20-mL syringe.
13. Push the syringe to remove the air inside.
14. Reconnect the output silicone tube to the 20-mL syringe.
15. Connect the 20-mL syringe to a syringe pump Model 11 Plus.
16. Close the air valve of the vacuum.
17. Remove the lid of the cell culture dish from the flow chamber attached with the gasket carefully (*see* **Note 17**).
18. Remove the medium from the cell monolayer dish and place it at the center of the microscope stage.
19. Place the flow chamber apparatus attached to the gasket into the cell monolayer dish.

20. Open the air valve of the vacuum to allow the flow chamber apparatus and gasket to attach tightly to the inner surface of the cell monolayer dish.
21. Remove the paperclips from the input and output silicone tubes.
22. Introduce the cell suspension into the flow chamber at a wall shear stress of 2, 1.5, 1, and 0.5 dyn/cm^2 using the syringe pump Model 11 Plus (*see* **Notes 18–20**).
23. Record images with a model ADT-40S CCD camera equipped on a CKX41 inverted microscope using a 20× objective lens according to the manufacturer's instructions (*see* **Note 21**).
24. Count the number of rolling leukocytes on the CHO transfectants per 30 s (*see* **Note 22**).

4 Notes

1. These cells express CD34 modified with 6-sulfo sLe^x on both the core 2 and the extended core 1 structures of its *O*-glycan moieties, which serves as an L-selectin ligand.
2. FBS should be heat-inactivated for 30 min at 56 °C.
3. Set the temperature at −19 °C to cut PLN sections. Check the sections under a microscope. If the tissue morphology is not sufficient, change the temperature and re-cut the sections.
4. Be careful not to make a circle on the liquid remaining on the glass slide.
5. This step is performed to block non-specific cell binding sites on the tissues.
6. As a typical example, the PLN sections are incubated with or without 10 μg/mL mAb (monoclonal antibody) S2, which recognizes L-selectin ligand carbohydrates expressed in HEVs [15], or with control mouse IgM in the experiment shown in Figs. 1 and 2.
7. This **step 7** is performed to selectively lyse red blood cells. Leukocytes are resistant to this ACK solution treatment.
8. CellTracker Orange CMTMR fluorescent probes freely pass through cell membranes and are converted to cell-impermeable fluorescent products. Other fluorescent probes with similar properties may also be used in this step.
9. After taking off one slide from the shaker platform, proceed with **steps 5** through **8** without pausing. After that, the second slide should be processed in the same way. Keep the shaker running until the last slide is removed.

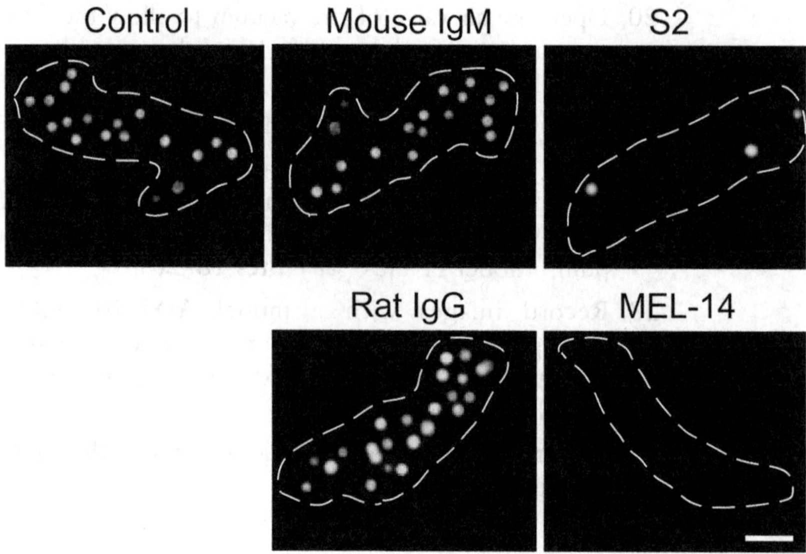

Fig. 1 Typical photomicrographs of leukocyte binding to HEVs in the modified Stamper–Woodruff cell-binding assay. Binding of CMTMR-labeled leukocytes incubated with or without (*Control*) 10 μg/mL MEL-14 or control rat IgG to PLN tissue sections incubated with or without 10 μg/mL S2 or control mouse IgM. *Dotted line*, outline of HEV. Bar, 40 μm. Modified from Hirakawa et al. [15]

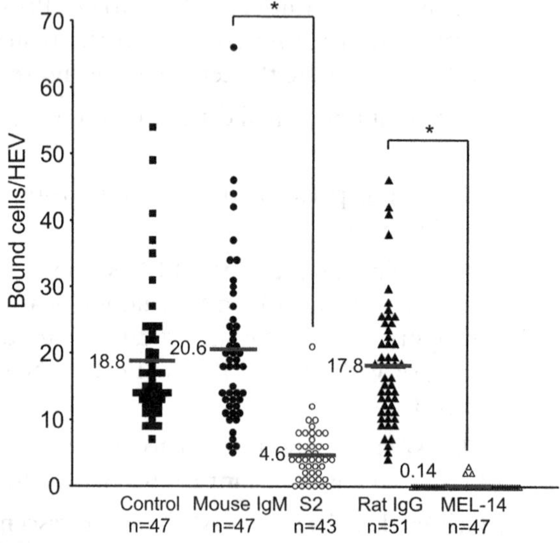

Fig. 2 Number of bound leukocytes per HEV in the modified Stamper–Woodruff cell-binding assay. The number of leukocytes bound to an HEV is plotted. The number of HEVs analyzed per sample (*n*) is indicated at the *bottom. Horizontal gray lines* and *numbers* next to them represent the average number of leukocytes bound per HEV. *NS* not significant; $*P < 0.001$. Modified from Hirakawa et al. [15]

10. It is important to dip the glass slides very gently and slowly into the tubes in these steps. If this step is not performed gently enough, HEV-bound cells may detach from the tissue.
11. Typical results of a modified Stamper–Woodruff cell-binding assay are shown in Figs. 1 and 2. The results indicate that mAb S2 significantly inhibits the binding of fluorescently labeled leukocytes to the HEVs in PLNs. The binding observed in this assay is dependent on L-selectin because the anti-L-selectin mAb, MEL-14 [14], completely blocks the cell binding. In Fig. 2, the number of leukocytes bound to HEVs is counted.
12. As a typical example, cells are incubated with or without 10 μg/mL mAb S2 in DME/F-12 medium supplemented with 1 % fetal bovine serum for 10 min at room temperature in the experiment shown in Fig. 4.
13. Prepare unlabeled leukocytes from the spleens of WT mice according to the procedures described in Subheading 3.1.2, except that **steps 11–14** should be omitted.
14. For the L-selectin inhibition experiment, leukocytes are preincubated with 10 μg/mL mAb MEL-14 for 10 min at 4 °C before introducing the cells into the flow chamber as described below.
15. A typical setup of the system is schematically shown in Fig. 3.

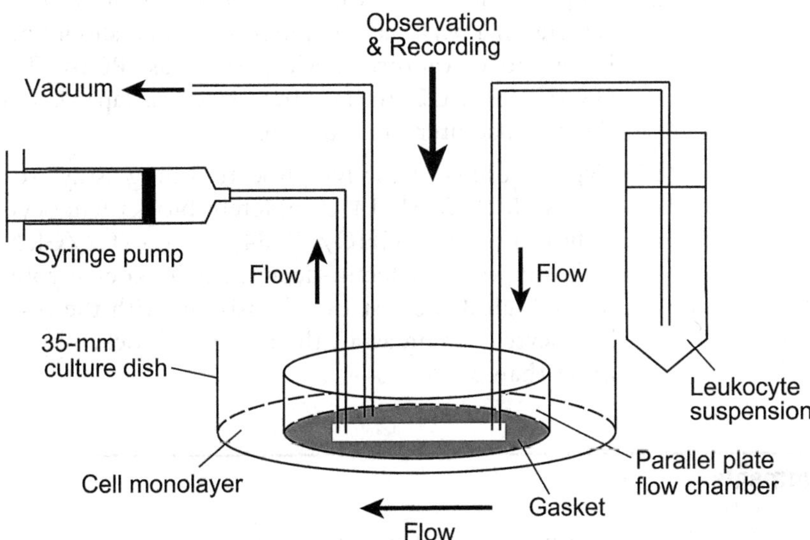

Fig. 3 Schematic representation of the experimental setup for the leukocyte rolling assay. Place the parallel plate flow chamber on the microscope stage with the input and output tubes to the *right* and *left*, respectively, and the vacuum tube to the back. The flow chamber apparatus and gasket are attached tightly to the inner surface of the cell monolayer dish by vacuum. The leukocyte suspension is introduced into the parallel plate flow chamber using a syringe pump. Leukocyte rolling is recorded with a CCD camera attached to an inverted microscope

16. It is important to remove air bubbles in the silicone tubes carefully in this step to avoid air bubbles stripping off cell monolayers from the culture dish in **step 22**.
17. Make sure that the gasket is attached underneath the parallel plate flow chamber after this step.
18. The equation relating wall shear stress to volumetric flow rate through the parallel flow chamber is given as follows:

    $\tau_w = \mu\gamma = 6\mu Q/a^2 b$ (*see* **Note 19**),

    where

    τ_w = wall shear stress, dyn/cm^2.

    γ = shear rate, 1/s.

    μ = apparent viscosity of the media, P (poise) (*see* **Note 20**).

    a = channel height (gasket thickness), cm.

    b = channel width (gasket flow path width), cm.

    Q = volumetric flow rate, ml/s.
19. Calculate the volumetric flow rate to attain the wall shear stress in the flow chamber of 2, 1.5, 1, and 0.5 dyn/cm^2. Based on the calculation, adjust the volumetric flow rate using a syringe pump Model 11 Plus.
20. 1 P = 1 dyn·s/cm^2. The analogous unit in the International System of Units (SI) is Pa·s (pascal second): 1 Pa·s = 10 P. For H_2O, μ = 0.89 mPa·s = 0.0089 P at 25 °C.
21. To playback a video captured by the model ADT-40S CCD camera on a personal computer, the video should be digitized by a video capture device, such as PCast TV capture (PC-SMP2E/U2, Buffalo Inc., Nagoya, Japan), according to the manufacturer's instructions.
22. Typical results of the lymphocyte rolling assay are shown in Fig. 4. Mab MEL-14 completely blocks leukocyte rolling, indicating that CHO/CD34/F7/C1/C2/GlcNAc6ST-2 cells support L-selectin-dependent leukocyte rolling under physiological shear stress. Consistent with the results of the leukocyte homing assay, the rolling is blocked by mAb S2 by more than 80 %.

Acknowledgments

I would like to thank Dr. Jotaro Hirakawa, Ms. Kaori Sato, and Mr. Ryuji Matsumura (University of Shizuoka, School of Pharmaceutical Sciences) for collaboration. This work was supported in part by Grants-in-Aid for Scientific Research, Category (B)

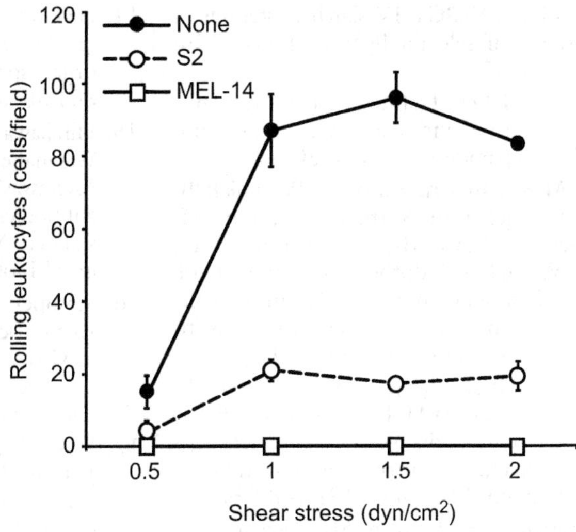

Fig. 4 Typical results of the leukocyte rolling assay. The number of rolling leukocytes per 30 s in a field observed using a 20× objective lens on CHO transfectants stably expressing 6-sulfo-sialyl Lewis X structure (CHO/CD34/F7/C1/C2/GlcNAc6ST-2), pretreated with or without 10 μg/mL mAb S2. Lymphocytes applied to the flow chamber were treated with or without 10 μg/mL mAb MEL-14. Modified from Hirakawa et al. [15]

from the Ministry of Education, Culture, Sports, Science, and Technology, Japan (21390023 and 24390018), and by the Institute for Fermentation, Osaka, Japan.

References

1. Rosen SD, Bertozzi CR (1994) The selectins and their ligands. Curr Opin Cell Biol 6: 663–673
2. McEver RP (2002) Selectins: lectins that initiate cell adhesion under flow. Curr Opin Cell Biol 14:581–586
3. Ley K, Kansas GS (2004) Selectins in T-cell recruitment to non-lymphoid tissues and sites of inflammation. Nat Rev Immunol 4:325–335
4. Arbones ML, Ord DC, Ley K, Ratech H, Maynard-Curry C, Otten G, Capon DJ, Tedder TF (1994) Lymphocyte homing and leukocyte rolling and migration are impaired in L-selectin-deficient mice. Immunity 1:247–260
5. Girard JP, Moussion C, Forster R (2012) HEVs, lymphatics and homeostatic immune cell trafficking in lymph nodes. Nat Rev Immunol 12:762–773
6. Kawashima H, Petryniak B, Hiraoka N, Mitoma J, Huckaby V, Nakayama J, Uchimura K, Kadomatsu K, Muramatsu T, Lowe JB, Fukuda M (2005) *N*-Acetylglucosamine-6-*O*-sulfotransferases 1 and 2 cooperatively control lymphocyte homing through L-selectin ligand biosynthesis in high endothelial venules. Nat Immunol 6:1096–1104
7. Uchimura K, Gauguet JM, Singer MS, Tsay D, Kannagi R, Muramatsu T, von Andrian UH, Rosen SD (2005) A major class of L-selectin ligands is eliminated in mice deficient in two sulfotransferases expressed in high endothelial venules. Nat Immunol 6:1105–1113
8. Homeister JW, Thall AD, Petryniak B, Maly P, Rogers CE, Smith PL, Kelly RJ, Gersten KM, Askari SW, Cheng G, Smithson G, Marks RM, Misra AK, Hindsgaul O, von Andrian UH, Lowe JB (2001) The α(1,3)fucosyltransferases FucT-IV and FucT-VII exert collaborative control over selectin-dependent leukocyte recruitment and lymphocyte homing. Immunity 15:115–126
9. Yang WH, Nussbaum C, Grewal PK, Marth JD, Sperandio M (2012) Coordinated roles of

ST3Gal-VI and ST3Gal-IV sialyltransferases in the synthesis of selectin ligands. Blood 120: 1015–1026

10. Springer TA (1994) Traffic signals for lymphocyte recirculation and leukocyte emigration: the multistep paradigm. Cell 76:301–314
11. Bao X, Moseman EA, Saito H, Petryniak B, Thiriot A, Hatakeyama S, Ito Y, Kawashima H, Yamaguchi Y, Lowe JB, von Andrian UH, Fukuda M (2010) Endothelial heparan sulfate controls chemokine presentation in recruitment of lymphocytes and dendritic cells to lymph nodes. Immunity 33:817–829
12. Tsuboi K, Hirakawa J, Seki E, Imai Y, Yamaguchi Y, Fukuda M, Kawashima H (2013) Role of high endothelial venule-expressed heparan sulfate in chemokine presentation and lymphocyte homing. J Immunol 191:448–455
13. von Andrian UH, Hasslen SR, Nelson RD, Erlandsen SL, Butcher EC (1995) A central role for microvillous receptor presentation in leukocyte adhesion under flow. Cell 82:989–999
14. Gallatin WM, Weissman IL, Butcher EC (1983) A cell-surface molecule involved in organ-specific homing of lymphocytes. Nature 304:30–34
15. Hirakawa J, Tsuboi K, Sato K, Kobayashi M, Watanabe S, Takakura A, Imai Y, Ito Y, Fukuda M, Kawashima H (2010) Novel anti-carbohydrate antibodies reveal the cooperative function of sulfated *N*- and *O*-glycans in lymphocyte homing. J Biol Chem 285:40864–40878
16. Stamper HB Jr, Woodruff JJ (1976) Lymphocyte homing into lymph nodes: in vitro demonstration of the selective affinity of recirculating lymphocytes for high-endothelial venules. J Exp Med 144:828–833
17. Yeh JC, Hiraoka N, Petryniak B, Nakayama J, Ellies LG, Rabuka D, Hindsgaul O, Marth JD, Lowe JB, Fukuda M (2001) Novel sulfated lymphocyte homing receptors and their control by a Core1 extension β1, 3-N-acetylglucosaminyltransferase. Cell 105: 957–969

Chapter 36

Assessment of Weak Sugar-Binding Ability Using Lectin Tetramer and Membrane-Based Glycans

Kazuo Yamamoto

Abstract

To consider biological significance of glycosylation of proteins, it is necessary to evaluate the importance of sugar-recognition processes mediated by lectins. Though the interaction between sugars and proteins, especially animal lectins, is quite weak with K_d approximately 10^{-4} M, cellular and molecular recognitions mediated via sugar–protein interaction increase their avidity by 1–3 orders of magnitude by the self-association of both receptors and their ligands on cell surfaces. To assess the weak interaction between lectins and their sugar ligands, we established lectin tetramer binding to cell surface glycans using flow cytometry. This strategy is highly sensitive, and useful to determine whether or not a putative lectin domain may have sugar-binding ability.

Key words Lectin tetramer, Specificity, Biotinylation, Cell surface glycan, Flow cytometry

1 Introduction

Newly synthesized proteins are posttranslationally modified with glycans in the endoplasmic reticulum (ER) and the Golgi apparatus. Glycosylation of proteins is a most complicated process associated with more than hundred kinds of glycosyltransferases and glycosidases [1]. Especially in case of animal lectins, the interaction between carbohydrate and lectins is very weak [2], thus insight into the role of sugar recognition in physiological processes is limited using conventional methods. To address this difficulty, we established a highly sensitive method to investigate the specificities and other significant characteristics of lectins. In this chapter, we describe the method preparing soluble lectin tetramer (*see* Fig. 1). By using the lectin tetramer, binding of the lectin to cell surface glycans were quantitatively monitored by flow cytometry. In combination with the cells treated with several sugar-processing inhibitors or endo- and exo-glycosidases, this method could access the weak sugar-binding ability of lectins without purification of both native lectins and their native ligands in large scale [3].

Jun Hirabayashi (ed.), *Lectins: Methods and Protocols*, Methods in Molecular Biology, vol. 1200,
DOI 10.1007/978-1-4939-1292-6_36, © Springer Science+Business Media New York 2014

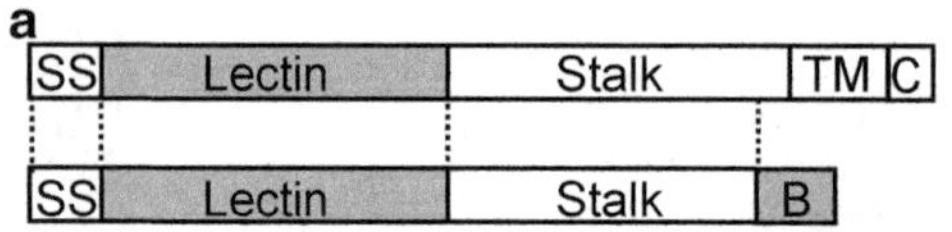

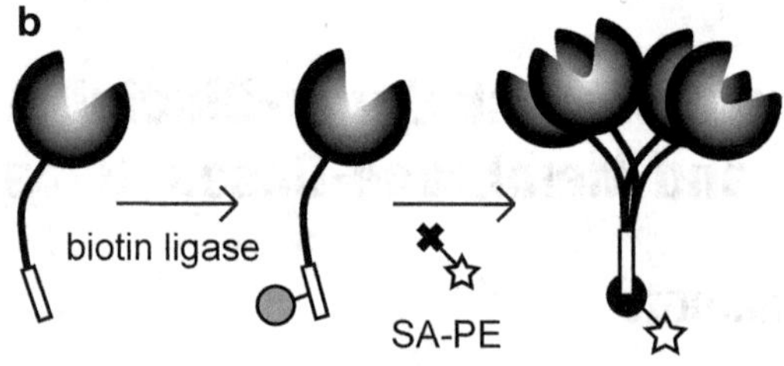

Fig. 1 Strategies for preparing PE-labeled lectin tetramer. (**a**) Preparation of plasmids encoding a soluble lectin domain with a biotinylation tag. (**b**) The purified recombinant soluble lectin domain with a C-terminal biotinylation tag is enzymatically biotinylated with biotin-protein ligase, BirA, and then tetramerized with PE-labeled streptavidin (PE-SA)

2 Materials

2.1 Instrumentation

1. FACSCalibur flow cytometer (BD Biosciences, San Jose, CA).
2. CELLQuest data analysis software (BD Biosciences).
3. Bath sonicator, e.g., Branson, ED-40 (Danbury, CT) (*see* **Note 1**).
4. Microcentrifuge for 1.5 mL-microcentrifuge tubes (*see* **Note 2**).
5. Vacuum manifold for a 96-well filter plate (Millipore, Bedford, MA) (*see* **Note 3**).
6. Shaker (*see* **Note 4**).
7. CO_2 incubator.

2.2 Preparation of Biotinylated Soluble Lectins

1. BL21(DE3)pLysS *Escherichia coli* cells.
2. Plasmid harboring cDNA encoding soluble lectin with biotinylation tag (*see* **Note 1**).
3. pET-3c vector (New England Biolabs, Ipswich, MA).
4. Isopropyl β-thiogalactopyranoside (IPTG).
5. Polyethyleneglycol 20,000.
6. Biotin-protein ligase, BirA (Avidity, Aurora, CO).
7. LB/Amp medium: LB broth containing 0.1 mg/mL ampicillin.
8. TBS: 20 mM Tris–HCl, pH 8.0, containing 0.15 M NaCl.
9. HBS: 20 mM HEPES–NaOH, pH 7.4, containing 150 mM NaCl.

10. Lysis buffer: 20 mM Tris–HCl, pH 8.0, 0.15 M NaCl, 0.5 % (v/v) Triton X-100, 1 mM phenylmethanesulfonyl fluoride (PMSF, Sigma-Aldrich, St. Louis, MO) (*see* **Note 2**).
11. Wash buffer: 20 mM Tris–HCl, pH 8.0, 0.15 M NaCl, 0.5 % (v/v) Triton X-100, 1 mM ethylenediaminetetraacetic acid (EDTA), 1 mM dithiothreitol (DTT).
12. Solubilization buffer: 50 mM Tris–HCl, pH 8.0, 6 M guanidine, 1 mM DTT, and 0.1 mM EDTA.
13. Refolding buffer: 100 mM Tris–HCl, pH 7.5, 0.4 M arginine, 5 mM reduced glutathione, 0.5 mM oxidized glutathione, and 0.5 mM phenylmethanesulfonyl fluoride (PMSF) (*see* **Notes 3** and **4**).
14. Dialysis buffer: 20 mM Tris–HCl, pH 7.5, 25 mM NaCl, and 0.1 mM EDTA.
15. UNO Q-6 ion-exchange column (12 mm × 53 mm; Bio-Rad, Hercules, CA).
16. Superdex-75 10/300 GL column (10 mm × 300 mm; GE Healthcare, Backinghamshire, UK).

2.3 Binding Assay for PE-Labeled Lectin Tetramer Using Flow Cytometry

1. *R*-phycoerythrin (PE)-conjugated streptavidin (BD Biosciences).
2. Propidium iodide (PI).
3. HeLaS3 cells (American Type Culture Collection, Manassas, VA).
4. Cell culture medium: RPMI1640 medium (Sigma-Aldrich) supplemented with 10 % (v/v) heat-inactivated fetal calf serum, 2 mM glutamine, 100 U/mL penicillin, 100 μg/mL streptomycin, and 50 μM 2-mercaptoethanol.
5. Phosphate buffered saline (PBS): 10 mM sodium phosphate, pH 7.4, containing 0.15 M NaCl.
6. PBS–EDTA: PBS containing 1 mM EDTA.
7. HEPES buffered saline (HBS): 20 mM HEPES–NaOH, pH 7.4, 0.15 M NaCl, and 1 mM EDTA.
8. BSA-containing HBS (HBSB): HBS containing 0.1 % NaN_3 and 0.1 % bovine serum albumin.
9. U bottom 96-well plate (Millipore).
10. 10-cm culture dish.

3 Methods

3.1 Preparation of Biotinylated Soluble Lectins

1. Prepare BL21(DE3)pLysS cells transfected with expression plasmid harboring cDNA encoding soluble lectin with biotinylation tag.
2. Grow BL21(DE3)pLysS cells overnight in 3 mL of LB/Amp medium at 37 °C in a shaker (*see* **Note 5**).

3. Add 1 mL of the overnight culture to 200 mL LB/Amp medium in a 0.5-L flask.
4. Grow at 37 °C with vigorous shaking until OD_{600} reaches 0.5.
5. Add 0.2 mL of 1 M IPTG (1 mM final) and grow the cells for another 4 h.
6. Harvest the cells by centrifugation at 4,000 × *g* for 15 min.
7. Resuspend cell pellet in 30 mL of TBS by pipetting.
8. Centrifuge at 3,000 × *g* for 10 min.
9. Repeat **steps** 7 and **8** twice.
10. Resuspend cells in 10 mL of lysis buffer.
11. Disrupt the cells by sonication for 15 min using a microtip (*see* **Note 6**).
12. Centrifuge cell lysate at 9,000 × *g* for 20 min.
13. Resuspend the pellet in 30 mL of wash buffer by pipetting (*see* **Note** 7).
14. Centrifuge at 9,000 × *g* for 20 min.
15. Repeat **steps 13** and **14** three times.
16. Add 10 mL of solubilization buffer and incubate with gentle stirring at 4 °C.
17. Add solubilized solution drop by drop into 500 mL of refolding buffer while stirring.
18. Allow to stand at 4 °C for 72 h.
19. Transfer to dialysis tubing and dialyze twice against TBS.
20. Concentrate the protein solution by polyethyleneglycol 20,000 at 4 °C.
21. Dialyze three times against 20 mM Tris–HCl, pH 7.5, containing 25 mM NaCl.
22. Transfer all material in the dialysis bag and centrifuge at 25,000 × *g* for 1 h to remove insoluble material.
23. Load supernatant onto a UNO Q-6 column equilibrated with 20 mM Tris–HCl, pH 7.5, containing 25 mM NaCl.
24. Elute protein with 18 mL of a linear gradient of NaCl from 25 to 500 mM in 20 mM Tris–HCl, pH 7.5. Collect 1-mL fractions.
25. Analyze fractions by sodium dodecyl sulfate–polyacrylamide gel electrophoresis (SDS-PAGE).
26. Mix purified proteins fused to a C-terminal biotinylation tag with a biotin-protein ligase, Bir A in 50 mM Tris–HCl, pH 8.3, containing 10 mM ATP, 10 mM manganese acetate, 50 μM biotin at 30 °C for 1 h.
27. Remove the remaining free biotin by gel filtration on a Superdex-75 10/300 GL column equilibrated with TBS (*see* **Note 8**).

3.2 Binding Assay for PE-Labeled Lectin Tetramer Using Flow Cytometry

1. Mix biotinylated protein with PE-conjugated streptavidin at a molar ratio of 4:1 for 1 h on ice to prepare the *R*-phycoerythrin (PE)-labeled soluble lectin tetramer.
2. Culture HeLaS3 cells in a 10-cm culture dish at 37 °C under 5 % CO_2 condition.
3. Aspirate the culture medium and wash the culture dish with 5 mL of PBS.
4. Add 1 mL of trypsin-EDTA solution into the dish and incubate at 37 °C for 5 min.
5. Harvest cells with 5 mL of cell culture medium by pipetting and centrifuge at 230 × *g* for 10 min.
6. Suspend the harvested cells with 10 mL of HBS and centrifuge at 230 × *g* for 10 min.
7. Suspend the cells in HBSB at a concentration of 2×10^7 cells/ml (*see* **Note 9**).
8. Transfer cells at 2×10^5 cells/well into a 96-well U bottom plate.
9. Add PE-labeled lectin tetramer into the well at a concentration 10–100 μg/mL and incubate at 25 °C for 30 min in the dark (*see* **Note 10**).
10. Add 200 μL of HBS and spin down the cells at 510 × *g* for 3 min.
11. Discard the supernatant.
12. Repeat **steps 10** and **11**.
13. Suspended the cells in 200 μL of HBS containing 1 μg/mL propidium iodide (PI).
14. Analyze the binding of PE-labeled lectin tetramer to the cells with FACSCalibur (*see* **Notes 11** and **12**).

4 Notes

1. A cDNA encoding the enzymatic biotinylation sequence GGGLNDIFEAQKIEWHE is introduced at the 3′-end of a cDNA by ligation with a synthetic DNA of 5′-ggaattccatatgga attcccgggggcggtctgaacgacatcttcgaagctcagaaaatcgaatggcac-gaataaggatccgcg-3′ [4].
2. PMSF is easily solubilized in methanol and then added to the buffer.
3. The pH condition of refolding is critical to obtain large amounts of recombinant proteins, which depends on the sample.
4. Instead of EDTA, several metal ions are occasionally added to the refolding and dialysis buffers to enhance the correct folding of denatured polypeptides.

5. Proper aeration is essential to achieve vigorous growth.
6. Place the sample on ice during sonication.
7. Most of induced recombinant proteins are recovered in an insoluble form. Check for the presence of the recombinant protein by SDS-PAGE.
8. Biotinylation of protein is confirmed with a gel-shift assay using polyacrylamide gels as described [5].
9. To determine the divalent cation dependency of the tetramer binding, 1 mM of metal ions was added to HBSB instead of 1 mM EDTA.
10. To test the effect of exogenous monosaccharides and oligosaccharides on the binding of PE-labeled lectin tetramer to the cells, tetramer is preincubated with various concentrations of monosaccharides or oligosaccharides at 25 °C for 30 min before being added to the cells.
11. The cell-surface fluorescence at 575 nm associated with PE is recorded. In total, 10^4 live cells gated by forward and side scattering and PI exclusion were acquired for analysis.
12. Remodeling cell surface glycans by glycosylation inhibitors is a useful method (*see* Chapter 32).

References

1. Narimatsu H (2004) Construction of a human glycogene library and comprehensive functional analysis. Glycoconj J 21:17–24
2. Knibbs RN, Takagaki M, Blake DA, Goldstein IJ (1988) The role of valence on the high-affinity binding of *Griffonia simplicifolia*. Biochemistry 37:16952–16957
3. Yamamoto K, Kawasaki N (2010) Detection of weak-binding sugar activity using membrane-based carbohydrates. Methods Enzymol 478: 233–240
4. Wada H, Matsumoto N, Maenaka K, Suzuki K, Yamamoto K (2004) The inhibitory NK cell receptor CD94/NKG2A and the activating receptor CD94/NKG2C bind the top of HLA-E through mostly shared but partly distinct sets of HLA-E residues. Eur J Immunol 34:81–90
5. Kawasaki N, Matsuo I, Totani K, Nawa D, Suzuki N, Yamaguchi D, Matsumoto N, Ito Y, Yamamoto K (2007) Detection of weak sugar binding activity of VIP36 using VIP36-streptavidin complex and membrane-based sugar chains. J Biochem 141:221–229

Part IV

Structural Biology and Engineering of Lectins

Chapter 37

Perspectives in Glycomics and Lectin Engineering

Jan Tkac, Tomas Bertok, Jozef Nahalka, and Peter Gemeiner

Abstract

This chapter would like to provide a short survey of the most promising concepts applied recently in analysis of glycoproteins based on lectins. The first part describes the most exciting analytical approaches used in the field of glycoprofiling based on integration of nanoparticles, nanowires, nanotubes, or nanochannels or using novel transducing platforms allowing to detect very low levels of glycoproteins in a label-free mode of operation. The second part describes application of recombinant lectins containing several tags applied for oriented and ordered immobilization of lectins. Besides already established concepts of glycoprofiling several novel aspects, which we think will be taken into account for future, more robust glycan analysis, are described including modified lectins, peptide lectin aptamers, and DNA aptamers with lectin-like specificity introduced by modified nucleotides. The last part of the chapter describes a novel concept of a glycocodon, which can lead to a better understanding of glycan–lectin interaction and for design of novel lectins with unknown specificities and/or better affinities toward glycan target or for rational design of peptide lectin aptamers or DNA aptamers.

Key words Biosensors, Glycomics, Lectins, Nanoparticles, DNA aptamers, Lectin peptide aptamers, Recombinant lectins

1 Introduction

Since the introduction of DNA biochips in 1995 [1], the technology has been intensively applied in assays of genome-wide expression to seek information about possible functions of novel or poorly characterized genes [2] and for diagnostic purposes, as well [3]. Even though DNA microarray technology has shed light on many physiological functions of genes by determination of expression of gene clusters, there is quite often only a very low correlation between RNA and protein abundance detected in single-cell organisms [4] and in higher ones, including humans [5]. Since quantitative analysis of proteins is central to proteomics with a focus on design of novel drugs, diagnostics of diseases, and their therapeutic applications, protein microarrays were successfully launched to address these issues [6].

Jun Hirabayashi (ed.), *Lectins: Methods and Protocols*, Methods in Molecular Biology, vol. 1200, DOI 10.1007/978-1-4939-1292-6_37, © Springer Science+Business Media New York 2014

Analysis of finely tuned posttranslational modifications (PTMs) of proteins is an additional challenge for current analytical technology. Glycosylation is a highly abundant form of PTM of proteins and it is estimated that 70–80 % of human proteins are glycosylated [7]. Importance of glycans can be further highlighted by the fact that 70 % of all therapeutic proteins are glycosylated [8]. Glycan-mediated recognition plays an important role in many different cell's processes such as fertilization, immune response, differentiation of cells, cell–matrix interaction, cell–cell adhesion, etc. [9, 10]. Glycans present on the surface of cells are naturally involved in pathological processes including viral and bacterial infections, in neurological disorder and in tumor growth and metastasis [3, 11–16]. Thus, better understanding of glycan-mediated pathogenesis is essential in order to establish a "policy" to develop efficient routes for disease treatment with several recent studies as good examples, e.g., "neutralization" of various forms of viruses [17, 18] or more efficient vaccines against various diseases [19, 20]. A changed glycosylation on a protein backbone can be effectively applied in early stage diagnostics of several diseases, including different forms of cancer with known glycan-based biomarkers [21–23]. Moreover, many previously established and even commercially successful strategies used to treat diseases are currently being revisited in light of glycan recognition in order to lower side effects, enhance serum half-life, or decrease cellular toxicity [3, 24, 25]. Recently, the first glyco-engineered antibody was approved to the market, which was called by the authors "a triumph for glyco-engineering" [26].

Glycomics focuses on revealing finely tuned reading mechanisms in the cell orchestra based on graded affinity, avidity, and multivalency of glycans (i.e., sugar chains covalently attached to proteins and lipids) [27]. Glycans are information-rich molecules applicable in coding tools of the cell since they can form enormous number of possible unique sequences from basic building units [28]. It is estimated that the size of the cellular glycome can be up to 500,000 glycan modified biomolecules (proteins and lipids) formed from 7,000 unique glycan sequences [29]. Thus, it is not a surprise the glycome is sometimes referred to as the "third alphabet" in biology, after genetics and proteomics [30]. A huge glycan variation can explain human complexity in light of a paradoxically small genome. This glycan complexity together with similar physicochemical properties of glycans is the main reason why the progress in the field of glycomics has been behind advances in genomics and proteomics [31].

Traditional glycoprofiling protocols rely on glycan release from a biomolecule with a subsequent quantification by an array of techniques including capillary electrophoresis, liquid chromatography, and mass spectrometry [30, 32–37]. There is an alternative way for glycoprofiling by application of lectins, natural glycan recognizing proteins [28, 38, 39] in combination with various transducing protocols [12, 40, 41]. The most powerful glycoprofiling tool relies on lectins arrayed on solid surfaces for direct analysis of glycoproteins, glycolipids, membranes, and even glycans on the surface of intact cells [11, 12, 42, 43]. Even though lectin microarrays offer high-throughput assay protocols with a minute consumption of samples and reagents, there are some drawbacks such as a need to fluorescently label the sample or the lectin, which negatively affects the performance of detection [11, 12], relatively high detection limits, and quite narrow working concentration ranges. Thus, the ideal detection platform should be based on protocols without a need to label a glycoprotein or a lectin, in a way similar to natural processes occurring within a cell [30].

Lectins (lat. *legere* = to choose) are proteins able to recognize and reversibly bind to free or bound mono- and oligosaccharides [44]. They are not usually catalytically active, do not participate in the immune response of higher organisms, and can be found in viruses, bacteria, fungi, plants, and animals. They are therefore a relatively heterogeneous group of oligomeric proteins belonging to distinct families with similar sequences and are considered as natural glycocode decipherers [28]. Lectins, unlike antibodies, have a low specificity and affinity with K_d ranged from 10^{-3} to 10^{-7} M [30] and lectins with a new specificity cannot be raised in a way similar to antibodies.

In the following sections we will focus on ways how to improve glycan detection either by application of novel nanoscale-controlled patterning protocols, nanoengineered devices or by application of novel, recombinant lectins, lectin-like aptamers, or lectin peptide aptamers. The final part of this book chapter will focus on a completely novel area in the glycomics—the idea of a glycocodon.

2 Perspectives of Novel Formats of Analysis Applicable in Glycomics

The use of nanotechnology, sophisticated nanoscale patterning protocols, and advanced detection platforms can help to overcome the drawbacks of lectin microarray technology allowing it to work in a label-free mode of operation, with a high sensitivity, low detection limits, a wide concentration window and in some cases, real-time analysis of a binding event is possible [9, 30, 45–51]. These devices can differ in their mode of signal transduction compared to traditional methods and will be divided into three categories according to their mode of action. Various traditional analytical techniques based on lectins (i.e., surface plasmon resonance and quartz crystal microbalance) are covered by different chapters accompanied this one within this book and are not discussed here.

2.1 Mechanical Platforms

Microcantilever biochips offer a novel approach for detection of a molecular binding based on a change in mass accumulated on the surface of a cantilever during biorecognition. It is a label-free technique allowing to monitor biospecific interaction in a real time,

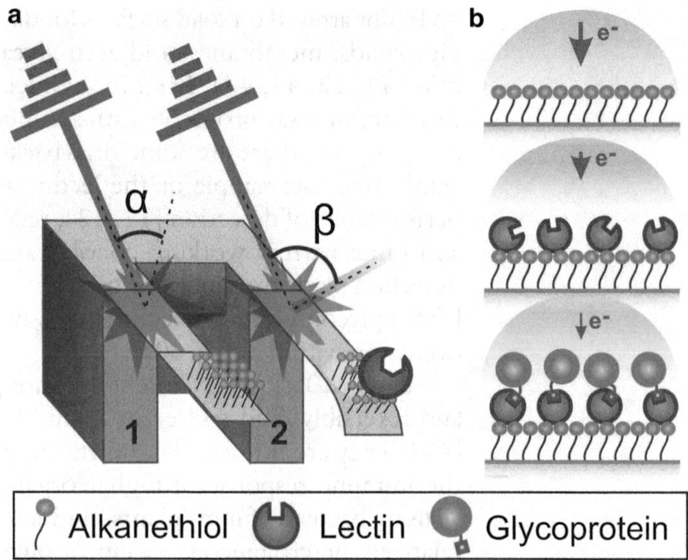

Fig. 1 (**a**) A mechanical platform of detection based on an array of microcantilever biochips with a cantilever bent after a biorecognition took place. (**b**) An electronic detection platform of analysis based on an electrochemical impedance spectroscopy (EIS) with an increase in overall resistance of a soluble redox probe, represented by a *red arrow*, to the interface after a biomolecular interaction

thus, affinity constants of the interaction can be acquired. When a biorecognition takes place, a particular cantilever bends, which results in the shift of a laser beam angle, allowing for a direct detection of the binding event (Fig. 1a).

The device was prepared with variations in the density and composition of a glycan determinant immobilized via a thiol-gold surface chemistry on a cantilever surface. Namely galactose, trimannose, and nonamannose were attached on the surface and probed with two different lectins—cyanovirin A and Concanavalin A (Con A). The later was successfully detected on a surface with optimal glycan composition down to a nM range [52]. The sensitivity is not impressive, but comparable to traditional surface plasmon resonance and quartz crystal microbalance lectin-based biosensors. The Seeberger's group later extended this concept for analysis of several *Escherichia coli* strains on microcantilever biochips functionalized with different mannosides with a specific and reproducible detection with an amount of detectable *E. coli* cells over four orders of magnitude [53].

2.2 Electrical Platforms

Electrical/electrochemical detection is quite often utilized in combination with other techniques in the field of glycomics for some time [54]. Electrical platforms of detection of a biorecognition event are primarily based on changes in the electrical signals such as resistance, impedance, capacitance, conductance, potential, and current [55]. These analytical techniques are usually nondestructive and extremely sensitive, offering quite a wide concentration working range with a possibility to work in an array format of analysis [30].

2.2.1 Electrochemical Impedance Spectroscopy (EIS)

The most frequently used label-free electrochemical technique is EIS, which is based on an electric perturbation of a thin layer on the conductive surface by small alternating current amplitude with ability to provide characteristics of this interface utilizable in sensing. EIS results are typically transformed into a complex plane Nyquist plot vectors, which by application of an equivalent circuit can provide information about electron transfer resistance of a soluble redox probe in a direct way (Fig. 1b). When a biorecognition took place, an electrode interface is modified and a subtle change in interfacial layer characteristics can be used for detection. EIS investigation is most frequently performed in the presence of a redox probe with detection of a change of resistance of the interface used for a signal generation. EIS is extensively used as a nondestructive technique for reliable analysis of surface conditions and allows complex biorecognition events to be probed in a simple, sensitive, and label-free manner and is being increasingly popular to develop electrochemical lectin-based biosensors for glycan determination [9, 30].

Initial efforts to detect glycoproteins by EIS were launched by the group of Prof. Joshi with sialic acid binding *Sambucus nigra* agglutinin (SNA) and a galactose binding peanut agglutinin covalently immobilized on printed circuit board electrodes [56]. The assays were really quick with a response time of 80 s and with sensitivity of glycoprotein detection down to 10 pg/mL (e.g., 150 fM), while using a cost-effective electrode material [56]. A group of Prof. Oliveira put a substantial effort to use lectin modified surfaces with EIS detection for discrimination between healthy human samples and samples from patients infected by a mosquito-borne Dengue virus (breakbone fever) with a high mortality rate [9]. Their device with two different lectins immobilized on gold nanoparticles offered a detection limit in the low nM range [57]. Another EIS-based biosensor was built on a surface of the silicon chip with an array of gold electrodes interfaced with nanoporous alumina membrane with high density of nanowells [58]. The biosensor offered a high reliability of assays and a good agreement with enzyme-linked lectin assays (ELLA). The detection limit of a biosensor for its analyte was five orders of magnitude lower compared to ELLA (i.e., 20 fM vs. 4.6 nM). An assay time for the biosensor of 15 min was much shorter compared to 4 h needed for ELLA. Moreover, a minute amount of sample (10 μL) was sufficient for the analysis by the biosensor [58].

In our recent work we focused on the development of ultrasensitive impedimetric lectin biosensors with detection limits down to a single-molecule level based on controlled architecture at the nanoscale [59–61]. In the first study the biosensor was able to detect a glycoprotein in a concentration window spanning seven orders of magnitude with a detection limit for the glycoprotein down to 0.3 fM, which was the lowest glycoprotein concentration detected [59]. In the following study an incorporation of gold nanoparticles offered even lower and unprecedented detection limit of 0.5 aM with quite a wide dynamic concentration range covered [61]. In our last study the EIS-biosensors were constructed with three different lectins to be able to detect changes on immunoglobulins with progression of a rheumatoid arthritis in humans. The biosensor with improved antifouling properties offered a detection limit in the fM range and worked properly even with 1,000× diluted human plasma. The biosensor performance was directly compared to the state-of-the-art glycoprofiling tool based on fluorescent lectin microarrays with a detection limit in the nM level [60]. Moreover, a sandwich configuration offered a detection limit down to aM concentration [60]. A detection limit down to fM range for analysis of alpha-fetoprotein (a biomarker for hepatocellular carcinoma) was recently observed on a device modified by arrays of single-walled carbon nanotubes and wheat-germ agglutinin with EIS as a transducing mechanism [62].

2.2.2 Nanotube Field Effect Transistor (NTFET) Sensors

In NTFETs, semiconducting nanotubes or nanowires act as a channel between two metal electrodes (source and drain) while the two electrodes are held at a constant bias voltage using a so-called gate electrode (Fig. 2a) [30, 49]. When the device with an

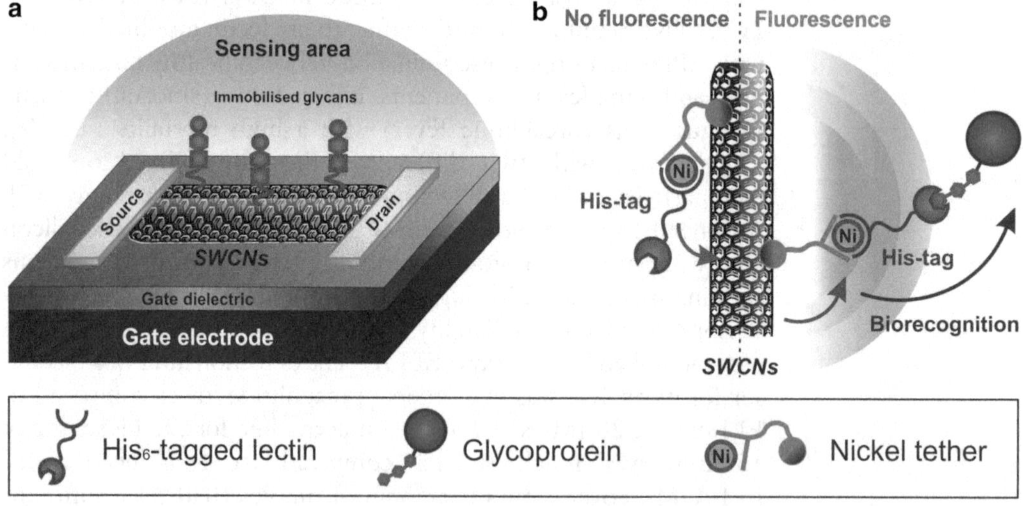

Fig. 2 (**a**) A field-effect transistor (FET) sensing based on a changed conductivity of a single-walled carbon nanotube (SWCNT) positioned in between a source and a drain. (**b**) An optical detection platform based on quenching of an intrinsic fluorescence of a SWCNT by Ni-tether employed for lectin immobilization via His_6 tag. A fluorescence of a SWCNT is partly restored after biorecognition since Ni-tether is pushed away from the SWCNT surface

immobilized biorecognition element is exposed to the sample containing its binding partner, a change of the device conductivity can be applied for quantification of the analyte. The application of the FET devices in the field of glycobiology was pioneered by Star's group [63, 64]. In the initial study carbon nanotubes were employed as a channel when glycoconjugate was immobilized on a surface of the device and an analyte lectin down to 2 nM concentration could be detected [63]. A forthcoming study confirmed that a carbon nanotube biosensor for detection of a lectin outperformed a device based on graphene [64]. However, semiconducting carbon nanotubes with a high purity are required to achieve better signal quality as a further research goal. Silicon nanowires were applied as an FET channel more effectively compared to carbon nanotubes and graphene, since a detection limit for a lectin down to 100 fg/mL (≈fM level) was achieved [65]. Even though such a remarkable concentration of lectin with a glycan modified FET device was detected, analysis of glycoproteins on a lectin immobilized surface can be more problematic since the device is able to detect changes in a close proximity to the surface and a biorecognition lectin-glycoprotein can be too far from the surface to be detected. The solution, however, can lie in an application of ultra-diluted (100× or 1,000×) phosphate buffers allowing to detect biorecognition event at distances 7.5–23.9 nm from the surface, but for analysis of protein levels in serum, serum has to be desalted prior to detection [66].

Another interesting approach applied in glycoassays was based on immobilization of mannose inside a nanochannel and changes in the nanochannel conductance were after binding of Con A detected in a concentration window from 10 to 1,500 nM [67]. The question how sensitive analysis of glycoproteins with lectins immobilized within a nanochannel can be has to be still answered.

2.3 Optical Platforms

There are two different optical sensing mechanisms applied in label-free glycoanalysis. The first is based on an intrinsic fluorescence of carbon nanotubes and the second one on a localized surface plasmon resonance detected on gold nanoislands. Both concepts have advantage since there is no necessity for an electronic interfacing, which is a problematic aspect of FET devices, and the nanoscale sensors require only a minute amount of sample for analysis.

2.3.1 Quenching of an Intrinsic Carbon Nanotube Fluorescence

This platform of detection employs fluorescent carbon nanotubes with a flexible NTA-nickel tether attached, modulating fluorescent intensity of carbon nanotubes on one side and being applied as a coupling agent for His_6-tagged lectins. When the glycoprotein interacts with an immobilized lectin, a nickel ion moves away from the carbon nanotube surface, partially restoring a quenched fluorescence of carbon nanotubes (Fig. 2b). An increase in the fluorescence output can be applied not only for quantification of a glycoprotein level, but for monitoring of the interaction in a real time, providing kinetic and affinity constants, as well. The absolute detection limit of the device for the glycoprotein was not that impressive (2 μg, i.e., 670 nM), but authors believe the device has a room for improvement (i.e., by using high quality nanotube sensors) [47, 68]. In a recent study authors extended this initial study for glycoprofiling of different forms of IgGs [69].

2.3.2 A Localized Surface Plasmon Resonance

Noble metal nanostructures exhibiting a localized surface plasmon resonance, sensitive to changes in the refractive index near the nanostructures, can be integrated into a biosensor device. There is only one report on application of such a device in glycoassays (Fig. 3). In this case mannose was immobilized on the surface of Au nanoislands and sensitivity toward a lectin was probed under stationary or flow conditions. Moreover, kinetic parameters of lectin interaction were obtained in an agreement with traditional techniques. Mannose-coated transducers offered an excellent selectivity toward Con A down to concentration of 5 nM in the presence of a large excess of bovine serum albumin (BSA) [70].

In summary, it can be concluded that from all novel nanoengineered devices offering label-free mode of detection EIS-based biosensors have a great potential for glycan analysis since they can clearly outperform the state-of-the-art tool in a glycoprofiling, lectin microarrays, in terms of a detection limit achieved and a dynamic concentration range of analysis offered. EIS lectin biosensors were successfully applied in analysis of complex samples such as human serum even at dilution of 1,000×. Only such sensitive devices can really detect ultralow concentration of disease markers directly in human serum, a feature important for early stage prognosis of a particular disease. Moreover, biosensor devices with a detection limit down to single molecule level have a potential to be applied for identification of novel biomarkers, which can be present in human body liquids at concentrations not detectable by other analytical platforms of detection involving lectins. Other analytical tools based on mechanical, FET, and optical signal transduction mechanisms have to prove their analytical potential in glycoprofiling with lectins immobilized on surfaces of such devices.

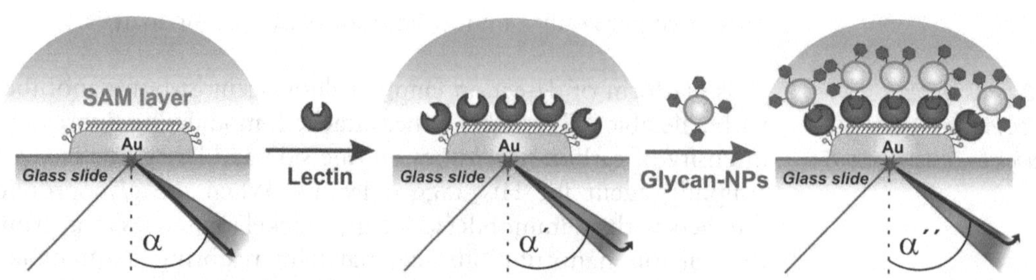

Fig. 3 A localized surface plasmon resonance (LSPR) employed for a label-free recognition based on a shift of reflected light as in case of traditional SPR technique

3 Perspectives in Lectin Engineering

The glycan binding sites of lectins are usually a shallow groove or a pocket present at the protein surface, or at the interface of oligomers [7]. Four main amino acids are part of an affinity site including asparagine, aspartic acid, glycine (arginine in Con A), and an aromatic residue for interaction with glycan via hydrogen bonds and hydrophobic interactions [71]. Ionic interactions are especially involved in recognition of negatively charged glycans containing sialic acids. Lectin-monosaccharide binding is relatively weak, which is why several approaches were applied to enhance practical utility of lectins in glycoprofiling [7].

Recombinant DNA technology for producing lectins was traditionally applied to establish primary structure; to study genetics, evolution, and biosynthesis; to elucidate the role of amino acids in recognition; to produce lectins with altered specificity and/or affinity; and to study their function in the organism of origin [72]. Novel trend is to apply this technology for producing lectins to be utilized for construction of various lectin-based biodevices. Recombinant lectin technology can significantly reduce drawbacks of traditional lectin isolation such as a long processing time, often quite a low yield, and batch-to-batch variation of the product quality depending on the source, with presence of various contaminants or different lectin isoforms [11, 73]. Moreover, recombinant technology offers to produce lectins either without any glycosylation, which can in many cases complicate glycoprofiling, by expression in prokaryotic hosts and to introduce various tags (His_6-tag, glutathione transferase), which can be effectively utilized not only for one-step purification process, but more importantly for an oriented immobilization of lectins on various surfaces [38]. Although lectin peptide aptamers have not been produced yet, it is a question of time, when such artificial glycan binding proteins emerge as an efficient tool in the area of glycobiology. It is estimated that another player in the area of glycoprofiling will make a substantial fingerprint i.e. lectin aptamers based on expanded genetic alphabet by introduction of modified nucleotides. Other concepts based on modified lectins in glycoprofiling with added value are finally described.

3.1 Oriented Immobilization of Recombinant Lectins

A controlled immobilization of lectins on a diverse range of surfaces can have a detrimental effect on the sensitivity of assays, since lectins can be attached in a way a biorecognition site is directly exposed to the solution phase for an efficient biorecognition. As a result almost 100 % of immobilized lectin molecules can have a proper orientation with an increased chance for catching its analyte (Fig. 4a). Moreover the presence of a linker, which attaches tag to the protein backbone, can significantly lower possible interaction of the protein with the surface, which can eventually lead to a denaturation of a protein [74].

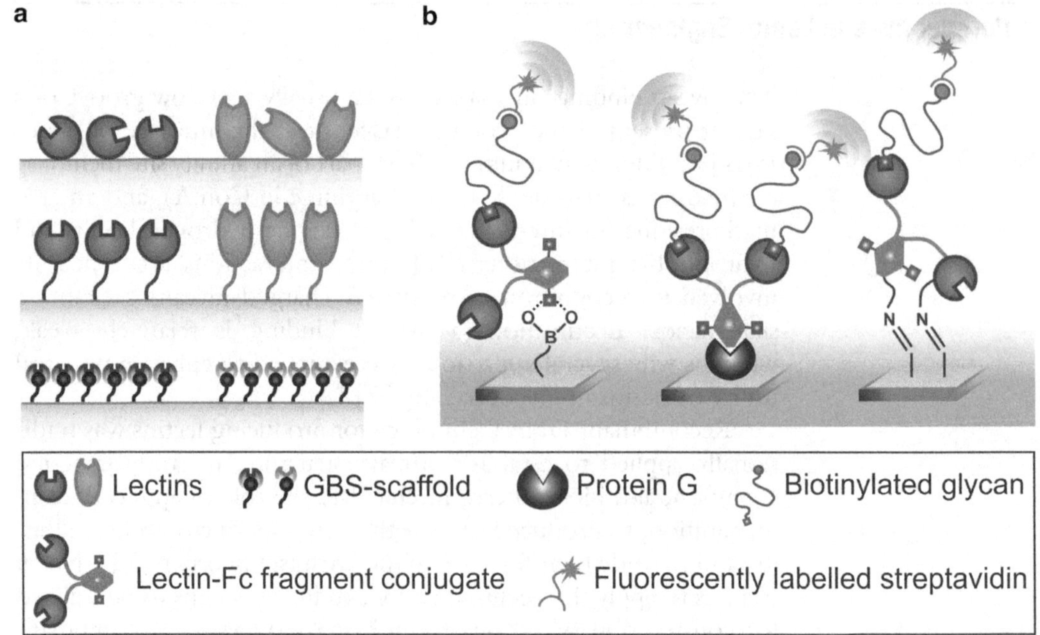

Fig. 4 (**a**) A schematic representation of possibilities to control uniformity of glycan binding proteins with a random amine coupling (*upper image*), oriented immobilization via introduced purification tag (image in the *middle*), and immobilization of uniform lectin peptide aptamers differing only in peptide insert providing a biorecognition element; *GBS* glycan binding site. (**b**) Various ways for immobilization of Fc-fused lectin on interfaces based on boronate affinity toward glycans present in Fc fragment (*left*), on affinity of protein G toward Fc fragment (*middle*) or a random amine coupling (*right*)

In a pilot study seven bacterial lectins, having a His_6-GST tag, were expressed in *E. coli* and subsequently applied for construction of a complete recombinant lectin microarray, which was utilized to probe differences between several tumor cells (ACHN, TK10, SK-MEL-5, and M14 cancer cell lines) [75]. For that purpose isolated membrane micelle from the tumor cell lines was employed. Such a procedure avoids use of proteases, which can change composition of samples containing glycoproteins, and at the same time there is no need to work with whole cells, allowing to work with a small spot sizes. The results showed distinct variations between tumor cell lines expressing different glycan moieties. In order to have a control spot on a lectin microarray a mutated form of one lectin was introduced, allowing to quantify specificity of interaction. Moreover, it was found out that in the presence of monosaccharides during a lectin printing process better resolved spot morphology and lectin activity were achieved [75]. In a next study of the Mahal's group, the effect of oriented immobilization of lectins on the sensitivity of lectin microarray assays was quantified. Oriented immobilization of lectins offered a detection limit of approx. 12 ng/mL (ca. 640 pM protein) [76], a significantly lower level compared to a detection limit of 10 μg/mL achieved on a lectin microarray with a random immobilization [33]. In a next study the group developed oriented immobilization of recombinant lectins in a single step deposition of lectins together with a glutathione to an activated chip surface. Such an approach simplifies an overall immobilization process because the surface does not need to be modified by glutathione prior to lectin immobilization [77].

Another group developed an oriented immobilization of recombinant lectins produced with a fused Fc-fragment. Such a fragment has an affinity toward protein G (expressed in *Streptococcus* sp., much like a protein A) or a carbohydrate moiety of Fc fragment has an affinity for boronate derivatives (Fig. 4b). Thus, a surface modified by a boronate derivatives or a protein G was effectively applied for oriented immobilization of a recombinant lectin via a fused Fc fragment [78]. Although boronate immobilization approach showed the highest sensitivity of detection, presence of boronate functionalities on the chip surface induced nonspecific interactions with glycoproteins and thus dextran blocking was introduced to minimize unwanted glycoprotein interactions. Additional drawback of such approach can be expected by introduction of Fc fragment having glycan entities, which can interact with glycan-binding proteins, which might be present in complex samples.

3.2 Perspectives for Peptide Lectin Aptamers

An alternative to production of mutated forms of lectins or glycosidases [7] for subsequent application in glycoprofiling in a future might be a preparation of novel forms of glycan-recognizing proteins termed here as peptide lectin aptamers (PLA). Such proteins will be peptide aptamers with a lectin-like affinity to recognize various forms of glycans. The term peptide aptamer was coined by Colas et al. in 1996 [79] and is defined as a combinatorial protein molecule having a variable peptide sequence, with an affinity for a given target protein, displayed on an inert, constant scaffold protein [80]. Construction of various bioanalytical devices such as PLA microarrays can benefit from such biorecognition elements since it would be possible to generate "army of terracotta soldiers" looking at the molecular level almost identical besides distinct "facial" feature of each entity provided by a unique peptide sequence. Even though lectin peptide aptamers have not been prepared yet, in our opinion, it is only a question of time, when such recognition elements will be prepared.

A beneficial feature of such a protein will be high solubility, a small, uniform size of a scaffold protein with an extended chemical and thermal stability, and a possibility to express such proteins in prokaryotic expression systems, which is a cost-effective process [81]. Moreover, when a small PLA will be immobilized on the surface of various bioanalytical devices, a higher density of biorecognition element can enhance sensitivity of detection, while suppressing nonspecific interactions and lowering background signals [80]. Peptide aptamers are produced by protein engineering from high-complexity combinatorial libraries with appropriate isolation/selection methods [80–82]. Thus, a need to have knowledge of the protein structure and the mechanism behind binding is not necessary. There are however some requirements for the scaffold protein to posses such as lack of a biological activity and ability to accommodate a wide range of peptides without changing a 3-D structure [83].

Currently there are over 50 proteins described as potential affinity scaffolds, but only quite a few of them have been applied for bioanalytical purposes [81]. Scaffold proteins were constructed from a diverse range of proteins differing in origin, size, structure, engineering protocols, mode of interaction, and applicability and typically have from 58 up to 166 amino acids (Fig. 5) [81, 84]. For example peptide aptamers based on a Stefin A protein (a cysteine protease inhibitor) are working well in an immobilized state on gold and modified gold surfaces with K_d of a peptide aptamer for its analyte down to nM range [85, 86]. Moreover, the scaffold based on Stefin A can accommodate and tolerate more than one peptide insert, which can dramatically widen practical application of such peptide aptamers [87]. It is possible that in case of lectin peptide aptamers a restricted range of amino acids enriched in four amino acids involved in glycan recognition could be possible, a concept which was successfully implemented for recognition of a maltose binding protein in the past [88].

3.3 Perspectives for Novel Lectin-Like Aptamers

The name aptamer is derived from the Latin expression "aptus" (to fit) and the Greek word "meros" (part) was coined in 1990 by Ellington and Szostak in order to introduce artificial RNA molecules binding to a small organic dye [89]. Aptamers are single-stranded oligonucleotides selectively binding small molecules, macromolecules, or whole cells, generally with a size of 15–60 nt (i.e., 5–20 kDa) (Fig. 6) [80]. Additional advantages of aptamers include relative simplicity of a chemical modification (introduction of a biotin or a fluorescent label), simple regeneration/reusability, and stability at a high temperature and/or at a high salt concentration [90]. Due to a small size of aptamers, they can be effectively attached to the interfaces with high densities, a feature important for construction of various robust and sensitive bioanalytical devices [91, 92]. Either DNA or RNA has to be chosen for preparation of aptamers keeping in mind a final application. For example RNA is structurally more flexible compared to DNA and thus, theoretically such aptamers can be raised against a wider range of analytes [81]. Contrary, a major limitation of using RNA is their susceptibility to chemical and/or enzymatic degradation. Furthermore, selection of RNA aptamers is a time-consuming process requiring additional enzymatic steps. Modifications of the

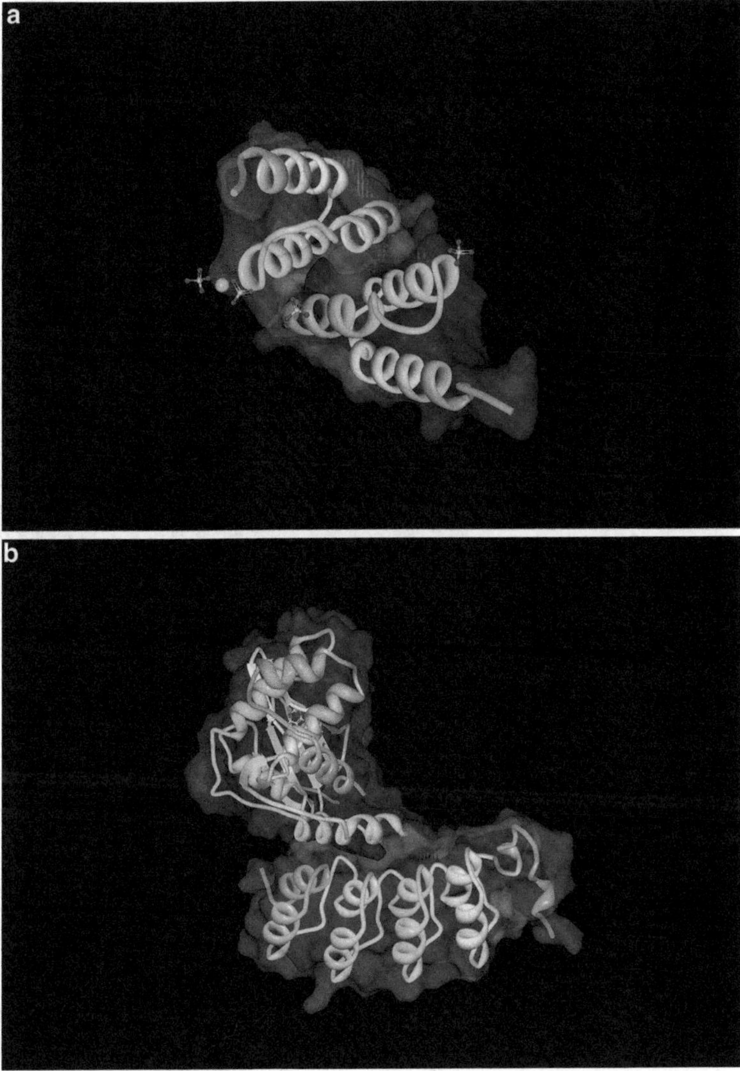

Fig. 5 Peptide aptamers based on an affibody (58 AA, PDB code 1LP1, on *left*) or a DARPin (166 AA, PDB code 2BKK, on *right*) scaffold in complex with its analyte. A peptide aptamer is in both cases at the bottom part of a figure, while its analyte is above the peptide aptamer

DNA or RNA backbone or introduction of modified nucleotides can produce aptamers more resistant to degradation [93].

When aptamers interacts with its analyte, usually a conformational change occurs creating a specific binding site for the target. Aptamers for proteins generally exhibit quite a high affinity in nM or sub-nM level due to presence of large complex areas with structures rich in hydrogen-bond donors and acceptors [80]. A relatively high affinity of aptamer makes such oligonucleotides an attractive alternative to lectins or antibodies as detection reagents for carbohydrate

Fig. 6 An RNA aptamer (*purple*) bound to its analyte peptide (*white-magenta* chain) (a PDB code 1EXY)

antigens. In order to increase palette of analytes being recognized by aptamers, modified nucleotides were introduced.

A rational approach for preparation of aptamers with a high affinity binding of glycoproteins by an extending library of nucleotides modified by incorporation of a boronic acid moiety was recently introduced by Wang's lab [94]. The study showed that affinity with K_d of 6–17 nM for fibrinogen using boronate modified DNA aptamers was higher compared to the affinity with K_d of 64–122 nM for the same analyte using DNA aptamers with natural pool of nucleotides. The fact that for the interaction between a glycoprotein and boronate modified DNA aptamer it is important an interaction between boronate moiety and glycan of fibrinogen was confirmed by analysis of a deglycosylated fibrinogen with a decreased affinity (K_d of 87–390 nM) [94].

An interesting approach for preparation of a wider library of nucleotides was recently introduced by incorporation of six new 5-position modified dUTP derivatives with five derivatives containing an aromatic ring [95]. DNA aptamers based on an extended pool of nucleotides were able to bind a necrosis factor receptor superfamily member 9 (TNFRSF9) with a high affinity of K_d = 4–6 nM for the first time. Interestingly two new derivatives containing either indole derivative or a benzene ring were the best TNFRSF9 binders. This fact is quite interesting since TNFRSF9 is a glycoprotein [96] and we can only speculate that these two aromatic derivatives of dUTP were involved in recognition of

TNFRSF9 via a glycan interaction. In a recent and similar study a derivative of imidazole (7-(2-thienyl)imidazo[4,5-*b*]pyridine, Ds) containing nucleotides was applied for generation of novel DNA aptamers binding to two glycoproteins vascular endothelial cell growth factor-165 (VEGF-165) and interferon-γ (IFN-γ) with an enhanced affinity [97]. The study revealed K_d down to 0.65 pM with DNA aptamers based on Ds nucleotides, while the best K_d of 57 pM for the DNA aptamers containing natural nucleotides was found for VEGF-165 [97]. Similarly, DNA aptamers based on Ds nucleotides offered much lower K_d of 0.038 nM compared to DNA aptamers based on natural nucleotides with K_d of 9.1 nM for IFN-γ [97]. Here we can again only speculate if the role of Ds nucleotides in enhanced affinity for two glycoproteins is in interaction of Ds modified nucleotides with the glycan moiety of glycoproteins.

3.4 Other Novel Forms of Lectins

There are several very interesting strategies as to how to enhance analytical applicability of lectins by their simple modifications, which can dramatically influence the field of glycoprofiling in a future.

The first study focused on application of multimers of eight different lectins prepared by incubation of biotinylated lectins with streptavidin. A wheat germ agglutinin (WGA) multimers integrated into lectin microarrays showed 4–40 times better sensitivity in analysis of glycans in human plasma and much better performance in glycoprofiling of samples from people having pancreatic cancer compared to utilization of WGA lectin monomer [98]. Authors of the study suggested that such lectin multimers with an enhanced affinity toward glycans can broaden the range of glycans, which can be detected. Moreover, according to authors lectin multimers might provide a fundamentally new biorecognition information not achievable by lectin monomers [98]. The second study described attachment of a boronate functionality to two different lectins in order to enhance affinity 2- to 60-fold for a particular glycan binding [99]. Such modified lectins were tested in a whole cell lysate with an excellent specificity for analysis of 295 N-linked glycopeptides. These results revealed that application of boronate modified lectins can facilitate identification of glycans present on the surface of low-abundant glycoproteins [99]. The third study indicated that by preparation of a lectin mutant with artificially introduced cysteine into lectin *Galanthus nivalis* agglutinin it was possible to prepare lectin dimmers via a disulfide linkage between two lectin mutants [100]. Agglutination activity of a lectin dimmer increased 16-fold compared to a lectin monomer and interestingly a transformation monomer/dimmer can be redox-switchable by addition of mild oxidation or reducing agents [100].

It can be summed up there are very exciting concepts already introduced in the field of lectin glycoengineering such as integration of deglycosylated forms of recombinant lectins into lectin microarrays with enhanced sensitivity of analysis and with lower detection limits achieved. It is only a question of time, when a wider range of recombinant lectins with purification/immobilization tags will be applied for oriented immobilization of lectins combined with various transducers or devices. Application of lectins modified by boronate derivatives or in a form of multimers/dimmers is a promising way for analysis of low abundant glycoproteins and possibly for analysis of glycoproteins, which cannot be detected by unmodified lectins.

We propose a future application of lectin peptide aptamers, which can even further enhance overall order of an immobilization process compared to immobilization of recombinant lectins with different tags for construction of various devices applicable in glycoprofiling. Although there is one rational approach for designing DNA aptamers with enhanced affinity toward glycoproteins by introduction of a boronate moiety into nucleotides there are two other reports focused on enhanced nucleotide alphabet created for generation of novel high affinity DNA aptamers against important targets. Interestingly in these two studies described especially glycoproteins were the main targets for such novel DNA aptamers based on an extended nucleotide alphabet, since such glycoproteins could not be recognized by "natural" DNA aptamers consisted of only natural nucleotides. Moreover, nucleotides in such novel DNA aptamers were modified mainly by aromatic amino acids, which are usually involved in glycan recognition by lectins. Thus, it is necessary to prove in the future if such modified nucleotides are really involved in glycan recognition.

4 A Glycocodon Hypothesis

A codon ($n_1n_2n_3$) is a sequence of three DNA or RNA nucleotides that corresponds to a specific amino acid or stop signal during protein synthesis, and the full set of codons is called a genetic code. The current state-of-the-art knowledge about the origin of the genetic code still remains as one of unsolved problems, and enormous number of theories can be divided to RNA world theories, protein world theories, co-evolution theories, and stereochemical theories. Integration of these theories leads us to the conclusion that a system of four codons ("gnc," n = a—adenine, g—guanine, c—cytosine, u—uracil) and four amino acids (G—glycine, A—alanine, V—valine, D—aspartic acid) could be the original genetic code [101–105]. Research on a selection of particular RNA sequences with an amino acid binding activity, and a relation of those activities to the genetic code, have revealed an evidence that there is a highly robust connection between the genetic code and RNA-amino acid binding affinity [106]. It seems that the main part of the genetic code is influenced by a stereochemical prebiotic selection during the first polymerization of G, A, V, and D amino acids and g, c, u, and a nucleic acids; however, only first two nucleotides of codons (n_1n_2) are directly related to amino acid stereoselectivity [107].

Recently, a similar evolution process was proposed for "the glycocode" [108]. The bioinformatics quantification of "GAVD-dipeptides" in monosaccharide-specific proteins revealed that the amino acid triplets, the glycocodons ($aa_1aa_2aa_3$), can be deduced for each glycan letter (monosaccharide). The glycocodons are composed from one polar amino acid, interacting with sugar –OH groups, and one specific dipeptide, usually detecting C–C hydrophobic patch (see Glc, Gal, and GlcNAc binding, Fig. 7a, c). Figure 7 depicts a quantification spectra of "GAVD-dipeptides" in glucose (Glc), galactose (Gal), mannose (Man), fucose (Fuc), *N*-acetylglucosamine (GlcNAc), and *N*-acetylgalactosamine (GalNAc) specific proteins. In the case of Glc, Gal, and GalNAc the maximal values of incidence of "GAVD-dipeptides" were taken for coding; AA for Glc, GA plus AG for Gal, and DD plus VV for GalNAc. In the case of Man and GlcNAc GD plus DG and GV plus VG from "GAVD-dipeptide pool" were taken for coding, because maximal values have been already taken by previous monosaccharides. During evolution, the GAVD-glycocodons were transformed to novel glycocodons by a positive selection for the increased diversity and functionality of a "sugar–protein language" that can be made with a larger amino acid alphabet. Nevertheless, evolution process holds hydropathic similarity; amino acids in the glycocodons are substituted by amino acids with similar polar properties, which minimizes errors in established sugar–protein interactions. The bioinformatics quantification of dipeptides composed from all 20 amino acids revealed that GA plus AG for Gal were substituted mainly with SW and WS, AA for Glc can be substituted with MF, GD plus DG for Man can be substituted with AY plus YA, GV plus VG for GlcNAc can be substituted with SF plus FS, and DD plus VV for GalNAc can be substituted with QD plus LF. AV plus VA from a "GAVD-dipeptide pool" were selected for NAc-group sensing, in the case of GalNAc, they were transformed to MS plus SM and IT plus TI dipeptides. Figure 7d shows how the GalNAc glycocodons are used during N-glycosylation by bacterial oligosaccharyltransferase (*Campylobacter lari*, PGlB). PGlB accepts different oligosaccharides from a lipid carrier requiring an acetamido group at the C2 carbon of the first monosaccharide (GalNAc is the best), or even a "monosaccharide" *N*-acetylgalactosamine-diphosphoundecaprenyl is a good substrate [109, 110]. PGlB connects the C1 carbon in the first saccharide moiety (*N*-acetylgalactosamine) with the amide nitrogen of the acceptor (sequon) asparagine. DQNATF peptide has been recognized as an optimal acceptor sequence for PGlB [111]. According to the glycocodon theory, DQ dipeptide plus the acceptor asparagine makes the glycocodon for GalNAc. The next ATF sequence of the sequon peptide is inserted into the catalytic center in such a way that the KTI and HLF glycocodons are formed. This process shows a basic difference

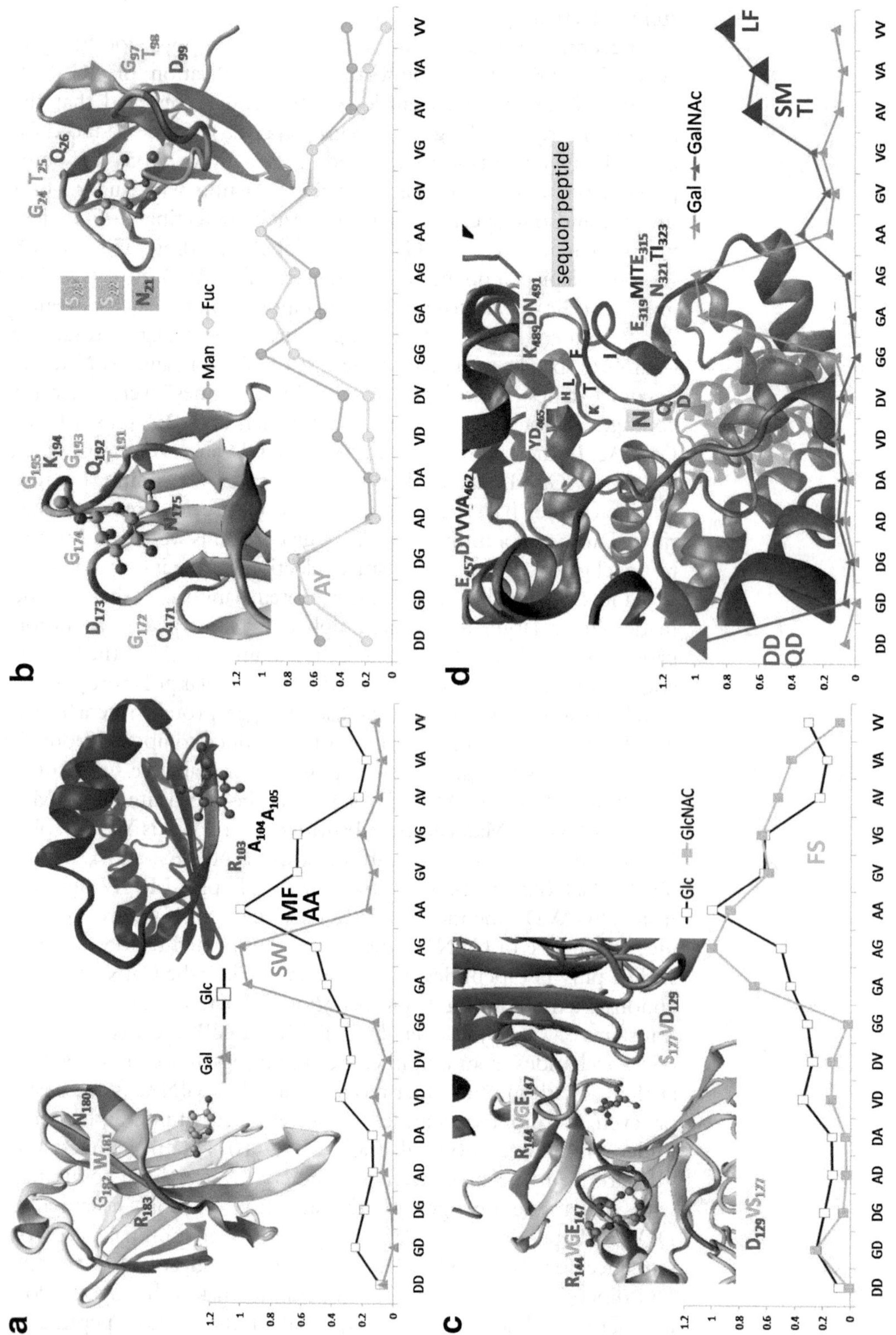
a
b
c
d
Gal
Glc
SW
MF
AA
Man
Fuc
AY
GlcNAC
FS
GalNAc
sequon peptide
DD
QD
SM
TI
LF
DD GD DG AD DA VD DV GG GA AG AA GV VG AV VA VV
1.2
1
0.8
0.6
0.4
0.2
0

Fig. 7 A glycocodon theory—G (glycine), A (alanine), V (valine), and D (aspartic acid) are elementary amino acids and the first primordial interactions between GAVD-peptides and sugars were evolutionary conserved and used in the glycocodons; the full set of glycocodons is proposed to call—the glycocode [108]. (**a**) A distribution of GAVD-dipeptides in galactose (Gal) and glucose (Glc) specific proteins [108]. In human Galectin-3 (a yellow protein structure, 1KJL), Gal is sensed by two overlapping glycocodons—NWG plus WGR that are derived from ancient GA—AG specific dipeptides; in perchloric acid-soluble protein from Pseudomonas syringae (a blue protein structure, 3K0T), Glc is sensed by the RAA glycocodon—ancient AA specific dipeptide can be today transformed to MF. (**b**) A distribution of GAVD-dipeptides in mannose (Man) and fucose (Fuc) specific proteins [108]. In bacteriocin from *Pseudomonas* sp. complexed with Met-mannose (a cyan protein structure, 3M7J), Man is sensed by two overlapping glycocodons—QGD plus DGN—ancient GD—DG specific dipeptides can be transformed today to AY—YA dipeptides; in a PA-IIL lectin from *P. aeruginosa* (an orange protein structure, 2JDK), Fuc is sensed by the specific glycocodon NSS and by two GTQ and GTD glycocodons derived from ancient GA dipeptide specific for Gal (Fuc is actually 6-deoxy-L-galactose). (**c**) A distribution of GAVD-dipeptides in *N*-acetylglucosamine (GlcNAc) and glucose (Glc) specific proteins [108]. In human L-ficolin (an assembly of two monomers is shown, 2J3O), GlcNAc is sensed by two overlapping glycocodons—RVG plus VGE—ancient VG specific dipeptide can be transformed today to FS dipeptide. (**d**) A distribution of GAVD-dipeptides in *N*-acetylgalactosamine (GalNAc) and galactose (Gal) specific proteins [108]. In oligosaccharyltransferase from Campylobacter lari (3RCE), GalNAc (a first moiety of the oligosaccharide) is sensed by two overlapping glycocodons—EMI plus ITE derived from ancient VV and VA dipeptides specific for GalNAc. GalNAc is transferred to asparagines of the DQN glycocodon—DQ dipeptide is derived from ancient DD dipeptide specific for GalNAc

between the codons and the glycocodons. When the codons are read from the nucleotide sequence, they are read in succession and do not overlap with one another. On the contrary, the glycocodons are used in a way of a "key-lock" principle—three different protein chains can make the glycocodon in 3D space and the glycocodons frequently overlap (Fig. 7).

It should be emphasized that the glycocodons were theoretically deduced by a bioinformatics study and it will be necessary to perform a study in the laboratory to establish the strongest correlation between the monosaccharides and the glycocodons and to determine the shortest peptides for the recognition of the specific monosaccharide. However, the glycocodon theory represents a tool showing how the PLA or novel DNA aptamers based on nucleotide derivatives should be organized and programmed.

5 Conclusions

This chapter described various tools, which have been recently applied in order to extend analytical usefulness of lectin-based devices in glycoprofiling. A positive aspect of recent effort in the field is utilization of a great potential nanotechnology can bring into quite a complex and challenging analysis of glycans. Such approaches proved analysis of glycans by different biosensors can be extremely sensitive with a concentration range spanning few orders of magnitude, which are features essential in analysis of low-abundant glycoproteins. Control of immobilization of lectins is other important issue, which was successfully addressed in a pilot study showing that attachment of recombinant lectins on surfaces via various tags present in recombinant lectin improved sensitivity of glycan analysis. The book chapter described also future prospect of PLA in order to increase sensitivity and stability of analysis, while suppressing nonspecific interactions. Additional issue to focus on in the future is investigation if modified nucleotides can be successfully applied in preparation of novel DNA aptamers targeting glycoproteins with high affinity and selectivity. The final part of the chapter describes the concept of a glycocodon, which can lead to a better understanding of glycan–lectin interaction and for design of novel lectins with unknown specificities and/or better affinities toward glycan target or for rational design of PLA or DNA aptamers.

Acknowledgement

The financial support from the Slovak research and development agency APVV 0282-11, from VEGA 2/0127/10 and 2/0162/14 is acknowledged. The research leading to these results has received partly funding from the European Research Council under the

European Union's Seventh Framework Programme (FP/2007-2013)/ERC Grant Agreement n. 311532, from the European Union's Seventh Framework Programme for research, technological development and demonstration under Grant agreement n. 317420 and from Qatar Foundation under Project n. 6-381-1-078. This contribution was partly supported by the project: Centre of excellence for white-green biotechnology, ITMS 26220120054, supported by the Research & Development Operational Programme funded by the ERDF.

References

1. Schena M, Shalon D, Davis RW, Brown PO (1995) Quantitative monitoring of gene expression patterns with a complementary DNA microarray. Science 270:467–470
2. Eisen MB, Spellman PT, Brown PO, Botstein D (1998) Cluster analysis and display of genome-wide expression patterns. Proc Natl Acad Sci 95:14863–14868
3. Alizadeh AA, Eisen MB, Davis RE, Ma C, Lossos IS, Rosenwald A, Boldrick JC, Sabet H, Tran T, Yu X, Powell JI, Yang L, Marti GE, Moore T, Hudson J, Lu L, Lewis DB, Tibshirani R, Sherlock G, Chan WC, Greiner TC, Weisenburger DD, Armitage JO, Warnke R, Levy R, Wilson W, Grever MR, Byrd JC, Botstein D, Brown PO, Staudt LM (2000) Distinct types of diffuse large b-cell lymphoma identified by gene expression profiling. Nature 403:503–511
4. Gygi SP, Rochon Y, Franza BR, Aebersold R (1999) Correlation between protein and mRNA abundance in yeast. Mol Cell Biol 19:1720–1730
5. Gry M, Rimini R, Stromberg S, Asplund A, Ponten F, Uhlen M, Nilsson P (2009) Correlations between RNA and protein expression profiles in 23 human cell lines. BMC Genomics 10:365
6. Lee J-R, Magee DM, Gaster RS, LaBaer J, Wang SX (2013) Emerging protein array technologies for proteomics. Expert Rev Proteome 10:65–75
7. Arnaud J, Audfray A, Imberty A (2013) Binding sugars: from natural lectins to synthetic receptors and engineered neolectins. Chem Soc Rev 42:4798–4813
8. Baker JL, Çelik E, DeLisa MP (2013) Expanding the glycoengineering toolbox: the rise of bacterial *N*-linked protein glycosylation. Trends Biotechnol 31:313–323
9. Bertók T, Katrlík J, Gemeiner P, Tkac J (2013) Electrochemical lectin based biosensors as a label-free tool in glycomics. Microchim Acta 180:1–13
10. Varki A et al (2009) Essentials of glycobiology, 2nd edn. Cold Spring Harbor Laboratory Press, Cold Spring Harbor, NY
11. Gemeiner P, Mislovicová D, Tkác J, Svitel J, Pätoprsty V, Hrabárová E, Kogan G, Kozár T (2009) Lectinomics II: a highway to biomedical/clinical diagnostics. Biotechnol Adv 27:1–15
12. Katrlík J, Švitel J, Gemeiner P, Kožár T, Tkac J (2010) Glycan and lectin microarrays for glycomics and medicinal applications. Med Res Rev 30:394–418
13. Krishnamoorthy L, Bess JW, Preston AB, Nagashima K, Mahal LK (2009) HIV-1 and microvesicles from T cells share a common glycome, arguing for a common origin. Nat Chem Biol 5:244–250
14. Schauer R, Kamerling JP (2011) The chemistry and biology of trypanosomal trans-sialidases: virulence factors in chagas disease and sleeping sickness. ChemBioChem 12:2246–2264
15. Song X, Lasanajak Y, Xia B, Heimburg-Molinaro J, Rhea JM, Ju H, Zhao C, Molinaro RJ, Cummings RD, Smith DF (2011) Shotgun glycomics: a microarray strategy for functional glycomics. Nat Methods 8:85–90
16. Vaishnava S, Yamamoto M, Severson KM, Ruhn KA, Yu X, Koren O, Ley R, Wakeland EK, Hooper LV (2011) The antibacterial lectin regIIIγ promotes the spatial segregation of microbiota and host in the intestine. Science 334:255–258
17. Burton DR, Poignard P, Stanfield RL, Wilson IA (2012) Broadly neutralizing antibodies present new prospects to counter highly antigenically diverse viruses. Science 337: 183–186
18. Pejchal R, Doores KJ, Walker LM, Khayat R, Huang P-S, Wang S-K, Stanfield RL, Julien J-P, Ramos A, Crispin M, Depetris R, Katpally U, Marozsan A, Cupo A, Maloveste S, Liu Y, McBride R, Ito Y, Sanders RW, Ogohara C, Paulson JC, Feizi T, Scanlan CN, Wong C-H, Moore JP, Olson WC, Ward AB, Poignard P,

Schief WR, Burton DR, Wilson IA (2011) A potent and broad neutralizing antibody recognizes and penetrates the HIV glycan shield. Science 334:1097–1103

19. Kim J-H, Resende R, Wennekes T, Chen H-M, Bance N, Buchini S, Watts AG, Pilling P, Streltsov VA, Petric M, Liggins R, Barrett S, McKimm-Breschkin JL, Niikura M, Withers SG (2013) Mechanism-based covalent neuraminidase inhibitors with broad-spectrum influenza antiviral activity. Science 340:71–75
20. Klein F, Halper-Stromberg A, Horwitz JA, Gruell H, Scheid JF, Bournazos S, Mouquet H, Spatz LA, Diskin R, Abadir A, Zang T, Dorner M, Billerbeck E, Labitt RN, Gaebler C, Marcovecchio PM, Incesu R-B, Eisenreich TR, Bieniasz PD, Seaman MS, Bjorkman PJ, Ravetch JV, Ploss A, Nussenzweig MC (2012) HIV therapy by a combination of broadly neutralizing antibodies in humanized mice. Nature 492:118–122
21. Chandler KB, Goldman R (2013) Glycoprotein disease markers and single protein-omics. Mol Cell Proteomics 12:836–845
22. Ferens-Sieczkowska M, Kowalska B, Kratz EM (2013) Seminal plasma glycoproteins in male infertility and prostate diseases: is there a chance for glyco-biomarkers? Biomarkers 18:10–22
23. Gilgunn S, Conroy PJ, Saldova R, Rudd PM, O'Kennedy RJ (2013) Aberrant PSA glycosylation – a sweet predictor of prostate cancer. Nat Rev Urol 10:99–107
24. Schmaltz RM, Hanson SR, Wong C-H (2011) Enzymes in the synthesis of glycoconjugates. Chem Rev 111:4259–4307
25. van Bueren JJL, Rispens T, Verploegen S, van der Palen-Merkus T, Stapel S, Workman LJ, James H, van Berkel PHC, van de Winkel JGJ, Platts-Mills TAE, Parren PWHI (2011) Anti-galactose-[alpha]-1,3-galactose ige from allergic patients does not bind [alpha]-galactosylated glycans on intact therapeutic antibody fc domains. Nat Biotechnol 29:574–576
26. Beck A, Reichert JM (2012) Marketing approval of mogamulizumab: a triumph for glyco-engineering. MAbs 4:419–425
27. Raman R, Raguram S, Venkataraman G, Paulson JC, Sasisekharan R (2005) Glycomics: an integrated systems approach to structure–function relationships of glycans. Nat Methods 2:817–824
28. Gabius H-J, André S, Jiménez-Barbero J, Romero A, Solís D (2011) From lectin structure to functional glycomics: principles of the sugar code. Trends Biochem Sci 36:298–313
29. Cummings RD (2009) The repertoire of glycan determinants in the human glycome. Mol Biosyst 5:1087–1104
30. Reuel NF, Mu B, Zhang J, Hinckley A, Strano MS (2012) Nanoengineered glycan sensors enabling native glycoprofiling for medicinal applications: towards profiling glycoproteins without labeling or liberation steps. Chem Soc Rev 41:5744–5779
31. Bertozzi CR, Kiessling LL (2001) Chemical glycobiology. Science 291:2357–2364
32. Furukawa JI, Fujitani N, Shinohara Y (2013) Recent advances in cellular glycomic analyses. Biomolecules 3:198–225
33. Rakus JF, Mahal LK (2011) New technologies for glycomic analysis: toward a systematic understanding of the glycome. Annu Rev Anal Chem 4:367–392
34. Smith DF, Cummings RD (2013) Application of microarrays to deciphering the structure and function of the human glycome. Mol Cell Proteomics 12:902–912
35. Alley WR, Mann BF, Novotny MV (2013) High-sensitivity analytical approaches for the structural characterization of glycoproteins. Chem Rev 113:2668–2732
36. Lazar IM, Lee W, Lazar AC (2013) Glycoproteomics on the rise: established methods, advanced techniques, sophisticated biological applications. Electrophoresis 34:113–125
37. Novotny M, Alley W Jr, Mann B (2013) Analytical glycobiology at high sensitivity: current approaches and directions. Glycoconj J 30:89–117
38. Oliveira C, Teixeira JA, Domingues L (2013) Recombinant lectins: an array of tailor-made glycan-interaction biosynthetic tools. Crit Rev Biotechnol 33:66–80
39. Murphy P, André S, Gabius H-J (2013) The third dimension of reading the sugar code by lectins: design of glycoclusters with cyclic scaffolds as tools with the aim to define correlations between spatial presentation and activity. Molecules 18:4026–4053
40. Mislovičová D, Katrlík J, Paulovičová E, Gemeiner P, Tkac J (2012) Comparison of three distinct ella protocols for determination of apparent affinity constants between con a and glycoproteins. Colloids Surf B Biointerfaces 94:163–169
41. Mislovičová D, Gemeiner P, Kozarova A, Kožár T (2009) Lectinomics i. Relevance of exogenous plant lectins in biomedical diagnostics. Biologia 64:1–19
42. Hirabayashi J, Yamada M, Kuno A, Tateno H (2013) Lectin microarrays: concept, principle and applications. Chem Soc Rev 42: 4443–4458
43. Krishnamoorthy L, Mahal LK (2009) Glycomic analysis: an array of technologies. ACS Chem Biol 4:715–732

44. Lis H, Sharon N (1998) Lectins: carbohydrate-specific proteins that mediate cellular recognition. Chem Rev 98:637–674
45. Cunningham S, Gerlach JQ, Kane M, Joshi L (2010) Glyco-biosensors: recent advances and applications for the detection of free and bound carbohydrates. Analyst 135:2471–2480
46. Gerlach JQ, Cunningham S, Kane M, Joshi L (2010) Glycobiomimics and glycobiosensors. Biochem Soc Trans 38:1333–1336
47. Reuel NF, Ahn J-H, Kim J-H, Zhang J, Boghossian AA, Mahal LK, Strano MS (2011) Transduction of glycan–lectin binding using near-infrared fluorescent single-walled carbon nanotubes for glycan profiling. J Am Chem Soc 133:17923–17933
48. Sanchez-Pomales G, Zangmeister RA (2011) Recent advances in electrochemical glycobiosensing. Int J Electrochem 2011
49. Tkac J, Davis JJ (2009) Label-free field effect protein sensing. In: Davis JJ (ed) Engineering the bioelectronic interface: applications to analyte biosensing and protein detection. Royal Society of Chemistry, Cambridge, pp 193–224, doi:10.1039/9781847559777
50. Reichardt NC, Martin-Lomas M, Penades S (2013) Glyconanotechnology. Chem Soc Rev 42:4358–4376
51. Zeng X, Andrade CAS, Oliveira MDL, Sun X-L (2012) Carbohydrate–protein interactions and their biosensing applications. Anal Bioanal Chem 402:3161–3176
52. Gruber K, Horlacher T, Castelli R, Mader A, Seeberger PH, Hermann BA (2011) Cantilever array sensors detect specific carbohydrate-protein interactions with picomolar sensitivity. ACS Nano 5:3670–3678
53. Mader A, Gruber K, Castelli R, Hermann BA, Seeberger PH, Radler JO, Leisner M (2012) Discrimination of escherichia coli strains using glycan cantilever array sensors. Nano Lett 12:420–423
54. Jelinek R, Kolusheva S (2004) Carbohydrate biosensors. Chem Rev 104:5987–6015
55. Luo X, Davis JJ (2013) Electrical biosensors and the label free detection of protein disease biomarkers. Chem Soc Rev 42:5944–5962
56. La Belle JT, Gerlach JQ, Svarovsky S, Joshi L (2007) Label-free impedimetric detection of glycan–lectin interactions. Anal Chem 79:6959–6964
57. Oliveira MDL, Correia MTS, Coelho LCBB, Diniz FB (2008) Electrochemical evaluation of lectin–sugar interaction on gold electrode modified with colloidal gold and polyvinyl butyral. Colloids Surf B Biointerfaces 66:13–19
58. Nagaraj VJ, Aithal S, Eaton S, Bothara M, Wiktor P, Prasad S (2010) Nanomonitor: a miniature electronic biosensor for glycan biomarker detection. Nanomedicine 5:369–378
59. Bertok T, Gemeiner P, Mikula M, Gemeiner P, Tkac J (2013) Ultrasensitive impedimetric lectin based biosensor for glycoproteins containing sialic acid. Microchim Acta 180:151–159
60. Bertok T, Klukova L, Sediva A, Kasák P, Semak V, Micusik M, Omastova M, Chovanová L, Vlček M, Imrich R, Vikartovska A, Tkac J (2013) Ultrasensitive impedimetric lectin biosensors with efficient antifouling properties applied in glycoprofiling of human serum samples. Anal Chem 85:7324–7332
61. Bertok T, Sediva A, Katrlik J, Gemeiner P, Mikula M, Nosko M, Tkac J (2013) Label-free detection of glycoproteins by the lectin biosensor down to attomolar level using gold nanoparticles. Talanta 108:11–18
62. Yang H, Li Z, Wei X, Huang R, Qi H, Gao Q, Li C, Zhang C (2013) Detection and discrimination of alpha-fetoprotein with a label-free electrochemical impedance spectroscopy biosensor array based on lectin functionalized carbon nanotubes. Talanta 111:62–68
63. Vedala H, Chen Y, Cecioni S, Imberty A, Vidal S, Star A (2011) Nanoelectronic detection of lectin-carbohydrate interactions using carbon nanotubes. Nano Lett 11:170–175
64. Chen YN, Vedala H, Kotchey GP, Audfray A, Cecioni S, Imberty A, Vidal S, Star A (2012) Electronic detection of lectins using carbohydrate-functionalized nanostructures: graphene versus carbon nanotubes. ACS Nano 6:760–770
65. Zhang GJ, Huang MJ, Ang JJ, Yao QF, Ning Y (2013) Label-free detection of carbohydrate-protein interactions using nanoscale field-effect transistor biosensors. Anal Chem 85: 4392–4397
66. Huang Y-W, Wu C-S, Chuang C-K, Pang S-T, Pan T-M, Yang Y-S, Ko F-H (2013) Real-time and label-free detection of the prostate-specific antigen in human serum by a polycrystalline silicon nanowire field-effect transistor biosensor. Anal Chem 85:7912–7918
67. Ali M, Nasir S, Ramirez P, Cervera J, Mafe S, Ensinger W (2013) Carbohydrate-mediated biomolecular recognition and gating of synthetic ion channels. J Phys Chem C 117:18234–18242
68. Kruss S, Hilmer AJ, Zhang J, Reuel NF, Mu B, Strano MS (2013) Carbon nanotubes as optical biomedical sensors. Adv Drug Deliv Rev. doi:10.1016/j.addr.2013.07.015
69. Reuel NF, Grassbaugh B, Kruss S, Mundy JZ, Opel C, Ogunniyi AO, Egodage K, Wahl R, Helk B, Zhang J, Kalcioglu ZI, Tvrdy K, Bellisario DO, Mu B, Blake SS, Van Vliet KJ, Love JC, Wittrup KD, Strano MS (2013) Emergent properties of nanosensor arrays: applications for monitoring igg affinity distributions, weakly affined hypermannosylation,

and colony selection for biomanufacturing. ACS Nano 7:7472–7482

70. Bellapadrona G, Tesler AB, Grunstein D, Hossain LH, Kikkeri R, Seeberger PH, Vaskevich A, Rubinstein I (2012) Optimization of localized surface plasmon resonance transducers for studying carbohydrate-protein interactions. Anal Chem 84:232–240
71. Jin S, Cheng Y, Reid S, Li M, Wang B (2010) Carbohydrate recognition by boronolectins, small molecules, and lectins. Med Res Rev 30:171–257
72. Streicher H, Sharon N (2003) Recombinant plant lectins and their mutants. Methods Enzymol 363:47–77, In: Yuan CL, Reiko TL (eds) doi:10.1016/S0076-6879(03)01043-7
73. Geisler C, Jarvis DL (2011) Letter to the glyco-forum: effective glycoanalysis with maackia amurensis lectins requires a clear understanding of their binding specificities. Glycobiology 21:988–993
74. Alava T, Mann JA, Théodore C, Benitez JJ, Dichtel WR, Parpia JM, Craighead HG (2013) Control of the graphene–protein interface is required to preserve adsorbed protein function. Anal Chem 85:2754–2759
75. Hsu K-L, Gildersleeve JC, Mahal LK (2008) A simple strategy for the creation of a recombinant lectin microarray. Mol Biosyst 4: 654–662
76. Propheter DC, Hsu K-L, Mahal LK (2010) Fabrication of an oriented lectin microarray. ChemBioChem 11:1203–1207
77. Propheter DC, Mahal LK (2011) Orientation of gst-tagged lectins via in situ surface modification to create an expanded lectin microarray for glycomic analysis. Mol Biosyst 7: 2114–2117
78. Chen M-L, Adak AK, Yeh N-C, Yang W-B, Chuang Y-J, Wong C-H, Hwang K-C, Hwu J-RR, Hsieh S-L, Lin C-C (2008) Fabrication of an oriented fc-fused lectin microarray through boronate formation. Angew Chem Int Ed 47:8627–8630
79. Colas P, Cohen B, Jessen T, Grishina I, McCoy J, Brent R (1996) Genetic selection of peptide aptamers that recognize and inhibit cyclin-dependent kinase 2. Nature 380: 548–550
80. Mascini M, Palchetti I, Tombelli S (2012) Nucleic acid and peptide aptamers: fundamentals and bioanalytical aspects. Angew Chem Int Ed 51:1316–1332
81. Ruigrok VJB, Levisson M, Eppink MHM, Smidt H, van der Oost J (2011) Alternative affinity tools: more attractive than antibodies? Biochem J 436:1–13
82. Ståhl S, Kronqvist N, Jonsson A, Löfblom J (2013) Affinity proteins and their generation. J Chem Technol Biotechnol 88:25–38
83. Woodman R, Yeh JTH, Laurenson S, Ferrigno PK (2005) Design and validation of a neutral protein scaffold for the presentation of peptide aptamers. J Mol Biol 352:1118–1133
84. Gebauer M, Skerra A (2009) Engineered protein scaffolds as next-generation antibody therapeutics. Curr Opin Chem Biol 13:245–255
85. Davis JJ, Tkac J, Humphreys R, Buxton AT, Lee TA, Ko Ferrigno P (2009) Peptide aptamers in label-free protein detection: 2. Chemical optimization and detection of distinct protein isoforms. Anal Chem 81: 3314–3320
86. Davis JJ, Tkac J, Laurenson S, Ferrigno PK (2007) Peptide aptamers in label-free protein detection: 1. Characterization of the immobilized scaffold. Anal Chem 79:1089–1096
87. Stadler LKJ, Hoffmann T, Tomlinson DC, Song QF, Lee T, Busby M, Nyathi Y, Gendra E, Tiede C, Flanagan K, Cockell SJ, Wipat A, Harwood C, Wagner SD, Knowles MA, Davis JJ, Keegan N, Ferrigno PK (2011) Structurefunction studies of an engineered scaffold protein derived from stefin A. II: development and applications of the sqt variant. Protein Eng Des Sel 24:751–763
88. Koide A, Gilbreth RN, Esaki K, Tereshko V, Koide S (2007) High-affinity single-domain binding proteins with a binary-code interface. Proc Natl Acad Sci 104:6632–6637
89. Ellington AD, Szostak JW (1990) In vitro selection of rna molecules that bind specific ligands. Nature 346:818–822
90. Keefe AD, Pai S, Ellington A (2010) Aptamers as therapeutics. Nat Rev Drug Discov 9: 537–550
91. Iliuk AB, Hu L, Tao WA (2011) Aptamer in bioanalytical applications. Anal Chem 83: 4440–4452
92. Liu J, Cao Z, Lu Y (2009) Functional nucleic acid sensors. Chem Rev 109:1948–1998
93. Tolle F, Mayer G (2013) Dressed for success – applying chemistry to modulate aptamer functionality. Chem Sci 4:60–67
94. Li M, Lin N, Huang Z, Du L, Altier C, Fang H, Wang B (2008) Selecting aptamers for a glycoprotein through the incorporation of the boronic acid moiety. J Am Chem Soc 130:12636–12638
95. Vaught JD, Bock C, Carter J, Fitzwater T, Otis M, Schneider D, Rolando J, Waugh S, Wilcox SK, Eaton BE (2010) Expanding the chemistry of DNA for in vitro selection. J Am Chem Soc 132:4141–4151
96. Eckstrum K, Bany B (2011) Tumor necrosis factor receptor subfamily 9 (tnfrsf9) gene is expressed in distinct cell populations in mouse uterus and conceptus during implantation period of pregnancy. Cell Tissue Res 344:567–576

97. Kimoto M, Yamashige R, Matsunaga K-I, Yokoyama S, Hirao I (2013) Generation of high-affinity DNA aptamers using an expanded genetic alphabet. Nat Biotechnol 31:453–457
98. Cao Z, Partyka K, McDonald M, Brouhard E, Hincapie M, Brand RE, Hancock WS, Haab BB (2013) Modulation of glycan detection on specific glycoproteins by lectin multimerization. Anal Chem 85:1689–1698
99. Lu Y-W, Chien C-W, Lin P-C, Huang L-D, Chen C-Y, Wu S-W, Han C-L, Khoo K-H, Lin C-C, Chen Y-J (2013) Bad-lectins: boronic acid-decorated lectins with enhanced binding affinity for the selective enrichment of glycoproteins. Anal Chem 85:8268–8276
100. McDonald RE, Hughes DJ, Davis BG (2004) Modular control of lectin function: redox-switchable agglutination. Angew Chem Int Ed 43:3025–3029
101. Eigen M, Schuster P (1978) The hypercycle. Naturwissenschaften 65:341–369
102. Ikehara K, Omori Y, Arai R, Hirose A (2002) A novel theory on the origin of the genetic code: a gnc-sns hypothesis. J Mol Evol 54:530–538
103. Ikehara K (2005) Possible steps to the emergence of life: the [GADV]-protein world hypothesis. Chem Rec 5:107–118
104. Di Giulio M (2008) An extension of the coevolution theory of the origin of the genetic code. Biol Direct 3:1–21
105. Lehmann J, Cibils M, Libchaber A (2009) Emergence of a code in the polymerization of amino acids along RNA templates. PLOS ONE 4
106. Yarus M, Widmann J, Knight R (2009) RNA–amino acid binding: a stereochemical era for the genetic code. J Mol Evol 69:406–429
107. Nahalka J (2011) Quantification of peptide bond types in human proteome indicates how DNA codons were assembled at prebiotic conditions. J Proteome Bioinf 4:153–159
108. Nahalka J (2012) Glycocodon theory—the first table of glycocodons. J Theor Biol 307:193–204
109. Wacker M, Feldman MF, Callewaert N, Kowarik M, Clarke BR, Pohl NL, Hernandez M, Vines ED, Valvano MA, Whitfield C, Aebi M (2006) Substrate specificity of bacterial oligosaccharyltransferase suggests a common transfer mechanism for the bacterial and eukaryotic systems. Proc Natl Acad Sci 103:7088–7093
110. Li L, Woodward R, Ding Y, Liu X-W, Yi W, Bhatt VS, Chen M, Zhang L-W, Wang PG (2010) Overexpression and topology of bacterial oligosaccharyltransferase pglb. Biochem Biophys Res Commun 394:1069–1074
111. Schwarz F, Huang W, Li C, Schulz BL, Lizak C, Palumbo A, Numao S, Neri D, Aebi M, Wang L-X (2010) A combined method for producing homogeneous glycoproteins with eukaryotic n-glycosylation. Nat Chem Biol 6:264–266

Chapter 38

Molecular Basis of a Pandemic of Avian-Type Influenza Virus

Nongluk Sriwilaijaroen and Yasuo Suzuki

Abstract

Despite heroic efforts to prevent the emergence of an influenza pandemic, avian influenza A virus has prevailed by crossing the species barriers to infect humans worldwide, occasionally with morbidity and mortality at unprecedented levels, and the virus later usually continues circulation in humans as a seasonal influenza virus, resulting in health-social-economic problems each year. Here, we review current knowledge of influenza viruses, their life cycle, interspecies transmission, and past pandemics and discuss the molecular basis of pandemic acquisition, notably of hemagglutinin (lectin) acting as a key contributor to change in host specificity in viral infection.

Key words Influenza, Replication, Transmission, Host range, Pandemic, Hemagglutinin, Sialylglycoconjugate

1 Introduction

An influenza pandemic is grim as it is unpredictable, rapidly spreads throughout the world, and is mostly associated with severe clinical disease and death in humans, leading to the serious socioeconomic problems. Influenza is an infectious respiratory illness epidemically caused by human influenza A, B, and C viruses (classified on the basis of serologic responses to matrix proteins and nucleoproteins), which are single-stranded, negative-sense RNA viruses of the family *Orthomyxoviridae* [1]. Type A viruses have the greatest genetic diversity, harboring numerous antigenically distinct subtypes of the two main viral surface glycoproteins; so far, 18 hemagglutinin (HA) and 11 neuraminidase (NA) subtypes have been identified. All possible combined 16×9 subtypes, except for H17–18 and N10–11 subtypes, which are detected only in bats (mammals) and exhibit functions completely different from those of the other subtypes [2–4], are maintained in waterfowls, mainly in wild ducks. These avian viruses are occasionally transmitted to other species that are immunologically naïve, in which they may only cause

Jun Hirabayashi (ed.), *Lectins: Methods and Protocols*, Methods in Molecular Biology, vol. 1200,
DOI 10.1007/978-1-4939-1292-6_38, © Springer Science+Business Media New York 2014

outbreaks or may acquire mutations so as to be efficiently transmitted between new hosts, and they can lead to a pandemic in human populations. An influenza pandemic has continued to circulate as an epidemic with an antigenic variant each year, and two human influenza A subtypes, H3N2 Hong Kong/68 and H1N1 2009 variants, have been seasonally found among humans.

The unexpectedly rapid emergence of A/H1N1 2009 swine pandemic recently, ongoing outbreaks of avian influenza viruses in humans, circulating avian/human/swine influenza viruses in pigs, and the emergence of influenza A viruses that are resistant to currently available anti-influenza virus drugs have raised a great concern that a pandemic could spread rapidly without time to prepare a public health response to stop the illness spread and could threaten human health and life throughout the world, becoming a major impediment to socioeconomic development. We have collected available information in the influenza A virus field, especially factors playing important roles in determining viral transmission, in order to know how best to perform surveillance, prevent, slow, or limit a future pandemic.

2 Influenza A Virus Infection and Replication

Influenza A virus contains eight (–) ssRNA genomic segments that encode at least ten proteins; nine are structural proteins and 1–4 depending on the virus strain and host species are nonstructural proteins (*see* Fig. 1; also *see* Table 1). Once inside the host, the virus is able to escape the host's innate immune responses in two ways: mainly by viral nonstructural protein 1 (NS1) attacking multiple steps of the type I IFN system, resulting in evasion of type I IFN responses [20], and by viral NA removing decoy receptors on mucins (*see* Figs. 2 and 3), cilia, and cellular glycocalyx and preventing self-aggregation of virus particles [22]. Also, the virus is capable of evasion of adaptive immune responses: evasion of the preexisting humoral or neutralizing antibodies and seasonal vaccines by antigenic variation in HA and NA antigens [23], and evasion of cellular immune response by amino acid substitutions in cytotoxic T-lymphocyte (CTL) epitopes of viral proteins, resulting in a decrease in CTL response [24]. Furthermore, although the precise functions of PB1-F2 remain unclear and are virus strain-specific and host-specific, it has been thought that PB1-F2 plays roles in both innate and adaptive immune responses in order to support influenza virus infection [25].

The replication cycle of an influenza A virus (*see* Fig. 2) starts from attachment of viral HAs to sialic acid (Sia, 5-amino-3,5-dideoxy-D-*glycero*-D-*galacto*-2-nonulosonic acid or neuraminic acid (Neu)) receptors on the host cell surface (*see* Subheading 5.3). This attachment mediates internalization of the virus into the cell by receptor-mediated endocytosis. While an early endosome gradually matures, the acidity in the endosome gradually increases. The low pH activates the integral membrane protein M2 of influenza virus, which is a pH-gated proton channel in the viral lipid

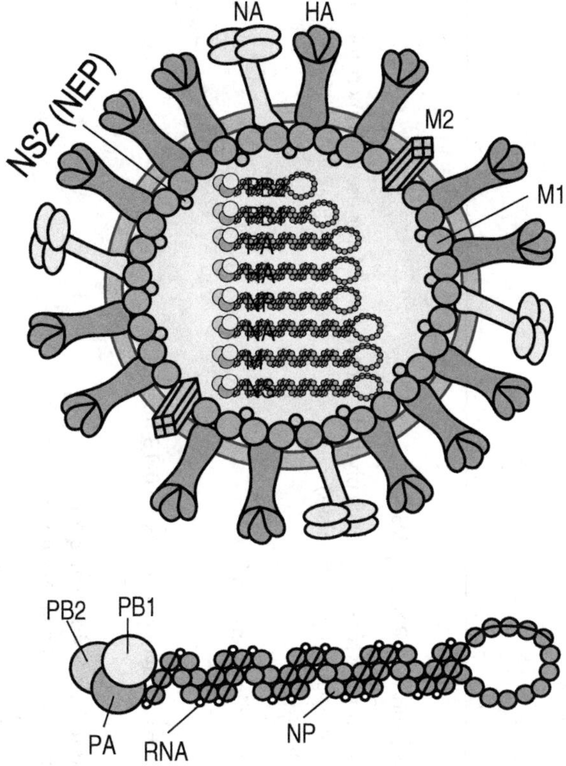

Fig. 1 Influenza A virus structure. Influenza A virus particles are roughly spherical in shape with sizes ranging from 80 to 120 nm in diameter, and each particle is enveloped by a lipid bilayer derived from the host cell membrane. Inside the envelope, each of eight (–) ssRNA genomic segments are wrapped with multiple nucleoproteins (NPs) and bound to RNA polymerases (PB1, PB2, PA) forming the vRNPs. The inner layer of the lipid envelope is attached to M1 molecules bound to vRNPs and to NS2. The outer layer of the lipid envelope is spiked with HA, NA, and M2 molecules with a ratio of about 5/2/1. *See* color figure in the online version

envelope, conducting protons into the virion interior. Acidification of the virus interior causes weakening of electrostatic interaction, leading to dissociation of M1 proteins from the viral RNP complexes (unpacking of the viral genome). The low pH in late endosomes also triggers a conformational change in HAs, resulting in exposure of their fusion peptides that immediately bind hydrophobically to the endosomal membrane (fusion), followed by release of vRNPs into the cytoplasm. During the course of the endocytic pathway, sialidase of NAs has been shown to be active [26], possibly in order to promote HA-mediated fusion [27]; however, further studies are needed to determine the exact mechanisms of this NA function. It should be noted that HA fusion will not occur if the HA protein (HA0) is not cleaved to form HA1 and HA2 by a membrane-bound host protease either before or during the release of progeny virions or with incoming viruses prior to endocytosis at the cell surface (*see* Subheading 5.2).

Table 1
Influenza A virus proteins encoded by each viral RNA segment and their functions

RNA segment (no. of nucleotides)[a]	Gene product (no. of amino acids)[a]	Molecules per virion	Function
	Structural proteins		
	Polymerase complex		Viral mRNA transcription and viral RNA replication
1 (2,341)	PB2 (759)	30–60	1. Recognition of caps of host mRNAs, endonucleolytically cleaved by PA for use as primers for viral mRNA transcription (cap-snatching mechanism) – A host range determinant [5] 2. Nuclear import 3. Inhibition of expression of interferon-β [6]
2 (2,341)	PB1 (757)	30–60	1. Nucleotide addition
3 (2,233)	PA (716)	30–60	1. Endonuclease activity that cleaves host mRNA 10-13 nucleotides
4 (1,778)	Hemagglutinin (566)	500	1. Major antigen 2. Proteolytically cleaved to be a fusion-active form – A pathogenic determinant (dependent on virus and host) 3. Sialic acid and linkage binding – A major determinant of host range and tissue tropism 4. Fusion
5 (1,565)	Nucleoprotein (498)	1,000	1. Nucleocapsid protein (viral RNA coating) 2. Intracellular trafficking of viral genome, viral transcription/replication and packaging [7] 3. Induction of apoptosis [8]
6 (1,413)	Neuraminidase (454)	100	1. Sialidase activity that prevents virus aggregation and facilitates viral entry into and budding from the host cell – Variations in the stalk lengths of NAs between residues 36-90 (N2 numbering) may play roles in pathogenesis, transmission, and host range – H274Y mutation in N1 subtype, E119V in N1/N2, N294S in N1/N2, and R292K in N2/N9 are associated with decrease in NAI susceptibility 2. Antigen

7 (1,027)	Matrix protein M1 (252)	3,000	1. Matrix protein that controls direction of viral RNPs transport [9] (dissociated from incoming vRNPs allowing their import into host nucleus and associated with newly assembled viral RNPs promoting their export from the nucleus) 2. Binding to vRNPs, M1 inhibits viral RNA polymerase activity [10] 3. Required for virus assembly and budding [11]
Spliced	Matrix protein M2 (97)	20–60	1. Ion channel to modulate pH of the virion during viral entry allowing M1/NP dissociation and that of the Golgi during transport of viral integral membrane proteins – S31N mutation decreases adamantane susceptibility
8 (890) Spliced	Nonstructural protein2 (NS2) (121)	130–200	1. Nuclear export protein that is involved in the nuclear export of viral RNPs
	Nonstructural proteins		
2 (2,341) Ribosomal leaky scanning	PB1-F2 (up to 90 amino acids depending on virus strains)	–	1. Pro-apoptotic protein [12] 2. Inhibition of induction of type I interferon synthesis – N66S mutation increases virulence [13]
	N40 (lack of 40 residues in N-terminal)		1. Interacts with the viral polymerase complex with an unclear function
Note: Both proteins are nonessential for virus replication, but their expression level changes are detrimental to virus replication [14]			
8 (890) Spliced	NS1 (strain specific length of 230–237 amino acids) [15]	–	1. Inhibition of the nuclear export of cellular mRNAs to promote cap-snatching 2. Inhibition of splicing of cellular mRNAs is also for use of host splicing apparatus for viral splicing [16] 3. Inhibition of the nuclear export of viral mRNAs until the appropriate time for their expression 4. Inhibition of host innate and adaptive immune responses – P42S mutation increases in virulence [17] 5. Anti-apoptotic and pro-apoptotic protein (control of host lifespan in order to complete its replication cycle) (After 13 h.p.i., pro-apoptotic signaling≫anti-apoptotic signaling) [15, 18]
	NS3[b] (a deletion of NS1 amino acids 126–168)	–	Unknown [19]

[a]The number of nucleotides and that of amino acids may vary depending on the host and strain of influenza A virus
[b]NS3 was experimentally found during the adaptation of a human virus within a mouse host [19]

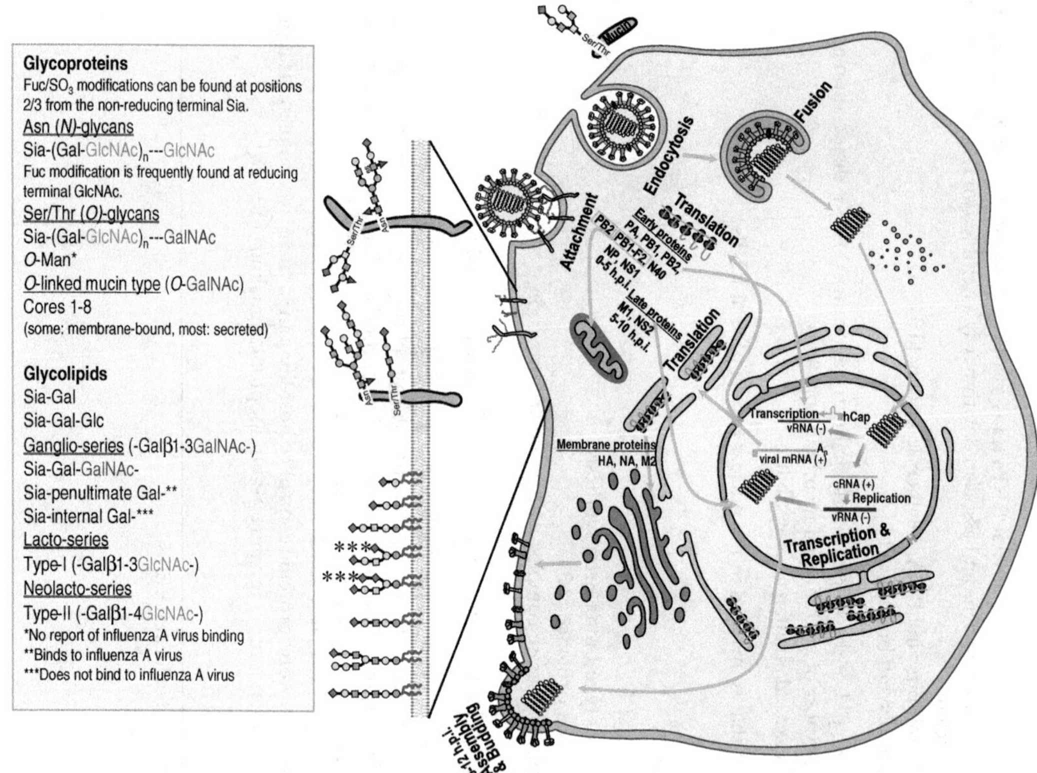

Fig. 2 Replication cycle of influenza A virus. The viral HAs can bind to sialylglycoconjugated proteins or lipids (with cartoon representations in the *left panel*; symbol and text nomenclature used according to the Nomenclature Committee of the Consortium for Functional Glycomics) including mucin (an *O*-GalNAc glycoprotein with Siaα2-3Gal linkage consisting of mucus shielding the epithelial surface for cellular protection from both physical and chemical damage and pathogen infection). The virus particles adsorbed to mucins can be released by viral NAs that preferentially cleave sialic acid moieties with preference to Siaα2-3Gal linkage over Siaα2-6Gal linkage [21]. *Right panel*: Schematic of replication cycle of influenza A virus, which can be divided into six distinct parts: (1) attachment, (2) receptor-mediated endocytosis, (3) fusion, (4) transcription and replication, (5) translation (protein synthesis), and (6) assembly, budding, and release. *See* the text for details. *See* color figure in the online version

The vRNPs in the cytoplasm are immediately imported into the nucleus most probably by nuclear localization signals in proteins composed of vRNPs, and the viral RNA polymerase transcribes the (−) vRNAs primed with 5′-capped RNA fragments, which are derived from cellular mRNAs by a cap-snatching mechanism, to viral mRNAs and replicates the unprimed (−) vRNAs to complementary RNAs, (+) cRNAs, used as templates to generate (−) vRNAs (transcription and replication). The viral mRNAs are subsequently exported to the cytoplasm for translation into viral proteins by the cellular protein-synthesizing machinery. Viral proteins needed for viral replication and transcription are transported back to the nucleus. The newly synthesized vRNPs are exported from the nucleus to the plasma membrane, mediated by M1 and NS2

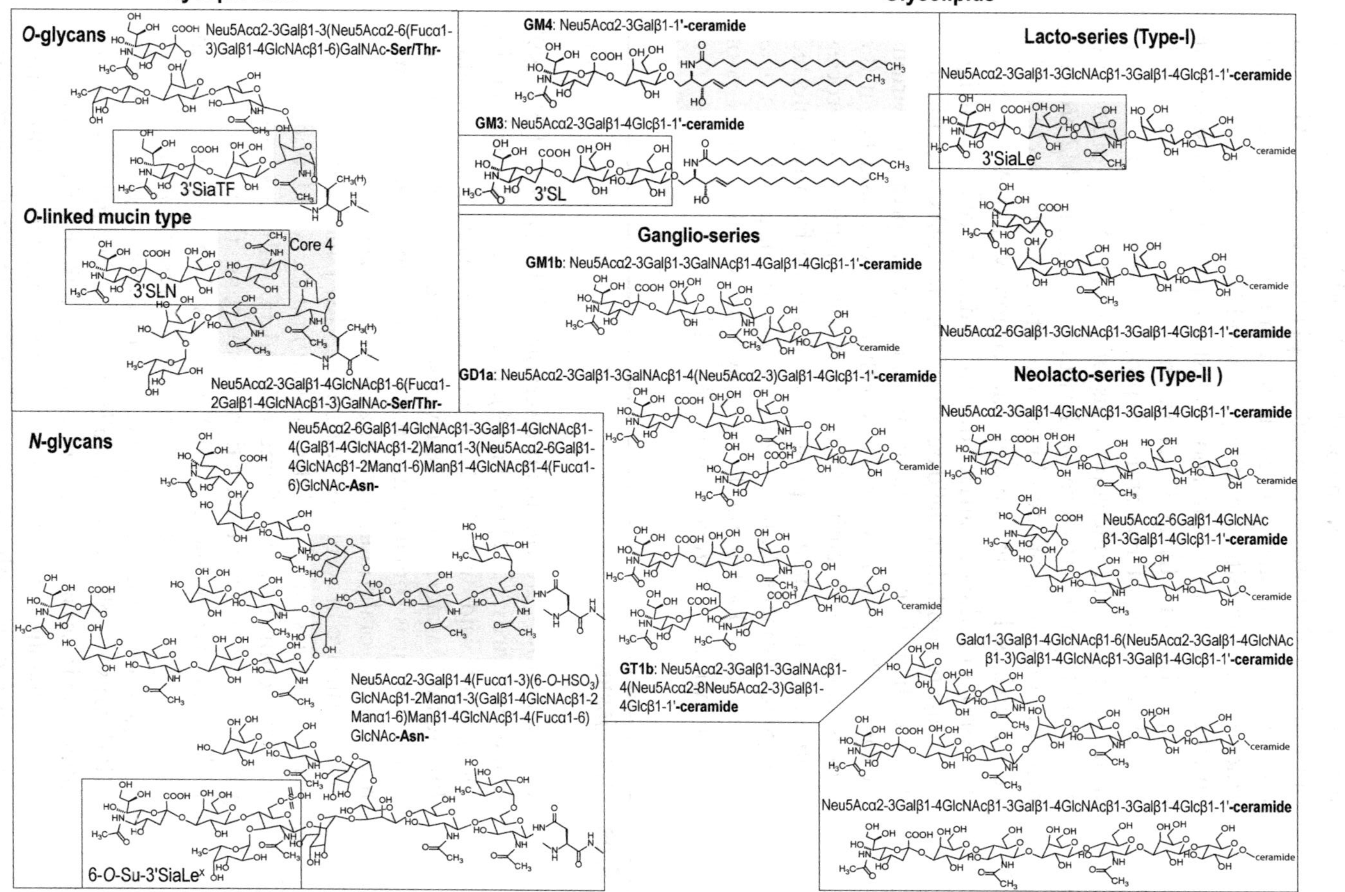

Fig. 3 Examples of chemical structures of glycans recognized by influenza A viruses. Glycoproteins are classified into *N*-linked glycans attached to Asn-X-Ser/Thr, X being any amino acid except proline, and *O*-linked glycans attached to Ser/Thr. Glycolipids are divided into simple gangliosides GM3, gala-series (GM4), globo-series (not shown here), ganglio-series, lacto-series, and neolacto-series. Core structures are highlighted by *green shading*. Ceramide structure is highlighted by *pink shading*. *Rectangles* show terminal trisaccharides with abbreviation names: 3′SiaTF for Neu5Acα2-3Galβ1-3GalNAc, 3′SLN for Neu5Acα2-3Galβ1-4GlcNAc, 6-*O*-Su-3′SiaLe^X for Neu5Acα2-3Galβ1-4(Fucα1-3)(6-*O*-HSO_3)GlcNAc, 3′SL for Neu5Acα2-3Galβ1-4Glc, and 3′SiaLe^c for Neu5Acα2-3Galβ1-3GlcNAc. *See* color figure in the online version

(NEP). Viral HA, M2, and NA proteins are synthesized and glycosylated in the rough endoplasmic reticulum (RER) and the Golgi apparatus and are transported to the cell surface via the trans-Golgi network (TGN). Within the acidic TGN, M2 transports H^+ ions out of the TGN lumen to equilibrate pH between the TGN and the host cytoplasm in order to maintain an HA metastable configuration [28]. At the site of budding, HA and NA cluster in lipid rafts enriched in sphingomyelin and cholesterol to provide a sufficient concentration of HA and NA in the budding virus [29, 30]. Cytoplasmic tails of HA and NA proteins bind to M1 proteins, which interact with nucleoproteins of vRNPs. M1 proteins also attach to M2 proteins to form viral particles (assembly). M2 proteins mediate alteration of membrane curvature at the neck of the budding virus, leading to membrane scission. Viral HAs retain the new virions on the cell surface due to binding to cellular sialylglycoconjugates, and thus sialidase activity of viral NAs is required to destroy these bonds, resulting in release and spread of virions (release). The viral HAs attach to other host cells and begin the process anew. The entire influenza A virus replication process (from viral attachment to progeny viruses burst from the infected cell) generally takes about 5–12 h, resulting in the production of as many as 100,000–1,000,000 progeny viruses; only about 1 % of progeny viruses can infect other cells [31].

3 Influenza A Virus Transmission

Wild waterfowls are the main reservoir of H1–H16 and N1–N9 influenza A viruses [32] (*see* Fig. 4). Influenza A viruses replicate in the gut of wild waterfowls, which are usually asymptomatic. Infected wild waterfowls excrete viruses in feces and spread viruses mainly via virus-contaminated water and fomites (fecal–contaminated–water–oral route). Influenza A viruses have been thought to able to survive in water for several days [33] and thus migratory wild waterfowls can spread the virus around the world, normally in a north–south direction. Some H5 and H7 subtypes crossing to poultry have acquired mutations converting them into highly pathogenic avian influenza (HPAI) viruses. Low pathogenic avian influenza (LPAI) viruses replicate mainly in respiratory and intestinal organs of poultry and cause epidemics of mild disease, whereas HPAI viruses replicate systemically and cause fatal influenza. Avian influenza viruses may be directly transmitted from infected birds or virus-contaminated environments or indirectly transmitted through mixing with another virus(es) in an intermediate host, such as pigs, to mammals. Influenza viruses typically replicate in the respiratory system of mammals, including humans, pigs, and horses, usually entering through the eyes, nose, mouth, throat, bronchi, and lungs, and are transmitted through the air by coughs or sneezes or through secretions or fomites.

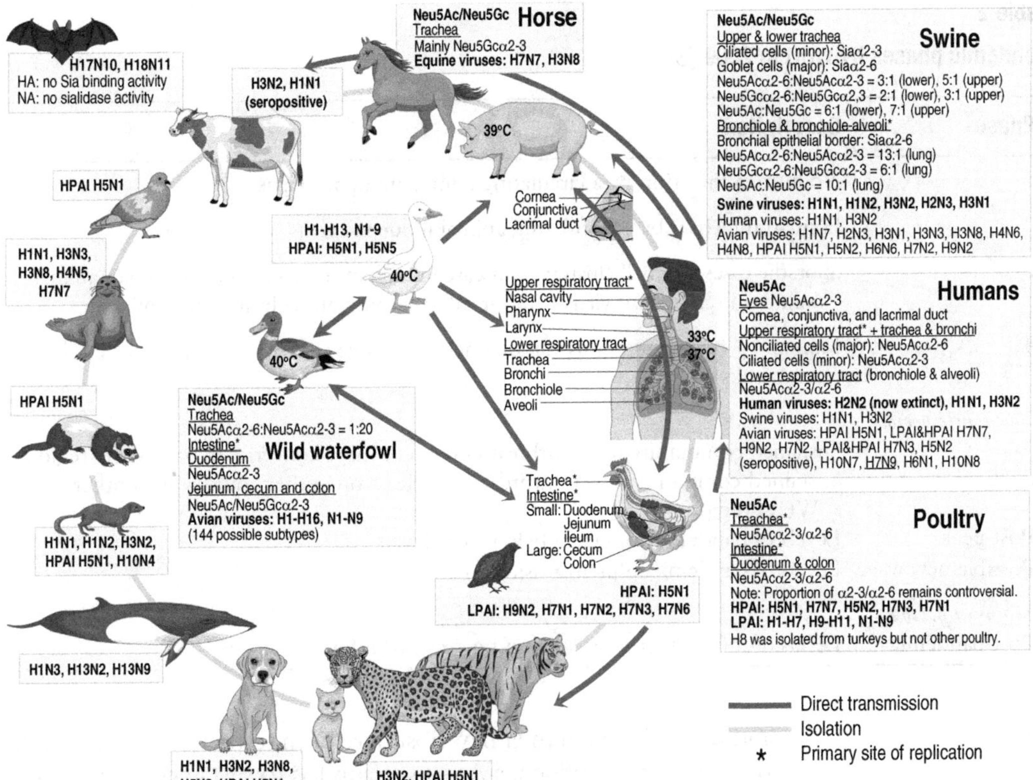

Fig. 4 Host range of influenza A viruses. Wild waterfowls are the natural hosts of influenza A viruses of H1–H16 and N1–N9 subtypes. Influenza A viruses from one host are sometimes transmitted to and continue circulating in other host species if they can change and adapt to the new hosts, and once adapted to the new host, they usually lose the capacity to circulate in the previous host. Influenza viruses from wild waterfowl are often transmitted to and from the domestic ducks using the same aquatic areas, and infected domestic ducks spread the viruses to others, including poultry, pigs, and farmers, in a local area. Influenza A viruses have been isolated from various animals as shown in the figure, indicating their capacity to cross the species barrier. Cell surface receptors with Neu5Ac/Neu5Gcα2-6/α2-3 specifically recognized by viral HAs, which initiate viral infection, being a major determinant restricting the host range of influenza A viruses are shown. *See* color figure in the online version

4 Emergence of Influenza Pandemics

Having (1) numerous wild waterfowl species as natural reservoirs, (2) various animal hosts, (3) RNA polymerase without proofreading, and (4) segmented genome (3 and 4 causing a high mutation rate), influenza A viruses have been difficult to control and/or eradicate. Efforts for prevention of the next pandemic, either by minimization of cross-infection between species or rapid identification of novel strains, constitute an essential and primary step for preventing influenza infection in human beings. Interspecies transmission of influenza A viruses between animal hosts including pigs, horses, and birds, as well as humans, has occasionally been detected, but successful propagation

Table 2

Pandemic phases by WHO in 2009 [34]

Phase	Description
1	No human infection by a circulating animal influenza virus
2	Human infection by a circulating animal influenza virus
3	Sporadic cases or small clusters of disease in humans by an animal or human-animal influenza reassortant virus without sufficient human-to-human transmission
4	Human-to-human transmission of a new virus able to sustain community-level outbreaks
Pandemic period	
5	Sustained community-level outbreaks in two or more countries in one WHO region
6	Sustained community-level outbreaks in at least one other country in another WHO region
Post-peak	Levels of pandemic influenza below peak levels
Possible new wave	Levels of pandemic influenza rising again
Seasonal period	
Post-pandemic	Levels of influenza activity as seen for seasonal influenza

and transmission in their new host have been restricted. In the past 95 years, only four influenza A virus strains led to sustained outbreaks in human populations and started pandemics (*see* Table 2; also *see* Fig. 5).

4.1 Past Pandemics

4.1.1 H1N1 Spanish Influenza Pandemic (1918–1919): The Greatest Parental Influenza

The Spanish influenza pandemic resulted from an avian-descended H1N1 virus. It killed at least 40 million people globally in 1918–1919, with almost 50 % of the deaths occurring in healthy young adults of 20–40 years of age, although its clinical symptoms and pathological manifestations were mainly in the respiratory tract [45]. This could be due to a too-strong and damaging response to the infection by healthy immune systems [46]; it is by far the most devastating influenza pandemic. The pandemic apparently originated during World War I in the USA at the beginning of 1918 before it appeared in France and then in Spain at the end of 1918 and thus being named Spanish influenza. Despite extensive

Fig. 5 (continued) between human-avian-swine viruses (2009 pandemic), (2) recurrence of a previous pandemic virus in new group of populations from a frozen refrigerator (an H1N1 Russian flu in 1977), (3) seasonal viral evolution by intrasubtypic reassortment: for example, A/Fujian/411/2002 (H3N2) having a major antigenic variant due to reassortment between two distinct clades of co-circulating H3N2 viruses [35] (not shown in this chart), and (4) seasonal viral evolution by adaptation associated with point mutations. The left panel shows the timeline of direct transmission of the first reported avian influenza A virus subtypes (H7N7 1996 [36], HPAI H5N1 1997 [37, 38], H9N2 1999 [36], H7N2 2002 (only serologic evidence)—2003 [36], HPAI H7N7 2003 [39], LPAI and HPAI H7N3 2004 [36, 40], H10N7 2010 [41], and H7N9 [42], H6N1 [43], and H10N8 [44] 2013) from avians to humans. Human infections with some of these avian influenza virus subtypes, especially HPAI H5N1, LPAI H9N2, and LPAI H7N9, have occasionally continued to be reported until now (January 2014) [36, 44]. *See* color figure in the online version

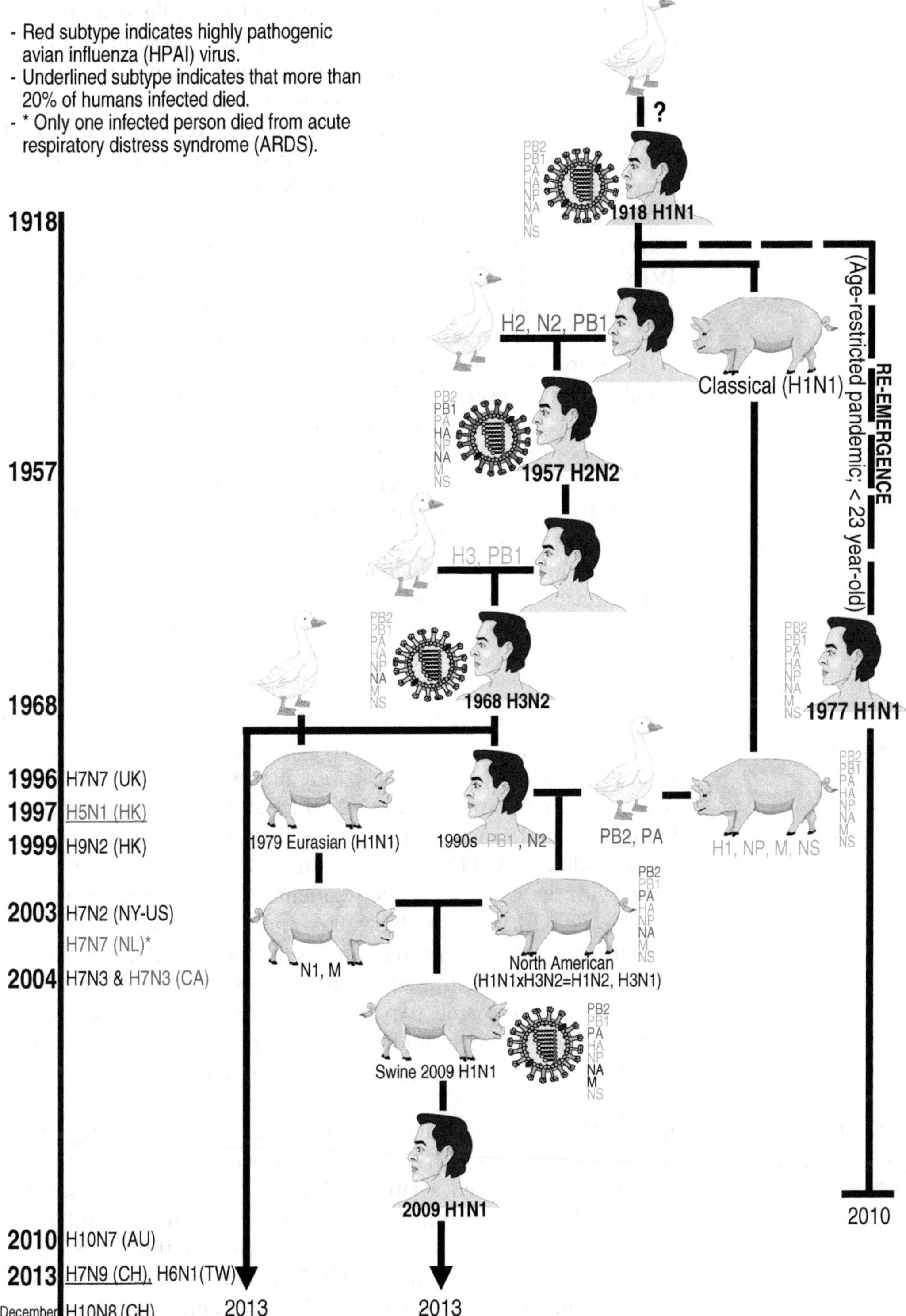

Fig. 5 A family tree of human influenza A viruses. Based on influenza history, the 1918 H1N1 virus provides descendants of influenza A viruses that have continued to circulate in human populations by (1) generation of a novel pandemic virus by reassortment between human-avian viruses (1957 and 1968 pandemics) or

investigations but a lack of available pre-1918 influenza samples, the emergence pathway of the Spanish pandemic strain either by direct avian viral mutations alone or by reassortment with another virus(es) in an intermediate host is still a mystery.

The pandemic virus induced humans to develop immunity that provides selective pressure and drives the virus to evolve their antigenicity. The resulting 1918-derived H1N1 virus, which had antigenic change annually, caused an epidemic with lower death rates and triggered human immunity to the virus over time. Somehow the 1918-derived H1N1 virus underwent dramatic genetic change with acquisition of three novel gene segments, avian-like H2, N2, and PB1 gene segments, resulting in emergence of the H2N2 pandemic in 1957 and disappearance of the 1918-derived H1N1 virus from circulation. However, the 1918-derived H1N1 virus from the pre-1957 period reappeared in 1977, its reemergence believed to be from a laboratory in Russia or Northern China, and caused a (low-grade) pandemic mostly affecting young people less than 20 years of age due to immunological memory to the virus in most elderly people. After that, H1N1 Russian/77 variant became epidemic yearly until the 2009 H1N1 pandemic emerged, and the virus disappeared from humans. The disappearance of the preexisting H1N1 seasonal virus was suggested to be due in part to stalk-specific antibodies boosted from infection with the 2009 H1N1 pandemic virus [47].

4.1.2 H2N2 Asian Influenza Pandemic (1957–1958)

In early 1956, an influenza outbreak of a new H2N2 strain occurred in China and spread worldwide, resulting in a pandemic in 1957. It has been believed that the pandemic H2N2 virus evolved via reassortant between avian H2N2 strain and the preexisting circulating human 1918 H1N1 strain; it consisted of three gene segments coding HA (H2), NA (N2), and PB1 derived from an avian virus, with the other five gene segments derived from a previously circulating human virus. New HA and NA surface antigens to human immunity for protection resulted in the Asian influenza pandemic virus infecting an estimated 1–3 million people worldwide with approximately two million deaths. The virus became seasonally endemic and sporadic and it disappeared from the human population after the next pandemic appeared in 1968 [48].

4.1.3 H3N2 Hong Kong Influenza Pandemic (1968–1969)

In July 1968, a new influenza A virus was detected in Hong Kong. It was identified as H3N2, which was believed to be a result of reassortment between avian and human influenza A viruses: the HA and PB1 gene segments were derived from avian influenza virus and the other six gene segments, including the NA N2 gene segment, were derived from the 1957 H2N2 virus. The virus killed up to one million humans varying widely depending on the source, less deaths than those in previous pandemics. The relatively small number of death is thought to be due to exposure of people to the 1957 virus, who apparently retained anti-N2 antibodies, which did not prevent 1968 infection but limited virus replication and reduced the duration and severity of illness. Although the next pandemic occurred in 2009, the virus is still in circulation globally (as of 2013) as a seasonal influenza strain.

4.1.4 H1N1 Swine Influenza Pandemic (2009–2010)

In April 2009, a widespread outbreak of a new strain of influenza A/H1N1 subtype referred to as swine flu was reported in Mexico. By June 2009, the virus had spread worldwide, starting the first influenza pandemic of the twenty-first century. In August 2010, the influenza activity returned to a normal level as seen for seasonal influenza; at least 18,000 laboratory-confirmed deaths from the pandemic 2009 were reported, but infection with the virus resulted in mild disease with no requirement of hospitalization [49]. To date, the influenza A viruses identified among people are 1968 H3N2 and 2009 H1N1 viruses.

Genetic composition of the pandemic 2009 viruses isolated from initial cases indicated that the viruses are composed of PB2 and PA gene segments of North American avian virus origin, PB1 segment of human H3N2 virus origin, HA (H1), NP and NS segments of classical swine virus origin (their genes having been found to circulate in pigs since 1997/1998 known as avian/human/swine triple reassortant H1N2 swine viruses), and NA (N1) and M segments of Eurasian avian-like swine H1N1 virus origin (which emerged in European pigs in 1979); hence, their original description was "quadruple" reassortants (*see* Fig. 5). Similarity between the HA sequence patterns of the 1918 pandemic and 2009 pandemic viruses and their cross-antibody neutralization [50–53] suggested that H1 HA of the 2009 pandemic virus may have originated from or derived from the same origin of the 1918 pandemic virus; probably, the 1918-like pandemic H1N1 virus was transmitted and established in domestic pigs between 1918 and 1920 referred to as the classical swine lineage that circulated continuously in pigs in the USA. Domestic pigs having a short lifespan (4- to 6-month-old pigs are killed for their flesh, and female breeding pigs (sows) remain in the farm until the age of 4–5 years and then taken to be slaughtered for sausages and bacon) are a frozen-like source for the influenza virus due to a lack of selection pressure in pigs. Consequently, the HA sequence of the 2009 swine origin derived from the classical swine lineage is not much different from its origin.

4.2 Pandemic Mechanisms

The past pandemics and ongoing direct transmission of avian influenza A viruses into humans suggest two plausible mechanisms that would permit influenza A viruses to overcome selective pressure and subsequently become established in human populations (*see* Fig. 5; also *see* Subheading 5.3.3). As shown in Fig. 6, one mechanism is an adaptation mechanism, in which a nonhuman virus acquires a mutation(s) (a mutation(s) in the HA gene to recognize the Neu5Acα2-6Gal receptor considered to be an essential prerequisite for the beginning of a pandemic) during adaptation to

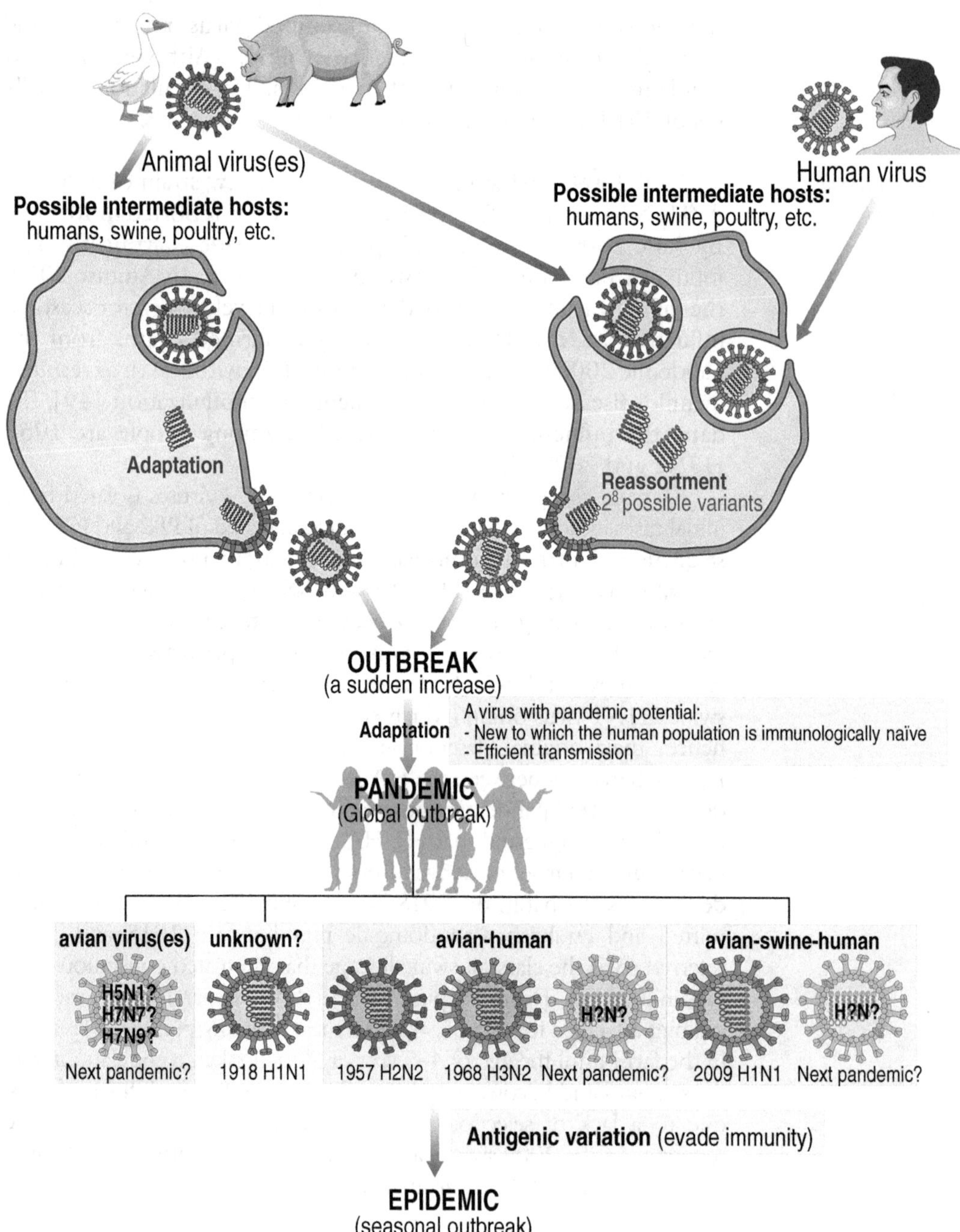

Fig. 6 Proposed mechanisms of emergence of an influenza pandemic. The *left panel* is an adaptation mechanism and the *right panel* is a reassortment mechanism. *See* text for details. *See* color figure in the online version

an intermediate host leading to sufficient human-to-human transmission. The other mechanism is a reassortment mechanism, in which a nonhuman virus reassorts with a nonhuman virus(es) and/or a human virus(es) in an intermediate host producing 2^8 possible reassortants (if two viruses are reassorted), by which the reassortant with ability leading to sufficient human-to-human transmission becomes dominant in the intermediate host. Then the adapted/reassorted virus can cause influenza outbreaks in immunologically naïve humans, allowing it to adapt (fine tune) for efficient and sustained human-to-human transmission and finally causing a global outbreak (pandemic). The first recorded H1N1 pandemic in 1918 is of an unknown origin, the H2N2 pandemic in 1957 and the H3N2 pandemic in 1968 arose from genetic reassortment between avian virus and human virus with an unknown intermediate host, and the latest H1N1 pandemic in 2009 arose from genetic reassortment among avian virus, swine virus, and human virus with another swine virus in pigs as intermediate hosts. It is not known when and how a future pandemic will emerge, by an adaptation or reassortment mechanism, and it is also impossible to predict which virus subtype will be the next pandemic. In the post-pandemic period, most people develop immunity to the pandemic strain and thus the pandemic virus can continue to cause seasonal outbreaks (epidemics) if it can change its surface antigens (antigenic variation) to evade host immunity.

5 Hemagglutinin

HA was so named due to its ability to agglutinate red blood cells via binding to Sia on red blood cells. Phylogenetically, 18 HA subtypes are classified into group 1 and group 2 (*see* Fig. 7a). Notable differences in structure between the two groups of HAs are in the region involved in HA conformational change required for membrane fusion; group 1 HAs contain an additional turn of the helix at residues 56–58, blocking accessibility of *tert*-butyl hydroquinone, which is accessible to group 2 [54]. HA is encoded by the fourth segment of the influenza A viral genome and is assembled as a homotrimeric precursor (HA0) (*see* Fig. 7b). HA is a major target of neutralizing antibodies, plays a pivotal role in avian influenza virus pathogenicity, and is a major determinant of host range restriction. It is a lectin that contains one or more carbohydrate recognition domains that determine host specificity [55] and plays a crucial role in fusion of the viral envelope and cellular endosomal membrane for release of the viral genome into host cells (*see* Fig. 7c–f).

5.1 HA1 Is a Major Target of Neutralizing Antibodies

Change in HA antigen is responsible for epidemics and pandemics. Change in HA antigen either by accumulation of mutations in HA1 of five proposed antigenic sites (based on amino acid sequence comparison among viruses isolated from different years or among variants grown in the presence of mouse monoclonal antibodies) as shown in Fig. 7c [56–59] or by intrasubtypic reassortment between distinct clades of co-circulating influenza A viruses [35] is responsible for evasion of recognition by the host antibodies and thus

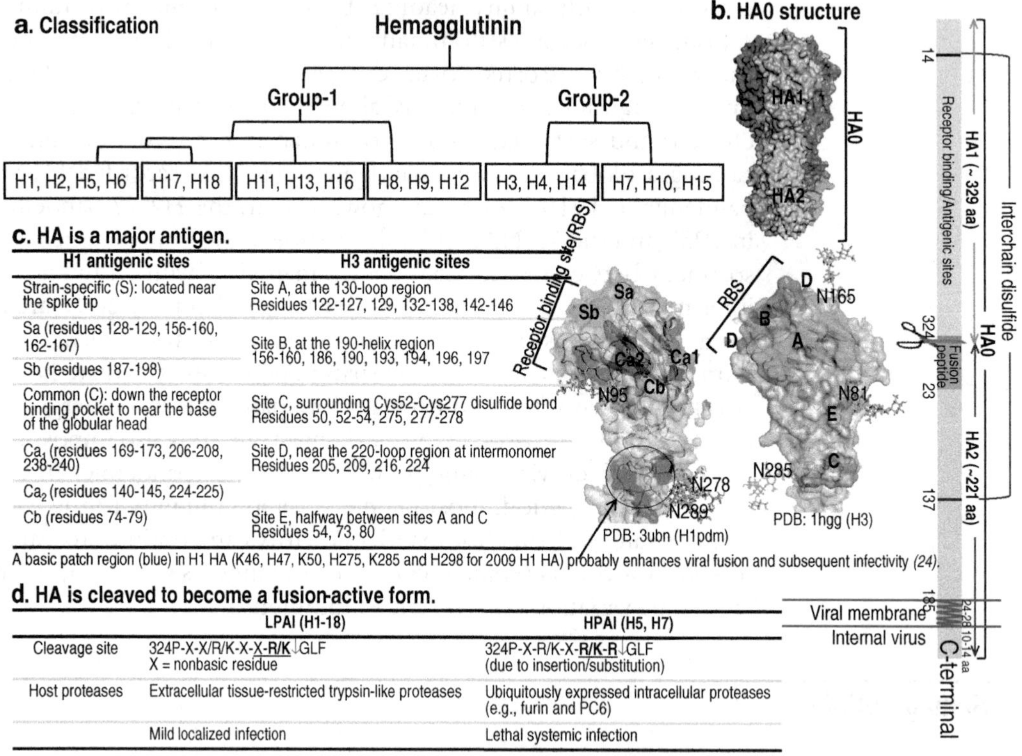

H1 antigenic sites	H3 antigenic sites
Strain-specific (S): located near the spike tip	Site A, at the 130-loop region Residues 122-127, 129, 132-138, 142-146
Sa (residues 128-129, 156-160, 162-167)	Site B, at the 190-helix region 156-160, 186, 190, 193, 194, 196, 197
Sb (residues 187-198)	
Common (C): down the receptor binding pocket to near the base of the globular head	Site C, surrounding Cys52-Cys277 disulfide bond Residues 50, 52-54, 275, 277-278
Ca_1 (residues 169-173, 206-208, 238-240)	Site D, near the 220-loop region at intermonomer Residues 205, 209, 216, 224
Ca_2 (residues 140-145, 224-225)	
Cb (residues 74-79)	Site E, halfway between sites A and C Residues 54, 73, 80

A basic patch region (blue) in H1 HA (K46, H47, K50, H275, K285 and H298 for 2009 H1 HA) probably enhances viral fusion and subsequent infectivity *(24)*.

	LPAI (H1-18)	HPAI (H5, H7)
Cleavage site	324P-X-X/R/K-X-**X-R/K**↓GLF X = nonbasic residue	324P-X-R/K-X-**R/K-R**↓GLF (due to insertion/substitution)
Host proteases	Extracellular tissue-restricted trypsin-like proteases	Ubiquitously expressed intracellular proteases (e.g., furin and PC6)
	Mild localized infection	Lethal systemic infection

Fig. 7 HA structure and its functions. (**a**) 18 HA subtypes are phylogenetically classified into six clades segregated into two groups as indicated in the chart. (**b**) HA forms a homo-trimeric precursor (HA0) by which each monomer consists of HA1 and HA2 (PDB: 1rd8) and the C-terminus of HA2 anchors each monomer in the viral membrane. It provides four events important for the entry of influenza virus into the host cell. (**c**) Antigenicity. HA1 is a major viral antigen with five important antigenic sites as indicated. (**d**) Cleavage. HA0 with a single basic residue is cleaved by extracellular proteases into HA1 (~329 amino acids) and HA2 (~221 amino acids). HA0 with multiple basic amino acids at the cleavage site (residues 324–329) in HPAI viruses is cleaved intracellularly in the trans-Golgi compartment by ubiquitous proteases. Two amino acids in the cleavage sites that are critical for the recognition by proteases are *underlined*. (**e**) Attachment. HA1 carries a receptor binding site at the membrane-distal tip, which is formed by 190-helix, 130-loop, and 220-loop. Amino acid substitutions in this binding site and virus-receptor binding assays revealed two amino acid residues critical for determinants of virus binding preference: at positions 190 and 225 for H1 virus (E190 and G225 HA preferring α2-3 receptors, and D190 and D225 HA favoring α2-6 receptors) and at positions 226 and 228 for H2 and H3 viruses (Q226 and G228 HA preferring α2-3 receptors, and L190 and S225 HA preferring α2-6 receptors). *Left panel*: Crystal structures of HAs of A/California/04/2009 (H1N1) interactions with 3′SLN (PDB: 3ubq) and with 6′SLN (PDB: 3ubn) and those of A/Anhui/1/2013 (H7N9) interactions with 3′SLN (PDB: 4BSD) and with 6′SLN (PDB: 4BSC). *Circular broken lines* indicate amino acids that interact with Neu5Ac and are highly conserved among different H1–H16 HA subtypes, except position 155 (V/I for increased binding to Neu5Gc). The *other broken lines* indicate amino acids involved in interactions with internal sugars of the glycan receptor. (**f**) Fusion. At acidic pH, the cleaved HA in HA1 (*gray*)-S-S-HA2 (*green*) form is conformationally changed. The N-terminal HA2 fusion peptide (residues 1–23) buried inside the interior of the HA molecule at neutral pH pops out and is quickly inserted into the host endosomal membrane, leading to membrane fusion (modeled in *right panel*). *See* color figure in the online version

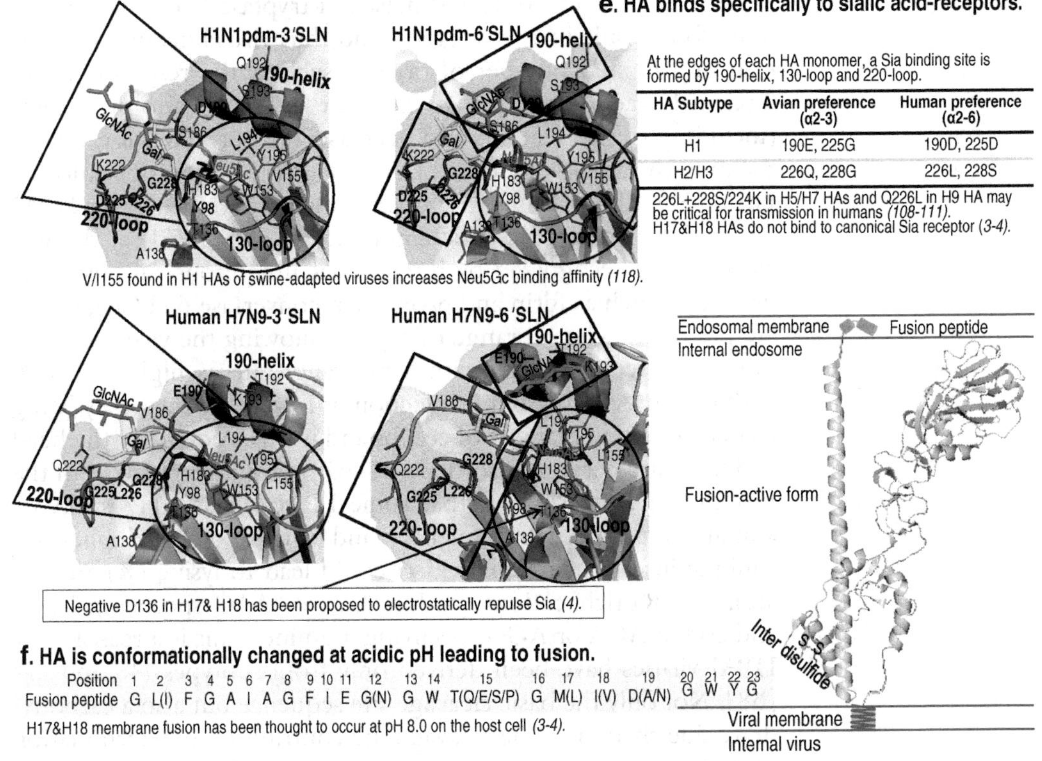

Fig. 7 (continued)

continuous circulation of the virus in host populations. Amino acid sequencing studies of HAs of avian and animal viruses isolated from different periods of time have shown that the HAs of avians and animals that have shorter lifespans have higher conservation of amino acid sequences than that of human virus isolates, suggesting that avian/animal viruses are subjected to little immune pressure, resulting in less antigenic variation than that of human virus strains [60, 61].

Introduction of a novel HA antigen, resulting from genetic reassortment during mixed infection, from direct introduction of a nonhuman influenza virus, or from reintroduction of human influenza viruses that had disappeared from circulation, into immunologically naïve human populations is a key factor of an influenza virus with pandemic potential.

5.2 HA Cleavage Is a Critical Determinant of Pathogenicity in Gallinaceous Poultry

To enable HA conformational changes that lead to membrane fusion, which is critical for viral infectivity and dissemination, HA0 must be cleaved by a host cell protease into subunits HA1 and HA2; thereby, the host protease is a determinant of tissue tropism of the virus. Influenza virus HA0 usually contains a monobasic cleavage site (*see* Fig. 7d), which is recognized by extracellular trypsin-like proteases, such as tryptase Clara from rat bronchiolar epithelial Clara cells [62] and mast cell tryptase from the porcine lung, found only in a few organs, and thus virus infection is localized in a limited number of organs, such as the respiratory and intestinal tracts, resulting in mild or asymptomatic symptoms (including ruffled feathers and decreased egg production); hence, the causative viruses are called LPAI viruses [63]. Multiplication of H5 or H7 LPAI viruses in chickens and turkeys generates HPAI viruses having multiple basic amino acids (*see* Fig. 7d) at the HA cleavage site recognized by intracellular ubiquitous subtilisin-like proteases, such as furin and proprotein convertase 6 (PC6), which are present in a broad range of organs, allowing the virus to infect multiple internal organs with a mortality rate as high as 100 % within 48 h (lethal systemic infection or fowl plaque typically being characterized by cyanosis of combs and wattles, edema of the head and face, and nervous disorders [64–67]). Sequence analysis of the HA cleavage site showed that some LPAI H5 and H7 subtypes contain a purine-rich sequence, and thus a direct duplication (unique insertion) in this region could lead to lysine (K) and/or arginine (R)-rich codons (codon AAA or AAG specifying lysine and codon AGA or AGG specifying arginine); this is a reason why HPAI viruses have been derived only from subtypes H5 and H7 [68]. Not only the basic cleavage site sequence but also a carbohydrate side chain near the cleavage site contributes to determination of pathogenicity (virulence) if it interferes with the host protease accessibility.

Since the first report of a HPAI H5N1 progenitor strain in 1996 from a farmed goose in Guangdong Province, China (A/gs/Guangdong/1/96 designated as Clade 0), the world has intermittently experienced HPAI virus outbreaks, both recurrence and new HPAI virus outbreaks, in domestic birds classified into groups or clades (20 clades having been recognized at present) and subdivided into subclades and lineages based on their phylogenetic divergence as the virus continues to evolve rapidly [69, 70]. Although they are generally restricted to domestic poultry on farms with high mortality rates and substantial economic losses, HPAI H5N1 viruses have occasionally been isolated from some species of wild waterfowls, including wood ducks and laughing gulls, with varying degrees of severity [71–75]; thus, migration of susceptible waterfowls could spread HPAI H5N1 viruses over long distances, leading to difficulties for avian influenza control. The significant species-related variation in susceptibility to and clinical disease caused by H5N1 virus infection has been determined not only in wild birds but also in other animals. For example, pigs can be infected with HPAI H5N1 viruses, but they have almost no or very weak disease symptoms or only slight respiratory illness. Without influenza-like symptoms, the virus may adapt to mammalian hosts in the respiratory tract of this potential intermediate host [76], which contains gradual increases in Neu5Acα2-6Gal, a human receptor, over Neu5Acα2-3Gal, an avian receptor, from upper and lower parts of the porcine trachea towards the porcine lung, a primary target organ for swine-adapted virus replication [77]. Humans can be infected with HPAI H5N1 virus (first report in 1997) with severe disease and high death rate. The ecological success of this virus in crossing the species barrier from poultry to infect diverse species including wild migratory birds and other mammals including pigs, cats, and dogs with sporadic infections in humans often with fatal outcomes [78, 79] highlighted the possibility of HPAI H5N1 development to a pandemic strain either by gradual modification of existing structures or rapid modification by reassortment with a human epidemic strain. Although the world has been at phase 3 in WHO's six phases of pandemic alert since 2006 [80], HPAI H5N1 viruses should not be neglected in efforts to perform surveillance and health management planning.

In addition to HPAI H5N1 viruses, other avian influenza viruses including LPAI H5N1, HPAI and LPAI H7N7, HPAI H7N3, and LPAI H9N2 viruses have occasionally crossed the species barrier to infect other mammals including humans (*see* Fig. 4) and have caused generally mild disease (with conjunctivitis or mild respiratory symptoms) in humans. Recently, LPAI H7N9 virus (a novel avian–avian reassortant virus: HA from wild-duck H7N3 virus, NA from wild-bird H7N9 virus, PA, PB1, PB2, NP, and M from chicken H9N2 virus, and NS from another chicken H9N2 virus) identified in humans in February 2013 in China has killed 45 (due to severe pneumonia) of the 139 laboratory-confirmed cases (case-fatality ratio of about 32 %) according to WHO data in November 2013 [42, 81]. This evidence indicated that HPAI viruses primarily infect poultry and cause severe illness and high death rates in poultry and that they occasionally infect other non-poultry species with variation in severity depending on the virus strain and host. LPAI viruses spread silently in poultry and occasionally spread to other non-poultry species and often cause mild illness but are capable of causing severe disease, such as disease caused by LPAI H7N9 virus infection in humans. Thus, more studies are needed to understand differences in pathogeneses of these viral infections.

5.3 Receptor Specificity Is Responsible for the Host Range Restriction of Influenza Virus

Influenza viruses enter the body and search for cells among the host cells in which they can replicate and grow. Influenza virus homing is triggered by interactions between viral HA spikes and sialylglycoconjugates on host cell surface [82], which play roles in a wide variety of host biological processes, including cell proliferation, apoptosis, and differentiation [83]. More than 50 types of sialic acids are found in nature, with *N*-acetylneuraminic acid (Neu5Ac) and *N*-glycolylneuraminic acid (Neu5Gc) being the most prevalent forms. Not only sialic acid type but also glycosidic linkage type (the most common terminal linkages being α2-3 and α2-6 linkages), substructure (such as GalNAc or GlcNAc), and other modifications (such as fucosylation and sulfation) cause diversity in sialylglycoconjugates (*see* Figs. 2, *left panel* and 3) [84, 85]. The sialic acid type and the glycosidic linkage type on the host cell surface are the principal determinants of host range restriction of influenza viruses, although other glycan modifications may be involved in the virus-receptor binding preference. Therefore, the distribution of sialylglycoconjugates among animal species and tissues, a crucial factor for influenza A infection and transmission, has been extensively investigated either by lectin histochemical analysis with *Maackia amurensis* agglutinin (MAA-I specific for Siaα2-3Galβ1-4GlcNAc-, MAA-II for Siaα2-3Galβ1-3GalNAc) and *Sambucus nigra* agglutinin (SNA specific for Siaα2-6Galβ1-4GlcNAc-) or by structural characterization using sequential glycosidase digestion in combination with HPLC and mass spectrometry. Figure 4 shows sialic acid-containing receptors in main target organs in important host species of influenza A viruses.

5.3.1 Wild Waterfowls

So far (2013), all H1–H16 and N1–N9 avian influenza viruses have been reported in 12 bird orders, most having been isolated from the order Anseriformes, especially in the family Anatidae (ducks, swans, and geese), and the order Charadriiformes (shore birds) in the family Laridae (gulls, terns, and relatives). It should be noted that the newest H17N10 and H18N11 viruses recognized in 2012 and 2013, respectively, were found only in bats, the little yellow-shouldered bat *Sturnira lilium* for H17N10 and the flat-faced fruit bat *Artibeus planirostris* for H18N11, in the family Phyllostomidae, a family of frugivorous bats that are abundant in Central and South America [2, 3]. Ducks in the Anatinae subfamily belonging to the family Anatidae are the most common source of influenza A virus isolation and risk for virus transmission [86]. Almost all ducks are naturally attracted to aquatic areas including wetlands, lakes, and ponds for resting, feeding, and breeding in their course of migration, allowing influenza viruses to be transmitted to and from domestic duck populations. Infected domestic ducks spread the virus to other avian species in a local area [73]. The duck tracheal and intestinal epithelium was shown to predominantly express Siaα2-3Gal oligosaccharides (the ratio of Siaα2-6Gal to Siaα2-3Gal in the duck trachea being approximately 1:20) [87, 88]. Not only Neu5Acα2-3Gal but also Neu5Gcα2-3Gal (not found in chickens) glycans are present in the epithelium of the duck jejunum, cecum, and colon [89]. Correlated with the duck hosts, the duck-isolated influenza viruses preferentially bind to Neu5Ac/Neu5Gcα2-3 receptors (avian receptors) [82, 89–91]. This also agrees with the finding that avian influenza virus isolates replicate efficiently in chorioallantoic cells of 10-day-old chicken embryonated eggs that contain *N*-glycans, which are essential for entry into host cells of influenza virus infection [92], with molar percents of α2-3 linkage and α2-6 linkage of 27.2 and 8.3, respectively [93].

Studies on sialic acid substructure binding specificity of influenza viruses revealed that although most avian viruses share their preferential binding to terminal Neu5Acα2-3Gal, duck-isolated influenza viruses prefer the β1-3 linkage between Neu5Acα2-3Gal and the next sugar residue such as 3′SiaLe^c and 3′SiaTF, whereas gull-isolated influenza viruses show high affinity for the β1-4 linkage such as 3′SLN, for fucosylated receptors such as 3′SiaLe^x, and for sulfated receptors such as Neu5Acα2-3Galβ1-4(6-*O*-HSO_3)GlcNAc (6-*O*-Su-3′SLN) and 6-*O*-Su-3′SiaLe^X (*see* Fig. 3) [90]. These receptor-binding specificity data of influenza viruses are correlated well with intestinal epithelial staining with SNA and MAA lectins showing that the duck intestinal epithelium expressed a high level of Siaα2-3Galβ1-3GalNAc-moieties (preferential to MAA-II), whereas the gull intestinal epithelium dominantly expressed Siaα2-3Galβ1-4GlcNAc-moieties (preferential to MAA-I).

Screening using a virus-receptor binding assay together with molecular modeling revealed that gull-viral HAs with 193R/K displayed increased affinity for 6-*O*-Su-3′SLN and 6-*O*-Su-3′SiaLe^X due to favorable electrostatic interactions of the sulfate group of the receptor and positively charged side chain of 193R/K [94, 95]. The gull-viral HAs with 222Q exhibited binding affinity for the fucosylated receptor 3′SiaLe^x similar to binding affinity for the nonfucosylated counterpart 3′SLN, while duck influenza viruses showed inefficient binding to the fucosylated receptor due to steric interference between its bulky 222K on the HA and the fucose moiety of the receptor [94, 95]. Only some gull-isolated influenza viruses have potential to infect ducks, indicating that there is a host-range restriction between avian species [96].

5.3.2 Poultry

Several influenza A viruses including H1–H13 and N1–N9 subtypes have been isolated from domesticated poultry in the family Phasianidae of the order Galliformes, including turkeys, chickens, quails, and guinea fowls [97, 98]. Adapted avian influenza viruses in domestic poultry can be divided into two main forms according to their capacity to cause low or high virulence in the infected poultry (*see* Subheading 5.2). Both forms of avian isolates from poultry before 2002 mainly bind to α2-3 sialyl linkages using either synthetic sialyloligosaccharides or erythrocytes as molecular probes for influenza virus binding specificity [99–102], but since 2002, some of the isolates have shown an increase in binding to α2-6 sialyl linkages (*see* Subheading 5.3.3). Tissue staining with avian and human influenza viruses and with MAA and SNA lectins has shown the presence of Siaα2-3Gal- and Siaα2-6Gal-terminated sialyloligosaccharides in respiratory and intestinal epithelia of gallinaceous poultry, including chickens and quails [87, 103–108]. However, the proportion of α2-3 and α2-6 sialyl linkages in respiratory and intestinal epithelia of the poultry is still controversial [87, 107, 109]. Thus, more investigations of the structure and distribution of receptors in the replication sites of influenza A viruses are needed to understand the basis of viral infection and transmission.

5.3.3 Human Beings

Human-adapted influenza A viruses that possess efficient human-to-human transmission ability mainly target the human upper respiratory tract, where they can be readily spread with a sneeze or cough. Lectin histochemistry of human respiratory tissues demonstrated that epithelial cells in the upper respiratory tract (nose-larynx) and in the upper part of the lower respiratory tract (trachea and bronchi) are enriched in α2-6 sialylated glycan receptors with a small proportion of α2-3 sialylated glycans [110]; using human airway epithelium (HAE) cells, lectin staining indicated that α2-6-linked sialylated receptors are dominantly present on the surface of nonciliated cells, while α2-3-linked sialylated receptors are present on ciliated cells [111, 112]. In the lower part of the lower respiratory tract (lung), α2-6-sialylated glycans can be found on epithelial cells of the bronchioles and alveolar type-I cells; α2-3-sialylated glycans can be found on nonciliated cuboidal bronchiolar cells and alveolar type-II cells [110, 113]. Recent mass spectromic analysis of glycan structures of human respiratory tract tissues showed that both Sia α2-3 and α2-6 glycans are present in the lung and bronchus [114]. The pattern of lectin localization correlated with the pattern of virus binding and infection: human-adapted viruses bound extensively to bronchial epithelial cells but intensively to alveolar cells, and the opposite results were found for avian viruses [110]; human-adapted viruses and avian viruses preferentially infected nonciliated cells and ciliated cells in the HAE, respectively [111, 112]. Clinically, seasonal influenza viruses mainly infect the upper respiratory tract [115]; however, pulmonary complications of influenza virus infection related to secondary bacterial pneumonia (such as by *Staphylococcus aureus* infection) rather than primary influenza pneumonia can occur, especially in children less than 2 years of age, adults more than 65 years of age, pregnant women, and people with comorbid illnesses/poor nutrition [115, 116]. The 2009 H1N1pdm viruses mostly attack the upper respiratory tract, resulting in subclinical infections or mild upper airway illness, but some are able to replicate in the lower respiratory tract as seen from diffuse alveolar damage in autopsy tissue samples from patients who died from the 2009 H1N1pdm virus. The viruses probably acquire D222G in HAs, leading to dual receptor specificity for α2-3- and α2-6-linked sialic acids [117], and more than 25 % of samples were co-infected with bacteria [118, 119]. Either HPAI H5 or H7 infection in terrestrial poultry spreads rapidly and causes damage throughout the avian body [120], but HPAI H5N1 infection in humans seems to be restricted to the respiratory tract and intestine, and H5N1 virus mainly replicates in pneumocytes, frequently causing death with acute respiratory distress syndrome (ARDS) (fatality rate of about 60 %) [121]. Either HPAI or LPAI H7 infection or LPAI H9, H10, and H6 infection in humans can result in disease in both ocular tissues that predominantly express α2-3-linked Sia receptors [122] and the respiratory tract with uncomplicated influenza-like illness [40, 41, 123, 124], but one veterinary doctor who was infected with HPAI H7N7 virus, in an outbreak in the Netherlands in 2003, died with ARDS [39] and approximately 32 % of people infected with a novel reassortant LPAI (H7N9) virus died from severe pneumonia and breathing difficulties (dyspnea) [42, 81, 125]. The first human case of avian influenza A (H10N8) virus has recently been detected in China in a 73-year-old immunocompromised female, who visited a live bird market and was hospitalized on November 30, 2013 and died of severe pneumonia on December 6, 2013 [44]. Not only epidemiologic surveillance but also molecular surveillance of influenza virus infection has become strengthened for rapid response to outbreaks of influenza virus having potentially unpredictable changes.

In general, influenza viruses evolve with changes in their environment, such as immune response and receptors, until achieving optimal viral fitness. Since 2002, some H5 and H7 poultry isolates, including A/Ck/Egypt/RIMD12-3/2008 (H5N1) of sublineage A [101], A/Tky/VA/4529/02 (H7N2), A/Ck/Conn/260413-2/03 (H7N2) of the North American lineage, and A/Laughing gull/DE/22/02 (H7N3) of the Eurasian lineage, have displayed significantly increased binding to α2-6 sialyl glycans [95, 126]. It should be noted that after the first outbreak of HPAI H5N1 virus in Egypt in 2006 [127], the virus has continued to undergo mutations, resulting in sublineages A–D; at present (2013), sublineages B and D are dominant in Egypt, while sublineage A has not been detected. Hemagglutination of an H9 human isolate, A/Hong Kong/1073/99 (H9N2), with guinea pig erythrocytes was shown to be inhibited by both α2-3 and α2-6-linked sialic acid containing polymers [128]. This characteristic of avian H5, H7, and H9 viruses highlights the possibility of the potential of these avian influenza viruses for development to infect and spread among humans in the future. Similar to H2 and H3 HAs of pandemic H2N2 in 1957 and H3N2 in 1968 (*see* Fig. 7e), Q226L and G228S/N224K mutations in H5 HA [129, 130], Q226L and G228S mutations in H7 HA [131], and Q226L mutation in H9 HA [132] have been experimentally shown to be associated with preferential binding to the α2-6 human-type receptor. Notably, the H5 virus harboring either Q226L-G228S [129] or Q226L-N224K [130] mutation in combination with loss of the 158–161 glycosylation site (N158D or T160A) near the receptor binding pocket and T318I or H107Y substitution in the stalk region (believed to increase the stability of the HA variant) has been shown to have preferential binding to Siaα2-6Gal, efficient respiratory droplet transmission in ferrets, and viral attachment to human tracheal epithelia. Nonetheless, another viral factor(s) has been believed to be involved for avian viruses to gain efficient human-to-human transmission (*see* Subheadings 6 and 7).

5.3.4 Pigs

Pigs serve as intermediate hosts for pandemic generation due to being mixing reservoirs of influenza A viruses, allowing genetic reassortment [133]. Indeed, interspecies transmission of avian and human viruses to pigs and vice versa has been documented in nature [134–136], and the recent pandemic H1N1 2009 has been confirmed to be of swine origin [137]. Lectin staining demonstrated high levels of α2-3 and α2-6 Sia expressed in the porcine respiratory epithelium [88, 133], and HPLC and matrix-assisted laser desorption/ionization time-of-flight mass spectrometry (MALDI-TOF-MS) analyses showed gradually increased molar ratios of α2-6/α2-3-linked sialyl glycans of 3.2-, 4.9-, and 13.2-fold for Neu5Ac and 1.8-, 2.7-, and 5.9-fold for Neu5Gc from the upper trachea and the lower trachea towards the lungs (the major replication site of swine-adapted influenza viruses) of a pig, respectively [77]. Neu5Ac/Neu5Gc ratios are 24.8/4.3 in the swine upper trachea, 27.1/4.1 in the swine lower trachea, 40.5/4.2 in the swine lung [77], and 98/2 in the duck intestine [89], whereas normal human tissues carrying nonfunctional hydroxylase to produce Neu5Gc [138] possess only Neu5Ac if Neu5Gc-containing food such as pork has not been eaten. Most duck-derived and swine-derived influenza A viruses displayed marked binding to Neu5Gc, related to the presence of V/I155 in H1 swine-adapted HAs [139], but they preserved preferential binding to Neu5Ac glycoconjugates, whereas human-adapted influenza viruses showed preferential binding to only Neu5Ac glycoconjugates [89, 140]. The swine-origin pandemic H1N1 2009 virus containing V155 rapidly spreads worldwide. Either T155Y or E158G mutation generated by a reverse genetics system in human H3 HA facilitates virus binding to Neu5Gc but retains strong binding affinity to Neu5Ac [141]. HAlo virus (A/Vietnam/1203/04 (H5N1) virus with removal of the multibasic cleavage site, responsible for high pathogenicity) with Y161A mutation generated by a reverse genetics system showed change of preferential binding from Neu5Ac to Neu5Gc with a five- to tenfold growth defect on MDCK cells [142]. It is still uncertain whether different ratios of Neu5Ac/Neu5Gc among animal species affect potential infection of influenza A viruses. Clearly, avian viruses with α2-3 binding preference would not overcome the interspecies barrier for efficient transmission in humans unless its binding preference is switched to α2-6. Thus, findings that classical swine influenza A viruses bind preferentially to Neu5Acα2-6Gal [88, 118, 143, 144] and that avian-like swine viruses acquired higher binding affinity for Neu5Acα2-6Gal over time [88, 118, 143] suggest that pigs provide a great source of natural selection of virus variants with α2-6 receptor-binding HAs, a prerequisite for a human pandemic.

5.3.5 Other Animals

Epithelial cells of the horse trachea showed prevalence of Siaα2-3Gal using lectin staining with Neu5Gc accounting for more than 90 % of Sia by HPLC analysis. Although most equine influenza viruses display high recognition of Neu5Gcα2-3Gal, they still prefer binding to Neu5Acα2-3Gal [82].

Seal and whale lung cells contain predominately Siaα2-3Gal over Siaα2-6Gal by lectin staining, and both seal and whale viruses prefer to recognize Siaα2-3Gal [145].

6 PB2

Changes in amino acid(s) in the RNA polymerase PB2 subunit resulting in different surface shape and/or charge affecting its protein's interaction with cellular factors have been thought to contribute to efficient transmission of influenza viruses in humans, a characteristic of an influenza virus in a pandemic outbreak. T271A plays roles in (1) acquisition of HA mutation conferring recognition of a human-type receptor and (2) efficient respiratory droplet transmission [146, 147]. E627K/Q591R/D701N facilitates (3) efficient influenza virus replication in the upper respiratory tract of humans and (4) efficient influenza virus replication at 33 °C in the human upper part airway [102].

7 Other Influenza Virus Proteins

Changes in amino acids in other viral proteins, such as PA and NS1, interacting with cellular factors could contribute to the emergence of an influenza pandemic, and further studies are therefore needed to clarify viral factors involved in generation of a potential pandemic virus.

8 Concluding Remarks

Of the three types of influenza viruses, only type A can lead to a pandemic, possibly due to the variety of subtypes originating from wild water fowls that harmoniously interact with the virus in cooperation with the virus's ability to cross the species barrier to infect a variety of animals (*see* Fig. 8). A virus crossing the species barrier to infect a new host species must experience a new environment in the host body including cellular receptors, host factors supporting/against virus replication, and local temperature, and thus is limited unless there is transmission evolution to surmount the species barrier. Of the influenza A viruses crossing into and establishing in

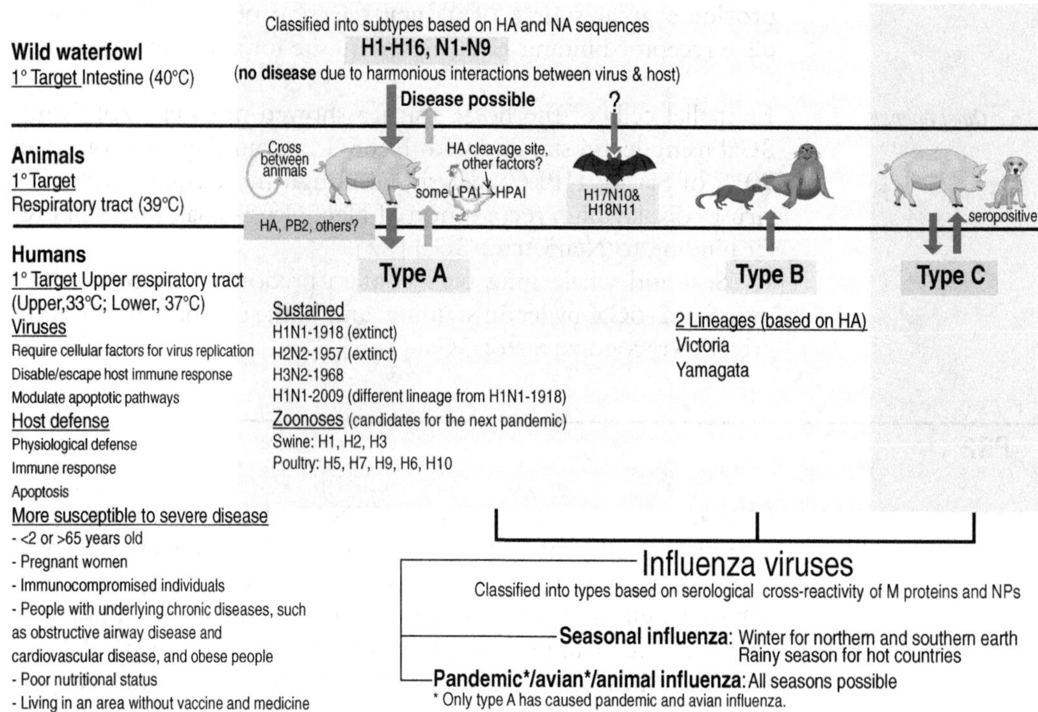

Fig. 8 A summary of human infection with influenza virus and emergence of an influenza pandemic. Airborne influenza in humans is caused by type A, B, or C, prevalent in the rainy season in several tropical regions, such as Thailand, Vietnam, and Brazil, and in winter (November–April for the Northern Hemisphere and May–October for the Southern Hemisphere), due to virus stability and host vulnerability to infection. Types A and B cause annual epidemic influenza typically due to virus antigenic variation (minor change); generally, they do not cause severe disease except in people under conditions as indicated [115, 116]. Type C usually causes only mild illness in humans [148]. Type B is restricted to humans, though occasionally found in seals [149] and ferrets [150], type C can be isolated from humans and pigs and is found in seropositive dogs [151], and type A viruses have natural reservoirs in the intestinal tract (40 °C) of wild birds (harmonious interactions with the virus allowing the production of variety of subtypes without selective pressure) and are transmitted via feces to domestic birds (intestinal tract, 40 °C) and from domestic birds to other animals (respiratory tract of pigs, 39 °C) if they are able to fine-tune for transmission to and replication in the new host. So far, only H1N1 Spanish/18, H2N2 Japan/57, H3N2 Hong Kong/68, and H1N1 Swine/09 have been successfully established in humans. Currently only H3N2 Hong Kong/68 and H1N1/09 variants are circulating in humans. Epizootic influenza A viruses still cross to humans occasionally, and thus qualitative surveillance of the next pandemic zoonoses is needed. *See* color figure in the online version

terrestrial poultry, some of the LPAI H5 and H7 subtypes have evolved into HPAI viruses with a universal pathogenic marker of a multibasic cleavage site causing systemic infection with a mortality rate as high as 100 % in poultry [152]. It was virtually unknown what factors in poultry drive the virus to acquire an HPAI property and why the HPAI viruses have continued to circulate in poultry despite the fact that a rapid and high fatality rate due to the HPAI property could result in a dead end for virus transmission. These questions challenge researchers to unravel the ultimate selection parameter for survival of the fittest during virus-host co-evolution, each of which has evolved to prevail with the host attempting to escape from or restrict the infection by various means including host immunity and apoptosis and the virus evolving strategies to block/evade host clearance mechanisms and to support its propagation including entry to the host cell and use of the host cellular machinery for each step of its life cycle [153]. LPAI and HPAI viruses can sometimes infect other animals but with limited transmission, including wild birds [154], pigs [75], and humans [155], and cause mild to severe and fatal diseases (assessed by the number of severe cases and deaths) depending on the followings: (1) environmental factors, such as cold weather facilitating viral infection, (2) susceptibility and response of each host, which are accounted for by host genetics and other factors including age and health status, and (3) virus strain, with each virus strain having distinct pathogenic profiles in different host species: more research is needed to understand viral pathogenesis. For efficient transmission in humans, a nonhuman virus acquires mutations through either an adaptation or reassortment mechanism for sufficient human-to-human transmission, providing a chance for the virus to further evolve until it can achieve suitable interactions with host factors for efficient replication and transmission in human populations and eventually leading to a pandemic. Viral factors believed to play key roles in generating an influenza virus with pandemic potential are HA and PB2. Homotrimeric HA carries the followings: (1) antigenic sites, which if new, are not recognized by the host immune system, (2) receptor binding sites, which if they have 190D and 225D in H1 HA, 226L together with 228S (or 224K in H5 HA) in H2, H3, H5, or H7 HA, and 226L in H9 HA, are believed to confer virus preferential binding to a human-type receptor, (3) glycosylation sites, which if there is loss of glycosylation at 158–160, are believed to enhance HPAI H5N1 virus binding to a human-type receptor, and (4) stalk domains with T318A or H107Y, possibly being acquired for sustained and efficient human-to-human transmission of the H5 HA variants. Viral PB2 with 271A together with 627K/591R/701N enhances HA binding to human-type receptors, enhances respiratory droplet transmission, supports viral growth in a mammalian host at 33 °C, and facilitates efficient viral replication in the human upper respiratory tract. Investigation should be continued to identify other factors involved in efficient influenza virus replication and transmission in humans for use as viral genetic markers for early surveillance of the emergence of a pandemic.

Although there has been an accumulation of information on influenza; (1) new variants have emerged, (2) avian viruses have occasionally infected other animals including intermediate hosts carrying both avian-type and human-type receptors, such as pheasants, turkeys, quails, and guinea fowls, with pigs in particularly tending to drive the virus binding to α2-6 human-type receptors, and (3) an influenza pandemic is still unpredictable. HPAI H5N1 viruses have caused sporadic human infections since 1997, and the pandemic phase is currently at level 3 (*see* Table 2, small clusters of disease in people). A new LPAI H7N9 virus reassorted between avian viruses in poultry just crossed species to infect humans in February 2013 and is now (January 2014) at pandemic level 3. Except for the pandemic 1918 virus, past pandemics emerged from an existing human strain picking up new genes (new to human immunity but efficient replication in the human upper respiratory tract) from an avian and/or swine virus(es). This evidence suggested that new variants that have emerged via reassortment acquire major change in their genetic materials fitting with a new host condition more easily and faster than do variants that have emerged via point mutations alone, which have gradual changes in their genetic materials. Thus, avoiding intermediate host infection with more than one influenza virus strain should be important for preventing/delaying the next pandemic. More knowledge of the molecular requirements of reassortment at levels of viral and host factors could lead to a better understanding of how appropriate viruses emerge, leading to strategies for efficient prevention and antiviral interventions. Identifying viral and host factors, especially knowledge gained from their interaction structures, required for efficient replication in each host species may be a key for understanding virus–host determinants and surveillance of viral host jumps and pathogenesis of influenza virus infection leading to the disease.

Available data have suggested that an influenza pandemic has never emerged through direct viral mutations alone. However, highly mutable avian influenza A viruses that have sporadically continued direct transmission to and infection in humans have raised concerns for pandemic potential with unpredictable pathogenesis (depending on virus–host interactions). HPAI H5N1 viruses have continued to infect humans with high morbidity and mortality rates, some isolates showing increased binding to α2-6 human-type receptors, and they are able to infect a variety of animals including wild birds and pigeons, which are responsible for introduction of the viruses they carry into different areas, pigs, which are mixing vessels driving the virus to bind to human-type receptors, and cats, which are in close contact with human beings [156]. Also, novel reassortant LPAI (invisible disease in domestic poultry) H7N9 viruses contain some mammalian flu adaptations, PB2-627K and 226L, and they target upper and lower respiratory tracts of infected primates [157] and cause severe illness with a high death rate in humans. In addition to surveillance of human infection, extensive surveillance of infection of these viruses to other animals and back to migratory birds should be carried out since control of the viral spread into other regions relies on early recognition. Continuing surveillance is important for understanding how a pandemic emerges and establishing strategies for efficient control and treatment if a pandemic arises as well as for prevention and control of the next pandemic. The best way for preventing influenza spread and a pandemic is to avoid direct contact with materials having suspected contamination as well as hygiene in healthcare for both animals and farmers, especially in mixed duck–poultry–pig farms.

References

1. Smith FI, Palese P (1989) Variation in influenza virus genes: epidemiological, pathogenic, and evolutionary consequences. In: Krug RM (ed) The influenza viruses. Plenum, New York, pp 319–359
2. Tong S, Li Y, Rivailler P et al (2012) A distinct lineage of influenza A virus from bats. Proc Natl Acad Sci U S A 109:4269–4274
3. Tong S, Zhu X, Li Y et al (2013) New world bats harbor diverse influenza a viruses. PLoS Pathog 9:e1003657
4. Sun X, Shi Y, Lu X et al (2013) Bat-derived influenza hemagglutinin H17 does not bind canonical avian or human receptors and most likely uses a unique entry mechanism. Cell Rep 3:769–778
5. Foeglein A, Loucaides EM, Mura M et al (2011) Influence of PB2 host-range determinants on the intranuclear mobility of the influenza A virus polymerase. J Gen Virol 92: 1650–1661
6. Graef KM, Vreede FT, Lau YF et al (2010) The PB2 subunit of the influenza virus RNA polymerase affects virulence by interacting with the mitochondrial antiviral signaling protein and inhibiting expression of beta interferon. J Virol 84:8433–8445
7. Portela A, Digard P (2002) The influenza virus nucleoprotein: a multifunctional RNA-binding protein pivotal to virus replication. J Gen Virol 83:723–734
8. Tripathi S, Batra J, Cao W et al (2013) Influenza A virus nucleoprotein induces apoptosis in human airway epithelial cells: implications of a novel interaction between nucleoprotein and host protein Clusterin. Cell Death Dis 4:e562
9. Martin K, Helenius A (1991) Nuclear transport of influenza virus ribonucleoproteins: the viral matrix protein (M1) promotes export and inhibits import. Cell 67:117–130
10. Watanabe K, Handa H, Mizumoto K et al (1996) Mechanism for inhibition of influenza virus RNA polymerase activity by matrix protein. J Virol 70:241–247
11. Ali A, Avalos RT, Ponimaskin E et al (2000) Influenza virus assembly: effect of influenza virus glycoproteins on the membrane association of M1 protein. J Virol 74:8709–8719
12. Zamarin D, Garcia-Sastre A, Xiao X et al (2005) Influenza virus PB1-F2 protein induces cell death through mitochondrial ANT3 and VDAC1. PLoS Pathog 1:e4
13. Varga ZT, Grant A, Manicassamy B et al (2012) Influenza virus protein PB1-F2 inhibits the induction of type I interferon by binding to MAVS and decreasing mitochondrial membrane potential. J Virol 86:8359–8366
14. Wise HM, Foeglein A, Sun J et al (2009) A complicated message: Identification of a novel PB1-related protein translated from influenza A virus segment 2 mRNA. J Virol 83:8021–8031
15. Hale BG, Randall RE, Ortin J et al (2008) The multifunctional NS1 protein of influenza A viruses. J Gen Virol 89:2359–2376
16. Tsai PL, Chiou NT, Kuss S et al (2013) Cellular RNA binding proteins NS1-BP and hnRNP K regulate influenza A virus RNA splicing. PLoS Pathog 9:e1003460
17. Jiao P, Tian G, Li Y et al (2008) A single-amino-acid substitution in the NS1 protein changes the pathogenicity of H5N1 avian influenza viruses in mice. J Virol 82:1146–1154
18. Zhirnov OP, Klenk HD (2007) Control of apoptosis in influenza virus-infected cells by up-regulation of Akt and p53 signaling. Apoptosis 12:1419–1432
19. Selman M, Dankar SK, Forbes NE et al (2012) Adaptive mutation in influenza A virus non-structural gene is linked to host switching and induces a novel protein by alternative splicing. Emerg Microbes Infect 1:e42
20. Garcia-Sastre A (2011) Induction and evasion of type I interferon responses by influenza viruses. Virus Res 162:12–18
21. Kobasa D, Kodihalli S, Luo M et al (1999) Amino acid residues contributing to the substrate specificity of the influenza A virus neuraminidase. J Virol 73:6743–6751
22. Matrosovich MN, Matrosovich TY, Gray T et al (2004) Neuraminidase is important for the initiation of influenza virus infection in human airway epithelium. J Virol 78: 12665–12667
23. Schmolke M, Garcia-Sastre A (2010) Evasion of innate and adaptive immune responses by influenza A virus. Cell Microbiol 12:873–880

24. Rimmelzwaan GF, Berkhoff EG, Nieuwkoop NJ et al (2004) Functional compensation of a detrimental amino acid substitution in a cytotoxic-T-lymphocyte epitope of influenza a viruses by comutations. J Virol 78: 8946–8949
25. Chakrabarti AK, Pasricha G (2013) An insight into the PB1F2 protein and its multifunctional role in enhancing the pathogenicity of the influenza A viruses. Virology 440:97–104
26. Suzuki T, Takahashi T, Guo CT et al (2005) Sialidase activity of influenza A virus in an endocytic pathway enhances viral replication. J Virol 79:11705–11715
27. Su B, Wurtzer S, Rameix-Welti MA et al (2009) Enhancement of the influenza A hemagglutinin (HA)-mediated cell-cell fusion and virus entry by the viral neuraminidase (NA). PLoS One 4:e8495
28. Schnell JR, Chou JJ (2008) Structure and mechanism of the M2 proton channel of influenza A virus. Nature 451:591–595
29. Takeda M, Leser GP, Russell CJ et al (2003) Influenza virus hemagglutinin concentrates in lipid raft microdomains for efficient viral fusion. Proc Natl Acad Sci U S A 100: 14610–14617
30. Zhang J, Pekosz A, Lamb RA (2000) Influenza virus assembly and lipid raft microdomains: a role for the cytoplasmic tails of the spike glycoproteins. J Virol 74:4634–4644
31. Barry JM (2004) The great influenza: the epic story of the deadliest plague in history. Viking, New York
32. Ito T, Kawaoka Y (2000) Host-range barrier of influenza A viruses. Vet Microbiol 74:71–75
33. Webster RG, Yakhno M, Hinshaw VS et al (1978) Intestinal influenza: replication and characterization of influenza viruses in ducks. Virology 84:268–278
34. WHO website. Current WHO phase of pandemic alert for Pandemic (H1N1) 2009. http://www.who.int/csr/disease/swineflu/phase/en/. Accessed 26 Nov 2013
35. Holmes EC, Ghedin E, Miller N et al (2005) Whole-genome analysis of human influenza A virus reveals multiple persistent lineages and reassortment among recent H3N2 viruses. PLoS Biol 3:e300
36. CDC website. Avian influenza A virus infections of humans. http://www.cdc.gov/flu/avian/gen-info/avian-flu-humans.htm. Accessed 26 Nov 26 2013
37. Claas EC, Osterhaus AD, van Beek R et al (1998) Human influenza A H5N1 virus related to a highly pathogenic avian influenza virus. Lancet 351:472–477
38. Subbarao K, Klimov A, Katz J et al (1998) Characterization of an avian influenza A (H5N1) virus isolated from a child with a fatal respiratory illness. Science 279:393–396
39. Belser JA, Zeng H, Katz JM et al (2011) Infection with highly pathogenic H7 influenza viruses results in an attenuated proinflammatory cytokine and chemokine response early after infection. J Infect Dis 203:40–48
40. Hirst M, Astell CR, Griffith M et al (2004) Novel avian influenza H7N3 strain outbreak, British Columbia. Emerg Infect Dis 10: 2192–2195
41. Arzey GG, Kirkland PD, Arzey KE et al (2012) Influenza virus A (H10N7) in chickens and poultry abattoir workers, Australia. Emerg Infect Dis 18:814–816
42. Gao R, Cao B, Hu Y et al (2013) Human infection with a novel avian-origin influenza A (H7N9) virus. N Engl J Med 368: 1888–1897
43. Yuan J, Zhang L, Kan X et al (2013) Origin and molecular characteristics of a novel 2013 avian influenza A(H6N1) virus causing human infection in Taiwan. Clin Infect Dis 57:1367–1368
44. WHO website. Emerging disease surveillance and response (avian influenza). http://www.wpro.who.int/emerging_diseases/AvianInfluenza/en/. Accessed 9 Jan 2014
45. Horimoto T, Kawaoka Y (2005) Influenza: lessons from past pandemics, warnings from current incidents. Nat Rev Microbiol 3:591–600
46. Palese P (2004) Influenza: old and new threats. Nat Med 10:S82–S87
47. Pica N, Hai R, Krammer F et al (2012) Hemagglutinin stalk antibodies elicited by the 2009 pandemic influenza virus as a mechanism for the extinction of seasonal H1N1 viruses. Proc Natl Acad Sci U S A 109:2573–2578
48. Simonsen L, Clarke MJ, Schonberger LB et al (1998) Pandemic versus epidemic influenza mortality: a pattern of changing age distribution. J Infect Dis 178:53–60
49. Dawood FS, Jain S, Finelli L et al (2009) Emergence of a novel swine-origin influenza A (H1N1) virus in humans. N Engl J Med 360:2605–2615
50. Xu R, Ekiert DC, Krause JC et al (2010) Structural basis of preexisting immunity to the 2009 H1N1 pandemic influenza virus. Science 328:357–360
51. Sriwilaijaroen N, Suzuki Y (2012) Molecular basis of the structure and function of H1 hemagglutinin of influenza virus. Proc Jpn Acad Ser B Phys Biol Sci 88:226–249
52. Manicassamy B, Medina RA, Hai R et al (2010) Protection of mice against lethal challenge with 2009 H1N1 influenza A virus by

1918-like and classical swine H1N1 based vaccines. PLoS Pathog 6:e1000745
53. Cohen J (2010) Swine flu pandemic. What's old is new: 1918 virus matches 2009 H1N1 strain. Science 327:1563–1564
54. Russell RJ, Kerry PS, Stevens DJ et al (2008) Structure of influenza hemagglutinin in complex with an inhibitor of membrane fusion. Proc Natl Acad Sci U S A 105:17736–17741
55. Sauter NK, Glick GD, Crowther RL et al (1992) Crystallographic detection of a second ligand binding site in influenza virus hemagglutinin. Proc Natl Acad Sci U S A 89:324–328
56. Raymond FL, Caton AJ, Cox NJ et al (1986) The antigenicity and evolution of influenza H1 haemagglutinin, from 1950–1957 and 1977–1983: two pathways from one gene. Virology 148:275–287
57. Stray SJ, Pittman LB (2012) Subtype- and antigenic site-specific differences in biophysical influences on evolution of influenza virus hemagglutinin. Virol J 9:91
58. Wiley DC, Wilson IA, Skehel JJ (1981) Structural identification of the antibody-binding sites of Hong Kong influenza haemagglutinin and their involvement in antigenic variation. Nature 289:373–378
59. Wilson IA, Cox NJ (1990) Structural basis of immune recognition of influenza virus hemagglutinin. Annu Rev Immunol 8:737–771
60. Kawaoka Y, Bean WJ, Webster RG (1989) Evolution of the hemagglutinin of equine H3 influenza viruses. Virology 169:283–292
61. Luoh SM, McGregor MW, Hinshaw VS (1992) Hemagglutinin mutations related to antigenic variation in H1 swine influenza viruses. J Virol 66:1066–1073
62. Sakai K, Kawaguchi Y, Kishino Y et al (1993) Electron immunohistochemical localization in rat bronchiolar epithelial cells of tryptase Clara, which determines the pneumotropism and pathogenicity of Sendai virus and influenza virus. J Histochem Cytochem 41:89–93
63. Webster RG, Bean WJ, Gorman OT et al (1992) Evolution and ecology of influenza A viruses. Microbiol Rev 56:152–179
64. Woo GH, Kim HY, Bae YC et al (2011) Comparative histopathological characteristics of highly pathogenic avian influenza (HPAI) in chickens and domestic ducks in 2008 Korea. Histol Histopathol 26:167–175
65. Banks J, Speidel ES, Moore E et al (2001) Changes in the haemagglutinin and the neuraminidase genes prior to the emergence of highly pathogenic H7N1 avian influenza viruses in Italy. Arch Virol 146:963–973
66. Webster RG, Rott R (1987) Influenza virus A pathogenicity: the pivotal role of hemagglutinin. Cell 50:665–666
67. Ito T, Goto H, Yamamoto E et al (2001) Generation of a highly pathogenic avian influenza A virus from an avirulent field isolate by passaging in chickens. J Virol 75:4439–4443
68. Perdue ML, Garcia M, Senne D et al (1997) Virulence-associated sequence duplication at the hemagglutinin cleavage site of avian influenza viruses. Virus Res 49:173–186
69. Abdel-Ghafar AN, Chotpitayasunondh T, Gao Z et al (2008) Update on avian influenza A (H5N1) virus infection in humans. N Engl J Med 358:261–273
70. WHO website. Updated unified nomenclature system for the highly pathogenic H5N1 avian influenza viruses. http://www.who.int/influenza/gisrs_laboratory/h5n1_nomenclature/en/. Accessed 1 Dec 1 2013
71. Keawcharoen J, van Riel D, van Amerongen G et al (2008) Wild ducks as long-distance vectors of highly pathogenic avian influenza virus (H5N1). Emerg Infect Dis 14:600–607
72. Brown JD, Stallknecht DE, Beck JR et al (2006) Susceptibility of North American ducks and gulls to H5N1 highly pathogenic avian influenza viruses. Emerg Infect Dis 12:1663–1670
73. Kim JK, Negovetich NJ, Forrest HL et al (2009) Ducks: the "Trojan horses" of H5N1 influenza. Influenza Other Respir Viruses 3:121–128
74. Cui Z, Hu J, He L et al (2013) Differential immune response of mallard duck peripheral blood mononuclear cells to two highly pathogenic avian influenza H5N1 viruses with distinct pathogenicity in mallard ducks. Arch Virol 159:339–343
75. Chen H, Smith GJ, Zhang SY et al (2005) Avian flu: H5N1 virus outbreak in migratory waterfowl. Nature 436:191–192
76. Nidom CA, Takano R, Yamada S et al (2010) Influenza A (H5N1) viruses from pigs, Indonesia. Emerg Infect Dis 16:1515–1523
77. Sriwilaijaroen N, Kondo S, Yagi H et al (2011) *N*-Glycans from porcine trachea and lung: predominant NeuAcα2-6Gal could be a selective pressure for influenza variants in favor of human-type receptor. PLoS One 6:e16302
78. Vijaykrishna D, Bahl J, Riley S et al (2008) Evolutionary dynamics and emergence of panzootic H5N1 influenza viruses. PLoS Pathog 4:e1000161
79. CDC website. Highly pathogenic avian influenza A (H5N1) in birds and other animals. http://www.cdc.gov/flu/avianflu/h5n1-animals.htm. Accessed 26 Nov 2013
80. WHO website. Current WHO phase of pandemic alert for avian influenza H5N1. http://apps.who.int/csr/disease/avian_influenza/phase/en/index.html. Accessed 26 Nov 2013

81. WHO website. Human infection with avian influenza A(H7N9) virus – update. http://www.who.int/csr/don/2013_11_06/en/index.html?utm_source=twitterfeed&utm_medium=twitter. Accessed 26 Nov 2013
82. Suzuki Y, Ito T, Suzuki T et al (2000) Sialic acid species as a determinant of the host range of influenza A viruses. J Virol 74:11825–11831
83. Wang B, Brand-Miller J (2003) The role and potential of sialic acid in human nutrition. Eur J Clin Nutr 57:1351–1369
84. Suzuki Y (2005) Sialobiology of influenza: molecular mechanism of host range variation of influenza viruses. Biol Pharm Bull 28:399–408
85. Paulson JC, de Vries RP (2013) H5N1 receptor specificity as a factor in pandemic risk. Virus Res 178:99–113
86. Gilbert M, Xiao X, Domenech J et al (2006) Anatidae migration in the western Palearctic and spread of highly pathogenic avian influenza H5N1 virus. Emerg Infect Dis 12:1650–1656
87. Kuchipudi SV, Nelli R, White GA et al (2009) Differences in influenza virus receptors in chickens and ducks: implications for interspecies transmission. J Mol Genet Med 3:143–151
88. Ito T, Couceiro JN, Kelm S et al (1998) Molecular basis for the generation in pigs of influenza A viruses with pandemic potential. J Virol 72:7367–7373
89. Ito T, Suzuki Y, Suzuki T et al (2000) Recognition of *N*-glycolylneuraminic acid linked to galactose by the α2,3 linkage is associated with intestinal replication of influenza A virus in ducks. J Virol 74:9300–9305
90. Gambaryan A, Yamnikova S, Lvov D et al (2005) Receptor specificity of influenza viruses from birds and mammals: new data on involvement of the inner fragments of the carbohydrate chain. Virology 334:276–283
91. Masuda H, Suzuki T, Sugiyama Y et al (1999) Substitution of amino acid residue in influenza A virus hemagglutinin affects recognition of sialyl-oligosaccharides containing *N*-glycolylneuraminic acid. FEBS Lett 464: 71–74
92. Chu VC, Whittaker GR (2004) Influenza virus entry and infection require host cell *N*-linked glycoprotein. Proc Natl Acad Sci U S A 101:18153–18158
93. Sriwilaijaroen N, Kondo S, Yagi H et al (2009) Analysis of *N*-glycans in embryonated chicken egg chorioallantoic and amniotic cells responsible for binding and adaptation of human and avian influenza viruses. Glycoconj J 26:433–443
94. Gambaryan A, Tuzikov A, Pazynina G et al (2006) Evolution of the receptor binding phenotype of influenza A (H5) viruses. Virology 344:432–438
95. Gambaryan AS, Matrosovich TY, Philipp J et al (2012) Receptor-binding profiles of H7 subtype influenza viruses in different host species. J Virol 86:4370–4379
96. Kawaoka Y, Chambers TM, Sladen WL et al (1988) Is the gene pool of influenza viruses in shorebirds and gulls different from that in wild ducks? Virology 163:247–250
97. Senne DA (2003) Avian influenza in the Western Hemisphere including the Pacific Islands and Australia. Avian Dis 47:798–805
98. Chen H, Bu Z, Wang J (2008) Epidemiology and control of H5N1 avian influenza in China. In: Klenk H-D, Matrosovich MN, Stech J (eds) Avian influenza (monographs in virology), vol 27. Karger, Basel, pp 27–40
99. Lu X, Qi J, Shi Y et al (2013) Structure and receptor binding specificity of hemagglutinin H13 from avian influenza A virus H13N6. J Virol 87:9077–9085
100. Petersen H, Matrosovich M, Pleschka S et al (2012) Replication and adaptive mutations of low pathogenic avian influenza viruses in tracheal organ cultures of different avian species. PLoS One 7:e42260
101. Watanabe Y, Ibrahim MS, Ellakany HF et al (2011) Acquisition of human-type receptor binding specificity by new H5N1 influenza virus sublineages during their emergence in birds in Egypt. PLoS Pathog 7:e1002068
102. Yamada S, Hatta M, Staker BL et al (2010) Biological and structural characterization of a host-adapting amino acid in influenza virus. PLoS Pathog 6:e1001034
103. Gambaryan A, Webster R, Matrosovich M (2002) Differences between influenza virus receptors on target cells of duck and chicken. Arch Virol 147:1197–1208
104. Gambaryan AS, Tuzikov AB, Bovin NV et al (2003) Differences between influenza virus receptors on target cells of duck and chicken and receptor specificity of the 1997 H5N1 chicken and human influenza viruses from Hong Kong. Avian Dis 47:1154–1160
105. Guo CT, Takahashi N, Yagi H et al (2007) The quail and chicken intestine have sialylgalactose sugar chains responsible for the binding of influenza A viruses to human type receptors. Glycobiology 17:713–724
106. Pillai SP, Lee CW (2010) Species and age related differences in the type and distribution of influenza virus receptors in different tissues of chickens, ducks and turkeys. Virol J 7:5
107. Wan H, Perez DR (2006) Quail carry sialic acid receptors compatible with binding of avian and human influenza viruses. Virology 346:278–286

108. Liu Y, Han C, Wang X et al (2009) Influenza A virus receptors in the respiratory and intestinal tracts of pigeons. Avian Pathol 38:263–266
109. Kim JA, Ryu SY, Seo SH (2005) Cells in the respiratory and intestinal tracts of chickens have different proportions of both human and avian influenza virus receptors. J Microbiol 43:366–369
110. Shinya K, Ebina M, Yamada S et al (2006) Avian flu: influenza virus receptors in the human airway. Nature 440:435–436
111. Matrosovich MN, Matrosovich TY, Gray T et al (2004) Human and avian influenza viruses target different cell types in cultures of human airway epithelium. Proc Natl Acad Sci U S A 101:4620–4624
112. Thompson CI, Barclay WS, Zambon MC et al (2006) Infection of human airway epithelium by human and avian strains of influenza a virus. J Virol 80:8060–8068
113. van Riel D, Munster VJ, de Wit E et al (2007) Human and avian influenza viruses target different cells in the lower respiratory tract of humans and other mammals. Am J Pathol 171:1215–1223
114. Walther T, Karamanska R, Chan RW et al (2013) Glycomic analysis of human respiratory tract tissues and correlation with influenza virus infection. PLoS Pathog 9:e1003223
115. Bouvier NM, Lowen AC (2010) Animal models for influenza virus pathogenesis and transmission. Viruses 2:1530–1563
116. Rothberg MB, Haessler SD, Brown RB (2008) Complications of viral influenza. Am J Med 121:258–264
117. Chutinimitkul S, Herfst S, Steel J et al (2010) Virulence-associated substitution D222G in the hemagglutinin of 2009 pandemic influenza A(H1N1) virus affects receptor binding. J Virol 84:11802–11813
118. Childs RA, Palma AS, Wharton S et al (2009) Receptor-binding specificity of pandemic influenza A (H1N1) 2009 virus determined by carbohydrate microarray. Nat Biotechnol 27:797–799
119. Shieh WJ, Blau DM, Denison AM et al (2010) 2009 Pandemic influenza A (H1N1): pathology and pathogenesis of 100 fatal cases in the United States. Am J Pathol 177: 166–175
120. Peiris JS, de Jong MD, Guan Y (2007) Avian influenza virus (H5N1): a threat to human health. Clin Microbiol Rev 20:243–267
121. Uiprasertkul M, Puthavathana P, Sangsiriwut K et al (2005) Influenza A H5N1 replication sites in humans. Emerg Infect Dis 11:1036–1041
122. Olofsson S, Kumlin U, Dimock K et al (2005) Avian influenza and sialic acid receptors: more than meets the eye? Lancet Infect Dis 5:184–188
123. Nguyen-Van-Tam JS, Nair P, Acheson P et al (2006) Outbreak of low pathogenicity H7N3 avian influenza in UK, including associated case of human conjunctivitis. Euro Surveill 11:E060504.2
124. Editorialteam (2007) Avian influenza A/(H7N2) outbreak in the United Kingdom. Euro Surveill 12:E070531.2
125. Morens DM, Taubenberger JK, Fauci AS (2013) Pandemic influenza viruses—hoping for the road not taken. N Engl J Med 368:2345–2348
126. Belser JA, Blixt O, Chen LM et al (2008) Contemporary North American influenza H7 viruses possess human receptor specificity: Implications for virus transmissibility. Proc Natl Acad Sci U S A 105:7558–7563
127. Balicer RD, Reznikovich S, Berman E et al (2007) Multifocal avian influenza (H5N1) outbreak. Emerg Infect Dis 13:1601–1603
128. Saito T, Lim W, Suzuki T et al (2001) Characterization of a human H9N2 influenza virus isolated in Hong Kong. Vaccine 20: 125–133
129. Herfst S, Schrauwen EJ, Linster M et al (2012) Airborne transmission of influenza A/H5N1 virus between ferrets. Science 336:1534–1541
130. Imai M, Watanabe T, Hatta M et al (2012) Experimental adaptation of an influenza H5 HA confers respiratory droplet transmission to a reassortant H5 HA/H1N1 virus in ferrets. Nature 486:420–428
131. Srinivasan K, Raman R, Jayaraman A et al (2013) Quantitative description of glycan-receptor binding of influenza A virus H7 hemagglutinin. PLoS One 8:e49597
132. Wan H, Perez DR (2007) Amino acid 226 in the hemagglutinin of H9N2 influenza viruses determines cell tropism and replication in human airway epithelial cells. J Virol 81:5181–5191
133. Kida H, Ito T, Yasuda J et al (1994) Potential for transmission of avian influenza viruses to pigs. J Gen Virol 75(Pt 9):2183–2188
134. Brown IH (2000) The epidemiology and evolution of influenza viruses in pigs. Vet Microbiol 74:29–46
135. Pensaert M, Ottis K, Vandeputte J et al (1981) Evidence for the natural transmission of influenza A virus from wild ducts to swine and its potential importance for man. Bull World Health Organ 59:75–78
136. Ottis K, Sidoli L, Bachmann PA et al (1982) Human influenza A viruses in pigs: isolation of a H3N2 strain antigenically related to A/England/42/72 and evidence for continuous circulation of human viruses in the pig population. Arch Virol 73:103–108
137. Smith GJ, Vijaykrishna D, Bahl J et al (2009) Origins and evolutionary genomics of the

2009 swine-origin H1N1 influenza A epidemic. Nature 459:1122–1125
138. Irie A, Koyama S, Kozutsumi Y et al (1998) The molecular basis for the absence of *N*-glycolylneuraminic acid in humans. J Biol Chem 273:15866–15871
139. Matrosovich MN, Klenk HD, Kawaoka Y (2006) Receptor specificity, host-range, and pathogenicity of influenza viruses. In: Kawaoka Y (ed) Influenza virology: current topics. Caister Academic Press, Wymondham, pp 95–137
140. Suzuki T, Horiike G, Yamazaki Y et al (1997) Swine influenza virus strains recognize sialylsugar chains containing the molecular species of sialic acid predominantly present in the swine tracheal epithelium. FEBS Lett 404: 192–196
141. Takahashi T, Hashimoto A, Maruyama M et al (2009) Identification of amino acid residues of influenza A virus H3 HA contributing to the recognition of molecular species of sialic acid. FEBS Lett 583:3171–3174
142. Wang M, Tscherne DM, McCullough C et al (2012) Residue Y161 of influenza virus hemagglutinin is involved in viral recognition of sialylated complexes from different hosts. J Virol 86:4455–4462
143. Matrosovich M, Tuzikov A, Bovin N et al (2000) Early alterations of the receptor-binding properties of H1, H2, and H3 avian influenza virus hemagglutinins after their introduction into mammals. J Virol 74: 8502–8512
144. Gambaryan AS, Karasin AI, Tuzikov AB et al (2005) Receptor-binding properties of swine influenza viruses isolated and propagated in MDCK cells. Virus Res 114:15–22
145. Ito T, Kawaoka Y, Nomura A et al (1999) Receptor specificity of influenza A viruses from sea mammals correlates with lung sialyloligosaccharides in these animals. J Vet Med Sci 61:955–958
146. Bussey KA, Bousse TL, Desmet EA et al (2010) PB2 residue 271 plays a key role in enhanced polymerase activity of influenza A viruses in mammalian host cells. J Virol 84:4395–4406
147. Zhang Y, Zhang Q, Gao Y et al (2012) Key molecular factors in hemagglutinin and PB2 contribute to efficient transmission of the 2009 H1N1 pandemic influenza virus. J Virol 86:9666–9674
148. Kauppila J, Ronkko E, Juvonen R et al (2013) Influenza C virus infection in military recruits-symptoms and clinical manifestation. J Med Virol 86:879–885
149. Bodewes R, Morick D, de Mutsert G et al (2013) Recurring influenza B virus infections in seals. Emerg Infect Dis 19:511–512
150. Francis T Jr (1940) A new type of virus from epidemic influenza. Science 92:405–408
151. Crescenzo-Chaigne B, van der Werf S (2007) Rescue of influenza C virus from recombinant DNA. J Virol 81:11282–11289
152. Alexander DJ (2000) A review of avian influenza in different bird species. Vet Microbiol 74:3–13
153. Konstantinov K, Foisner R, Byrd D et al (1995) Integral membrane proteins associated with the nuclear lamina are novel autoimmune antigens of the nuclear envelope. Clin Immunol Immunopathol 74:89–99
154. Kim HR, Lee YJ, Park CK et al (2012) Highly pathogenic avian influenza (H5N1) outbreaks in wild birds and poultry, South Korea. Emerg Infect Dis 18:480–483
155. Peiris JS (2009) Avian influenza viruses in humans. Rev Sci Tech 28:161–173
156. WHO website. Influenza at the human-animal interface (HAI). http://www.who.int/influenza/human_animal_interface/en/. Accessed 28 Nov 2013
157. Watanabe T, Kiso M, Fukuyama S et al (2013) Characterization of H7N9 influenza A viruses isolated from humans. Nature 501:551–555

Chapter 39

Basic Procedure of X-Ray Crystallography for Analysis of Lectin–Sugar Interactions

Zui Fujimoto

Abstract

Most three-dimensional structures of lectins have been determined by X-ray crystallography. This method determines the molecular structure using X-ray diffraction of a crystal, thereby providing structural information at the atomic level. In this chapter, an overview of the method for protein crystallography is briefly introduced, including a description of several techniques for analysis of the molecular and sugar-binding structure of lectins.

Key words Carbohydrate-binding, Crystal structure, Crystallization, Protein data bank, X-ray crystallography

1 Introduction

Lectin–sugar interactions can be studied by mutagenesis and three-dimensional structural analyses. The three-dimensional structure of lectins provides a clear insight into a protein fold, carbohydrate-binding structure, and illustrates the mechanism of sugar-recognizing specificity. The first three-dimensional structure of a lectin was determined for concanavalin A by X-ray crystallography [1, 2]. With progress in structural biology and development of analytical techniques, the number of lectin structures has been increasing year by year and by now thousands of entries for lectin and lectin-like proteins have accumulated in the Protein Data Bank (PDB, http://www.rcsb.org/pdb/). The structure of a protein is determined primarily by three methods: X-ray crystallography, NMR spectroscopy, and electron microscopy. Most of the three-dimensional structures of lectins have been determined by X-ray crystallography because it is a powerful and direct method of determining molecular structure at the atomic level, if a crystal of good quality is obtained. Each method has not only advantages and specialties but also disadvantages. Electron microscopy reveals the overall shape of the molecule, but it is difficult to obtain ligand-binding structure at the atomic level.

Jun Hirabayashi (ed.), *Lectins: Methods and Protocols*, Methods in Molecular Biology, vol. 1200,
DOI 10.1007/978-1-4939-1292-6_39, © Springer Science+Business Media New York 2014

Using NMR spectroscopy, it is difficult to determine the macromolecular structure of large molecular mass and complicated carbohydrate structures. A disadvantage of X-ray crystallography is that the target protein needs to be crystallized. However, most proteins can now be crystallized owing to advances in methodology and molecular biology. Furthermore, the latest technique, X-ray-free electron laser, is expected to obtain X-ray diffractions from a sample that is not crystallized or is crystallized into very small crystals and to determine the structure of proteins, which remain undetermined so far. X-ray crystallography currently is the leading method of analysis of the three-dimensional structure of proteins.

In this chapter, an overview of the traditional method of protein crystallography is briefly introduced, focusing on structural analysis of lectin–sugar interactions. For details on principles and methodology of protein crystallography, please refer to the range of excellent textbooks available elsewhere.

2 Protein Preparation

As a first step to determine the crystal structure of a lectin, it is necessary to obtain at least a few milligrams of pure protein sample. For this purpose, a protein sample is usually produced using the *Escherichia coli* expression system. This makes it easy to obtain large amounts of recombinant protein and to purify the sample in a few steps using an inserted affinity tag such as GST tag or histidine tag. It also enables the introduction of mutations into the protein to analyze functional amino acids or domains. However, some proteins occasionally misfold into inclusion bodies and refolding is difficult to achieve, and in this case, another expression system, such as yeast, insect, or mammalian cell system or a cell-free system, becomes necessary. Otherwise proteins can be isolated directly from their source, if sufficient quantities are available. Because the natural protein is occasionally glycosylated, removal of glycan by enzyme treatment may be necessary to retain homogeneity of the sample.

To obtain crystals, high purity and homogeneity of the sample are necessary. The sample should be confirmed to show the presence of a single band on SDS-PAGE. Solubility, stability, and ligand-binding activity are also necessary that can be indicators of proper protein folding. The sample is preferably dissolved in distilled water, but some buffer reagents up to 10 mM, or some additives such as glycerol, metal ions, dithiothreitol, detergent, or protease inhibitor, can be used to maintain solubility and stability of the sample. Phosphate buffer is usually avoided because it occasionally results in salt crystal along with calcium or various other divalent metal ions. Carbohydrate ligands can be used to stabilize the sample because lectin usually forms stable protein–sugar complexes.

However, complicated sugars that aggregate the protein sample should be avoided. It may take from several days to several months for crystals to grow, and the sample needs to remain stable through this period. It is important to identify the correct conditions for stable dissolution of the sample before crystallization.

3 Crystallization

Before crystallization trials, the sample needs to be concentrated to 5–15 mg/mL. Centrifugal concentration is often used and it enables simultaneous desalting or buffer transfer. Lectins are usually composed of a small domain with molecular mass of 20,000 or less and show high solubility in water; they may need to be concentrated to 20 mg/mL or more. To determine the appropriate protein concentration for crystallization screening, a test kit is available from Hampton Research (Hampton Research, Aliso Viejo, California, USA), and it is recommended to use this for any sample, which is to be crystallized for the first time. If a lectin, such as C-type lectin and calreticulin, requires calcium ions for sugar binding, this should be added into the sample solution up to a concentration of 2 mM before the crystallization setup.

Crystallization trials are set up using the hanging-drop or sitting-drop vapor diffusion method with sparse-matrix crystal screening kits (Fig. 1; also *see* ref. 3). These are commercially available from several biochemical companies (such as Hampton Research; Emerald Biostructures, Bainbridge Island, Washington, USA; Molecular Dimensions, Newmarket, Suffolk, UK; Qiagen, Hilden, Germany), and contain 96 different precipitant solutions per kit. Crystallization solutions consist of precipitant, salt, and buffer reagent. Various types of salt (such as ammonium sulfate, lithium sulfate, sodium citrate or magnesium sulfate), organic solvents (such as *2*-methyl-*2,4*-pentanediol or *2*-propanol), or polyethylene glycol of various degrees of polymerization are often used as protein precipitants. Crystallization plates can be set up either manually or by robotic dispenser. For manual setup, 24-well tissue culture trays are generally used, with 96-well sitting plates for robotic setup (Fig. 1). Normally 1–5 screening plates (300–500 conditions) are set up for the first screening trials. An aliquot (approximately 1 μL) of protein sample is combined with the same volume of precipitant solution and equilibrated against a large volume of precipitant solution, from 50 μL for sitting-drop to 1,000 μL for hanging-drop setup. The well is covered with a cover glass sealed with grease for the hanging-drop setup, or sealed with clear film for the sitting-drop setup, and is incubated in a temperature-controlled environment, typically used at 20 °C, and occasionally at 4 °C. Because the concentration of the precipitant in the reservoir is higher than in the drop, the water in the sample

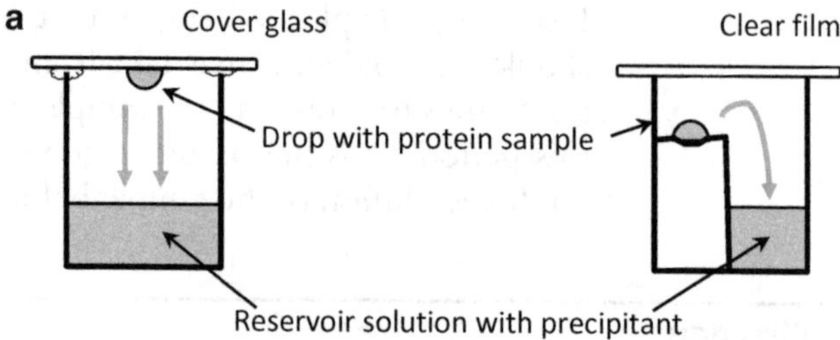

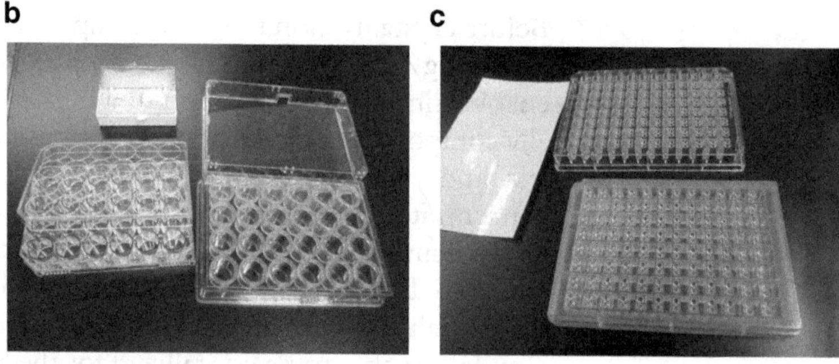

Fig. 1 (**a**) Illustration of hanging- (*left*) and sitting-drop (*right*) vapor-diffusion crystallization method setups. (**b**) 24-Well hanging-drop crystallization plates with cover glasses. (**c**) 96-Well sitting-drop crystallization plates with a sealing film

drop will evaporate towards the reservoir. The protein and precipitant concentrations of the drop increase, resulting in a supersaturated state of the protein, which causes it to crystallize. After several days or weeks, the appearance of crystals in sample drops is observed using a microscope. The sitting-drop vapor-diffusion protocol is described later.

If a precipitate appears in most of the drops, then the concentration of the sample should be reduced for another screening. In contrast, if most of the drops remain clear for several days, then the concentration of the sample should be increased. If one or more hits are observed in the screens, then refinement of the crystallization conditions is necessary to obtain one single large crystal in a drop (Fig. 2). Refinement is conducted by varying the concentrations of some components in the crystallization solution, changing pH of the buffer, varying the protein concentrations, or changing the protein-precipitant solution ratio. Using additives, changing to similar buffers or precipitants or changing the incubation temperature may also be considered.

To analyze the lectin–sugar interaction, it is necessary to make crystals of the sugar complex. For this purpose, co-crystallization and soaking methods are used. For co-crystallization, the protein sample and sugar ligands are mixed before crystallization. The molar concentration of the sugar is usually set at a value two- to tenfold higher than that of the protein, and the mixed sample is

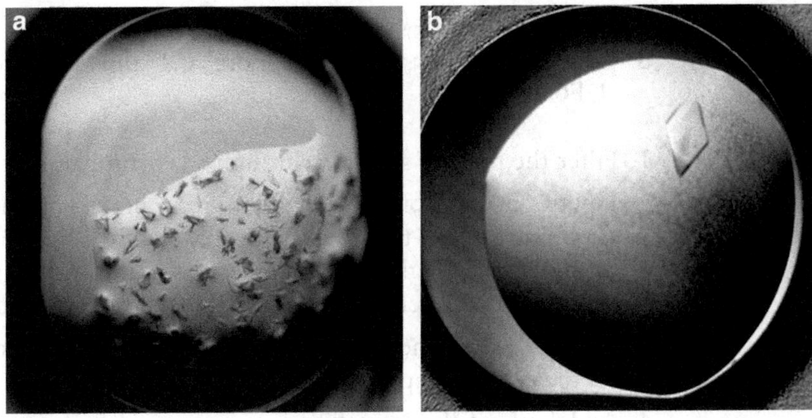

Fig. 2 (**a**) The appearance of crystals of EW29 lectin C-half domain at the initial crystallization screening. (**b**) Single crystal grown under optimized crystallization conditions

incubated for one or a few days and then subjected to crystallization trials. For soaking, the crystal of the protein itself is first prepared and then sugar solution is added to the drop containing crystals or the crystal is soaked in the precipitant solution containing the sugar ligands, before performing the X-ray diffraction experiment. A sugar concentration of 1–20 % is used, with a soaking time of a few seconds to a few days. When crystallization is set up using only the protein sample, the lectin sugar-binding site is occasionally involved in the crystal packing interface, and the ligand cannot be docked in the crystal or the crystal breaks to give no diffractions. In such a case, the co-crystallization method must be selected for further crystallization. However, the piece of the broken crystal occasionally diffracts to yield sufficient data. In our case involving analysis of the C-half domain of EW29 galactose-binding lectin, the protein itself did not crystallize, but two sugar complex samples crystallized in different conditions [4]. Because lectin usually forms stable complexes with sugars, it is recommended to try crystallization screening using the sugar-complexed sample at the first trial.

If a crystal is not obtained, then it is recommended that the protein expression or purification step should be repeated. Increasing the purity and stability of the protein, or changing the construct, transferring the purification tag from the N-terminus to the C-terminus or deleting some domains or regions, may be considered.

3.1 Sitting-Drop Vapor Diffusion Method

3.1.1 Materials

1. Protein solution at 10–20 mg/mL in distilled water or 2–10 mM buffer solution.
2. Crystal screening kit, e.g., crystal screen HT, Hampton Research (HR2-130).
3. 96-Well crystallization sitting drop plate (HR3-192).

4. Clear sealing film (HR3-609).
5. Centrifugal filter, e.g., Millipore 0.1 μm Ultrafree Filter (UFC30VV00).

3.1.2 Procedure

1. Filter the protein solution through a centrifugal filter.
2. Pipette 50 μL of crystallization reagents from the crystal screening kit into the reservoir deep well of the 96-well sitting-drop plate. An eight-channel pipette is useful. Add reagent A1 to the A1 deep well, and repeat for the 96 reservoir deep wells.
3. Pipette 1 μL of the filtered protein solution into the small well on the sitting drop pedestal. Repeat for all 96 wells.
4. Add 1 μL of the crystallization reagent from the reservoir to the sample drop. Repeat for all 96 wells.
5. Seal the entire plate with a clear sealing film. Confirm that all 96 wells are firmly sealed.
6. Place the plate at room temperature, preferably under the temperature-controlled conditions.
7. Observe the sample drops under a microscope at 20–100× magnification after 1, 3, and 7 days (and if necessary after 1 and 2 weeks, and 1, 3, and 6 months).
8. If crystals appear, crystallization conditions should be optimized to obtain a crystal suitable for X-ray diffraction analysis.

4 X-Ray Diffraction Analysis

In X-ray diffraction, X-rays are scattered by the electrons of the molecules examined, and the X-rays scattered from the molecules in the crystal give diffractions. To collect X-ray diffraction data, specialized instrumentations, such as laboratory X-ray generator or synchrotron facility, are used as an X-ray source, and a charge-coupled device (CCD) or an imaging plate area detector is used for detection of the diffraction. X-rays produced at synchrotron beamlines produce X-rays of high intensity and this enables sufficient diffraction data to be obtained from very small crystals. Over 20 synchrotron facilities are available worldwide for protein crystallography.

Because X-rays damage the crystals, data collection needs to be conducted under low temperatures, particularly at synchrotron beamlines. The crystal is usually mounted in a cryo-loop and placed under streaming nitrogen gas cooled to around 100 K. Some crystallization solutions can be cooled without freezing, but cryoprotectants, such as glycerol, trehalose, or ethylene glycol, are necessary to be added to some solutions to prevent it from freezing and to keep crystals diffractable.

Whether a crystal will provide adequate diffraction is unknown until the X-ray experiment. If a crystal does not diffract or diffraction produces insufficient resolution, then the crystal screening must be repeated to give crystals of a larger size and higher quality or produced under different conditions. When a crystal produces sufficient diffractions, a complete dataset is collected. The mounted crystal is rotated and the X-ray diffraction patterns are recorded for a small angle per image, and collected for less than or equal to 360°. Collection of one dataset takes a few days using a laboratory X-ray generator but can be completed within an hour by synchrotron radiation. At a synchrotron facility, the wavelength of the X-rays is usually set at approximately 1 Å, but the wavelength around the absorption edge of a heavy atom will need to be altered as described below for multi-wavelength anomalous dispersion (MAD) or single-wavelength anomalous dispersion (SAD) experiments, where the use of synchrotron radiation is indispensable. Protein crystals give thousands of diffraction spots on one image, and a complete dataset contains from tens of thousands to millions of intensities.

The dataset obtained is then processed using computer software connected with the X-ray facilities. Programs HKL2000, mosfilm, and XDS are often used [5–7]. These programs determine the crystal space group, the unit cell parameters, and the other crystallographic parameters, and provide the dataset applicable to structure determination after integration, merging, and scaling procedures have been performed on the diffraction images.

The resolution of the data implies the number of diffractions obtained from a crystal during each protein crystallography experiment and higher (smaller number) resolution data contains more diffractions, such as diffraction spots, that are observed in the outer region of the image (Fig. 3). The resolution is a reflection of the quality of the electron density maps. Higher resolution of the data implies higher resolution of the electron density maps, which in turn results in greater accuracy of the positions of the atoms in the structure, making model building easier. For example, the hole in an aromatic ring may be visible in a 1.5-Å resolution map, whereas a side chain appears to be a bulky projection at 2.5 Å resolution (Fig. 3).

5 Structure Determination

Structure determination can be performed on general personal computers. The collected data comprised more than several tens of thousands of merged reflections, and each has intensity with a standard deviation, which is first converted to a structural factor. To calculate the electron density map of a molecule, structural factor and phase of every reflection are necessary, but at this stage the dataset does not contain phase values, and phases need to be determined. For this phase calculation, several methods have been developed; however, it occasionally makes protein crystallography difficult.

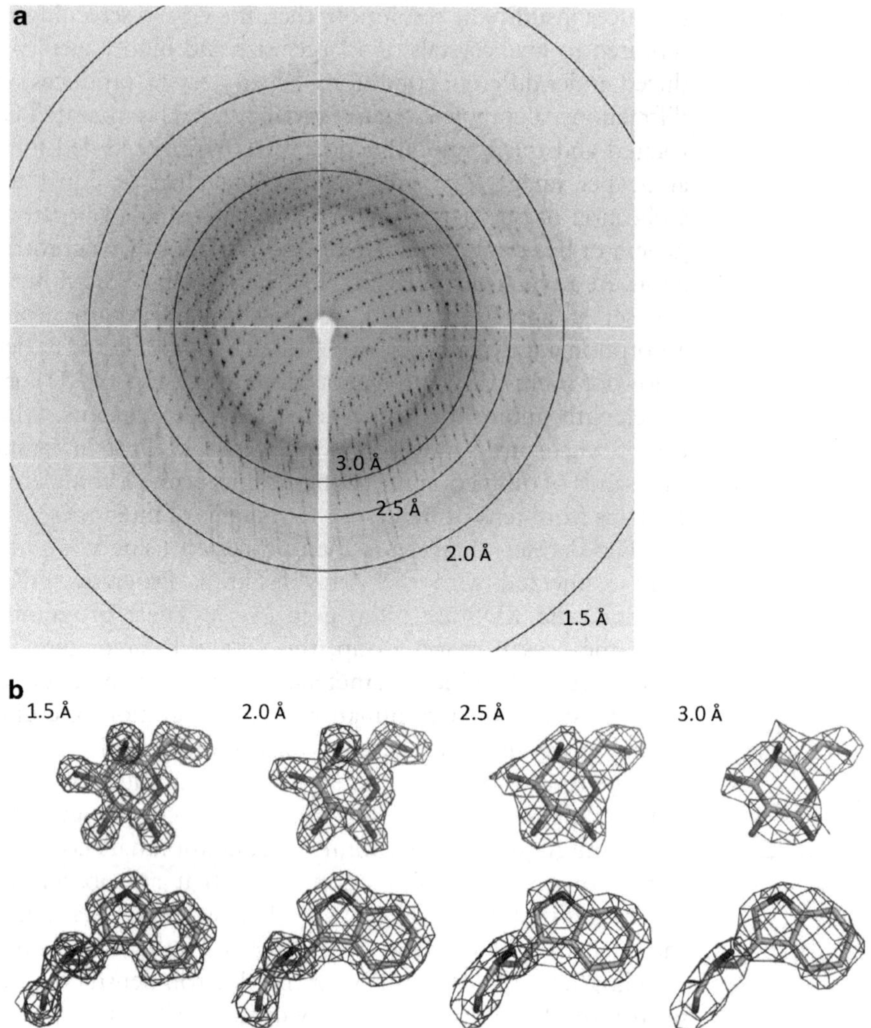

Fig. 3 (**a**) X-ray diffraction image of a protein crystal. *Circles* indicate resolutions. (**b**) Electron density map calculated from the data of 1.5, 2.0, 2.5, and 3.0 Å resolution around the tryptophan and the bound galactose of rice α-galactosidase contoured at 1.0σ

If the crystal structure of the same protein or a similar protein has been solved, the molecular replacement method can be applied. After rotation and translation, phases can be calculated from the reference model and the electron density can be calculated. For successful phase calculation, the reference structure usually needs to have amino acid sequence identity of over 30 % throughout the molecule, but in some cases good results can be achieved with a structure of 25 % identity using high-resolution diffraction data. Crystal structures of lectins have been widely analyzed and accumulated in PDB, and they are introduced by Z. Fujimoto (*see* Chapter 46). To find out whether a similar structure is available the sequence search at the PDB website can be used.

If there is no reference model or the molecular replacement method does not work, then phases are determined using heavy-atom derivative data. If a protein sample is prepared using the *E. coli* expression system, then MAD or SAD is a very effective method using selenomethionine-derivative crystals ([8]; also *see* Chapter 40). The selenomethionyl proteins, in which methionines are replaced by selenomethionine, can be expressed by *E. coli* strains auxotrophic for methionine in minimal media supplemented with selenomethionine instead of methionine and usually crystallized under the same conditions as the native protein. Anomalous scattering data can be obtained by collecting diffractions around the X-ray absorption edge of the selenium atom (0.978 Å) at synchrotron facilities, and they enable phase calculation.

When a selenomethionyl protein is unavailable, other heavy-atom derivative crystals, prepared by soaking heavy-atom compounds into protein crystals, can be used for MAD or SAD or for the multiple isomorphous replacement (MIR) method. Platinum, gold, and mercury derivatives are often used because they are sufficiently heavy, appropriate types of chemical reagents are available, and anomalous scattering data can be obtained because their absorption edges are in the changeable wavelength range for normal use. However, difficulties are occasionally encountered in seeking derivatives because heavy atoms do not always soak into the crystal or they may occasionally break the crystals into pieces. Because lectin shows binding specificity for particular sugars, selenium-containing sugar ligands would be applicable for use in phase determination methods instead of heavy-atom compounds, and this method is mentioned by H. Makyio and R. Kato (*see* Chapter 40).

When the phases are determined, the electron density map can be calculated, from which the molecular model can be constructed. The modelling program Coot is often used [9]. The quality of the map depends on the quality and resolution of the diffraction data. If the resolution of the data is better than 3.0 Å, individual peptides can be traced. Auto-model building program, such as arp/warp, can be applied on 2.0 Å resolution or better data [10]. First, peptide tracing is conducted along with the amino acid sequence information. Recalculation of the phases using the built model improves the electron density map, and then manual model rebuilding is performed on the newly produced map. Model refinement using a macromolecular refinement program followed by manual rebuilding is repeated several times. Water and other molecules around the protein are introduced into the model. For the sugar-complex structure, the bound ligand can be modeled at this stage. If the ligands are successfully complexed with the protein, the electron density for the bound ligand can be observed.

The accuracy of the constructed model is confirmed by the crystallographic R-factor, which indicated the error between the calculated and observed amplitudes [11]. The R-factor of the final macromolecular model usually decreases to 20 % or lower.

Stereochemistry of the model is also checked by programs such as PROCHECK or RAMPAGE [12, 13]. Ramachandran plot is one index, in which most of the main-chain torsion angles should be plotted in the restricted areas [14]. Many validation programs are used to check the structure, until the investigator is satisfied, and then the structure can be deposited in PDB.

Most of the work listed above on structure determination can be performed by computer programs. CCP4 and PHENIX program packages are widely used [15, 16], and they can be obtained through their website free of charge for academic researchers. Owing to recent progress in molecular biology, computer technology, the development of powerful X-ray sources, and crystallographic methodology, it is now relatively easy to try to determine the crystal structure of a macromolecule. The three-dimensional structure allows us to understand how lectin interacts with sugars at an atomic level. Novel lectins have been consecutively discovered from the genome analysis of several organisms. How about trying to start the three-dimensional structural analysis of a lectin in passing, when you can successfully express the protein for characterization?

References

1. Edelman GM, Cunningham BA, Reeke GN Jr, Becker JW, Waxdal MJ, Wang JL (1972) The covalent and three-dimensional structure of concanavalin A. Proc Natl Acad Sci USA 69:2580–2584
2. Hardman KD, Ainsworth CF (1972) Structure of concanavalin A at 2.4-Å resolution. Biochemistry 11:4910–4919
3. Newman J, Egan D, Walter TS et al (2005) Towards rationalization of crystallization screening for small- to medium-sized academic laboratories: the PACT/JCSG+ strategy. Acta Crystallogr D Biol Crystallogr 61:1426–1431
4. Suzuki R, Kuno A, Hasegawa T et al (2009) Sugar-complex structures of the C-half domain of the galactose-binding lectin EW29 from the earthworm *Lumbricus terrestris*. Acta Crystallogr D Biol Crystallogr 65:49–57
5. Otwinowski Z, Minor W (1997) Processing of X-ray diffraction data collected in oscillation mode. Methods Enzymol 276:307–326
6. Battye TG, Kontogiannis L, Johnson O, Powell HR, Leslie AG (2011) iMOSFLM: a new graphical interface for diffraction-image processing with MOSFLM. Acta Crystallogr D Biol Crystallogr 67:271–281
7. Kabsch W (2010) XDS. Acta Crystallogr D Biol Crystallogr 66:125–132
8. Hendrickson WA, Horton JR, LeMaster DM (1990) Selenomethionyl proteins produced for analysis by multiwavelength anomalous diffraction (MAD): a vehicle for direct determination of three-dimensional structure. EMBO J 9:1665–1672
9. Emsley P, Lohkamp B, Scott WG, Cowtan K (2010) Features and development of Coot. Acta Crystallogr D Biol Crystallogr 66: 486–501
10. Cohen SX, Morris RJ, Fernandez FJ et al (2004) Towards complete validated models in the next generation of ARP/wARP. Acta Crystallogr D Biol Crystallogr 60:2222–2229
11. Kleywegt GJ, Jones TA (1997) Model building and refinement practice. Methods Enzymol 277:208–230
12. Laskowski RA, MacArthur MW, Moss DS, Thornton JM (1993) PROCHECK: a program to check the stereochemical quality of protein structures. J Appl Crystallogr 26:283–291
13. Lovell SC, Davis IW, Arendall WB 3rd et al (2003) Structure validation by Cα geometry: φ, ψ and Cβ deviation. Proteins 50:437–450
14. Ramachandran GN, Sasisekharan V (1968) Conformation of polypeptides and proteins. Adv Protein Chem 23:283–437
15. Winn MD, Ballard CC, Cowtan KD et al (2011) Overview of the CCP4 suite and current developments. J Appl Crystallogr 67:235–242
16. Adams PD, Afonine PV, Bunkoczi G et al (2010) PHENIX: a comprehensive Python-based system for macromolecular structure solution. J Appl Crystallogr 66:213–221

Chapter 40

A New Structure Determination Method of Lectins Using a Selenium-Containing Sugar Ligand

Hisayoshi Makyio and Ryuichi Kato

Abstract

Phase determination is essential for solving X-ray crystal structures of proteins and their complexes. Conventional phase-determination methods using heavy atoms (Pt, Au, Hg, etc.) or the selenium (Se) atom are routinely utilized in structure determination of protein crystals. Here, we describe an alternative phase-determination method for proteins such as lectins in which a Se-containing glycan is used as a ligand. In this technique, the Se atoms are simply introduced into the protein crystal as a complex, and the phase of the protein can be determined using anomalous signals from the Se-containing sugar.

Key words X-ray crystallography, Phase determination, Anomalous dispersion, Glycan synthesis, Lectin

1 Introduction

Although sugar-binding proteins play important roles in the immune response and other cellular processes, their molecular mechanisms have not been extensively studied to date [1]. Three-dimensional structures of sugar-binding proteins will provide information that will allow elucidation of these mechanisms [2]. From that viewpoint, it is very important to reveal the structures of sugar-binding proteins; therefore, new methods for structure determination would be useful.

In order to determine a protein structure by X-ray crystallography, the phase problem must be solved as described by Z. Fujimoto (*see* Chap. 39). Several conventional methods are used to determine phase [3, 4]. One such method, multiple isomorphous replacement (MIR), uses heavy atoms (Pt, Au, Hg, etc.) for phase determination of new proteins. In addition to the native crystals, the MIR method also requires crystals that contain heavy atoms; however, isomorphism among the crystals can make it difficult to determine the phase of the protein. By contrast, the multi-wavelength anomalous dispersion (MAD) and single-wavelength

Jun Hirabayashi (ed.), *Lectins: Methods and Protocols*, Methods in Molecular Biology, vol. 1200,
DOI 10.1007/978-1-4939-1292-6_40, © Springer Science+Business Media New York 2014

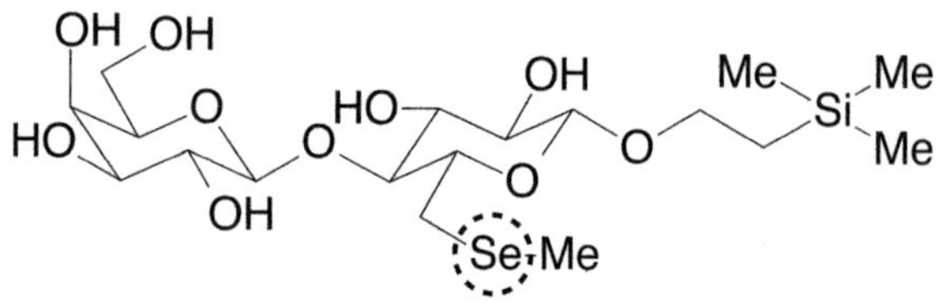

Fig. 1 Structural formula of Se-containing sugar GSC819. GSC819 was used for phase determination of human galectin-9 NCRD. The selenium atom is enclosed by a *dotted circle*

anomalous dispersion (SAD) methods do not require multiple crystals. In these techniques, a single crystal can provide enough information to determine the phase of a protein by exploiting the anomalous dispersion effect of atoms (typically selenium or heavy atoms). Both the MAD and SAD methods require selenomethionine-(Se-Met) substituted or heavy atom-bound protein crystals. Expression of Se-Met protein is sometimes difficult for host cells because of its toxicity to the cells. When heavy atoms are used for the MAD and SAD methods, there are no reliable ways to determine whether the heavy atoms are bound to the protein or not. Thus, a phase-determination method that does not depend on the expression level of a substituted protein or that provides the specific binding of heavy atoms to a protein would be a useful tool in solving the structures of novel proteins.

Here, we describe an alternative phase-determination method that uses Se-containing sugar (*see* Fig. 1). This method does not depend on the expression level of a substituted protein or specific binding of heavy atoms. As an illustrative example, we applied this method to solving the structure of human galectin-9 N-terminal carbohydrate recognition domain (hGal-9 NCRD), which has been well studied, and for which the crystal structure has already been determined [5–7]. The structure of hGal-9 NCRD was successfully determined using the Se-sugar method (*see* Fig. 2). We believe that this method will be applicable to the determination of X-ray crystal structures of other sugar-binding proteins.

2 Materials

All solutions (buffers) should be prepared using ultrapure water (Milli-Q water).

2.1 Preparation of Protein and Se-Containing Sugar

1. hGal-9 NCRD fused to glutathione S-transferase on plasmid vector pGEX-4 T1 (GE Healthcare Life Science) (*see* **Note 1**).
2. *Escherichia coli* strain BL21(DE3) cells (Merck KGaA).

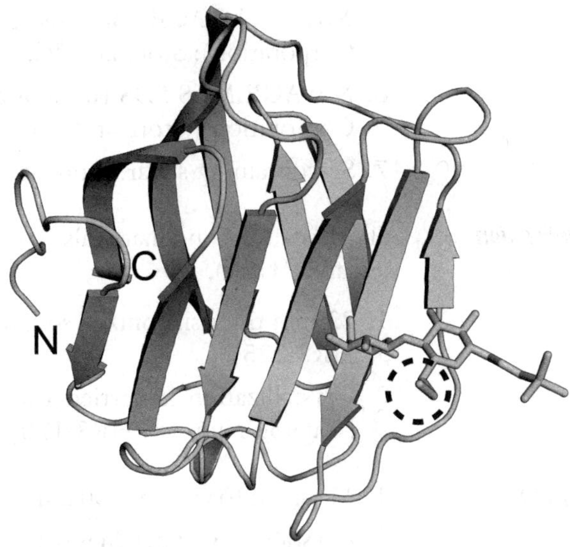

Fig. 2 The overall structure of human galectin-9 NCRD with Se-containing sugar. Human galectin-9 NCRD with Se-containing sugar is represented as a ribbon model. The carbon atoms of the Se-containing sugar are represented as sticks. The selenium atom is enclosed by a *dotted circle*

3. LB medium (1 L): Dissolve 10 g tryptone, 5 g yeast extract, and 10 g NaCl in 1 L deionized water. After sterilization by autoclaving, add ampicillin to a final concentration of 100 μg/mL.
4. LB-agar: Add agar into LB medium at a final concentration of 1.5 %.
5. Isopropyl-β-D-thiogalactopyranoside (IPTG): 1 M aqueous solution as a stock.
6. Purification buffer: 50 mM Tris–HCl buffer (pH 8.0), 500 mM NaCl, and 1 mM dithiothreitol (DTT).
7. Glutathione-Sepharose 4B resin (GE Healthcare Life Science).
8. Econo-Pac chromatography column (1.5 × 12 cm, Bio-Rad Laboratories, Inc.).
9. Glutathione (reduced form).
10. Thrombin protease (GE Healthcare Life Science).
11. Benzamidine Sepharose (GE Healthcare Life Science).
12. Amicon Ultra-15 Centrifugal Filter Units (Millipore Corporation).
13. Gel filtration buffer: 10 mM Tris–HCl buffer (pH 8.0), 100 mM NaCl, and 1 mM dithiothreitol (DTT).
14. Superdex 75 10/300 GL column (GE Healthcare Life Science).

15. Novex NuPAGE Bis-Tris precast gel (Life Technologies Corporation): Store at 4 °C.
16. NuPAGE MES SDS running buffer (20×) (Life Technologies Corporation): Store at 25 °C.
17. Se-containing sugar: Store at −80 °C (*see* **Note 2**).

2.2 Crystallization

All crystallization materials were purchased from Hampton Research (USA).

1. 22 mm thick siliconized square cover slides (Catalog Number: HR3-225).
2. Crystallization is carried out using VDX Plate with sealant (Catalog Number: HR3-170).

2.3 X-Ray Data Collection

1. Mounted CryoLoop (20 μm) (Hampton Research) (*see* **Note 3**).
2. CrystalCap Copper Magnetic (Catalog Number: HR4-745).
3. Cryo canes (aluminum cryogenic vial storage holder) (Thermo Fisher Scientific Inc., USA) (Catalog Number: 5015-0001).

3 Methods

3.1 Expression of Protein

1. Introduce hGal-9 NCRD on plasmid vector pGEX-4T1 into *E. coli* strain BL21(DE3) cells. Select transformed *E. coli* cells selected on LB-agar plates (+ampicillin at final concentration of 50 μg/mL) at 37 °C for 14–18 h.
2. Grow selected *E. coli* cells in LB medium (10 mL) at 37 °C for 14–18 h (preculture).
3. Add preculture (1–5 mL) to LB medium (1 L) in a flask (5 L). Grow cells at 37 °C in shaking incubator.
4. Check the turbidity of cells by measuring absorbance at 600 nm (OD_{600}). When the turbidity of cells reaches OD_{600} = 0.4–0.6, add IPTG to cells at a final concentration of 0.1 mM to induce expression of hGal-9 NCRD.
5. Shift the temperature from 37 to 25 °C and grow cells for an additional 20–22 h in a shaking incubator.
6. Transfer the cells from flasks to bottles. Centrifuge the bottles at 11,800 × *g* for 20 min. Wash pelleted cells with purification buffer. Transfer washed cells from bottles to plastic tubes (50 mL) and harvest cells by centrifugation at 8,400 × *g*. Store harvested cells at −80 °C before purification.

3.2 Purification of Protein

1. Resuspend stored cells in purification buffer.
2. Disrupt suspended cells by sonication in a water bath with crushed ice.

3. After disruption of the cells, centrifuge the solution at 34,500 × *g* for 20 min at 4 °C.
4. Transfer the supernatant into a 50-mL plastic tube containing glutathione-Sepharose 4B resin (3 mL).
5. Incubate the 50-mL tube at 4 °C for 1–2 h to bind hGal-9 NCRD protein to glutathione-Sepharose 4B resin.
6. Transfer the glutathione-Sepharose 4B resin into an Econo-Pac chromatography column at 4 °C.
7. Wash the column with purification buffer (50 mL).
8. Elute the bound hGal-9 NCRD from the resin with purification buffer (20 mL) containing 15 mM glutathione (reduced form).
9. Remove glutathione S-transferase from the fusion protein by cleavage with thrombin protease (4 units/mL) at 22 °C for 12–18 h.
10. Pass cleaved protein through benzamidine-Sepharose resin to remove thrombin protease.
11. Concentrate the protein solution using Amicon Ultra-15 Centrifugal Filter Units (10,000 MW cutoff) by centrifugation at 2,000 × *g* at 4 °C.
12. Load the concentrated protein solution onto a Superdex 75 10/300 GL column.
13. Collect fractions corresponding to hGal-9 NCRD protein and check them by SDS-PAGE (Novex NuPAGE Bis-Tris Precast Gel with NuPAGE MES SDS Running Buffer (1×)).
14. Concentrate fractions containing purified hGal-9 NCRD protein by centrifugation (2,000 × *g*) at 4 °C using Amicon Ultra-15 Centrifugal Filter Units (10,000 MW cutoff). Final protein concentration should be 3–4 mg/mL.
15. Divide concentrated hGal-9 NCRD protein solution into small aliquots (50–100 μL) and store at −80 °C before use.

3.3 Crystallization with Se-Containing Sugar

1. Se-containing sugar solution: Dissolve the Se-containing sugar (*see* Fig. 1) in Milli-Q water to make a 100 mM stock solution. Divide the stock solution into small aliquots (10–20 μL) and store at −80 °C before use.
2. Protein solution: Take out an aliquot of hGal-9 NCRD protein from storage at −80 °C and thaw at 4 °C. Centrifuge the dissolved protein solution at 18,000 × *g* for 20 min at 4 °C. After centrifugation, use the supernatant of the protein solution for crystallization.
3. Crystallization screening: Mix protein solution and Se-containing sugar solution at final concentrations of 5–10 mg/mL and 10 mM, respectively. Incubate the mixed

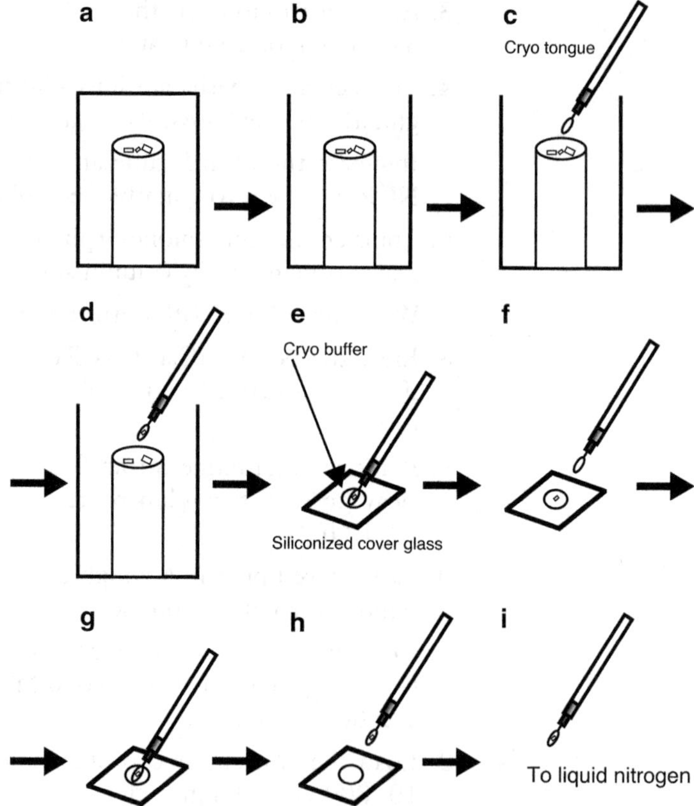

Fig. 3 Procedure for cryoprotecting for crystals. (**a**) Crystals of protein with Se-containing sugar appear in the crystallization wells (in this case, the schematic drawing depicts the sitting-drop method). (**b**) The seal at the top of crystallization well is cut and opened. (**c**, **d**). The protein crystal is picked up into the cryo-loop using CryoTongs under the microscope. (**e**) A drop of cryo-buffer is prepared on a siliconized cover glass, and the crystal that was picked up using CryoTongs is transferred into the cryo-buffer drop. (**f**) The crystal stays in the cryo-buffer for several seconds (if the cryo-buffer contains glycerol, the crystal is left for 30 s). (**g**, **h**) The crystal is picked up again using the CryoTongs. (**i**) Finally, the crystal is frozen and stored in liquid nitrogen until collection of X-ray data

solution on ice for 5 min before mixing in the crystallization precipitant solution. Perform crystallization screening by the vapor-diffusion method; place the mixed solution on a plastic plate or siliconized glass as a droplet (1 μL).

4. Cryoprotection of crystals: When crystals appear, transfer them to cryoprotectant buffer (*see* **Note 4**). Incubate crystals in cryoprotectant buffer for 1 min at 20 °C. Directly transfer crystals from cryoprotectant buffer to liquid nitrogen using a CryoLoop on a CrystalCap Copper Magnetic. Place CryoLoops into cryo canes and store them in liquid nitrogen prior to X-ray data collection (*see* Fig. 3).

3.4 X-Ray Data Collection and Structure Determination

Principle of X-ray scattering experiment and its analysis is summarized in Subheadings 4 and 5 of Chap. 39. In this subheading and the following Subheading 4, more specific procedures are described.

1. Transfer crystals from the cryo cane stored in liquid nitrogen to the goniometer on a beamline at the synchrotron radiation facility (*see* **Note 5**).
2. Align the crystal on the CryoLoop with the center position of the X-ray beam.
3. Check the quality of the crystals using the X-ray beam (*see* **Note 6**).
4. Check the anomalous signals from the crystals using fluorescence XAFS (X-ray absorption fine structure) measurements.
5. Determine how to collect the reflection data, and collect the diffraction data set (*see* **Note** 7).
6. Process the data using the appropriate software (*see* **Note 8**).
7. Scale and merge the processed data (*see* **Note 9**).
8. Convert the scaled data from MTZ format to SCA format (*see* **Note 10**).
9. If the phase of the protein is determined correctly, the structure of the protein with Se-containing sugar can be automatically solved by model-building software (*see* **Note 11**).
10. Further refine the structure using Coot [8] and Refmac [9] in CCP4 [10].
11. Check the final quality of refinement using Molprobity [11] and sfcheck [10].

4 Notes

1. The expression system depends on the protein.
2. The Se-containing sugar is synthesized and provided by our collaborator, but not commercially available [12].
3. The size of the CryoLoop depends on the crystal size. CryoLoops of 0.1–0.2 mm (Catalog Number: HR4-955) or 0.2–0.3 mm (HR4-957) are usually suitable for our experiments.
4. The cryoprotectant buffer depends on the crystals. In our trials, crystallization buffer with 20–25 % (w/v) glycerol or 20–25 % (v/v) ethylene glycol was used for the initial trials of our experiments. In order to reduce background noise for X-ray data collection, the cryoprotectant buffer should not contain the Se-containing sugar.

5. Robots that can automatically perform this step are installed at modern synchrotron radiation facilities (e.g., the Photon Factory (Japan) [13], the Advanced Photon Source (USA), and the European Synchrotron Radiation Facility (France)).
6. "Quality" refers to the X-ray diffraction pattern. For example, if there are few diffraction spots at lower resolution and strong spots at high resolution near 1.0 Å, the crystal is likely to be a salt crystal. When there are no diffraction spots from the crystal, you could apply more X-ray beam to the crystals in order to check diffraction spots at lower resolution.
7. Data-processing software (e.g., HKL2000 [14], XDS [15], imosflm [16]) provide the "strategy" for data collection.
8. We usually use XDS [15] for reflection data processing. For more detail, see the XDS website (http://xds.mpimf-heidelberg.mpg.de/).
9. The Pointless and Aimless software in CCP4 [10] are used for scaling and merging of reflection data.
10. After scaling with Pointless and Aimless, the reflection file is in MTZ format (*see* http://www.ccp4.ac.uk/html/mtzformat.html). For phase determination using the PHENIX program packages [17, 18], it is convenient to convert the file format from MTZ to SCA. The command "phenix.reflection_file_converter" in the PHENIX program packages is useful for this conversion (*see* http://www.phenix-online.org/documentation/reflection_file_tools.htm).
11. Our standard protocol for the SAD or the MAD method uses the PHENIX program packages [17, 18] from the command line. Two example scripts for use with the PHENIX program packages are as follows:
    (a) For finding three Se sites and determination of the initial phase: "phenix.autosol w1.sca seq.dat 3 Se lambda=0.9798"
    (b) For further model building: "phenix.autobuild after_autosol semet=false". For more detail, see the PHENIX program packages manual on the website (http://www.phenix-online.org/).

Acknowledgement

The Se-containing sugars were provided by Dr. Ando Hiromune (Institute for Integrated Cell-Material Sciences, Kyoto University; Department of Applied Life Science, Faculty of Applied Biological Sciences, Gifu University).

References

1. Berg JM, Tymoczko JL, Stryer L. Lectins are specific carbohydrate-binding proteins. http://www.ncbi.nlm.nih.gov/books/NBK22545/
2. Quiocho FA (1986) Carbohydrate-binding proteins: tertiary structures and protein-sugar interactions. Annu Rev Biochem 55:287–315
3. Taylor G (2003) The phase problem. Acta Crystallogr D Biol Crystallogr 59:1881–1890
4. Taylor GL (2010) Introduction to phasing. Acta Crystallogr D Biol Crystallogr 66:325–338
5. Nagae M, Nishi N, Murata T et al (2006) Crystal structure of the galectin-9 N-terminal carbohydrate recognition domain from *Mus musculus* reveals the basic mechanism of carbohydrate recognition. J Biol Chem 281:35884–35893
6. Nagae M, Nishi N, Nakamura-Tsuruta S et al (2008) Structural analysis of the human galectin-9 N-terminal carbohydrate recognition domain reveals unexpected properties that differ from the mouse orthologue. J Mol Biol 375:119–135
7. Nagae M, Nishi N, Murata T et al (2008) Structural analysis of the recognition mechanism of poly-*N*-acetyllactosamine by the human galectin-9 N-terminal carbohydrate recognition domain. Glycobiology 19:112–117
8. Emsley P, Lohkamp B, Scott WG et al (2010) Features and development of Coot. Acta Crystallogr D Biol Crystallogr 66:486–501
9. Murshudov GN, Skubák P, Lebedev AA et al (2011) REFMAC 5 for the refinement of macromolecular crystal structures. Acta Crystallogr D Biol Crystallogr 67:355–367
10. Winn MD, Ballard CC, Cowtan KD et al (2011) Overview of the CCP4 suite and current developments. Acta Crystallogr D Biol Crystallogr 67:235–242
11. Chen VB, Arendall WB, Headd JJ et al (2009) MolProbity: all-atom structure validation for macromolecular crystallography. Acta Crystallogr D Biol Crystallogr 66:12–21
12. Suzuki T, Makyio H, Ando H et al (2014) Expanded potential of seleno-carbohydrates as a molecular tool for X-ray structural determination of a carbohydrate-protein complex with single/multi-wavelength anomalous dispersion phasing. Bioorg Med Chem 22:2090–2101
13. Hiraki M, Watanabe S, Phonda N et al (2008) High-throughput operation of sample-exchange robots with double tongs at the Photon Factory beamlines. J Synchrotron Radiat 15:300–303
14. Otwinowski Z, Minor W (1997) Processing of X-ray diffraction data collected in oscillation mode. Methods Enzymol 276:307–326
15. Kabsch W (2010) XDS. Acta Crystallogr D Biol Crystallogr 66:125–132
16. Battye TGG, Kontogiannis L, Johnson O et al (2011) iMOSFLM: a new graphical interface for diffraction-image processing with MOSFLM. Acta Crystallogr D Biol Crystallogr 67:271–281
17. Adams PD, Afonine PV, Bunkóczi G et al (2010) PHENIX: a comprehensive Python-based system for macromolecular structure solution. Acta Crystallogr D Biol Crystallogr 66:213–221
18. Terwilliger TC, Adams PD, Read RJ et al (2009) Decision-making in structure solution using Bayesian estimates of map quality: the PHENIX AutoSol wizard. Acta Crystallogr D Biol Crystallogr 65:582–601

Chapter 41

NMR Analysis of Carbohydrate-Binding Interactions in Solution: An Approach Using Analysis of Saturation Transfer Difference NMR Spectroscopy

Hikaru Hemmi

Abstract

One of the most commonly used ligand-based NMR methods for detecting ligand binding is saturation transfer difference (STD) nuclear magnetic resonance (NMR) spectroscopy. The STD NMR method is an invaluable technique for assessing carbohydrate–lectin interactions in solution, because STD NMR can be used to detect weak ligand binding (K_d ca. 10^{-3}–10^{-8} M). STD NMR spectra identify the binding epitope of a carbohydrate ligand when bound to lectin. Further, the STD NMR method uses ^{1}H-detected NMR spectra of only the carbohydrate, and so only small quantities of non-labeled lectin are required. In this chapter, I describe a protocol for the STD NMR method, including the experimental procedures used to acquire, process, and analyze STD NMR data, using STD NMR studies for methyl-β-D-galactopyranoside (β-Me-Gal) binding to the C-terminal domain of an R-type lectin from earthworm (EW29Ch) as an example.

Key words Saturation transfer difference, Nuclear magnetic resonance spectroscopy, Lectin, Carbohydrate–lectin interaction, Epitope, Methyl-β-D-galactopyranoside, R-type lectin

1 Introduction

Lectins are proteins that bind specifically to carbohydrates. They exist in organisms ranging from viruses and plants to humans and serve to mediate biological recognition events. Lectin-carbohydrate interactions play crucial roles in numerous biological processes. The binding of an individual lectin site (monovalent binding) to a monosaccharide is extremely weak, with K_d values typically ranging from 0.1 to 10 mM [1–3]. The approaches used to characterize protein-protein or protein-nucleic acid interactions are often therefore limited for analyzing protein–carbohydrate interactions. As such, the interactions between lectin and carbohydrates are often analyzed using frontal affinity chromatography (FAC; *see* Chapter 21), isothermal titration calorimetry (ITC; *see* Chapter 18), nuclear magnetic resonance (NMR) spectroscopy, and X-ray crystallography (*see* Chapters 39 and 40). NMR spectroscopy as a protein-based

Jun Hirabayashi (ed.), *Lectins: Methods and Protocols*, Methods in Molecular Biology, vol. 1200,
DOI 10.1007/978-1-4939-1292-6_41,

NMR approach is a powerful tool that generates high-resolution data not only about the tertiary structure of lectin but also about the dynamics and binding interactions at the atomic level in the solution state. Protein-based NMR approaches for characterizing a carbohydrate-protein interaction are necessary to uniformly label the protein with ^{13}C and ^{15}N stable isotopes, and to assign the chemical shifts of the protein using multinuclear and multidimensional NMR spectroscopy, in addition to the chemical shifts of carbohydrate [4]. When ligand-based NMR approaches are used, ^{1}H-detected NMR spectra of only the carbohydrate are analyzed. When using ligand-based approaches, it is therefore not necessary to obtain large quantities (mg quantities) of labeled protein and to assign the chemical shifts of the protein [4]. One of the most commonly used ligand-based NMR methods for detecting ligand binding is saturation transfer difference (STD) NMR spectroscopy [5, 6]. The STD NMR method is commonly used to assess the binding of a ligand to target proteins. It can also be an excellent technique for determining the binding epitope of a ligand. The STD NMR method is based on saturation transfer from the protein to a bound ligand. The initial step in STD NMR is the saturation of protons on the protein, which is selectively excited in a spectral region that contains signals from the protein, but not from the ligand (Fig. 1). The saturation of the protein protons is then transferred to the adjacent protons on the ligand by intermolecular ^{1}H-^{1}H cross-relaxation. The ligand protons nearest to the protein will therefore have the strongest signal in the STD NMR spectrum, and the ligand protons that are not in close contact with the protein will have weak or no signal (Fig. 1). STD NMR can be used to detect weak ligand binding, with K_d values ranging from ~10^{-3} to 10^{-8} M [7], and is therefore invaluable for studying carbohydrate–lectin interactions.

Here, the author describes a protocol for the STD NMR method, including the experimental procedures used to acquire, process, and analyze STD NMR data, using STD NMR studies for methyl-β-D-galactopyranoside (β-Me-Gal) binding to the C-terminal domain of an R-type lectin from earthworm [8] as an example.

2 Materials

2.1 NMR Sample Preparation

1. Purified lectin, EW29Ch, solution containing a protease inhibitor cocktail (*see* **Note 1**).
2. Phosphate buffer: 0.5 M KH_2PO_4, 0.5 M Na_2HPO_4, pH 6.1.
3. Deuterium oxide (D_2O) (99.9 %) (Isotech Labs, Inc.).
4. The carbohydrate ligand methyl-β-D-galactopyranoside (β-Me-Gal, Seikagaku Co., Tokyo, Japan) dissolved in distilled water.
5. 5-mm ultra-precision NMR sample tubes, and 5-mm NMR tube caps (Shigemi Inc.).
6. Distilled water.

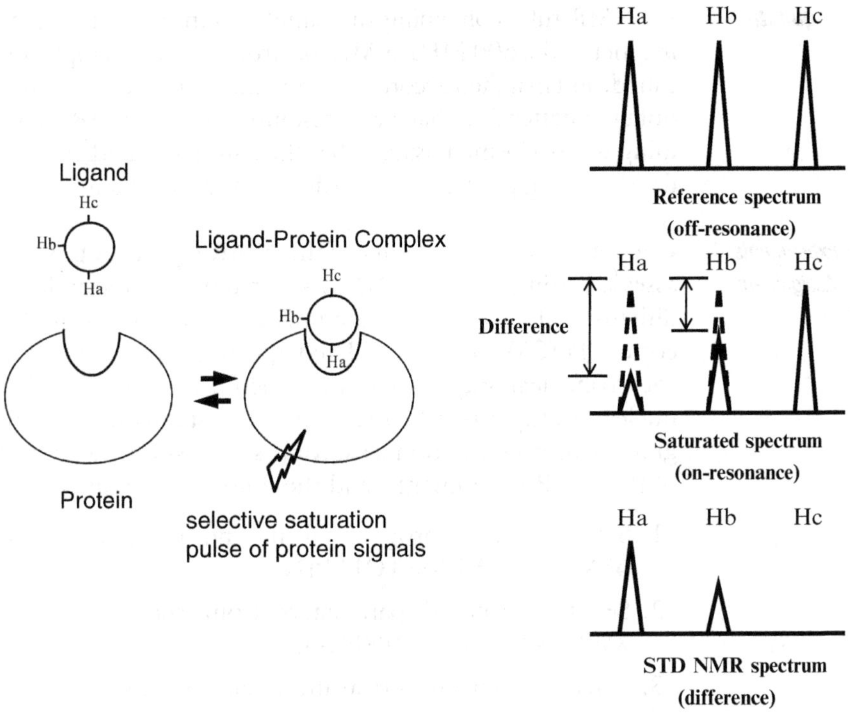

Fig. 1 A schematic of the STD NMR method for a ligand in fast exchange between the bound and free states. Irradiation of a selective pulse of the protein at a resonance where ligand signals are absent leads to a very efficient saturation of the entire protein by spin diffusion. Saturation is transferred to the ligand in a distance-dependent manner by intermolecular saturation transfer. The protons of the ligand in closest proximity to the protein will therefore experience the largest effects, and the ligand protons that are not in close contact with the protein will experience weak or no effects. Subtraction of the on- and off-resonance spectra gives a difference spectrum (STD), where the peak intensities are proportional to the distance of the ligand protons from the protein

2.2 NMR Hardware and Software

1. Bruker Avance 600 MHz spectrometer.
2. 5-mm TXI probe with *xyz*-axis gradient (*see* **Note 2**).
3. XWINNMR software (Bruker) running on Silicon Graphics O_2, or TOPSPIN software (Bruker) running on a PC (Linux) for data acquisition and processing.
4. The temperature unit was set at 298 K (*see* **Note 3**).

3 Methods

3.1 NMR Sample Preparation

1. A 600-μL solution containing 50 μM EW29Ch, 50 mM β-Me-Gal (protein:ligand ratio of 1:100, *see* **Note 4**), and 50 mM phosphate buffer was prepared using 0.5 M phosphate buffer and distilled water.
2. The solution was lyophilized, and then the lyophilizate was redissolved in D_2O. The final volume was adjusted to 600 μL.

3.2 NMR Acquisition

An NMR tube containing the sample solution was inserted into the magnet of the 600 MHz NMR spectrometer. The sample was locked, tuned, and matched according to the manufacturer's user manual. To obtain a better line shape and resolution of the NMR signals, shimming was performed using either the automatic gradient or the manual shimming procedure, according to the user manual.

3.2.1 NMR Experiments for Spectrum Assignment of β-Me-Gal

Chemical shift assignments of the NMR signals of β-Me-Gal were assessed using a 1D ^{1}H NMR spectrum, and the following two-dimensional (2D) NMR experiments: COSY (correlated spectroscopy), TOCSY (total correlated spectroscopy), [^{1}H-^{13}C] HSQC (heteronuclear single-quantum coherence), and HMBC (heteronuclear multiple-bond correlation). The standard NMR pulse program with presaturation as solvent suppression was used for each of the NMR experiments, and the following parameters were set:

1. Set the pulse program from the pulse program library of XWINNMR (or TOPSPIN).
2. Set the standard parameters from the parameter files of XWINNMR (or TOPSPIN).
3. Check and change a relaxation delay of ~2 s.
4. Set the number of scans (ns) to 32 (depending on the sample concentration and the performance of the NMR probe).
5. Set the spectral width to 6,000 Hz.
6. Check and adjust the pulse width and the power level (dB) for the pulse width.
7. Set the receiver gain using the command "rga".
8. Collect the NMR data using the command "zg".
9. Process (Fourier transform, and a phase and baseline correction) the NMR data, following the instructions in the user manual.

Chemical shifts obtained from the NMR spectra can be assigned by homonuclear through-bond correlations, heteronuclear through-bond correlations [9], and comparisons with published NMR data and NMR chemical shift databases, such as the online Glycosciences.de.

3.2.2 STD NMR Experiments

In the first instance, a ^{1}H NMR spectrum was acquired as a reference spectrum (*see* Subheading 3.2.1), or the off-resonance spectrum that was extracted from a pseudo-2D STD NMR experiment (*see* Subheading 3.2.3) was used. The standard pseudo-2D STD NMR pulse program (Fig. 2) and the parameters for the STD NMR experiments were set according to the following steps:

1. Set the pulse program from the pulse program library of TOPSPIN.
2. Set the standard parameters from the parameter files of TOPSPIN.

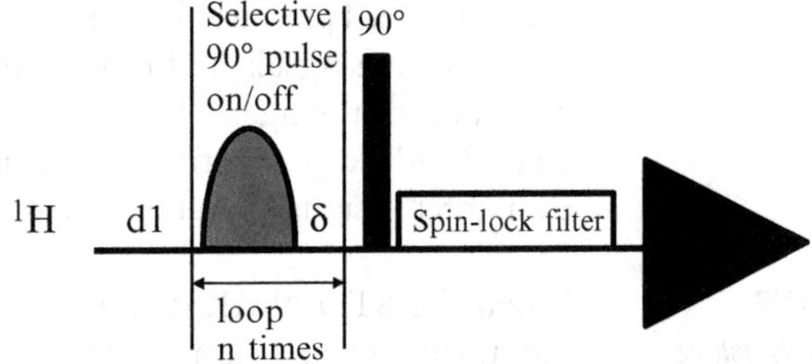

Fig. 2 NMR pulse sequence for the 1D STD NMR experiment for D_2O samples [6]. The length of selective 90° Gaussian pulse (bandwidth = 86 Hz) was set to 50 ms and the delay δ between the selective pulses is 1 ms (δ = 4 μs in the Bruker pulse program); the duration of the presaturation period is adjusted by the number of loops n (typically $n = 40$). The additional spin-lock pulse with a bandwidth = 4,960 Hz after 90° high-power pulse was set to normally 30 ms to remove unwanted protein background. The STD NMR experiment was set up as a pseudo-2D by switching the on- and off-resonance frequencies of the selective pulse between −1 and 1 ppm and ~30 ppm after every scan. The FIDs of on- and off-resonance spectra are stored and processed separately. Subtracting the on- and off-resonance spectra results in the STD NMR spectrum

3. Check and change the relaxation delay to 3–4 s to ensure complete relaxation of the ligand.
4. Set the number of scans (ns) to 256–2,048, depending on signal intensity.
5. Define the frequency list (fq2list) with an on- and off-resonance frequency (*see* **Note 5**).
6. Set the number of loop times of a train of Gaussian-shaped pulses with a 50-ms duration for a total saturation time of 2 s to 40 (*see* **Note 6**).
7. Set the time for a spin-lock filter to 10–30 ms, with a strength of 5 kHz to suppress the residual protein background.
8. Check and adjust the pulse width and the power level (dB) for the pulse width.
9. Set the receiver gain using the command "rga".
10. Collect the NMR data using the command "zg". Free induction decay values with on- and off-resonance protein saturation are recorded in an alternative fashion as a pseudo-2D experiment (*see* **Note 7**).

For samples with higher concentrations of H_2O, the standard STD NMR pulse program with the Watergate 3-9-19 pulse train is employed, with gradients gpz1 and gpz2 of 30 % strength for each. For more complex carbohydrates, 2D STD NMR experiments such as STD-TOCSY and STD-HSQC are used.

It is recommended that several control experiments are carried out to ensure that the STD NMR experiment is set up properly on the spectrometer being used. One of such experiments is an STD NMR experiment using a ligand-only sample in the same buffer to ensure no excitation of the ligand by itself and no STD signals. In addition, an STD NMR experiment using a ligand-protein sample known to exhibit STD signals can be used as a positive standard control.

3.2.3 STD NMR Processing and Analysis

When the STD NMR data were recorded as a pseudo-2D experiment, only the F2 dimension of a 2D data set was processed. The on- and off-resonance spectra were then extracted from the pseudo-2D NMR data. The difference spectrum representing the STD NMR spectrum could then be obtained by subtracting the on-resonance spectrum from the off-resonance spectrum.

The steps for processing and subtraction using TOPSPIN are the following:

1. Process only the F2 dimension of the pseudo-2D STD NMR data using the command "xf2".
2. Perform phase correction according to the user manual, and then (if needed) apply baseline correction using the command "abs2".
3. Extract the on- and off-resonance spectrum from the pseudo-2D spectrum using the Bruker AU program "split2D". The on- and off-resonance spectra are saved separately into new experiment-processing directories.
4. Load the on- and off-resonance spectra in the same experiment-processing directory using multiple display, and subtract the on-resonance from off-resonance spectrum by clicking the icon "Δ". The difference spectrum (STD NMR spectrum) is then saved separately.

To create the binding epitope map of a ligand, the relative STD effects were calculated by determining the individual signal intensities (integral values) in the STD (I_{STD}) and reference spectra (I_0). The ratios of the intensities I_{STD}/I_0 (relative STD effects) were normalized using the largest STD effect that was set to 100 % [6]. Figure 3a shows the reference ^{1}H NMR spectrum of the mixture of β-Me-Gal and EW29Ch (upper spectrum), and the corresponding STD NMR spectrum (lower spectrum). The STD NMR spectrum was acquired using in-house pulse program, and not the Bruker pulse program. The STD epitope map of β-Me-Gal binding to EW29Ch was obtained by normalizing the largest STD effect of the H4 proton to 100 % (Fig. 3b). The H3, H5, H6a, and H6b protons showed similar STD effects (40–53 %), and the H1 proton and protons of the O-methyl group exhibited the lowest STD effects of 10 % and 11 %, respectively. These results indicate that EW29Ch mainly recognized the region from H3 to H6a/H6b. The epitopes of galactose residue for binding to EW29Ch from this STD NMR experiment therefore correlate well with those obtained from the crystal structure of EW29Ch in complex with lactose [8].

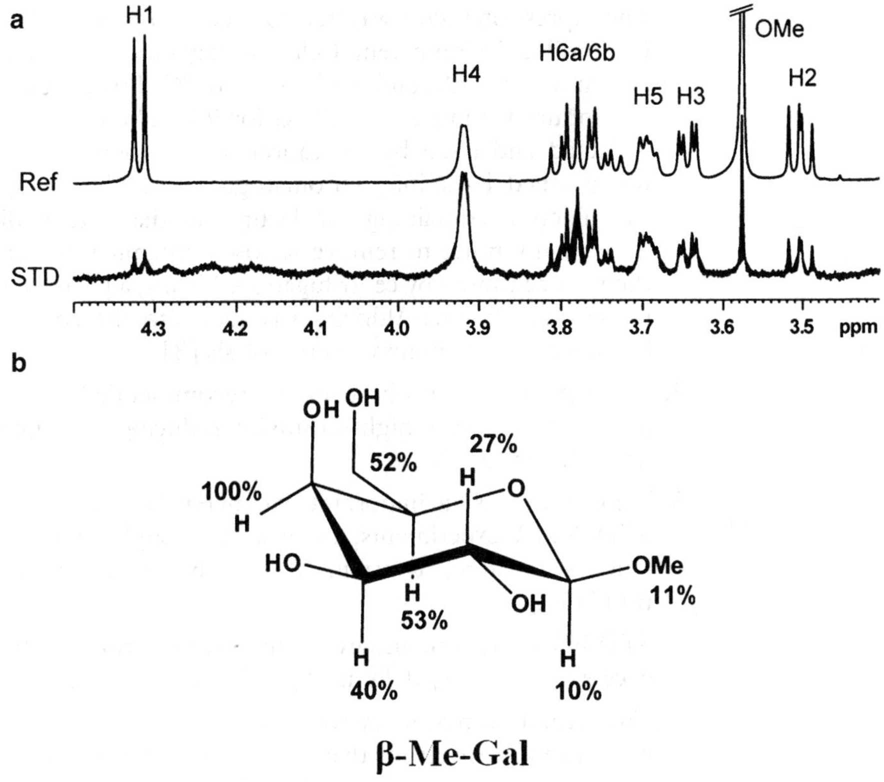

Fig. 3 (**a**) 1D ^{1}H reference NMR spectrum and STD NMR spectrum for the mixture of β-Me-Gal (5 mM) and EW29Ch (50 μM) [8]. Saturation of the protein was achieved with a train of Gaussian pulse, resulting in a total saturation time of 2 s. On-resonance irradiation of the protein was performed at −0.4 ppm and an off-resonance at 30 ppm where no ligand and protein signals were present. (**b**) The structure shows the relative STD NMR effects (%) of β-Me-Gal. Values were calculated by determining the individual signal intensities in the STD NMR spectrum (I_{STD}), and the reference spectrum (I_0) [6]. The ratios of the intensities I_{STD}/I_0 were normalized to the largest STD effect of the H4 proton of β-Me-Gal that was set to 100 %

Yan et al. revealed that T_1 relaxation of ligand protons has a severe interference on the epitope map derived from an STD NMR experiment [10]. Thus, for more accurate epitope mapping, the total STD values calculated from the initial slopes of the STD buildup curves are used (*see* **Note 6**). In summary, STD NMR spectroscopy is a very useful method for identifying ligands that interact with a target protein and characterizing the epitopes of the ligands. Knowledge of this interaction is expected to provide valuable information for the design of new lectins, forming critical biomarkers.

4 Notes

1. To overexpress EW29Ch in *Escherichia coli*, the expression vector was constructed by incorporating the open reading frame of the EW29Ch gene into the pET21 plasmid.

The expression vector was transformed into *E. coli* cells, BL21-CodonPlus™ Competent Cell (Stratagene), which was then grown in Luria–Bertani medium at 37 °C. Protein expression was induced using 1 mM IPTG for 9 h. The cells were then collected and lysed by sonication, and recombinant protein was purified by affinity chromatography on lactose-agarose. The fraction containing the lectin was dialyzed in distilled water many times to remove lactose contamination, and was then concentrated by centrifugation. Finally, a protease inhibitor cocktail (Sigma-Aldrich) was added to the concentrated lectin solution to minimize proteolysis [8].

2. The use of a cryoprobe is highly recommended because the probe delivers very high sensitivity, reducing the number of necessary scans.
3. Temperature is an important parameter for the sensitivity of STD NMR experiments, because it strongly influences the kinetics and affinity constant of ligand–protein complex formation [11].
4. STD NMR experiments are commonly performed with a large excess of ligand, typically at a ligand:protein ratio of 50–100:1.
5. The saturation frequency (on-resonance) should be set at the region of the spectrum that contains protein, but not ligand resonances (normally around −1.0 to 1.0 ppm). The off-resonance frequency is normally set at 30–40 ppm, which is the region where protein or ligand signals are absent.
6. The STD signals depend on the saturation time; the longer the saturation time, the stronger the STD signal. Generally, a saturation time of 2 s gives good results. For more accurate epitope mapping, the STD NMR experiment acquires data at different saturation times (normally from 0.5 to 4.0 s). This obtains total STD values that are calculated from the initial slopes of the STD buildup curves for the STD amplification factor ($= I_{STD}/I_0 \times$ ligand excess) as a function of saturation time, because the total STD values neglect the overestimation of STD effects of ligands from the T_1 bias [12, 13].
7. The STD NMR data are recorded as a pseudo-2D NMR experiment by switching the on- and off-resonance frequencies of the selective pulse because of instability due to long data collection times and low S/N.

Acknowledgement

This work was supported in part by a Grant-in-Aid for Scientific Research (KAKENHI) (C) (20580373 and 24580500) from the Japan Society for the Promotion of Science (JSPS).

References

1. Weis WI, Drickamer K (1996) Structural basis of lectin-carbohydrate recognition. Annu Rev Biochem 65:441–473
2. Rini JM (1995) Lectin structure. Annu Rev Biophys Biomol Struct 24:551–577
3. Weis WI (1997) Cell-surface carbohydrate recognition by animal and viral lectins. Curr Opin Struct Biol 7:624–630
4. Bewley CA, Shahzad-ul-Hussan S (2013) Characterizing carbohydrate-protein interactions by nuclear magnetic resonance spectroscopy. Biopolymers 99:796–806
5. Mayer M, Meyer B (1999) Characterization of ligand binding by saturation transfer difference NMR spectroscopy. Angew Chem Int Ed 38:1784–1788
6. Mayer M, Meyer B (2001) Group epitope mapping by saturation transfer difference NMR to identify segments of a ligand in direct contact with a protein receptor. J Am Chem Soc 123:6108–6117
7. Meyer B, Peters T (2003) NMR spectroscopy techniques for screening and identifying ligand binding to protein receptors. Angew Chem Int Ed 42:864–890
8. Hemmi H, Kuno A, Ito S, Suzuki R, Hasegawa T, Hirabayashi J (2009) FEBS J 276:2095–2105
9. Bubb WA (2003) NMR spectroscopy in the study of carbohydrates: characterizing the structural complexity. Concepts Magn Reson 19(A):1–19
10. Yan J, Kline AD, Mo H, Shapiro M, Zartler ER (2003) The effect of relaxation on the epitope mapping by saturation transfer difference NMR. J Magn Reson 163:270–276
11. Groves P, Kövér KE, André S, Bandorowicz-Pikula J, Batta G, Bruix M, Buchet R, Canales A, Cañada FJ, Gabius H-J, Laurents DV, Naranjo JR, Palczewska M, Pikula S, Rial E, Strzelecka-Kiliszek A, Jiménez-Barbero J (2007) Temperature dependence of ligand-protein complex formation as reflected by saturation transfer difference NMR experiments. Magn Reson Chem 45:745–748
12. Mayer M, James TL (2004) NMR-based characterization of phenothiazines as a RNA binding scaffold. J Am Chem Soc 126: 4453–4460
13. Bhunia A, Bhattachariya S (2010) Mapping residue-specific contacts of polymyxin B with lipopolysaccharide by saturation transfer difference NMR: insights into outer-membrane disruption and endotoxin neutralization. Biopolymers 96:273–287

Chapter 42

Small-Angle X-Ray Scattering to Obtain Models of Multivalent Lectin–Glycan Complexes

Stephen D. Weeks and Julie Bouckaert

Abstract

Recent advances in small-angle X-ray scattering (SAXS) have led to the ability to model the glycans on glycoproteins and to obtain the low-resolution solution structures of complexes of lectins bound to multivalent glycan-presenting scaffolds. This progress in SAXS can respond to the increasing interest in the biological action of glycoproteins and lectins and in the design of multivalent glycan-based antagonists. Carbohydrates make up a significant part of the X-ray scattering content in SAXS and should be included in the model together with the protein, whose structure is most often based on a crystal structure or NMR ensemble, to give a far-improved fit with the experimental data. The modeling of the spatial positioning of glycans on proteins or in the architecture of lectin–glycan complexes delivers low-resolution structural information hitherto unmatched by any other method. SAXS data on the bacterial lectin FimH, strongly bound to heptyl α-D-mannose on a sevenfold derivatized β-cyclodextrin, permitted determination of the stoichiometry of the complex and the geometry of the lectin deposition on the multivalent β-cyclodextrin. The SAXS methods can be applied to larger complexes as the technique imposes no limit on the size of the macromolecular assembly in solution.

Key words Small-angle X-ray scattering, ATSAS, Lectin, Glycan, FimH, Multivalency, Heptyl α-D-mannose, β-Cyclodextrin

1 Introduction

In the era of genomics, the concept of epi-genomics quickly gained access when it became increasingly evident that the functioning of organisms could not be explained by their DNA content only. Life heavily depends on the rapid responses that cells and macromolecules give in an ever-changing environment. Environmental changes could be an infection, aging, metabolic imbalance, and the confrontation with allergens, antigens, or injuries and are usually handled well by the immune system. The speed and specificity of cellular responses to such molecular stimuli are manifested in the well-regulated activity of housekeeping proteins [1]. Posttranslational modifications affect the cellular localization and appearance of housekeeping proteins in the cell membrane, thus

Jun Hirabayashi (ed.), *Lectins: Methods and Protocols*, Methods in Molecular Biology, vol. 1200,
DOI 10.1007/978-1-4939-1292-6_42,

allowing to vary the perception of these proteins by potential interaction partners [2]. Bacteria mimic or adopt these communication mechanisms to disguise themselves and settle without rejection in the host. Thus, biological recognition depends to a great extent on intercellular communication via the recognition of certain proteins on one cell by receptors on the other. Such rapidly adaptable biological recognition is adopted in lectin–glycan interactions because of the enormous encoding potential of glycans [3]. Lectins are proteins that exhibit a finely tuned specificity for their glycosylated receptors. Moreover, lectin specificity extends beyond the three-dimensional structure of the carbohydrate-binding pocket and its compatibility with possible conformers of the recognized glycan sequences, into a fourth dimension created in lectin oligomerization. Clusters of lectin-carbohydrate complexes act as molecular signaling entities by phosphorylation at the cytoplasmic side of the receptor, such as uroplakin-IIIa upon recognition of the high-mannose glycan on uroplakin-Ia by the *Escherichia coli* FimH adhesin [4]. *E. coli* adhesion to epithelial linings in the human body, mediated by the type-1 fimbrial lectin FimH, can be inhibited using mannosidic substances, with affinity of up to 5 nM for a small monomeric compound heptyl α-D-mannose (HM) [5].

Multivalent interactions creating avidity for cellular receptors are the most common way of biological interaction in cell adhesion events, whether between cells of the same organism or in host-pathogen communications [6, 7]. Nevertheless, multivalent interactions are perhaps the least well understood due to their complexity [8]. What is well known is that the nature of these evolutionarily optimized multivalent biological interactions allows a highly fine-tuned regulation of the frequency of their occurrence, of their affinity, and consequently of the cellular signals in host and pathogen [7]. We would like to increase our understanding of the intricacies of common multivalent biological interactions by exposing the involved molecules, in particular bacterial lectins, to multivalent synthetic probes composed of the glycan receptor molecules. In doing so, one may elucidate principles of affinity and cellular response regulation through stoichiometry, geometry, and cooperativity [9]. This in turn will provide information on how to move forward the best possible designs of multivalent, polymeric inhibitors and neutralizers of bacterial adhesion [10].

The dynamic nature of polysaccharide chains on glycoproteins, and the often large size of lectin-glycoconjugate complexes, has typically precluded the structural analyses of these biologically important species by either X-ray crystallography or nuclear magnetic resonance (NMR). To overcome these problems, small-angle X-ray scattering (SAXS) has become an increasingly popular technique in the structural biologist toolkit, complementing the traditional high-resolution methods. Crucially for this group of proteins, SAXS permits the characterization of the protein in a native state in solution with no size limitations [11]. Examples of lectins that have been studied using SAXS, in ensembles that were not possible to be studied using crystallography, are lectins that oligomerize upon glycan recognition into pore-forming toxins [12–14], oligomer conformers regulatory for lectin activity of mannan-binding protein [15, 16], and the architecture for the dual specificity of the BC2L-C lectin from *Burkholderia cenocepacia* [17].

Although an old technique, SAXS has undergone somewhat of renaissance in the past decade [11]. The principal developments have occurred methodologically, and include the ability to build ab initio bead models of a protein [18] or to employ rigid body refinement with available high-resolution structures to generate complex structures [19]. In the specific field of glycoprotein structural analysis, computational methods have recently been created to model the carbohydrate chain and include it in structural models used to fit SAXS data [20, 21]. These software developments have occurred in parallel with improvements in the synchrotron beamline hardware setup, in particular in facilitating sample handling whilst using lower amounts of material. This, plus the increasing number of commissioned beamlines, has permitted a much wider base of users.

In the following sections we shall describe the application of SAXS to study the interaction of a lectin with a multivalent ligand [10]. Whilst the example is specific, the methods used can be readily applied to other proteins of interest. We do not elaborate much on the theoretical framework of this technique and we recommend the interested user to a number of excellent treatises on the topic [22–24]. In terms of the computational tools used for data analysis, we have restricted ourselves to the ATSAS package [25]. This is principally because it is the suite of programs most commonly employed and thus has good user support. Additionally, a number of computational intensive programs within this suite can be run remotely using the online ATSAS server (http://www.embl-hamburg.de/biosaxs/). Even so the user should be aware that other programs exist and are currently being developed (*see* references within ref. 25).

2 Materials

1. The buffer solution (hereafter called buffer): 25 mM HEPES (*see* **Note 1**), 150 mM NaCl, 0.5 mM EDTA, 5 mM DTT (*see* **Note 2**).
2. The purified, recombinant lectin domain of FimH [26].
3. Compound β-cyclodextrin sevenfold conjugated to heptyl α-D-mannopyrannoside (hereafter called HM) [10].

4. Access to a synchrotron beamline equipped for SAXS experiments (*see* **Note 3**).
5. Bovine serum albumin (BSA; SIGMA ultrapure) resuspended in buffer at 5 mg/mL, for placing intensity measurements on a relative scale for mass determination.

3 Methods

3.1 Sample Preparation

1. FimH lectin domain is mixed in a sevenfold molar excess with the β-cyclodextrin sevenfold conjugated to HM.
2. The resultant complexes FimH are purified using size-exclusion chromatography (SEC) in the buffer.
3. Peak fractions are pooled and concentrated using a centrifugal concentration device (maximum 10-kDa molecular weight cutoff) that has been pre-rinsed with buffer. Taking samples as they concentrate reduces the risk of sample aggregation (*see* **Note 4**).
4. As a contrast-based method (*see* Subheading 3.2, **step 5**), it is essential for SAXS data collection that the buffer control sample is of exactly the same composition as in the sample of the lectin-ligand complex. For this purpose, the flow-through of the concentration device is kept and used as a blank for reference data collection.
5. The concentrating samples are recuperated at the appropriate times from the device at concentrations of approximately 1, 2, 5, and 10 mg/mL. The concentration of the FimH lectin domain is determined using an absorption coefficient at 280 nm of 26,025 M/cm (*see* **Note 5**).
6. The stoichiometry of the multivalent complex had previously been determined to be three FimH lectin domains for one β-cyclodextrin, using isothermal titration calorimetry (ITC) with reverse titration (FimH injected into the measurement cell containing the ligand) [10]. Knowledge of the stoichiometry and thus molar mass of the particles are crucial for normalization of SAXS data and can also be determined using refractometry (*see* **Note 6**).

3.2 Data Collection and Processing

1. SAXS curves are measured at 10 °C in the range of momentum transfer 0.008 $Å^{-1} < q < 0.60$ $Å^{-1}$ (where $q = 4\pi \sin(\theta)/\lambda$; 2θ is the scattering angle and $\lambda = 1.5$ Å is the X-ray wavelength). For data collection 2-min exposures, with 15-s collection windows, are used to minimize radiation damage (*see* **Note 7**).
2. Following integration and radial averaging of the individual detector frames, the processed curves are averaged together—at this point curves showing radiation damage which is detected

by the telltale sign of aggregation in the Guinier region can be discarded. Visual inspection of the data is performed using PRIMUS [27] found in the ATSAS software suite. This program also allows for simple manipulation of the data including averaging and subtraction. Alternatively data manipulation and analysis can be performed via the command line using DATtools, a group of applications also found in the ATSAS suite of programs.

3. The two buffer samples are further averaged together and then subtracted from the data of the target sample, again using either PRIMUS or DATtools, to yield the SAXS profile of the sample alone. This process is repeated for the lectin-ligand sample at the various known concentrations (*see* **Note 8**).
4. The data is then finally placed on the same scale by dividing the scattering intensity of each sample by the predetermined concentration (Fig. 1a). In this regard, the determination of the sample concentration is cardinal (*see* **Note 5**) and should equally be applicable also to glycoprotein [21] and lectin–glycan complexes [10] (*see* **Note 6**).

3.3 Data Analysis and Merging

1. Fixed sample beamlines have the benefit of permitting the study of the effect of sample concentration on solution properties. In particular the scattering profile is sensitive to non-ideal solute behavior [22]. This may arise either from inter-particle association or repulsion. The sample properties can be readily distinguished by examination of the scaled data in PRIMUS (Fig. 1a). Non-ideal behavior can also be observed in the plot of the radius of gyration (R_g), or the forward scattering value ($I(0)$), versus sample concentration (Fig. 1b). Both of these values can be calculated using the Guinier approximation (Eq. 1):

$$I(q) = I(0)e^{\left(-q^2.R_g^2/3\right)} \quad (1)$$

or in a more familiar expansion comparable to that of straight line (Eq. 2):

$$\ln I(q) = \ln I(0) - \left(q^2.R_g^2\right)/3 \quad (2)$$

This approximation assumes linearity for a globular protein in the part of the SAXS curve that meets the condition $q \times R_g < 1.3$ known as the Guinier region [22]. $I(0)$ and R_g are thus calculated from a fit of the plot $ln\ I(q)$ versus q^2 within this region (Fig. 1c). Plotting and fitting of the data can be performed in automated fashion using the ATSAS program AUTORG [25], running the application either from the command line or called from within PRIMUS. In the latter case the graphical interface permits manipulation and optimization of the fitting region.

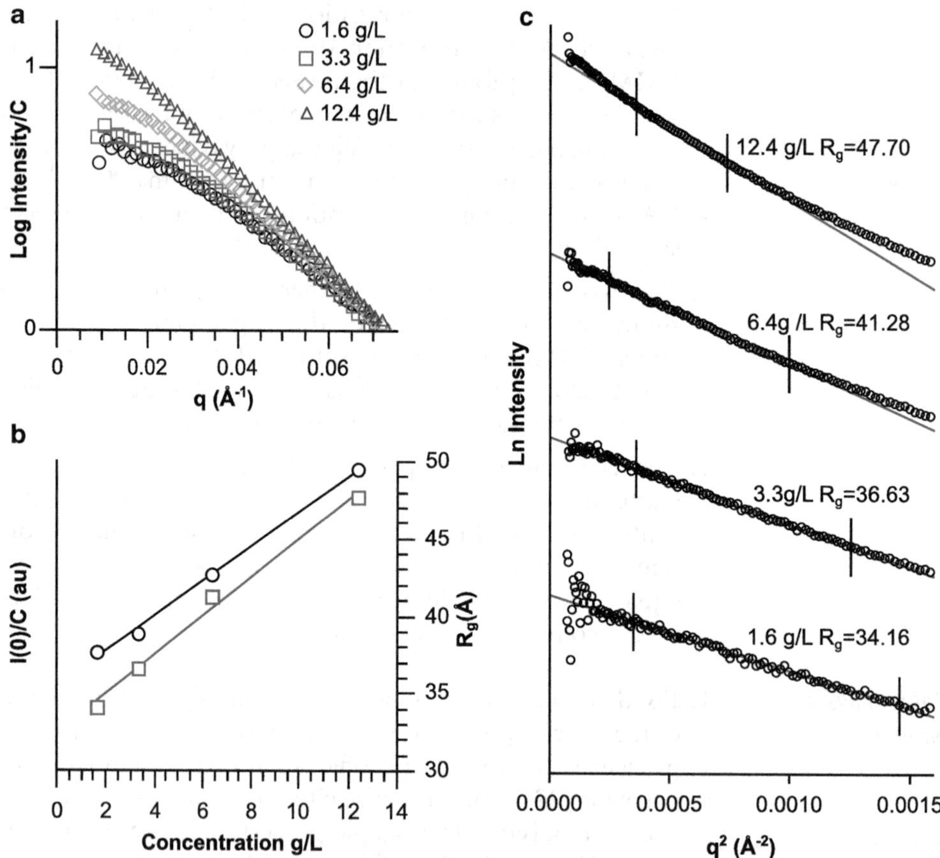

Fig. 1 Scaled data from PRIMUS to evaluate inter-particle interference and Guinier analysis for the obtention of R_g. (**a**) SAXS data were measured at four different concentrations of FimH premixed with the multivalent ligand. The logarithm of the intensity was normalized through division by the sample concentration (**c**). At low angle ($q < 0.02$ Å) and for the two highest concentrations, there are clear indications for inter-particle association. (**b**) Plot showing variation of the forward scattering value $I(0)$ normalized against sample concentration (*left axis, black hollow circles*) and the radius of gyration R_g (*right axis, red hollow squares*) with sample concentration (*horizontal axis*). (**c**) Both values $I(0)$ and R_g can be calculated using the Guinier approximation, from a fit of the plot $ln\ I(q)$ versus q^2 within the linear region. Guinier plots were generated using PyXPlot (http://www.pyxplot.org.uk) where each curve is offset by a factor of 2 to the curve below it. The *vertical lines* on each plot correspond to the lower and upper limits of the q^2 values, predetermined using AUTORG, between which linear least squares fitting was performed

2. In the case of an ideal behaving sample, best practice would be to use the scattering data of the highest concentration sample for further steps as this has the lowest signal-to-noise ratio (*see* **Note 8**). Even so it is common in the field to merge data from the low q-range of a SAXS curve of a dilute sample with the wider q-range data of the highest concentration sample—to minimize the effect of any overlooked non-ideal behavior. This approach though typically transfers the higher error of the low concentration data to the final curve and results in overly generous χ^2 values. Those values are used to qualify model

fitting in subsequent modeling steps. Software has been, and is being, developed to perform this step without bias. In the simplest case, the program ALMERGE [28] scales the lowest concentration data to the highest and identifies the best overlap between the two data sets. It outputs a single merged file taking the low *q*-range intensity values from the low concentration data, followed by the averaged intensity data encompassing the overlapping region of the two files, and the remainder of the curve from the high-concentration data. Importantly, it transplants to this file the error from the highest concentration data.

3. For further modeling steps, it is crucial to know the molar mass of the scattering species (*see* **Note 6**). A number of methods exist for extrapolating this information from SAXS data [22]. In the case of glycoproteins, a comparative analysis of the different methods suggests that the most accurate approach is based on the forward scattering value [20]. This value, which cannot be directly measured, is determined either by the Guinier approximation (Eq. 2) or from the pair distribution function (*see* **step 4**). The mass can be extracted from this value either by calibration of the data on an absolute scale with water or by comparison to the value determined for a protein of known molecular weight, such as BSA, collected during the same data collection period as the sample of interest [29]. In the latter case the mass of the sample under study is calculated by Eq. 3:

$$\mathrm{Mass}_{\mathrm{Sample}} = \frac{I(0)_{\mathrm{sample}}}{I(0)_{\mathrm{BSA}}} \times \mathrm{Mass}_{\mathrm{BSA}} \tag{3}$$

where $I(0)_{\mathrm{BSA}}$ is the calculated forward scattering intensity of BSA (divided by the concentration of the measured preparation) and $\mathrm{Mass}_{\mathrm{BSA}}$ is 66 kDa.

4. The pair distribution function, $P(r)$, of the scattering data provides an intuitive "real space" profile of the interatomic distances of the scattering lectin complex [22]. Calculation of this function not only generates values for $I(0)$ and R_g (using information from the whole scattering curve rather than the Guinier region alone) but it also yields the maximum dimension (D_{max}) of the complex as well as provides information related to the overall shape of the particle [23]. In addition, the regularized scattering curve calculated from this function is used for ab initio model building (*see* Subheading 3.4). Calculation of this function is performed by the indirect-transform method implemented in the ATSAS program GNOM [30]. GNOM can be automatically run on a scattering profile run via the command line using DATGNOM [31]. Alternatively, GNOM is integrated into PRIMUS. This arrangement allows the user to view the interatomic distance profile and fit of the experimental data whilst dynamically modulating the amount of the SAXS curve that is used and adjusting the D_{max} to optimize the result.

3.4 Ab Initio Model Building

1. Dummy atom models (DAM) are generated using the program DAMMIN [18]. This program can be run locally or via the ATSAS online server. The program only requires the final output file from GNOM [30] to run. The angular units must be selected correctly, but for the remaining questions such as symmetry or shape anisotropy the default options (P1 and unknown, resp.) are used (*see* **Note 10**).
2. Typically at least ten instances of the program are run (Fig. 2a). If the online server is employed ensure that the random sequence is initialized from a different seed value by waiting approximately for 1 min between submissions. This number can be checked in the log file—if two instances are the same they will result in the same final model.
3. Following dummy atom model generation the most typical models are identified by DAMAVER [32]. This set of programs identifies the bead model showing the lowest difference to the other structures, superposes all similar structures to the first, and then generates an average model (Fig. 2b). It is simplest to run this in automatic mode using the command

```
>$ damaver -a dammin_model*.pdb
```

Alongside the software-specific logfiles the program generates a new pdb files of the superposed bead models (having the suffix "r" after the filename), the average model file (default filename "damaver.pdb"), a filtered model where the central core of the aligned beads are shown (default filename "damfilt.pdb"), and a model file equivalent to the average file where the beads are flagged (default filename "damstart.pdb").

4. At this point, in terms of the quality of the fit to the SAXS data, the bead model with the lowest discrepancy to the other models can be used for analysis. Alternatively the average model can be resubmitted to DAMMIN inputting the "damstart.pdb" file when requested for an initial dummy atom model (DAM) (Fig. 2c). As of writing this feature is currently not available in the online DAMMIN version and has to be implemented using a local installation of the software.

3.5 Modeling with Available Crystal Structures

1. When atomic structures are available for a lectin moiety these can be modeled against the processed SAXS data with rigid body refinement using SASREF [19] (Fig. 3). Like DAMMIN this can be run either locally or by using the ATSAS online server. This process requires knowledge of the stoichiometry of the interaction; thus the earlier steps in Subheading 3.3 describing molecular mass determination should be first performed.
2. To begin modeling, coordinates of an atomic structure are necessary for both the ligand and the lectin. For the latter case, the structure most similar to the lectin under study must be

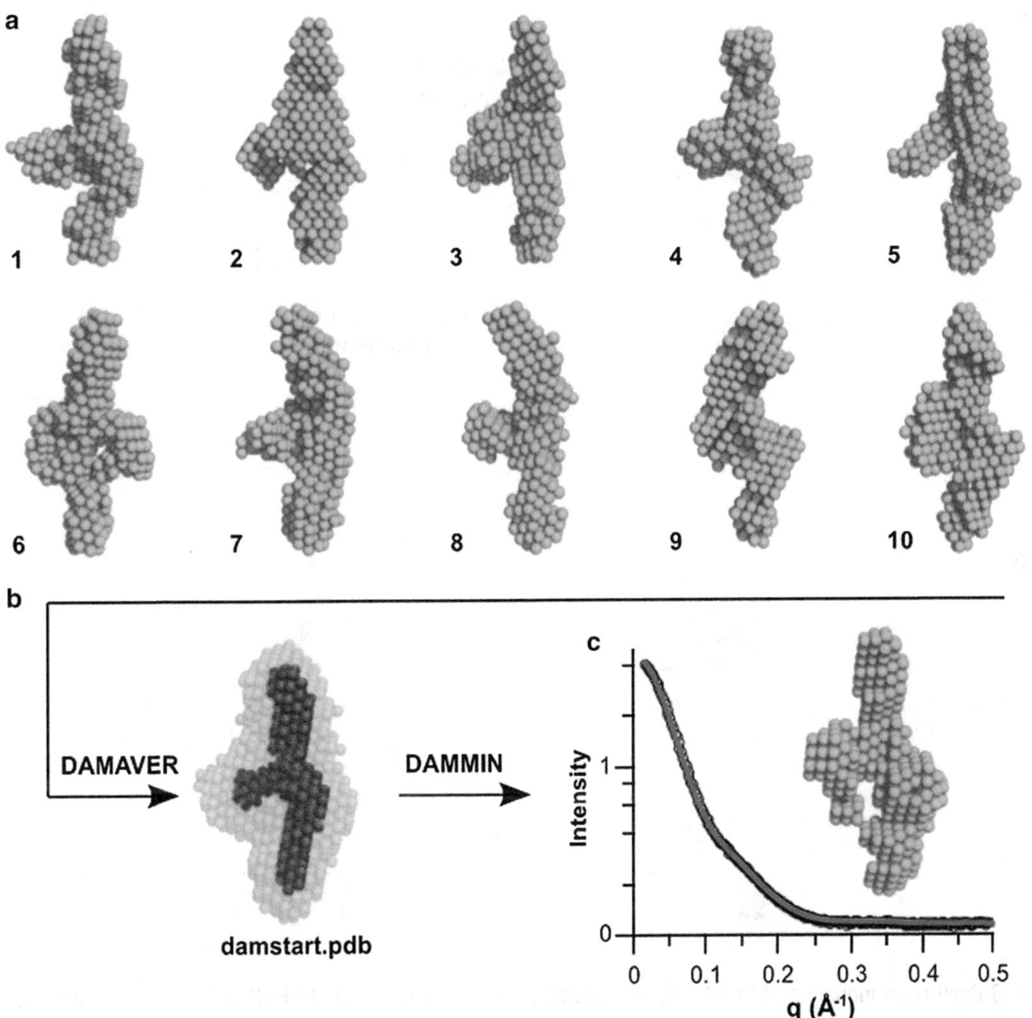

Fig. 2 Dummy atom model (DAM) generation of a complex between the lectin-binding domain of FimH and a cyclic heptavalent ligand. Models show a preference to bind only three moieties of the lectin in agreement with the determined molecular mass. (**a**) Ten instances of DAMMIN were run using the ATSAS online server. The resultant models have χ^2 values ranging from 1.48 to 1.51 when compared to the experimental data. DAMAVER identifies Model 1 ($\chi^2 = 1.49$) as having the lowest average normalized spatial discrepancy (NSD) when compared to the other structures. (**b**) DAMAVER generates an average model identifying a core structure with high atom occupancy (*dark spheres*). The presence of the surrounding atoms (*transparent spheres*) is refined by repeating the DAMMIN using the model as an input. (**c**) The final refined DAM structure and the corresponding fit (*red line* $\chi^2 = 1.49$) to the experimental data (*grey circles*). The model clearly shows three "arms" likely corresponding to three FimH lectin domains

identified in the protein data bank (PDB, www.rcsb.org (*see* **Note 11**). The simplest way of performing this is by sequence alignment using the sequence search tab on the front page of the PDB site. Clicking on the "options" button permits the opportunity to change the search parameters.

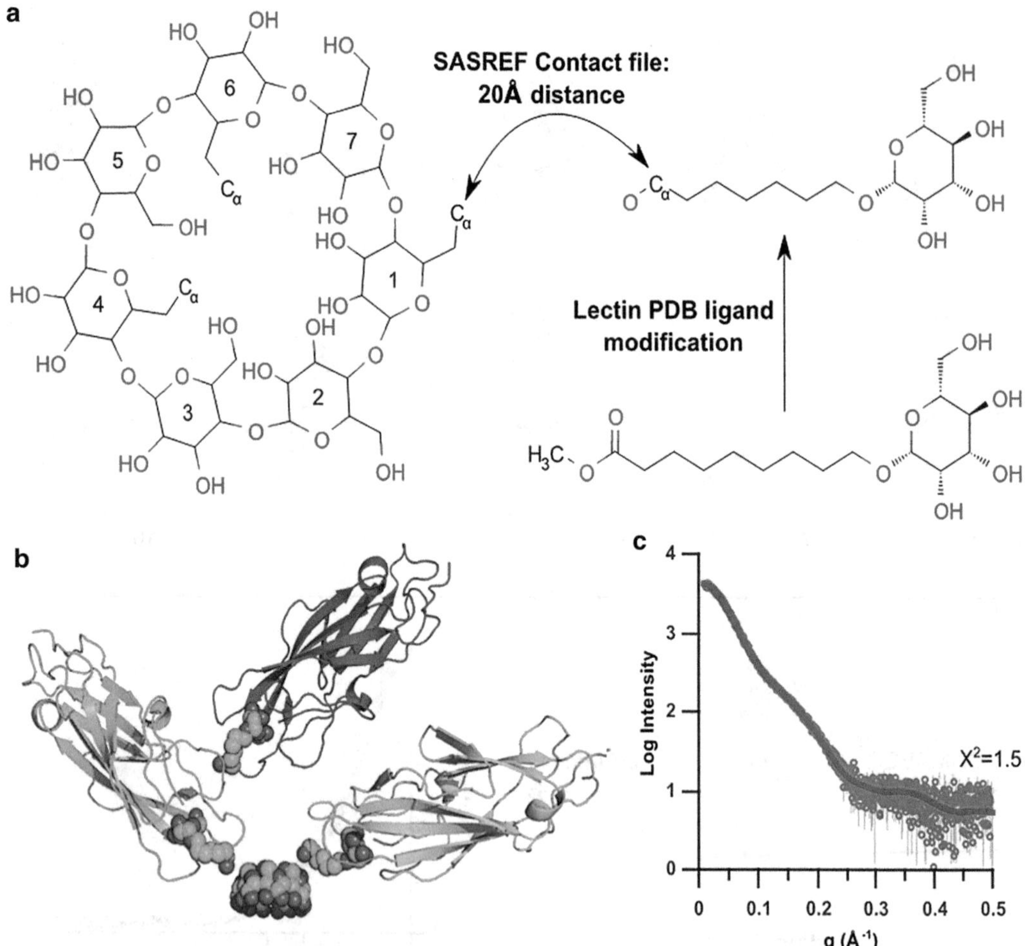

Fig. 3 Rigid body modeling of SAXS data using SASREF. (**a**) Preparation of the PDB and the contact files: for defining a constraint between two components, SASREF only recognizes $C\alpha$ atoms as anchor atoms. With nonprotein-based ligands these have to be manually introduced into the PDB text file. *Left panel*: Three hydroxyl groups of β-cyclodextrin are renamed as CA in the PDB. Modification of a ligand bound to the lectin domain of FimH (PDB entry 4AVI, *see* Ref. 35), the methyl ester group is removed and carbon 7 of the octyl chain is renamed as CA. (**b**) Structure following rigid body refinement of the FimH lectin domains around a heptavalent β-cyclodextrin scaffold. (**c**) Calculated fit of the model (*red line*) compared to the processed SAXS data (*grey circles*, $\chi^2 = 1.5$, with standard deviations, *thin grey lines*)

3. Carefully examine the search hits to identify structures containing the bound carbohydrate ligand. The corresponding PDB text file can then be downloaded.
4. For the multivalent ligand, a model of the scaffold is required. In the simplest case a suitable coordinate file can be identified by web search. Alternatively a model can be generated using one of a number of freely available tools, such as the electronic Ligand Builder and Optimization Workbench (eLBOW) which is part of the PHENIX software suite [33].

The eLBOW program accepts a description of the molecule in a SMILES string format. A PDB file is generated as output, wherein the geometry of the structure has been optimized whilst fulfilling basic chemical restraints [34].

5. To improve convergence of the solution, as well as generating biologically relevant structures SASREF [19] accepts a restraints file. This file, alternatively described as the contact conditions by the program authors, defines a minimal distance between two single or two groups of atoms. Crucially though, SASREF only recognizes $C\alpha$ atoms in the coordinate files when defining the distance restraint. As this atom description is reserved for protein it has to be manually introduced into both the ligand PDB file and into the bound carbohydrate moiety of the lectin structure. To do this, the PDB file is opened with a basic text editor. An appropriate atom is selected (typically one that would be attached to a linker group) and the atom name, in Column 3 of the PDB file, is substituted to CA, for example

```
ATOM    75  O6  GLC    4    -25.629 -29.269 -18.610
```

is converted to

```
ATOM    75  CA  GLC    4    -25.629 -29.269 -18.610
```

In the above example the O6 atom of a glucose residue 4 has been changed to a $C\alpha$ atom. It is important to ensure that the same spacing between the columns in the PDB file is maintained; otherwise it will not be read correctly.

6. The contact conditions file takes on the following format:

```
dist 20
1 1 1 2 162 162
dist 20
1 4 4 3 162 162
dist 20
1 6 6 4 162 162
```

In the above file, used to describe contacts between a heptavalent ligand and three copies of the same lectin, a minimum distance of 20 Å is set between atoms 1, 4, and 6 of the ligand (Column 1) with atom 162 of the three different lectin molecules (Column 4, numbers 2, 3, and 4). As stated above only CA atoms are recognized by SASREF, and the atom number in the restraints file corresponds to the residue number in the PDB file. In the example given in **step 5**, this value is found in Column 5 of the PDB file and would accordingly be 4, rather than the actual atom number 75 given in Column 2. To facilitate preparation of the contact conditions file, the ATSAS online server provides a simple utility where the user can upload the appropriate PDB files and generate the atom selections by a drop-down menu. In the case above, where three lectin structures are bound to the same ligand, a separate PDB file (simply duplicated and renamed) has to be generated for each search copy (Fig. 3b).

7. Once the PDB and contact conditions files have been prepared these are loaded, alongside the processed SAXS data, by the SASREF software whether running it locally or on the online server. Successful completion of the program results in the output of a PDB model and a fit file where the final calculated scattering of the refined model is compared to the experimental data (Fig. 3c).

4 Notes

1. The authors have experienced that phosphate-based buffers accelerate radiation-induced aggregation of the proteins and should be avoided.
2. DTT serves to protect cysteines against oxidation, preventing unwanted covalent inter-particle association that would interfere with the monodispersity, and thus the SAXS profile, of the sample. DTT is to be added fresh to the buffer throughout the purification of the protein, at a concentration of 5–10 mM depending on the cysteine content (higher for more cysteines). Following the final concentration step, no further addition of the reductant should occur to ensure that the protein and buffer samples match exactly in terms of solvent content. DTT also acts as a free radical scavenger protecting the sample against X-ray radiation-induced damage. In the case where the protein contains disulfide bonds, and the reducing agent cannot be used, an alternative scavenger such as glycerol can be employed instead.
3. Samples can be loaded directly into sample chamber at fixed concentrations or, alternatively, a number of beamlines provide the opportunity to collect scattering data of a sample as it elutes from a gel filtration column. Whilst the latter setup provides a number of advantages for simplicity, we focus on handling of fixed concentration data. Typically, the fixed concentration setup requires less protein sample and it is easier to collect data at a higher sample concentration (absence of dilution effects occurring during SEC), resulting in data with a higher signal-to-noise ratio. However, many of the methods applied in the analysis of the data can apply to both fixed and continuous concentration approaches.
4. For SAXS, sample homogeneity and monodispersity are essential conditions; it is therefore necessary to confirm these properties by other methods. Dynamic light scattering (DLS)

allows evaluation of the presence of polydispersity or aggregation in a sample as well as provides a measure of the hydrodynamic radius of a monodisperse particle [10]. A polydispersity value of 20 % or less signifies a very-well-behaved sample. Alternative buffers and additives can be tested to increase monodispersity.

5. The concentration should be determined just prior to SAXS data collection for accurate scaling of the processed data. If wet-lab facilities are available to the user at the synchrotron facility, then this step of sample preparation and concentration determination can be performed in a short time window. Otherwise individual samples are prepared a priori, and the concentration re-measured at the beamline.

6. Glycoproteins and lectin-glycoconjugate complexes contain the glycan mass that is not ultraviolet absorbent, but that has to be accounted for in the scaling of the SAXS data exactly because of its X-ray scattering power. For glycoproteins, the glycan mass, percentage incorporation, and site(s) of modification may be determined by mass spectrometry (MS). For lectin-glycoconjugate complexes, the glycan mass is typically known but information related to stoichiometry of such assemblies is essential. Accurate molar mass of the sample can be achieved using static light scattering (SLS). Alternatively, the combination of SEC or composite gradients (CG) with modern laboratory refractometry such as multi-angle laser light scattering (MALS) also allows measurement of the molar mass of assemblies that are non-covalently associated.

7. At some beamlines, such as P12 on Petra III (Hamburg, Germany) **steps 2–4** of the data processing are performed in a fully automated fashion during data collection [11, 28].

8. On beamlines where the sample is directly injected into the sample cell, SAXS measurements are typically collected for the buffer control both before and after the lectin-ligand sample. Multiple short exposures of each sample are performed to minimize radiation damage during data collection. In theory, this can be optimized for a particular sample but the majority of beamlines provide suitable suggestions for both the length and the number of exposures that can be applied generically to all samples.

9. With reversible non-ideal behavior—either weak self-association or repulsion—the scaled SAXS curves can be examined to ensure that there are no gross changes in the profile (Fig. 1a). In this case, it then may be possible to extrapolate the curve to infinite dilution using ALMERGE. For this to work effectively, the non-ideal behavior must have a linear concentration dependence as observed by plotting R_g or $I(0)$ for the different measured samples (Fig. 1b). Ultimately though, non-ideal behavior and aggregation should be avoided.

This may be achieved by changing the buffer conditions for example by varying either the pH or the salt concentration or by adding stabilizers.

10. Unless a particle had high icosahedral symmetry, DAMMIN cannot generate models where a central region is absent in protein. This typically results in too much mass in the center in the lectin multivalent ligand models. Even so the quality of the map provides excellent hints to the arrangement of the lectin moiety around the carbohydrate cluster.
11. When no structure with high sequence similarity is available, a suitable template can be identified using the HHPRED server [36]. This server simplifies the task of identifying the best structure matching the query sequence and provides a simple pipeline for generating a complete model of the protein of interest using MODELLER.

Acknowledgements

This work was carried out with financial support to JB from the French Agence Nationale de la Recherche (ANR-12-BSV5-0016-01), the Centre National de la Recherche Scientifique, and the Ministeře de l'Enseignement Supeřieur et de la Recherche in France. SDW acknowledges the support of Marie-Curie Reintegration Grant. Measurements were made at beamlines X33 at the DORIS ring in Hamburg, Germany, and SWING at the French synchrotron SOLEIL, in Saint-Aubin.

References

1. Lau KS, Partridge EA, Grigorian A, Silvescu CI, Reinhold VN, Demetriou M, Dennis JW (2007) Complex *N*-glycan number and degree of branching cooperate to regulate cell proliferation and differentiation. Cell 129:123–134
2. Haga Y, Ishii K, Hibino K, Sako Y, Ito Y, Taniguchi N, Suzuki T (2012) Visualizing specific protein glycoforms by transmembrane fluorescence resonance energy transfer. Nature Commun 3:907
3. Gabius H-J, Andre S, Jimenez-Barbero J, Romero A, Solis D (2011) From lectin structure to functional glycomics: principles of the sugar code. Trends Biochem Sci 36:298–313
4. Thumbikat P, Berry RE, Zhou G, Billips BK, Yaggie RE, Zaichuk T, Sun TT, Schaeffer AJ, Klumpp DJ (2009) Bacteria-induced uroplakin signaling mediates bladder response to infection. PLoS Pathog 5:e1000415
5. Bouckaert J, Berglund J, Schembri M, De Genst E, Cools L, Wuhrer M, Hung CS, Pinkner J, Slattegard R, Zavialov A, Choudhury D, Langermann S, Hultgren SJ, Wyns L, Klemm P, Oscarson S, Knight SD, De Greve H (2005) Receptor binding studies disclose a novel class of high-affinity inhibitors of the Escherichia coli FimH adhesin. Mol Microbiol 55:441–455
6. Dam TK, Brewer CF (2010) Lectins as pattern recognition molecules: the effects of epitope density in innate immunity. Glycobiology 20:270–279
7. Sacchettini JC, Baum LG, Brewer CF (2001) Multivalent protein-carbohydrate interactions. A new paradigm for supermolecular assembly and signal transduction Biochemistry 40:3009–3015
8. Turnbull WB (2011) Multivalent interactions: a hop, skip and jump. Nature Chem 3:267–268
9. Badjic JD, Nelson A, Cantrill SJ, Turnbull WB, Stoddart JF (2005) Multivalency and cooperativity in supramolecular chemistry. Acc Chem Res 38:723–732

10. Bouckaert J, Li Z, Xavier C, Almant M, Caveliers V, Lahoutte T, Weeks SD, Kovensky J, Gouin SG (2013) Heptyl α-D-mannosides grafted on a β-cyclodextrin core to interfere with *Escherichia coli* adhesion: an in vivo multivalent effect. Chemistry 19:7847–7855
11. Graewert MA, Svergun DI (2013) Impact and progress in small and wide angle X-ray scattering (SAXS and WAXS). Curr Opin Struct Biol 23:1–7
12. Fujisawa T, Kuwahara H, Hiromasa Y, Niidome T, Aoyagi H, Hatakeyama T (1997) Small-angle X-ray scattering study on CEL-III, a hemolytic lectin from Holothuroidea *Cucumaria echinata*, and its oligomer induced by the binding of specific carbohydrate. FEBS Lett 414:79–83
13. Goda S, Sadakata H, Unno H, Hatakeyama T (2013) Effects of detergents on the oligomeric structures of hemolytic lectin CEL-III as determined by small-angle X-ray scattering. Biosci Biotechnol Biochem 77:679–681
14. Feil SC, Lawrence S, Mulhern TD, Holien JK, Hotze EM, Farrand S, Tweten RK, Parker MW (2012) Structure of the lectin regulatory domain of the cholesterol-dependent cytolysin lectinolysin reveals the basis for its Lewis antigen specificity. Structure 20:248–258
15. Tabarani G, Thepaut M, Stroebel D, Ebel C, Vives C, Vachette P, Durand D, Fieschi F (2009) DC-SIGN neck domain is a pH-sensor controlling oligomerization: SAXS and hydrodynamic studies of extracellular domain. J Biol Chem 284:21229–21240
16. Dong M, Xu S, Oliveira CL, Pedersen JS, Thiel S, Besenbacher F, Vorup-Jensen T (2007) Conformational changes in mannan-binding lectin bound to ligand surfaces. J Immunol 178:3016–3022
17. Sulak O, Cioci G, Lameignere E, Balloy V, Round A, Gutsche I, Malinovska L, Chignard M, Kosma P, Aubert DF, Marolda CL, Valvano MA, Wimmerova M, Imberty A (2011) *Burkholderia cenocepacia* BC2L-C is a super lectin with dual specificity and proinflammatory activity. PLoS Pathog 7:e1002238
18. Svergun DI (1999) Restoring low resolution structure of biological macromolecules from solution scattering using simulated annealing. Biophys J 76:2879–2886
19. Petoukhov MV, Svergun DI (2005) Global rigid body modelling of macromolecular complexes against small-angle scattering data. Biophys J 89:1237–1250
20. Guttman M, Weinkam P, Sali A, Lee KK (2013) All-atom ensemble modeling to analyze small-angle x-ray scattering of glycosylated proteins. Structure 21:321–331
21. Felix J, Elegheert J, Gutsche I, Shkumatov AV, Wen Y, Bracke N, Pannecoucke E, Vandenberghe I, Devreese B, Svergun DI, Pauwels E, Vergauwen B, Savvides SN (2013) Human IL-34 and CSF-1 establish structurally similar extracellular assemblies with their common hematopoietic receptor. Structure 21:528–539
22. Putnam CD, Hammel M, Hura GL, Tainer JA (2007) X-ray solution scattering (SAXS) combined with crystallography and computation: defining accurate macromolecular structures, conformations and assemblies in solution. Quart Rev Biophys 40:191–285
23. Koch MHJ, Vachette P, Svergun DI (2003) Small-angle scattering: a view on the properties, structures and structural changes of biological macromolecules in solution. Quart Rev Biophys 34:147–227
24. Jacques DA, Guss JM, Trewhella J (2012) Reliable structural interpretation of small-angle scattering data from bio-molecules in solution-the importance of quality control and a standard reporting framework. BMC Struct Biol 12:9
25. Petoukhov MV, Franke D, Shkumatov AV, Tria G, Kikhney AG, Gajda M, Gorba C, Mertens HDT, Konarev PV, Svergun DI (2012) New developments in the ATSAS program package for small-angle scattering data analysis. J Appl Crystallogr 45:342–350
26. Wellens A, Garofalo C, Nguyen H, Van Gerven N, Slattegard R, Hernalsteens JP, Wyns L, Oscarson S, De Greve H, Hultgren S, Bouckaert J (2008) Intervening with urinary tract infections using anti-adhesives based on the crystal structure of the FimH-oligomannose-3 complex. PLoS One 3:e2040
27. Konarev PV, Volkov VV, Sokolova AV, Koch MHJ, Svergun DI (2003) PRIMUS—a Windows-PC based system for small-angle scattering data analysis. J Appl Crystallogr 36:1277–1282
28. Franke D, Kikhney AG, Svergun DI (2012) Automated acquisition and analysis of small angle X-ray scattering data. Nucl Instrum Methods Phys Res, Sect A 689:52–59
29. Mylonas E, Svergun DI (2007) Accuracy of molecular mass determination of proteins in solution by small-angle X-ray scattering. J Appl Crystallogr 40:s245–s249
30. Svergun DI (1992) Determination of the regularization parameter in indirect-transform methods using perceptual criteria. J Appl Crystallogr 25:495–503
31. Petoukhov MV, Konarev PV, Kikhney AG, Svergun DI (2007) ATSAS 2.1—towards automated and web-supported small-angle scattering data analysis. J Appl Cryst 40:s223–s228

32. Volkov VV, Svergun DI (2003) Uniqueness of ab-initio shape determination in small-angle scattering. J Appl Crystallogr 36:860–864
33. Adams PD, Afonine PV, Bunkoczi G, Chen VB, Davis IW, Echols N, Headd JJ, Hung LW, Kapral GJ, Grosse-Kunstleve RW, Mccoy AJ, Moriarty NW, Oeffner R, Read RJ, Richardson DC, Richardson JS, Terwilliger TC, Zwart PH (2010) PHENIX: a comprehensive Python-based system for macromolecular structure solution. Acta Crystallogr, Sect D: Biol Crystallogr 66:213–221
34. Moriarty NW, Grosse-Kunstleve RW, Adams PD (2009) Electronic ligand builder and optimization workbench (eLBOW): a tool for ligand coordinate and restraint generation. Acta Crystallogr, Sect D: Biol Crystallogr 65: 1074–1080
35. Wellens A, Lahmann M, Touaibia M, Vaucher J, Oscarson S, Roy R, Remaut H, Bouckaert J (2012) The tyrosine gate as a potential entropic lever in the receptor-binding site of the bacterial adhesin FimH. Biochemistry 51: 4790–4799
36. Soding J, Biegert A, Lupas AN (2013) The HHpred interactive server for protein homology detection and structure prediction. Nucleic Acids Res 33:W244–W248

Chapter 43

Directed Evolution of Lectins by an Improved Error-Prone PCR and Ribosome Display Method

Dan Hu, Hiroaki Tateno, and Jun Hirabayashi

Abstract

Lectins are useful reagents for the structural characterization of glycans. However, currently available lectins have an apparent drawback in their "repertoire," lacking some critical probes, such as those for sulfated glycans. Thus, engineering lectins with novel specificity would be of great practical value. Here, we describe a directed evolution strategy to tailor novel lectins for novel specificity or biological functions. Our strategy uses a reinforced ribosome display-based selection combined with error-prone PCR to isolate mutants with target specificity and an evanescent-field fluorescence-assisted glycoconjugate microarray to rapidly evaluate the specificity of selected mutants. A successful case of screening a lectin, which has acquired an ability to recognize 6-sulfo-galactose-terminated glycans, is described.

Key words Glycans, Directed evolution, Ribosome display, Error-prone PCR, Lectin engineering, Glycoconjugate microarray

1 Introduction

Carbohydrates are intricate information-carrying biopolymers that are generating increasing interest in the post-genomic era. Dramatic changes in glycosylation features have been observed in various key biological events such as embryogenesis, differentiation, and tumorigenesis [1, 2]. The systematic analysis of glycan structures, i.e., glycomics, is therefore critical to understand the roles of carbohydrates and to identify potential biomarkers [3, 4]. Lectins, defined as sugar-binding proteins, are useful reagents for the structural characterization of glycans. Until now, lectins have been extensively used in cell typing, histochemical staining, and glycoprotein fractionation. Recently, a novel technique called lectin microarray, in which a panel of well-defined lectins is immobilized onto a solid support, has been successfully used for high-throughput analysis of complex carbohydrate structures included in serum glycoproteins and whole cells [5–7]. Despite these achievements, several issues remain to be addressed in order to make lectins more

Jun Hirabayashi (ed.), *Lectins: Methods and Protocols*, Methods in Molecular Biology, vol. 1200,
DOI 10.1007/978-1-4939-1292-6_43, © Springer Science+Business Media New York 2014

meaningful tools for glycomics [8]: (1) Most of the lectins used at present are of plant origin, having many inherent problems such as inconsistent activity and unreliable availability. (2) A currently available lectin set has an apparent drawback in its "repertoire," lacking some critical probes, such as those for Siaα2-3-linked or sulfated glycans. (3) Lectins often have broad specificity, which complicates the glycomic interpretation of complex samples. Therefore, engineering lectins having much improved properties would be of great practical value.

Here, we describe a representative protocol to engineer a novel lectin, which acquired unique specificity for 6-sulfo-galactose from a galactose-binding lectin, EW29Ch, by error-prone PCR combined with ribosome display-based selection. Selected lectin mutants can be rapidly tested for their capacity to bind 6-sulfo-galactose-terminated glycans by an evanescent-field fluorescence-assisted glycoconjugate microarray. An outline of the total procedure is shown in Fig. 1.

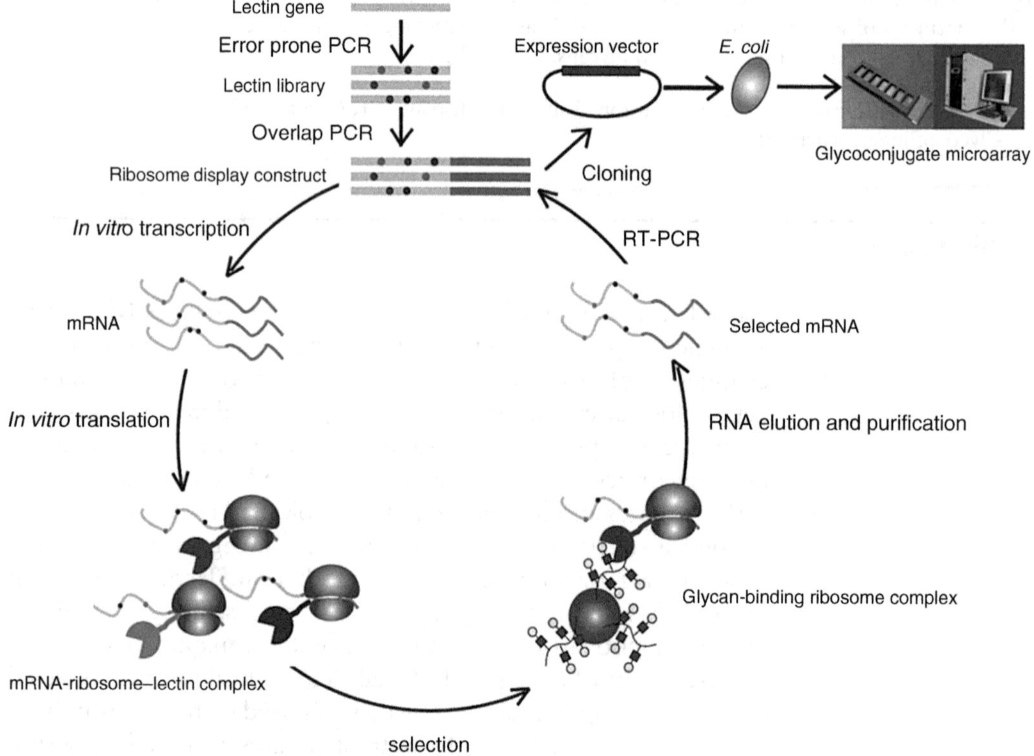

Fig. 1 An outline of the total procedure for PCR-based mutagenesis and selection by a reinforced ribosome display method

2 Materials

2.1 Construction of EW29Ch Library for Ribosome Display by Error-Prone PCR

1. DNA-modifying enzymes: *Nde*I, *Hin*dIII, *Xho*I, *Nhe*I, *Pst*I, *Nhe*I, and *Dpn*I (New England Biolabs).
2. Agarose (Takara).
3. QIAquick Gel Extraction Kit, QIAGEN Plasmid Min Kit, Minelute PCR Purification Kit (Qiagen).
4. PCR amplification: Either PCR Master Mix (Promega, 2× PCR Master Mix) or LA Taq PCR (Takara, 10× LA PCR buffer, 25 mM $MgCl_2$, 2.5 mM dNTPs, and LA Taq DNA polymerase) is used.
5. T4 DNA Polymerase (Takara, 10× T4 DNA polymerase buffer, 0.1 % bovine serum albumin (BSA), 2.5 mM dNTP, 1 unit/μL T4 DNA polymerase).
6. 10 mM $MnCl_2$.
7. Sterile deionized water.
8. 1× Tris-acetate electrophoresis buffer (1× TAE buffer): Prepare a stock solution of 50× TAE, and dilute it 1:50 with H_2O before use.
9. 50× TAE buffer: 242 g/L Tris base, 57.1 mL/L glacial acetic acid, 100 mL/L of 0.5 M ethylenediamine tetraacetic acid (EDTA, pH 8.0).
10. 0.5 M EDTA (pH 8.0).
11. Ethidium bromide (10 mg/mL).

2.2 Selection of Target Mutants by Ribosome Display

1. T7 RiboMAX™ Express Large Scale RNA Production System (Promega) containing Enzyme Mix; RQ1 RNase-Free DNase; RiboMAX™ Express T7 2× Buffer.
2. RNeasy Mini Kit (Qiagen).
3. Biotinylated multivalent carbohydrate polymers (mannose-polyacrylamide-biotin (Man-PAA), and 6′-sulfo-N-Acetyllactosamine-polyacrylamide-biotin (6′-sulfo-LN-PAA)) (Glycotech), which are obtained from Glycotech (http://www.glycotech.com/probes/multivalbio.html). They are dissolved in PBS at a concentration of 1 mg/mL, and stored at −80 °C until use.
4. Binding buffer: PBS containing 50 mM $MgCl_2$, and 1 % (v/v) Tween 20.
5. Dynabeads M-280 streptavidin (Invitrogen).
6. *E. coli* S30 Extract System for linear template (Promega).
7. RNasin® Ribonuclease Inhibitor (Promega).
8. Blocking buffer: 125 μL blocking buffer with lactose containing 82.5 μL of binding buffer, 25 μL of 10 % BSA, 5 μL of

RNasin® Ribonuclease Inhibitor, and 12.5 μL of 100 mM lactose. 125 μL blocking buffer without lactose containing 95 μL of binding buffer, 25 μL of 10 % BSA, and 5 μL of RNasin® Ribonuclease Inhibitor.

9. Elution buffer: 100 μL of PBS containing 50 mM EDTA, and 2 μL of RNasin® Ribonuclease Inhibitor.
10. One-step RT-PCR Kit (Takara).

2.3 Rapid Verification of the Specificity of the Selected Mutants by Glycoconjugate Microarray

1. TBS: 25 mM Tris–HCl, pH 7.4 containing 0.8 % NaCl.
2. 1 % BSA.
3. Isopropyl-β-D-1-thiogalactopyranoside (IPTG).
4. BugBuster® HT Protein Extraction Reagent (Novagen)
5. Probing buffer: TBS containing 2.7 mM KCl, 1 mM $CaCl_2$, 1 mM $MnCl_2$, and 1 % Triton X-100.
6. Glycostation Reader 1200 (GP Biosciences Ltd).
7. Spotting solution (Matsunami Glass Ind., Ltd).
8. Epoxy-coated glass slide (Schott AG).
9. Noncontact microarray printing robot (MicroSys 4000; Genomic Solutions, Inc., MI).
10. *E. coli* DH5α competent cells.
11. Quick Ligation™ Kit (New England Biolabs).
12. QIAGEN Plasmid Mini Kit.
13. ABI 3130xl Genetic Sequencer.
14. Electrocompetent BL21-CondonPlus (DE3)-RIL.
15. Anti-Flag M2 antibody (Sigma).
16. Cy3-labeled goat anti-mouse IgG (H+L) antibody (Life Technologies).

3 Methods

3.1 Construction of EW29Ch Library for Ribosome Display by Error-Prone PCR

1. The plasmid construct pRARE_EW29Ch is schematically shown in Fig. 2. Prepare pRARE_EW29Ch by inserting three regions into the pET27b: insert the coding sequence of the target gene EW29Ch (GeneBank accession NO: AB010783) without a stop codon between *Nde*I and *Hin*dIII sites; insert the coding sequence of spacer gene III between *Xho*I and *Nhe*I sites; insert the rare condon linker between *Pst*I and *Nhe*I sites. The sequence of the rare codon cluster: *CTGCAG*GTTAAG AATTCA**CGGGGACGG**GTT**CGG**GTT*GCTAGC* (*Pst*I and *Nhe*I sites are shown in italic, and codons corresponding to deficiency of cognate aminoacyl-tRNAs are indicated by underline and bold) (*see* **Note 1**).

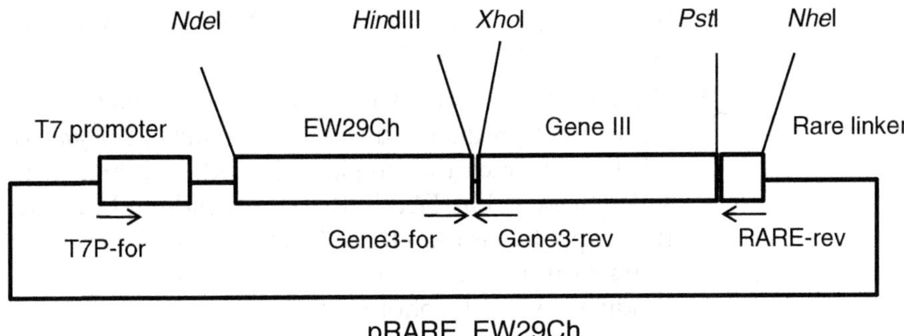

Fig. 2 Diagram of plasmid pRARE_EW29Ch for ribosome display. Sites for relevant restriction enzymes and primers are shown

Table 1
Oligonucleotides used for error-prone PCR amplification to construct EW29Ch libraries in the developed ribosome display method

Primer name	Sequence
T7P-for	5′-TAATACGACTCACTATAGGGGAATTGTG-3′
gene3-rev	5′-CCACCTCCTCCTCCGCCGGTAGAAGATATC-3′
gene3-for	5′-CTCGAGGGGATATCTTCTACCGGCGGAGGA-3′
RARE-rev	5′-GCTAGCAACCCGAACCCGT-3′
Flag-EW29Ch	5′-ATGGACTACAAAGACGATGACGACAAGCATATGAAGCCGAAG TTCTTC-3′, the Flag tag is underlined and Nde 1 site is shown in italic
EW29Ch-end	5′-TCACTCGAGTGCGGCCGCAAGCTT-3, Xho 1 site is shown in italic

2. Generate the random mutant library of EW29Ch by error-prone PCR. Set up one error-prone PCR amplification tube of 50 μL total volume with 25 μL of 2× PCR Master buffer, 7 μL of 25 mM $MgCl_2$, 2.5 μL of 10 mM $MnCl_2$, 1 μL of 10 μM T7P-for, 1 μL of 10 μM gene3-rev (Table 1), and 5 ng pRARE_EW29Ch plasmid as template (*see* **Note 2**).
3. Proceed to thermal cycling according to the following conditions: one cycle at 98 °C for 2 min, followed by 30 cycles at 98 °C for 0.5 min, 65 °C for 0.5 min, and 72 °C for 0.75 min, then one cycle at 72 °C for 5 min, and hold at 4 °C.
4. Use the primers gene3-for and RARE-rev (Table 1) to amplify the gene III spacer fragment by PCR. Set up one PCR amplification tube of 50 μL total volume with 5 μL of 10× LA PCR buffer, 5 μL of 25 mM $MgCl_2$, 4 μL of 2.5 mM dNTPs, 1 μL of 10 μM gene3-for, 1 μL of 10 μM RARE-rev (*see* Table 1), 0.5 μL LA Taq DNA polymerase, and 5 ng pRARE_EW29Ch plasmid as template. Cycle with the following conditions: one cycle at 98 °C for 2 min, followed by 30 cycles at 98 °C for

0.5 min, 65 °C for 0.5 min, and 72 °C for 0.75 min, then one cycle at 72 °C for 5 min, and hold at 4 °C.

5. Treat both the EW29Ch library PCR product and gene III spacer PCR product with 1 μL of 20 unit/μL *Dpn*I for 1 h at 37 °C to degrade the template plasmid, and then analyze on a 1 % agarose gel in TAE buffer with ethidium bromide. Excise the proper bands from the gel and purify using a QIAquick Gel Extraction Kit. Determine the concentrations of the PCR products by spectrophotometric measurement at 260 nm.
6. Treat the purified PCR products described as above with T4 DNA polymerase to generate blunt-ended PCR products. Prepare the reaction mixture in a microcentrifuge tube to the total volume of 29 μL with 1 μg of PCR products, 3 μL of 10× T4 DNA polymerase buffer, 3 μL of 0.1 % BSA, and 2 μL of 2.5 mM dNTP. Heat the reaction mixture at 70 °C for 5 min and then place at 37 °C incubation bath. After that, add 1 μL of 1 unit/μL T4 DNA polymerase and incubate for 5 min. Purify the blunt-ended PCR products using Minelute PCR Purification Kit and elute with 15 μL H_2O. Determine the concentrations of the blunt-ended PCR products by spectrophotometric measurement at 260 nm (*see* **Note 3**).
7. Assemble the blunt-ended EW29Ch library and gene III spacer products by overlap PCR to generate EW29Ch-gene III spacer construct for ribosome display. To do so, set up five overlap PCR tubes of 50 μL volume with 25 ng EW29Ch library PCR product, 25 ng gene3 spacer PCR product, 5 μL of 10× LA PCR buffer, 5 μL of 25 mM $MgCl_2$, 4 μL of 2.5 mM dNTPs, and 0.5 μL LA Taq DNA polymerase. The conditions for cycling are as follows: one cycle at 98 °C for 5 min, followed by five cycles at 98 °C for 0.5 min, 60 °C for 0.5 min, and 72 °C for 1 min 10 s, and then one cycle at 72 °C for 5 min. In the first five cycles, no primers are added. After that, add 1 μL mixture of 50 μM T7P-for and 50 μM RARE-rev and continue the reaction for another 12 cycles. The PCR conditions are set as follows: one cycle at 98 °C for 5 min, followed by 12 cycles at 98 °C for 0.5 min, 65 °C for 0.5 min, and 72 °C for 1 min 10 s, then one cycle at 72 °C for 5 min, and hold at 4 °C (*see* **Note 4**).
8. Purify the overlap PCR products using Minelute PCR purification kit and elute with 15 μL H_2O. Determine the concentrations of the overlap PCR products by spectrophotometric measurement at 260 nm.

3.2 Selection of Target Mutants by Ribosome Display

1. Use the purified overlap PCR products as template for in vitro transcription. Set up 20 μL reactions containing 10 μL of RiboMAX™ Express T7 2× Buffer, 1 μg overlap PCR products, and 2 μL enzyme mix. Carry out the reaction at 37 °C for 30 min.

Add 1 μL of 1 unit/μL RQ1 RNase-free DNase and incubate at 37 °C for 15 min to degrade the DNA template. Purify the in vitro-transcripted RNA using RNeasy Mini Kit. Determine the concentrations of the purified RNA by spectrophotometric measurement at 260 nm and store at –80 °C until use.

2. Preparation of carbohydrate polymer-coated beads. Dissolve the biotinylated multivalent carbohydrate polymers (http://www.glycotech.com/probes/multivalbio.html) including the nontarget glycan mannose-polyacrylamide-biotin (Man-PAA) and the target glycan 6′-sulfo-*N*-acetyllactosamine-polyacrylamide-biotin (6′-sulfo-LN-PAA) in PBS at a concentration of 1 mg/mL. Dilute them to 100 μg/mL with binding buffer before use. Transfer 10 μL of Dynabeads M-280 streptavidin to a 1.5 mL tube, and wash beads with binding buffer three times. Add 10 μL of 100 μg/mL carbohydrate polymers to the tube, and incubate at 4 °C for 30 min on a shaker. After that, remove the unbound carbohydrate polymers by washing beads three times with binding buffer (*see* **Note 5**).
3. Carry out the in vitro translation with *E. coli* S30 extract system for linear template according to the manufacturer's protocol. Set up 50 μL reactions containing 2 μg purified mRNA, 20 μL of S30 premix, 5 μL of 1 mM amino acid mixture minus methionine, 5 μL of 1 mM amino acid mixture minus leucine, 15 μL of S30 extract, and 2 μL of RNasin® Ribonuclease Inhibitor. Carry out the reaction at 37 °C for 10 min. Stop the reaction by adding 200 μL ice-cold binding buffer and place on ice for 5 min. Centrifuge the translation mix (250 μL) at 10,000 × *g* for 5 min and transfer the supernatant to a new ice-cold tube.
4. Transfer 125 μL of the translation mix to a tube containing 125 μL blocking buffer with or without 10 mM lactose. Gently shake the tube on a mixer (1,100 rpm) in an ice-cold room for 1 h (*see* **Note 6**).
5. Transfer 250 μL of the blocked translation mix to a tube containing the carbohydrate polymer-coating magnetic Dynabeads. Gently shake the tube on a mixer (1,100 rpm) in an ice-cold room for 1 h.
6. Wash the carbohydrate polymer-coated Dynabeads with binding buffer four times to remove unbound ribosome complexes.
7. To elute the mRNA, add 100 μL ice-cold elution buffer containing 2 μL RNase inhibitor and incubate for 10 min on ice with gentle shaking. Purify the eluted mRNA immediately using an RNeasy Minelute Cleanup Kit according to the manufacturer's instructions (*see* **Note 7**).
8. Elute the mRNA in 12 μL of RNase-free water and subject to one-step RT-PCR. Set up 25 μL reactions containing 12 μL of mRNA elute, 2.5 μL of 10× one-step PCR buffer, 5 μL of

25 mM $MgCl_2$, 2.5 μL of 10 mM dNTPs, 0.5 μL of RNase inhibitor, 0.5 μL of AMV RTase XL, 0.5 μL of AMV optimized Taq DNA polymerase, 1 μL of 10 μM T7P-for, and 1 μL of 10 μM RARE-rev. Set the RT-PCR conditions as follows: one cycle at 50 °C for 30 min and one cycle at 98 °C for 2 min, followed by 40 cycles at 98 °C for 0.5 min and 68 °C for 1.5 min, then one cycle at 72 °C for 5 min, and hold at 4 °C.

9. Analyze the RT-PCR product on a 1 % agarose gel, and isolate the proper bands by gel extraction. The purified RT-PCR products are ready for the next round of selection by repeating **steps 1–8**.

3.3 Rapid Verification of the Specificity of the Selected Mutants by Glycoconjugate Microarray

After two rounds of selection against 6′-sulfo-LN-PAA, the reduction of RT-PCR products in the presence of 10 mM lactose is no longer observed (Fig. 3), suggesting that EW29Ch mutants with target specificity (6′-sulfo-LN) are enriched.

1. Preparation of glycoconjugate microarray is described previously [9]. Dissolve glycoproteins and glycoside-PAA conjugates in a spotting solution at a final concentration of 0.5 and 0.1 mg/mL, respectively.

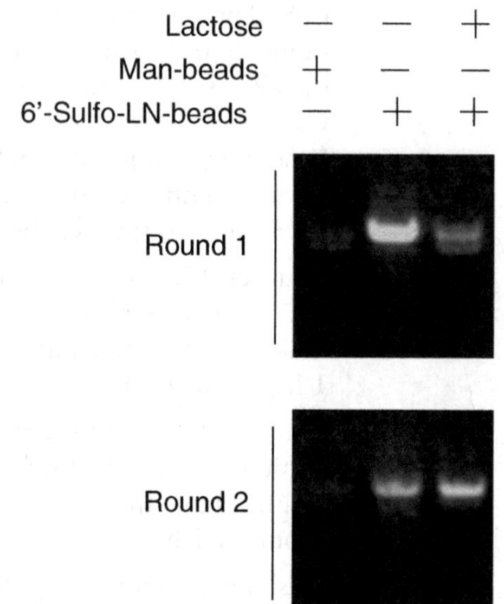

Fig. 3 Selection of 6′-sulfo-LN-binding lectins by pretreatment of the translation mixture with excessive lactose. EW29Ch library generated by error-prone PCR was selected with 6′-sulfo-LN-coated beads with or without pretreatment of the translation mixture with lactose. The precipitated mRNAs under individual conditions were isolated and reverse transcribed to cDNA by one-step RT-PCR. A part of cDNA was used for analysis by agarose-gel electrophoresis; the rest of the cDNAs selected in the presence of lactose were subsequently used for the second round of ribosome display selection. In all experiments, Man-PAA was used as a negative control

2. Filter using 0.22 μm pore size filter to remove insoluble particles, and spot glycoprotein and glycoside-PAA on a microarray-grade epoxy-coated glass slide attached with a silicone rubber sheet with 14 chambers, using a noncontact microarray printing robot with a spot diameter size of 220 μm spaced at a 260 μm interval.
3. Incubate the glass slide in a humidity-controlled incubator at 25 °C for 3 h to allow immobilization. After that, wash the glass slide with the probing buffer to remove the excess amounts of non-immobilized materials and block with 100 μL of TBS containing 1 % BSA at 20 °C for 1 h. The glycoconjugate microarray is stored is 4 °C and now ready for use.
4. To clone the selected EW29Ch mutants into an expression vector containing *N*-terminal Flag tag, restriction sites need to be created. In our vector systems, *Nde*I and *Xho*I sites are introduced to the expression vector of pET27b_Flag_EW29Ch, which contains an *N*-terminal Flag tag followed by *Nde*I site, EW29Ch coding sequence, and *Xho*I site.
5. Set up 50 μL reactions containing 5 μL of 10× LA PCR buffer, 5 μL of 25 mM $MgCl_2$, 4 μL of 2.5 mM dNTPs, 1 μL of 10 μM phosphorylated Flag-EW29Ch, 1 μL of 10 μM phosphorylated EW29Ch-end (*see* Table 1), 0.5 μL LA Taq DNA polymerase, and 5 ng pRARE_EW29Ch plasmid as template. The conditions for cycling are as follows: one cycle at 98 °C for 5 min, followed by 30 cycles at 98 °C for 0.5 min, 60 °C for 0.5 min, and 72 °C for 1 min, and then one cycle at 72 °C for 5 min.
6. Purify the PCR products and insert the fragments into the pET27b which was digested with *Nde*I/*Xho*I and further blunt-ended with T4 DNA polymerase as described above. The ligated products were transformed into *E. coli* DH5α competent cells and the sequence of pET27b_Flag_EW29Ch was confirmed.
7. To clone the selected EW29Ch mutants to the Flag-tagged expression vector, the RT-PCR product after two rounds of selection was amplified by PCR using primers, Flag-EW29Ch and Gene3-rev. Set up 50 μL reactions containing 5 μL of 10× LA PCR buffer, 5 μL of 25 mM $MgCl_2$, 4 μL of 2.5 mM dNTPs, 1 μL of 10 μM Flag-EW29Ch, 1 μL of 10 μM gene3-rev (*see* Table 1), 0.5 μL LA Taq DNA polymerase, and 1 μL purified RT-PCR product as template. The conditions for cycling are as follows: one cycle at 98 °C for 5 min, followed by 30 cycles at 98 °C for 0.5 min, 60 °C for 0.5 min, and 72 °C for 1 min, and then one cycle at 72 °C for 5 min.
8. Purify the PCR products by agarose gel electrophoresis and digest with *Nde*I and *Xho*I. Purify the released EW29Ch

mutant fragments, and ligate with *Nde*I- and *Xho*I-treated pET27b_Flag_EW29Ch using Quick Ligation™ Kit. Transform *E. coli* DH5α competent cells with the ligated products according to the heat-shock transformation protocol and grow the transformed *E. coli* cells on the LB plate.

9. Randomly select 20 clones and isolate the plasmids from randomly picked clones using QIAGEN Plasmid Mini Kit. Analyze the sequence of the plasmids using ABI 3130xl Genetic Sequencer. These plasmids were transformed into Electrocompetent BL21-CondonPlus (DE3)-RIL.
10. After overnight growth, scrape each colony from the petri dish into 2 mL of LB medium with a sterile scraper, and incubate at 37 °C overnight. Add IPTG to each culture at a final concentration of 1 mM and shake each culture for 24 h at 20 °C.
11. Harvest the cells by centrifugation at 3,500 × *g* for 10 min at 4 °C and wash each pellet with PBS (−) three times. Lyse the cells by suspension in 1/5 culture volume of BugBuster® HT Protein Extraction Reagent. Pellet intact cells and debris by centrifugation at 10,000 × *g* for 5 min. Recover each supernatant and dilute 20 μL of supernatant in 100 μL probing buffer. Add 1 μL of 0.45 mg/ml anti-Flag M2 antibody and incubate at 4 °C for 20 min followed by incubation with 1 μL of 0.1 mg/mL Cy3-labeled goat anti-mouse IgG antibody for another 20 min. The labeled solution was applied onto a glycoconjugate microarray. After incubation at 20 °C for 3 h, binding was detected by an evanescent-type scanner, Glycostation Reader 1200 under Cy3 mode (*see* **Note 8**).

4 Notes

1. Ribosome display method is a very powerful method for in vitro selection of functional proteins from large libraries. The system is based on a simple principle to achieve linkage between mRNA and a protein in a manner mediated by ribosome under the following conditions: removal of the stop codon, presence of magnesium, and low temperature. This method has been used successfully in engineering antibodies with high affinity (K_d, ~10^{-9} M). However, since the interaction between lectin and glycan is much weaker than that between antibody and antigen, a longer time is required for the selection of lectins, during which a key mRNA–ribosome–protein interaction is often disrupted. Thus, an improved stabilization of the complex is necessary for the lectin selection. In our study, we observed that incorporation of rare codon sequence at 3′ of the ribosome display construct could significantly increase the stability of mRNA–ribosome–protein complex

based on the basis of principle of ribosome-stalling machinery, which should be a key contributor to our success in engineering of lectins with target specificity [10].

2. The mutation rate of error-prone PCR is proportional to the concentration of Mn^{2+} and Mg^{2+}. However, the frequency of insertion and deletion is also increased at higher concentrations of Mn^{2+} and Mg^{2+}, which will significantly affect the quality of generated libraries. Determining optimal concentrations of Mn^{2+} and Mg^{2+} should be performed first. In addition, the template used also has an effect on the optimal concentrations of Mn^{2+} and Mg^{2+}.
3. Since PCR using Taq DNA polymerase often results in a single, 3′-adenine overhang to each end of the PCR products, the adenine at 3′ end should be removed by T4 DNA polymerase as this will hinder the subsequent overlap PCR.
4. The maximal amount of DNA including blunt-ended EW29Ch library and gene III spacer products used in overlap PCR is about 50 ng/50 μL. Adding an excessive amount of DNA will greatly decrease the amount of overlap PCR products.
5. Because there is a high protein concentration in the in vitro translation solution, nonspecific interactions can easily take place. In fact, we observed that the intact streptavidin-coated beads themselves could nonspecifically precipitate a large amount of EW29Ch. However, the nonspecific binding of EW29Ch to the beads was substantially reduced as the increment of immobilized nontarget Man-PAA. Lower concentrations than 100 μg/mL are considered to cause the nonspecific interactions by the streptavidin-coated beads, while higher concentrations than 100 μg/mL will aimlessly exceed the binding capacity of streptavidin-coated beads. Thus, in the present study, we decided to use the concentration of 100 μg/mL throughout the experiments.
6. In our study, specific enrichment of target mutants was achieved by pre-blocking with 10 mM lactose to prevent the undesired selection of wild types or those preserving similar activity (neutral mutants) before interaction with carbohydrate polymer-coated beads.
7. RNeasy Minelute Cleanup Kit is very effective in recovering small amount of selected RNA.
8. In our study, we have introduced the advanced technology of evanescent-field fluorescence-assisted glycoconjugate microarray into our evaluation system in place of FAC to facilitate screening lines of mutants derived by the selection procedures. Since the microarray technique provides us with a highly sensitive and high-throughput screening, it greatly facilitated analysis of the sugar-binding activity of a series of candidate clones even in the form of crude extracts.

References

1. Hedlund M, Ng E, Varki A, Varki NM (2008) alpha 2-6-Linked sialic acids on *N*-glycans modulate carcinoma differentiation in vivo. Cancer Res 68:388–394
2. Tateno H, Toyota M, Saito S, Onuma Y, Ito Y, Hiemori K, Fukumura M, Matsushima A, Nakanishi M, Ohnuma K, Akutsu H, Umezawa A, Horimoto K, Hirabayashi J, Asashima M (2011) Glycome diagnosis of human induced pluripotent stem cells using lectin microarray. J Biol Chem 286:20345–20353
3. Hirabayashi J (2004) Lectin-based structural glycomics: glycoproteomics and glycan profiling. Glycoconj J 21:35–40
4. Hirabayashi J, Kuno A, Tateno H (2011) Lectin-based structural glycomics: a practical approach to complex glycans. Electrophoresis 32:1118–1128
5. Kuno A, Uchiyama N, Koseki-Kuno S, Ebe Y, Takashima S, Yamada M, Hirabayashi J (2005) Evanescent-field fluorescence-assisted lectin microarray: a new strategy for glycan profiling. Nat Methods 2:851–856
6. Pilobello KT, Slawek DE, Mahal LK (2007) A ratiometric lectin microarray approach to analysis of the dynamic mammalian glycome. Proc Natl Acad Sci U S A 104:11534–11539
7. Tateno H, Uchiyama N, Kuno A, Togayachi A, Sato T, Narimatsu H, Hirabayashi J (2007) A novel strategy for mammalian cell surface glycome profiling using lectin microarray. Glycobiology 17:1138–1146
8. Hsu KL, Gildersleeve JC, Mahal LK (2008) A simple strategy for the creation of a recombinant lectin microarray. Mol Biosyst 4:654–662
9. Tateno H, Mori A, Uchiyama N, Yabe R, Iwaki J, Shikanai T, Angata T, Narimatsu H, Hirabayashi J (2008) Glycoconjugate microarray based on an evanescent-field fluorescence-assisted detection principle for investigation of glycan-binding proteins. Glycobiology 18:789–798
10. Yabe R, Suzuki R, Kuno A, Fujimoto Z, Jigami Y, Hirabayashi J (2007) Tailoring a novel sialic acid-binding lectin from a ricin-B chain-like galactose-binding protein by natural evolution-mimicry. J Biochem 141:389–399

Chapter 44

Tracing Ancestral Specificity of Lectins: Ancestral Sequence Reconstruction Method as a New Approach in Protein Engineering

Tomohisa Ogawa and Tsuyoshi Shirai

Abstract

Protein evolution is a process of molecular design leading to the diversity of functional proteins found in nature. Recent advances in bioinformatics and structural biology, in addition to recombinant protein expression techniques, enable us to analyze more directly the molecular evolution of proteins by a new method using ancestral sequence reconstruction (ASR), the so-called experimental molecular archaeology. ASR has been used to reveal molecular properties and structures correlating with changing geology, ecology, and physiology, and to identify the structure elements important to changing physiological functions to fill substantial gaps in the processes of protein evolution. In this chapter, we describe ASR as a new method of protein engineering studies, and their application to analyzing lectins, of which evolutionary processes and structural features contributing to molecular stability, specificity, and unique functions have been elucidated. Experimental molecular archeology using ASR and crystal structures of full-length ancestral proteins is useful in understanding the evolutionary process of the functional and structural diversified lectins by tracing ancestral specificities.

Key words Ancestral sequence reconstruction, Galectin, Lectin, Molecular phylogeny, Protein engineering, Bioinformatics, Protein evolution, Recombinant expression

1 Introduction

Lectins are group of sugar-binding proteins except for antibodies and enzymes that recognize specific carbohydrate structures, resulting in the specific regulation of various cellular functions. They are widely distributed in almost all taxa from microbial organisms, plant, and animal. Based on the structural similarity of carbohydrate recognition domain (CRD) of lectins and their characteristics, for example, animal lectins are classified into several categories; C-type lectins (CTLs), galectins, I-type, F-type, P-type lectins, pentraxins, tachylectins, SUEL/RBL in addition to the Ricin-type, Lily-type, 6× β-propeller/tectonin-type lectins [1]. Furthermore, recent studies including genome-wide screening

Jun Hirabayashi (ed.), *Lectins: Methods and Protocols*, Methods in Molecular Biology, vol. 1200,
DOI 10.1007/978-1-4939-1292-6_44,

revealed that the CRDs of lectins have been spread in genome as distinct structural motifs. For example, C-type lectin domain (CTLD) superfamily is well known as a large group of extracellular proteins with conserved CRD sequences but different functions including more than a thousand identified members, which has been classified into 17 groups (I–XVII) [2–4].

SUEL/RBL family lectins showed the specific binding activities to L-rhamnose or D-galactose, and mainly isolated from eggs and ovary cells of fishes and invertebrates [5–11]. Most RBLs have two or three tandem repeated CRDs, which consist of 95–105 amino acid residues and share the conserved topology of four disulfide bonds, while SUEL forms a homodimer composed of two identical subunits including single CRD *via* inter-subunit disulfide bond, resulting in the hemmaglutinating activity with bivalent binding properties. RBL homologues have been also reported as integrated domains involved in the ligand binding in membrane receptors such as polycystic kidney disdase-1-like (PKD-1), axon guidance receptor EVA-1, HuC21orf63, and the adhesion-class G-protein-coupled receptor latrophilin (LPHN) [5, 12–15].

Galectins are family of carbohydrate-binding proteins defined by their Ca^{2+}-independent affinity for β-galactoside sugar, and can be classified into three types; prototype, tandem repeat type, and chimera type [16]. Recently, the homologue of galectin-related inter fiber protein (Grifin) has been identified in zebrafish as a lens crystallin protein [17]. Furthermore, novel galectin-related protein, CvGal, which contains four canonical galectin CRDs, has been discovered from the hemocytes of eastern oyster, *Crassostrea virginica* [18]. Unique domain architecture for genes/proteins consist of galectin-CRDs with an N-terminal GlyXY domain that can form a collagen triple helix was also found in nematogalectin isolated from freshwater hydrozoan *Hydra* and marine hydrozoan *Clytia*. Nematogalectin is a major component of the nematocyst tubule, and is transcribed by namatocyte-specific alternative splicing [19]. Thus, the lectin family proteins are diversified by unique evolutionary process.

To identify the structural elements for protein functions and to develop the useful proteins, of which functions and properties have been improved, the protein engineering technique is one of the most efficient and widely used systems. It includes the rational design of proteins and directed evolution as efficient strategies. Directed evolution is a method to harness the power of in vitro natural selection (mimics Darwinian evolution) using repeated rounds of random mutagenesis and selection/screening to evolve and produce novel proteins/genes towards a well-defined goal [20]. On the other hand, rational design usually requires both the information of the structures or sequences of proteins and knowledge about the relationships between sequence, structure, and mechanism/function. In the sequence-based rational design, a popular approach to more effectively identify the structural elements of functionality in protein sequence has been the use of evolutionary information. Multiple sequence alignments and phylogenetic analyses are powerful tools for the exploration of amino acid conservation and relationships among groups of homologous proteins.

More recently, a new protein engineering method using ancestral sequence reconstruction (ASR) has been reported [21–30]. ASR is thought to be one of expanded methods for sequence-based rational design to reveal the relationships between molecular properties and structures correlating with changing geology, ecology, and physiology, and to fill substantial gaps in the processes of protein evolution. In this chapter, we describe protocols of the new protein engineering method, ASR, and its application to analyzing conger eel galectins, of which evolutionary processes and unique structural features contributing to molecular stability and specificity have been elucidated [26, 28, 30]. Conger eel galectins, Con I and Con II, function as biodefense molecules in the skin mucus and frontier organs [31–35]. They are prototype galectins, of which 3D structure form jellyroll motif including twofold symmetric homodimers with 5- and 6-stranded β-sheets in each subunit. They have different stabilities and carbohydrate-binding specificities, although they do have the conserved carbohydrate recognition domain (CRD) common to other galectins. Previous studies of Con I and Con II, based on molecular evolutionary and X-ray crystallography analyses, revealed that these proteins have evolved *via* accelerated substitutions under natural selection pressure [32, 36, 37]. To understand the rapid adaptive differentiation of Con I and Con II, ASR analysis using the reconstructed ancestral mutants, Con-anc and Con-anc,' was adapted [26, 28, 30].

Experimental molecular archeology using ASR and crystal structures of full-length ancestral proteins is useful in understanding the evolutionary process of the functional and structural diversified lectins by tracing ancestral specificities.

2 Materials

The DNA and amino acids sequences of extant genes used for ancestral sequence reconstruction were retrieved from the International Nucleotides Sequence Database Collaboration (INSDC) (DDBJ/GenBank/ENA, http://www.insdc.org/) and UniProtKB/SwissProt (http://www.uniprot.org/) databases, respectively. The 3D structures were from RCSB PDB database (http://www.rcsb.org/pdb/home/home.do). For example, galectins' data used in our studies [26, 30] are as follows: ConI (*Conger myriaster* congerin I, AB010276.1), ConII (*C. myriaster* congerin II, AB010277.1), *Anguilla japonica* galectin-1 (AJL1, AB098064.1),

Hippoglossus hippoglossus galectin (AHA1, DQ993254.1), *Paralichthys olivaceus* galectin (PoGal, AF220550.1), *Tetraodon nigroviridis* galectin (TnGal, CR649222.2), *Danio rerio* galectin-like lectin lgals1l1 (DrGal1_L1, BC164225.1), *D. rerio* galectin-like lectin Gal1-L2 (DrGal1_L2, AY421704.1), *D. rerio* Galectin-like lectin lgals1l3 (DrGal1_L3, BC165230.1), *Ictalurus punctatus* galectin (IpGal, CF261531), *Bos taurus* galectin-1 (BTG1, BC103156.1), *Homo sapiens* galectin-1 (HSG1, AK312161.1), *Mus musculus* galectin-1 (MMG1, BC099479.1), *Cricetulus* sp. galectin L-14 (CRG1, M96676.1), *Xenopus laevis* galectin-1 (XLG1, AF170341.1), and *H. sapiens* galectin-2 (HSG2, BC059782.1).

3 Methods

3.1 Ancestral Sequence Prediction

Sequence alignment: Because the ancestral sequence inference is executed on codon bases, the gene sequences of interest are aligned as follows. For example, the alignment of amino acid sequences of the extant galectins, as shown above, is first prepared by using the XCED program packages (*see* **Note 1**), which include the MAFFT multiple alignment program [38] (*see* **Note 2**), and then alignment of the nucleotide sequences, which are retrieved from the DDBJ database [39], is made in accordance with the amino acid sequence alignment.

1. Prepare the sequence data (DNA/RNA and protein sequences, respectively) with FASTA format, and save it as a text file. For example, copy & paste from the DDBJ database or type sequences. In the case for nucleotide sequences, the untranslated (noncoding) sequences are trimmed. Only sequences of coding regions are used.
2. For Web version (ver. 7), paste the sequence data into "Input" box or upload a text file. The type of sequences (DNA or protein) is automatically recognized.
3. Select some formats for output file (uppercase or lowercase, direction of sequence, order). Select some settings for strategy, parameters.
4. Crick "Submit" button, then you obtain the aligned sequences with CLUSTAL format.

Phylogenetic analysis: The tree topology is based on the amino acid sequences of the extant proteins with the NJ (neighbor-joining) method. The tree is rooted by using the sequence of farthest species as the outgroup.

1. On the Web site at MAFFT (ver. 7), following sequence alignment, crick "Phylogenetic tree" button.

2. After selecting some settings including method (NJ or UPGMA, etc.), substitution model, heterogeneity among sites, and bootstrap, crick "go" button.

Ancestral sequence prediction: The PAML (Phylogenetic Analysis by Maximum Likelihood) application (*see* **Note 3**) is employed to infer the last common ancestral sequence of target proteins from the phylogeny and the sequence alignment [40]. An F1X4 matrix is used for the codon substitution model with the universal codon table. The free *dN/dS* ratio with M8 (beta & omega) model is adapted. The ancestral gene sequence can be also evaluated with the MCMC Bayesian method by using the MrBayes application (*see* **Note 4**) in order to verify the sequence inferred with PAML [41]. The universal codon table, the GTR model with gamma-distributed rate variation across sites and proportion of invariable sites, and M3 model for omega variation are used for the simulation. The ancestral sequence with MrBayes is determined by averaging over the posterior probabilities of amino acid residues in the sequences of top 10 % likelihood during a total of 5,000 generations.

After downloading and compiling the PAML programs, open a command terminal and run the programs from the terminal as follows:

1. Prepare the control file (for example, galectins-aln.ctl as shown in Fig. 1) and the files of sequence data in the PHYLIP format (seqfile) and tree structure (treefile), for example, galectins-aln.nuc and galectins-aln.tree, respectively (Fig. 2). In the control file, the names of seqfile, treefile, and outfile (results dump file), and the parameters need to be rewritten.
2. For Mac OSX version, start X11 by double-clicking the "X11" icon, and open a terminal window.
3. On terminal window, change directory (Unix command cd) to the paml folder, copy control file to the same folder (Unix command cp), and start PAML by typing "codeml" with control file (Fig. 1) as follows

   % ./codeml galectins-aln.ctl

   The "codeml" is a part of the PAML package processing for the codon-based (seqtype = 1, codonml) and the amino acid-based (seqtype = 2, aaml) analyses.
4. You obtain some output files, lnf, rst, rub.
5. The main result file (for example, outfile name: galectins-aln) contains the codon usage in sequences, dN/dS (dN, dS) values for each pairwise and each branch (Fig. 3).
6. The "rst" file includes the following supplemental results for codeml containing ancestral sequences at each node (Fig. 4).
   (a) Number of codon sites with 0, 1, 2, 3 position differences.
   (b) Ancestral reconstruction by codonml.

```
seqfile = galectins-aln.nuc  * sequence data file name
treefile = galectins-aln.tree * tree structure file name
 outfile = galectins-aln     * main result file name
      noisy = 9  * 0,1,2,3,9: how much rubbish on the screen
    verbose = 0  * 0: concise; 1: detailed, 2: too much
    runmode = 0  * 0: user tree;  1: semi-automatic;  2: automatic
               * 3: StepwiseAddition; (4,5):PerturbationNNI; -2: pairwise
    seqtype = 1  * 1:codons; 2:AAs; 3:codons-->AAs
  CodonFreq = 1  * 0:1/61 each, 1:F1X4, 2:F3X4, 3:codon table
     aaDist = 0  * 0:equal, +:geometric; -:linear, 1-6:G1974,Miyata,c,p,v,a
 aaRatefile = wag.dat * only used for aa seqs with model=empirical(_F)
               * dayhoff.dat, jones.dat, wag.dat, mtmam.dat, or your own
      model = 1
               * models for codons:
                 * 0:one, 1:b, 2:2 or more dN/dS ratios for branches
               * models for AAs or codon-translated AAs:
                 * 0:poisson, 1:proportional, 2:Empirical, 3:Empirical+F
                 * 6:FromCodon, 7:AAClasses, 8:REVaa_0, 9:REVaa(nr=189)
    NSsites = 0  * 0:one w;1:neutral;2:selection; 3:discrete;4:freqs;
               * 5:gamma;6:2gamma;7:beta;8:beta&w;9:beta&gamma;
               * 10:beta&gamma+1; 11:beta&normal>1; 12:0&2normal>1;
               * 13:3normal>0
      icode = 0  * 0:universal code; 1:mammalian mt; 2-10:see below
      Mgene = 0  * 0:rates, 1:separate;
  fix_kappa = 0  * 1: kappa fixed, 0: kappa to be estimated
      kappa = 2  * 2 initial or fixed kappa
  fix_omega = 0  * 0 1: omega or omega_1 fixed, 0: estimate
      omega = .4 * .4 initial or fixed omega, for codons or codon-based Aas
  fix_alpha = 1  * 0: estimate gamma shape parameter; 1: fix it at alpha
      alpha = 0. * initial or fixed alpha, 0:infinity (constant rate)
     Malpha = 0  * different alphas for genes
      ncatG = 8  * # of categories in dG of NSsites models
      clock = 0   * 0:no clock, 1:global clock; 2:local clock; 3:TipDate
      getSE = 0  * 0: don't want them, 1: want S.E.s of estimates
RateAncestor = 1  * (0,1,2): rates (alpha>0) or ancestral states (1 or 2)
  Small_Diff = .5e-6
*   cleandata = 0  * remove sites with ambiguity data (1:yes, 0:no)?
*       ndata = 10
*  fix_blength = -1  * 0: ignore, -1: random, 1: initial, 2: fixed
      method = 0   * 0: simultaneous; 1: one branch at a time
* Genetic codes: 0:universal, 1:mammalian mt., 2:yeast mt., 3:mold mt.,
* 4: invertebrate mt., 5: ciliate nuclear, 6: echinoderm mt.,
* 7: euplotid mt., 8: alternative yeast nu. 9: ascidian mt.,
* 10: blepharisma nu.
* These codes correspond to transl_table 1 to 11 of GENEBANK.
```

Fig. 1 The example of codeml control file, galectin-aln.ctl, for PAML analysis

(c) Probability of best state at each node.

(d) Summary of changes along branches.

(e) List of extant and reconstructed sequences.

(f) Overall accuracy of the ancestral sequences.

Assessment of predicted sequences: ASR method has some issues with ambiguity, depending on the choice of evaluation method, evolutionary model, and sequence sets. It is often difficult to obtain a complete, highly accurate sequence, as molecular evolution is believed to be a highly stochastic process and there is no guarantee that ancestral sequences can be identified without errors, because the probability that the sequence as a whole is accurate is only ~0.37

a

```
 14   438 S

BTG1
atg------gcttgtggtctggtcgccagcaacctgaatctcaaacctggggagtgcctcagagtgcggggcgaggtggccgca
---------gacgccaagagcttcttgctgaacctgggcaaagacgacaac---------aacctgtgcctccacttcaaccct
cgtttcaacgcgcatggggacgtcaacaccatcgtgtgtaacagcaaggacgctggg------gcctgggggccgagcagagg
gaatctgccttccccttccagcctggaagtgtcgtggaggtatgcatctccttcaaccagacggacctaaccatcaagctgcct
gatggatacgaattcaagttccccaaccgcctc---aacctggaggccatcaactacctgtctgcaggtggtgacttcaagatc
aagtgtgtggcctttgag

CONI
atg------agtggaggacttcaggtcaaaaactttgacttcactgtcggaaaattcttgactgtcggaggtttcatcaacaat
---------tctccacaacgtttctcggtcaatgtgggcgaatccatgaat---------tcactttcattgcacctcgaccat
cgtttcaactatggtgcggaccaaaataccatcgtcatgaactccacgcttaagggcgataatggttgggagacggagcagcgg
agcacaaacttcaccctcagtgcagggcagtatttgagatcaccctctcat atgacatcaacaagttttacattgatatactt
gatggtcccaatttggagttccccaaccgctat---tcaaaggaattcttgcccttcctttccctggcaggagatgctagactc
acgcttgtgaaactagaa

CONII
atg------agtgatagagctgaggtgagaaacattcccttcaagttaggaatgtacttgactgtcggaggtgtcgtcaactcc
---------aatgcaactcgtttctcgatcaatgtgggcgaatccaccgat---------tcaattgcaatgcacatggaccat
cggttcagctatggtgcggaccaaaatgtcctcgtcttaaactccctggttcacaatgtt---ggttggcagcaggaggagcgg
tccaagaagttccccttcactaaagggatcatttcagacaaccatcacatttgacacccacacgttttatattcagctaagt
aatggtgagacagtggagttccccaaccgcaat---aaagatgcagccttcaacttaatttacctggcaggagatgctagactc
acgtttgtgagactagaa

AHA1
atg---------aaagacatgatggtaaagaacatgtccttcaaggtcggacagaccctgacccttgttggagttgccaaacct
---------gatgcgacaaatttcgcattgaatattggctcctctgaccag---------gacattgtgatgcacatcaaccct
cgtttcaacgcccacggcgatgagaacgcagtggtgtgcaactcttacatcggagga------cagtggtgtgaggagctccgt
gagggaggctttcctttccagctaggacaggagttcaagatcaccattgaattcacccctcaggagttcctggtgactttatcc
gatggctccaacatccacttccccaaccgcatc---ggggcggagaagtactccttcatgagctttgaggggaggctcgcatc
aggagcatcgagatcaag

AJL1
atg---------gatttcgtggaggtgaaaaacctgatcatgaagtcaggaatggaactgaaggtcaacggtgtcttcaacgcc
---------aatccagaacgtttctctatcaatgtgggccactctaccgaa---------gaaattgcggtgcacgttgacgtg
cgtttcagctatttaagtgacaaacgccaattgatcataaaccacaagaccggcgac------gcctggcaagaagaacagaga
gatgctagattccccttcacagcagggcaggcatttcaggtgtccgttgtcttcaactttgatacttttgacatttatctgcca
gatggccaggtggcgcacttcaccaaccatctg---ggtgcccaggaatacaaatacatttctttgtggggatgccacagtc
aaaaacataagtgtgaat
...........................
```

b

```
(((((BTG1,HSG1),XLG1),HSG2),((CONI,CONII),AJL1)),(((AHA1,PoGal),TnGal),(DrGal1_L1,Dr
Gal1_L2)),(DrGal1_L3,IpGal));
```

Fig. 2 The examples of the file format for the sequence data (**a**) and tree structure files (**b**)

(i.e., 0.99^{100}) even if each residue of a protein made up of 100 residues, is identified with posterior probability of 0.99 (i.e., 99% are expected to be correct). The inference methods such as MP (maximum parsimony), ML (maximum likelihood), and BI (Bayesian inference) can lead to errors in predicted ancestral sequences, resulting in potentially misleading estimates of the properties of the ancestral protein. This is a major problem in ancestor reconstruction studies, and considerable efforts have been made.

To avoid false conclusions as a result of such ambiguity, the accuracy of reconstructed ancestral sequence is critical for such studies, and the last common ancestral sequence of target proteins is verified by two methods. *First*, the amino acid sequences translated from the inferred ancestral nucleotide sequences are used to

a

```
TREE # 1:  (((((1, 6), 8), 7), ((2, 3), 5)), (((4, 13), 14), (9, 10)), (11, 12));   MP score: -1
lnL(ntime: 25  np: 51):  -4852.708216    +0.000000
  15..16   16..17   17..18   18..19   19..1    19..6    18..8    17..7    16..20   20..21   21..2    21..3    20..5    15..22   22..23   23..24   24..4    24..13
 23..14   22..25   25..9    25..10   15..26   26..11   26..12

(((((BTG1: 0.313743, HSG1: 0.088418): 0.842558, XLG1: 1.644521): 0.287241, HSG2: 1.312904): 0.203588, ((CONI: 0.605267, CONII: 0.409499): 0.696706, AJL1: 1.065473): 0.568119): 0.175155, (((AHA1: 0.056423, PoGal: 0.182931): 0.752255, TnGal: 0.556775): 0.436797, (DrGall_L1: 0.230649, DrGall_L2: 0.093904): 1.892932): 0.387274, (DrGall_L3: 0.774008, IpGal: 0.846101): 0.514001);
```

b

```
Nei & Gojobori 1986. dN/dS (dN, dS)(Pairwise deletion)(Note: This matrix is not used in later m.l. analysis.Use runmode = -2 for ML pairwise comparison.)
BTG1
CONI        0.4735 (0.6996  1.4775)
CONII       0.5569 (0.7934  1.4248) 2.7860 (0.3959  0.1421)
AHA1        0.3350 (0.6433  1.9203) 0.5137 (0.7308  1.4225) 0.3218 (0.6963  2.1639)
AJL1        0.3317 (0.7660  2.3093) 0.7038 (0.6969  0.9901) 0.5736 (0.5495  0.9580) 0.5589 (0.6791  1.2151)
HSG1        0.1756 (0.0637  0.3628) 0.4915 (0.7315  1.4882) 0.5839 (0.8004  1.3707) 0.4952 (0.6414  1.2953) 0.3602 (0.7645  2.1228)
HSG2        0.3329 (0.5916  1.7771) 0.3715 (0.7020  1.8898) 0.3463 (0.7823  2.2588) 0.4853 (0.7622  1.5708) 0.5988 (0.7471  1.2476) 0.4708 (0.5648  1.1996)
XLG1        0.2817 (0.4981  1.7682) 0.5242 (0.7869  1.5011) 0.4341 (0.8199  1.8886) 0.4652 (0.7077  1.5214) 0.3413 (0.7910  2.3176) 0.1863 (0.4467  2.3974)  0.2949 (0.6399  2.1698)
DrGallL1    0.1951 (0.7006  3.5916)-1.0000 (0.7846 -1.0000)-1.0000 (0.7119 -1.0000)-1.0000 (0.3521 -1.0000)-1.0000 (0.6657 -1.0000)-1.0000 (0.7108 -1.0000) -1.0000 (0.6786 -1.0000)
DrGallL2   -1.0000 (0.6402 -1.0000)-1.0000 (0.7129 -1.0000)-1.0000 (0.6683 -1.0000)-1.0000 (0.2954 -1.0000)-1.0000 (0.6390 -1.0000) 0.1806 (0.6239  3.4542) -1.0000 (0.5985 -1.0000)
DrGallL3    0.3992 (0.5169  1.2948) 0.4813 (0.7083  1.4717) 0.3194 (0.6679  2.0914) 0.2220 (0.5199  2.3417) 0.1829 (0.6477  3.5408) 0.4586 (0.5256  1.1461)  0.2841 (0.5174  1.8211)
IpGal       0.5184 (0.6560  1.2654)-1.0000 (0.7046 -1.0000) 0.1750 (0.6700  3.8295) 0.2927 (0.5974  2.0409) 0.1483 (0.6074  4.0953) 0.4426 (0.6559  1.4821) -1.0000 (0.5902 -1.0000)
PoGal      -1.0000 (0.6769 -1.0000) 0.6045 (0.8216  1.3592) 0.3552 (0.7284  2.0508) 0.2488 (0.0460  0.1850) 0.5389 (0.7045  1.3074) 0.3764 (0.6722  1.7858)  0.3600 (0.7678  2.1330)
TnGal       0.3776 (0.5916  1.5670) 0.4501 (0.7510  1.6685)-1.0000 (0.7289 -1.0000) 0.1964 (0.2108  1.0732) 0.3645 (0.6641  1.8217) 0.4434 (0.5910  1.3330)  0.4155 (0.6802  1.6368)

0.2300 (0.7365 3.2028)
0.1820 (0.7077 3.8886) 1.1956 (0.1135  0.0950)
0.3981 (0.6898 1.7325)-1.0000 (0.4987 -1.0000) -1.0000 (0.4457 -1.0000)
0.3159 (0.6987 2.2114)-1.0000 (0.4698 -1.0000) -1.0000 (0.4326 -1.0000) 0.1631 (0.2516 1.5423)
0.3592 (0.7078 1.9706) 0.0766 (0.3599  4.6960) -1.0000 (0.2989 -1.0000) 0.2356 (0.4941 2.0976) 0.2675 (0.5813 2.1727)
0.2487 (0.6231 2.5056) 0.1945 (0.3822  1.9655)  0.1043 (0.2995  2.8711) 0.3078 (0.4664 1.5155) 0.3831 (0.5335 1.3927) 0.2140 (0.2195 1.0256)


dN & dS for each branch
 branch          t         N         S     dN/dS       dN       dS   N*dN   S*dS
  15..16     0.175     324.7     113.3    0.9311   0.0573   0.0615   18.6    7.0
  16..17     0.204     324.7     113.3  999.0000   0.0915   0.0001   29.7    0.0
  17..18     0.287     324.7     113.3  999.0000   0.1291   0.0001   41.9    0.0
  18..19     0.843     324.7     113.3    0.1558   0.1169   0.7505   38.0   85.0
  19..1      0.314     324.7     113.3    0.2491   0.0588   0.2359   19.1   26.7
  19..6      0.088     324.7     113.3    0.0464   0.0047   0.1005    1.5   11.4
  18..8      1.645     324.7     113.3    0.2188   0.2849   1.3024   92.5  147.6
  17..7      1.313     324.7     113.3    0.2747   0.2600   0.9465   84.4  107.3
  16..20     0.568     324.7     113.3    0.6389   0.1652   0.2586   53.6   29.3
  20..21     0.697     324.7     113.3    0.3576   0.1586   0.4434   51.5   50.2
  21..2      0.605     324.7     113.3  999.0000   0.2721   0.0003   88.3    0.0
  21..3      0.409     324.7     113.3    1.8468   0.1549   0.0839   50.3    9.5
  20..5      1.065     324.7     113.3    0.5409   0.2912   0.5384   94.6   61.0
  15..22     0.387     324.7     113.3    1.5342   0.1419   0.0925   46.1   10.5
  22..23     0.437     324.7     113.3    0.2764   0.0868   0.3140   28.2   35.6
  23..24     0.752     324.7     113.3    0.1254   0.0894   0.7130   29.0   80.8
  24..4      0.056     324.7     113.3  999.0000   0.0254   0.0000    8.2    0.0
  24..13     0.183     324.7     113.3    0.1323   0.0226   0.1709    7.3   19.4
  23..14     0.557     324.7     113.3    0.1817   0.0857   0.4717   27.8   53.5
  22..25     1.893     324.7     113.3    0.0395   0.0865   2.1911   28.1  248.3
  25..9      0.231     324.7     113.3  998.8254   0.1037   0.0001   33.7    0.0
  25..10     0.094     324.7     113.3    0.1194   0.0108   0.0901    3.5   10.2
  15..26     0.514     324.7     113.3    0.2145   0.0880   0.4102   28.6   46.5
  26..11     0.774     324.7     113.3    0.1330   0.0960   0.7221   31.2   81.8
  26..12     0.846     324.7     113.3    0.2056   0.1410   0.6861   45.8   77.7
```

Fig. 3 PAML results of phylogenetic tree (**a**) and *dN/dS* values (**b**) for galectins

```
Number of codon sites with 0,1,2,3 position differences

   2 vs.   1     23    44    46    22   0.4735 (0.6996 1.4775)
   3 vs.   1     21    41    48    25   0.5569 (0.7934 1.4248)
   3 vs.   2     61    45    26     4   2.7860 (0.3959 0.1421)
... ... ... ...

List of extant and reconstructed sequences
  26   438
BTG1              ATG --- --- GCT TGT GGT CTG GTC GCC AGC AAC CTG AAT CTC AAA CCT GGG GAG TGC CTC AGA GTG CGG GGC GAG GTG GCC GCA --- --- --- GAC
GCC AAG AGC TTC TTG CTG AAC CTG GGC AAA GAC GAC AAC --- --- --- AAC CTG TGC CTC CAC TTC AAC CCT CGT TTC AAC GCG CAT GGG GAC GTC AAC ACC ATC GTG
TGT AAC AGC AAG GAC GCT GGG --- --- GCC TGG GGG GCC GAG CAG AGG GAA TCT GCC TTC CCC TTC CAG CCT GGA AGT GTC GTG GAG GTA TGC ATC TCC TTC AAC CAG
ACG GAC CTA ACC ATC AAG CTG CCT GAT GGA TAC GAA TTC AAG TTC CCC AAC CGC CTC --- AAC CTG GAG GCC ATC AAC TAC CTG TCT GCA GGT GGT GAC TTC AAG ATC
AAG TGT GTG GCC TTT GAG

CONI              ATG --- --- AGT GGA GGA CTT CAG GTC AAA AAC TTT GAC TTC ACT GTC GGA AAA TTC TTG ACT GTC GGA GGT TTC ATC AAC AAT --- --- --- TCT
CCA CAA CGT TTC TCG GTC AAT GTG GGC GAA TCC ATG AAT --- --- --- TCA CTT TCA TTG CAC CTC GAC CAT CGT TTC AAC TAT GGT GCG GAC CAA AAT ACC ATC GTC
ATG AAC TCC ACG CTT AAG GGC GAT AAT GGT TGG GAG ACG GAG CAG CGG AGC ACA AAC TTC ACC CTC AGT GCA GGG CAG TAT TTT GAG ATC ACC CTC TCA TAT GAC ATC
AAC AAG TTT TAC ATT GAT ATA CTT GAT GGT CCC AAT TTG GAG TTC CCC AAC CGC TAT --- TCA AAG GAA TTC TTG CCC TTC CTT TCC CTG GCA GGA GAT GCT AGA CTC
ACG CTT GTG AAA CTA GAA

... ... ... ...

node #15           ATG ATG CAG ACT GGT GGA CTG ATG GTC AAG AAC ATG ACC TTC AAG GCC GGA CAG GAC CTG ACG ATC ACA GGT GTC CTC AAG CCT AAA AAA AAA GAC
TCT AAT AGC TTC TCA ATC AAT ATT GGC CAC GAC GCC GAC AAA AAA AAA GAC ATT GCC CTG CAC TTC AAC CCT CGT TTC AAC GCC CAC GGC GAC AAA AAC ACC ATC GTC
TGC AAC TCC AAG CAG GGC GGC AAC AAT GAC TGG GGG CAG GAG CAG CGG GAA GAT AGC TTC CCC TTC CAG CAA GGA GAG GAG TTC AAG GTC ACC ATC ACC TTC AAC AAT
GAC AAG TTC TAC GTC AAG CTG CCT GAT GGC CCC GTG ATG AAC TTC CCC AAC CGC CTC AAA GGC GAC GAG GAC TTC ACC TAC ATG TAC GTT GAA GGA GAT GTC AAG ATC
AAG GGC GTC AAG ATA AAA

node #16           ATG ATG CAG ACT GGT GGA CTG GTG GTC AAG AAC ATG ACC TTC AAG GCC GGA CAG GAC CTG ACG ATC ACA GGT GTC GTC AAC CCT AAA AAA AAA GAC
GCT AAT AGC TTC TCA ATC AAT ATG GGC CAA GAC ACC GAC AAA AAA AAA GAC ATT GCC CTG CAC TTC AAC CCT CGT TTC AAC GAC CAC GGC GAC AAA AAC ACC ATC GTC
TGC AAC TCC AAG GAC GGC GGC AAT AAT GAC TGG GGG CAG GAG CAG CGG GAA GAT AGC TTC CCC TTC CAG CAA GGG GAG GAG TTC CAG GTC ACC ATC ACC TTC AAC AAT
GAC AAG TTT TAC GTC AAG CTG CCT GAT GGC CAC GAG TTG AAC TTC CCC AAC CGC CTC AAA GGC GAC GAG GAC TTC AAC TAC ATG TAC GTG GAA GGA GAT GTC AAG ATC
AAG TGT GTC AAG TTA AAA

... ... ... ...


Overall accuracy of the 12 ancestral sequences:
  0.64179  0.63593  0.62765  0.64326  0.86087  0.58283  0.74937  0.65872  0.69126  0.92051  0.90535  0.63513
for a site.

Amino acid sequences inferred by codonml.

Results unreliable for sites with alignment gaps.
Node #15           MIQTGGLVVK NMTFKAGQDL TITGVLKPRR RDSNSFSINI GHDADRRRDI ALHFNPRFNA HGDKNTIVCN SKQGGNNDWG QEQREDSFPF QQGEEFKVTI TFNNDKFYVK
LPDGPVMNFP NRLRGDEDFT YMYVEGDVKI KGVKIK
Node #16           MIQTGGLVVK NMTFKAGQDL TITGVVNPRR RDANSFSINM GQDTDRRRDI ALHFNPRFND HGDKNTIVCN SKDGGNNDWG QEQREDSFPF QQGEEFQVTI TFNNDKFYVK
LPDGHELNFP NRLRGDEDFN YMYVEGDVKI KCVKLK
... ... ... ...
```

Fig. 4 Ancestral sequences of galectins and their accuracy estimated by PAML analysis

reconstruct molecular phylogenies to determine if an ancestor is connected to the node of proteins with zero distance, which is the necessary condition for true ancestor. *Second*, the robustness of the ancestral sequence is tested by reconstructions with a reduced number of genes. For example, the galectin genes, except for that of ConI and ConII, are excluded one by one from the phylogeny, and the ConI-ConII node ancestor sequence is inferred with the same parameters to verify that each amino acid/codon is reproducible. The genes are randomly excluded for each reconstruction. A total of 250 reconstructions are executed, and the differences between reconstructed sequences and between reconstructed sequence and predicted ancestral sequence are plotted against the number of omitted genes. In order to indentify fragile amino acid residues in the ancestral sequence, the reproduction rate of the residues among sequences, for which one of the extant genes is excluded, can be defined as (number of times the same amino acid of ancestor has been obtained)/(maximum number of genes omitted from in original phylogeny).

Furthermore, ASR method is a more reliable method because all possible ancestral mutants, in which ambiguous amino acid sites are replaced by equally probable candidates individually or in combination, are reproducible and the biological and physicochemical properties and 3D structures of the molecules can be assessed. Indeed, when ancestral congerins were reconstructed based on insufficient sequence information lacking recently determined fish galectin genes, the ancestral Con-anc protein was shown to have a strand-swapped structure resembling ConI, indicating that Con-anc was more likely to be an intermediate mutant of the ancestor to ConI, and that the revised Con-anc' or Con-anc'-N28K are more appropriate ancestors. The accuracy of ASR can be assessed by analysis of protein activities, stabilities, specificities, and even 3D structures in the laboratory using biochemical or biophysical methods. In many cases, ancestral sequences cannot be unambiguously determined, and several amino acids might be assigned to a residue site with almost equal probabilities.

3.2 Preparation and Expression of Recombinant Ancestral Mutants

The cDNAs encoding ancestral mutants were constructed by PCR, and ligation of fragments at suitable restriction sites as shown in Fig. 5. For example of Con-anc mutant [26], four cDNA fragments, namely, Con-ancN, Con-ancMa, Con-ancMb, and Con-ancC, were prepared by PCR using ancN100S, ancMb90S, and ancC100S as templates, and ancN25S/ancN20AS, ancMb21S/ancMb21AS, and ancC21S/ancC48AS as primer sets (Fig. 5). The two sets of PCR products—Con-ancN and Con-ancMa, and Con-ancMb, and Con-ancC—are ligated to each other after digestion with *Acl*I and *Sac*I, respectively. Next, a second PCR is carried out using the ligation products Con-ancN/Con-ancMa and Con-ancMb/Con-ancC as templates and ancN25S/ancMa54AS and

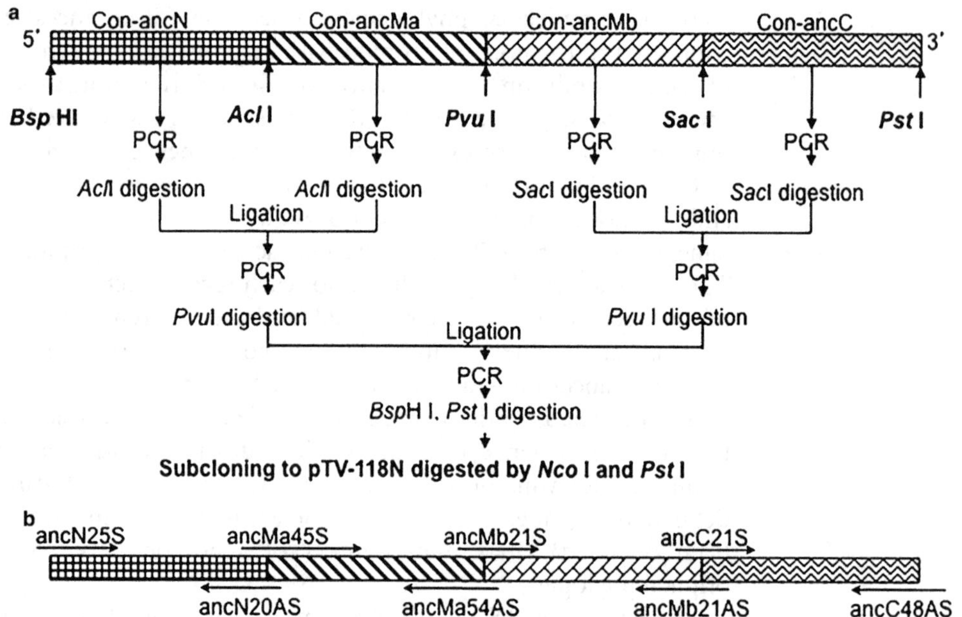

Fig. 5 Schematic structure of Con-anc DNA construct. (**a**) Scheme for construction of Con-anc expression vector. (**b**) Primer sites using for construction of Con-anc DNA. Each *boxes* are indicate the same pattern boxes in Fig. 5a

ancMb21S/ancC48AS as external primers. Subsequently, these fragments are ligated at the *Pvu*I site and used for the third PCR as the template. Finally, the PCR product encoding the complete sequence of Con-anc is ligated into the pTV-118N vector at the *Nco*I and *Pst*I sites after digestion with *Bsp*HI and *Pst*I, thereby resulting in pTV-Con-anc (*see* **Note 5**).

Expression of recombinant ancestral mutants is carried out by using pTV vector and *E. coli* JM109 system [26, 42]. In brief, each mutant is purified by affinity chromatography on an HCl-treated Sepharose 4B column (GE Healthcare, UK), followed by anion-exchange chromatography on a 5-mL HiTrap Q column (GE Healthcare). The purity of each mutant was confirmed by sodium dodecyl sulfate-polyacrylamide gel electrophoresis (SDS-PAGE).

3.3 Crystal Structures of Full-Length Ancestral Proteins

The crystal structures of full length Con-anc', Con-anc'-N28K (the low reliable site, N28, was replaced with the second best amino acid Lys), and Con-anc (the ancestor previously inferred with fewer genes) were determined by X-ray crystallography.

The crystals are grown by the hanging-drop vapor-diffusion method. The diffraction data are collected by using synchrotron radiation sources, BL38B1 of SPring-8 (Hyogo, Japan). The crystal structures are solved by the molecular replacement method using the known structures as a search model with MOLREP (molecular replacement program for protein X-ray crystallography)

[43], for example ConI (for Con-anc) or ConII (for Con-anc' and Con-anc'-N28K). The structure refinements are executed with COOT and REFMAC5 (manual and automatic structure refinement programs against X-ray diffraction data) [44, 45] (*see* **Note 6**).

4 Notes

1. The XCED programs (Linux/x86, Solaris/SPARC, and Mac OSX versions) are freely available for academic use by downloading from XCED home page (http://align.bmr.kyushu-u.ac.jp/xced/).
2. MAFFT program (ver. 7) is also able to download directly from the Web site at http://mafft.cbrc.jp/alignment/software/index.html, or the online version is available at http://mafft.cbrc.jp/alignment/server/. The quick guide manuals and tips are also available from each website.
3. PAML software (current version 4.6 (March, 2012); Windows, UNIX/Linux, and Mac OSX versions, respectively) and its manual are freely available from the Web site at http://abacus.gene.ucl.ac.uk/software/paml.html.
4. MrBayes software (current version 3.2.2 (September, 2013); Windows, UNIX/Linux and Mac OSX versions, respectively) and its manual are freely available from the Web site at http://mrbayes.sourceforge.net/.
5. The artificial synthetic genes encoding ancestral mutants have become commercially available from some biotechnology companies such as Invitrogen, Life technologies (Carlsbad, CA, USA) and Eurofins MWG Operon (Ebersberg, Germany).
6. These programs (current version 6.3.0.2 (September, 2013); Windows, UNIX/Linux and Mac OSX versions, respectively) and manuals are freely available as a part of or a supplement of the CCP4 (Collaborative Computational Project No.4) package from the Web site at http://www.ccp4.ac.uk/download/index.php.

References

1. Gabius H-J (1997) Animal lectins. Eur J Biochem 243:543–576
2. Drickamer K (1993) Ca^{2+}-dependent carbohydrate-recognition domains in animal proteins. Curr Opin Struct Biol 3:393–400
3. Drickamer K, Fadden AJ (2003) Genomic analysis of C-type lectins. Biochem Soc Symp 69:59–72
4. Zelensky AN, Gready JE (2005) The C-type lectin-like domain superfamily. FEBS J 272: 6179–6217
5. Ogawa T, Watanabe M, Naganuma T, Muramoto K (2011) Diversified carbohydrate-binding lectins from marine resources. J Amino Acids 2011:838914. doi:10.4061/2011/838914
6. Ozeki Y, Matsui T, Suzuki M, Titani K (1991) Amino acid sequence and molecular characterization of a D-galactoside-specific lectin purified from sea urchin (*Anthocidaris crassispina*) eggs. Biochemistry 30:2391–2394
7. Tateno H, Saneyoshi A, Ogawa T, Muramoto K, Kamiya H, Saneyoshi M (1998) Isolation

and characterization of rhamnose-binding lectins from eggs of steelhead trout (*Oncorhynchus mykiss*) homologous to low density lipoprotein receptor superfamily. J Biol Chem 273: 19190–19197

8. Tateno H, Ogawa T, Muramoto K, Kamiya H, Saneyoshi M (2002) Distribution and molecular evolution of rhamnose-binding lectins in Salmonidae: isolation and characterization of two lectins from white-spotted charr (*Salvelinus leucomaenis*) eggs. Biosci Biotechnol Biochem 66:1356–1365
9. Hosono M, Ishikawa K, Mineki R, Murayama K, Numata C, Ogawa Y, Takayanagi Y, Nitta K (2002) Tandem repeat structure of rhamnose-binding lectin from catfish (*Silurus asotus*) egg. Biochim Biophys Acta 1472:668–675
10. Naganuma T, Ogawa T, Hirabayashi J, Kasai K, Kamiya H, Muramoto K (2006) Isolation, characterization and molecular evolution of a novel pearl shell lectin from a marine bivalve, *Pteria penguin*. Mol Divers 10:607–618
11. Shirai T, Watanabe Y, Lee M-S, Ogawa T, Muramoto K (2009) Structure of rhamnose-binding lectin CSL3: unique pseudo-tetrameric architecture of a pattern recognition protein. J Mol Biol 391:390–403
12. Weston BS, Malhas AN, Price RG (2003) Structure-function relationships of the extracellular domain of the autosomal dominant polycystic kidney disease-associated protein, polycystin-1. FEBS Lett 538:8–13
13. Fujisawa K, Wrana JL, Culotti JG (2007) The slit receptor EVA-1 coactivates a SAX-3/robo-mediated guidance signal in *C. elegans*. Science 317:1934–1938
14. Mitsunaga K, Harada-Itadani J, Shikanai T, Tateno H, Ikehara Y, Hirabayashi J, Narimatsu H, Angata T (2009) Human C21orf63 is a heparin-binding protein. J Biochem 146:369–373
15. Lelianova VG, Davletov BA, Sterling A, Rahman MA, Grishin EV, Totty NF, Ushkaryov YA (1997) α-Latrotoxin receptor, latrophilin, is a novel member of the secretin family of G protein-coupled receptors. J Biol Chem 272: 21504–21508
16. Kasai K, Hirabayashi J (1996) Galectins. A family of animal lectins that decipher glycocodes. J Biochem 119:1–8
17. Ahmed H, Vasta GR (2008) Unlike mammalian GRIFIN, the zebrafish homologue (DrGRIFIN) represents a functional carbohydrate-binding galectin. Biochem Biophys Res Commun 371:350–355
18. Tasumi S, Vasta GR (2007) A galectin of unique domain organization from hemocytes of the eastern oyster (*Crassostrea virginica*) is a receptor for the protistan parasite *Perkinsus marinus*. J Immunol 179:3086–3098
19. Hwang JS, Takaku Y, Momose T, Adamczyk P, Özbek S, Ikeo K, Khalturin K, Hemmrich G, Bosch TCG, Holstein TW, David CN, Gojobori T (2010) Nematogalectin, a nematocyst protein with GlyXY and galectin domains, demonstrates nematocyte-specific alternative splicing in Hydra. Proc Natl Acad Sci U S A 107: 18539–18544
20. Bloom JD, Arnold FH (2009) In the light of directed evolution: pathways of adaptive protein evolution. Proc Natl Acad Sci U S A 106:9995–10000
21. Thornton JW (2001) Evolution of vertebrate steroid receptors from an ancestral estrogen receptor by ligand exploitation and serial genome expansions. Proc Natl Acad Sci U S A 98:5671–5676
22. Thornton JW, Need E, Crews D (2003) Resurrecting the ancestral steroid receptor: ancient origin of estrogen signaling. Science 301(5640):1714–1717
23. Thornton JW (2004) Resurrecting ancient genes: experimental analysis of extinct molecules. Nat Rev Genet 5:366–375
24. Chang BSW, Ugalde JA, Matz MV (2005) Applications of ancestral protein reconstruction in understanding protein function: GFP-like proteins. Methods Enzymol 395: 652–670
25. Bridgham JT, Carroll SM, Thornton JW (2006) Evolution of hormone-receptor complexity by molecular exploitation. Science 312: 97–101
26. Konno A, Ogawa T, Shirai T, Muramoto K (2007) Reconstruction of a probable ancestral form of conger eel galectins revealed their rapid adaptive evolution process for specific carbohydrate recognition. Mol Biol Evol 24: 2504–2514
27. Ortlund EA, Bridgham JT, Redinbo MR, Thornton JW (2007) Crystal structure of an ancient protein: evolution by conformational epistasis. Science 317:1544–1548
28. Konno A, Yonemaru S, Kitagawa A, Muramoto K, Shirai T, Ogawa T (2010) Protein engineering of conger eel galectins by tracing of molecular evolution using probable ancestral mutants. BMC Evol Biol 10:43
29. Field SF, Matz MV (2010) Retracing evolution of red fluorescence in GFP-like proteins from faviina corals. Mol Biol Evol 27:225–233
30. Konno A, Kitagawa A, Watanabe M, Ogawa T, Shirai T (2011) Tracing protein evolution through ancestral structures of fish galectin. Structure 19:711–721

31. Muramoto K, Goto R, Kamiya H (1988) Purification and properties of agglutinins from conger eel, *Conger Myriaster* (Brevoort), skin mucus. Dev Comp Immunol 12:309–318
32. Ogawa T, Ishii C, Kagawa D, Muramoto K, Kamiya H (1999) Accelerated evolution in the protein-coding region of galectin cDNAs, congerin I and congerin II, from skin mucus of conger eel (*Conger myriaster*). Biosci Biotechnol Biochem 63:1203–1208
33. Muramoto K, Kagawa D, Sato T, Ogawa T, Nishida Y, Kamiya H (1999) Functional and structural characterization of multiple galectins from the skin mucus of conger eel, *Conger myriaster*. Comp Biochem Physiol B Biochem Mol Biol 123:33–45
34. Nakamura O, Watanabe T, Kamiya H, Muramoto K (2001) Galectin containing cells in the skin and mucosal tissues in Japanese conger eel, *Conger myriaster*: an immunohistochemical study. Develop Comp Immunol 25:431–437
35. Nakamura, O., Matsuoka, H., Ogawa, T., Muramoto, K., Kamiya, H., and Watanabe, T. (2006) Opsonic effect of congerin, a mucosal galectin of the Japanese conger, *Conger myriaster* (Brevoort)" *Fish & Shellfish Immun* 20, 433–435.
36. Shirai T, Mitsuyama C, Niwa Y, Matsui Y, Hotta H, Yamane T, Kamiya H, Ishii C, Ogawa T, Muramoto K (1999) High-resolution structure of conger eel galectin, congerin I, in lactose-liganded and ligand-free forms: emergence of a new structure class by accelerated evolution. Structure 7:1223–1233
37. Shirai T, Matsui Y, Mitsuyama C, Yamane T, Kamiya H, Ishii C, Ogawa T, Muramoto K (2002) Crystal structure of a conger eel galectin (congerin II) at 1.45 Å resolution: implication for the accelerated evolution of a new ligand-binding site following gene duplication. J Mol Biol 321:879–889
38. Katoh K, Misawa K, Kuma K, Miyata T (2002) MAFFT: a novel method for rapid multiple sequence alignment based on fast Fourier transform. Nucleic Acids Res 30:3059–3066
39. Kaminuma E, Mashima J, Kodama Y, Gojobori T, Ogasawara O, Okubo K, Takagi T, Nakamura Y (2010) DDBJ launches a new archive database with analytical tools for next-generation sequence data. Nucleic Acids Res 38:D33–D38
40. Winn MD, Ballard CC, Cowtan KD, Dodson EJ, Emsley P, Evans PR, Keegan RM, Krissinel EB, Leslie AG, McCoy A, McNicholas SJ, Murshudov GN, Pannu NS, Potterton EA, Powell HR, Read RJ, Vagin A, Wilson KS (2011) Overview of the CCP4 suite and current developments. Acta Crystallogr D Biol Crystallogr 67:235–242
41. Ronquist F, Teslenko M, van der Mark P, Ayres DL, Darling A, Hohna S, Larget B, Liu L, Suchard MA, Huelsenbeck JP (2012) MrBayes 3.2: efficient Bayesian phylogenetic inference and model choice across a large model space. Syst Biol 61:539–542
42. Ogawa T, Ishii C, Suda Y, Kamiya H, Muramoto K (2002) High-level expression and characterization of fully active recombinant conger eel galectins in *Eschericia coli*. Biosci Biotechnol Biochem 66:476–480
43. Collaborative Computational Project, Number 4 (1994) The CCP4 suite: programs for protein crystallography. Acta Crystallogr D Biol Crystallogr 50:760–763
44. Emsley P, Lohkamp B, Scott WG, Cowtan K (2004) Features and development of Coot. Acta Crystallogr D Biol Crystallogr 66: 486–501
45. Murshudov GN, Vagin AA, Dodson EJ (1997) Refinement of macromolecular structures by the maximum-likelihood method. Acta Crystallogr D Biol Crystallogr 53:240–255

Part V

Comprehensive Lists

Chapter 45

Comprehensive List of Lectins: Origins, Natures, and Carbohydrate Specificities

Yuka Kobayashi, Hiroaki Tateno, Haruko Ogawa, Kazuo Yamamoto, and Jun Hirabayashi

Abstract

More than 100 years have passed since the first lectin ricin was discovered. Since then, a wide variety of lectins (*lect* means "select" in Latin) have been isolated from plants, animals, fungi, bacteria, as well as viruses, and their structures and properties have been characterized. At present, as many as 48 protein scaffolds have been identified as functional lectins from the viewpoint of three-dimensional structures as described in this chapter. In this chapter, representative 53 lectins are selected, and their major properties that include hemagglutinating activity, mitogen activity, blood group specificity, molecular weight, metal requirement, and sugar specificities are summarized as a comprehensive table. The list will provide a practically useful, comprehensive list for not only experienced lectin users but also many other non-expert researchers, who are not familiar to lectins and, therefore, have no access to advanced lectin biotechnologies described in other chapters.

Key words Lectin family, Hemagglutinating activity, Mitogen activity, Blood group specificity, Metal requirement, Oligosaccharide specificity

1 Introduction

It is generally believed that the earliest description of a lectin was given by Peter Hermann Stillmark in 1888 at the University of Dorpat as described by Els van Damme (*see* Chapter 1). Stillmark isolated ricin, an extremely toxic hemagglutinin, from seeds of the castor plant, *Ricinus communis*. However, the first lectin to be purified on a large scale and made commercially available was concanavalin A (Con A), which is now the most used lectin for characterization and purification of sugar-containing molecules and cells. The legume lectins including Con A are probably the

Jun Hirabayashi (ed.), *Lectins: Methods and Protocols*, Methods in Molecular Biology, vol. 1200, DOI 10.1007/978-1-4939-1292-6_45, © Springer Science+Business Media New York 2014

most thoroughly studied lectins. As defined in 1980 by Irwin J Goldstein et al., "Lectins are multivalent carbohydrate-binding proteins or glycoproteins except for enzymes and antibodies" [1, 2]. Lectins occur ubiquitously in nature. They may bind to a soluble carbohydrate or to a carbohydrate moiety, which is a part of a glycoprotein or a glycolipid. They typically agglutinate certain animal cells and/or precipitate glycoconjugates. In this chapter, the authors attempt at providing a comprehensive list of so far identified lectins with focus on commercially available 53 lectins, of which properties will be of useful information for readers of this book from both basic and practical viewpoints. The list also enables linkage to structural information, i.e., into what protein family (Pfam) or scaffold each lectin is classified, which Z. Fujimoto et al. summarize (*see* Chapter 46).

2 Criteria: How 50 Lectins and Their Properties Are Selected

As described, plant lectins have a long history of investigation. In fact, they have been contributed as useful means or bioactive proteins to analyze animal cells in diverse areas of biomedical and biological researches. A comprehensive yet concise overview of the biochemical, carbohydrate-binding specificity, biological activities, and applications of most of the currently known plant lectins was provided in a previous handbook [3]. The handbook lists approximately 200 lectin entries in alphabetical order (common name, taxonomy, classification code, group, and specificity).

On the other hand, animal lectins undoubtedly fulfill a variety of functions of glycoconjugates. Many of them have been implicated extensive functions as recognition molecules, e.g., in the first line of defense against pathogens as well as in roles concerned with cell trafficking, immune regulation, and prevention of autoimmunity. Thus, the information of animal lectins is a highly valuable resource not only for basic studies in lectin biology but also for advanced research in pursuit of several applications in biotechnology, immunology, and clinical practice [4–6]. However, resources of many animal lectins are restricted, and a more practical difficulty of them is instability compared with plant lectins and other classes (e.g., bacterial and fungal) of lectins.

So far, many researchers have attempted summarizing lectins for the review. Considering such a background, we have created a lectin list with the concept as described below:

1. For the first place, lectins must be commercially available so that researchers can access these lectins for immediate use for their studies.

2. The lectins should better be well evaluated (and thus, well known), having been investigated in multiple laboratories so that objective, consistent data are available.
3. Uniform and comprehensive methods should be used to assess the binding specificity of each lectin toward a panel of oligosaccharides, hopefully with quantitative data in terms of dissociation constant (K_d).

With the criteria above, the authors constructed Table 1, which lists 53 commercially available lectins that have been reported in multiple academic papers along with individual characteristics, including hemagglutinating activity (HA), mitogen activity, blood group specificity, molecular weight, metal-ion requirement, and sugar specificity. Advanced frontal affinity chromatography (FAC) described by C. Sato (*see* Chapter 22) revealed detailed oligosaccharide specificities of most of these lectins quantitatively. For the analyses, pyridylaminated (PA) glycans are extensively used. For reference, Fig. 1 lists representative 100 oligosaccharides. For both researchers, who know lectins very well and those who are about to use lectins for the first time, this list is an easy-to-read table that emphasizes key characteristics. Please refer to the cited references for more information about each lectin.

There are also cases in Table 1, where properties of two or three names apply to the same lectin because the names include the abbreviations “A” for agglutinin, “L” for lectin, or “M” for mitogen (e.g., ECA/ECL, MAM/MAL). Table 1 also provides information of commercial sources, thereby allowing users to access these lectins for their immediate use.

Table 1
List of lectin specifications

No.	Common name (abbreviation)	Origin common name	HA activity[a] (blood group specificity)	Mitogen Activity	MW in nature (oligomer)	Family (Pfam ID)	Metal requirement (if any)	Monosaccharide specificity	Key structures for recognition	PA-oligos in Fig. 1[b]	Commercial source[c]	Ref.
1	AAL	*Aleuria aurantia* Orange peel fungus, fruiting body	Yes (nonspecific)	No	67 kDa (33 kDa×2)	AAL-like (PF07938)	None	Fuc	Fucα1-6, 1-2, 1-3	15, 41, 97	JOM (rec) Vec	[7–9]
2	ABA ABL	*Agaricus bisporus* common mushroom, fruiting body	Yes (nonspecific)	No	64 kDa (16 kDa×4)	Fungal fruit body/ Actinoporin-like/ ABL-like (PF07367)	None	Gal, GlcNAc (2 binding sites)	Core1	17, 99, 100 (pNP)	EY JOM Sigma Wako (rec)	[10–12]
3	ACG	*Agrocybe cylindracea* Poplar mushroom, fruiting body	Yes (nonspecific)	Yes	34 kDa (17 kDa×2)	Galectin (PF00337)	None	Gal	Siaα2-3Galβ1-4GlcNAc	58, 69, 88	JOM[f]	[13–18]
4	ACL ACA	*Amaranthus caudatus* velvet flower, seed	Yes (nonspecific)	Yes	62 kDa (31 kDa×2)	ACA-like (PF07468)	None	GalNAc	Core1 and Core2	99 (pNP), 100 (pNP)	JOM[f] Vec	[19–23]
5	AOL	*Aspergillus oryzae* fungus	Yes (nonspecific)		35 kDa (35 kDa×1)	AAL-like (PF07938)	None	Fuc	Fucα1-6GlcNAc (core Fuc) Fucα1-2Galβ1-4GlcNAc (H-type2)	39, 41, 73	JOM (rec) TCI	[24–26]
6	BanLec	*Musa acuminate* banana, pulp	Yes (nonspecific)	Yes	30 kDa (15 kDa×2)	Jacalin related (PF01419)	Cd, Zn	Man	Manα1-3-Man, Glcβ1-3	8, 10, 11	EY	[27–29]

7	BC2LCN AiLec-S1	*Burkholderia cenocepacia* Gram-negative bacterium	Yes (weak)	N/A	32 kDa (16 kDa×2)	BC2LCN (PF00229)	None	Fuc	Fucα1-2Galβ1-3GlcNAc (H-type1), Fucα1-2Galβ1-3(Fucα1-3) GlcNAc (Le^b) Fucα1-2Galβ1-3GalNAc (H-type3)	Not included in Fig. 1	Elicityl (rec) Wako (rec)	[30–34]
8	BPL BPA	*Bauhinia purpurea* orchid tree, seed	Yes (nonspecific)	Yes	195 kDa (47 kDa×4)	L-type (PF00139)	None	Gal/GalNAc	Galβ1-3GalNAc	36, 51, 62	EY JOM[f] Sigma Vec	[35–44]
9	Calsepa	*Calystegia sepium* hedge bindweed, rhizome	Yes (nonspecific)	Yes	65 kDa (18 kDa×4)	Jacalin-related (PF01419)		Man	Man, Maltose	18, 30, 33	EY	[45–48]
10	Con A	*Canavalia ensiformis* Jack bean, seed	Yes (nonspecific)	Yes	104 kDa (26 kDa×4)	L-type (PF00139)	Ca, Mn	Man Glc	Biantennary hybrid type (N-linked) high mannose type (N-linked) complex type (N-linked)	7, 10, 13	EY JOM Sigma Vec	[49–54]
11	DBA	*Dolichos biflorus* horse gram (plant), seed	Yes (A1)	No	110 kDa (27 kDa×4)	L-type (PF00139)	Cal, Mn, Zn	GalNAc	GalNAcβ1-3GalNAc	64, 74	EY JOM Vec	[55–60]
12	DSA	*Datura stramonium* Jimson weed, seed	Yes (nonspecific)	Yes	86 kDa (A: 40 kDa + B: 46 kDa)	Hevein/Chitin binding (PF00187)	N/A	GlcNAc	N-linked (Galβ1-4GlcNAc-)n	37, 41, 49	EY JOM Sigma Vec	[61–64]
13	ECA ECL	*Erythrina cristagalli* coral tree, seed	Yes (nonspecific)	Yes	57 kDa (28 kDa×2)	L-type (PF00139)	Ca, Mn	Gal	Galβ1-4GlcNAc->Lac>GalNAc>Gal	37, 45, 49	EY JOM Sigma Vec	[65–68]
14	EEA EEL	*Euonymus europaeus* spindle tree, seed	Yes (B, O(H), A2)	N/A	57 kDa (?)		Ca, Zn		Galα1-3Galβ(1–3, 1–4)GlcNAc Fucα1-2Galβ (1–3, 1–4) GlcNAc	75	EY Vec	[69, 70]

(continued)

Table 1
(continued)

No.	Common name (abbreviation)	Origin common name	HA activity[a] (blood group specificity)	Mitogen Activity	MW in nature (oligomer)	Family (Pfam ID)	Metal requirement (if any)	Monosaccharide specificity	Key structures for recognition	PA-oligos in Fig. 1[b]	Commercial source[c]	Ref.
15	GNA GNL	*Galanthus nivalis*, bulb	Yes (nonspecific)	No	50 kDa (13 kDa×4)	Monocot (PF01453)	None	Man	High mannose type (*N*-linked)	2, 14, 41	EY JOM[f] Sigma Vec	[71–74]
16	GSL-I GS-I BSA-I	*Griffonia simplicifolia Bandeiraea simplicifolia* Bandeiraea plant (shrub), seed	Yes (B)	No	114-115 kDa (A: 32 kDa/B: 33 kDa×4)	L-type (PF00139)	Ca	Gal (B subunit), GalNAc (A subunit)	Gal-α/GalNAc-α	72 (GSA-IA4) 75, 89 (GSL-IB4)	EY JOM[f] Sigma Vec	[75–82]
17	GSL-II GS-II	*Griffonia simplicifolia* Bandeiraea plant (shrub), seed	Yes (nonspecific)	No	113 kDa (30 kDa×4)	L-type (PF00139)	Ca	GlcNAc	Agalacto *N*-linked, GlcNAc(β1-4GlcNAc)n	19, 26, 94	EY JOM[f] Vec	[75–82]
18	HHA HHL AL	*Hippeastrum hybrid* Amaryllis, bulb	Yes (nonspecific)	No	50 kDa (12.5 kDa×4)	Monocot (PF01453)	None	Man	High mannose type (*N*-linked)	2, 14, 23	Vec JOM[f] Vec	[83–86]
19	HPA HPL	*Helix pomatia* edible snail, albumin gland	Yes (A)	No	79 kDa (13 kDa×6)	H-type (PF09458)	None	GalNAc	GalNAcα1-3GalNAc (Forssman-antigen), GalNAcα-Ser (Tn-antigen)	61, 72, 74	EY Sigma	[87–93]
20	Jacalin	*Artocarpas integliforia* jack fruit, seed	Yes (nonspecific)	Yes	50-65 kDa (14-18 kDa×4)	Jacalin-related (PF01419)	None	GalNAc	Core1 and Core2	99 (pNP), 100 (pNP)	EY Sigma Vec	[94–107]
21	LCA LcH	*Lens culinaris* lentil (bean), seed	Yes (nonspecific)	Yes	46 kDa (α:×2+β:×2)	L-type (PF00139)	Ca, Mn	Man, Glc	Fucose-linked α(1–6) to core GlcNAc of *N*-linked glycopeptides	24, 25, 40	EY JOM Sigma Vec	[108–114]

22	LEL	*Lycipersicon esculentum* tomato, fruit	Yes	No	100 kDa (100 kDa × 1)	Hevein/ Chitin-binding (PF00187)	N/A	GlcNAc	(GlcNAcβ1-4)n, (Galβ1-4GlcNAc)n (polylactosamine)	84, 93, 95	EY Sigma Vec	[115–118]
23	LTA LTL	*Lotus tetragonolobus* Winged or asparagus pea, awws	Yes (O)	No	54 kDa (A:/B:/C: × 4)	L-type (PF00139)	Ca, Mn	Fuc	LewisX Galβ1-4(Fucα1-3) GlcNAc (Lewis X)	73, 97	EY JOM Vec	[119, 120]
24	MAM MAL MAL-I	*Maackia amurensis* Maackia plant, seed	Yes (nonspecific)	Yes	140 kDa (35 kDa × 4)	L-type (PF00139)	Ca, Mn	Sia	Siaα2-3 Galβ1-4GlcNAc	No FAC data	EY (mix)[d] JOM Sigma Vec	[121–124]
25	MAH MAL-II	*Maackia amurensis* Maackia plant, seed	Yes	No	140 kDa (35 kDa × 4)	L-type (PF00139)	Ca, Mn	Sia	Siaα2-3 Galβ1-3GalNAc	No FAC data	EY (mix)[d] JOM Vec	[121–132]
26	MOA MOL	*Marasmius oreades* fairy ring mushroom	Yes (B>>A)	N/A	16, 907 Da	R-type (PF00652)	None	Gal	Galα1-3Gal, Galα1-3Galβ1-4GlcNAc/Glc	75, 78, 85	EY	[125–132]
27	MPL MPA	*Maclura pomifera* osage orange, seed	Yes (nonspecific)	Yes	40–46 kDa (11 kDa × 4)	Jacalin-related (PF01419)	None	Gal	Gal-α/GalNAc-α/ Core1	99 (pNP) GalNAcα (pNP) pNP-Gala	EY JOM[f] Vec	[133–142]
28	NPA	*Narcissus pseudonarcissus* daffodil, bulb	Yes	No	26 kDa (13 kDa × 2)	Monocot (PF01453)	None	Man	High-Man including Manα1-6Man	2, 14, 41	EY (mix)[d] Vec	[84, 85, 143]
29	PA-IL	*Pseudomonas aeruginosa* Gram-negative bacterium	Yes (nonspecific)	N/A	26 kDa (12.7 kDa × 4)	PA-IL (PF07828)	Ca, Mn	Gal	Melibiose > methyl αGal > Gal > methyl βGal > GalNAc	70, 78	Elicityl Sigma	[144–155]
30	PA-IIL	*Pseudomonas aeruginosa* Gram-negative bacterium	Yes (nonspecific)	N/A	26 kDa (11.7 kDa × 2)	PA-IIL-like/ fucolectin (PF07472)		Fuc	Lea Galβ1-4(Fucα1-3) GlcNAc (Lewis X) > Galβ1-3(Fucα1-4) GlcNAc (Lewis a) > Lex	80	Elicityl	[144–155]

(continued)

Table 1
(continued)

No.	Common name (abbreviation)	Origin common name	HA activity[a] (blood group specificity)	Mitogen Activity	MW in nature (oligomer)	Family (Pfam ID)	Metal requirement (if any)	Monosaccharide specificity	Key structures for recognition	PA-oligos in Fig. 1[b]	Commercial source[c]	Ref.
31	PHA-E4 E-PHA	*Phaseolus vulgaris* red kidney bean, seed	Yes (O > A, B)	No	128 kDa (32 kDa × 4)	L-type (PF00139)	Ca, Mn	GalNAc	*N*-glycans containing bisecting GlcNAc	43	EY JOM Vec	[156–162]
32	PHA-L4 L-PHA	*Phaseolus vulgaris* red kidney bean, seed	Yes (weak) (non-specific)	Yes	126 kDa (32 kDa × 4)	L-type (PF00139)	Ca, Mn	GalNAc	*N*-glycans containing GlcNAcβ1-6Man branch	21, 37, 49	EY JOM Vec	[156–162]
33	PHA-P PHA-M	*Phaseolus vulgaris* red kidney bean, seed	Yes (non-specific)	Yes	126–128 kDa (32 kDa × 4)	L-type (PF00139)		GalNAc	–	21, 37, 49	EY (mix)[d] JOM	[156–162]
34	PhoSL	*Pholiota squarrosa* shaggy scalycap mushroom, fruiting body	Yes (non-specific)	N/A	14 kDa	Fungal fruit body/ actinoporin-like/ ABL-like (PF07367)	None	Fuc	*N*-glycans containing core α1-6Fuc	19, 23, 24	JOM[f]	[163]
35	PNA	*Arachis hypogaea* peanut, seed	Yes (non-specific)	No	98 kDa (25 kDa × 4)	L-type (PF00139)	Ca, Mn	Gal	Galβ1-3GalNAc of *N*-glycans and glycolipids	62, 65, 99	EY JOM Sigma Vec	[164–167]
36	PSA	*Pisum sativum* garden pea, seed	Yes (non-specific)	Yes	50 kDa (a: × 2 + b: × 2)	L-type (PF00139)	Ca, Mn	Man, Glc	High-Man-type *N*-glycans biantennary *N*-glycans containing core α1-6Fuc	14, 39, 42	EY JOM Sigma Vec	[114, 168–170]
37	PSL	*Polyporus squamosus* pheasant's back mushroom, fruiting body	Yes (B > A, O)	N/A	56 kDa (28 kDa × 2)	R-type (PF00652)		Sia	Neu5Acα2-6Galβ1-4GlcNAc	54, 55, 56	EY	[171–174]

38	PTL-I PTA WBA-1	*Psophocarpus tetragonolobus* winged bean, seed	Yes (A, weak)	No	58 kDa (29 kDa × 2)	L-type (PF00139)	N/A	Gal	A-type sugar GalNAcα > Galα	74, 75, 76	EY JOM[f] Sigma Vec	[175–183]
39	PVL	*Psathyrella velutina* Weeping widow (mushroom), fruiting body	Yes		40 kDa (40 kDa × 1)	PVL-like (PF13517)	None	GlcNAc, NeuAc	GlcNAc(β1-4GlcNAc)n Siaa2-3Gal	17, 19, 94	Wako	[184–189]
40	PWM	*Phytolacca americana* pokeweed, root	Yes (nonspecific)	Yes	0.9 kDa × 1	Hevein/ Chitin-binding (PF00187)	None	N/A	(Galβ1-4GlcNAc-)n	83, 84, 95	EY	[190–196]
41	RCA120 RCA-I	*Ricinus communis* caster bean, seed	Yes (nonspecific)	No	120 kDa (30 kDa × 4)	R-type (PF00652)	None	Gal	LacNAcβ, GalNAcβ, Galβ,Lacβ of *N*- and *O*-glycan	37, 45, 49	Not available[e] Sigma Vec	[197–202]
42	RSL	*Ralstonia solanacearum* Gram-negative bacterium	Yes (nonspecific)	No data	Approx. 140 kDa (9.9 kDa × >10)	AAL-like (PF07938)	Ca	Fuc	Fucα1-2Gal, Fucα1-6GlcNAc	14, 39, 73	Elicityl (rec)	[203–205]
43	SBA	*Glycine max* soybean, seed	Yes (A1 > A2 >> B)	No	120 kDa (30 kDa × 4)	L-type (PF00139)	Ca, Mn	GalNAc	GalNAcα1-3Gal of *O*-linked glycopeptides	61, 70, 72	EY JOM Sigma Vec	[156, 206–210]
44	SNA-I SNA	*Sambucus nigra* elderberry (plant), bark	Yes	No	140 kDa	R-type (PF00652)	None	Sia	Siaa2-6Gal/ GalNAc (Tn)	54, 55, 56	EY JOM[f] Sigma Vec	[211–214]
45	SSA	*Sambucus sieboldiana* Japanese elderberry (plant), bark	Yes	No	160 kDa	R-type (PF00652)	None	Sia	Siaa2-6Gal/ GalNAc (Tn)	54, 55, 56	JOM	[215, 216]

(continued)

Table 1 (continued)

No.	Common name (abbreviation)	Origin common name	HA activity[a] (blood group specificity)	Mitogen Activity	MW in nature (oligomer)	Family (Pfam ID)	Metal requirement (if any)	Monosaccharide specificity	Key structures for recognition	PA-oligos in Fig. 1[b]	Commercial source[c]	Ref.
46	STA	*Solanum tuberosum* potato, root tubers	Yes (nonspecific)	No	(34 kDa)	Hevein/ Chitin-binding (PF00187)	None	GlcNAc	(GlcNAc-)n, (GlcNAc-4MurNAc)n (peptidoglycan backbone)	95	EY Sigma Vec	[217–220]
47	TJA-I	*Trichosanthes japonica* Japanese snake gourd, root tubers	Yes		64 KDa	R-type (PF00652)	None	Fuc	Siaα2-6Gal/ GalNAc	52, 54, 55	Not available[g]	[221–225]
48	TxLc-I	*Tulipa gesneriana* tulip, bulb	Yes (nonspecific)	Yes	120/60 kDa (28 kDa × 4 or 2)	Monocot (PF01453)	N/A	GalNAc	Man1-3 core, bi- and tri-antennary *N*-glycans, GalNAc	41, 42, 45	EY JOM[f]	[226–231]
49	UDA	*Urtica dioica* stinging nettle (plant), rhizome	Yes (nonspecific)		(9 kDa)	Hevein/ Chitin-binding (PF00187)	Zn	GlcNAc	GlcNAcβ1-4GlcNAc, Man5–Man9	13, 95	EY	[232–234]
50	UEA-I	*Ulex europaeus* gorse (plant), seed	Yes (O>>A1, A2)	No	53 KDa	L-type (PF00139)	Ca, Mn	Fuc	Fucα1-2Galβ1-4GlcNAc-	73, 80	EY JOM Sigma Vec	[235, 236]

51	VVA-G VVA VVL	*Vicia villosa* hairy vetch (bean), seed	Yes (A1)	No	110 kDa	L-type (PF00139)	Ca, Mn	GalNAc	GalNAc	61, 64, 71	EY (Mix)[d] JOM[f] Sigma Vec (Mix)[d]	[43, 237–241]
52	WFL WFA	*Wisteria floribunda* Japanese wisteria (plant), seed	Yes (nonspecific)	Yes	75 kDa (37 kDa×2)	L-type (PF00139)	N/A	GalNAc	GalNAcβ> GalNAcα>>> Gal-α,β	61, 64	EY JOM[f] Sigma Vec	[242–244]
53	WGA	*Triticum vulgaris* wheat germ	Yes (nonspecific)	No	43 kDa (21 kDa×2)	Hevein/ Chitin-binding (PF00187)	None	GlcNAc Sia	GlcNAc(β1-4GlcNAc)n	95	EY JOM Vec	[245–250]

N/A Not assessed

[a]Activity when assayed using trypsin-treated rabbit erythrocytes

[b]Up to 3 PA-oligosaccharides are ranked in the order of affinity assessed by frontal affinity chromatography

[c]Manufacturer's abbreviations are EY (EY Laboratories, Inc.; http://www.eylabs.com/), Vec (Vector Laboratories, Inc; https://www.vectorlabs.com/), Elicityl (http://www.elicityl-oligotech.com/), JOM (J-Oil Mills, Inc; http://www.j-oil.com/), TCI (Tokyo Chemical Industry Co. Ltd; http://www.tcichemicals.com/), Wako (Wako Pure Chemical Industries, Ltd; http://www.wako-chem.co.jp/english/), Elicityl (http://www.elicityl-oligotech.com/)

[d]These commercial products are a substantial mixture of few isolectins

[e]Production of this lectin has been discontinued because of concern about impurity of isolectin, RCA60, or RCA I, a highly poisonous toxin

[f]Inquired (build-to-order manufacturing)

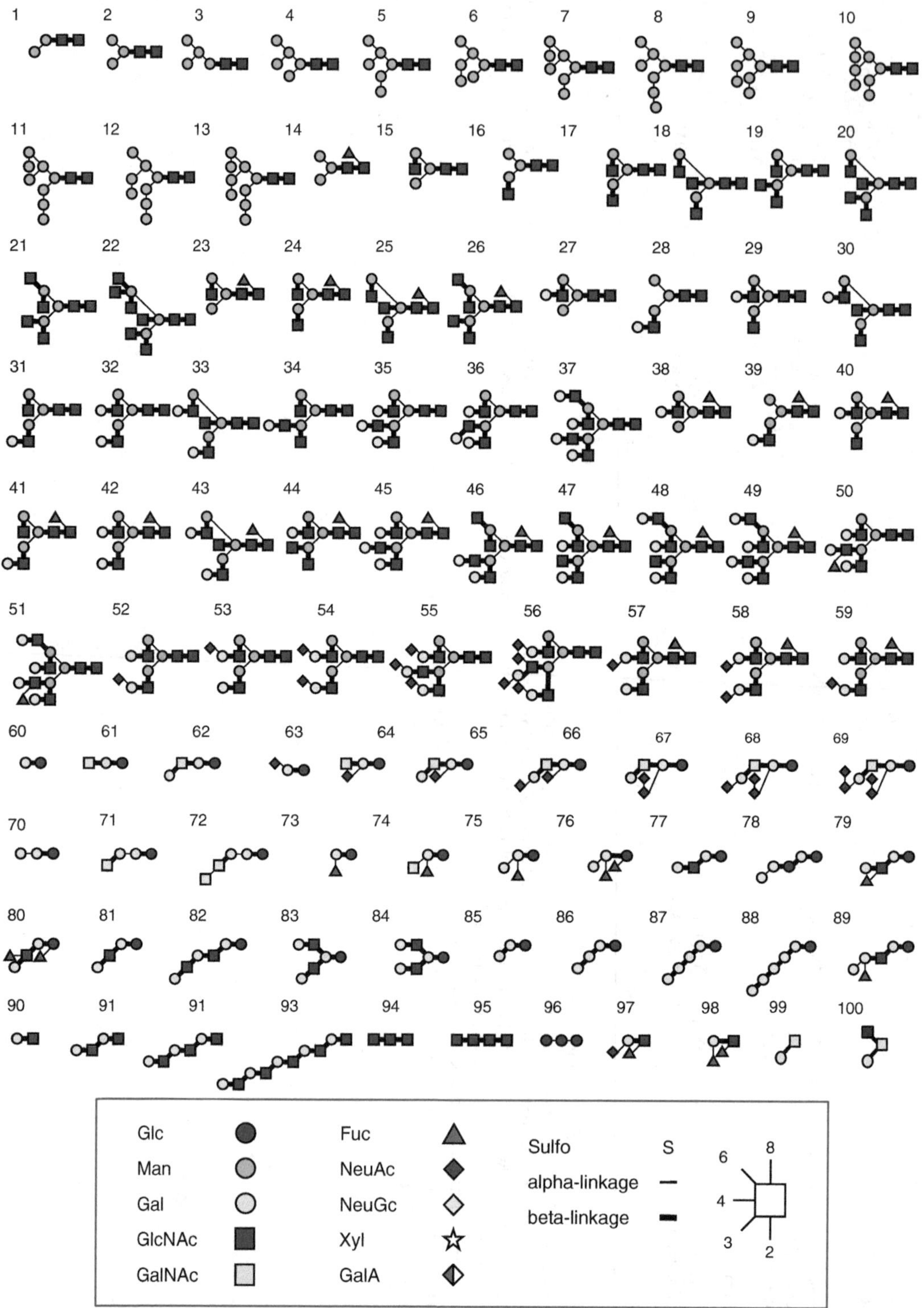

Fig. 1 Schematic representation of the PA-oligosaccharides used for FAC analysis. Note that the reducing terminal is pyridylaminated for FAC analysis

References

1. Goldstein IJ, Hughes RC, Monsigny M et al (1986) What should be called a lectin. Nature 285:66–66
2. Sharon N, Lis H (2003) Lectins. Springer, Dordrecht
3. Van Damme EJM, Peumans WJ, Pusztai A et al (1998) Handbook of plant lectins: properties and biomedical applications. Wiley, Chichester
4. Gabius HJ, Andre S, Kaltner H et al (2002) The sugar code: functional lectinomics. Biochim Biophys Acta 1572:165–177
5. Kilpatrick DC (2002) Animal lectins: a historical introduction and overview. Biochim Biophys Acta 1572:187–197
6. Kumar D, Mittal Y (2011) AnimalLectinDb: An integrated animal lectin database. Bioinformation 6:134–136
7. Fukumori F, Takeuchi N, Hagiwara T et al (1990) Primary structure of a fucose-specific lectin obtained from a mushroom, *Aleuria aurantia*. J Biochem 107:190–196
8. Harada H, Kamei M, Tokumoto Y et al (1987) Systematic fractionation of oligosaccharides of human immunoglobulin G by serial affinity chromatography on immobilized lectin columns. Anal Biochem 164:374–381
9. Kochibe N, Furukawa K (1980) Purification and properties of a novel fucose-specific hemagglutinin of *Aleuria aurantia*. Biochemistry 19:2841–2846
10. Presant C A, and Kornfeld S (1972) Characterization of the cell surface receptor for the Agaricus bisporus hemagglutinin. J Biol Chem 247(21):6937–6945
11. Nakamura-Tsuruta S, Kominami J, Kuno A, et al (2006) Evidence that *Agaricus bisporus* agglutinin (ABA) has dual sugar-binding specificity. Biochem Biophys Res Commun 347:215–220
12. Yu L, Fernig DG, Smith JA et al (1993) Reversible inhibition of proliferation of epithelial cell lines by *Agaricus bisporus* (edible mushroom) lectin. Cancer Res 53: 4627–4632
13. Ban M, Yoon HJ, Demirkan E et al (2005) Structural basis of a fungal galectin from *Agrocybe cylindracea* for recognizing sialoconjugate. J Mol Biol 351:695–706
14. Imamura K, Takeuchi H, Yabe R et al (2011) Engineering of the glycan-binding specificity of *Agrocybe cylindracea* galectin towards alpha(2,3)-linked sialic acid by saturation mutagenesis. J Biochem 150:545–552
15. Liu C, Zhao X, Xu XC et al (2008) Hemagglutinating activity and conformation of a lactose-binding lectin from mushroom *Agrocybe cylindracea*. Int J Biol Macromol 42: 138–144
16. Ngai PH, Zhao Z, Ng TB (2005) Agrocybin, an antifungal peptide from the edible mushroom *Agrocybe cylindracea*. Peptides 26:191–196
17. Wang H, Ng TB, Liu Q (2002) Isolation of a new heterodimeric lectin with mitogenic activity from fruiting bodies of the mushroom *Agrocybe cylindracea*. Life Sci 70:877–885
18. Yagi F, Hiroyama H, Kodama S (2001) *Agrocybe cylindracea* lectin is a member of the galectin family. Glycoconj J 18:745–749
19. Akiyoshi H, Sugii S, Nahid MA et al (2011) Detection of chromogranin A in the adrenal gland extracts of different animal species by an enzyme-linked immunosorbent assay using Thomsen-Friedenreich antigen-specific *Amaranthus caudatus* lectin. Vet Immunol Immunopathol 144:255–258
20. Boland CR, Chen YF, Rinderle SJ et al (1991) Use of the lectin from *Amaranthus caudatus* as a histochemical probe of proliferating colonic epithelial cells. Cancer Res 51:657–665
21. Rinderle SJ, Goldstein IJ, Matta KL et al (1989) Isolation and characterization of amaranthin, a lectin present in the seeds of *Amaranthus caudatus*, that recognizes the T- (or cryptic T)-antigen. J Biol Chem 264: 16123–16131
22. Rinderle SJ, Goldstein IJ, Remsen EE (1990) Physicochemical properties of amaranthin, the lectin from *Amaranthus caudatus* seeds. Biochemistry 29:10555–10561
23. Wu AM, Wu JH, Yang Z et al (2008) Differential contributions of recognition factors of two plant lectins—*Amaranthus caudatus* lectin and *Arachis hypogea* agglutinin, reacting with Thomsen-Friedenreich disaccharide (Galbeta1-3GalNAcalpha1-Ser/Thr). Biochimie 90:1769–1780
24. Ishida H, Hata Y, Kawato A et al (2004) Isolation of a novel promoter for efficient protein production in *Aspergillus oryzae*. Biosci Biotechnol Biochem 68:1849–1857
25. Matsumura K, Higashida K, Ishida H et al (2007) Carbohydrate binding specificity of a fucose-specific lectin from *Aspergillus oryzae*: a novel probe for core fucose. J Biol Chem 282:15700–15708
26. Tateno H, Nakamura-Tsuruta S, Hirabayashi J (2009) Comparative analysis of core-fucose-

binding lectins from *Lens culinaris* and *Pisum sativum* using frontal affinity chromatography. Glycobiology 19:527–536
27. Jin ZQ, Zhang DY, Xu BY (2004) Cloning and developmental and tissue-specific expression of banana (*Musa acuminate* AAA) lectin gene. Yi Chuan Xue Bao 31:508–512
28. Meagher JL, Winter HC, Ezell P et al (2005) Crystal structure of banana lectin reveals a novel second sugar binding site. Glycobiology 15:1033–1042
29. Peumans WJ, Zhang W, Barre A et al (2000) Fruit-specific lectins from banana and plantain. Planta 211:546–554
30. Onuma Y, Tateno H, Hirabayashi J et al (2013) rBC2LCN, a new probe for live cell imaging of human pluripotent stem cells. Biochem Biophys Res Commun 431:524–529
31. Sulak O, Cioci G, Delia M et al (2010) A TNF-like trimeric lectin domain from *Burkholderia cenocepacia* with specificity for fucosylated human histo-blood group antigens. Structure 18:59–72
32. Sulak O, Cioci G, Lameignere E et al (2011) *Burkholderia cenocepacia* BC2L-C is a super lectin with dual specificity and proinflammatory activity. PLoS Pathog 7:e1002238
33. Tateno H, Matsushima A, Hiemori K et al (2013) Podocalyxin is a glycoprotein ligand of the human pluripotent stem cell-specific probe rBC2LCN. Stem Cells Trans Med 2:265–273
34. Tateno H, Toyota M, Saito S et al (2011) Glycome diagnosis of human induced pluripotent stem cells using lectin microarray. J Biol Chem 286:20345–20353
35. Imai Y, Osawa T (1983) Enrichment of IL-2-producer T cells from mouse spleen by use of *Bauhinia purpurea* lectin. Scand J Immunol 18:217–224
36. Kasper M, Schuh D, Muller M (1994) *Bauhinia purpurea* lectin (BPA) binding of rat type I pneumocytes: alveolar epithelial alterations after radiation-induced lung injury. Exp Toxicol Pathol 46:361–367
37. Kusui K, Yamamoto K, Konami Y et al (1991) cDNA cloning and expression of *Bauhinia purpurea* lectin. J Biochem 109:899–903
38. Osawa T, Irimura T, Kawaguchi T (1978) *Bauhinia purpurea* agglutinin. Methods Enzymol 50:367–372
39. Sarker AB, Akagi T, Teramoto N et al (1994) *Bauhinia purpurea* (BPA) binding to normal and neoplastic thyroid glands. Pathol Res Pract 190:1005–1011
40. Sarker AB, Koirala TR, Aftabuddin et al (1994) Lectin histochemistry of normal lung and pulmonary carcinoma. Indian J Pathol Microbiol 37:29–38
41. Sarker AB, Koirala TR, Murakami I (1994) *Bauhinia purpurea* agglutinin (BPA) binding sites in human gastrointestinal tract. Indian J Pathol Microbiol 37:21–28
42. Sarker AB, Koirala TR, Murakami I et al (1995) *Bauhinia purpurea* and *Pisum sativum* lectin binding in human breast. Indian J Pathol Microbiol 38:261–265
43. Shue GL, Kawa S, Kato M et al (1993) Expression of glycoconjugates in pancreatic, gastric, and colonic tissue by *Bauhinia purpurea*, *Vicia villosa*, and peanut lectins. Scand J Gastroenterol 28:599–604
44. Yamamoto K, Konami Y, Osawa T et al (1992) Alteration of the carbohydrate-binding specificity of the *Bauhinia purpurea* lectin through the preparation of a chimeric lectin. J Biochem 111:87–90
45. Bourne Y, Roig-Zamboni V, Barre A et al (2004) The crystal structure of the *Calystegia sepium* agglutinin reveals a novel quaternary arrangement of lectin subunits with a beta-prism fold. J Biol Chem 279:527–533
46. Nakamura-Tsuruta S, Uchiyama N, Peumans WJ et al (2008) Analysis of the sugar-binding specificity of mannose-binding-type Jacalin-related lectins by frontal affinity chromatography—an approach to functional classification. FEBS J 275:1227–1239
47. Peumans WJ, Winter HC, Bemer V et al (1997) Isolation of a novel plant lectin with an unusual specificity from *Calystegia sepium*. Glycoconj J 14:259–265
48. Van Damme EJ, Barre A, Verhaert P et al (1996) Molecular cloning of the mitogenic mannose/maltose-specific rhizome lectin from *Calystegia sepium*. FEBS Lett 397:352–356
49. Carrington DM, Auffret A, Hanke DE (1985) Polypeptide ligation occurs during post-translational modification of concanavalin A. Nature 313:64–67
50. Derewenda Z, Yariv J, Helliwell JR et al (1989) The structure of the saccharide-binding site of concanavalin A. EMBO J 8:2189–2193
51. Goldstein IJ, Reichert CM, Misaki A (1974) Interaction of concanavalin A with model substrates. Ann N Y Acad Sci 234:283–296
52. Kornfeld R, Ferris C (1975) Interaction of immunoglobulin glycopeptides with concanavalin A. J Biol Chem 250:2614–2619
53. Mega T, Oku H, Hase S (1992) Characterization of carbohydrate-binding specificity of concanavalin A by competitive binding of

pyridylamino sugar chains. J Biochem 111: 396–400

54. Ohyama Y, Kasai K, Nomoto H et al (1985) Frontal affinity chromatography of ovalbumin glycoasparagines on a concanavalin A-sepharose column. A quantitative study of the binding specificity of the lectin. J Biol Chem 260: 6882–6887
55. Casset F, Peters T, Etzler M et al (1996) Conformational analysis of blood group A trisaccharide in solution and in the binding site of *Dolichos biflorus* lectin using transient and transferred nuclear Overhauser enhancement (NOE) and rotating-frame NOE experiments. Eur J Biochem 239:710–719
56. Etzler ME, Gupta S, Borrebaeck C (1981) Carbohydrate binding properties of the *Dolichos biflorus* lectin and its subunits. J Biol Chem 256:2367–2370
57. Etzler ME, Kabat EA (1970) Purification and characterization of a lectin (plant hemagglutinin) with blood group A specificity from *Dolichos biflorus*. Biochemistry 9:869–877
58. Hamelryck TW, Loris R, Bouckaert J et al (1999) Carbohydrate binding, quaternary structure and a novel hydrophobic binding site in two legume lectin oligomers from *Dolichos biflorus*. J Mol Biol 286:1161–1177
59. Hammarstrom S, Murphy LA, Goldstein IJ et al (1977) Carbohydrate binding specificity of four *N*-acetyl-D-galactosamine-"specific" lectins: *Helix pomatia* A hemagglutinin, soy bean agglutinin, lima bean lectin, and *Dolichos biflorus* lectin. Biochemistry 16:2750–2755
60. Schnell DJ, Etzler ME (1987) Primary structure of the *Dolichos biflorus* seed lectin. J Biol Chem 262:7220–7225
61. Crowley JF, Goldstein IJ (1981) *Datura stramonium* lectin: isolation and characterization of the homogeneous lectin. FEBS Lett 130:149
62. Cummings RD, Kornfeld S (1984) The distribution of repeating [Gal beta 1,4GlcNAc beta 1,3] sequences in asparagine-linked oligosaccharides of the mouse lymphoma cell lines BW5147 and PHAR 2.1. J Biol Chem 259: 6253–6260
63. Endo T, Iino K, Nozawa S et al (1988) Immobilized *Datura stramonium* agglutinin column chromatography, a novel method to discriminate the urinary hCGs of patients with invasive mole and choriocarcinoma from those of normal pregnant women and patients with hydatidiform mole. Jpn J Cancer Res 79: 160–164
64. Yamashita K, Totani K, Ohkura T et al (1987) Carbohydrate binding properties of complex-type oligosaccharides on immobilized *Datura stramonium* lectin. J Biol Chem 262: 1602–1607
65. De Boeck H, Loontiens FG, Lis H et al (1984) Binding of simple carbohydrates and some *N*-acetyllactosamine-containing oligosaccharides to *Erythrina cristagalli* agglutinin as followed with a fluorescent indicator ligand. Arch Biochem Biophys 234:297–304
66. Iglesias JL, Lis H, Sharon N (1982) Purification and properties of a D-galactose/*N*-acetyl-D-galactosamine-specific lectin from *Erythrina cristagalli*. Eur J Biochem 123:247–252
67. Kaladas PM, Kabat EA, Iglesias JL et al (1982) Immunochemical studies on the combining site of the D-galactose/*N*-acetyl-D-galactosamine specific lectin from *Erythrina cristagalli* seeds. Arch Biochem Biophys 217: 624–637
68. Teneberg S, Angstrom J, Jovall PA et al (1994) Characterization of binding of Gal beta 4GlcNAc-specific lectins from *Erythrina cristagalli* and *Erythrina corallodendron* to glycosphingolipids. Detection, isolation, and characterization of a novel glycosphingolipid of bovine buttermilk. J Biol Chem 269: 8554–8563
69. Fouquaert E, Peumans WJ, Smith DF et al (2008) The "old" *Euonymus europaeus* agglutinin represents a novel family of ubiquitous plant proteins. Plant Physiol 147:1316–1324
70. Roussel F, Dalion J, Wissocq JC (1992) *Euonymus europaeus* lectin as an endothelial and epithelial marker in canine tissues. Lab Anim 26:114–121
71. Hester G, Kaku H, Goldstein IJ et al (1995) Structure of mannose-specific snowdrop (*Galanthus nivalis*) lectin is representative of a new plant lectin family. Nat Struct Biol 2: 472–479
72. Van Damme EJ, De Clercq N, Claessens F et al (1991) Molecular cloning and characterization of multiple isoforms of the snowdrop (*Galanthus nivalis* L.) lectin. Planta 186: 35–43
73. Van Damme EJ, Kaku H, Perini F et al (1991) Biosynthesis, primary structure and molecular cloning of snowdrop (*Galanthus nivalis* L.) lectin. Eur J Biochem 202:23–30
74. Van Damme EJ, Peumans WJ (1988) Biosynthesis of the snowdrop (*Galanthus nivalis*) lectin in ripening ovaries. Plant Physiol 86: 922–926
75. Eckhardt AE, Malone BN, Goldstein IJ (1982) Inhibition of Ehrlich ascites tumor cell growth by *Griffonia simplicifolia* I lectin in vivo. Cancer Res 42:2977–2979

76. Edge AS, Spiro RG (1984) Presence of sulfate in *N*-glycosidically linked carbohydrate units of calf thyroid plasma membrane glycoproteins. J Biol Chem 259:4710–4713
77. Knibbs RN, Perini F, Goldstein IJ (1989) Structure of the major concanavalin A reactive oligosaccharides of the extracellular matrix component laminin. Biochemistry 28:6379–6392
78. Maddox DE, Goldstein IJ, Lobuglio AF (1982) *Griffonia simplicifolia* I lectin mediates macrophage-induced cytotoxicity against Ehrlich ascites tumor. Cell Immunol 71:202–207
79. Maddox DE, Shibata S, Goldstein IJ (1982) Stimulated macrophages express a new glycoprotein receptor reactive with *Griffonia simplicifolia* I-B4 isolectin. Proc Natl Acad Sci U S A 79:166–170
80. Murphy LA, Goldstein IJ (1977) Five alpha-D-galactopyranosyl-binding isolectins from *Bandeiraea simplicifolia* seeds. J Biol Chem 252:4739–4742
81. Peters BP, Goldstein IJ (1979) The use of fluorescein-conjugated *Bandeiraea simplicifolia* B4-isolectin as a histochemical reagent for the detection of alpha-D-galactopyranosyl groups. Their occurrence in basement membranes. Exp Cell Res 120:321–334
82. Shibata S, Goldstein IJ, Baker DA (1982) Isolation and characterization of a Lewis b-active lectin from *Griffonia simplicifolia* seeds. J Biol Chem 257:9324–9329
83. Balzarini J, Hatse S, Vermeire K et al (2004) Mannose-specific plant lectins from the Amaryllidaceae family qualify as efficient microbicides for prevention of human immunodeficiency virus infection. Antimicrob Agents Chemother 48:3858–3870
84. Balzarini J, Schols D, Neyts J et al (1991) Alpha-(1-3)- and alpha-(1-6)-D-mannose-specific plant lectins are markedly inhibitory to human immunodeficiency virus and cytomegalovirus infections in vitro. Antimicrob Agents Chemother 35:410–416
85. Kaku H, Van Damme EJ, Peumans WJ et al (1990) Carbohydrate-binding specificity of the daffodil (*Narcissus pseudonarcissus*) and amaryllis (*Hippeastrum hybr.*) bulb lectins. Arch Biochem Biophys 279:298–304
86. Lisowska E, Duk M (1972) The reaction of products of sequential periodate oxidation of human erythrocyte glycoproteins with hemagglutinin from *Helix pomatia*. Arch Immunol Ther Exp (Warsz) 20:869–875
87. Axelsson B, Kimura A, Hammarstrom S et al (1978) *Helix pomatia* A hemagglutinin: selectivity of binding to lymphocyte surface glycoproteins on T cells and certain B cells. Eur J Immunol 8:757–764
88. Brostrom H, Hellstrom U, Hammarstrom S et al (1985) A new surface marker on equine peripheral blood lymphocytes. I. Subpopulations of lymphocytes with receptors for *Helix pomatia* A hemagglutinin (HP). Vet Immunol Immunopathol 8:35–46
89. Dillner ML, Hammarstrom S, Perlmann P (1975) The lack of mitogenic response of neuraminidase-treated and untreated human blood lymphocytes to divalent, hexavalent, or insoluble *Helix pomatia* a hemagglutinin. Exp Cell Res 96:374–382
90. Hellstrom U, Hammarstrom ML, Hammarstrom S et al (1984) Fractionation of human lymphocytes on *Helix pomatia* a hemagglutinin-sepharose and wheat germ agglutinin-sepharose. Methods Enzymol 108:153–168
91. Hellstrom U, Hammarstrom S, Klein G (1978) Enrichment of *Helix pomatia* (HP) lectin binding variant from the TA3St mouse ascites tumor by repeated column selection. Eur J Cancer 14:1265–1272
92. Morein B, Hellstrom U, Axelsson L et al (1979) *Helix pomatia* a hemagglutinin, a surface marker for bovine T-lymphocytes. Vet Immunol Immunopathol 1:27–36
93. Poros A, Ahrlund-Richter L, Klein E et al (1983) Expression of *Helix pomatia* (HP) haemagglutinin receptors on cytolytic lymphocytes activated in mixed cultures. J Immunol Methods 57:9–19
94. Antony L, Basu D, Appukuttan PS (1989) Alpha-galactoside-binding isolectins from wild jack fruit seed (*Artocarpus hirsuta*): purification and properties. Indian J Biochem Biophys 26:361–366
95. Appukuttan PS, Kumar GS, Basu D (1984) Polysaccharide precipitation as a model to study sugar binding by lectins: jack fruit seed lectin interaction with galactomannan. Indian J Biochem Biophys 21:353–356
96. Arslan MI, Chulavatnatol M (2000) Characterisation of Jack fruit lectin. Bangladesh Med Res Counc Bull 26:23–26
97. Basu D, Delucas L, Parks EH et al (1988) Preliminary crystallographic study of the alpha-D-galactose-specific lectin from jack fruit (*Artocarpus integra*) seeds. J Mol Biol 201: 661–662
98. Bourne Y, Astoul CH, Zamboni V et al (2002) Structural basis for the unusual carbohydrate-binding specificity of jacalin towards galactose and mannose. Biochem J 364:173–180
99. Jeyaprakash AA, Geetha Rani P, Banuprakash Reddy G et al (2002) Crystal structure of the jacalin-T-antigen complex and a comparative study of lectin-T-antigen complexes. J Mol Biol 321:637–645

100. Komath SS, Bhanu K, Maiya BG et al (2000) Binding of porphyrins by the tumor-specific lectin, jacalin [Jack fruit (*Artocarpus integrifolia*) agglutinin]. Biosci Rep 20:265–276

101. Mahanta SK, Sanker S, Rao NV et al (1992) Primary structure of a Thomsen-Friedenreich-antigen-specific lectin, jacalin [*Artocarpus integrifolia* (jack fruit) agglutinin]. Evidence for the presence of an internal repeat. Biochem J 284(Pt 1):95–101

102. Namjuntra P, Muanwongyathi P, Chulavatnatol M (1985) A sperm-agglutinating lectin from seeds of Jack fruit (*Artocarpus heterophyllus*). Biochem Biophys Res Commun 128:833–839

103. Pratap JV, Jeyaprakash AA, Rani PG et al (2002) Crystal structures of artocarpin, a Moraceae lectin with mannose specificity, and its complex with methyl-alpha-D-mannose: implications to the generation of carbohydrate specificity. J Mol Biol 317:237–247

104. Remani P, Augustine J, Vijayan KK et al (1989) Jack fruit lectin binding pattern in benign and malignant lesions of the breast. In Vivo 3:275–278

105. Remani P, Joy A, Vijayan KK et al (1990) Jack fruit lectin binding pattern in carcinoma of the uterine cervix. J Exp Pathol 5:89–96

106. Remani P, Nair RA, Sreelekha TT et al (2000) Altered expression of jack fruit lectin specific glycoconjugates in benign and malignant human colorectum. J Exp Clin Cancer Res 19:519–523

107. Restum-Miguel N, Prouvost-Danon A (1985) Effects of multiple oral dosing on IgE synthesis in mice: oral sensitization by albumin extracts from seeds of Jack fruit (*Artocarpus integrifolia*) containing lectins. Immunology 54:497–504

108. Casset F, Hamelryck T, Loris R et al (1995) NMR, molecular modeling, and crystallographic studies of lentil lectin-sucrose interaction. J Biol Chem 270:25619–25628

109. Kaifu R, Osawa T, Jeanloz RW (1975) Synthesis of 2-*O*-(2-acetamido-2-deoxy-beta-D-glucopyranosyl)-D-mannose, and its interaction with D-mannose-specific lectins. Carbohydr Res 40:111–117

110. Loris R, Casset F, Bouckaert J et al (1994) The monosaccharide binding site of lentil lectin: an X-ray and molecular modelling study. Glycoconj J 11:507–517

111. Loris R, Steyaert J, Maes D et al (1993) Crystal structure determination and refinement at 2.3-A resolution of the lentil lectin. Biochemistry 32:8772–8781

112. Roth J, Neupert G, Thoss K (1975) Interaction of *Lens culinaris* lectin, concanavalin A, *Ricinus communis* agglutinin and wheat germ agglutinin with the cell surface of normal and transformed rat liver cells. Exp Pathol (Jena) 10:309–317

113. Schwarz FP, Misquith S, Surolia A (1996) Effect of substituent on the thermodynamics of D-glucopyranoside binding to concanavalin A, pea (*Pisum sativum*) lectin and lentil (*Lens culinaris*) lectin. Biochem J 316(Pt 1): 123–129

114. Schwarz FP, Puri KD, Bhat RG et al (1993) Thermodynamics of monosaccharide binding to concanavalin A, pea (*Pisum sativum*) lectin, and lentil (*Lens culinaris*) lectin. J Biol Chem 268:7668–7677

115. Kilpatrick DC, Graham C, Urbaniak SJ (1986) Inhibition of human lymphocyte transformation by tomato lectin. Scand J Immunol 24:11–19

116. Kilpatrick DC, Graham C, Urbaniak SJ et al (1984) A comparison of tomato (*Lycopersicon esculentum*) lectin with its deglycosylated derivative. Biochem J 220:843–847

117. Oguri S, Amano K, Nakashita H et al (2008) Molecular structure and properties of lectin from tomato fruit. Biosci Biotechnol Biochem 72:2640–2650

118. Peumans WJ, Rouge P, Van Damme EJ (2003) The tomato lectin consists of two homologous chitin-binding modules separated by an extensin-like linker. Biochem J 376:717–724

119. Konami Y, Yamamoto K, Osawa T (1990) The primary structure of the *Lotus tetragonolobus* seed lectin. FEBS Lett 268:281–286

120. Moreno FB, de Oliveira TM, Martil DE et al (2008) Identification of a new quaternary association for legume lectins. J Struct Biol 161:133–143

121. Imberty A, Gautier C, Lescar J et al (2000) An unusual carbohydrate binding site revealed by the structures of two *Maackia amurensis* lectins complexed with sialic acid-containing oligosaccharides. J Biol Chem 275: 17541–17548

122. Knibbs RN, Goldstein IJ, Ratcliffe RM et al (1991) Characterization of the carbohydrate binding specificity of the leukoagglutinating lectin from *Maackia amurensis*. Comparison with other sialic acid-specific lectins. J Biol Chem 266:83–88

123. Konami Y, Ishida C, Yamamoto K et al (1994) A unique amino acid sequence involved in the putative carbohydrate-binding domain of a legume lectin specific for sialylated carbohydrate chains: primary sequence determination of *Maackia amurensis* hemagglutinin (MAH). J Biochem 115:767–777

124. Yamamoto K, Konami Y, Irimura T (1997) Sialic acid-binding motif of *Maackia amurensis* lectins. J Biochem 121:756–761
125. Grahn E, Askarieh G, Holmner A et al (2007) Crystal structure of the *Marasmius oreades* mushroom lectin in complex with a xenotransplantation epitope. J Mol Biol 369: 710–721
126. Grahn E, Holmner A, Cronet C et al (2004) Crystallization and preliminary X-ray crystallographic studies of a lectin from the mushroom *Marasmius oreades*. Acta Crystallogr D Biol Crystallogr 60:2038–2039
127. Grahn EM, Winter HC, Tateno H et al (2009) Structural characterization of a lectin from the mushroom *Marasmius oreades* in complex with the blood group B trisaccharide and calcium. J Mol Biol 390:457–466
128. Kirkeby S, Winter HC, Goldstein IJ (2004) Comparison of the binding properties of the mushroom *Marasmius oreades* lectin and *Griffonia simplicifolia* I-B isolectin to alphagalactosyl carbohydrate antigens in the surface phase. Xenotransplantation 11:254–261
129. Loganathan D, Winter HC, Judd WJ et al (2003) Immobilized Marasmius oreades agglutinin: use for binding and isolation of glycoproteins containing the xenotransplantation or human type B epitopes. Glycobiology 13:955–960
130. Rempel BP, Winter HC, Goldstein IJ et al (2002) Characterization of the recognition of blood group B trisaccharide derivatives by the lectin from *Marasmius oreades* using frontal affinity chromatography-mass spectrometry. Glycoconj J 19:175–180
131. Tateno H, Goldstein IJ (2004) Partial identification of carbohydrate-binding sites of a Galalpha1,3Galbeta1,4GlcNAc-specific lectin from the mushroom *Marasmius oreades* by site-directed mutagenesis. Arch Biochem Biophys 427:101–109
132. Teneberg S, Alsen B, Angstrom J et al (2003) Studies on Galalpha3-binding proteins: comparison of the glycosphingolipid binding specificities of *Marasmius oreades* lectin and *Euonymus europaeus* lectin. Glycobiology 13: 479–486
133. Allen PZ, Connelly MC, Apicella MA (1980) Interaction of lectins with *Neisseria gonorrhoeae*. Can J Microbiol 26:468–474
134. Barton RW (1982) The binding of *Maclura pomifera* lectin to cells of the T-lymphocyte lineage in the rat. Cell Immunol 67:101–111
135. Bausch JN, Poretz RD (1977) Purification and properties of the hemagglutinin from *Maclura pomifera* seeds. Biochemistry 16: 5790–5794
136. Bausch JN, Richey J, Poretz RD (1981) Five structurally related proteins from affinity-purified *Maclura pomifera* lectin. Biochemistry 20:2618–2620
137. Hearn MT, Smith PK, Mallia AK (1982) Isolation of the *Maclura pomifera* hemagglutinin on a deoxymelibiotol affinity support and preliminary characterization by buffer electrofocusing and high-performance liquid chromatography. Biosci Rep 2:247–255
138. Jirgensons B (1980) Circular dichroism tests on the effect of alkali on conformation of lectins. Biochim Biophys Acta 625:193–201
139. Jones JM, Soderberg F (1979) Cytotoxicity of lymphoid cells induced by *Maclura pomifera* (MP) lectin. Cell Immunol 42: 319–326
140. Reano A, Faure M, Jacques Y et al (1982) Lectins as markers of human epidermal cell differentiation. Differentiation 22:205–210
141. Sarkar M, Wu AM, Kabat EA (1981) Immunochemical studies on the carbohydrate specificity of *Maclura pomifera* lectin. Arch Biochem Biophys 209:204–218
142. Ulevitch RJ, Jones JM, Feldman JD (1974) Isolation and characterization of *Maclura pomifera* (MP) lectin. Prep Biochem 4: 273–281
143. Van Damme EJ, Allen AK, Peumans WJ (1987) Leaves of the orchid twayblade (*Listera ovata*) contain a mannose-specific lectin. Plant Physiol 85:566–569
144. Avichezer D, Gilboa-Garber N, Garber NC et al (1994) *Pseudomonas aeruginosa* PA-I lectin gene molecular analysis and expression in *Escherichia coli*. Biochim Biophys Acta 1218:11–20
145. Avichezer D, Katcoff DJ, Garber NC et al (1992) Analysis of the amino acid sequence of the *Pseudomonas aeruginosa* galactophilic PA-I lectin. J Biol Chem 267:23023–23027
146. Blanchard B, Nurisso A, Hollville E et al (2008) Structural basis of the preferential binding for globo-series glycosphingolipids displayed by *Pseudomonas aeruginosa* lectin I. J Mol Biol 383:837–853
147. Chen CP, Song SC, Gilboa-Garber N et al (1998) Studies on the binding site of the galactose-specific agglutinin PA-IL from *Pseudomonas aeruginosa*. Glycobiology 8:7–16
148. Cioci G, Mitchell EP, Gautier C et al (2003) Structural basis of calcium and galactose recognition by the lectin PA-IL of *Pseudomonas aeruginosa*. FEBS Lett 555:297–301

149. Garber N, Guempel U, Belz A et al (1992) On the specificity of the D-galactose-binding lectin (PA-I) of *Pseudomonas aeruginosa* and its strong binding to hydrophobic derivatives of D-galactose and thiogalactose. Biochim Biophys Acta 1116:331–333
150. Gilboa-Garber N (1972) Purification and properties of hemagglutinin from *Pseudomonas aeruginosa* and its reaction with human blood cells. Biochim Biophys Acta 273: 165–173
151. Gilboa-Garber N (1982) *Pseudomonas aeruginosa* lectins. Methods Enzymol 83:378–385
152. Gilboa-Garber N, Katcoff DJ, Garber NC (2000) Identification and characterization of *Pseudomonas aeruginosa* PA-IIL lectin gene and protein compared to PA-IL. FEMS Immunol Med Microbiol 29:53–57
153. Gilboa-Garber N, Mizrahi L (1979) Interaction of the mannosephilic lectins of *Pseudomonas aeruginosa* with luminous species of marine enterobacteria. Microbios 26: 31–36
154. Gilboa-Garber N, Sudakevitz D, Sheffi M et al (1994) PA-I and PA-II lectin interactions with the ABO(H) and P blood group glycosphingolipid antigens may contribute to the broad spectrum adherence of *Pseudomonas aeruginosa* to human tissues in secondary infections. Glycoconj J 11:414–417
155. Nurisso A, Blanchard B, Audfray A et al (2010) Role of water molecules in structure and energetics of *Pseudomonas aeruginosa* lectin I interacting with disaccharides. J Biol Chem 285:20316–20327
156. Buts L, Dao-Thi MH, Loris R et al (2001) Weak protein-protein interactions in lectins: the crystal structure of a vegetative lectin from the legume *Dolichos biflorus.* J Mol Biol 309:193–201
157. Dahlgren K, Porath J, Lindahl-Kiessling K (1970) On the purification of phytohemagglutinins from *Phaseolus vulgaris* seeds. Arch Biochem Biophys 137:306–314
158. Dupuis G, Leclair B (1982) Studies on *Phaseolus vulgaris* phytohemagglutinin. Structural requirements for simple sugars to inhibit the agglutination of human group A erythrocytes. FEBS Lett 144:29–32
159. Faye L, Sturm A, Bollini R et al (1986) The position of the oligosaccharide side-chains of phytohemagglutinin and their accessibility to glycosidases determines their subsequent processing in the Golgi. Eur J Biochem 158: 655–661
160. Borberg H, Yesner I, Gesner B, Silber R (1968) The effect of *N* acetyl C-galactosamine and other sugars on the mitogenic activity and attachment of PHA to tonsil cells. Blood 31:747–757
161. Hoglund S, Dahlgren K (1970) On the morphology of phytohemagglutinin from *Phaseolus vulgaris* seeds. Eur J Biochem 17:23–26
162. Serafini-Cessi F, Franceschi C, Sperti S (1979) Specific interaction of human Tamm-Horsfall gylcoprotein with leucoagglutinin, a lectin from *Phaseolus vulgaris* (red kidney bean). Biochem J 183:381–388
163. Kobayashi Y, Tateno H, Dohra H et al (2012) A novel core fucose-specific lectin from the mushroom *Pholiota squarrosa.* J Biol Chem 287:33973–33982
164. Banerjee R, Das K, Ravishankar R et al (1996) Conformation, protein-carbohydrate interactions and a novel subunit association in the refined structure of peanut lectin-lactose complex. J Mol Biol 259:281–296
165. Pereira ME, Kabat EA, Lotan R et al (1976) Immunochemical studies on the specificity of the peanut (*Arachis hypogaea*) agglutinin. Carbohydr Res 51:107–118
166. Ravishankar R, Suguna K, Surolia A et al (1999) Structures of the complexes of peanut lectin with methyl-beta-galactose and *N*-acetyllactosamine and a comparative study of carbohydrate binding in Gal/GalNAc-specific legume lectins. Acta Crystallogr D Biol Crystallogr 55:1375–1382
167. Young NM, Johnston RA, Watson DC (1991) The amino acid sequence of peanut agglutinin. Eur J Biochem 196:631–637
168. Higgins TJ, Chandler PM, Zurawski G et al (1983) The biosynthesis and primary structure of pea seed lectin. J Biol Chem 258: 9544–9549
169. Kornfeld K, Reitman ML, Kornfeld R (1981) The carbohydrate-binding specificity of pea and lentil lectins. Fucose is an important determinant. J Biol Chem 256:6633–6640
170. Rini JM, Hardman KD, Einspahr H et al (1993) X-ray crystal structure of a pea lectin-trimannoside complex at 2.6 A resolution. J Biol Chem 268:10126–10132
171. Arigi E, Singh S, Kahlili AH et al (2007) Characterization of neutral and acidic glycosphingolipids from the lectin-producing mushroom, *Polyporus squamosus.* Glycobiology 17: 754–766
172. Mo H, Winter HC, Goldstein IJ (2000) Purification and characterization of a Neu5A

calpha2-6Galbeta1-4Glc/GlcNAc-specific lectin from the fruiting body of the polypore mushroom *Polyporus squamosus*. J Biol Chem 275:10623–10629

173. Tateno H, Winter HC, Goldstein IJ (2004) Cloning, expression in *Escherichia coli* and characterization of the recombinant Neu5 Acalpha2,6Galbeta1,4GlcNAc-specific high-affinity lectin and its mutants from the mushroom *Polyporus squamosus*. Biochem J 382: 667–675
174. Zhang B, Palcic MM, Mo H et al (2001) Rapid determination of the binding affinity and specificity of the mushroom *Polyporus squamosus* lectin using frontal affinity chromatography coupled to electrospray mass spectrometry. Glycobiology 11:141–147
175. Appukuttan PS, Basu D (1981) Isolation of an *N*-acetyl-D-galactosamine-binding protein from winged bean (*Psophocarpus tetragonolobus*). Anal Biochem 113:253–255
176. Kirkeby S, Singha NC, Surolia A (1997) Localized agglutinin staining in muscle capillaries from normal and very old atrophic human muscle using winged bean (*Psophocarpus tetragonolobus*) lectin. Histochem Cell Biol 107:31–37
177. Kortt AA (1984) Purification and properties of the basic lectins from winged bean seed [*Psophocarpus tetragonolobus* (L.) DC]. Eur J Biochem 138:519–525
178. Kortt AA (1985) Characterization of the acidic lectins from winged bean seed (*Psophocarpus tetragonolobus*(L.)DC). Arch Biochem Biophys 236:544–554
179. Matsuda T, Kabat EA, Surolia A (1989) Carbohydrate binding specificity of the basic lectin from winged bean (*Psophocarpus tetragonolobus*). Mol Immunol 26:189–195
180. Patanjali SR, Sajjan SU, Surolia A (1988) Erythrocyte-binding studies on an acidic lectin from winged bean (*Psophocarpus tetragonolobus*). Biochem J 252:625–631
181. Pueppke SG (1979) Purification and characterization of a lectin from seeds of the winged bean, *Psophocarpus tetragonolobus* (L.)DC. Biochim Biophys Acta 581:63–70
182. Shet MS, Madaiah M (1988) Chemical modification studies on a lectin from winged-bean [*Psophocarpus tetragonolobus* (L.) DC] tubers. Biochem J 254:351–357
183. Shet MS, Murugiswamy B, Madaiah M (1985) A lectin from winged bean (*Psophocarpus tetragonolobus*) tubers. Indian J Biochem Biophys 22:313–315
184. Cioci G, Mitchell EP, Chazalet V et al (2006) Beta-propeller crystal structure of *Psathyrella velutina* lectin: an integrin-like fungal protein interacting with monosaccharides and calcium. J Mol Biol 357:1575–1591
185. Endo T, Ohbayashi H, Kanazawa K et al (1992) Carbohydrate binding specificity of immobilized *Psathyrella velutina* lectin. J Biol Chem 267:707–713
186. Kobata A, Kochibe N, Endo T (1994) Affinity chromatography of oligosaccharides on *Psathyrella velutina* lectin column. Methods Enzymol 247:228–237
187. Kochibe N, Matta KL (1989) Purification and properties of an *N*-acetylglucosamine-specific lectin from *Psathyrella velutina* mushroom. J Biol Chem 264:173–177
188. Ueda H, Kojima K, Saitoh T et al (1999) Interaction of a lectin from *Psathyrella velutina* mushroom with *N*-acetylneuraminic acid. FEBS Lett 448:75–80
189. Ueda H, Takahashi N, Ogawa H (2003) Psathyrella velutina lectin as a specific probe for *N*-acetylneuraminic acid in glycoconjugates. Methods Enzymol 363:77–90
190. Farnes P, Barker BE, Brownhill LE et al (1964) Mitogenic activity in *Phytolacca americana* (Pokeweed). Lancet 2:1100–1101
191. Fujii T, Hayashida M, Hamasu M et al (2004) Structures of two lectins from the roots of pokeweed (*Phytolacca americana*). Acta Crystallogr D Biol Crystallogr 60:665–673
192. Hayashida M, Fujii T, Hamasu M et al (2003) Crystallization and preliminary X-ray analysis of lectin C from the roots of pokeweed (*Phytolacca americana*). Acta Crystallogr D Biol Crystallogr 59:1249–1252
193. Hayashida M, Fujii T, Hamasu M et al (2003) Similarity between protein-protein and protein-carbohydrate interactions, revealed by two crystal structures of lectins from the roots of pokeweed. J Mol Biol 334:551–565
194. Kino M, Yamaguchi K, Umekawa H et al (1995) Purification and characterization of three mitogenic lectins from the roots of pokeweed (*Phytolacca americana*). Biosci Biotechnol Biochem 59:683–688
195. Reisfeld RA, Borjeson J, Chessin LN et al (1967) Isolation and characterization of a mitogen from pokeweek (*Phytolacca americana*). Proc Natl Acad Sci U S A 58:2020–2027
196. Yokoyama K, Terao T, Osawa T (1978) Carbohydrate-binding specificity of pokeweed mitogens. Biochim Biophys Acta 538: 384–396

197. Olsnes S, Pappenheimer AM Jr, Meren R (1974) Lectins from *Abrus precatorius* and *Ricinus communis*. II Hybrid toxins and their interaction with chain-specific antibodies. J Immunol 113:842–847
198. Olsnes S, Refsnes K, Pihl A (1974) Mechanism of action of the toxic lectins abrin and ricin. Nature 249:627–631
199. Olsnes S, Saltvedt E, Pihl A (1974) Isolation and comparison of galactose-binding lectins from *Abrus precatorius* and *Ricinus communis*. J Biol Chem 249:803–810
200. Pappenheimer AM Jr, Olsnes S, Harper AA (1974) Lectins from *Abrus precatorius* and *Ricinus communis*. I. Immunochemical relationships between toxins and agglutinins. J Immunol 113:835–841
201. Saltvedt E (1976) Structure and toxicity of pure ricinus agglutinin. Biochim Biophys Acta 451:536–548
202. Yamamoto T, Iwasaki Y, Konno H et al (1985) Primary degeneration of motor neurons by toxic lectins conveyed from the peripheral nerve. J Neurol Sci 70:327–337
203. Kostlanova N, Mitchell EP, Lortat-Jacob H et al (2005) The fucose-binding lectin from *Ralstonia solanacearum*. A new type of beta-propeller architecture formed by oligomerization and interacting with fucoside, fucosyllactose, and plant xyloglucan. J Biol Chem 280:27839–27849
204. Sudakevitz D, Imberty A, Gilboa-Garber N (2002) Production, properties and specificity of a new bacterial L-fucose- and D-arabinose-binding lectin of the plant aggressive pathogen *Ralstonia solanacearum*, and its comparison to related plant and microbial lectins. J Biochem 132:353–358
205. Sudakevitz D, Kostlanova N, Blatman-Jan G et al (2004) A new *Ralstonia solanacearum* high-affinity mannose-binding lectin RS-IIL structurally resembling the *Pseudomonas aeruginosa* fucose-specific lectin PA-IIL. Mol Microbiol 52:691–700
206. Dorland L, van Halbeek H, Vleigenthart JF et al (1981) Primary structure of the carbohydrate chain of soybean agglutinin. A reinvestigation by high resolution 1H NMR spectroscopy. J Biol Chem 256:7708–7711
207. Galbraith W, Goldstein IJ (1970) Phytohemagglutinins: a new class of metalloproteins. Isolation, purification, and some properties of the lectin from *Phaseolus lunatus*. FEBS Lett 9:197–201
208. Jaffe CL, Ehrlich-Rogozinski S, Lis H et al (1977) Transition metal requirements of soybean agglutin. FEBS Lett 82:191–196
209. Lis H, Sharon N (1978) Soybean agglutinin–a plant glycoprotein. Structure of the carboxydrate unit. J Biol Chem 253:3468–3476
210. Vodkin LO, Rhodes PR, Goldberg RB (1983) cA lectin gene insertion has the structural features of a transposable element. Cell 34: 1023–1031
211. Broekaert WF, Nsimba-Lubaki M, Peeters B et al (1984) A lectin from elder (*Sambucus nigra* L.) bark. Biochem J 221:163–169
212. Roth J, Taatjes DJ, Weinstein J et al (1986) Differential subcompartmentation of terminal glycosylation in the Golgi apparatus of intestinal absorptive and goblet cells. J Biol Chem 261:14307–14312
213. Shibuya N, Goldstein IJ, Broekaert WF et al (1987) The elderberry (*Sambucus nigra* L.) bark lectin recognizes the Neu5Ac(alpha 2-6) Gal/GalNAc sequence. J Biol Chem 262: 1596–1601
214. Shibuya N, Goldstein IJ, Broekaert WF et al (1987) Fractionation of sialylated oligosaccharides, glycopeptides, and glycoproteins on immobilized elderberry (*Sambucus nigra* L.) bark lectin. Arch Biochem Biophys 254:1–8
215. Kaku H, Tanaka Y, Tazaki K et al (1996) Sialylated oligosaccharide-specific plant lectin from Japanese elderberry (*Sambucus sieboldiana*) bark tissue has a homologous structure to type II ribosome-inactivating proteins, ricin and abrin. cDNA cloning and molecular modeling study. J Biol Chem 271:1480–1485
216. Shibuya N, Tazaki K, Song ZW et al (1989) A comparative study of bark lectins from three elderberry (*Sambucus*) species. J Biochem 106:1098–1103
217. Ashford D, Allen AK, Neuberger A (1982) The production and properties of an antiserum to potato (*Solanum tuberosum*) lectin. Biochem J 201:641–645
218. Ashford D, Desai NN, Allen AK et al (1982) Structural studies of the carbohydrate moieties of lectins from potato (*Solanum tuberosum*) tubers and thorn-apple (*Datura stramonium*) seeds. Biochem J 201:199–208
219. Doi A, Matsumoto I, Seno N (1983) Fluorescence spectral studies on the specific interaction between sulfated glycosaminoglycans and potato lectin. J Biochem 93: 771–775
220. Kawashima H, Sueyoshi S, Li H et al (1990) Carbohydrate binding specificities of several poly-*N*-acetyllactosamine-binding lectins. Glycoconj J 7:323–334
221. Fukushima K, Hada T, Higashino K et al (1998) Elevated serum levels of *Trichosanthes*

japonica agglutinin-I binding alkaline phosphatase in relation to high-risk groups for hepatocellular carcinomas. Clin Cancer Res 4:2771–2777

222. Yamashita K, Fukushima K, Sakiyama T et al (1995) Expression of Sia alpha 2→6Gal beta 1→4GlcNAc residues on sugar chains of glycoproteins including carcinoembryonic antigens in human colon adenocarcinoma: applications of *Trichosanthes japonica* agglutinin I for early diagnosis. Cancer Res 55: 1675–1679
223. Yamashita K, Ohkura T, Umetsu K et al (1992) Purification and characterization of a Fuc alpha 1→2Gal beta 1→and GalNAc beta 1→-specific lectin in root tubers of *Trichosanthes japonica*. J Biol Chem 267: 25414–25422
224. Yamashita K, Umetsu K, Suzuki T et al (1992) Purification and characterization of a Neu5Ac alpha 2→6Gal beta 1→4GlcNAc and HSO3 (-)→6Gal beta 1→GlcNAc specific lectin in tuberous roots of Trichosanthes japonica. Biochemistry 31:11647–11650
225. Yoshikawa K, Umetsu K, Shinzawa H et al (1999) Determination of carbohydrate-deficient transferrin separated by lectin affinity chromatography for detecting chronic alcohol abuse. FEBS Lett 458:112–116
226. Cammue BP, Peeters B, Peumans WJ (1986) A new lectin from tulip (*Tulipa*) bulbs. Planta 169:583–588
227. Oda Y, Ichida S, Aonuma S et al (1989) Studies on chemical modification of *Tulipa gesneriana* lectin. Chem Pharm Bull (Tokyo) 37:2170–2173
228. Oda Y, Minami K (1986) Isolation and characterization of a lectin from tulip bulbs, *Tulipa gesneriana*. Eur J Biochem 159: 239–245
229. Oda Y, Minami K, Ichida S et al (1987) A new agglutinin from the *Tulipa gesneriana* bulbs. Eur J Biochem 165:297–302
230. Van Damme EJ, Brike F, Winter HC et al (1996) Molecular cloning of two different mannose-binding lectins from tulip bulbs. Eur J Biochem 236:419–427
231. Van Damme EJ, Peumans WJ (1989) Developmental changes and tissue distribution of lectin in *Tulipa*. Planta 178:10–18
232. Beintema JJ, Peumans WJ (1992) The primary structure of stinging nettle (*Urtica dioica*) agglutinin. A two-domain member of the hevein family. FEBS Lett 299:131–134
233. Le Moal MA, Truffa-Bachi P (1988) *Urtica dioica* agglutinin, a new mitogen for murine T lymphocytes: unaltered interleukin-1 production but late interleukin 2-mediated proliferation. Cell Immunol 115:24–35
234. Shibuya N, Goldstein IJ, Shafer JA et al (1986) Carbohydrate binding properties of the stinging nettle (*Urtica dioica*) rhizome lectin. Arch Biochem Biophys 249:215–224
235. Kurimura Y, Tsuji Y, Yamamoto K et al (1995) Efficient production and purification of extracellular 1,2-alpha-L-fucosidase of *Bacillus* sp. K40T. Biosci Biotechnol Biochem 59: 589–594
236. Matsumoto I, Osawa T (1969) Purification and characterization of an anti-H(O) phytohemagglutinin of *Ulex europeus*. Biochim Biophys Acta 194:180–189
237. Bausch SB, Chavkin C (1996) *Vicia villosa* agglutinin labels a subset of neurons coexpressing both the mu opioid receptor and parvalbumin in the developing rat subiculum. Brain Res Dev Brain Res 97:169–177
238. Ervasti JM, Burwell AL, Geissler AL (1997) Tissue-specific heterogeneity in alpha-dystroglycan sialoglycosylation. Skeletal muscle alpha-dystroglycan is a latent receptor for *Vicia villosa* agglutinin b4 masked by sialic acid modification. J Biol Chem 272: 22315–22321
239. Grubhoffer L, Ticha M, Kocourek J (1981) Isolation and properties of a lectin from the seeds of hairy vetch (*Vicia villosa* Roth). Biochem J 195:623–626
240. Kimura A, Wigzell H, Holmquist G et al (1979) Selective affinity fractionation of murine cytotoxic T lymphocytes (CTL). Unique lectin specific binding of the CTL associated surface glycoprotein, T 145. J Exp Med 149:473–484
241. Murakami T, Ohtsuka A, Ono K (1996) Neurons with perineuronal sulfated proteoglycans in the mouse brain and spinal cord: their distribution and reactions to lectin *Vicia villosa* agglutinin and Golgi's silver nitrate. Arch Histol Cytol 59:219–231
242. Kurokawa T, Tsuda M, Sugino Y (1976) Purification and characterization of a lectin from *Wistaria floribunda* seeds. J Biol Chem 251:5686–5693
243. Torres BV, McCrumb DK, Smith DF (1988) Glycolipid-lectin interactions: reactivity of lectins from Helix pomatia, *Wisteria floribunda*, and *Dolichos biflorus* with glycolipids containing *N*-acetylgalactosamine. Arch Biochem Biophys 262:1–11
244. Toyoshima S, Akiyama Y, Nakano K et al (1971) A phytomitogen from *Wistaria floribunda* seeds and its interaction with human peripheral lymphocytes. Biochemistry 10: 4457–4463

245. Kronis KA, Carver JP (1982) Specificity of isolectins of wheat germ agglutinin for sialyloligosaccharides: a 360-MHz proton nuclear magnetic resonance binding study. Biochemistry 21:3050–3057

246. Matsumoto I, Koyama T, Kitagaki-Ogawa H et al (1987) Separation of isolectins by high-performance hydrophobic interaction chromatography. J Chromatogr 400:77–81

247. Nagata Y, Burger MM (1974) Wheat germ agglutinin. Molecular characteristics and specificity for sugar binding. J Biol Chem 249:3116–3122

248. Nagata Y, Goldberg AR, Burger MM (1974) The isolation and purification of wheat germ and other agglutinins. Methods Enzymol 32: 611–615

249. Wright CS, Keith C, Langridge R et al (1974) A preliminary crystallographic study of wheat germ agglutinin. J Mol Biol 87: 843–846

250. Yamamoto K, Tsuji T, Matsumoto I et al (1981) Structural requirements for the binding of oligosaccharides and glycopeptides to immobilized wheat germ agglutinin. Biochemistry 20:5894–5899

Chapter 46

Lectin Structures: Classification Based on the 3-D Structures

Zui Fujimoto, Hiroaki Tateno, and Jun Hirabayashi

Abstract

Recent progress in structural biology has elucidated the three-dimensional structures and carbohydrate-binding mechanisms of most lectin families. Lectins are classified into 48 families based on their three-dimensional structures. A ribbon drawing gallery of the crystal and solution structures of representative lectins or lectin-like proteins is appended and may help to convey the diversity of lectin families, the similarity and differences between lectin families, as well as the carbohydrate-binding architectures of lectins.

Key words Carbohydrate-binding, Crystal structure, Protein Data Bank, Protein family, Protein fold, Solution structure

1 Introduction

Lectins are distributed throughout natural organisms, found in animals, plants, fungi, bacteria, and virus. Lectins are often categorized by their natural sources, e.g., animal lectins, plant lectins, fungal, or bacterial lectins. In other cases, lectins are occasionally divided according to their carbohydrate-binding specificities, such as galactose-binding, fucose-binding, *N*-glycan-binding. Recent progress in molecular biology has made it easy to classify lectins by amino acid sequences as described by E. Van Damme in Chap. 1, and this shed light on new findings that some lectins distributed in several kingdoms and different lectins share the same protein motifs. Furthermore, progress in structural biology has elucidated the three-dimensional structures and carbohydrate-binding mechanisms of most lectin families, enabling the fine classification of lectins. Considering these facts, it is necessary to create a new cross-species classification.

In this chapter, lectins are classified into 48 families based on their three-dimensional structure determined by X-ray crystallography or NMR analysis, and the representative protein fold from each lectin family is visually introduced. Structural insights into lectins

Jun Hirabayashi (ed.), *Lectins: Methods and Protocols*, Methods in Molecular Biology, vol. 1200, DOI 10.1007/978-1-4939-1292-6_46, © Springer Science+Business Media New York 2014

will provide an aid to understand the carbohydrate-recognition mechanisms and ligand-specificities of lectins. Furthermore, they will provide a structural platform in protein engineering and glycobiology of lectins and lectin-like domains for industrial, medical, and pharmaceutical applications.

2 Classification

Three-dimensional structures of lectins were extracted from the Protein Data Bank (PDB, http://www.rcsb.org/pdb/), and the structures of the lectin carbohydrate-recognition domains were classified into families using the entries in the Pfam protein families database (http://pfam.sanger.ac.uk/; also *see* Ref. 1) and InterPro: protein sequence analysis and classification (http://www.ebi.ac.uk/interpro/; also *see* Ref. 2) and are summarized in Table 1. Families were roughly listed in the order of animal, plant, algal, fungal, bacterial, and viral lectins, but families having closely related structures were consecutively listed. In principle, one lectin family is assumed to belong to one Pfam family. As exceptions, L-type, R-type, AB5 toxin, or viral capsid protein family is treated as a single family that contains several Pfam entries. L-type or R-type lectin family has been generally established and the concerned Pfam structures are closely related. AB5 toxins or viral capsid proteins are found in several Pfam families, but they are gathered into one category here because their structures appear to have similarities. The numbers of PDB entries related to the Pfam families are listed in Table 1, which will be an index of how much structural information is available for the lectin family. The large number of I-type lectins is because of the structural entries for human immunoglobulin. Distribution, basic protein fold, and assembly of the lectin are also listed in Table 1.

Lectin families, such as X-lectin, Nictaba-like lectin, and *Euonymus europaeus* lectin, whose members' structures have not been determined, are not included in Table 1. Nevertheless, the amino acid sequences suggest that X-lectin is related to Ficolins [3]. *Oscillatoria agardhii*, *Pseudomonas fluorescens*, and *Myxococcus xanthus* lectins have a similar ten-stranded β-barrel structure [4, 5] but the Pfam family has not been assigned and these structures are not included in Table 1. Similarly, serine-rich repeat adhesin of *Streptococcus gordonii* has an immunoglobulin-like β-sandwich structure and Salmonella phage tail spike protein has a right-handed parallel β-helix structure, but the Pfam family has not been assigned for their structures and are not included in Table 1. Structures of several other lectin-like proteins have been determined, e.g., *Vibrio cholerae* neuraminidase insertion domain [6], CD11b/CD18 adhesin [7] or *Flammulina velutipes* fruiting body protein [8], and these structures have the potential to create a new lectin structural family.

Table 1
Lectin families classified based on the three-dimensional structures

No	Family	Lectin component	Fold	Assembly	Distribution: Animal	Plant	Fungi	Bacteria	Virus	Pfam clan[a]	Pfam family[a,b] Number of PDB entries	Interpro[c]	CAZy[d]	Chap. Fig
1 a	L-type	Legume intelectin, concanavarin-A, Lentil lectin, Peanut agglutinin, Hairy vetch lectin (VVL), *Maackia amurensis* leukoagglutinin (MAL), *Ulex europaeus* lectin (UEA), *Canavalia maritima* lectin (ConM)	Concanavalin A-like β-sandwich (jellyroll)	Dimer of 2-chain fold	•	•	•			CL0004 Concanavalin	PF00139 Legume lectin domain (247)	IPR001220 Legume lectin domain		§11 1A
b	L-type-like	VIP36, VERGIC-53, LMAN1, Emp46p, Emp47p	Concanavalin A-like β-sandwich (jellyroll)	Monomer	•	•	•			CL0004 Concanavalin	PF03388 Legume-like lectin family (21)	IPR005052 Legume-like lectin		
2	Galectin	Galectin	Concanavalin A-like β-sandwich (jellyroll)	Dimer, monmer	•		•		•	CL0004 Concanavalin	PF00337 Galactoside-binding lectin (162)	IPR001079 Galectin, carbohydrate recognition domain		1B

(continued)

Table 1
(continued)

No	Family	Lectin component	Fold	Assembly	Distribution					Pfam clan[a]	Pfam family[a,b] Number of PDB entries	Interpro[c]	CAZy[d]	Chap. Fig
					Animal	Plant	Fungi	Bacteria	Virus					
3	Pentaxin	C-reactive protein (CRP), serum amyloid P component protein (SAP), female protein (FP)	Concanavalin A-like β-sandwich (jellyroll)	Pentamer	•					CL0004 Concanavalin	PF00354 Pentaxin family (22)	IPR001759 Pentaxin		1C
4	I-type	CD33 and related siglecs, sialoadhesin, CD22, myelin-associated glycoprotein (MAG)	Immunoglobulin-like β-sandwich	Linked to C-set domain	•					CL0011 Immunoglobulin superfamily	PF07686 Immunoglobulin V-set domain (2407)	IPR013106 Immunoglobulin V-set domain		1D
5	C-type	Mannose-binding protein, selectin, collectin	C-type α/β-fold	Monomer, dimer, trimer, linked to different domain	•					CL0056 C-type lectin-like superfamily	PF00059 Lectin C-type domain (248)	IPR001304 C-type lectin		1E
6	Hyaladherin	Hyaluronan-binding proteins, CD44, hyaladherins	C-type α/β-fold	Linked to different domain	•					CL0056 C-type lectin-like superfamily	PF00193 Extracellular link domain (16)	IPR000538 Link		1F
7	Chitinase-like	YKL-40, Ym1, oviductin	$(\beta/\alpha)_8$-barrel	Monomer	•	•	•	•	•	CL0058 Tim barrel glycosyl hydrolase superfamily	PF00704 Glycosyl hydrolases family 18 (206)	IPR001223 Glycoside hydrolase, family 18, catalytic domain	GH18	2A

8		M-type [e]	ER-associated degradation-enhancing α-mannosidase-like protein	$(\alpha/\alpha)_7$-barrel	Monomer	•	•	•	•		CL0059 Six-hairpin glycosidase superfamily	PF01532 Glycosyl hydrolase family 47 (17)	IPR001382 Glycoside hydrolase, family 47	GH47	2B
9	a	R-type	Plant toxin ricin, abrin, Elderberry lectin, CEL-III. EW29, Actinohivin	β-Trefoil	Tandem repeat, linked to enzyme	•	•	•	•	•	CL0066 β-Trefoil superfamily	PF00652 Ricin-type β-trefoil lectin domain (120)	IPR000772 Ricin B lectin domain	CBM13	2C
	b	R-type-like	HA33, *Marasmius oreades* lectin (MOA), *Sclerotinia sclerotiorum* agglutinin (SSA)	β-Trefoil	Linked to different domain			•	•	•	CL0066 β-Trefoil superfamily	PF14200 Ricin-type β-trefoil lectin domain-like (62)	IPR000772 Ricin B lectin domain	CBM13	
10		ACA-like	*Amaranthus caudatus* agglutinin	β-Trefoil	Dimer of tandem repeat		•				CL0066 β-Trefoil superfamily	PF07468 Agglutinin (2)	IPR008998 Agglutinin		2D
11		Botulinum neurotoxin-like	Botulinum toxin, Tetanus toxin	β-Trefoil	Linked to different domain				•		CL0066 β-Trefoil superfamily	PF07951 *Clostridium* neurotoxin, C-terminal receptor binding (63)	IPR013104 *Clostridium* neurotoxin, receptor-binding C-terminal		2E
12		F-box	Ubiquitin ligase E3	β-Sandwich (jellyroll)	Linked to different domain	•					CL0202 Galactose-binding domain-like superfamily	PF04300 F-box associated region (6)	IPR007397 F-box associated (FBA) domain		2F

(continued)

Table 1
(continued)

No	Family	Lectin component	Fold	Assembly	Distribution					Pfam clan[a]	Pfam family[a,b] Number of PDB entries	Interpro[c]	CAZy[d]	Chap. Fig
					Animal	Plant	Fungi	Bacteria	Virus					
13	F-type[f] (AAA-like, Eel-lectin, fucolectins)	*Anguilla anguilla* agglutinin (AAA)	β-Sandwich (jellyroll) with Ca ion	Monomer, tandem repeat, linked to different domain	•	•	•	•		CL0202 Galactose-binding domain-like superfamily	PF00754 F5/8 type C domain (119)	IPR000421 Coagulation factor 5/8 C-terminal type domain	CBM32, CBM47	3A
14	PA-IL-like	*Pseudomonas aeruginosa* PA-IL	β-Sandwich (jellyroll)	Dimer				•		CL0202 galactose-binding domain-like superfamily	PF07828 PA-IL-like protein (16)	IPR012905 PA-IL-like		3B
15	P-type	Cation-dependent mannose 6-phosphate receptor	P-type α/β-fold	Dimer	•					CL0226 Mannose 6-phosphate receptor	PF02157 Mannose-6-phosphate receptor (11)	–		3C
16	Ficolins	Ficolin, tachylectins, tenascin, TL-5A, TL-5B, slug *Limax flavus* lectin (LFA)	Fibrinogen-like	Trimer	•					CL0422 Fibrinogen C-terminal domain-like	PF00147 Fibrinogen β and γ chains, C-terminal globular domain (17)	IPR002181 Fibrinogen, α/β/γ chain, C-terminal globular domain		3D
17	Malectin	Malectin	β-Sandwich (jellyroll)	Monomer	•					CL0468 Malectin-like	PF11721 Di-glucose binding within endoplasmic reticulum (3)	IPR021720 Malectin		3E

18	Calnexin	Calnexin, calreticulin	β-Sandwich (jellyroll)	Monomer	•		–	PF00262 Calreticulin family (11)	IPR001580 Calreticulin/ calnexin		3F
19	Tachylectin2-like[g]	Tachylectin2	5-Bladed β-propeller	Monomer	•		–	Not assigned (1)	IPR023294 Tachylectin 2		4A
20	Tachycitin-like	Tachycitin	Tachycitin type β-sheet-cystine fold	Monomer	•		CL0155 Carbohydrate binding domain 14/19 clan	PF01607 Chitin binding Peritrophin-A domain (1)	IPR002557 Chitin binding domain	CBM14	4B
21	Hevein	*Urtica dioica* agglutinin, wheat germ agglutinin isolectin 3 (WGA3), hevein	Hevein-like cystine knot motif	Dimer	•	•	–	PF00187 Chitin recognition protein (36)	IPR001002 Chitin-binding, type 1	CBM18	4C
22	Jacalin-related	Jacalin, griffithsin	β-Prism I	Tetramer of 2-chain fold	•	•	–	PF01419 Jacalin-like lectin domain (67)	IPR001229 Mannose-binding lectin		4D
23	SUEL-related (CSL3-like)	Sea urchin egg lectin (SUEL), L-rhamnose-binding lectin CSL3, latrophilin-1 GPCR Gal lectin domain	α/β-Fold with two long structured loop	Dimer, linked to different domain	•		–	PF02140 Galactose binding lectin domain (7)	IPR000922 D-galactoside/ L-rhamnose binding SUEL lectin domain		4E
24	H-type[h]	Helix pomatia agglutinin (HPA), Discoidin	H-type six-stranded antiparallel β-sandwich	Hexamer	•	•	–	PF09458 H-type lectin domain (12)	IPR019019 H-type lectin domain		4F
25	Cystine knot (SHL-like)	*Selenocosmia huwena* lectin-I (SHL-I)	Cystine knot motif		•		CL0083 Omega toxin-like	PF07740 Ion channel inhibitory toxin (23)	IPR011696 Ion channel inhibitory toxin		5A

(continued)

Table 1
(continued)

No	Family	Lectin component	Fold	Assembly	Distribution					Pfam clan[a]	Pfam family[a,b] Number of PDB entries	Interpro[c]	CAZy[d]	Chap. Fig
					Animal	Plant	Fungi	Bacteria	Virus					
26	TgMIC4	*Toxoplasma gondii* micronemal protein 4(TgMIC4), EtMIC5, *Sarcocystis muris* lectin (SML-2)	Apple-like α/β-fold	Tandem repeat with PAN_1 domains	•					CL0168 PAN-like	PF00024 PAN domain (28)	IPR003014 PAN-1 domain		5B
27	TgMIC1	*Toxoplasma gondii* micronemal protein 1 (TgMIC1)	Micronemal sialic-acid binding protein	Linked to different domain	•					–	PF10564 Sialic-acid binding micronemal adhesive repeat (7)	IPR019562 Micronemal adhesive repeat, sialic-acid binding		5C
28	LysM	Effector Ecp6, chitin elicitor receptor kinase-1	LysM βααβ-fold (lysin motif)	Triple repeat	•	•	•	•		CL0187 LysM-like domain	PF01476 LysM domain (14)	IPR018392 Peptidoglycan-binding lysin domain	CBM50	5D
29	LNP-type (N-type)	Lectin nucleotide phospho-hydrolase (LNP)	RNAseH-like α/β-fold	Monomer	•	•		•		CL0108 Actin-like ATPase superfamily	PF01150 GDA1/CD39 (nucleoside phosphatase) family (45)	IPR000407 Nucleoside phosphatase GDA1/CD39		5E

30	Monocot (bulb-type lectin)	Snowdrop lectin (GNA), bacteriocin LLPA, monocot mannose-binding lectin (MMBL), *Polygonatum cyrtonema* lectin (PCL)	β-Prism II	Monomer, dimer, tetramer	•		•	–	PF01453 D-Mannose binding lectin (26)	IPR001480 Bulb-type lectin domain	5F
31	Fungal fruit-body (Actinoporin-like, ABL-like)	*Xerocomus chrysenteron* lectin (XCL), *Sclerotium rolfsii* lectin (SRL), *Agaricus bisporus* lectin (ABL), *Boletus edulis* lectin	α/β-Sandwich (actinoporin-like)	Tetramer, dimer	•	•		–	PF07367 Fungal fruit body lectin (17)	IPR009960 Fungal fruit body lectin	6A
32	CV-N	Cyanovirin-N	Two homologous motifs with 3-stranded β-sheet and β-hairpins	Monomer	•	•	•	–	PF08881 CVNH domain (CyanoVirin-N homology domain) (35)	IPR011058 Cyanovirin-N	6B
33	PVL-like	*Psathyrella velutina* lectin (PVL)	7-Bladed β-propeller	Monomer		•		CL0186 β-Propeller clan	PF13517 Repeat domain in *Vibrio*, *Colwellia*, *Bradyrhizobium*, and *Shewanella* (8)	–	6C

(continued)

Table 1
(continued)

No	Family	Lectin component	Fold	Assembly	Distribution: Animal	Plant	Fungi	Bacteria	Virus	Pfam clan[a]	Pfam family[a,b] Number of PDB entries	Interpro[c]	CAZy[d]	Chap. Fig
34	AAL-like	Aleuria aurantia lectin (AAL), *Ralstonia solanacearum* lectin (RSL), *Burkholderia ambifaria* lectin	6-Bladed β-propeller	Monomer			•	•		–	PF07938 Fungal fucose-specific lectin (17)	IPR012475 Fungal fucose-specific lectin		6D
35	Flocculins	Flocculins	PA14 β-sandwich	Monomer			•	•		CL0301 PA14 superfamily	PF07691 PA14 domain (44)	IPR011658 PA14		6E
36	PCL-like[i]	*Pleurotus cornucopiae* lectin (PCL)	β-Sandwich (jellyroll) with Ca ion	Tandem repeat			•			CL0202 Galactose-binding domain-like superfamily	PF08531 α-L-rhamnosidase N-terminal domain (3)	IPR013737 Bacterial α-L-rhamnosidase N-terminal	CBM67	6F
37	BC2LCN[j]	*Burkholderia cenocepacia* lectin BC2L-C N-terminal domain	TNFa β-sandwich (jellyroll)	Trimer				•		(CL0100 C1q and TNF superfamily)	(PF00229 TNF (tumour necrosis factor) family (86))	IPR006052 Tumour necrosis factor		7A
38	Staphylococcal toxin	Staphylococcal enterotoxin B	5-Stranded β-barrel OB-fold and β-grasp	Monomer				•		–	PF02876 Staphylococcal/ Streptococcal toxin, β-grasp domain (115)	IPR006173 Staphylococcal/ Streptococcal toxin, OB-fold		7B

39	a	AB5 toxin[k] (Cholera toxin B)	Cholera toxin B subunit, citrobacter toxin B subunit	AB5 toxin α/β OB-fold	AB5	•		–	PF01376 Heat-labile enterotoxin beta chain (50)	IPR001835 Heat-labile enterotoxin, B chain	7C
	b	AB5 toxin (ADP-ribosylating toxin, shiga-like toxin B)	Shiga-like toxin I B subunit	AB5 toxin α/β OB-fold	AB5	•	•	–	PF02258 Shiga-like toxin β subunit (20)	IPR003189 Shiga-like toxin, β subunit	
	c	AB5 toxin (*Bordetella pertussis* exotoxin B)	*Bordetella pertussis* exotoxin Subunit 2 and 3 C-terminal domain	C-type α/β-fold with AB5 toxin α/β OB fold	AB5	•		–	PF02918 Pertussis toxin, subunit 2 and 3, C-terminal domain (4)	IPR020063 *Bordetella pertussis* toxin B, subunit 2/3, C-terminal	
	d	AB5 toxin (LT-IIb-B)	*Escherichia coli* type II heat-labile enterotoxin (LT-IIb-B)	AB5 toxin α/β OB-fold	AB5	•		–	PF06453 Type II heat-labile enterotoxin , B subunit (LT-IIB) (6)	IPR010503 Type II heat-labile enterotoxin, B subunit	
40		PA-IIL-like (fucolectin)	RS-IIL, PA-IIL , BC2L-A, CV-IIL	2-Ca 9-stranded antiparallel β-sandwich	Dimer	•		–	PF07472 Fucose-binding lectin II (PA-IIL) (32)	IPR010907 Calcium-mediated lectin	7D
41		MVL	*Microcystis viridis* protein MVL	MVL α/β-fold	Dimer	•		–	PF12151 Mannan-binding protein (2)	IPR021992 Mannan-binding protein	7E
42		PapG (P-pili adhesin)	P pilus adhesin PapG	Immunoglobulin-like β-sandwich divided into 2-domain	Linked to different domain	•		CL0204 Bacterial adhesin superfamily	PF03627 PapG carbohydrate binding domain (2)	IPR005310 PapG, carbohydrate-binding	7F

(continued)

Table 1
(continued)

No	Family	Lectin component	Fold	Assembly	Distribution					Pfam clan[a]	Pfam family[a,b] Number of PDB entries	Interpro[c]	CAZy[d]	Chap. Fig
					Animal	Plant	Fungi	Bacteria	Virus					
43	FimH	FimH	11-Stranded β-sandwich with an immunoglobulin fold	Linked to different domain				•		(CL0204 Bacterial adhesin superfamily)	PF09160 FimH, mannose binding (26)	IPR015243 FimH, mannose-binding		8A
44	F17-G	Fimbrial adhesin F17-G	Immunoglobulin-like β-sandwich	Linked to different domain				•		(CL0204 Bacterial adhesin superfamily)	PF09222 Fimbrial adhesin F17-AG, lectin domain (14)	IPR015303 Fimbrial adhesin F17-AG, lectin domain		8B
45	Hemagglutinin	Influenza virus hemagglutinin, measles virus hemagglutinin	Hemagglutinin β-sandwich (jellyroll)	Trimer of 2-chain fold					•	–	PF00509 Haemagglutinin (255)	IPR001364 Haemagglutinin, influenzavirus A/B		8C
46	RotavirusVP4	Rotavirus VP4 VP8-domain,	Concanavalin A-like β-sandwich (jellyroll)	Virus capsid					•	–	PF00426 Outer capsid protein VP4 (Hemagglutinin) (19)	IPR000416 Haemagglutinin outer capsid protein VP4		8D

47	a	Viral protein (picornavirusl VP1)	Equine rhinitis A virus VP1, foot-and-mouth disease virus VP1	Viral 8-stranded β-sandwich	Virus capsid		•	CL0055 Positive stranded ssRNA viruses coat protein	PF00073 Picornavirus capsid protein (162)	IPR001676 Picornavirus capsid		8E
	b	Viral protein (Norovirus P-domain)	Norovirus P-domain	EF-Tu2-like β-sandwich	Virus capsid		•	CL0055 Positive stranded ssRNA viruses coat protein	PF00915 Calicivirus coat protein (200)	IPR004005 Calicivirus coat protein/ nonstructural polyprotein		
	c	Viral protein (polyomavirus VP1)	Polyomavirus VP1, Simian virus 40 VP1	Viral 8-stranded β-sandwich	Virus capsid		•	–	PF00718 Polyomavirus coat protein (29)	IPR000662 Capsid protein VP1, Polyomavirus		
48		Knob domain	Adenovirus knob domain	Viral 8-stranded β-sandwich (jellyroll)	Virus capsid		•	–	PF00541 Adenoviral fiber protein (knob domain) (36)	IPR000978 Adenoviral fiber protein, knob		8F

[a]http://pfam.sanger.ac.uk/

[b]Number in the parentheses indicates the number of PDB entries related to the Pfam family as of June 2014

[c]http://www.ebi.ac.uk/interpro/

[d]http://www.cazy.org/

[e]Structural information is available only for several α-1,2-mannosidases [11]

[f]F-type lectin family is originally designated for fucose-binding lectins such as *Anguilla anguilla* agglutinin (AAA) and striped bass lectin FBP32 [12, 13]

[g]Pfam family for tachylectin-2 has not been assigned [14], while Interpro has made a domain class "Tachylectin 2"

[h]H-type lectin family and Pfam family PF09458 are named after *Helix pomatia* agglutinin [15]

[i]Concerning PCL-like lectin, structure of *Pleurotus cornucopiae* lectin (PCL) is not available, but the crystal structure of its homologous protein, CBM67 rhamnose-binding domain from *Streptomyces avermitilis* α-L-rhamnosidase, has been determined [16]

[j]For family BC2LCN, structure of *Burkholderia cenocepacia* lectin BC2L-C N-terminal domain has been determined [17], but the Pfam family has not been assigned the structure has high similarity to the members of TNF, tumor necrosis factor, Pfam family PF00229

[k]Subtilase cytotoxin subunit B has a AB5 toxin α/β OB-fold structure, but Pfam family has not been assigned, and may be included in AB5 toxin family

Kocourek and Horejsi [9] designated lectins as proteins of non-immunoglobulin nature capable of specific recognition and reversible binding to carbohydrate moieties of complex carbohydrates without altering the covalent structure of any of the recognized glycosyl ligands. On the other hand, a lot of proteins that alter the covalent structure of the recognized glycosyl ligands, namely enzymes, have been reported. Some carbohydrate-active enzymes possess a modular structure, and the catalytic modules are occasionally attached to one or more noncatalytic lectin-like domains, designated as carbohydrate-binding modules (CBMs), for efficient catalysis. These enzymes and CBMs are well summarized in the CAZy database, http://www.cazy.org/; *see* Ref. 10). Some lectin families share the same protein folding with CBM or enzyme families. For example, the R-type lectin domain having a β-trefoil fold is shared in xylanases as a xylan-binding domain and is recognized as CBM family 13 [11]. The related CAZy families are listed in Table 1.

Ribbon drawings of the lectin structures are shown in Figs. 1, 2, 3, 4, 5, 6, 7, and 8. Figures are in order of the lectin families listed in Table 1. Some lectin structures are shown as a single domain and some as an assembly case by case. When a figure shows the assembled structure, one peptide chain capable of ligand binding is highlighted in dark grey. For the representatives, structures complexed with carbohydrate or a carbohydrate analogue and solved at the earlier stage are preferentially shown. These structural figures may help illustrate the diversity of lectin families, showing similarities and differences between lectin families, as well as the carbohydrate-binding architectures of lectin.

3 Topology of the β-Structures Observed in Lectins

β-Structures are majority in the structures of lectin, but they are topologically differentiated (Figs. 9 and 10). The most frequent fold in lectin families is the β-sandwich. β-sandwich contains several types of fold, including the Greek key, jellyroll, or immunoglobulin-like fold, and it basically comprises two antiparallel β-sheets, forming a globular structure. The Greek key is a motif where a β-hairpin turns into a four-stranded antiparallel β-sheet (Fig. 9a). The Greek key barrel is a fold consisting of two tandemly repeated Greek key motifs, whereas the jellyroll is a motif in which a β-hairpin makes three turns wrapping into a barrel shape (Fig. 9c, d). The immunoglobulin-like fold basically has three- and four-stranded antiparallel β-sheets folding into an oval structure (Fig. 9e, f).

The β-sandwich fold is occasionally included in a broad definition of β-barrel structure, but in a narrow sense, β-barrel consists of a single β-sheet, in which sequential β-strands are in contact with each other in order of appearance and the last strand contacts the

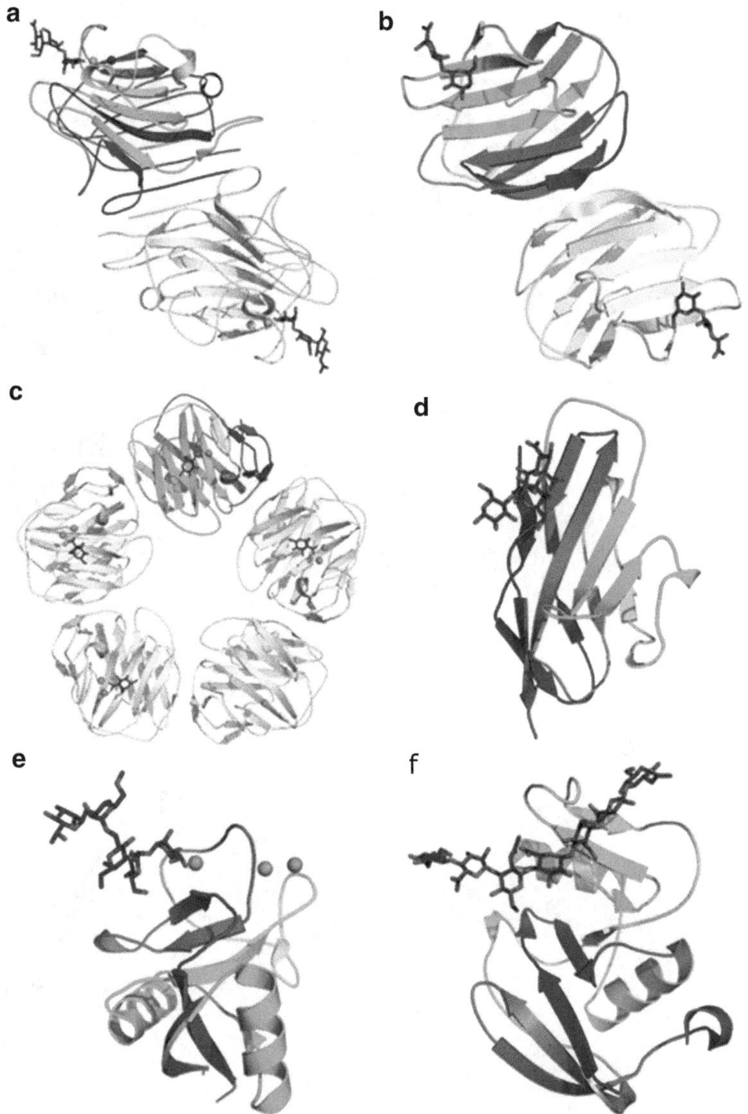

Fig. 1 Ribbon drawings of lectins from families 1–6 in Table 1. (**a**), 1 L-type. Crystal structure of the *Lathyrus ochrus* legume isolectin I complexed with *N*-acetyllactosamine-type trisaccharide (PDB code 1LOG; [19]). *Spheres* show calcium and manganese ions. Shown as a homodimer. (**b**), 2 Galectin. Crystal structure of the bovine galectin-1 complexed with *N*-acetyllactosamine (PDB code 1SLT; [20]). Shown as a homodimer. (**c**), 3 Pentaxin. Crystal structure of the human serum amyloid P component complexed with the 4,6-pyruvate acetal of β-D-galactose (PDB code 1GYK; [21]). *Spheres* show calcium ions. Shown as a homopentamer. (**d**), 4 I-type. Crystal structure of the human sialoadhesin N-terminal (V-set) domain complexed with 3′-sialyllactose (PDB code 1QFO and 1QFP; [22]). (**e**), 5 C-type. Crystal structure of the rat C-type mannose-binding protein complexed with manno-oligosaccharide (PDB code 2MSB; [23]). *Spheres* show calcium ions. (**f**), 6 Hyaladherin. Crystal structure of the cell surface receptor CD44 hyaluronan-binding domain complexed with hyaluronan (PDB code 2JCR; [24])

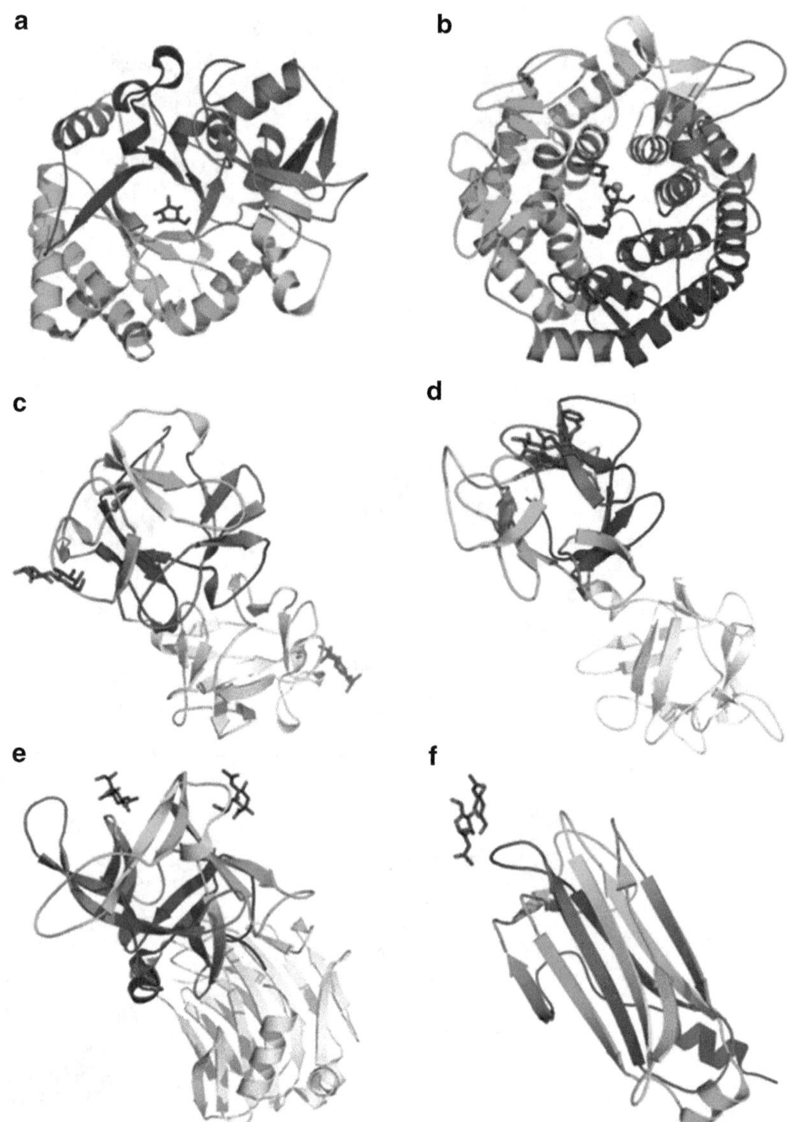

Fig. 2 Ribbon drawings of lectins from families 7–12 in Table 1. (**a**), Crystal structure of the mouse lectin Ym1 complexed with D-glucosamine (PDB code 1E9L; [25]). (**b**), Crystal structure of the human class I α-mannosidase complexed with mannosyl thiodisaccharide substrate analogue (PDB code 1X9D; [26]). *Sphere* shows the calcium ion. (**c**), Crystal structure of *Ricinus communis* plant toxin ricin B-chain complexed with lactose (PDB code 2AAI; [27]). Shown only for B-chain. Two β-trefoil domains are tandemly arranged. (**d**), Crystal structure of the *Amaranthus caudatus* agglutinin (ACA) complexed with T-antigen disaccharide analogue (PDB code 1JLX; [28]). Two β-trefoil domains are tandemly arranged. (**e**), Crystal structure of the botulinum neurotoxin C complexed with cell binding fragment complexed with sialic acid (PDB code 3R4S; [29]). β-Jerryroll and β-trefoil domains are tandemly arranged. (**f**), Crystal structure of the mouse ubiquitin ligase sugar-binding domain complexed with chitobiose (PDB code 1UMI; [30])

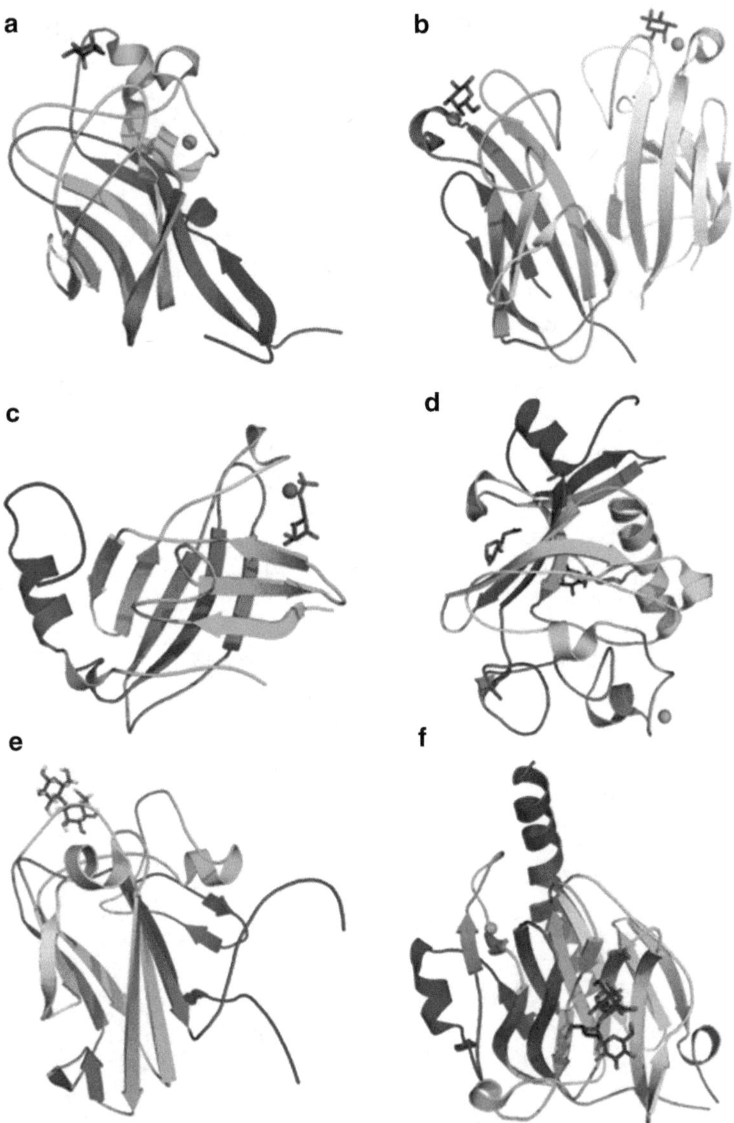

Fig. 3 Ribbon drawings of lectins from families 13–18 in Table 1. (**a**), Crystal structure of the *Anguilla anguilla* agglutinin (AAA), European eel lectin complexed with fucose (PDB code 1K12; [13]). *Sphere* shows the calcium ion. (**b**), Crystal structure of the *Pseudomonas aeruginosa* lectin I (PA-IL) complexed with galactose (PDB code 1OKO; [31]). Shown as a homodimer. (**c**), Crystal structure of the bovine cation-dependent mannose 6-phosphate receptor complexed with mannose 6-phosphate (PDB code 1M6P; [32]). *Sphere* shows the manganese ion. (**d**), Crystal structure of the human L-ficolin complexed with *N*-acetyl-D-glucosamine (PDB code 2J30; [33]). *Sphere* shows the calcium ion. (**e**), Solution structure of the *Xenopus* malectin complexed with maltose (PDB code 2KR2; [34]). (**f**), Crystal structure of the mouse calreticulin lectin domain complexed with Glc_1Man_3 tetrasaccharide (PDB code 3O0W; [35]). *Sphere* shows the calcium ion

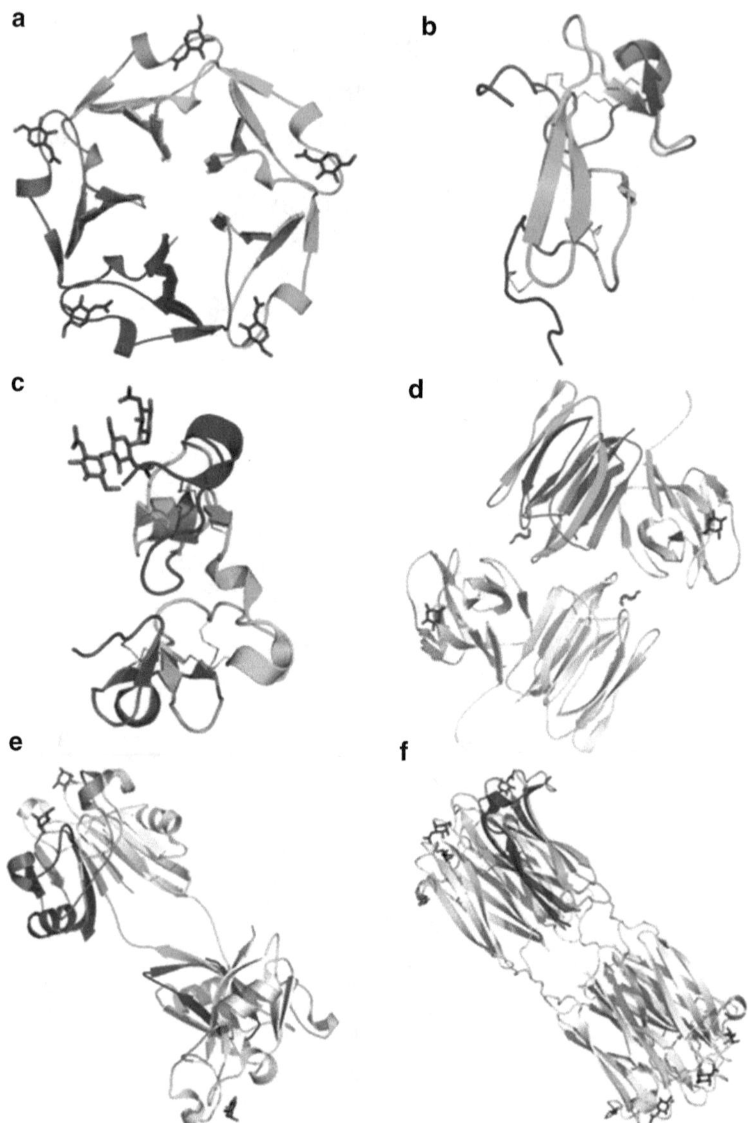

Fig. 4 Ribbon drawings of lectins from families 19–24 in Table 1. (**a**), Crystal structure of the Japanese horseshoe crab *Tachypleus tridentatus* Tachylectin-2 complexed with *N*-acetyl-D-glucosamine (PDB code 1TL2; [15]). (**b**), Solution structure of horseshoe crab Tachycitin (PDB code 1DQC; [36]). *Sphere* shows the calcium ion. (**c**), Crystal structure of the *Urtica dioica* agglutinin complexed with tri-*N*-acetylchitotriose (PDB code 1EHH; [37]). Disulfide bonds are shown as *lines*. (**d**), Crystal structure of the Jacalin, a tetrameric two-chain lectin from jackfruit seeds, complexed with methyl-α-D-galactose (PDB code 1JAC; [38]). (**e**), Solution structure of the salmon L-rhamnose-binding lectin CSL3 complexed with L-rhamnose (PDB code 2ZX2; [39]). Shown as a swapped homodimer. (**f**), Crystal structure of *Helix pomatia* agglutinin complexed with *N*-acetylgalactosamine (PDB code 3OOW; [16]). Interdomain disulfide bonds are shown as *lines*. Shown as a homohexamer

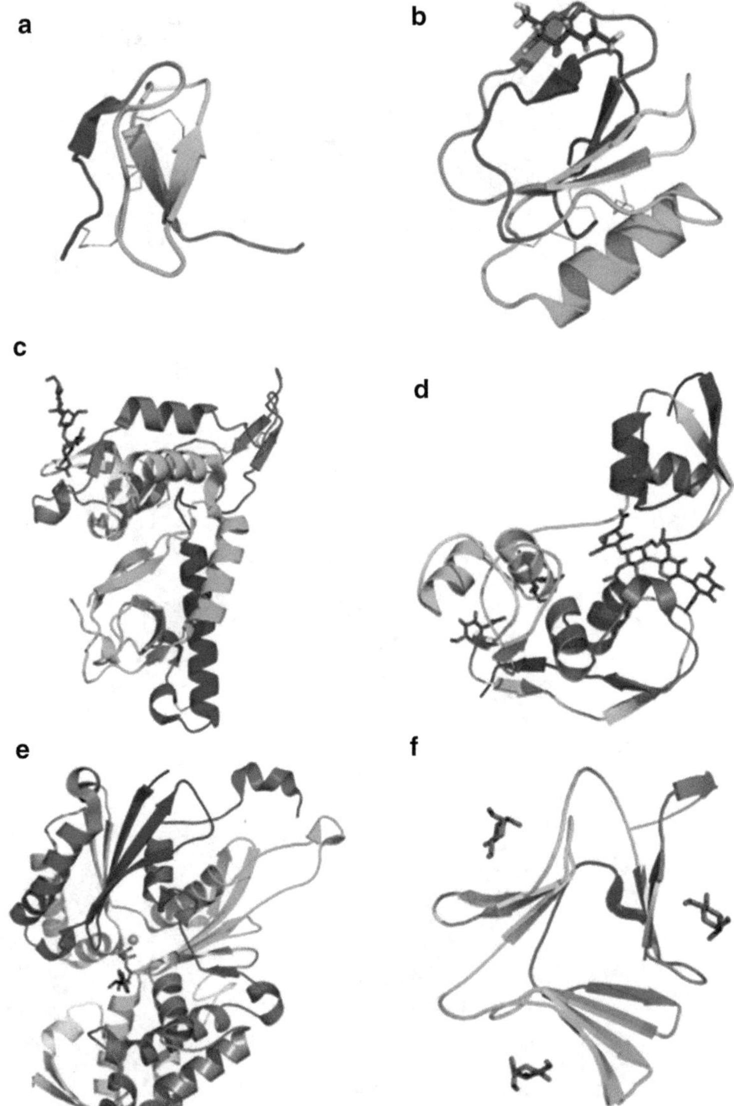

Fig. 5 Ribbon drawings of lectins from families 25–30 in Table 1. (**a**), Solution structure of Chinese bird spider lectin-I (SHL-I) (PDB code 2QK7; [40]). (**b**), Solution structure of *Toxosplasma gondii* TgMIC4 fifth PAN-domain complexed with lacto-*N*-biose (PDB code 2LL4; [41]). (**c**), Crystal structure of the *Toxoplasma gondii* adhesive protein TgMIC1 micronemal adhesive repeat-region in complex with sialyl-*N*-acetyllactosamine (PDB code 3F5A; [42]). Disulfide bonds are shown as *lines*. (**d**), Crystal structure of the fungal effector Ecp6 complexed with *N*-acetyl-D-glucosamine tetrasaccharide (PDB code 4B8V; [43]). (**e**), Crystal structure of the rat nucleoside triphosphate diphosphohydrolase (NTPDase) complexed with AMPPNP (PDB code 3CJA; [44]). *Sphere* shows the calcium ion. (**f**), Crystal structure of the snowdrop *Galanthus nivalis* lectin complexed with methylmannose (PDB code 1MSA; [45]*)*. Shown as a monomeric β-prism II domain

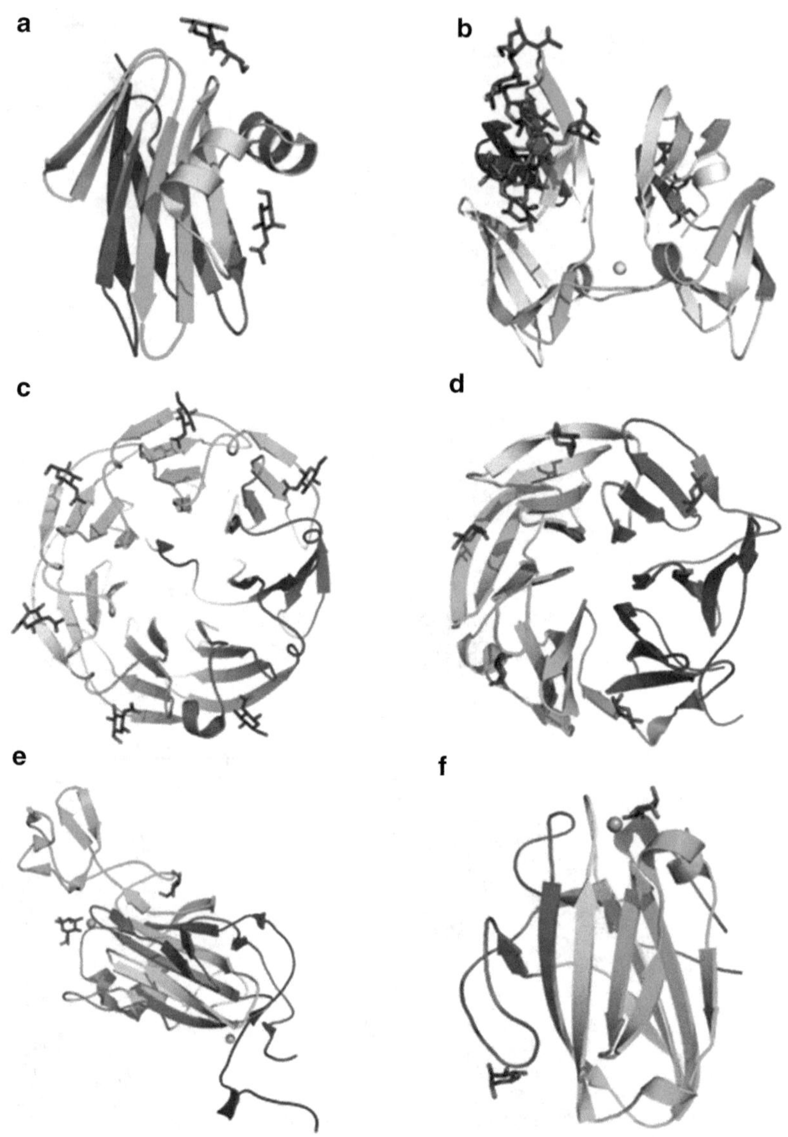

Fig. 6 Ribbon drawings of lectins from families 31–36 in Table 1. (**a**), Crystal structure of the mushroom *Agaricus bisporus* lectin complex with lacto-*N*-biose (PDB code 1Y2U; [46]). (**b**), Crystal structure of the cyanobacterium (blue-green algae) *Nostoc ellipsosporum* cyanovirin-N complexed with high mannose oligosaccharides (PDB code 3GXZ; [47]). *Sphere* shows the magnesium ion. Shown as a homodimer. (**c**), Crystal structure of the mushroom *Psathyrella velutina* lectin (PVL) complexed with *N*-acetylglucosamine (PDB code 2C4D; [48]). *Sphere* shows the calcium ion. (**d**) Crystal structure of the fungal *Aleuria aurantia* lectin (AAL) complexed with fucose (PDB code 1OFZ; [49]). (**e**), Crystal structure of the *Saccharomyces cerevisiae* flocculin complexed with mannose (PDB code 2XJP; [50]). *Spheres* show calcium ions. (**f**), Crystal structure of the rhamnose-binding domain of *Streptomyces avermitilis* α-L-rhamnosidase complexed with L-rhamnose (PDB code 3W5N; [16]). *Spheres* show calcium ions

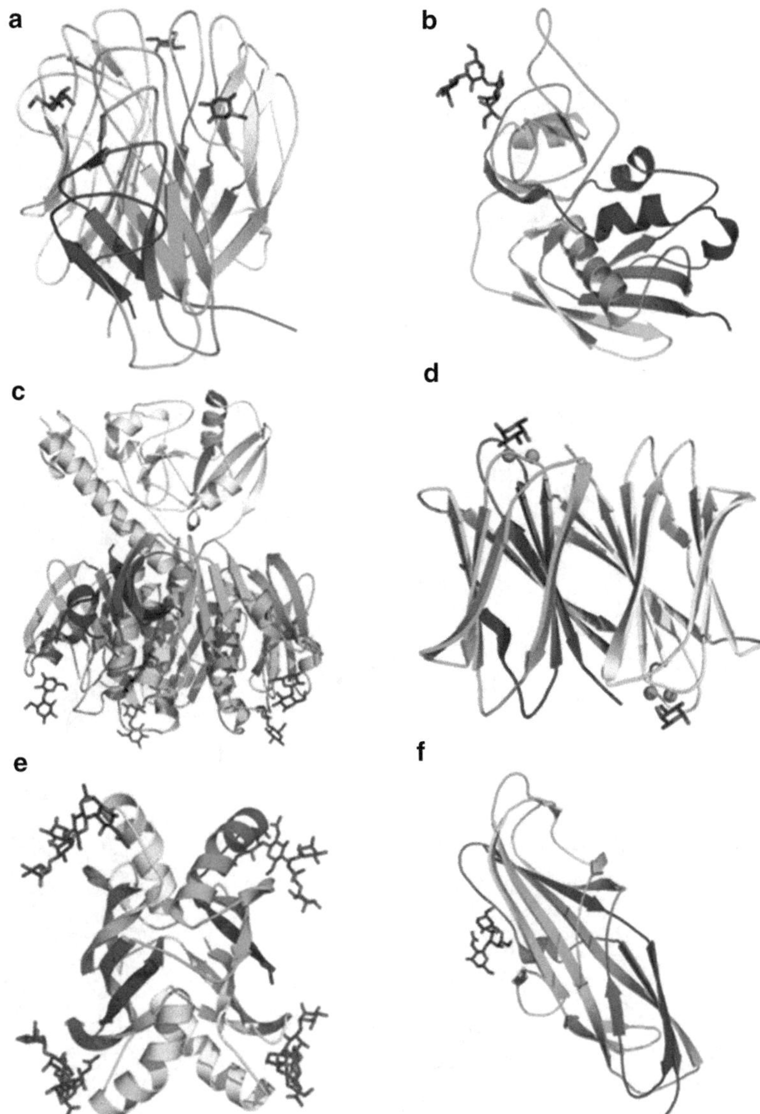

Fig. 7 Ribbon drawings of lectins from families 37–42 in Table 1. (**a**), Crystal structure of the N-terminal domain of opportunistic pathogen *Burkholderia cenocepacia* lectin BC2L-C complexed with methylseleno-α-L-fucopyranoside (PDB code 2WQ4; [18]). (**b**), Crystal structure of the *staphylococcal* enterotoxin B complexed with 3′-sialyllactose (PDB code 1SE3; [51]). (**c**), Crystal structure of the *Escherichia coli* heat-labile enterotoxin complexed with lactose (PDB code 1LTT; [52]). Shown as an $\alpha\beta_5$ complex. (**d**), Crystal structure of the *Pseudomonas aeruginosa* fucose-binding lectin (PA-IIL) complexed with fucose (PDB code 1GZT; [53]). Shown as a homodimer. (**e**), Crystal structure of the cyanobacterium *Microcystis viridis* protein MVL complexed with pentasaccharide $Man_3GlcNAc_2$ (PDB code 1ZHS; [54]). Shown as a homodimer of a two-domain chain. (**f**), Crystal structure of the receptor-binding N-domain of pyelonephritic *Escherichia coli* adhesin (PapG) complexed with globoside (GbO4) tetrasaccharide (PDB code 1J8R; [55])

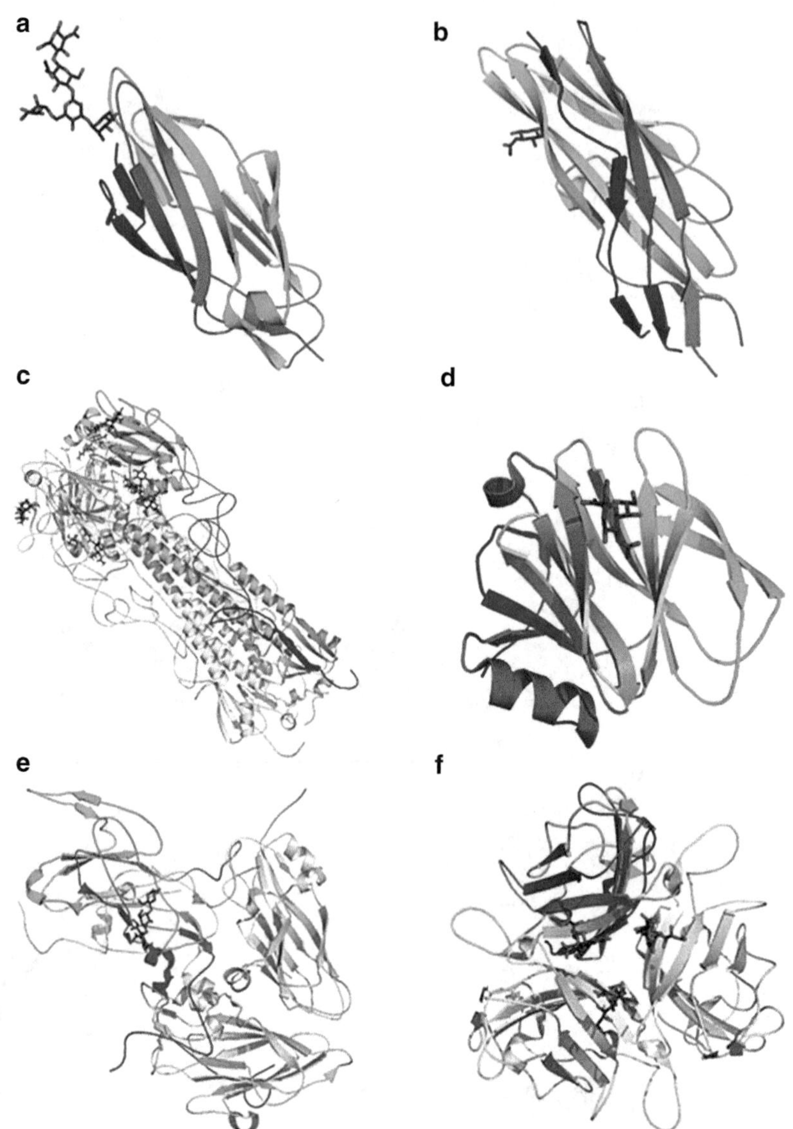

Fig. 8 Ribbon drawings of lectins from families 43–48 in Table 1. (**a**), Crystal structure of the receptor-binding domain of *Escherichia coli* FimH complexed with oligomannose-3 (PDB code 2VCO; [56]). (**b**), Crystal structure of the enterotoxigenic *Escherichia coli* F17-G adhesin complexed with *N*-acetyl-D-glucosamine (PDB code 1O9W; [57]). (**c**), Crystal structure of the influenza virus hemagglutinin complexed with sialyllactose (PDB code 1HGG; [58]). (**d**), Crystal structure of the VP8 N-terminal fragment of rotavirus outer capsid spike protein VP4 complexed with sialoside (PDB code 1KQR; [59]). (**e**), Crystal structure of the equine rhinitis A virus outer capsid protein VP1 complexed with 3′-sialyllactose (PDB code 2XBO; [60]). (**f**), Crystal structure of the Adenovirus serotype 37 knob domain complexed with 3′-sialyllactose (PDB code 1UXA; [61])

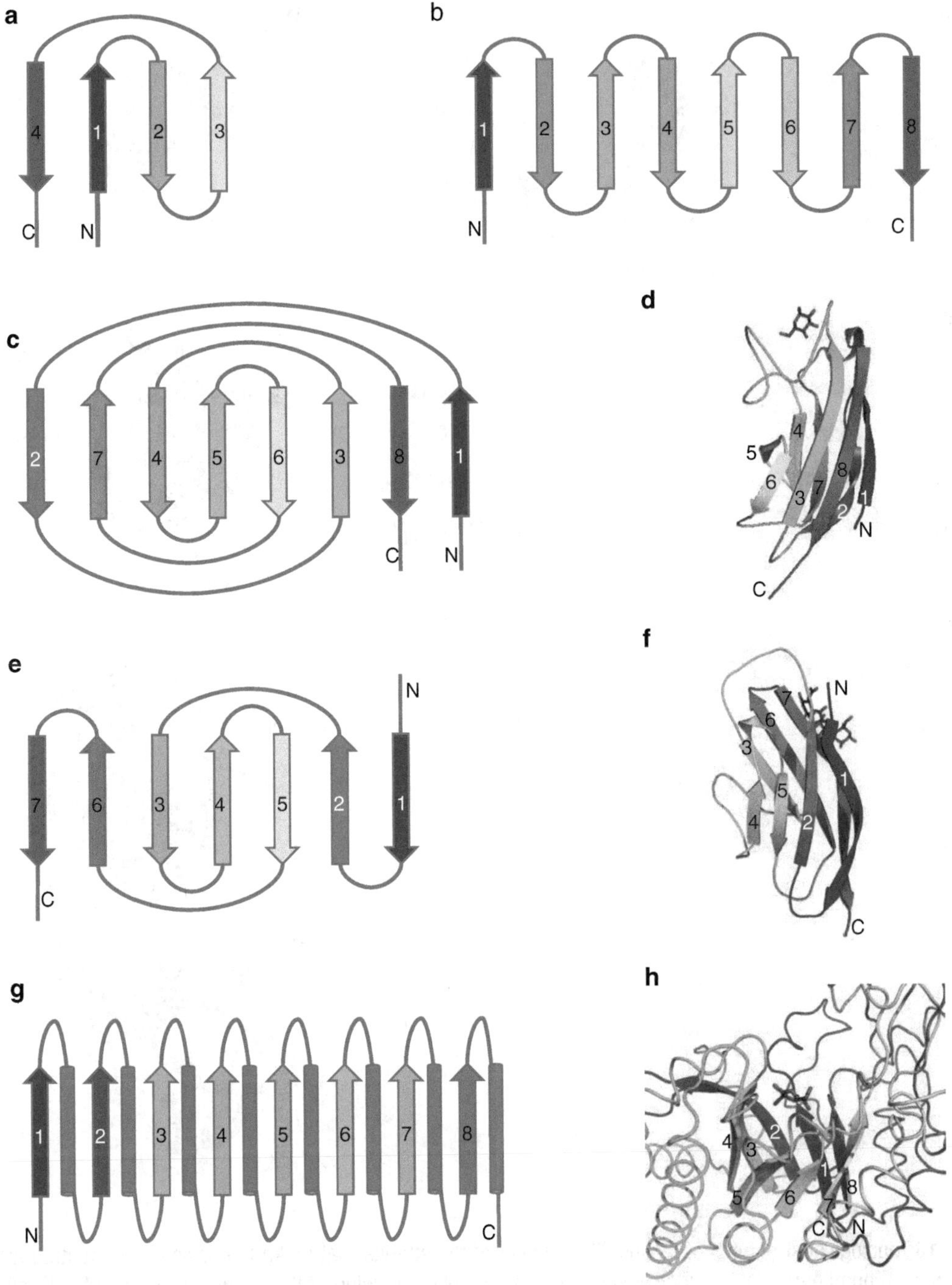

Fig. 9 Topological structures of lectins. The number of the strands that make the basic motif structure are shown and numbered. (**a**), Topological diagram of a Greek key motif. (**b**), Topological diagram of an eight-stranded up-and-down β-barrel fold. (**c**), Topological diagram and (**d**), ribbon drawing of the β-jerryroll fold for the *Anguilla anguilla* agglutinin (AAA) complexed with fucose (PDB code 1K12; [13]). (**e**), Topological diagram and (**f**), ribbon drawing of the immunoglobin-like β-sandwich fold for the human sialoadhesin N-terminal (V-set) domain complexed with 3′-sialyllactose (PDB code 1QFO and 1QFP; [22]). (**g**), Topological diagram and (**h**), ribbon drawing of the $(\beta/\alpha)_8$-barrel fold for the mouse lectin Ym1 complexed with D-glucosamine (PDB code 1E9L; [25])

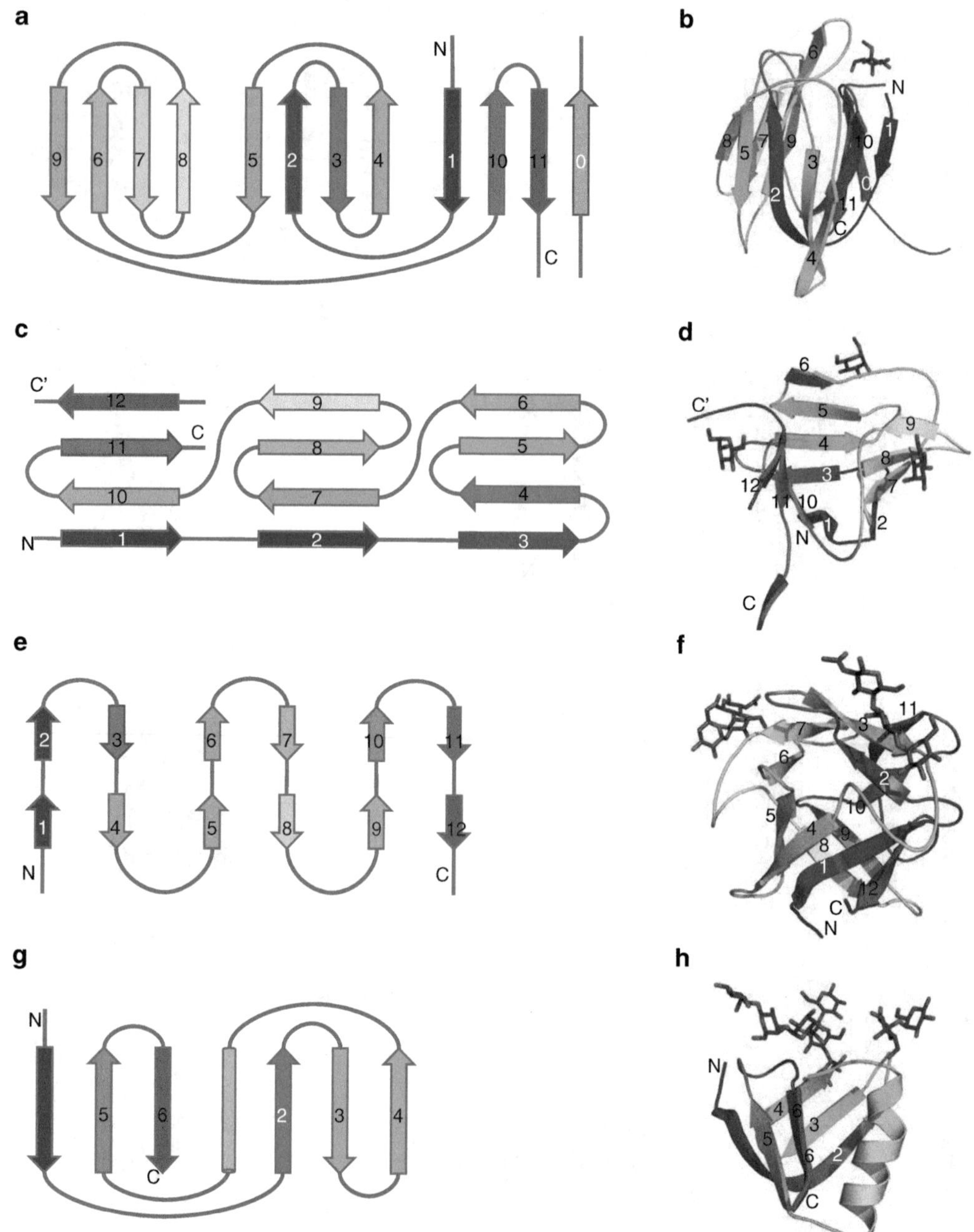

Fig. 10 Topological structures of lectins. The number of the strands that make the basic motif structure are shown and numbered. (**a**), Topological diagram and (**b**), ribbon drawing of the β-prism type-I fold for Jacalin complexed with methyl-α-D-galactose (PDB code 1JAC; [38]). (**c**), Topological diagram and (**d**), ribbon drawing of the β-prism type-II fold for snowdrop *Galanthus nivalis* lectin complexed with methylmannose (PDB code 1MSA; [45]). (**e**), Topological diagram and (**f**), ribbon drawing of the β-trefoil fold for the *Marasmius oreades* mushroom lectin in complex with trisaccharide Gal-β-1,3-Gal-β-1,4-GlcNAc. (PDB code 2IHO; [62]). (**g**), Topological diagram and (**h**), ribbon drawing of the OB-fold for the Shiga-like toxin I β-subunit complexed with its receptor Gb3 analogue (PDB code 1BOS; [52])

first strand, making a cylindrical structure. β-barrel is often observed as an antiparallel or up-and-down β-barrel (Fig. 9b), and a parallel β-barrel usually exists in a different structural fold, such as a $(\beta/\alpha)_8$-barrel that contains β_8-barrel at the center (Figs. 9g, h). Immunoglobulin-like and jellyroll folds are distinguished from β-barrel because the strands are not necessarily aligned in order.

The β-prism structure comprises three-repeated β-sheets, where three sheets are gathered around a pseudo-threefold axis, resulting in a triangular prism structure. β-Prism type I, observed in the lectins Jacalin and griffithsin, contains Greek key motifs for the sheet (Fig. 10a, b), whereas β-prism type II, observed in monocot lectins, has simple antiparallel β-sheets (Fig. 10c, d). A β-trefoil fold, observed in R-type lectins and *Amaranthus caudatus* agglutinin, contains three four-stranded Y-shaped β-hairpins, but the first and last β-strands (β1 and β4) of the three β-hairpins join to form a six-stranded β-barrel on one side of the protein, whereas the other two β-strands (β2 and β3) are gathered on the opposite side of the protein, folding into a globular structure (Fig. 10e, f).

The OB-fold (oligonucleotide/oligosaccharide-binding fold) is a type of β-sandwich with two three-stranded antiparallel β-sheets connecting one α-helix (Fig. 10g, h). It is observed in the β-subunits of AB-toxin. It is sometimes called five-stranded β-barrel, because the first and second β-strands appear to be one strand.

Lectins often share similar protein fold structures but their sugar-binding specificity is often determined by a few amino acid residues in the sugar-binding site. Therefore, modification of a pre-existing lectin by protein engineering is expected to be an effective way of generating new lectin ligand specificity. The structural information of the lectin–sugar complex will serve as a platform for the molecular engineering of lectin.

References

1. Finn RD, Mistry J, Tate J et al (2010) The Pfam protein families database. Nucleic Acids Res 38:D211–D222
2. Hunter S, Jones P, Mitchell A et al (2012) InterPro in 2011: new developments in the family and domain prediction database. Nucleic Acids Res 40:D306–D312
3. Lee JK, Baum LG, Moremen K, Pierce M (2004) The X-lectins: a new family with homology to the *Xenopus laevis* oocyte lectin XL-35. Glycoconj J 21:443–450
4. Koharudin LM, Gronenborn AM (2011) Structural basis of the anti-HIV activity of the cyanobacterial *Oscillatoria Agardhii* agglutinin. Structure 19:1170–1181
5. Koharudin LM, Kollipara S, Aiken C, Gronenborn AM (2012) Structural insights into the anti-HIV activity of the *Oscillatoria agardhii* agglutinin homolog lectin family. J Biol Chem 287:33796–33811
6. Moustafa I, Connaris H, Taylor M et al (2004) Sialic acid recognition by *Vibrio cholerae* neuraminidase. J Biol Chem 279:40819–40826
7. Lee JO, Rieu P, Arnaout MA, Liddington R (1995) Crystal structure of the A domain from the α subunit of integrin CR3 (CD11b/CD18). Cell 80:631–638
8. Paaventhan P, Joseph JS, Seow SV et al (2003) A 1.7A structure of Fve, a member of the new fungal immunomodulatory protein family. J Mol Biol 332:461–470
9. Kocourek J, Horejsi V (1983) Note on the recent discussion on definition of the term "lectin". In: Bog-Hansen TC, Spengler GA

(eds) Lectins: Biology, Biochemistry and Clinical Biochemistry, vol 3. Walter de Gruyter, Berlin and New York, pp 3–6

10. Cantarel BL, Coutinho PM, Rancurel C et al (2009) The Carbohydrate-Active EnZymes database (CAZy): an expert resource for Glycogenomics. Nucleic Acids Res 37: D233–238
11. Fujimoto Z, Kuno A, Kaneko S et al (2000) Crystal structure of *Streptomyces olivaceoviridis* E-86 β-xylanase containing xylan-binding domain. J Mol Biol 300:575–585
12. Vallee F, Lipari F, Yip P et al (2000) Crystal structure of a class I α1,2-mannosidase involved in *N*-glycan processing and endoplasmic reticulum quality control. EMBO J 19:581–588
13. Bianchet MA, Odom EW, Vasta GR, Amzel LM (2002) A novel fucose recognition fold involved in innate immunity. Nat Struct Biol 9:628–634
14. Odom EW, Vasta GR (2006) Characterization of a binary tandem domain F-type lectin from striped bass (*Morone saxatilis*). J Biol Chem 281:1698–1713
15. Beisel HG, Kawabata S, Iwanaga S, Huber R, Bode W (1999) Tachylectin-2: crystal structure of a specific GlcNAc/GalNAc-binding lectin involved in the innate immunity host defense of the Japanese horseshoe crab *Tachypleus tridentatus*. EMBO J 18:2313–2322
16. Sanchez JF, Lescar J, Chazalet V et al (2006) Biochemical and structural analysis of *Helix pomatia* agglutinin. A hexameric lectin with a novel fold J Biol Chem 281:20171–20180
17. Fujimoto Z, Jackson A, Michikawa M et al (2013) The structure of a *Streptomyces avermitilis* α-L-rhamnosidase reveals a novel carbohydrate-binding module CBM67 within the six-domain arrangement. J Biol Chem 288: 12376–12385
18. Sulák O, Cioci G, Delia M et al (2010) A TNF-like trimeric lectin domain from *Burkholderia cenocepacia* with specificity for fucosylated human histo-blood group antigens. Structure 18:59–72
19. Bourne Y, Rouge P, Cambillau C (1990) X-ray structure of a (α-Man(1-3)β-Man(1-4) GlcNAc)-lectin complex at 2.1-Å resolution. The role of water in sugar-lectin interaction. J Biol Chem 265:18161–18165
20. Liao D-I, Kapadia G, Ahmed H, Vasta GR, Herzberg O (1994) Structure of S-lectin, a developmentally regulated vertebrate β-galactoside-binding protein. Proc Natl Acad Sci U S A 91:1428–1432
21. Thompson D, Pepys MB, Tickle I, Wood S (2002) The structures of crystalline complexes of human serum amyloid P component with its carbohydrate ligand, the cyclic pyruvate acetal of galactose. J Mol Biol 320:1081–1086
22. May AP, Robinson RC, Vinson M, Crocker PR, Jones EY (1998) Crystal structure of the N-terminal domain of sialoadhesin in complex with 3′ sialyllactose at 1.85 Å resolution. Mol Cell 1:719–728
23. Weis WI, Drickamer K, Hendrickson WA (1992) Structure of a C-type mannose-binding protein complexed with an oligosaccharide. Nature 360:127–134
24. Banerji S, Wright AJ, Noble M et al (2007) Structures of the Cd44-hyaluronan complex provide insight into a fundamental carbohydrate-protein interaction. Nat Struct Mol Biol 14:234–239
25. Sun YJ, Chang NC, Hung SI et al (2001) The crystal structure of a novel mammalian lectin, Ym1, suggests a saccharide binding site. J Biol Chem 276:17507–17514
26. Karaveg K, Siriwardena A, Tempel W et al (2005) Mechanism of class 1 (glycosylhydrolase family 47) α-mannosidases involved in *N*-glycan processing and endoplasmic reticulum quality control. J Biol Chem 280:16197–16207
27. Rutenber E, Ready M, Robertus JD (1987) Structure and evolution of ricin B chain. Nature 326:624–626
28. Transue TR, Smith AK, Mo H, Goldstein IJ, Saper MA (1997) Structure of benzyl T-antigen disaccharide bound to *Amaranthus caudatus* agglutinin. Nat Struct Biol 4:779–783
29. Strotmeier J, Gu S, Jutzi S et al (2011) The biological activity of botulinum neurotoxin type C is dependent upon novel types of ganglioside binding sites. Mol Microbiol 81: 143–156
30. Mizushima T, Hirao T, Yoshida Y et al (2004) Structural basis of sugar-recognizing ubiquitin ligase. Nat Struct Mol Biol 11:365–370
31. Cioci G, Mitchell EP, Gautier C et al (2003) Structural basis of calcium and galactose recognition by the lectin PA-IL of *Pseudomonas aeruginosa*. FEBS Lett 555:297–301
32. Roberts DL, Weix DJ, Dahms NM, Kim JJ (1998) Molecular basis of lysosomal enzyme recognition: three-dimensional structure of the cation-dependent mannose 6-phosphate receptor. Cell 93:639–648
33. Garlatti V, Belloy N, Martin L et al (2007) Structural insights into the innate immune recognition specificities of L- and H-ficolins. EMBO J 26:623–633
34. Schallus T, Feher K, Sternberg U, Rybin V, Muhle-Goll C (2010) Analysis of the specific interactions between the lectin domain of

malectin and diglucosides. Glycobiology 20: 1010–1020

35. Kozlov G, Pocanschi CL, Rosenauer A et al (2010) Structural basis of carbohydrate recognition by calreticulin. J Biol Chem 285: 38612–38620
36. Suetake T, Tsuda S, Kawabata S et al (2000) Chitin-binding proteins in invertebrates and plants comprise a common chitin-binding structural motif. J Biol Chem 275:17929–17932
37. Harata K, Muraki M (2000) Crystal structures of *Urtica dioica* agglutinin and its complex with tri-*N*-acetylchitotriose. J Mol Biol 297: 673–681
38. Jeyaprakash AA, Srivastav A, Surolia A, Vijayan M (2004) Structural basis for the carbohydrate specificities of artocarpin: variation in the length of a loop as a strategy for generating ligand specificity. J Mol Biol 338:757–770
39. Shirai T, Watanabe Y, Lee MS, Ogawa T, Muramoto K (2009) Structure of rhamnose-binding lectin CSL3: unique pseudo-tetrameric architecture of a pattern recognition protein. J Mol Biol 391:390–403
40. Lü S, Liang S, Gu X (1999) Three-dimensional structure of *Selenocosmia huwena* lectin-I (SHL-I) from the venom of the spider *Selenocosmia huwena* by 2D-NMR. J Protein Chem 18:609–617
41. Marchant J, Cowper B, Liu Y et al (2012) Galactose recognition by the apicomplexan parasite *Toxoplasma gondii*. J Biol Chem 287: 16720–16733
42. Garnett JA, Liu Y, Leon E et al (2009) Detailed insights from microarray and crystallographic studies into carbohydrate recognition by microneme protein 1 (MIC1) of *Toxoplasma gondii*. Protein Sci 18:1935–1947
43. Sánchez-Vallet A, Saleem-Batcha R, Kombrink A et al (2013) Fungal effector Ecp6 outcompetes host immune receptor for chitin binding through intrachain LysM dimerization. Elife 2:e00790
44. Zebisch M, Strater N (2008) Structural insight into signal conversion and inactivation by NTPDase2 in purinergic signaling. Proc Natl Acad Sci U S A 105:6882–6887
45. Hester G, Kaku H, Goldstein IJ, Wright CS (1995) Structure of mannose-specific snowdrop (*Galanthus nivalis*) lectin is representative of a new plant lectin family. Nat Struct Biol 2:472–479
46. Carrizo ME, Capaldi S, Perduca M et al (2005) The antineoplastic lectin of the common edible mushroom (*Agaricus bisporus*) has two binding sites, each specific for a different configuration at a single epimeric hydroxyl. J Biol Chem 280: 10614–10623
47. Botos I, O'Keefe BR, Shenoy SR et al (2002) Structures of the complexes of a potent anti-HIV protein cyanovirin-N and high mannose oligosaccharides. J Biol Chem 277: 34336–34342
48. Cioci G, Mitchell EP, Chazalet V et al (2006) β-Propeller crystal structure of *Psathyrella velutina* lectin: an integrin-like fungal protein interacting with monosaccharides and calcium. J Mol Biol 357:1575–1591
49. Wimmerova M, Mitchell E, Sanchez JF, Gautier C, Imberty A (2003) Crystal structure of fungal lectin: six-bladed β-propeller fold and novel fucose recognition mode for *Aleuria aurantia* lectin. J Biol Chem 278:27059–27067
50. Veelders M, Bruckner S, Ott D et al (2010) Structural basis of flocculin-mediated social behavior in yeast. Proc Natl Acad Sci U S A 107:22511–22516
51. Swaminathan S, Furey W, Pletcher J, Sax M (1995) Residues defining V beta specificity in staphylococcal enterotoxins. Nat Struct Biol 2: 680–686
52. Ling H, Boodhoo A, Hazes B et al (1998) Structure of the Shiga-like toxin I B-pentamer complexed with an analogue of its receptor Gb3. Biochemistry 37:1777–1788
53. Mitchell E, Houles C, Sudakevitz D et al (2002) Structural basis for oligosaccharide-mediated adhesion of *Pseudomonas aeruginosa* in the lungs of cystic fibrosis patients. Nat Struct Biol 9:918–921
54. Williams DC Jr, Lee JY, Cai M, Bewley CA, Clore GM (2005) Crystal structures of the HIV-1 inhibitory cyanobacterial protein MVL free and bound to $Man_3GlcNAc_2$: structural basis for specificity and high-affinity binding to the core pentasaccharide from *N*-linked oligomannoside. J Biol Chem 280:29269–29276
55. Dodson KW, Pinkner JS, Rose T et al (2001) Structural basis of the interaction of the pyelonephritic *E. coli* adhesin to its human kidney receptor. Cell 105:733–743
56. Wellens A, Garofalo C, Nguyen H et al (2008) Intervening with urinary tract infections using anti-adhesives based on the crystal structure of the FimH-oligomannose-3 complex. PLoS One 3:e2040
57. Buts L, Wellens A, Van Molle I et al (2005) Impact of natural variation in bacterial F17G adhesins on crystallization behaviour. Acta Crystallogr D Biol Crystallogr 61:1149–1159
58. Sauter NK, Hanson JE, Glick GD et al (1992) Binding of influenza virus hemagglutinin to

analogs of its cell-surface receptor, sialic acid: analysis by proton nuclear magnetic resonance spectroscopy and X-ray crystallography. Biochemistry 31:9609–9621

59. Dormitzer PR, Sun ZY, Wagner G, Harrison SC (2002) The rhesus rotavirus VP4 sialic acid binding domain has a galectin fold with a novel carbohydrate binding site. EMBO J 21:885–897
60. Fry EE, Tuthill TJ, Harlos K et al (2010) Crystal structure of equine rhinitis A virus in complex with its sialic acid receptor. J Gen Virol 91:1971–1977
61. Burmeister WP, Guilligay D, Cusack S, Wadell G, Arnberg N (2004) Crystal structure of species D adenovirus fiber knobs and their sialic acid binding sites. J Virol 78:7727–7736
62. Grahn E, Askarieh G, Holmner A et al (2007) Crystal structure of the *Marasmius oreades* mushroom lectin in complex with a xenotransplantation epitope. J Mol Biol 369:710–721

ERRATUM

Small-Angle X-Ray Scattering to Obtain Models of Multivalent Lectin–Glycan Complexes

Stephen D. Weeks and Julie Bouckaert

Jun Hirabayashi (ed.), *Lectins: Methods and Protocols*, Methods in Molecular Biology, vol. 1200,
DOI 10.1007/978-1-4939-1292-6_42, © Springer Science+Business Media New York 2014

DOI 10.1007/978-1-4939-1292-6_47

The publisher regrets that the online and print versions of the above title were published with the wrong version of Figure 2 in Chapter 42.

The online version of the original chapter can be found at
http://dx.doi.org/10.1007/978-1-4939-1292-6_42

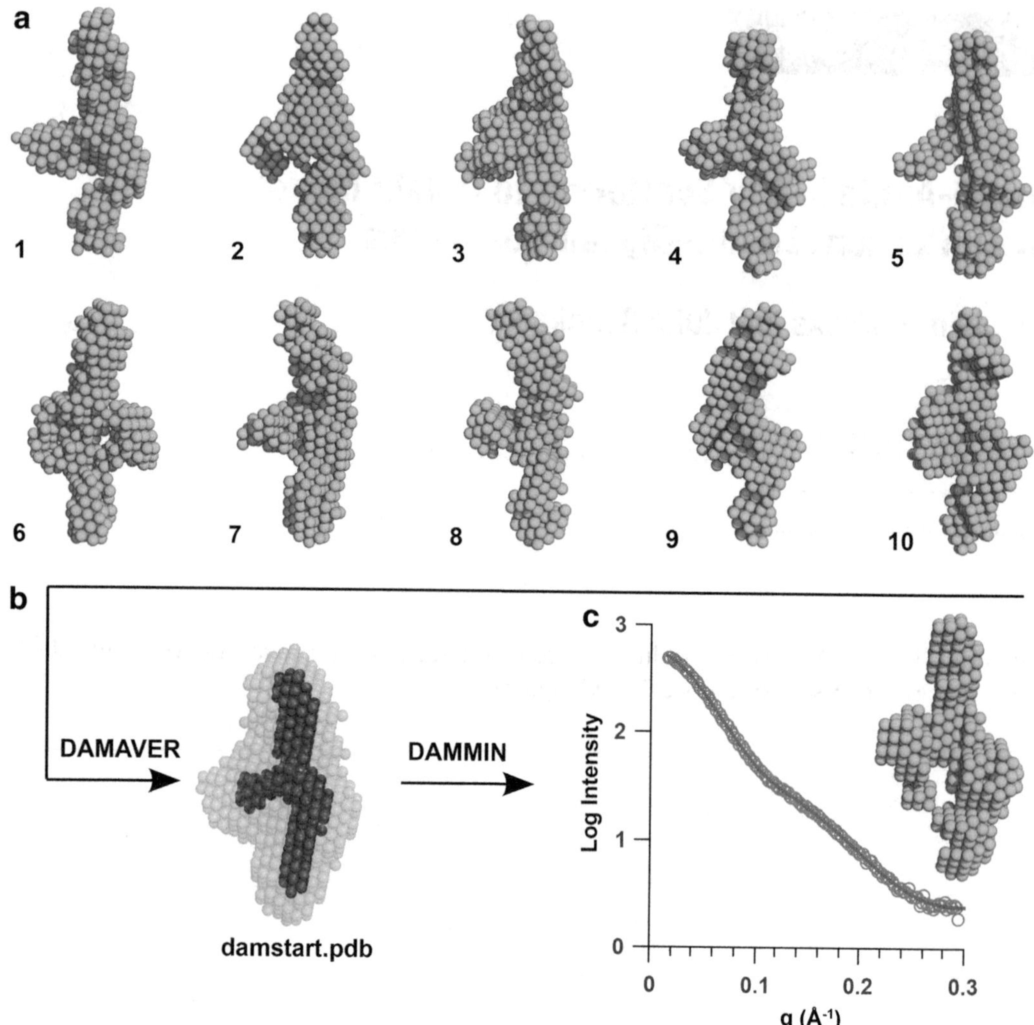

Fig. 2 Dummy atom model (DAM) generation of a complex between the lectin-binding domain of FimH and a cyclic heptavalent ligand. Models show a preference to bind only three moieties of the lectin in agreement with the determined molecular mass. (**a**) Ten instances of DAMMIN were run using the ATSAS online server. The resultant models have χ^2 values ranging from 1.48 to 1.51 when compared to the experimental data. DAMAVER identifies Model 1 ($\chi^2 = 1.49$) as having the lowest average normalized spatial discrepancy (NSD) when compared to the other structures. (**b**) DAMAVER generates an average model identifying a core structure with high atom occupancy (*dark spheres*). The presence of the surrounding atoms (*transparent spheres*) is refined by repeating the DAMMIN using the model as an input. (**c**) The final refined DAM structure and the corresponding fit (*red line* $\chi^2 = 1.49$) to the experimental data (*grey circles*). The model clearly shows three "arms" likely corresponding to three FimH lectin domains

Index

Jun Hirabayashi (ed.), *Lectins: Methods and Protocols*, Methods in Molecular Biology, vol. 1200,
DOI 10.1007/978-1-4939-1292-6, © Springer Science+Business Media New York 2014

S

T

U

W

X

Y